Radar Systems Principles

Harold R. Raemer

CRC Press
Boca Raton London New York Washington, D.C.

Acquiring Editor: *Norm Stanton*
Senior Project Editor: *Susan Fox*
Cover Design: *Dawn Boyd*
Illustrations: *Jim Houston, Visual Graphics*
Marketing Manager: *Susie Carlisle*
Direct Marketing Manager: *Becky McEldowney*

Library of Congress Cataloging-in-Publication Data

Raemer, Harold Roy, 1924-
Radar systems principles / by Harold R. Raemer
p. cm.
Includes bibliographical references and index.
ISBN 0-8493-9481-3 (alk. paper)
1. Radar. I. Title.
TK6575.R3 1996
621.3848—dc20

96-8565
CIP

International Standard Book Number 0-8493-9481-3
Library of Congress Card Number 96-8565
Printed in the United States of America 2 3 4 5 6 7 8 9 0
Printed on acid-free paper

PREFACE

This book emphasizes the basic theoretical concepts required for analysis, conceptual design, and performance evaluation of radar systems. Mathematical derivations of formulas used as "rules-of-thumb" by radar engineers are presented, with careful discussions of the assumptions behind these expressions and their ranges of validity. The derivations are followed by discussions of the applications of these ideas to a variety of radar scenarios. It includes chapters on the radar equation, constituents of a radar system, theory of radar signals in noise, radar detection theory, target signal statistics, radar waveform design, clutter and anti-clutter measures, radar antennas, propagation, and measurement of radar parameters. A set of problems for each chapter is provided at the end of the book. There are numerous figures to illustrate the principles and numerous references to well-known books on radar for coverage of practical design issues and specialized topics too detailed to cover in this book. The book would be suitable as a course textbook or a vehicle for self-study by engineers wishing to enhance their understanding of radar principles and their implications in actual systems.

The book is based on a set of notes developed for a course in Radar Systems Principles taught by the author at the Mitre Institute, Bedford, MA, in 1989–90. A significant portion of that course was also taught to graduate students by the author at Northeastern University, Boston, MA, a number of times during the 1970s and 1980s and once in the early 1990s. During each rendition of the course it was required to specify a textbook. The books that first came to mind were the standard texts on Radar Systems, e.g., M. Skolnick, *Introduction to Radar Systems* (McGraw-Hill, 1968, 2nd ed., 1980), which I used as the official text during the 1970s and most of the 1980s. When the course was given at Northeastern in 1988, I adopted a different text; Eaves and Reedy, *Principles of Modern Radar* (Van Nostrand, 1987). Both of those works have some excellent features, as do other well-known books on radar, such as Skolnick's *Radar Handbook* (McGraw-Hill, 1970, 2nd ed., 1986; also adopted as the text by Professor David McLaughlin at Northeastern during the early 1990s); D. Barton, *Modern Radar System Analysis* (Artech, 1988); and Nathanson, Reilly, and Cohen, *Radar Design Principles* (McGraw-Hill, 1991). However, for the most part, they are most suitable for engineering practitioners and less suitable as textbooks in an academic course in radar.

For the latter reason, I usually based my lectures on knowledge gained from a combination of an academic background in electromagnetic theory and statistical communication theory and experience trying to apply these disciplines to radar systems problems, and used the adopted textbook only in a supplementary way. I also sometimes used my own book, *Statistical Communication Theory and Applications* (Prentice-Hall, 1969) as a source of lecture material on some of the basic theoretical ideas behind radar system analysis.

Out of my experience teaching the radar course at Northeastern and Mitre and working on radar analysis problems in industrial laboratories and at the University, I developed the approach evident in this book. That approach is based on the premise that the detailed derivation from first principles of the mathematical relationships commonly used in radar engineering practice is a useful contribution to the process of learning about radar. Radar is an enormously broad subject, encompassing most of the EE subdisciplines. It is impossible to go back to the first principles on every facet of the subject and still cover sufficient material for a textbook. But a book that presents the standard formulas used by radar engineers with a minimum of derivation in order to achieve extensive coverage leaves the reader who is unfamiliar with these formulas with the choice of accepting them on faith without clear understanding of the limits of their applicability or spending considerable time with reference sources trying to understand where they came from.

The principal feature of this book is that it was designed to be primarily a textbook, not a handbook or reference book. Hence, the emphasis is on deriving a few general principles quite thoroughly and then applying them in the context of radar systems. The objective is to impart to the reader an in-depth understanding of some important theoretical principles of radar and thereby lay the groundwork for an ability to interpret the theory in a practical way. Planning and conceptual design of a radar system to achieve specified performance levels for target detection, imaging, or tracking requires that kind of understanding. Superficial knowledge of a few standard "rules of thumb" is not sufficient.

The extensive derivations and discussions of interpretation of the consequences of theory to be found in this book are achieved with a sometimes considerable sacrifice in breadth of coverage. An attempt is made to reach a compromise between depth of theoretical treatment characteristic of university textbooks and extensive coverage of real-world topics characteristic of industry-based reference books. Whether I have succeeded in that attempt is a matter of opinion, but it is hoped that this book might be useful as a text in a course for graduate or advanced undergraduate students *or* for practicing engineers without much specific experience in radar who want to learn about radar concepts from their core of basic knowledge of mathematics, physics, and engineering theory. For the latter group, it is understood that they would need to supplement this book with reference works such as those by Skolnick, Barton, Nathanson, and others in order to fill in details of practical implementation of radar systems.

The first chapter covers very briefly a few simple radar concepts such as the use of pulse-travel-time to determine range and the use of doppler to determine line-of-sight velocity, the radar frequency bands, a little radar history, and a few applications. Chapter 2 deals with derivation and some applications of the radar equation, which is the basis of radar system calculations and should be thoroughly understood by radar engineers. The presentation is a little more general than in most texts and includes bistatic radar. Chapter 3 breaks down into stages the process of transmitting a radar wave into the surrounding space and receiving the energy backscattered from the target. The transmitter, its antenna, the path to target, the target itself, the return path, receiving antenna, and the linear stages of the receiver constitute the stages of the process, and each of them is briefly discussed in Chapter 3.

Chapter 4 presents in some detail the idealized theory of "narrowband" signals in additive noise which forms the basis of an understanding of optimized detection of targets and extraction of target information from received radar signals.

For those familiar with communication theory, this chapter is a review. But for some readers it could serve as an introduction to the subject.

Chapter 5 provides extensive coverage of the theory of detection of radar signals in additive Gaussian noise, including mathematical details of detection and false-alarm probability calculations for radar pulses and pulse trains. This chapter borrows heavily from *Radar Detection,* by DiFranco and Rubin (McGraw-Hill, 1967 and Artech, 1980), the most thorough treatise on the subject. It summarizes some of the mathematical details in order to point out the simplifying assumptions behind the results and then presents these results in a form where they can be directly used by the reader in solving the assigned problems.

Chapter 6 covers most of the non-Rayleigh target signal amplitude probability density functions (PDF) that often fit observed data better than the Rayleigh PDF. This non-Rayleigh behavior may seriously affect detection probabilities and hence it is important in radar system analysis.

Chapter 7 covers the effect of the transmitted signal waveshape on the radar's ability to resolve targets in range and doppler, and the trade-off between range and doppler resolution. The ambiguity function, the vehicle for study of resolution, is covered in depth, and it is emphasized that it is a very close approximation to the output of a "matched" filter with unavoidable mismatches in range and/or doppler and hence can be used in the study of detection and parameter estimation.

Chapter 8 introduces the basic concepts of radar returns from extended targets, which include a large patch of illuminated terrain or a volume of precipitation. The latter are the major sources of radar clutter, which is the subject of Chapter 8. This chapter includes discussion of extraneous clutter returns due to range and doppler ambiguities caused by high and low PRFs respectively, a very important effect in modern radars. Chapter 9 emphasizes the use of doppler filtering to distinguish clutter from desired targets based on velocity differences between them.

Chapter 10 is a summary of some of the basic concepts of radar antennas, with emphasis on the theory behind the generation of the radiation pattern for rectangular and circular aperture antennas, the most commonly used classes of antenna used in radars. It includes discussions of antenna arrays and electronic scanning, both widely used in modern radars. Chapter 11 covers some important radar propagation topics which sometimes have an important effect on radar performance, but which are often neglected in simplified radar system analysis. Especially emphasized in this chapter is the "propagation factor" due to interference between the direct waves from radar-to-target and back and the earth-reflected paths. It also includes introductory discussions of refraction, diffraction, attenuation, and shadowing, all of which have important effects on radar performance.

Chapter 12 covers basic theoretical ideas behind the optimal measurement of range, frequency shift, and direction angles of radar signals in noise, using maximum likelihood estimation theory as the basis of the discussions. It concludes with an extensive discussion of the theory behind angle measurement techniques used in tracking radars, based on amplitude or phase comparison concepts.

The problem sets for each of the chapters, which are located at the end of the book, are designed to give the reader some practice in dealing with the ideas discussed in the text, essential to the learning process. Wherever possible, I have put the problems concerning particular theoretical ideas into the context of a hypothetical radar scenario with some degree of realism. That approach requires that I specify a set of radar system parameters and specific parameters for the radar-target-earth geometry. The reader who tries to solve the problems will not only learn the basic concepts but will be forced to deal repeatedly with the radar equation and hopefully will acquire some feeling for the expected range of numerical values of key parameters.

About the Author

Harold R. Raemer received his Ph.D. in Physics at Northwestern University, Evanston, IL, in 1959. He was a research engineer in industrial laboratories from 1952 to 1963, then joined the faculty at Northeastern University in Boston as an Associate Professor of Electrical Engineering. He became Professor in 1966, served as Chairman of the EE Department from 1967 to 1977, Acting Chair (1982–84), and Technical Director for RF Phenomena and Systems in the Center for Electromagnetics Research (1985–94). He retired in 1994, but still remains at the university as a Professor Emeritus and an active member of the Center's research staff. He has worked on many radar-related research projects during his career, both in industrial and government laboratories and at the university, and during the past 10 years has been working extensively on analytical modeling and computer simulation of radar clutter and multipath.

Acknowledgments

I want to express my appreciation to a number of people who helped in various ways in the preparation of this book. First, I want to mention some of the people at the Mitre Corporation in 1989–90 who were involved in the administration and coordination of the course that was the basis of this book, in particular, Charles Gager, Ronald Fante, George Randig, and Arthur Glazer. They provided many helpful suggestions on coverage of topics. I also extend my appreciation to Michael Silevitch, Director of the Center for Electromagnetics Research at Northeastern University, who, together with the management of the Mitre Corporation and Alex Schwartzkopf of NSF, provided the opportunity for me to spend a sabbatical year at Mitre, teaching this course, developing the notes that became this book, and working on advanced radar system analysis problems with a number of outstanding radar engineers, including those mentioned above, from whom I learned a great deal that helped in the development of the book. I also acknowledge the indirect contributions of a number of students in the radar courses at Northeastern and Mitre who, through questions and class discussions, influenced the distribution of topics in the notes and eventually in the book. Some of these students also helped to correct some typographical errors in the notes.

Special thanks are extended to Caroline Kent, who word-processed the entire manuscript with great skill, including the many equations that appear in the text. Thanks also are extended to Charles DiMarzio, who performed the computations for the curves of Figures 7.20, 7.21, 11.10, 11.11, and 11.12 in Mathematica and to Robert Bilotta, who performed the computations for Figure 10.12 in Matlab.

Finally, I want to thank my wife, Paulyne. Throughout my long professional career she has always been understanding and supportive of the various technical projects that often occupied my evenings at home. That was no less true of the most recent project, the completion of this book.

Dedication

This book is dedicated to my wife Paulyne, our children Dan, Liane, and Diane, their spouses, Wallis and Matt, and our grandchildren Ben, Zach, Jacob, Emily, Ross, Karly, and Maggie.

CONTENTS

Chapter 1

Introduction

CONTENTS

One possible definition of a radar system is "a system that attempts to infer information about a remotely located object from reflections of deliberately generated electromagnetic waves at radio frequencies." That definition would be considered too broad by some radar engineers. Without commenting about that issue, we will immediately pass to a specialized definition, that of the "monostatic" radar.

That word implies that the transmitter of the radiofrequency (RF) waves and the receiver that picks up the reflected waves from the object are colocated. In a "bistatic" radar, transmitter and receiver are at different locations. Bistatic radars are infrequently used. The overwhelming majority of radars in use today are monostatic. Unless explicitly indicated otherwise, "radar," in what follows, will mean a monostatic radar.

There are a number of books on radar in general and on specialized aspects of it in particular. In the process of teaching courses in radar systems during the past several years, the author adopted certain established radar texts as either the official textbook for the course or as a supplementary text. However, as indicated in the preface, most of the lecture material was derived by the author from very basic theoretical principles and these established texts were used for corroboration of the basic analytical tools or to refer the students to topics that there was insufficient time to cover in class. That is the way they are usually used in this book.

The "established texts" referred to above include those that are considered "classical" in the radar community, e.g., M.I. Skolnick's "Introduction to Radar Systems," whose first edition was published in 1962[1a] and whose second edition was published in 1980.[1b] Another text was written by D.K. Barton, "Radar System Analysis",[2a] first published in 1964 and whose sequel, "Modern Radar Systems Analysis,"[2b] was published in 1988. Still another was "Radar Design Principles," by F.E. Nathanson, the first edition of which was published in 1969[3a] and the second edition (co-authored by J.P. Reilly and M.N. Cohen) was published in 1991.[3b] In recent years, some other texts have emerged, in particular, a book edited by J.L. Eaves and E.K. Reedy and written by them and a number of colleagues at Georgia Tech Research Institute,[4] and others by S.A. Hovanessian,[5] N. Levanon,[6] L.V. Blake,[7] P. Rohan,[8] and N. Carpentier.[9] In addition to these, an extremely useful and complete source of information on virtually all of the important topics in radar is Skolnick's "Radar Handbook" (first edition, 1970[10a] and second edition, 1990.[10b]). Some other general radar texts predate Skolnick's first edition and the first edition of Barton's text. One of these is "Radar Systems Engineering," edited by L.H. Ridenour,[11] the first volume of the MIT Radiation Laboratory series.

Other general radar texts, including some recently published ones, are in the reference list.[12–18] Another useful reference source is an index edited by C. Gager[19], which lists references on radar published prior to 1990. Still another is the bibliography on pp. 681–785 of Nathanson et al.,[3b] which contains 785 references on radar and radar-related subjects published prior to 1992. In most of the chapters of this book the author has cited some of the above texts in connection with the topic covered in the chapter. In most cases, that is to provide the reader with additional readily available references on that topic to supplement the material covered in the chapter. Because most of the chapters of this book emphasize the derivations of the theoretical radar principles and their general interpretations, many details important in radar practice were not included in the chapter. An example of this is Chapter 10 on radar antennas, where texts such as Skolnick,[1a,1b] Barton,[2a,2b] and Eaves and Reedy[4] have entire chapters on radar antennas which contain many more practical details on the subject than Chapter 10 of this book. The same applies to Chapter 11 on radar propagation, Chapter 12 on measurement of radar parameters, Chapter 5 on target detection, Chapter 7 on resolution and ambiguities, and Chapter 8 on clutter. In other cases, a standard topic covered in this book was also covered, usually differently, in a number of these texts and the author corroborated some of the results he obtained from the theory with results given in other texts. In still other cases, the treatment of a topic in this book was inspired by the treatment of that topic in another text, always acknowledged by the author in the presentation of the topic.

1.1 COMPOSITION OF A RADAR SYSTEM

Figure 1.1 illustrates the basic monostatic radar. In Figure 1.1a the first phase of the radar's operation is shown, where the modulated RF signal generated within the transmitter (TRANS) feeds into a transmission line (TL) and in turn into the transmitting antenna (ANT) and finally into the surrounding space in the form of a wave propagating into all directions, but with directional preferences governed

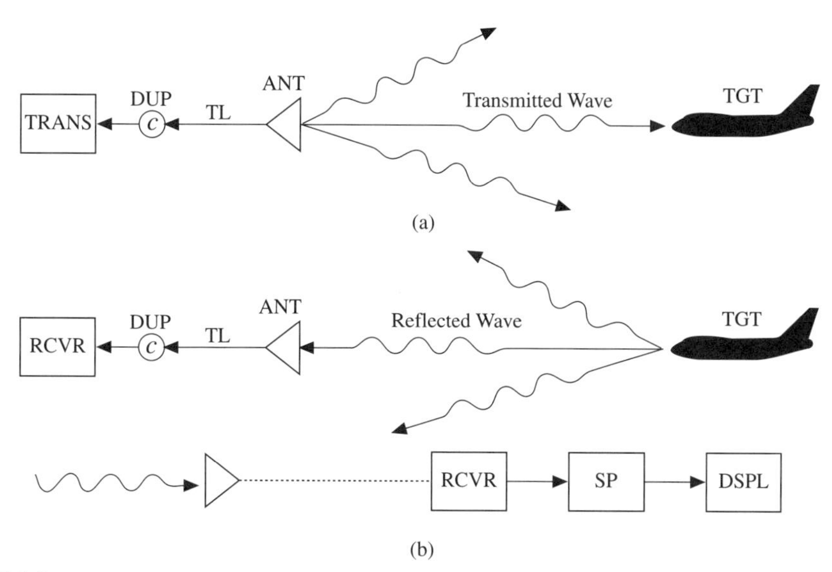

FIGURE 1.1
Basic monostatic radar.

by the design of the antenna. [The "duplexer" (DUP) allows the same antenna to be used for both transmission and reception. In most but not all radars that is the case.] Some of the wave energy travels in the direction of the object of interest, called a "target" in radar parlance and designated as TGT in the figure. Figure 1.1b shows the action of the target in receiving the transmitted wave energy and retransmitting or "scattering" it in all directions, but again with a directional distribution depending on its scattering properties. The wave energy scattered in the direction of the radar ("backscattered") is gathered in by the receiving antenna and routed through the transmission line and the duplexer into the receiver (RCVR) where it is converted into a voltage or current waveform. In Figure 1.1c the received signal is shown being routed into a signal processer (SP), where it is processed and fed into a display (DSPL).

A couple of points are worthy of mention in an introduction to radar. One is that the radio frequencies used in most radar systems tend toward the microwave region, whose low end is loosely defined as within the ultra-high-frequency (UHF) band (from 300 MHz to 1 GHz) and whose high end is somewhere in the millimeter band, above 30 GHz. It will become apparent later why these frequencies are favored for radars.

The other point is that radio waves travel in free space at the "speed of light," equal to 3×10^8 m/sec and usually denoted by c. Hence, all the processes depicted in Figure 1.1 take place within a period of less than a few milliseconds. Even for targets as far away as 300 km, if they can be detected, the entire propagation process from radar to target back to radar requires only 2 msec.

The well-known relationship between frequency f in Hertz and wavelength in meters is

$$\lambda = \frac{c}{f} = \frac{3(10^8)}{f} \tag{1.1}$$

so the low end of the radar spectrum has wavelength of the order of 1 m and the high end has wavelength of the order of a few millimeters.

1.2 HISTORICAL DEVELOPMENTS—1930S TO PRESENT

This topic will not be a major one in this book. However, it is useful in establishing perspective to mention that radar is a mature technology. It has been around since the 1930s. Some of the historical highlights of radar technology are summarized in Eaves and Reedy (table on p. 4)[4] and in Skolnick (pp. 8–12).[1b]

Among the highlights of early radar history as indicated in these sources are the demonstrations of reflection of radio waves from metallic and dielectric objects in general by Hertz in 1886, from ships by Hulsmeyer in 1903 and Tyler and Young at NRL in 1922, and from aircraft by Hyland in 1930. The demonstrations of the feasibility of pulsed microwave radar by Watson-Watt in Britain and Page at NRL between 1934 and 1936 began a period of intensive radar development in Britain and the United States between 1936 and 1940. In 1940, the MIT Radiation Laboratory was founded. The radar research and development activities conducted there resulted in the Radiation Laboratory Series of 28 books, which contained an enormous storehouse of information on radar theory and practice. Although the hardware implementation content of these books is now mostly obsolete, much of the fundamental theory content is still relevant and some of these books are still sometimes cited as references on certain radar topics (e.g., Volumes 12[20], 13,[21] and 24[22].)

A delineation of developments occurring during and after World War II that enhanced the development and improvements in the performance of radars from 1941 to the present time is given in Eaves and Reedy (p.5).[4] It includes the development of high-power klystrons, low-noise traveling wave tubes, and parametric amplifiers and masers, all of which enhanced the power generation and amplification capabilities of radar transmitters and receivers. Other developments listed are monopulse, short pulse techniques, phased arrays, synthetic aperture radar, and HF over-the-horizon radar. The later developments on that list are mostly related to the improvements in integrated circuit technology, e.g., very-high speed integrated circuits (VHSIC), all of which enhanced the digital signal-processing techniques available in radar systems.

Another very useful source on radar history is a journal article by D.K. Barton[23] entitled "A half-century of radar," which summarizes the history of radar from 1934 to 1984.

1.3 RADAR FREQUENCY BANDS

It is useful to know the frequency bands specifically designated by the radar community. They are given in Table 1.1, with the corresponding wavelength ranges. Most of the data in the table follows the IEEE Standard for Radar Frequency Bands.[24]*

1.4 APPLICATIONS OF RADAR

There are many applications of radar, but a few of the notable ones are delineated on p. 13 of Skolnik's, "Introduction to Radar Systems"[1b] and Table 1-1 on p. 5 of Barton's "Modern Radar Systems Analysis".[2b] The items listed in these

* Skolnick (Table 1.1 on p. 8)[1b], Barton (Table 1.2 on p. 6)[2b], and Nathanson et al. (Tables 1.1 and 1.2 on pp. 18 and 19)[3b] contain much of the same information as Table 1.1 in this text.

TABLE 1.1
Radar Bands

Designation	Frequency Range	Wavelength Range	Notes
HF	3–30 MHz	10–100 m	Used for "over-the-horizon radar"
VHF	30–300 MHz	1–10 m	
UHF	300 MHz–1 GHz	30 cm–1 m	Radio communication in general bounds UHF at 3 GHZ
L-band	1–2 GHz	15–30 cm	3–30 GHz-SHF (super-high frequency) includes part of S-band, C, X, Ku, K, and part of K_a-bands
S-band	2–4 GHz	7.5–15 cm	
C-band	4–8 GHz	3.75–7.5 cm	
X-band	8–12 GHz	2.5–3.75 cm	
Ku-band	12–18 GHz	1.67–2.5 cm	
K-band	18–27 GHz	1.11–1.67 cm	
K_a-band	27–40 GHz	7.5 mm–1.11 cm	30–300 GHz-EHF (extremely high frequency) includes part of K_a, V, W, and mm-band
V-band	40–75 GHz	4–7.5 cm	
W-band	75–110 GHz	2.7–4 mm	
mm-band	110–300 GHz	1–2.7 mm	

tabulations are military, transportation (for either military or civilian use), and scientific applications.

Among the specifically military applications are search and surveillance of enemy targets, navigation, control, and guidance of weapons, e.g., ballistic missiles, air-to-ground, ground-to-air, and air-to-air missiles used in combat, battlefield surveillance, and antiaircraft fire control.

Among the air transportation applications are aircraft navigation, avoidance of collisions with other aircraft or terrain obstacles, detection and avoidance of weather disturbances and clear-air turbulence, radar altimeters, air traffic control, including ground control approach, (GCA) airport taxiway control, and others.

Water-transportation applications include shore-based ship navigation, collision avoidance for ships and small boats, harbor and waterway control, and others.

Applications involving land transportation (existing or potential) include doppler radar for automobile speed measurement as used in law enforcement, or for speed measurement and tracking of military vehicles and collision avoidance radar for land vehicles.

Space applications include detection and tracking of satellites and radars for rendezvous and docking of space vehicles.

Scientific applications include remote sensing of the earth's environment from aircraft and satellites and radars for planetary observation.

Radars for some special applications include weather radars for study and monitoring of precipitation, clouds, and major weather disturbances (e.g., hurricanes and tornadoes), ground mapping radars, ground-penetrating radars for detection of buried objects, and radars for high-resolution imaging of objects and terrain (e.g., synthetic aperture imaging radars).

1.5 RADAR PULSES

A large majority of radars in use employ pulsed transmission. In a monostatic radar system, the most elementary measurement that can be made is that of the distance to target, or range, through the back-and-forth travel time of a pulse. This is illustrated in Figure 1.2.

The transmitted radio frequency (RF) pulse travels at the free-space velocity of light $c = 3 \times 10^8$ m/sec. The time required for the pulse to travel from transmitter to target is R/c sec. The return trip also requires R/c sec. The total travel time is

$$T = \frac{2R}{c} \tag{1.2}$$

Figure 1.2c shows an "A-scope" display, i.e., a display of amplitude of the received signal pulse vs. time, where the time scale is calibrated in distance units. Thus, if the center (or leading edge or trailing edge) of the transmitted pulse is at $t = 0$, corresponding to $R = 0$, and the center (or leading edge or trailing edge, which ever reference point was used for the transmitted pulse) of the return pulse is at $t = T$, corresponding to $R = cT/2$, then the target range is determined to be R.

The A-scope display shown in Figure 1.2c is highly idealized. First, it applies to transmission of a single pulse, where realistic radar transmissions are trains of pulses. Second, it assumes no noise and no extraneous targets ("clutter"). The former issue is very easily handled by superposing returns at $t = t_1$, $t = t_1 + Tr$, $t = t_1 + 2Tr$, etc., where Tr is the pulse repetition period (assuming of course the usual case where the pulses are equally spaced). The display shown in the figure still applies, but the amplitude is increased through addition of pulses. The second issue is more difficult to deal with, because extraneous targets might return signals of amplitude comparable to that of the desired targets and near its range and there are many

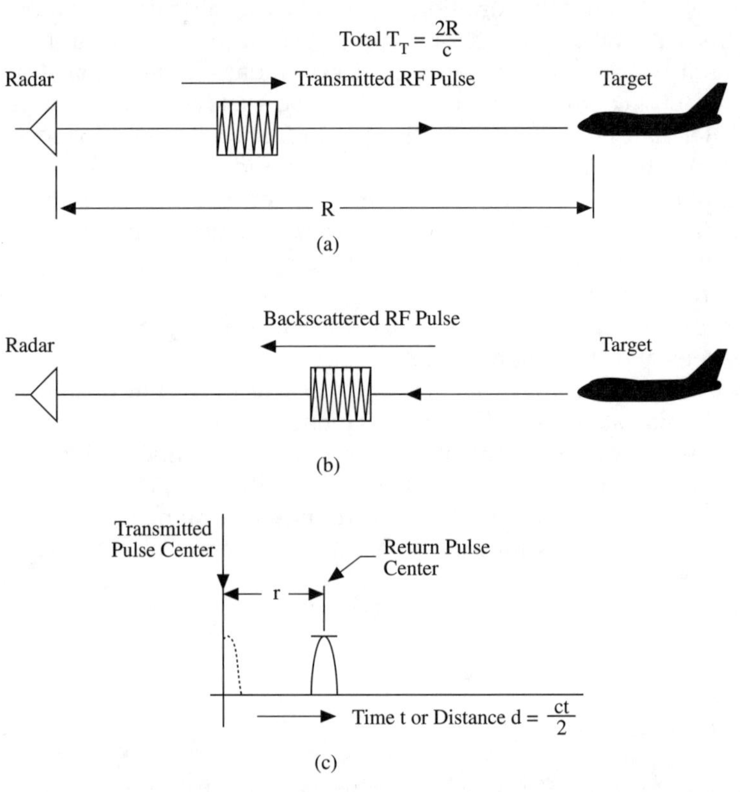

FIGURE 1.2
Range measurement through pulsing.

sources of noise that might corrupt the display and make it difficult to distinguish the desired target pulse from noise bursts and extraneous target pulses. Those topics will form a large portion of the material covered in this book.

To quantify the relationship between pulse travel time and range, we note (from 1.2) that 1 sec of time corresponds to $cT/2 = 1.5 \times 10^8$ m or 1.5×10^5 km. Pulse durations range from nanoseconds in high-resolution radars to the order of hundredths of milliseconds in conventional low-frequency radars. A pulse duration of 1 μsec corresponds to a pulse width calibrated in range units of 150 m. The pulse duration roughly corresponds to the minimum range difference we can resolve, as will be explained later. Thus, a range resolution of 1.5 m would require a pulse no longer than 10 nsec. A 5-μsec pulse will provide a range resolution of 0.75 km. "High-resolution" radar usually refers to pulse durations well below a microsecond, where ranges of a few meters or less can be resolved. Table 1.2 shows best attainable rough range resolution vs. pulse duration.

The reasons for the relationship between quantities proportional to pulse duration and range resolution, implied in the discussion above, will be explained in Chapter 7, Sections 7.2–7.4. There is also a relationship between other quantities proportional to pulse duration and the accuracy with which the range of a target can be measured in the presence of additive noise, which will be explained in Chapter 12, Sections 12.1.3 and 12.2.1.

There is a profound difference between range measurement and range resolution, although they may seem to be closely related. The fractional rms error in measurement of the target range is directly proportional to both the pulse duration and the voltage noise-to-signal ratio. The "range resolution" (Section 7.2 in Chapter 7) is the minimum range separation between two point targets (whose radar returns have the same amplitude) that allows those two targets to be seen by a radar as separate rather than as a single target. The range *measurement accuracy* (being inversely proportional to the fractional rms error in the measurement) is inversely proportional to the pulse duration for a given SNR. The range *resolution* is inversely proportional to pulse duration. Hence "high resolution" implies a pulse of short duration. A narrow pulse, in turn, also implies high range measurement accuracy. Hence these two measures of merit in a radar system, the ability to confine a target within a small span of possible ranges and the ability to resolve two targets whose ranges are very close together, are both enhanced by pulses of short duration. It will be brought out in discussions of range resolution in Chapter 7 and discussions of range measurement accuracy in Chapter 12 that both of these goals are related to large bandwidths.

TABLE 1.2

Pulse Duration vs. Range Resolution

Pulse Duration	Best Attainable Range Resolution (Roughly)
0.1 nsec	1.5 cm
1 nsec	15 cm
10 nsec	1.5 m
100 nsec	15 m
1 μsec	150 m
10 μsec	1500 m
100 μsec	15 km
1 msec	150 km

1.6 USE OF DOPPLER IN RADAR

The Doppler effect, familiar to students of elementary physics in connection with light waves, is also useful at microwave frequencies. If the relative velocity between target and radar has a component along the line of sight (LOS) between radar and target, a "Doppler shift" in the received signal frequency relative to the transmitted signal frequency arises. If the target is approaching the radar, the shift is upward. If it is receding from the radar, the shift is downward (see Fig. 1.3). In both cases, the shift is proportional to the magnitude of the LOS velocity component.

The Doppler shift for each leg of the wave's journey between radar and target is $-(f_0/c)\mathbf{V}\cdot\hat{r}$. Since there are two legs of that journey, the Doppler shift is [with the aid of (1.4)]

$$f_d = -\frac{2f_0}{c}\mathbf{V}\cdot\hat{r} = -\frac{2}{\lambda_0}\mathbf{V}\cdot\hat{r} \tag{1.3}$$

where λ_0 is the transmitted signal wavelength in meters, c is the speed of light, f_0 is the transmitted radar signal frequency in Hertz, $\mathbf{V}$ is the velocity vector of the target relative to the radar (i.e., the radar is assumed stationary), and $\hat{r}$ is a unit vector from the radar directed toward the target. In the approaching case, as shown in Figure 1.3a, Eq. (1.3) would give us

$$f_d = -\frac{2V}{\lambda_0}\cos\theta = +\frac{2V}{\lambda_0}\cos(\pi-\theta) = \frac{2V}{\lambda_0}|\cos\theta| > 0 \tag{1.4a}$$

where V is the relative speed between target and radar.

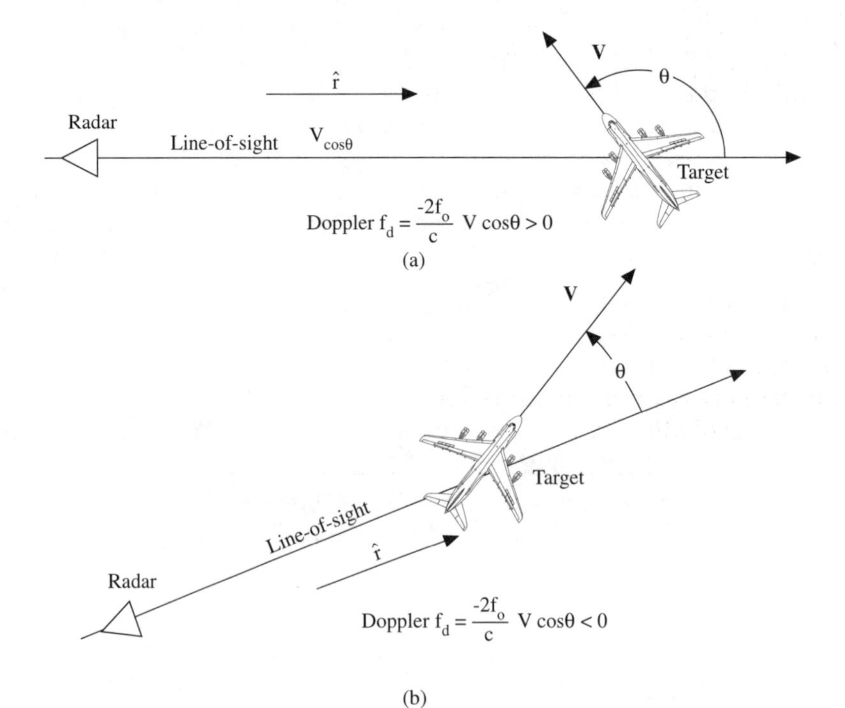

FIGURE 1.3
Use of Doppler effect in radar.

TABLE 1.3

Doppler Shifts at Centers of Radar Bands[a]

Band	Central Frequency	Wave-length	$2/\lambda$-m^{-1}	Doppler Shifts for Various V_{rel} (m/sec)				
				0.1 msec	1 msec	10 msec	100 msec	1 km/sec
HF	15 MHz	20 m	0.1	0.01 Hz	0.1 Hz	1 Hz	10 Hz	100 Hz
VHF	17 MHz	17.6 m	0.114	0.0114 Hz	0.114 Hz	1.14 Hz	11.4 Hz	114 Hz
UHF	165 MHz	1.8 m	1.11	0.111 Hz	1.11 Hz	11.1 Hz	111 Hz	1.11 kHz
L	1.5 GHz	20 cm	10	1 Hz	10 Hz	100 Hz	1 kHz	10 kHz
S	3 GHz	10 cm	20	2 Hz	20 Hz	200 Hz	2 kHz	20 kHz
C	6 GHz	5 cm	40	4 Hz	40 Hz	400 Hz	4 kHz	40 kHz
X	10 GHz	3.3 cm	67	6.7 Hz	67 Hz	670 Hz	6.7 kHz	67 kHz
Ku	15 GHz	2 cm	100	10 Hz	100 Hz	1 kHz	10 kHz	100 kHz
K	23 GHz	1.3 cm	154	15.4 Hz	154 Hz	1.54 kHz	15.4 kHz	154 kHz
K_a	34 GHz	0.8 cm	250	25 Hz	250 Hz	2.5 kHz	25 kHz	250 kHz
V	58 GHz	5 mm	400	40 Hz	400 Hz	4 kHz	40 kHz	400 kHz
W	93 GHz	3 mm	667	66.7 Hz	667 Hz	6.67 kHz	66.7 kHz	667 kHz
mm	205 GHz	1.46 mm	1370	137 Hz	1.37 kHz	13.7 kHz	137 kHz	1.37 MHz

[a] *Note*: 100 m/sec = 225 mph = 195.65 nmi/h.

In the receding case (Fig. 1.3b), application of (1.3) yields

$$f_d = -\frac{2V}{\lambda_0}\cos\theta = -\frac{2V}{\lambda_0}|\cos\theta| < 0 \qquad (1.4b)$$

To get some insight into the magnitude of Doppler shift that can be expected from typical targets at various frequencies, we refer back to Table 1.1, from which we can construct a table of Doppler shifts for the central frequencies of the standard radar bands and various values of the relative speed V_{rel} in m/sec, which appears in Table 1.3 (from Table 1.1).

Table 1.3 shows that Doppler is a useful measurement tool at the higher end of the radar bands, but at the lower end, the Doppler shift even for extremely high-speed targets is quite small and not easily detectable. However, since Doppler shows up as a fraction of the transmitted signal frequency, in principle, it is useful throughout the radar bands. Attainable frequency measurement accuracy relates to the size of the frequency being measured, as we will see in Chapter 12.

REFERENCES

1a. Skolnick, M. I., "Introduction to Radar Systems," 1st ed. McGraw-Hill, New York, 1962.
1b. Skolnick, M. I., "Introduction to Radar Systems," 2nd ed. McGraw-Hill, New York, 1980.
2a. Barton, D. K., "Radar Systems Analysis," Prentice-Hall, Englewood Cliffs, NJ, 1964.
2b. Barton D. K., "Modern Radar Systems Analysis." Artech, Norwood, MA, 1988.
3a. Nathanson, F. E., "Radar Design Principles." McGraw-Hill, New York, 1969.
3b. Nathanson, F. E., Reilly, J. P., and Cohen, M. N., "Radar Design Principles." McGraw-Hill, New York, 1991.
4. Eaves, J. L., and Reedy, E. K. (eds.), "Principles of Modern Radar." Van Nostrand Reinhold, New York, 1987.
5. Hovanessian, S. A., "Radar System Design and Analysis." Artech, Norwood, MA, 1984.
6. Levanon, N., "Radar Principles." Wiley, New York, 1988.
7. Blake, L. V., "Radar Range Performance Analysis." Artech, Norwood, MA, 1986.
8. Rohan, P., "Surveillance Radar Performance Prediction." Peter Peregrinus, Ltd., London, 1983.
9. Carpentier, M. H., "Principles of Modern Radar." Artech, Norwood, MA, 1988.

10a. Skolnick, M. I., "Radar Handbook," 1st ed. McGraw-Hill, New York, 1970.
10b. Skolnick, M. I., "Radar Handbook," 2nd ed. McGraw-Hill, New York, 1990.
11. Ridenour, L. H., "Radar Systems Engineering" (Vol. 1 of MIT Radiation Laboratories Series). McGraw-Hill, New York, 1947.
12. Povejsil, R. S., Raven, R. S., and Waterman, P., "Airbone Radar," McGraw-Hill, New York, 1970.
13. Rihaczek, "Principles of High Resolution Radar." Peninsula Press, Palo Alto, CA, 1977.
14. Brookner, E. (ed.), "Radar Technology," Artech, Norwood, MA, 1977.
15. Brookner, E. (ed.), "Aspects of Modern Radar." Artech, Norwood, MA, 1988.
16. Stimson, G. W., "Introduction to Airborne Radar," Hughes Aircraft Co., El Segundo, CA; Radar Systems Group, 1983.
17. Edde, B., "Radars." Prentice Hall, Englewood Cliffs, NJ, 1993.
18. Wehner, D. R., "High Resolution Radar." Artech, Norwood, MA, 1987, 1995.
19. Gager, C. (ed.), "Cumulative Index on Radar Systems for 1977–1984." IEEE Trans. AES, 25(4), 1–95, 1989.
20. Silver, S., "Microwave Antenna Theory and Design." McGraw-Hill, New York, 1949.
21. Kerr, D. E. (ed.), "Propagation of Short Radio Waves." McGraw-Hill, New York, 1951.
22. Lawson, J. L., and Uhlehbeck, G. E., "Threshold Signals." McGraw-Hill, New York, 1950.
23. Barton, D. K., "A Half-Century of Radar." IEEE Trans. MTT 32(9), 1161–1170, 1984.
24. IEEE Standard 521-1984, "IEEE Standard Letter Designations for Radar-Frequency Bands." New York, November 30, 1984.

Chapter 2

THE RADAR EQUATION

CONTENTS

The radar equation, sometimes called the "radar range equation," is fundamental to an understanding of radar systems.* It is really a very simple algebraic

* The radar equation is covered in virtually every standard text on radar, but the treatments are different. The references cited in this chapter[1–8] specify the location of this material in a few well-known general texts on radar.

relationship between various radar system parameters. No knowledge of electromagnetic wave theory is required to comprehend it, because it deals with power or energy and not directly with electromagnetic fields per se. It could just as well be applied to an active acoustic sensor system such as a sonar.

The derivation of the radar equation will be presented below starting with that of the "beacon equation," then proceeding to the monostatic and bistatic radar equations.

2.1 THE BEACON EQUATION: ANTENNA PROPERTIES

The first stage in the derivation of the radar equation is that of the "beacon equation" (Fig. 2.1), by which the power at a point q due to a transmitting source at point p can be determined. This is useful in introducing certain concepts to be used later in the radar context.

If total power P_{T} is radiated into the surrounding space by the transmitting source at p and it is radiated isotropically, i.e., with equal weighting in all directions over the unit sphere surrounding p, then the "power density" at q (defined as the power radiated into unit area of the sphere of radius r_{pq} centered at p) is

$$\left(\frac{dP_{\mathrm{T}}}{dA}\right)_q^{(\mathrm{iso})} = \frac{P_{\mathrm{T}}}{4\pi r_{pq}^2} \tag{2.1}$$

The assumption that p is an isotropic radiator is virtually never valid, so (2.1) must be modified by a quantity called the "gain" of the antenna in the direction of q relative to p. This direction is defined by two direction angles (θ_{pq}, ϕ_{pq}) in the p-centered spherical coordinate system (r_p, θ_p, ϕ_p). The second index q denotes a position coordinate of the point q with respect to the point p.

The gain of the transmitting antenna in the direction (θ_{pq}, ϕ_{pq}) is defined as

$$G_{\mathrm{T}}(\theta_{pq}, \phi_{pq}) = \frac{dP_{\mathrm{T}}}{dA}(\theta_{pq}, \phi_{pq}) \bigg/ \left(\frac{dP_{\mathrm{T}}}{dA}\right)_q^{(\mathrm{iso})} \tag{2.2}$$

where the denominator is the quantity given by (2.1). In words, the definition of the gain of the antenna in direction (θ_{pq}, ϕ_{pq}) is the ratio of the power density radiated into the direction (θ_{pq}, ϕ_{pq}) by the antenna to the power density radiated by an isotropic antenna that radiates the same total power into all of the surrounding space.

To be strictly correct about the concept of gain, it should be called "directivity," as it usually is in the antenna community. In the latter group, "gain" usually accounts for transmitter losses in the definition. The total power P_{T} in (2.1) is that actually generated within the transmitter in the definition of gain and that radiated into space in the definition of directivity. Thus, "gain" is proportional to "directivity," when the strict definitions of these terms are used.

In this book we will interpret the word "gain" when referred to an antenna to mean the "directivity" as defined in (2.2). It is a measure of the degree to which the antenna radiates power within a confined angular region. The smaller the angular region to which the radiated power is confined, the larger the gain. To illustrate the point, if the power radiated into the direction (θ_{pq}, ϕ_{pq}) is independent of θ_{pq} and ϕ_{pq}, i.e., if the radiated power density is the same in all directions, then (2.2) tells

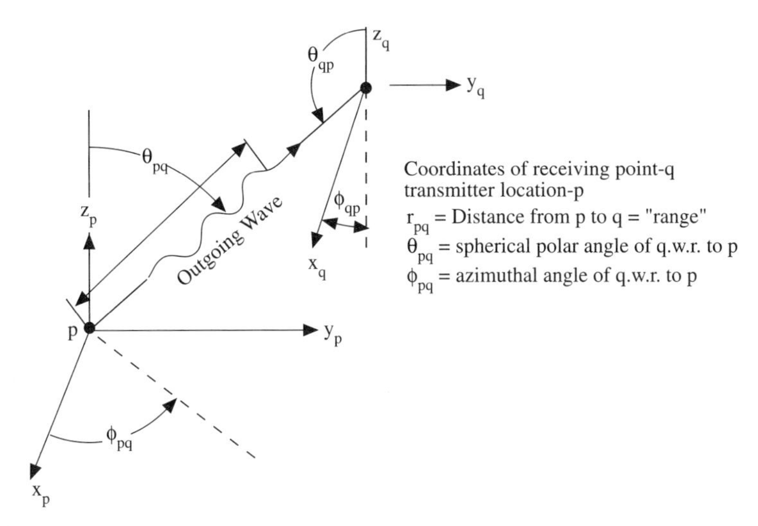

FIGURE 2.1
The beacon equation. One way transmission from p to q.

us that $G(\theta_{pq}, \phi_{pq}) = 1$ for all possible values of θ_{pq}, ϕ_{pq}. That condition describes an isotropic antenna, or an antenna with uniform angular coverage.

The single parameter usually used to describe the directional properties of an antenna is the "peak gain," defined as

$$G_0 = G(\theta_{pq}^{(0)}, \phi_{pq}^{(0)}) \tag{2.3}$$

where $(\theta_{pq}^{(0)}, \phi_{pq}^{(0)})$ are the direction angles in the direction where the gain has its maximum value.

The angular dependence of the gain can be described by a "relative gain" function or "antenna radiation pattern" function $g(\theta_{pq}, \phi_{pq})$ normalized to the peak gain G_0, i.e.,

$$g(\theta_{pq}, \phi_{pq}) = \frac{G(\theta_{pq}, \phi_{pq})}{G_0} \tag{2.4}$$

Using (2.3) and (2.4) in (2.1) we have for the power density at the point q as a function of the coordinates of q with respect to p

$$\frac{dP_{\mathrm{T}}}{dA}(r_{pq}, \theta_{pq}, \phi_{pq}) = \frac{P_{\mathrm{T}} G_{\mathrm{T0}}}{4\pi r_{pq}^2} g_{\mathrm{T}}(\theta_{pq}, \phi_{pq}) \tag{2.5}$$

To complete the beacon equation, we note that the power received by an antenna at the point q is

$$P_{\mathrm{R}}(r_{pq}, \theta_{pq}, \phi_{pq}) = \frac{dP_{\mathrm{T}}}{dA}(r_{pq}, \theta_{pq}, \phi_{pq})\, A_{\mathrm{eR}}^{(q)}(\theta_{qp}, \phi_{qp}) \tag{2.6}$$

where

$$A_{\mathrm{eR}}^{(q)}(\theta_{qp}, \phi_{qp})$$

is defined as the effective aperture area of a receiving antenna at q for radiation coming from a direction (θ_{qp}, ϕ_{qp}) where the reversal of the indices p and q indicates that the directional properties of the receiving antenna are referred to a coordinate system centered at the receiver location, i.e., (θ_{qp}, ϕ_{qp}) are the direction angles of the transmitter with respect to the receiver.

We will now write two equations analogous to (2.3) and (2.4) to define the "peak effective aperture area" or "peak receptivity" pattern as follows:

$$A_{\mathrm{eR0}}^{(q)} = A_{\mathrm{eR0}}^{(q)}\,(\theta_{qp}^{(0)}, \phi_{qp}^{(0)}) \qquad \text{(analogous to (2.3))} \tag{2.7}$$

where

$$(\theta_{qp}^{(0)}, \phi_{qp}^{(0)})$$

are the direction angles at which receptivity has its maximum value $A_{\mathrm{eR0}}^{(q)}$, and

$$g_{\mathrm{R}}^{(q)}(\theta_{qp}, \phi_{qp}) = \frac{A_{\mathrm{eR}}^{(q)}(\theta_{gp}, \phi_{gp})}{A_{\mathrm{eR0}}^{(q)}} \qquad \text{(analogous to (2.4))} \tag{2.8}$$

Incorporating (2.7) and (2.8) into (2.6) we obtain the most general form of the beacon equation, which gives the received power at q due to power transmitted at point p in terms of geometric parameters of the two sites p and q relative to each other and certain properties of the transmitting and receiving antennas;

$$P_{\mathrm{R}} = P_{\mathrm{R0}} g_{\mathrm{T}}^{(p)}(\theta_{pq}, \phi_{pq}) g_{\mathrm{R}}^{(q)}(\theta_{qp}, \phi_{qp}) \tag{2.9}$$

where

$$P_{\mathrm{R0}} = \frac{P_{\mathrm{T}} G_0^{(p)} A_{\mathrm{eR0}}^{(q)}}{4\pi r_{pq}^2} = \text{received power}$$

at q when q is at the peak of p's radiation pattern and p is at the peak of q's receptivity pattern.

We now assume that both antennas are "reciprocal," i.e., that the antennas have the same properties when used for transmission and reception. The implications of that assumption (stated without proof, which requires some electromagnetic theory to be done properly) are (given that the same antenna is used for transmission and reception)

$$g_{\mathrm{T}}^{(p)}(\theta_{pq}, \phi_{pq}) = g_{\mathrm{R}}^{(p)}(\theta_{pq}, \phi_{pq}) = g^{(p)}(\theta_{pq}, \phi_{pq}) \tag{2.10}$$

The implication of (2.10) is that the radiation pattern for transmission and the receptivity pattern when the antenna is in the receiving mode are the same for the same antenna [note that (2.10) cannot be used in (2.9) because the two antennas are different in general] and

$$G_0 = \frac{4\pi A_{\mathrm{e0}}}{\lambda^2} \tag{2.11}$$

where G_0, A_{e0}, and λ are the peak gain, peak effective aperture area, and radar wavelength, respectively, for an arbitrary reciprocal antenna.

Applying (2.11) to (2.9), we can write the beacon equation in more than one form, depending on whether G_0 or A_{e0} are the known parameters:

From (2.9) and (2.11) we can write

$$P_R = P_{R0} g_T^{(p)}(\theta_{pq}, \phi_{pq}) g_R^{(q)}(\theta_{qp}, \phi_{qp}) \tag{2.12}$$

where

$$P_{R0} = \frac{P_T G_{T0}^{(p)} A_{eR0}^{(q)}}{4\pi r_{pq}^2} = \frac{P_T G_{T0}^{(p)} G_{R0}^{(q)} \lambda^2}{(4\pi)^2 r_{pq}^2} = \frac{P_T A_{eT0}^{(p)} A_{eR0}^{(q)}}{\lambda^2 r_{pq}^2}$$

The beacon equation (2.12) is not used directly in radar unless we include beacon transponders in the definition of radar. But its derivation here has served to introduce such concepts as antenna gain and effective aperture area and to establish the groundwork for the derivation of the radar equation.

2.2 THE MONOSTATIC RADAR EQUATION: CONCEPT OF RCS

A monostatic radar is a radar in which the transmitter and receiver are colocated, as shown in Figure 2.2.

The radar "transceiver" (combined transmitter and receiver) is located at p. The target is located at q. The target coordinates with respect to transceiver are (r_{pq}, θ_{pq}, ϕ_{pq}) and the transceiver coordinates with respect to target are (r_{qp}, θ_{qp}, ϕ_{qp}). It is obvious that $r_{pq} = r_{qp}$. The relationship between (θ_{pq}, ϕ_{pq}) and (θ_{qp}, ϕ_{qp}) is determined by the definitions of the coordinates (x_p, y_p, z_p). z_q and z_p are pictured in the diagram as parallel, establishing a simple relationship between these two sets of angles. However, the reference coordinate system for the target at q is not usually parallel to that at p. Hence, we cannot specify in general what the relationship between the two sets of angles is without specifying the reference coordinate system for q with respect to that for p.

To derive the monostatic radar equation, we revert back to (2.5), which gives the power density at the target. We now consider the target as if it were a receiving

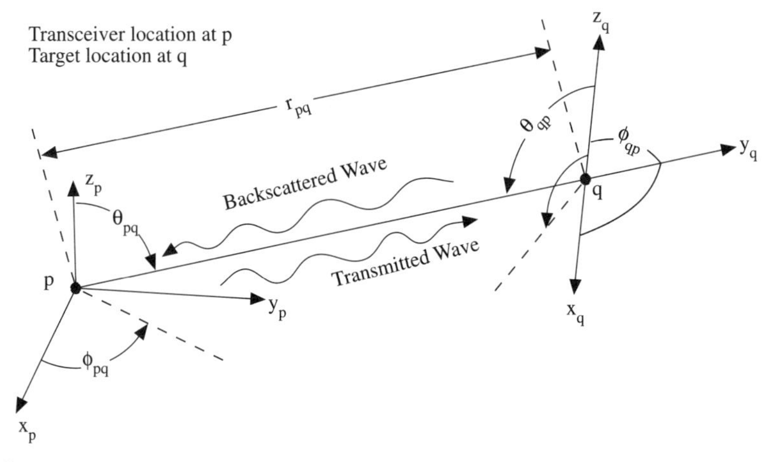

FIGURE 2.2
Monostatic radar geometry.

antenna, having a particular effective aperture area $A_{eT}^{(q)}(\theta_{qp}, \phi_{qp})$. The power "received" by the target is obtained from the beacon equation (2.12), where subscript R denoting "receiver" is replaced by t denoting "target." Thus from (2.12), the power received by the target at q is

$$(P_t)_q = \frac{P_T G_{T0}^{(p)} A_{et0}^{(q)}}{4\pi r_{pq}^2} g_T^{(p)}(\theta_{pq}, \phi_{pq}) g_t^{(q)}(\theta_{qp}, \phi_{qp}) \tag{2.13}$$

where $(P_t)_q$, $A_{et0}^{(q)}$, and $g_t^{(q)}(\theta_{qp}, \phi_{qp})$ are analogous to P_R, $A_{eR0}^{(q)}$, and $g_R^{(q)}(\theta_{qp}, \phi_{qp})$, respectively, where target t replaces receiver R in each case.

Still considering the target as an "antenna equivalent," we now assert that the target reradiates power into the space that surrounds it. The reradiated power is $(P_t)_q$, which now replaces P_T in (2.1), resulting in the power density at the transceiver at p due to reradiation by the target at q, given with the aid of (2.13) by

$$\left(\frac{dP_t}{dA}\right)_{qp} = \frac{(P_t)_q}{4\pi r_{pq}^2} = \frac{P_T G_{T0}^{(p)} A_{et0}^{(q)} G_{t0}^{(q)}}{(4\pi)^2 r_{pq}^4} g_T^{(p)}(\theta_{pq}, \phi_{pq}) g_t^{(q)}(\theta_{qp}, \phi_{qp}) g_t^{(q)}(\theta_{qp}, \phi_{qp}) \tag{2.14}$$

where $G_{t0}^{(q)}$ is the peak gain of the target considered as a transmitting antenna, $A_{et0}^{(q)}$ is the peak effective aperture area of the target considered as a receiving antenna, and $g_t^{(q)}(\theta_{qp}, \phi_{qp})$ is both the radiation pattern of the target for retransmission into the direction of the radar transceiver at p and the receptivity pattern of the target for reception from the direction of the radar at p. Inherent in the assumption that these two patterns are equivalent is the assumption that the target is equivalent to a reciprocal antenna.

To determine the received power at p from the power density as given by (2.14), we again invoke (2.6) as we did in Section 2.1, this time applying it to the power density of the primary radiation from the transmitter. The result is [with the aid of (2.8) modified for the new geometry]

$$P_R = \left(\frac{dP_t}{dA}\right)_{qp} A_{eR0}^{(p)} g_R^{(p)}(\theta_{pq}, \phi_{pq}) \tag{2.15}$$

where the effective aperture area of the receiver is now notated differently than in (2.6) reflecting the change in the receiver's location from q to p and the change in the reference angles from (θ_{pq}, ϕ_{pq}) to (θ_{qp}, ϕ_{qp}) (i.e., the transmitter at p has now become the target at q and the receiver at q has now become the receiver at p).

Applying (2.14) to (2.15), we obtain the final form of the monostatic radar equation, which we will express in three alternative forms with the aid of (2.11),

$$P_R = P_{R0}[g^{(p)}(\theta_{pq}, \phi_{pq})]^2 [g^{(q)}(\theta_{qp}, \phi_{qp})]^2 \tag{2.16}$$

where

$$P_{R0} = \frac{P_T G_0 A_{e0} \sigma_0}{(4\pi)^2 r^4} = \frac{P_T G_0^2 \lambda^2 \sigma_0}{(4\pi)^3 r^4} = \frac{P_T A_{e0}^2 \sigma_0}{4\pi \lambda^2 r^4} \tag{2.16$'$}$$

and where (assuming the same antenna is used for transmission and reception) the following notational changes have occurred and one new definition will be introduced, as follows:

$$A_{e0} = A_{eT0}^{(p)} = A_{eR0}^{(p)} \quad \text{(due to reciprocity of radar antenna)} \tag{2.16a}$$

$$G_0 = G_{T0}^{(p)} = G_{R0}^{(p)} \quad \text{(due to reciprocity of antenna)} \tag{2.16b}$$

$$r = r_{pq} \tag{2.16c}$$

$$g^{(p)}(\theta_{pq}, \phi_{pq}) = g_T^{(p)}(\theta_{pq}, \phi_{pq}) \tag{2.16d}$$

$$\begin{aligned} &= g_R^{(p)}(\theta_{pq}, \phi_{pq}) \quad \text{(due to antenna reciprocity)} \\ g^{(q)}(\theta_{qp}, \phi_{qp}) &= g_t^{(q)}(\theta_{qp}, \phi_{qp}) \quad \text{(due to target reciprocity)} \end{aligned} \tag{2.16e}$$

Note that the location index subscripts p have been deleted from A_{e0} and G_0, but not from $g(\theta, \phi)$. The new definition is

$$\sigma^{(q)}(\theta_{qp}, \phi_{qp}) = \sigma_0[g_t^{(q)}(\theta_{qp}, \phi_{qp})]^2 = \tag{2.17a}$$

backscattering cross section or "radar cross section" (abbreviated as RCS in radar practice) as a function of the incidence and scattering angles (one and the same in the backscattering case). Its peak value is

$$\sigma_0 = G_{t0}^{(q)} A_{et0}^{(q)} \tag{2.17b}$$

i.e., it is defined as the product of the peak effective aperture area for reception of the incident wave and the peak gain for retransmission of the scattered wave. The RCS has the dimensions of area [since $G_{t0}^{(q)}$ is dimensionless and $A_{et0}^{(q)}$ is an area] and can be viewed as proportional to the cross-sectional area of the target's projection normal to the wavefront. A perfect reflector would scatter back the transmitted power that it intercepts. For intuitive purposes we can view the RCS as closely related to its actual cross-sectional area, but that is a gross oversimplification, so one should not rely on that idea as a means of inferring RCS values from actual geometric dimensions.

A generalization of (2.16) to account for the case where different antennas are used for transmission and reception is as follows:

$$P_R = P_{R0} g_T^{(p)}(\theta_{pq}, \phi_{pq}) g_R^{(p)}(\theta_{pq}, \phi_{pq}) [g_t^{(q)}(\theta_{qp}, \phi_{qp})]^2 \tag{2.18}$$

where the subscripts T and R refer to transmitter and receiver, respectively, and where these subscripts are also reintroduced on G_0 and A_{e0}, leading to

$$P_{R0} = \frac{P_T G_{T0} A_{eR0} \sigma_0}{(4\pi)^2 r^4} = \frac{P_T G_{T0} G_{R0} \lambda^2 \sigma_0}{(4\pi)^3 r^4} = \frac{P_T A_{eT0} A_{eR0} \sigma_0}{4\pi \lambda^2 r^4} \tag{2.18$'$}$$

2.3 BISTATIC RADAR EQUATION

The bistatic radar (Fig. 2.3) has its transmitter and receiver at different locations. The transmitter and target are again located at p and q, respectively, but the receiver is now located at p'.

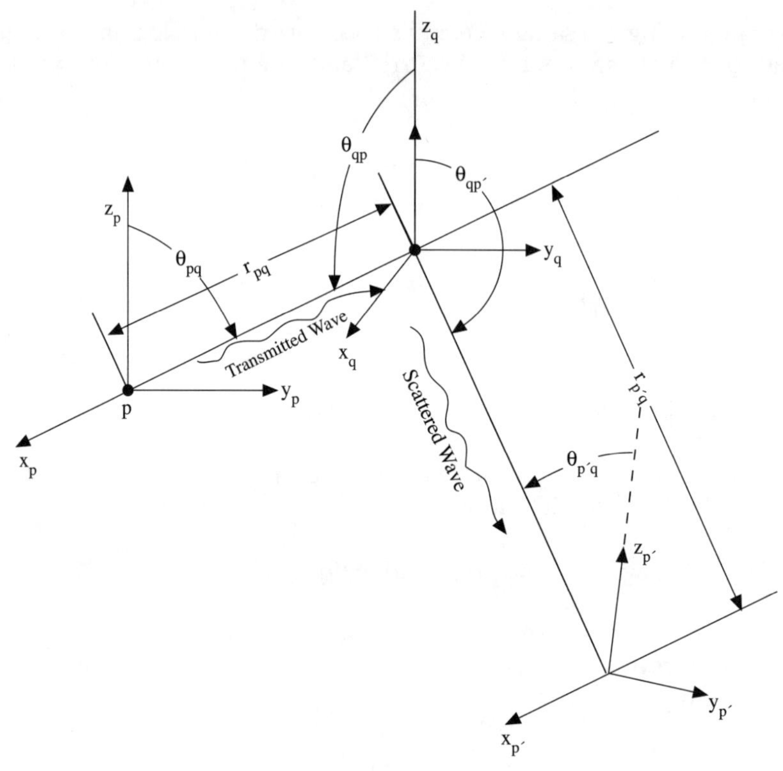

FIGURE 2.3
Bistatic radar geometry.

In Figure 2.3, the azimuth angles ϕ_{pq}, ϕ_{qp}, $\phi_{p'q}$, $\phi_{qp'}$ are not shown in order to keep the "clutter" in the diagram within bounds.

The scattering from the target toward the receiver is no longer backscattering but rather it is "oblique" or "bistatic" scattering, where the scattered wave direction can be arbitrary with respect to the direction of the incident wave, depending on transmitter and receiver location.

The first step in the derivation, determination of the power intercepted by the target and available for reradiation, is the same as that for the monostatic radar and culminates in (2.13). The next step, calculation of the power density at the receiver due to the wave scattered by the target in the direction of the receiver, takes us from (2.13) to a modified form of (2.14) wherein $g_t^{(q)}(\theta_{qp}, \phi_{qp})$ is replaced by $g_t^{(q)}(\theta_{qp'}, \phi_{qp'})$ and r_{pq}^4 is replaced by $r_{pq}^2 r_{p'q}^2$ to account for the fact that the scattering is now from q to p' rather than from q to p. The third and last step in the derivation, multiplication of the power density at the receiver by the effective aperture area of the receiving antenna , is identical to the step that took us from (2.14) to (2.15) except that in (2.15) p is replaced by p' to account for the new receiver location. The final results, then, are generalized versions of (2.18) and (2.18)′ as follows:

$$(P_\mathrm{R})_{p'} = P_{\mathrm{R}0} g_\mathrm{T}^{(p)}(\theta_{pq}, \phi_{pq}) g_\mathrm{R}^{(p')}(\theta_{p'q}, \phi_{p'q}) g_\mathrm{T}^{(q)}(\theta_{qp'}, \phi_{qp'}/\theta_{qp}, \phi_{qp}) \quad (2.19)$$

where

$$P_{\mathrm{R}0} = \frac{P_\mathrm{T} G_{\mathrm{T}0} A_{\mathrm{eR}0} \sigma_0}{(4\pi)^2 r^4} = \frac{P_\mathrm{T} G_{\mathrm{T}0} G_{\mathrm{R}0} \lambda^2 \sigma_0}{(4\pi)^3 r^4} = \frac{P_\mathrm{T} A_{\mathrm{eT}0} A_{\mathrm{eR}0} \sigma_0}{4\pi \lambda^2 r^4} \quad (2.19)'$$

and where

$$r = \sqrt{r_{pq}r_{p'q}} = \text{geometric mean of } r_{pq} \text{ and } r_{p'q} \tag{2.19)''}$$

If (2.19)″ is used, then we note that P_{R0} as given by (2.19)′ for bistatic radar appears identical to P_{R0} as given by (2.18)′ for monostatic radar. The apparent equivalence is due to the interpretation of r as the geometric mean of the distances from target to transmitter and from target to receiver.

The function on the extreme right in (2.19) is the product of the two functions $g_t^{(q)}(\theta_{qp}, \phi_{qp})$ and $g_t^{(q)}(\theta_{qp'}, \phi_{qp'})$. However, in writing that product, it is not indicated notationally that the reradiation pattern $g_t^{(q)}(\theta_{qp'}, \phi_{qp'})$ depends on the direction of the incident wave; hence, it is not correct to treat those two functions as if one were a function of one pair of angles and the other a function of the other pair of angles. Actually, the product itself depends in a possibly transcendental and highly complicated way on both pairs of angles, and these dependences are not separable. Hence we define the function as

$$g_t^{(q)}(\theta_{qp'}, \phi_{qp'}/\theta_{qp}, \phi_{qp}) = \text{scattering pattern} \tag{2.19)'''}$$

of the target at q in the direction of p' in response to a wave incident on the target from the direction (θ_{qp}, ϕ_{qp}).

Another point about (2.19)′ as compared with (2.18)′ is the difference in the meaning of σ_0 in the monostatic and bistatic case. In both cases, σ_0 is taken to mean the peak value of the RCS. However, in the monostatic case, that meaning is generally clear and unequivocal. It is the value of backscattering cross section in the direction in which that parameter is maximized. In the bistatic case, σ_0 also means the maximum scattering cross section (i.e., the cross section at the set of incidence and scattering angles where it is maximized), but that may or may not be the backscattering direction. It depends on the scatterer's orientation with respect to both the sets of angles, and it must be precisely defined to be meaningful.

2.4 CONSOLIDATION: THE RADAR EQUATION

We summarize below the results of the derivations in Sections (2.1), (2.2), and (2.3), as follows:

The general (monostatic or bistatic) radar equation [from (2.19)′]:

$$P_R = P_{R0}g_T^{(p)}(\theta_{pq}, \phi_{pq})g_R^{(p')}(\theta_{p'q}, \phi_{p'q})g_t^{(q)}(\theta_{qp'}, \phi_{qp'}/\theta_{qp}, \phi_{qp}) \tag{2.20}$$

where [from (2.19)′]

$$P_{R0} = \frac{P_T G_{T0} A_{eR0}\sigma_0}{(4\pi)^2 r^4} = \frac{P_T G_{T0} G_{R0}\lambda^2\sigma_0}{(4\pi)^3 r^4} = \frac{P_T A_{eT0} A_{eR0}\sigma_0}{4\pi\lambda^2 r^4} \tag{2.20)'}$$

and where

$$\sigma_0 = \text{peak value of target RCS in both cases}$$

[from (2.19)″]

$r = \sqrt{r_{qp}r_{qp'}}$ in the bistatic case
r_{qp} = in the monostatic case
P_T = total transmitted power
(θ_{pq}, ϕ_{pq}) = direction angles of target at q in the transmitter based coordinate system centered at p
$(\theta_{p'q}, \phi_{p'q})$ = direction angles of target at q in the receiver-based coordinate system centered at p'
(θ_{qp}, ϕ_{qp}) and $(\theta_{qp'}, \phi_{qp'})$ = direction angles of transmitter p and receiver at p' in the target-based coordinate system centered at q
G_{T0} = peak gain of transmitting antenna
G_{R0} = peak gain of receiving antenna
A_{eT0} = peak effective aperture area of transmitting antenna
A_{eR0} = peak effective aperture area of receiving antenna
$g_T^{(p)}(\theta_{pq}, \phi_{pq})$ = normalized radiation pattern (relative gain) of transmitting antenna at p in direction of target at q
$g_R^{(p')}(\theta_{p'q}, \phi_{p'q})$ = normalized radiation pattern (relative gain) of receiving antenna at p' in direction of target at q
$g_t^{(q)}(\theta_{qp'}, \phi_{qp'}/\theta_{qp}, \phi_{qp})$ = scattering pattern of target q in direction of receiver at p' in response to a wave incident from direction of p

In the monostatic case, we set $p' = p$ and then simplify the equations (2.20), but not much simplification occurs unless transmitting and receiving antennas are identical, in which case the simpler form of (2.20) is [see (2.16)]

$$P_R = P_{R0}[g^{(p)}(\theta_{pq}, \phi_{pq})]^2[g^{(q)}(\theta_{qp}, \phi_{qp})]^2 \tag{2.21}$$

where

$$P_{R0} = \frac{P_T G_0 A_{e0}\sigma_0}{(4\pi)^2 r^4} = \frac{P_T G_0^2 \lambda^2 \sigma_0}{(4\pi)^3 r^4} = \frac{P_T A_{e0}^2 \sigma_0}{4\pi^2 r^4} \tag{2.21$'$}$$

and where subscripts T and R can be omitted because properties of the antenna are the same for transmission and reception.

2.5 APPLICATIONS OF THE RADAR EQUATION

In general, the radar equation in a form based indirectly on one of the forms (2.20)′ or (2.21)′ is used to determine certain system parameters given other system parameters. A most important application is to find the maximum range of a radar for a given value of RCS. Noting the inverse fourth power dependence of the received power on range, we can determine the maximum range r_{max} by solving (2.20)′ or (2.21)′ for r_{max} for monostatic only as follows [where in (2.20)′ we should be reminded that "range" means "geometric mean of r_{pq} and $r_{p'q}$].

From (2.20)′

$$r_{max} = \sqrt[4]{\frac{P_T G_{T0} A_{eR0} \sigma_0}{(4\pi)^2 (P_{R0})_{min}}} = \sqrt[4]{\frac{P_T G_{T0} G_{R0} \lambda^2 \sigma_0}{(4\pi)^3 (P_{R0})_{min}}} = \sqrt[4]{\frac{P_T A_{eT0} A_{eR0} \sigma_0}{4\pi\lambda^2 (P_{R0})_{min}}} \tag{2.22}$$

or in the simpler case (2.21)′

$$r_{max} = \sqrt[4]{\frac{P_T G_0 A_{e0} \sigma_0}{(4\pi)^2 (P_{R0})_{min}}} = \sqrt[4]{\frac{P_T G_0^2 \lambda^2 \sigma_0}{(4\pi)^3 (P_{R0})_{min}}} = \sqrt[4]{\frac{P_T A_{e0}^2 \sigma_0}{4\pi\lambda^2 (P_{R0})_{min}}} \tag{2.23}$$

where in either case $(P_{R0})_{min}$ is usually referred to as the "minimum detectable signal power," and is generally proportional to the receiver noise level. However, that definition implies that pure detection of the target is the goal of the radar. I prefer that $P_{R0_{min}}$ be designated as the "minimum effective signal power," implying that it is the smallest value of received power that renders the radar effective in whatever function it is trying to perform, e.g., measurement of a particular target parameter or target tracking.

Another application of the radar equation is: given the range, to find the minimum required RCS $(\sigma_0)_{min}$ for detection or whatever other function is to be performed (e.g., location or tracking). To illustrate this, we will use the general form (2.20)′ or more specifically the first equation on the left in (2.20)′ or (2.21)′, resulting in

$$(\sigma_0)_{min} = \frac{(4\pi)^2 (P_{R0})_{min} r^4}{P_T G_{T0} A_{eR0}} \tag{2.24}$$

where $(P_{R0})_{min}$ has the same meaning as in (2.22) and (2.23).

We note that the two applications illustrated use only the primed forms of the equations (2.20)′ and (2.21)′, i.e., the radiation and scattering patterns are neglected. That is usually valid for finding r_{max} because the primed forms correspond to the best possible performance of the radar, which occurs when the target is at the peak of the radiation and scattering patterns.

If the target is "off-boresight" (the word "boresight" is usually used to mean the peak of the antenna beam), and if the target's scattering pattern is not at its peak in the direction of the receiver, then the unprimed equations (2.20) or (2.21) have to be used in (2.22), (2.23), or (2.24), i.e., they must include the radiation and scattering patterns. The key to this issue lies in the interpretation of the meanings of the quantities calculated with the radar equation. If r_{max} is interpretated to mean the largest attainable range, then (2.22) or (2.23) is adequate. If we ask for "the largest attainable range given that the target may be 3 or more degrees off boresight," then the calculation of r_{max} through (2.23) must account for the loss at an angle θ_{pq} due to the pattern function $[g^{(p)}(\theta_{pq}, \phi_{pq}]^2$ in (2.21). We would then have to include a factor

$$\sqrt[4]{|g^{(p)}(\theta_{pq}, \phi_{pq})|^2} = \sqrt{|g^{(p)}(\theta_{pq}, \phi_{pq})|}$$

on the right-hand side (RHS) of (2.23), where $\theta_{pq} = 3°$ (assuming the pattern were independent of ϕ_{pq}), which would reduce the maximum attainable range.

There are an infinite number of possibilities of this sort and discussion of these issues should be done in the context of specific examples. Before launching that discussion, it should be remarked that the radar equation can be written conveniently in decibels to conform to standard radar engineering practice. For example, the second form in (2.21)′ used in (2.21) could be written as

$$\begin{aligned}(P_R)_{dBw} = (P_T)_{dBw} &+ 2(G_0)_{dB} + 20\log_{10}\lambda + (\sigma_0)_{dBm^2} \\ &- 32.98 - 40\log_{10} r + 2|g^{(p)}(\theta_{pq}, \phi_{pq})|_{dB} \\ &+ 2|g^{(q)}(\theta_{qp}, \phi_{qp})|_{dB}\end{aligned} \tag{2.25}$$

where

$(P_{R0})_{dBw}$ = received power in dB over 1 W
$(P_T)_{dBw}$ = transmitted power in dB over 1 W
$(\sigma_0)_{dBm^2}$ = RCS in dB over 1 m^2
$|g \cdots|_{dB}$ = *one-way* power pattern in dB [must be doubled for two-way (radar) propagation]
$(G_0)_{dB}$ = peak *one-way* power gain in dB

It is particularly convenient at times to plot $(P_R)_{dB}$ vs. $\log_{10} r$ using (2.25) or one of its possible variations. This gives a straight line with negative slope of −40, as illustrated in Figure 2.4 for the case where P_T = 1 kW, λ = 3 cm [corresponding to frequency f = 10 GHz (X-band), G_0 = 100 (one way power gain = 20 dB], the RCS is 2 m^2 and both patterns are zero dB (i.e., target and antenna boresight and radar at peak of scattering pattern). Equation (2.25) then becomes

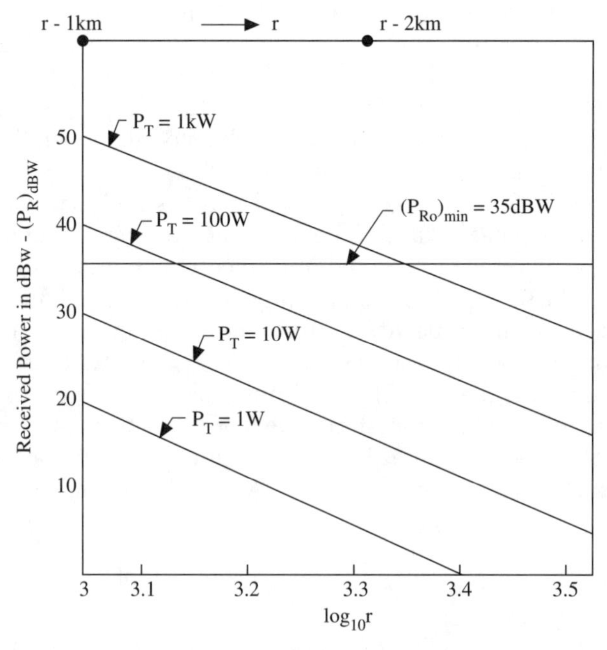

FIGURE 2.4
Illustration of use of radar equation in dB.

$$
\begin{aligned}
(P_R)_{dBw} &= 30 + 2(20) + 10.45 + 3 - 32.98 - 40 \log_{10} r \\
&= 50.47 - 40 \log_{10} r \\
&\approx 50 - 40 \log_{10} r
\end{aligned} \tag{2.26}
$$

The four linear plots shown in the figure serve to illustrate the utility of the radar equation when the received power is plotted in dB in showing how parameter variations can be studied graphically (such plots are usually on log scales, but Figure 2.4, which is on a linear scale, serves to illustrate the point). The highest plot shows the linear variation of received power in dBw with $\log_{10} r$ for the parameter values given above. The lower plots show 10 dB reductions in transmitted power, other parameter values remaining the same. Suppose the minimum effective received power is 35 dBw, as shown by the horizontal line at 35 dBw. By examining the intersections of the linear curves with that horizontal line, we can easily infer that the maximum range r_{max}, as given by (2.23), is about $10^{3.13} = 1.349$ km if transmitted power is 100 W and about $10^{3.39} = 2.455$ km if the transmitted power is 1 kw. If there were more curves and calibrations, r_{max} could be obtained very accurately for a much wider variation of values of transmitted power.

2.6 SOLID ANGLE OF BEAM COVERAGE

Another point will now be made about the relationship between peak gain of the antenna and the antenna beam pattern. Some people think of these two entities as if they were independent. The fact is that they are interdependent, in the sense that the peak gain [and through (2.11), the peak effective aperture area] is a "functional" of the pattern function. To see this, we invoke (2.5), which comes from the definition of gain (2.2) and the expression for power density (2.1). Dropping subscripts and superscripts p, given in (2.5), we can write it in the form

$$
G_0 g_T(\theta, \phi) = \frac{4\pi r^2 (dP_T/dA)(r, \theta, \phi)}{P_T} \tag{2.27a}
$$

We now note that the total transmitted power P_T can be obtained by integrating the power density over the entire sphere of radius r centered at the antenna. We first note that

$$
\frac{dP_T}{dA}(r, \theta, \phi) = \left[\frac{dP_T}{dA}\right]_0 g_T(\theta, \phi) \tag{2.27b}
$$

where $[dP_T/dA]_0$ is the peak value of the power density. Using (2.27b) in (2.27a) and cancelling out $g_T(\theta, \phi)$ on both sides of the resulting equation, (2.27a) takes the form

$$
G_0 = \frac{4\pi r^2 [dP_T/dA]_0}{r^2 \int_0^{\pi} d\theta \sin\theta \int_0^{2\pi} d\phi [dP_T/dA]_0 g_T(\theta, \phi)} = \frac{4\pi}{\int_0^{\pi} d\theta \sin\theta \int_0^{2\pi} d\phi\, g_T(\theta, \phi)} \tag{2.28}
$$

The denominator of the RHS of (2.28) is the integral of the normalized distribution of radiated power density over the unit sphere surrounding the antenna.

Equation (2.28) provides us with an alternative definition of peak antenna gain, as follows:

$$G_0 = \frac{4\pi}{\Omega_b} \tag{2.29}$$

where $\Omega_b = \int_0^{\pi} d\theta \sin\theta \int_0^{2\pi} d\phi\, g_T(\theta, \phi)$ is known as the "solid angle of beam coverage." The justification of this terminology can be explained through the artificial case of a pattern that has uniform coverage over a range of angles (θ, ϕ) and zero coverage elsewhere, such as the axially symmetric beam shown in Figure 2.5, which has the pattern function.

$$g_T(\theta, \phi) = \begin{cases} 1 & \text{if } \theta \le \theta_b \\ 0 & \text{if } \theta > \theta_b \end{cases} \tag{2.30}$$

As indicated in Figure 2.5, integration of the denominator of (2.28) yields

$$G_0 = \frac{2}{1 - \cos\theta_b} \tag{2.31}$$

If $\theta_b = \pi$, we now have uniform coverage over the entire sphere, i.e., the isotropic antenna, in which case $\cos\theta_b = \cos\pi = -1$ and $G_0 = 1$. That is the widest coverage and smallest peak gain. If the coverage is only over a hemisphere, then $\cos\theta_b = \cos\pi/2 = 0$, in which case the gain = 2, which is twice as high as the isotropic case, reflecting the fact that the solid angle covered, 2π steradians, is half that of the isotropic case. Finally, the case where $\theta_b = 60°$ gives us a solid angle of coverage equal to $2\pi(1 - \cos\pi/3) = \pi$ steradians and a peak gain of 4.

This example aptly illustrates the inverse tradeoff between angular coverage and peak gain although this kind of pattern could never exist in the real world.

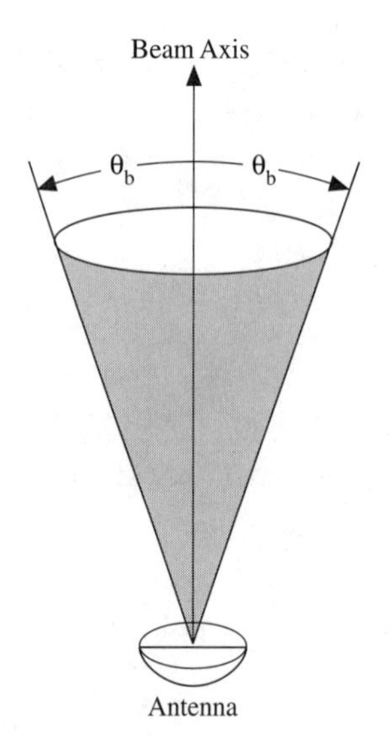

Shaded region is region of coverage

$g_T(\theta, \phi) = 1;\ \theta \le \theta_b$
$\qquad\qquad\ \ 0;\ \theta > \theta_b$

$$\Omega_b = \int_0^{\pi} d\theta \sin\theta \int_0^{2\pi} d\phi\, g_T(\theta, \phi)$$
$$= 2\pi \int_0^{\theta_b} d\theta \sin\theta$$
$$= 2\pi[1 - \cos\theta_b]$$
$$G_0 = \frac{4\pi}{\Omega_b} = \frac{2}{1 - \cos\theta_b}$$
$$\Omega_b = 4\pi \sin^2\left(\frac{\theta_b}{2}\right)$$

FIGURE 2.5
Solid angle of coverage.

Equation (2.28) or its equivalent (2.29) demonstrates that specification of the beam pattern automatically fixes that peak gain, but the reverse is not true; specification of the peak gain does not in any way restrict the beam pattern shape. Any pattern shape that fixes the value of the integral in (2.28) will suffice and an infinite number of such patterns exist.

In practice, one often sees what appears to be independent specification of G_0 and $g_T(\theta, \phi)$. The justification for that is that only a small part of the beam pattern, such as the portion for small value of θ, may be important in the system application. An approximation that is quite accurate for values of θ below 3° might be perfectly satisfactory for evaluation of system performance. But that approximation may exclude a great deal of sidelobe energy at large angles, which has a profound effect on the integral in the denominator of (2.28).

Calculating the gain through (2.28) with the approximate pattern might give highly erroneous results while the integration may be interactable with the true pattern, or the true pattern may not be known for larger angles.

In that case, the gain is often specified from experimental results or estimated from general experience.

2.7 FREQUENCY DEPENDENCE OF RECEIVED RADAR SIGNAL POWER

The radar equations as we have summarized them in Section 2.4 are expressed in terms of wavelength λ. Through (1.1) we can translate this into a dependence on frequency. Referring back to the three forms of P_{R0} in (2.20)′, it would appear superficially through (1.1) that we could express P_{R0} as (1) independent of frequency through the form on the left, (2) as dependent on the inverse-square of frequency through the form in the center, or (3) as proportional to the square of frequency through the form on the right. None of these frequency dependences is exactly correct, but the one closest to the truth is the third.

The reason for that is the fact that the third form is given in terms of the peak effective aperture areas rather than the peak gains. Since the effective aperture area is closely related to the true geometric area, it is more nearly independent of frequency than the gain. The latter, in other words, varies roughly as the square of the frequency through (1.1) and (2.11), where A_{e0} can be thought of as roughly constant. The truth is that both G_0 and A_{e0} have some degree of frequency dependence, which varies with the specific type of antenna, but approximating A_{e0} as frequency independent is a much more accurate assumption than approximating G_0 as frequency independent. The RCS σ_0 is also frequency dependent in general. By the reasoning we used in deriving the monostatic radar equation in Section 2.2, the peak value of RCS can be viewed as the product of G_{T0} and A_{eT0}, where these parameters refer to the target t as a reradiating antenna. Through (2.11) and (2.12) we can write, with the aid of (1.1)

$$\sigma_0 = G_{t0} A_{et0} = \frac{4\pi A_{et0}^2}{\lambda^2} = \left(\frac{4\pi}{c^2} A_{et0}^2\right) f^2 \tag{2.32}$$

A form of the radar equation for P_{R0} that accentuates the frequency dependence could be obtained from the third form of (2.20)′, (2.32), and (1.1), as follows:

$$P_{\mathrm{R0}} = \left(\frac{P_{\mathrm{T}}A_{\mathrm{eT0}}A_{\mathrm{eR0}}A_{\mathrm{et0}}^2}{c^4r^4}\right)f^4 \tag{2.33}$$

From (2.33) we conclude that, if the transmitted power and the effective aperture areas of the transmitting and receiving antennas and the target viewed as an antenna were truly frequency independent, then the received power at boresight and the angles of maximum RCS would vary as the fourth power of frequency. If we were able to fix the values of those parameters while varying the frequency, it would follow (see Table 1.2) that the return power would increase by the amounts shown in the informal tabulation below as we traveled in frequency space through the radar bands.

From the low end of VHF (30 MHz) to high end of VHF (300 MHz)-40 dB increase
From the low end of UHF (300 MHz) to high end of UHF (1 GHz) another 21 dB
From the low end of L-band (1 GHz) to high end of L-band (2 GHz) another 12 dB
From the low end of S-band (2 GHz) to high end of S-band (4 GHz) another 12 dB
From the low end of C-band (4 GHz) to high end of C-band (8 GHz) another 12 dB
From the low end of X-band (8 GHz) to high end of X-band (12 GHz) another 7 dB
From the low end of Ku-band (12 GHz) to high end of Ku-band (18 GHz) another 7 dB
From the low end of K-band (18 GHz) to high end of K-band (27 GHz) another 7 dB
From the low end of Ka-band (27 GHz) to high end of Ka-band (40 GHz) another 6.8 dB
From the low end of V-band (40 GHz) to high end of V-band (75 GHz) another 10.9 dB
From the low end of W-band (75 GHz) to high end of W-band (110 GHz) another 6.65 dB
From the low end of mm-band (110 GHz) to high end of mm-band (300 GHz) another 17 dB

Although these increases are not realistic because the effective aperture areas and power are not completely frequency independent, the above tabulation shows that even if the realistic frequency variation is closer to f^3 or f^2 than f^4, it still provides a powerful argument for using the highest frequencies available for radars, "other things being equal." However, "other things" are *not* really equal, and there are many reasons for choosing the lower frequency bands. These will be discussed later in the book when we have more knowledge of the departures from the idealized situation described by the simple radar equations we have covered.

2.8 RECEIVER NOISE: SNR

The internal noise power in a radio receiver is given by

$$P_{\mathrm{N}} = kT_0B_{\mathrm{R}}F_{\mathrm{N}} \quad W \tag{2.34}$$

where

k = Boltzmann's constant = 1.38×10^{-23} J/deg K
T_0 = absolute temperature—degrees K
B_R = Bandwidth of the receiver in Hertz
F_N = Noise figure in absolute units, equal to $(10^{F_{NdB}/10})$, where F_{NdB} is the noise figure in dB, the units in which it is usually given

The standard value of T_0 usually used in radar system analysis that does not focus on specific weather condition is 290°K but 300°K is often used because it is a "round number."

The noise figure F_N is defined as the *ratio of the total noise power at the output of the IF section of the receiver* (the last stage of the receiver that is approximately linear) *to the noise power that would exist at that point if no new noise were created within the RF and IF stages.* The denominator of that ratio is the noise at the receiver input multiplied by the gain of the RF and IF stages. All of that will be discussed in more detail when we get to the topic of radar receivers in Chapter 3.

It suffices to say at this point that the noise figure is a measure of the amount of noise created within the receiver. If the noise at the receiver input were merely amplified linearly in the RF and IF stages of the receiver, then F_N would be unity or equivalent to zero dB.

A typical value of receiver noise, with T_0 = 290°K, F_{NdB} = 3 dB, and B_R = 1 GHz [bandwidth = 10% of a central frequency of 10 GHz (X-band)], is $8(10^{-12})$ = 8 pW. We can usually think of receiver noise as being somewhere in the micromicrowatt or "picowatt" region of magnitudes.

Once we have the receiver noise power P_N, we can define the received signal-to-noise ratio, usually abbreviated as SNR. From any one of the forms (2.20) and (2.20)′ or the simpler forms (2.21) and (2.21)′, we can write the SNR, for example, as

$$\text{SNR} = \rho_{01} = \frac{P_R}{P_N} = \frac{P_T G_0 A_{e0} \sigma_0}{(4\pi)^2 r^4 k T_0 B_R F_N} \tag{2.35}$$

or any of the more general forms embedded in (2.20).

If we are using pulses, then there are situations in which we are interested in the received signal *energy* rather than *power*, and the ratio of that quantity to the quantity

$$N_0 = \frac{P_N}{B_R} = kT_0 F_N = \text{"noise power density"} \tag{2.36}$$

which is obtained from (2.34) by dividing the noise power by the receiver bandwidth. Receiver noise is spectrally distributed approximately uniformly across the passband of the receiver; hence (2.36) is the noise power per unit of frequency in any part of the receiver passband.

For a train of N_p rectangular pulses each of width τ_p, each with peak power P_T, the *energy* in the received signal is the sum of products of peak power and duration of each pulse, given by (2.21)′ as

$$E_R = N_p P_R \tau_p = \frac{P_T G_0 A_{e0} \sigma_0 [N_p \tau_p]}{(4\pi)^2 r^4} \tag{2.37}$$

The ratio of E_R to noise power density N_0 is a dimensionless ratio obtained from (2.36) and (2.37) as

$$\rho_{0N} = \frac{E_R}{N_0} = \frac{P_T G_0 A_{eo} \sigma_0 N_p \tau_p}{(4\pi)^2 r^4 k T_0 F_N} \tag{2.38}$$

comparing (2.38) with (2.35), we have

$$\frac{\rho_{0N}}{\rho_{01}} = N_p(B_R \tau_p) \tag{2.39}$$

The ratio ρ_{0N} of (2.38), as we will learn in Chapter 4, is the SNR at the output of a matched filter. With ideal matching, it will be found that the pulse width τ_p is the reciprocal of the receiver bandwidth, so that $B_R\tau_p = 1$. Then the improvement in SNR due to matched filtering will be found to be N_p, the number of pulses. That will be explained in detail in Section 4.

The parameter P_T has been understood to mean the *peak* transmitted power, hence P_R is the *peak* received power. If we wish to deal with *time-averaged* power in a pulse train, we note that it is given by (see Fig. 2.6)

$$(P_T)_{av} = P_T \frac{\tau_p}{T_r} = P_T \tau_p f_r \tag{2.40}$$

where T_r is the pulse repetition period and its reciprocal f_r is the pulse repetition frequency, usually abbreviated as PRF. The ratio (τ_p/T_r), equivalent to the product $(f_r\tau_p)$ is the "pulse duty cycle," defined as the fraction of the time during which pulse power is on.

It is obvious that if the transmitted power in the radar equation is the *average* power $(P_T)_{av}$, then the received power on the LHS of the equation is the *average* received power $(P_R)_{av}$. We must be careful to specify whether we are dealing with peak or average power in using radar equations for pulsed systems.

2.9 TRANSMITTER AND RECEIVER LOSSES

The transmitter losses have already been accounted for by using the *gain* G_0 rather than the *directivity* in the radar equation (see Section 2.1).

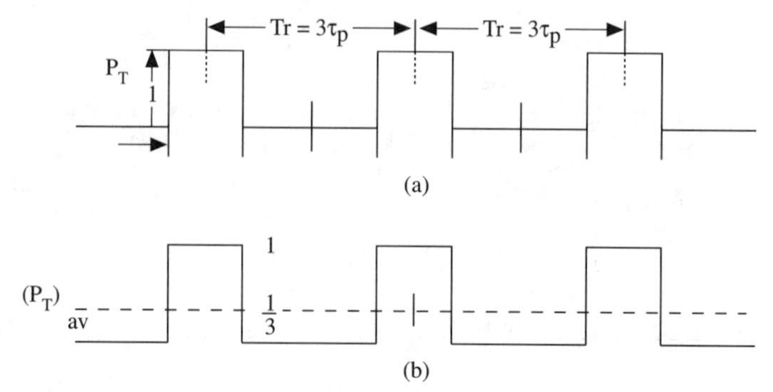

FIGURE 2.6
Peak and average power.

The receiver losses are usually lumped in with the noise figure F_N in defining SNR as in (2.34). To be more precise, we should include a factor $L_R = 10^{L_{rdB}/10}$ in the denominator of the radar equation, where L_{rdB} is the power loss between the receiver input and the output point. The noise figure F_N and the receiver loss L_R are not the same, but from an operational viewpoint we cannot separate them, because they both have the same effect, a degradation in SNR at the IF output point. Hence, it is satisfactory, for example, if receiver loss is 3 dB and noise figure is 3 dB, to characterize that situation as equivalent to a 6 dB effective noise figure.

2.10 A SPECIAL FORM OF THE RADAR EQUATION: THE POWER APERTURE FORMULA

The radar equation is derived and presented in widely different ways in different standard radar texts. It also has some specialized forms in which it is customarily used for certain applications. For example, in search radar, the antenna beam is usually scanned in azimuth and/or elevation angle, and as the beam scans past the target, it is only "on target" for a finite period of time, while the mainbeam is sweeping past it. That period is known as the "dwell-time." During the remaining portion of the scan, while the target is in the sidelobes, so little energy illuminates the target that for practical purposes there is negligible illumination and the target return descends below the noise level.

The total solid angle that must be searched is denoted by Ω_s. The azimuth and elevation beamwidths are $\Delta\theta_B$ and $\Delta\phi_B$, respectively. We conceptualize the process for analysis purposes as one where the beam is flat between elevation angles $\theta_0 - (\Delta\theta_B/2)$ and $\theta_0 + (\Delta\theta_B/2)$ and between azimuth angles $\phi_0 - (\Delta\phi_B/2)$ and $\phi_0 + (\Delta\phi_B/2)$ and zero outside those regions (not true in the real world, of course, but this kind of abstraction simplifies the concept of dwell-time. If the pattern is modeled realistically, sidelobe energy is present at all angles except those in the nulls and dwell-time cannot be defined precisely).

It is intuitively evident that the dwell-time will increase as the scan rate decreases and as the beamwidth in the plane of the angle being scanned increases, because a wide beam and a slow scan will keep the beam "on target" for a longer period of time. It is also evident that a larger average transmitted power will keep the target return above the noise level for a longer period for fixed values of scan rate and beamwidth and hence will effectively increase the dwell-time.

A specialized form of the radar equation often used for search radar highlights the "power-aperture" product. In this form, the radar equation is sometimes called the "power-aperture formula."[9,10]

To derive this form of the radar equation, we invoke (2.38), which gives us the SNR for a monostatic pulsed radar that uses the same antenna for transmission and reception and uses a matched filter to maximize SNR. (See Chapter 4, Section 4.1.)

The total solid angle covered by the beam at any one instant of time during the scan is denoted by Ω_b and approximated as $(\Delta\theta_B\Delta\phi_B)$. The total time available to search the entire solid angle Ω_s is t_s. The actual time "on-target," i.e., the dwell-time, is t_d. The fraction of the total solid angle covered by the beam instantaneously is (Ω_b/Ω_s). The fraction of the total available time that the radar is on-target is t_d/t_s. These two fractions are the same; hence

$$t_d = t_s \frac{\Omega_b}{\Omega_s} \tag{2.41}$$

Equation (2.38), applicable to a pulsed radar that receives N_P pulses during the dwell, is

$$\rho_0 = \frac{P_T G_0 A_{e0}\, \sigma_0 N_P \tau_P}{(4\pi)^2 r^4 k T_0 F_N} \tag{2.42}$$

From (2.29)

$$G_0 = \frac{4\pi}{\Omega_b} \tag{2.43a}$$

The quantity $N_P\tau_P$ in (2.42) can be expressed as

$$N_P\tau_P = fr\tau_P t_d = \left(\frac{\tau_P}{Tr}\right) t_d \tag{2.43b}$$

where fr is the PRF and Tr its reciprocal, the pulse repetition period.

From (2.40) and (2.43b), it follows that

$$P_T N_P \tau_P = P_T\left(\frac{\tau_P}{Tr}\right) t_d = (P_T)_{av} t_d \tag{2.43c}$$

Substitution of (2.41), (2.43a), and (2.43c) into (2.42) yields

$$\rho_{ON} = \frac{[(P_T)_{av} A_{e0}]\sigma_0}{4\pi k T_0 F_N r^4 R_s} \tag{2.44}$$

where $R_s = (\Omega_s/t_s)$ = rate of scanning over the solid angle Ω_s in steradians/second.

Equation (2.44) is a particular form of the "power-aperture formula," where the square-bracketed quantity is the power aperture. Its message is that the output SNR, the quantity that determines the probability of detection in the search process, depends only on the power aperture, the effective noise temperature (the product $F_N T_0$, which includes receiver losses in this form of the radar equation), the RCS, and the target range. Radar frequency is not directly involved except insofar as the RCS, $(P_T)_{av}$, A_{e0}, and F_N may be frequency dependent.

The form (2.44) has other implications for search radar that will not be discussed here.

2.11 ACCOUNTING FOR SYSTEM LOSSES IN THE RADAR EQUATION

There are a number of losses occurring in radar systems that are usually lumped into the radar equation. This was briefly mentioned in Section 2.10 in connection with transmitter and receiver losses. In the present section, some additional detail on system losses will be presented.

The forms of the radar equation given in (2.20) and (2.20)′ can be modified to include a "loss factor" denoted by L. The resulting form could be, for example

$$P_R = P_{R0} F_a L \tag{2.45a}$$

or

$$P_R = \frac{P_{R0} F_a}{L} \tag{2.45b}$$

where P_{R0} is given in (2.20)′ and F_a is the product $g_T^{(p)} g_R^{(p')} g_t^{(q)}$ contained in (2.20). The factor L as given in (2.45a) is a positive number ≤ 1 and as given in (2.45b) is a positive number ≥ 1. Both alternative forms can be found in the literature. If (2.45a,b) is expressed in dB, then the resulting form for either (2.45a) or (2.45b) is

$$(P_R)_{dB} = (P_{R0})_{dB} + (F_a)_{dB} - |L_{db}| \tag{2.45c}$$

where $|L_{dB}|$ is the total system loss in dB, which is always positive.

When the equation is written in absolute power units as in (2.45a,b), then the loss is expressed as a product of the various system losses, i.e., if there are M separable loss mechanisms

$$L = L_1 L_2 \cdots L_M \tag{2.45d}$$

If (2.45c) is used, then, of course, it is a sum of the form

$$|L_{dB}| = |L_{dB1}| + |L_{dB2} + \cdots + |L_{dBM}| \tag{2.45e}$$

Some of the important loss mechanisms are delineated by Skolnick (Section 2.12 on pp. 56–61).[1] The losses he lists are briefly discussed below:

1. Attenuation losses in transmission lines between transmitter output and antenna input terminals and reflection losses due to impedance mismatch at the junctions. From Figure 2.28 on page 57 of Skolnick,[1] these losses are usually low at frequencies below X-band (say a small fraction of a dB for a transmission line a few meters long), but in general increase with frequency and might be significant at the higher end of the radar spectrum (e.g., Ku or Ka band) for a transmission line of the same length.

These losses are often called "plumbing losses" and, by reciprocity, they are incurred both in the transmitter (source to antenna input path) and the receiver (antenna output to RF stage of the receiver).

If these losses in the transmitter are combined with those incurred in the source circuitry and in the transmitting antenna itself, they can be lumped together as "transmitter losses," meaning the ratio of the power radiated into space to the power generated. In the receiver, again the plumbing losses can be combined with losses in the receiving antenna and in the path from the input to the RF stage to the output of the IF stage where the signal processing operations are performed. The combination of all these losses can be lumped together as "receiver losses," meaning the ratio of the power at the input to the signal processing stage relative to that entering the receiving antenna from the surrounding space.

In Section 2.9, it was pointed out that the transmitter and receiver losses of this kind were already accounted for by using antenna "gain" rather than "directivity" in the radar equation. The meaning of that statement can be clarified by reading the discussion immediately following Eq. (2.2). The relationship between power generated P_{gen} and power radiated P_{rad} is

$$P_{rad} = P_{gen}L_T \tag{2.46a}$$

where $L_T = 10^{-(L_{TdB}/10)}$ is the "radiation efficiency," i.e., the loss between the power generated by the source and the power radiated into space, and L_{TdB} is that loss in dB.

Using (2.46a), the proper definitions of gain and directivity are

$$D_0 = \frac{4\pi r^2[dP_{rad}/dA]}{P_{rad}} \tag{2.46b}$$

$$G_0 = \frac{4\pi r^2[dP_{rad}/dA]}{P_{gen}} = D_0 L_T \leq D_0 \tag{2.46c}$$

From (2.46b,c), we see that the gain G_0 as used in the radar equation is really directivity D_0 if we do not include the loss factor L_T, which implies that the quantity P_T is being interpreted as the radiated power. But if P_T means the generated power, then the radar equation should contain the factor L_T. In fact, by reciprocity, if the same antenna is used for transmission and reception, then the factor should be L_T^2. If not, then the loss between the signal at the receiving antenna terminals and that at the input to the processing stage of the receiver can still be interpreted through reciprocity as the same as L_T. It can be renamed L_R for the receiving mode, in which case the correct factor on the radar equation is $L_T L_R$.

2. Losses due to the target's position in the antenna beam pattern, called beam-shape loss by Skolnick (p. 58).[1] These are already accounted for in our text by deriving the radar equation in its general form (2.20), which includes the angular functions due to the target's position in the beam pattern.

3. Losses due to limiting in the receiver, which are usually only a small fraction of a dB (see Skolnick, p. 59).[1]

4. Collapsing loss, a degradation due to the integration of noise pulses not containing target signal energy. This is covered in our text in Chapter 5.

5. Losses due to nonideal equipment, operator deficiencies, and degradations under field conditions relative to performance attainable under ideal laboratory conditions. Skolnick (pp. 60 and 61)[1] lists and briefly discusses these losses, but states that they are very difficult to quantify and very dependent on the specifics of the equipment, the operating personnel, and the field conditions. These losses are best treated in the radar equation by lumping them all in a reduction in received power of X dB, where X depends on the judgment and experience of the engineer.

6. Losses due to the propagation factor, e.g., cases where reflections from terrain interfere destructively with target reflections, or cases where atmospheric propagation losses, lens effects, or attenuation due to precipitation degrades the received signal. Those losses are discussed in Chapter 11. In the radar equation, they are accounted for by inserting a "propagation factor" on the received power, which is a function of the radar-target-earth geometry and acts to reduce the received power relative to its free-space value for certain geometries. (See Chapter 11, Sections 11.2 and 11.8.)

7. Straddling loss, a degradation due to the fact that the target is not in the center of a range gate or a Doppler filter, particularly when it might straddle two contiguous gates or filters. This loss is accounted for in Chapter 4 (see especially Sections 4.4, 4.5, 4.6, and 4.7), where SNR degradation factors due to mismatches in path delay, phase, and frequency in matched filters are defined and then related to

the concept of the ambiguity function. Such degradation factors can be incorporated as multiplicative quantities on the RHS of the radar equation and thereby account for straddling loss when banks of range gates or Doppler filters are used.

8. The losses due to "blind speeds" in nulling out or at least significantly reducing received power in MTI or pulse doppler radars. This is covered in Chapter 9.

9. The losses due to "blind ranges" in high PRF pulsed Doppler radars, also covered in Chapter 9.

Delineations of losses also appear in other well-known texts on radar, which cover those listed by Skolnick and still others. For example, Hovanessian (pp. 65–69)[3] lists the following additional losses not explicitly indicated (but sometimes implied), in Skolnick's listing.

10. Crossover loss, due to the target boresight angle in an angle-tracking radar being offset from the peak of the beam. A thorough quantitative discussion of that topic appears in our text in Chapter 12, Section 12.5.

11. Integration loss, due to noncoherent integration of a train of pulses. This is thoroughly covered in Chapter 5 of our text, and is explicitly discussed in Section 5.7.

12. Target fluctuation loss, which is the reduction in SNR due to the temporal fluctuation in target RCS, is covered in our text in Chapters 5 and 6. It can be accounted for in the radar equation by modeling the RCS as a random variable, using its peak value in the basic radar equation and then inserting a loss factor equal to the deepest fade that might be expected, according to some predetermined criterion, e.g., assuming that the RCS is above that level 98% of the time. This can be done quantitatively through a knowledge of the amplitude statistics (see Chapter 6). In detection theory, it is automatically accounted for in the detection probability calculations (see Chapter 5, Sections 5.4, 5.8, 5.9.4, and 5.9.6).

Coverage of these and other kinds of losses and how to quantify them in system analysis can also be found in other well-known radar texts in connection with discussions of specialized applications (e.g., search and surveillance radars or tracking radars). Some of these are indicated in the reference list at the end of this chapter.[11–13]

REFERENCES

1. Skolnick, M.I., "Introduction to Radar Systems," 2nd ed. McGraw-Hill, New York, 1980; Chapter 2, pp. 15–65.
2. Rohan, P., "Surveillance Radar Performance Prediction." Peter Peregrinus Ltd., London, 1983; Chapter 4, pp. 61–102.
3. Hovanessian, S.A., "Radar System Design and Analysis." Artech, Norwood, MA, 1984; pp. 4–13.
4. Blake, L.V., "Radar Range Performance Analysis." Artech, Norwood, MA, 1986; Chapter 1, pp. 1–31.
5. Eaves, J.L., and Reedy, E.K. (eds.), "Principles of Modern Radar." Van Nostrand Reinhold, New York, 1987; Section 1.4, pp. 5–21.
6. Barton, D.K., "Modern Radar System Analysis." Artech, Norwood, MA, 1988; Section 1.2, pp. 9–24.
7. Levanon, N., "Radar, Principles." Wiley, New York, 1988; pp. 4–13.
8. Nathanson, F.E., Reilly, J.P., and Cohen, M.N., "Radar Design Principles," 2nd ed. McGraw-Hill, New York, 1991; pp. 5–9, 49–60.
9. Barton, D.K., "Modern Radar System Analysis." Artech, Norwood, MA, 1988; Section 7.2, pp. 320–321 and Section 7.2, pp. 346–348.
10. Skolnick, M.I., "Radar Handbook," 1st ed. McGraw-Hill, New York, 1970; Sections 1-6 and 32-22.
11. Barton, D.K., "Modern Radar System Analysis." Artech, Norwood, MA, 1988; pp. 18–19.
12. Levanon, N., "Radar, Principles." Wiley, New York, 1988; p. 13.
13. Nathanson, F.E., Reilly, J.P., and Cohen, M.N., "Radar Design Principles," 2nd ed. McGraw-Hill, New York, 1991; pp. 50–52.

Chapter 3

Constituents of a Radar System: Signal Flow

CONTENTS

In this chapter, each major constituent of a radar system will be briefly indicated, with a view toward understanding how a radar signal is generated at the transmitter, propagates through a portion of the surrounding space, is scattered by a target, returns to the receiver, passes through the early receiver stages, and is converted to useful information and eventually displayed in some form. The constituents that will be delineated are

1. Transmitter
2. Transmitting antenna
3. Propagation path to target
4. Target

5. Propagation path-target to receiver
6. Receiving antenna
7. "Linear" stages of receiver (RF, IF)
8. Video stage of receiver
9. Signal processing block
10. Display

The symbol $s_k(t, \mathbf{\Omega}_k)$ denotes the radar signal waveform at the stage indicated in the diagram. The symbol $\mathbf{\Omega}_k$ represents a vector containing parameters such as amplitude, path delay, frequency, and phase shift due to causes other than path delay. We will begin with a brief discussion of the action of each constituent in changing the signal from $s_{k-1}(t, \mathbf{\Omega}_{k-1})$ to $s_k(t, \mathbf{\Omega}_k)$, where the index on $\mathbf{\Omega}$ refers to the numbered "stage" of the signal flow. These stages and their indices are indicated in Figure 3.1, which is a flowchart for a radar signal.

3.1 THE TRANSMITTER

The transmitter (discussed in some detail in Eaves and Reedy[1] in Chapter 5, "Radar Transmitters," by G.W. Ewell, in Skolnick's "Introduction to Radar Systems"[2] in Chapter 6, and Skolnick's "Radar Handbook"[3] in Chapter 4) consists essentially of an RF power source, a power supply, and a modulator. The details of these components will not be discussed here, except to summarize briefly some of the important considerations in choosing them for a radar application. Those interested in the "hardware" details of those devices can read about them in the indicated references or others.[4,5]

The RF power source requires an oscillator and an amplifier, the former to generate an RF signal at the required frequency and the latter to provide adequate power at the output. These two functions may not be separated within a power source. They are often interactive, in that a certain amount of DC power is supplied

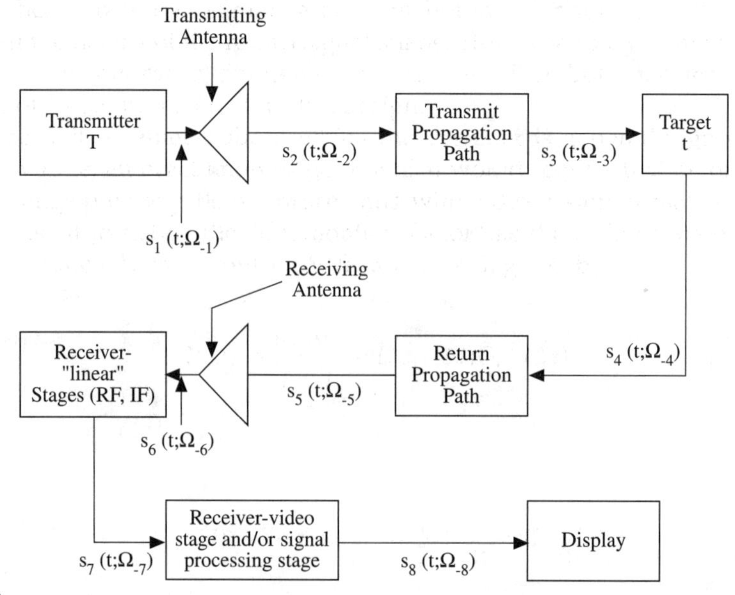

FIGURE 3.1
Flowchart for a radar signal.

to energize the signal. The oscillator–amplifier combination behaves as a linear circuit to produce an RF signal of a given amplitude at the output without a true physical separation between the functions of oscillation and amplification.

An excellent treatment of some of the details of microwave oscillators and amplifiers used in radar transmitters is presented in Eaves and Reedy.[1] The reader is referred to Table 5.1 on p. 116 of that reference for some of the important characteristics of commercially available magnetron oscillators, Table 5.3 on p. 124 for some characteristics of commercially available klystron amplifiers, Table 5.2 on p. 120 for characteristics of crossed-field amplifiers ("CFA"), and Table 5.5 on p. 128 for characteristics of traveling wave tube amplifiers (TWT); there is also material on other microwave devices that are important in radar transmitters, e.g., extended interaction oscillators (EIO), extended interaction amplifiers (EIA), twystron amplifiers, and gyrotron tubes (summarized in Table 5.7 on p. 134).

Comparisons between various types of RF sources with respect to peak power available, efficiency, instantaneous bandwidth, frequency range, and gain are summarized in Table 5.8 on p. 137 of Eaves and Reedy.[1]

The DC power supply, one of the other constituents of the transmitter, will also not be discussed in any detail here. The modulator, the third constituent, is simply a device that either interrupts the output of the RF source periodically to produce RF pulses *or* in effect varies the frequency of the RF source output to produce a frequency-modulated signal, *or* does both, in order to produce frequency-modulated ("chirp") pulses. Some of the properties and characteristics of modulators are discussed in Section 5.3 of Eaves and Reedy[1] on pp. 137–144. Since the modulator operates at frequencies well below the microwave region, it falls into the category of conventional low-frequency electronic devices. It would not be productive to discuss it in detail here. Instead we will allude to Figure 3.2, in which the block-diagram operation involved in generating the signal to be transmitted is shown.

It is indicated in the diagram that one of two basic configurations may be used in the RF power source working in conjunction with the modulator. Configuration (a) involves a master RF oscillator that generates an RF signal.

$$s'_{RF}(t;\omega_o) = a'_T \cos(\omega_o t + \psi'_T) \tag{3.1}$$

which then enters a power amplifier that applies a gain K and a possible phase shift $\delta\psi'_T$ (assuming of course that the power amplifier is linear), resulting in a signal

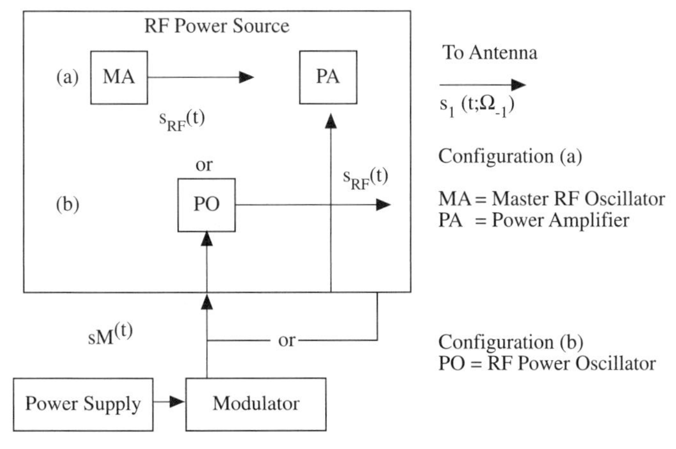

FIGURE 3.2
Producing the transmitted radar signal.

$$s_{RF}(t;\,\omega_o) = a_T \cos(\omega_o t + \psi_T) \tag{3.2}$$

where $a_T = Ka'_T$ and $\psi_T = \psi'_T + \delta\psi'_T$.

In this configuration, the modulator generates the low-frequency modulation signal $s_M(t)$ and applies it to the power amplifier, which has the effect of multiplying the signal $s_M(t)$ by the signal waveform of (3.2). A more sophisticated way to view the process is to view $s_M(t)$ and $s_{RF}(t;\,\omega_o)$ as being added coherently in the power amplifier; then the sum being squared (somewhat closer to what actually happens) and the product being the part that emerges at the output, i.e., the output of the process without consideration of the amplifier's passband is

$$\begin{aligned} s'_{out}(t) &= [s_{RF}(t;\,\omega_o) + s_M(t)]^2 \\ &= \left[\frac{a_T^2}{2} + \frac{a_T^2}{2}\cos(2\omega_o t + 2\psi_T)\right] + s_M^2(t) \\ &\quad + 2a_T s_M(t)\cos(\omega_o t + \psi_T) = 1 + 2 + 3 + 4 \end{aligned} \tag{3.3}$$

where the spectral locations of the terms are as follows: the first term is DC, the second term is at twice the RF, and the third term has the modulation spectrum squared, whose highest frequency is twice the highest modulation frequency, always very small compared with the RF. The only term of (3.3) that appears in the amplifier output is the fourth term, the product of the RF and modulation signal waveform. This is because terms 1, 2, and 3 are all far outside the amplifier passband. All of this is illustrated in Figure 3.3. In Figure 3.3a, the operation described above is shown schematically and the spectral components of the output are shown in Figure 3.3b.

In Figure 3.2b, the modulation signal $s_M(t)$ is applied directly to the RF power oscillator. There is no fundamental difference between these two configurations except that no amplification occurs, but that is because the power oscillator generates sufficient RF power without an amplifier. In this case Eq. (3.2) gives the oscillator output directly. The remainder of the discussion beyond Eq. (3.2) remains intact.

The result of this is that the signal $s_1(t;\,\mathbf{\Omega}_1)$ entering the transmission line on its way to the transmitting antenna has the generic form

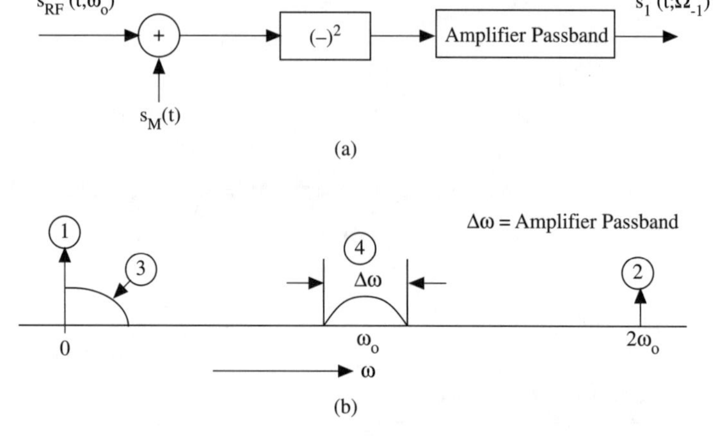

FIGURE 3.3
Generation of transmitted signal.

$$s_1(t;\, \mathbf{\Omega}_1) = A_T a_M(t) \cos[\omega_o t + \psi_T(t)] \tag{3.4}$$

where A_T is the amplitude of the signal, $a_M(t)$ is the waveform representing the normalized (to A_T) form of the time varying amplitude of the transmitted signal, and the function $\psi_T(t)$ is the phase shift induced by the modulation waveform plus the original RF phase. The time variations of $a_M(t)$ and $\psi_T(t)$ are slow compared with $\cos \omega_o t$ or $\sin \omega_o t$, or equivalently a waveform of the form (3.4) is known as a "narrowband" signal. This means that the modulation bandwidth is very small compared with the radar frequency ω_0. Alternatively we could represent (3.4) with two "quadrature" components $x_1(t;\, \mathbf{\Omega}_1)$ and $y_1(t;\, \mathbf{\Omega}_1)$ or

$$s_1(t;\, \mathbf{\Omega}_1) = x_1(t;\, \mathbf{\Omega}_1) \cos \omega_o t + y_1(t;\, \mathbf{\Omega}_1) \sin \omega_o t \tag{3.5}$$

where

$$x_1(t;\, \mathbf{\Omega}_1) = A_T a_M(t) \cos[\psi_T(t)]$$

$$y_1(t;\, \mathbf{\Omega}_1) = -A_T a_M(t) \sin[\psi_T(t)]$$

or in the form

$$s_1(t;\, \mathbf{\Omega}_1) = \mathrm{Re}[\hat{s}_1(t;\, \mathbf{\Omega}_1) e^{j\omega_o t}] \tag{3.6}$$

where

$$\hat{s}_1(t;\, \mathbf{\Omega}_1)) = A_T a_M(t) e^{j\psi_T(t)} = \text{"Complex envelope" of the signal } s_1(t;\, \mathbf{\Omega}_1)$$

and where it is recognized from (3.5) and (3.6) that

$$x_1(t;\, \mathbf{\Omega}_1) = \mathrm{Re}[\hat{s}_1(t;\, \mathbf{\Omega}_1)]$$

$$y_1(t;\, \mathbf{\Omega}_1) = -\mathrm{Im}[\hat{s}_1(t;\, \mathbf{\Omega}_1)]$$

3.2 THE TRANSMITTING ANTENNA

Since there will be somewhat detailed discussions of radar antennas in Chapter 10, the transmitting antenna will not be discussed here, except to say that the antenna is assumed to be *linear*. The transmission line leading from the output of the RF power source to the input terminals of the antenna, cascaded with the antenna itself, produces electric field components at the antenna aperture that have the form (see Fig. 3.1)

$$s_2(t;\, \mathbf{\Omega}_2) = a_2(t) \cos[\omega_o t + \psi_2(t)] \tag{3.7}$$

where

$$a_2(t) = K_{12} A_T a_M(t)$$

$$\psi_2(t) = \psi_T(t) + \Delta\psi_{12}$$

and where K_{12} is an amplitude scale factor due to the line and antenna (generally a loss, due to reflections at the termination and mismatch between the antenna impedance and that of free space) and $\Delta\psi_{12}$ is the phase shift introduced by the line-antenna network. Equation (3.7) indicates simply that the network, being linear, can change the amplitude scale and shift the phase of the signal entering it but will not change the frequency ω_o or distort the slowly varying functions $a_M(t)$ and $\psi_T(t)$.

3.3 THE PROPAGATION PATH TO TARGET

The next stage of the signal's journey is the propagation path from transmitter to target. It is assumed that the path is entirely in free space. That is the idealization that facilitates qualitative thinking about radar, and is the logical place to begin the analysis of a radar system. Deviations from it introduce phenomena such as ray bending (refraction), multipath due to scattering material along the path, attenuation due to fog, clouds, and rain, and other interesting propagation features that complicate the analysis. Most of these effects are discussed in Chapter 11.

If the path is entirely in "infinite" free space (i.e., no protrusions or boundaries anywhere in the environment) then the only effect of that path is to introduce a time delay $\tau_{d1} = r/c$ and a possible Doppler shift $\omega_{d1} = \omega_o V_{rel}/c$ where V_{rel} is the relative line-of-sight velocity component between transmitter and target. Thus the signal waveform appearing at the target is given by

$$s_3(t; \mathbf{\Omega}_3) = a_3(t - \tau_{d1}) \cos[(\omega_o + \omega_{d1})(t - \tau_{d1}) + \psi_3(t - \tau_{d1})] \tag{3.8}$$

where $a_3(t) = (K_{23}/r)a_2(t)$, $\psi_3(t - \tau_{d1}) = \psi_2(t - \tau_{d1})$, the phase shift incurred in the transmitter and its antenna, and where K_{23} is a scale factor due to the antenna pattern and the $1/r$ loss in the field strength (translating into a $1/r^2$ power loss) is accounted for.

3.4 THE TARGET

The target, which will be discussed in more detail in later chapters, is assumed to be a linear scatterer. That means that the transformation from $s_3(t; \mathbf{\Omega}_4)$, the signal voltage of the wave incident on the target, to $s_4(t; \mathbf{\Omega}_4)$, the signal voltage of the wave launched at the target due to the scattering process, consists of only a loss factor in the amplitude and a phase shift incurred in scattering, i.e.,

$$s_4(t; \mathbf{\Omega}_4) = a_4(t - \tau_d) \cos[(\omega_o + \omega_d)(t - \tau_d) + \psi_4(t - \tau_d)] \tag{3.9}$$

where

$$a_4(t) = K_{34}a_3(t) \quad \text{and} \quad \psi_4(t) = \psi_2(t) + \Delta\psi_{34}$$

$\Delta\psi_{34}$ being the phase shift due to the scattering and K_{34} being the amplitude loss due to scattering.

3.5 THE RETURN PROPAGATION PATH

Assuming monostatic radar, the return path performs the same transformation as the forward path discussed in 3.3, resulting in

$$s_5(t; \boldsymbol{\Omega}_5) = a_5(t - \tau_d)\cos[(\omega_o + \omega_d)(t - \tau_d) + \psi_5(t - \tau_d)] \tag{3.10}$$

where $\tau_d = 2\tau_{d1}$, $\omega_d = 2\omega_{d1}$, $a_5(t) = (K_{45}/r)a_4(t)$ and $\psi_5(t) = \psi_4(t) + \Delta\psi_{45}$, $\Delta\psi_{45}$ being the incremental phase shift. In the case of bistatic radar, if the forward and return paths are denoted by subscripts a and b, respectively, then $\tau_d = \tau_{d1a} + \tau_{d1b}$, $\omega_d = \omega_{d1a} + \omega_{d1b}$ and the factor K_{45}/r becomes K_{45b}/r_b on the return path and K_{23a}/r_a on the forward path covered in (3.3).

3.6 THE RECEIVING ANTENNA

The receiving antenna, like the transmitting antenna, performs a linear transformation on the signal impinging on it, i.e., an amplitude loss K_{56} and a fixed phase shift $\Delta\psi_{56}$. The result is

$$s_6(t; \boldsymbol{\Omega}_6) = a_6(t - \tau_d)\cos[(\omega_o + \omega_d)(t - \tau_d) + \psi_6(t - \tau_d)] \tag{3.11}$$

where

$$a_6(t) = K_{56}a_5(t)$$

$$\psi_6(t - \tau_d) = \psi_5(t - \tau_d) + \Delta\psi_{56}$$

3.7 "LINEAR" STAGES OF RECEIVER (RF, IF)

At this point there is an actual departure from pure linearity, but it is our objective to show that the nonlinear part of the signal is filtered out and the net result is that the linearity property is preserved mathematically in the signal waveform that emerges from these stages.

Before doing this, we will discuss some general principles of a radar receiver.[6,7] There are three general types of receiver configurations that can be used in a radar. The superregenerative receiver (Eaves and Reedy,[1] Section 7.2.1, pp. 184–185) is not included because of its prohibitively large noise. The three other generic types of radar receiver are

1. The "crystal video" receiver
2. The tuned radio frequency (TRF) receiver
3. The superheterodyne receiver

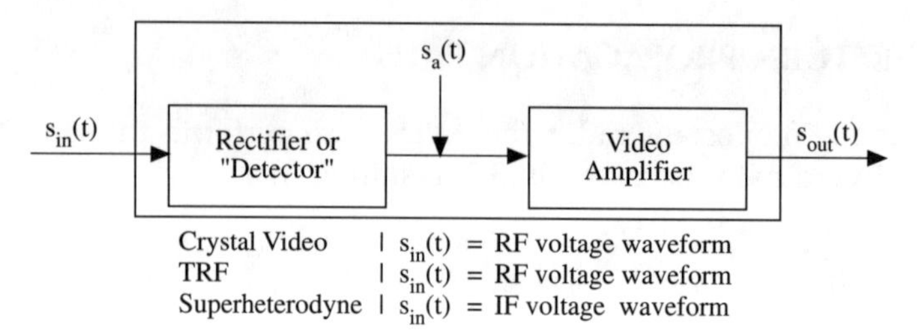

FIGURE 3.4
Last two stages of typical radar receiver.

The last two stages of any of these three receiver types are the rectifier (or "detector" as it is traditionally called) and the video amplifier, shown in Figure 3.4.

The input to these last two stages is a voltage waveform of the general class

$$s_{in}(t) = Ap(t)\cos[\omega_o t + \psi(t)] \tag{3.12}$$

where $p(t)$ is the pulse waveform and ω_0 is the radar frequency in the crystal video or TRF case and the "intermediate frequency" or "IF" in the superheterodyne case. The rectifier transforms the signal into a "video signal" by taking the square of (3.12) ("square law detector"), or its absolute value ("linear detector") or a sum of terms of various powers of the input signal. In any case, the rectifier performs a nonlinear operation on the input signal and produces a voltage waveform that is always positive. The example of a pulsed RF or IF signal at the input is shown in Figure 3.5, for the linear and quadratic rectifier (Fig. 3.5a and b, respectively).

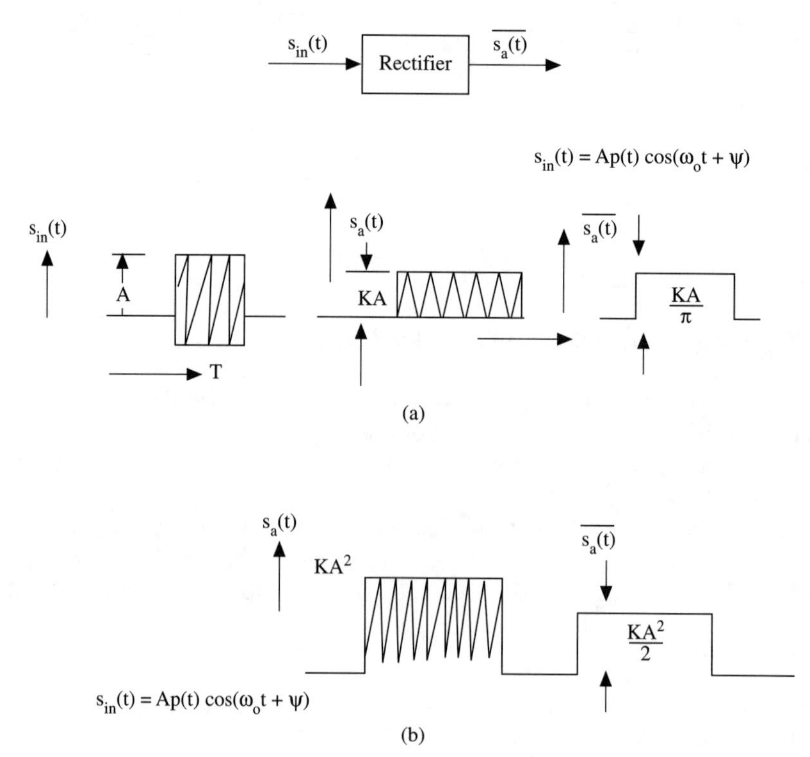

FIGURE 3.5
Rectifier operation.

For the linear full-wave rectifier, the transformation is

$$\overline{s_a(t)} = K'|\overline{s_{in}(t)}| = K'Ap(t)|\overline{\cos[\omega_o t + \psi(t)]}| \tag{3.13}$$

For the quadratic rectifier

$$\overline{s_a(t)} = K'[s_{in}(t)]^2 = \frac{K'A^2p^2(t)}{2} + \frac{K'A^2p^2(t)}{2}\overline{\cos\{2[\omega_o t + \psi(t)]\}} \tag{3.14}$$

The signal $s_a(t)$ given by (3.13) or (3.14) contains the effect of a lowpass filtering operation, which is part of the rectifier's action. That is represented by a Fourier series of the form (see Spiegel[8], 23.11, p. 132)

$$|\cos[\omega_o t + \psi(t)]| = \frac{2}{\pi} - \frac{4}{\pi}\left[\frac{\cos[2\omega_o t + 2\psi(t)]}{3} + \frac{\cos[4\omega_o t + 4\psi(t)]}{15} + \frac{\cos[6\omega_o t + 6\psi(t)]}{35} - \cdots\right] \tag{3.15}$$

The lowpass filtering operation, whose bandwidth is orders of magnitude below $(2\omega_o)$, will filter out all but the first term of (3.15), with the result

$$\overline{s_a(t)} = \frac{K'A}{\pi}|p(t)| \tag{3.16}$$

for the linear rectifier. The same filtering operation in the quadratic detector will eliminate all but the first term of (3.14), with the result

$$s_a(t) = \frac{K'A^2p^2(t)}{2} \tag{3.17}$$

Referring back to Figure 3.4, the video amplifier introduces gain G_v into the signal but otherwise does not affect the waveform (3.16) or (3.17), provided its passband is wide enough to accommodate all of the frequencies present in $|p(t)|$ or $p^2(t)$. It is no more than a lowpass filter with bandwidth in excess of that of $|p(t)|$ or $p^2(t)$ or whatever form the low frequency signal has when it emerges from the rectifier. The output of the video amplifier, then, is

$$s_{out}(t) = \frac{G_v K'A}{\pi}|p(t)| \quad \text{for a linear rectifier} \tag{3.18a}$$

$$= \frac{G_v K'A^2}{2}p^2(t) \quad \text{for a quadratic rectifier} \tag{3.18b}$$

where G_v is the gain of the video amplifier and the latter is assumed designed in such a manner as not to distort the modulation waveform.

We now return to the subject of this section, the "linear stages" of the receiver, i.e., those prior to those shown in Figure 3.4, culminating in the signal $s_{in}(t)$ given by (3.12). We designate the receiver portion shown in Figure 3.4 as nonlinear stages (NLS) and show the three basic forms of radar receiver in Figure 3.6.

The crystal video receiver, as shown in Figure 3.6a, uses no RF amplifier, but merely feeds the RF signal directly from the antenna into the NLS. The advantages are simplicity, low cost, and small size. The bandwidth is very large, being that of the RF input circuit [i.e., the antenna and the RF transmission line (waveguide) that carries the RF signal]. The noise level is very high, because there is nothing to limit the noise that naturally accompanies the RF signal. In TRF and superheterodyne receivers, the RF amplifier serves to reduce the noise level, as we will see in the discussion to follow. In a crystal video receiver, the basic RF noise that comes in with the signal is carried directly into the NLS, where still more noise is generated, resulting in a requirement for enormous video amplifier gain in order to get a sufficiently strong signal (because *all* of the gain is obtained from the video amplifier). Thus crystal video receivers have very low sensitivity and hence are rarely used in modern radars.

The TRF receiver, illustrated in Figure 3.6b, has an RF amplifier prior to NLS. This amplifier provides part of the gain and increases the SNR at the input to NLS, resulting in a smaller gain requirement for the video amplifier and further reduction in bandwidth (because the passband of the RF amplifier is smaller than that of the RF input circuitry). The latter has an additional SNR benefit because noise occupies the entire RF passband uniformly. If the RF amplifier reduces the effective passband, it reduces the noise level in proportion to its fractional passband reduction. The obvious limit on that reduction is the requirement that the RF amplifier pass the signal spectrum and hence not distort the signal.

The overall conclusion is that the TRF receiver has much higher sensitivity than the crystal video. However, with a small cost increase, still further improvement can be achieved through use of the superheterodyne receiver, shown in Figure 3.6c. In the latter, the local oscillator generates a signal

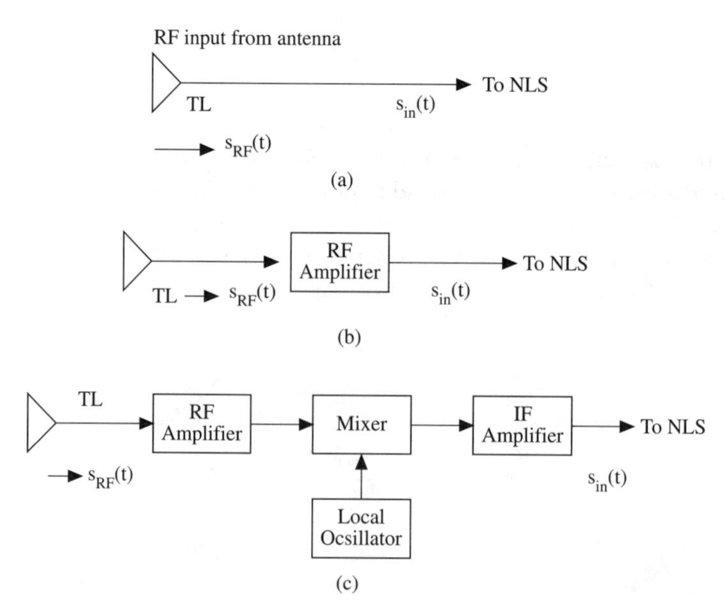

FIGURE 3.6
Linear stages of receiver.

$$s_{LO}(t) = A_{LO} \cos[(\omega_{RF} - \omega_{IF})t] \tag{3.19}$$

where ω_{IF}, the "intermediate frequency," is a frequency much lower than the RF (e.g., of the order of MHz or tens or hundreds of MHz if the RF is of the order of GHz). The output of the RF amplifier is combined with the local oscillator signal $s_{LO}(t)$ of (3.19) in a way that results in a signal at the IF whose slowly varying amplitude and phase are undistorted. That signal is much easier to amplify and to work with in general than the RF signal, because its frequency brings it within the region of "low-frequency" electronics. The end result is much higher sensitivity than can be obtained with TRF or crystal video with a minimum incremental cost. Hence the superheterodyne receiver is used in most modern radars.

The action of the mixer is basically an addition of the local oscillator signal and the RF amplifier output followed by a squaring operation. The RF amplifier output is

$$s_{RF}(t) = ap(t) \cos[\omega_{RF}t + \psi(t)] \tag{3.20}$$

where again $p(t)$ and $\psi(t)$ are slowly varying compared with $\cos \omega_o t$ or $\sin \omega_o t$, where ω_o denotes either ω_{RF} or ω_{IF}.

Adding $s_{RF}(t)$ as in (3.20) and $s_{LO}(t)$ as in (3.19) and squaring the result, we would obtain (noting that $\cos^2 \theta = 1/2 + 1/2 \cos 2\theta$, $\cos(A + B) = 1/2[\cos(A + B) + \cos(A - B)]$.

$$\begin{aligned}[s_{RF}(t) + s_{LO}(t)]^2 = {} & \frac{a^2p^2(t)}{2}\{1 + \cos[2\omega_{RF}t + 2\psi(t)]\} \\ & + \frac{A_{LO}^2}{2}\{1 + \cos[2(\omega_{RF} - \omega_{IF})t]\} \\ & + \frac{2A_{LO}ap(t)}{2}\{\cos[(2\omega_{RF} - \omega_{IF})t + \psi(t)] \\ & + \cos[\omega_{IF}t + \psi(t)]\} = s_1(t) + s_2(t) \\ & + s_3(t) + s_4(t) + s_5(t)\end{aligned} \tag{3.21}$$

where

$$\begin{aligned} s_1(t) &= \frac{a^2p^2(t)}{2} + \frac{A_{LO}^2}{2} \\ s_2(t) &= \frac{A_{LO}^2}{2}\cos[2(\omega_{RF} - \omega_{IF})t] \\ s_3(t) &= A_{LO}ap(t) \cos[(2\omega_{RF} - \omega_{IF})t + \psi(t)] \\ s_4(t) &= \frac{a^2p^2(t)}{2}\cos[2\omega_{RF}t + 2\psi(t)] \\ s_5(t) &= A_{LO}\, ap(t) \cos[\omega_{IF}t + \psi(t)] \end{aligned}$$

The five signal outputs in (3.21) can be illustrated in frequency space with Figure 3.7.

It is obvious from Figure 3.7 that the only one of the five mixer output signals that is at IF is $s_5(t)$. The spectra of the other four signals $s_1(t)$, $s_2(t)$, $s_3(t)$, and $s_4(t)$ are centered, respectively, at DC (far below IF), $2(\omega_{RF} - \omega_{IF})$ (far above IF, near RF), $(2\omega_{RF} - \omega_{IF})$ (still further above IF, near twice the RF), and $2\omega_{RF}$ (twice the RF, still further above IF). The net result of all this is that the IF amplifier, whose passband is centered at IF and whose bandwidth is generally about 10 to 20% of IF, passes only $s_5(t)$ and rejects all of the other five signals.

The true mixer operation is more complicated than indicated by (3.21). Actually it is a complicated nonlinear device whose output contains many harmonics and hence many intermodulation products. The true output is of the form (see Eq. (7.20) on p. 202 of Eaves and Reedy[6] and the discussion on pp. 202, 203, and 204 of that reference):

$$s_{mix}(t) = a_o + a_1(s_{RF} + s_{LO}) + a_2(s_{RF} + s_{LO})^2 + a_3(s_{RF} + s_{LO})^3 + \cdots + a_N(s_{RF} + s_{LO})^N \tag{3.22}$$

where N is an arbitrary positive integer. It is an interesting and important exercise to determine the "intermodulation products" in the output arising from the various terms in (3.22), some of which are near IF and hence passed by the IF amplifier. In addition to this there are noise terms in the general output (3.22) that arise from intermodulation between the input RF noise and the RF and/or LO signals. These can be studied by writing

$$v_{RF}(t) = s_{RF}(t) + n_{RF}(t) \tag{3.23}$$

where

$$n_{RF}(t) = a_n(t)\cos[\omega_{RF}t + \psi(t)]$$

and where $a_n(t)$ and $\psi_n(t)$ are random functions that vary slowly compared with $\cos\omega_o t$ or $\sin\omega_o t$, where ω_o can be ω_{RF} or ω_{IF}.

If $v_{RF}(t)$ is substituted for s_{RF} in (3.22), it is feasible to determine both the spurious signals within the IF passband *and* the additional noise components introduced by various harmonics. This will be assigned as a homework problem. Figure 7–9 of Eaves and Reedy[6] indicates spurious signals generated by this process.

To complete this section, we note that the idealized superheterodyne receiver has no spurious intermodulation processes and its output is essentially $s_5(t)$ in (3.21)

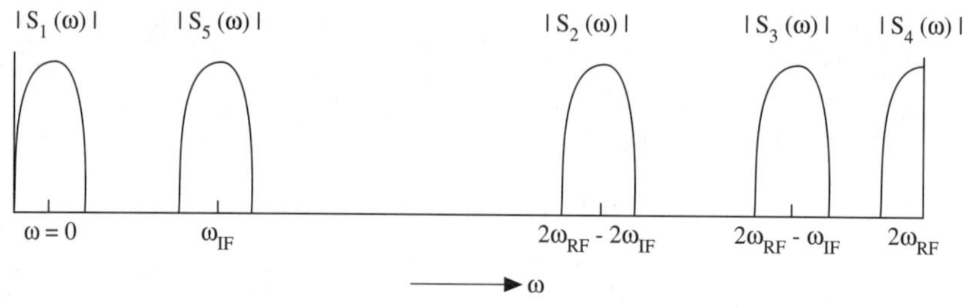

FIGURE 3.7
Outputs of the mixer in frequency space.

multiplied by the gain of the IF amplifier G_{IF}. Thus the input to the NLS is assumed to be

$$s_{in}(t) = G_{IF}A_{LO}ap(t)\cos[\omega_{IF}t + \psi(t)] \tag{3.24}$$

We can use this to obtain the signal $s_7(t; \mathbf{\Omega}_7)$ as given in Figure 3.1.

Returning to the notation used in Sections 3.1 through 3.6, we can invoke the results (3.4), (3.7), (3.8), (3.9), (3.10), (3.11), and (3.24) to produce the form of the signal waveform at the input to the nonlinear stages of the receiver. Before doing that, it should be remarked that the mixer, although highly *nonlinear*, in conjunction with the IF amplifier, which *is* linear, acts as a linear system if it is ideal, because it preserves the shape of the modulation waveform, both its amplitude and its phase. Also, the noise, additive to the signal at the receiver input, is *still* additive and uncoupled to the signal at the IF output.

The idealized waveform $s_7(t; \mathbf{\Omega}_7)$, from (3.4), (3.7)–(3.11), and (3.24), is

$$s_7(t; \mathbf{\Omega}_7) = Aa_M(t - \tau_d)\cos[(\omega_o + \omega_d)(t - \tau_d) + \psi_o + \psi_T(t)] \tag{3.25}$$

where

A = $A_TK_{12}K_{23}K_{34}K_{45}K_{56}K_{67}K'A_{LO}G_{RF}G_{IF}/r^2$
A_T = amplitude of transmitted signal
K_{mn} = loss factor on amplitude in passing from stage m to stage n (excluding $1/r^2$ loss)
K' = scale factor on rectifier law in receiver
A_{LO} = amplitude of local oscillator signal
G_{RF} = gain of RF amplifier
G_{IF} = gain of IF amplifier
r = range-to-target for monostatic radar and geometric mean of two ranges for bistatic radar
τ_d = two-way path delay
= $\frac{2r}{c}$ for monostatic radar
= $\frac{(r_{pq} + r_{qp'})}{c}$ for bistatic radar (where p, q, p' are transmitter, target, and receiver locations, respectively)
ω_d = two-way Doppler shift
= $(2/c)\, v\omega_{RF}$ for monostatic radar
= $\frac{(v_{pq} + v_{qp'})}{c}\omega_{RF}$ for bistatic radar
ψ_o = $\psi_T + \Delta\psi_{12} + \Delta\psi_{23} + \Delta\psi_{34} + \Delta\psi_{45} + \Delta\psi_{56} + \Delta\psi_{67}$
ψ_T = phase of transmitted signal
$\Delta\psi_{mn}$ = phase shift (excluding path delay) in passing from stage m to stage n

Note that no such phase shift exists for the propagation paths *if* the medium is infinite free space, where $\Delta\psi_{23}$ and $\Delta\psi_{45}$ are zero and where

ω_o = central frequency of signal at the input to the nonlinear stages

Note that $\omega_o = \omega_{RF}$ in a crystal video or TRF receiver and ω_{IF} in a superheterodyne receiver. Note also that the Doppler is defined in terms of ω_{RF}, not in terms of ω_o. This is because the Doppler shift was absorbed in the slowly varying phase $\psi(t)$ in (3.20) and the discussion following it. That portion of $\psi(t)$ is $\omega_d t$, which varies slowly compared with RF. In some treatments ω_{RF} might already contain the Doppler shift.

The other consideration with respect to the "linear" receiver stages is that of noise. As indicated above, the noise at the receiver input is additive and uncoupled to the signal. The RF amplifier provides a pure multiplicative gain, which acts the same with respect to signal and noise and hence does not disturb the noise statistics provided the input noise is Gaussian (which nearly all of it is). The same applies to the IF amplifier.

As indicated earlier, the mixer, if it is ideal and generates no spurious intermodulation products, does not disturb the Gaussian characteristic of the noise and does not introduce coupling between signal and noise. Hence for the idealized receiver system, the input to NLS can be considered as a signal of the form $s_7(t; \mathbf{\Omega}_7)$ plus a Gaussian noise waveform centered at the frequency ω_o.

The RF input circuitry, the RF amplifier, the mixer, and the IF amplifier all generate some new noise within the components that make up these subsystems. The concept of noise figure as a measure of *new* noise created within a device was introduced in Section 2.8 [Eq. (2.34)]. We now consider the noise figure for a cascade of two or more linear stages, e.g., RF amplifier, mixer, and IF amplifier. The noise power input to the cascade is P_{ni} and the output is P_{no}, as shown in Figure 3.8 for a three-stage cascade.

The output of the first stage, a pure power gain G_1, is

$$P_{n12} = P_{ni}G_1 + \Delta P_{n1} = P_{ni}G_1\left(1 + \frac{\Delta P_{n1}}{G_1 P_{ni}}\right) \tag{3.26a}$$

where ΔP_{nK} is the amount of noise power generated in the Kth stage. The power output of the second stage is

$$P_{n23} = P_{n12}G_2 + \Delta P_{n2} = P_{ni}G_1G_2\left(1 + \frac{\Delta P_{n1}}{P_{ni}G_1}\right)\left\{1 + \frac{\Delta P_{n2}}{P_{ni}G_1G_2[1 + (\Delta P_{n1}/P_{ni}G_1)]}\right\} \tag{3.26b}$$

The final output for the three-stage cascade is

$$\begin{aligned} P_{no} &= P_{n23}G_3 + \Delta P_{n3} = P_{ni}G_1G_2G_3\left(1 + \frac{\Delta P_{ni}}{P_{ni}G_1}\right) \\ &\times \left\{1 + \frac{\Delta P_{n2}}{P_{ni}G_1G_2[1 + (\Delta P_{n1}/P_{ni}G_1)]}\right\} \\ &\cdots \left\{1 + \frac{\Delta P_{n3}}{P_{ni}G_1G_2G_3\left(1 + \frac{\Delta P_{n1}}{P_{ni}G_1}\right)\left[1 + \frac{\Delta P_{n2}}{P_{ni}G_1G_2[1 + (\Delta P_{n1}/P_{ni}G_1)]}\right]}\right\} \end{aligned} \tag{3.26c}$$

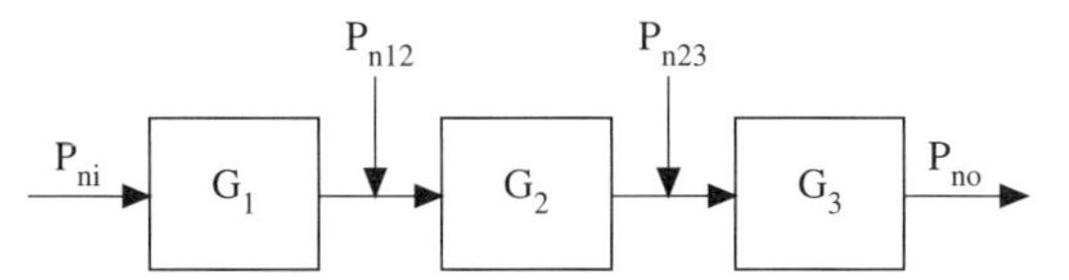

FIGURE 3.8
Illustration of noise figure vs. gain.

The noise figure for this cascade, defined as the ratio of the noise output to what *would* exist if no new noise were created within the devices, is

$$F_N = \frac{P_{no}}{P_{ni}G_1G_2G_3} = \left(1 + \frac{\Delta P_{n1}}{P_{ni}G_1}\right)\left\{1 + \frac{\Delta P_{n2}}{P_{ni}G_1G_2[1 + (\Delta P_{n1}/P_{ni}G_1)]}\right\}$$

$$\times \left\{1 + \frac{\Delta P_{n3}}{P_{ni}G_1G_2G_3\left(1 + \frac{\Delta P_{n1}}{P_{ni}G_1}\right)\left[1 + \frac{\Delta P_{n2}}{P_{ni}G_1G_2[1 + (\Delta P_{n1}/P_{ni}G_1)]}\right]}\right\} \tag{3.26d}$$

If the gains were made large enough so that

$$G_1 >> \frac{\Delta P_{n1}}{P_{ni}} \tag{3.27a}$$

$$G_1G_2 >> \frac{\Delta P_{n2}}{P_{ni}} \tag{3.27b}$$

$$G_1G_2G_3 >> \frac{\Delta P_{n3}}{P_{ni}} \tag{3.27c}$$

then it follows that

$$F_N \approx 1 \tag{3.28}$$

This illustrates that it is possible, in principle, to reduce noise in a cascade by providing high enough gains to offset the effects of noise created within the devices that make up the cascade. If the cascade is a single device, then condition (3.27a) would be sufficient to keep the noise figure near unity (0 dB). If it were a cascade of two devices, then conditions (3.27a) *and* (3.27b) would be required to meet that goal. In any case, high gains attained without increasing the amount of noise generated within the device would significantly reduce the noise figure. That is why an RF amplifier prior to mixing is helpful in reducing the noise input to the mixer and thereby reducing the overall receiver noise at the input to the NLS.

A point should be made about terminology. We have been designating F_N, an absolute number, as the "noise *figure*." It is common to call that quantity "noise *factor*," and its value in dB "noise *figure*." Throughout this book, F_N as defined in Chapter 1 will be called "noise figure" and its value will usually be expressed in dB.

REFERENCES

1. Eaves, J.L., and Reedy, E.K. (eds.), "Principles of Modern Radar." Van Nostrand Reinhold, New York, 1987, Chapter 5 by G.W. Ewell.
2. Skolnick, M.I., "Introduction to Radar Systems," 2nd ed. McGraw-Hill, New York, 1980, Chapter 6.
3. Skolnick, M.I. (ed.), "Radar Handbook," 2nd ed. McGraw-Hill, New York, 1990, Chapter 4, by T.A. Weil.
4. Ewell, G.W., "Radar Transmitters." McGraw-Hill, New York, 1981.
5. Nathanson, F.E., Reilly, J.P., and Cohen, M.N., "Radar Design Principles," 2nd ed. McGraw-Hill, New York, 1991, pp. 31–33 and 673–679.
6. Eaves, J.L., and Reedy, E.K. (eds.), "Principles of Modern Radar." Van Nostrand Reinhold, New York, 1987, Chapter 7 by T.L. Lane.
7. Skolnick, M.I., "Radar Handbook," 2nd ed. McGraw-Hill, New York, 1990, Chapter 3 by J.W. Taylor, Jr.
8. Spiegel, M.R., "Mathematical Handbook" (Schaum's Outline), McGraw-Hill, New York, 1968.

Chapter 4

Idealized Theory of Radar Signals in Additive Noise

CONTENTS

In this chapter we will discuss the highly idealized theory of signals in additive noise as it applies to radar. We will begin with the matched filter, a classical theoretical concept that is certainly not unique to radar, but very important in establishing a "frame of reference" for studies of radar signal detection and parameter estimation. A few general sources on the material covered in this chapter are given in the reference list.[1–5]

4.1 THEORY OF THE MATCHED FILTER

Consider a *deterministic* signal waveform $s_i(t)$ and a noise waveform $n_i(t)$ the latter being a sample function of a random process. The sum of signal and noise waveforms, denoted by $v_i(t)$, is

$$v_i(t) = s_i(t) + n_i(t) \tag{4.1}$$

The voltage or current waveform $v_i(t)$ is the input to a linear filter whose impulse response is $h(t)$. The output of the filter is

$$v_o(t) = s_o(t) + n_o(t) \tag{4.2}$$

where signal and noise outputs $s_o(t)$ and $n_o(t)$ are

$$s_o(t) = \int_{-\infty}^{\infty} d\tau h(\tau) s_i(t - \tau) \qquad (4.2a)'$$

$$n_o(t) = \int_{-\infty}^{\infty} d\tau h(\tau) n_i(t - \tau) \qquad (4.2b)'$$

and of course

$$v_o(t) = \int_{-\infty}^{\infty} d\tau h(\tau) v_i(t - \tau) \qquad (4.2c)'$$

This situation described above is depicted in Figure 4.1.

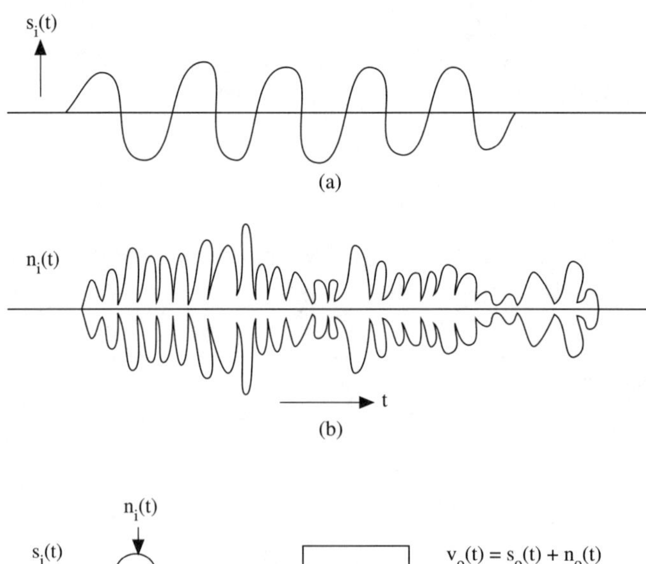

FIGURE 4.1
Signal-plus-noise through linear filter.

An alternative way to describe $s_i(t)$ and $n_i(t)$ is through their Fourier transforms $S_i(\omega)$ and $N_i(\omega)$, respectively, i.e.,

$$s_i(t) = \frac{1}{2\pi}\int_{-\infty}^{\infty} d\omega S_i(\omega)e^{j\omega t} \tag{4.3a}$$

$$n_i(t) = \frac{1}{2\pi}\int_{-\infty}^{\infty} d\omega N_i(\omega)e^{j\omega t} \tag{4.3b}$$

The impulse response $h(t)$ can also be described through its Fourier transform, the frequency response function, $H(\omega)$, i.e.,

$$h(t) = \frac{1}{2\pi}\int_{-\infty}^{\infty} d\omega H(\omega)e^{j\omega t} \tag{4.3c}$$

Using (4.3a,b,c) in (4.2a,b)′ and changing the order of integration, we have

$$s_o(t) = \frac{1}{2\pi}\int_{-\infty}^{\infty} d\omega_1 \int_{-\infty}^{\infty} d\omega_2 H(\omega_1)S_i(\omega_2)e^{j\omega_1 t}\left[\frac{1}{2\pi}\int_{-\infty}^{\infty} d\tau e^{-j(\omega_1-\omega_2)\tau}\right] \tag{4.4a}$$

$$n_o(t) = \frac{1}{2\pi}\int_{-\infty}^{\infty} d\omega_1 \int_{-\infty}^{\infty} d\omega_2 H(\omega_1)N_i(\omega_2)e^{j\omega_1 t}\left[\frac{1}{2\pi}\int_{-\infty}^{\infty} d\tau e^{-j(\omega_1-\omega_2)\tau}\right] \tag{4.4b}$$

The integral on τ in (4.4a,b), where the limits are changed to $+T/2$ and $-T/2$, is

$$I(\omega_2 - \omega_1) = \frac{1}{2\pi}\int_{-T/2}^{T/2} d\tau e^{j(\omega_2-\omega_1)\tau} = \frac{T}{2\pi}\left[\frac{e^{j(\omega_2-\omega_1)(T/2)} - e^{-j(\omega_2-\omega_1)(T/2)}}{2_j(\omega_2 - \omega_1)(T/2)}\right]$$

$$= \frac{T}{2\pi}\operatorname{sinc}\left[(\omega_2 - \omega_1)\frac{T}{2}\right] \tag{4.5}$$

If $\omega_2 = \omega_1$, $I(0) = T/2\pi$, which approaches infinity as T approaches infinity. If $\omega_2 \neq \omega_1$, then $I(\omega_2 - \omega_1)$ approaches zero as T approaches infinity. Also, if $x = (\omega_1 - \omega_2)(T/2)$, then

$$\frac{1}{2\pi}\int_{-\infty}^{\infty} dx \operatorname{sinc}(x) = \int_{-\infty}^{\infty} I\left(\frac{2x}{T}\right)dx = 1 \tag{4.6}$$

These properties are those of the unit impulse or delta function, from which it follows that

$$I(\omega_2 - \omega_1) = \frac{1}{2\pi}\int_{-\infty}^{\infty} d\tau e^{j(\omega_2-\omega_1)\tau} = \delta(\omega_2 - \omega_1) \tag{4.7}$$

Applying (4.7) to (4.4a,b), we obtain

$$s_o(t) = \frac{1}{2\pi}\int_{-\infty}^{\infty} d\omega H(\omega)S_i(\omega)e^{j\omega t} \tag{4.8a}$$

$$n_o(t) = \frac{1}{2\pi}\int_{-\infty}^{\infty} d\omega H(\omega)N_i(\omega)e^{j\omega t} \tag{4.8b}$$

To do matched filter theory in the simplest possible way, we leave (4.8a) intact. However, strictly speaking, random noise *does not have* a Fourier transform in the usual sense. To express the output noise power, take an ensemble average of its square, denoted by $\langle\text{--}\rangle$, i.e.,

$$\langle n_o^2(t)\rangle = \frac{1}{(2\pi)^2}\int_{-\infty}^{\infty} d\omega_1 \int_{-\infty}^{\infty} d\omega_2 H(\omega_1)H(\omega_2)e^{j(\omega_1+\omega_2)t}\langle N_i(\omega_1)N_i(\omega_2)\rangle \tag{4.9}$$

We then take a time-average from $t = -T/2$ to $t = +T/2$ inside the integral.

$$\frac{1}{T}\int_{-T/2}^{T/2} dt e^{j(\omega_1+\omega_2)t} = \frac{1}{T}\left\{T \operatorname{sinc}\left[(\omega_1 + \omega_2)\frac{T}{2}\right]\right\} \tag{4.10}$$

From the same arguments as those that led us from (4.5) to (4.7), we have, as T is allowed to become arbitrarily large [with the aid of (4.8b)]

$$\frac{1}{T}\int_{-T/2}^{T/2} dt\langle n_o^2(t)\rangle \rightarrow \frac{1}{2\pi}\int_{-\infty}^{\infty} d\omega H(\omega)H(-\omega)\frac{\langle N_i(\omega)N_i(-\omega)\rangle}{T} \tag{4.11}$$

as $T \rightarrow \infty$.

Denoting the operation of time averaging by an overline and noting that the Fourier transform of any *real* time waveform $x(t)$ has the property

$$X(-\omega) = \int_{-\infty}^{\infty} dt x(\tau)e^{j(-\omega)\tau} = X^*(\omega) \tag{4.12}$$

it follows from (4.9), with the aid of (4.10), (4.11), and (4.12), that

$$\overline{\langle n_o^2(t)\rangle} = \overline{\langle n_o^2\rangle} = \frac{1}{2\pi}\int_{-\infty}^{\infty} d\omega |H(\omega)|^2 G_n(\omega) \tag{4.13}$$

where it is explicitly indicated that $\overline{\langle n_o^2(t)\rangle}$ is independent of time (the time having been averaged out) and where

$$G_n(\omega) = \lim_{T\rightarrow\infty}\frac{\langle |N_i(\omega)|^2\rangle}{T} \tag{4.14}$$

is defined as the spectral density of the noise, or "noise power spectrum," the average noise power per unit of frequency (Hertz).

At this point we must delineate some assumptions about the signal and noise inputs to the filter, as follows:

1. The signal waveform $s_i(t)$ is deterministic, from which it follows that the output signal $s_o(t)$ is also deterministic.
2. The noise waveform $n_i(t)$ is a sample function of a stationary random process, which would seem to follow from the development above, $\langle n_o^2(t)\rangle$, the mean square of the output noise $n_o(t)$, is independent of time. However, there is a subtle point buried in that development in that we performed *both* time and ensemble averaging to obtain (4.13). A random function is defined as "ergodic" if the time and ensemble averages of itself, its square, or other functions of its waveform at a given time are equivalent and independent of that time. Ergodicity and stationarity are not exactly the same property, because the latter, strictly speaking, relates to its probability density functions (PDF) being independent of time. However, for practical purposes, we will consider ergodicity and stationarity as equivalent, and will use the concepts of stationarity to characterize random waveforms as having time-independent averages or PDFs.

   Returning to the point about averaging, $G_n(\omega)$ in (4.14) is an ensemble average, and (4.13) implies that the mean output noise power is independent of time. The mean *input* noise power $\langle n_i^2(t)\rangle$, will also be assumed independent of time. Since $\langle n_i^2(t)\rangle$ was not explicitly used in the development, it was not necessary to make that assumption earlier. It will be made now, for future use:

$$\langle n_i^2(t)\rangle = \overline{n_i^2(t)} = \langle n_i^2\rangle = \sigma_{n_i}^2 \tag{4.15}$$

3. The noise and signal are uncorrelated. Since the noise was assumed "stationary," time and ensemble averages are considered interchangeable, as stated above, and hence this assumption can be expressed as

$$\langle s_i(t_1)n_i(t_2)\rangle = \overline{s_i(t_1)n_i(t_2)} = 0 \tag{4.16}$$

   for arbitrary values of t_1 and t_2.
4. The input noise has zero mean, i.e.,

$$\langle n_i(t)\rangle = 0 \tag{4.17a}$$

   from which it follows through (4.2b)′ that

$$\langle n_o(t)\rangle = \int_{-\infty}^{\infty} d\tau h(\tau)\langle n_i(t-\tau)\rangle = 0 \tag{4.17b}$$

   i.e., the output noise $n_o(t)$ also has zero mean.

From (4.15) and (4.16) the mean-square input voltage is

$$\langle v_i^2(t)\rangle = s_i^2(t) + 2\langle s_i(t)n_i(t)\rangle + \langle n_i^2\rangle = s_i^2(t) + \langle n_i^2\rangle \tag{4.18a}$$

The mean square output voltage is

$$\langle v_o^2(t)\rangle = s_o^2(t) + 2\langle s_o(t)n_o(t)\rangle + \langle n_o^2\rangle \tag{4.18b}$$

where $s_o^2(t)$ is given by squaring (4.8a) [recognizing that $s_o^2(t) = s_o(t)s_o^*(t) = |s_o(t)|^2$, because $s_o(t)$ is real], $\langle n_o^2\rangle$ is obtained from (4.13), and $\langle s_o(t)n_o(t)\rangle$ can be obtained by taking the ensemble average of the product of (4.2a)′ and (4.2b)′, resulting in

$$\begin{aligned}\langle s_o(t)n_o(t)\rangle &= \int_{-\infty}^{\infty} d\tau_1 \int_{-\infty}^{\infty} d\tau_2 h(\tau_1)h(\tau_2)\langle s_i(t-\tau)n_i(t-\tau)\rangle \\ &= 0, \quad \text{with the aid of (4.16)}\end{aligned} \tag{4.19}$$

From (4.19), we conclude that the *output* signal and noise are uncorrelated as a consequence of the facts that (1) the *input* signal and noise are uncorrelated, and (2) the linear filtering operation does not change that property.

From (4.8a), (4.18b), (4.13), and (4.19) the output signal SNR of the filter at a *specific time* t_o, is given by

$$\rho_o(t_o) = \frac{|s_o(t_o)|^2}{\langle n_o^2 \rangle} = \frac{\left| \frac{1}{2\pi} \int_{-\infty}^{\infty} d\omega S_i(\omega) H(\omega) e^{j\omega t_o} \right|^2}{\frac{1}{2\pi} \int_{-\infty}^{\infty} d\omega |H(\omega)|^2 G_n(\omega)} \tag{4.20}$$

As a reference point, we would like to eventually relate ρ_o to the *input* SNR, defined as $\rho_i(t)$ and given by

$$\rho_i(t) = \frac{|s_i(t)|^2}{\langle n_i^2 \rangle} \tag{4.21}$$

Through the Schwartz inequality

$$\left| \int_{-\infty}^{\infty} dx f(x) g(x) \right|^2 \leq \int_{-\infty}^{\infty} dx |f(x)|^2 \int_{-\infty}^{\infty} dy |g(y)|^2 \tag{4.22}$$

for arbitrary functions $f(x)$ and $g(y)$, it is easily shown that the maximum possible value of the numerator of (4.20) is given by

$$\left| \int d\omega \left[\frac{S_i(\omega)}{\sqrt{2\pi}\sqrt{G_n(\omega)}} e^{j\omega t_o} \right] \frac{\sqrt{G_n(\omega)}}{\sqrt{2\pi}} H(\omega) \right|^2_{max}$$

$$= \left[\frac{1}{2\pi} \int_{-\infty}^{\infty} d\omega \frac{|S_i(\omega)|^2}{G_n(\omega)} \right] \left[\frac{1}{2\pi} \int_{-\infty}^{\infty} d\omega |H(\omega)|^2 G_n(\omega) \right] \tag{4.23}$$

where it is evident that, in (4.22),

$$f(\omega) = \frac{S_i(\omega) e^{j\omega t_o}}{\sqrt{2\pi G_n(\omega)}}, \qquad g(\omega) = \sqrt{\frac{G_n(\omega)}{2\pi}} H(\omega)$$

If the numerator of (4.20) has its maximum value, then the value of $\rho_o(t_o)$ is

$$[\rho_o(t_o)]^{max}_{num} = \frac{\left[\frac{1}{2\pi} \int_{-\infty}^{\infty} d\omega \frac{|S_i(\omega)|^2}{G_n(\omega)} \right] \left[\frac{1}{2\pi} \int_{-\infty}^{\infty} d\omega |H(\omega)|^2 G_n(\omega) \right]}{\frac{1}{2\pi} \int_{-\infty}^{\infty} d\omega |H(\omega)|^2 G_n(\omega)} = \frac{1}{2\pi} \int_{-\infty}^{\infty} d\omega \frac{|S_i(\omega)|^2}{G_n(\omega)} \tag{4.24}$$

The condition (4.24) is met if

$$H(\omega) = K \frac{S_i^*(\omega)}{G_n(\omega)} e^{-j\omega t_o} \tag{4.25}$$

(where K is an arbitrary scale factor) in which case we attain the maximum possible SNR at the filter output [since the denominator cancelled out in (4.24), the value of ρ_o given in (4.24) is the true maximum output SNR], which with the aid of (4.25) is

$$[\rho_o(t_o)]_{\max} = \rho_o^{(o)} = \frac{\left|\frac{1}{2\pi}\int_{-\infty}^{\infty} d\omega S_i(\omega)\left(K\frac{S_i^*(\omega)}{G_n(\omega)}e^{-j\omega t_o}\right)e^{j\omega t_o}\right|^2}{\frac{1}{2\pi}\int_{-\infty}^{\infty} d\omega G_n(\omega)\left|K\frac{S_i^*(\omega)}{G_n(\omega)}e^{-j\omega t_o}\right|^2}$$

$$= \frac{\left|\frac{1}{2\pi}\int_{-\infty}^{\infty} d\omega \frac{|S_i(\omega)|^2}{G_n(\omega)}\right|^2}{\frac{1}{2\pi}\int_{-\infty}^{\infty} d\omega \frac{|S_i(\omega)|^2}{G_n(\omega)}} = \frac{1}{2\pi}\int_{-\infty}^{\infty} d\omega \frac{|S_i(\omega)|^2}{G_n(\omega)} \tag{4.26}$$

which is consistent with (4.24).

The assumption we will now make, consistent with the usual radar applications, is that the noise spectrum is approximately flat over the signal band. To illustrate this, we show in Figure 4.2 the condition required to validate this assumption, which we will call the "white noise" assumption.

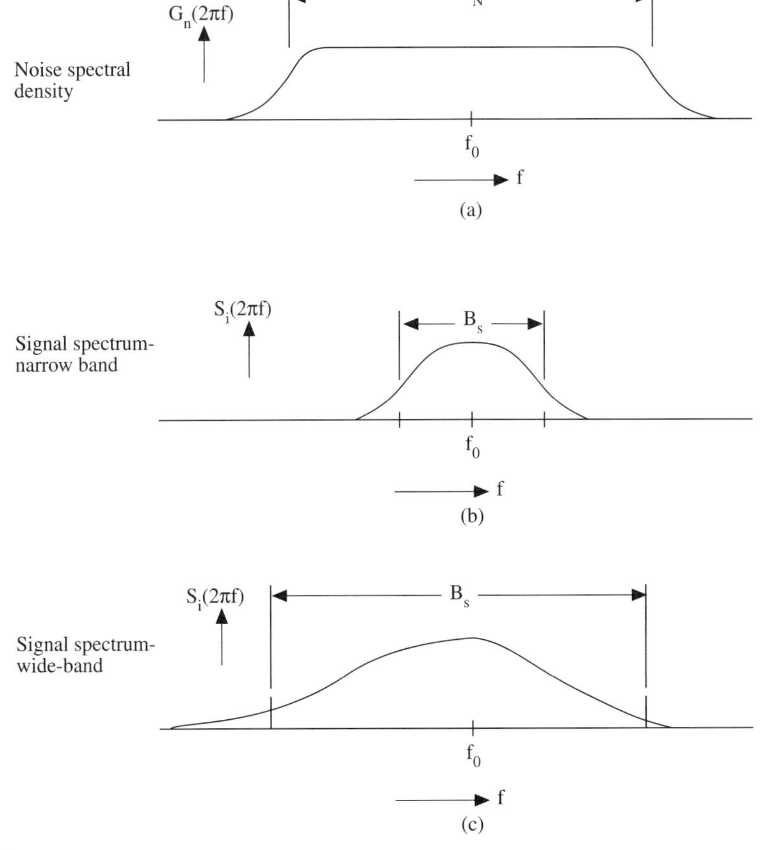

FIGURE 4.2
Illustration of the "white noise" assumption.

Figure 4.2b shows a case in which the signal spectrum falls off to negligible values well within the region where the noise spectrum is approximately flat. Figure 4.2c shows a case where there is significant signal power in the spectral regions where the noise spectrum is variable. The assumption in the case shown in Figure 4.2b is

$$B_s << B_n \tag{4.27a}$$

where B_s and B_n are signal and noise bandwidths, respectively. The assumption in Figure 4.2c is

$$B_s \approx B_n \tag{4.27b}$$

To approximate the noise as flat over the signal band, we require condition (4.27a). If that condition is not met, i.e., if (4.27b) holds, then strictly speaking we are dealing with the "colored noise" case, in which the "optimum filter" (defined as the filter that maximizes the output SNR at time t_o) is the filter whose frequency response function is the conjugate of the signal spectrum (with a delay factor $e^{-j\omega t_o}$) weighted by the reciprocal of the noise spectral density. This process "matches" the impulse response to the signal spectrum and preferentially passes those frequencies not contained in the noise spectrum while rejecting much of the spectral region that is heavy in noise. The latter process reduces the signal information contained in the output in order to maximally filter out noise.

If $G_n(\omega)$ can be approximated as flat, as in Figure 4.2a and b, then we have a true "matched" filter. In that case,

$$G_n(\omega) = N_o = \text{noise power density} \tag{4.28}$$

throughout the signal spectrum.

If the approximation (4.28) is applied in (4.25), then the "optimum filter" becomes

$$H(\omega) = K' S_i^*(\omega) e^{-j\omega t_o} \tag{4.29}$$

(where K' is an arbitrary scale factor) and the output SNR, from (4.26), is

$$[\rho_o(t_o)]_{max} = \frac{\left(1/2\pi \int_{-\infty}^{\infty} d\omega |S_i^*(\omega)|^2\right)}{N_o} = \rho_o^{(o)} \tag{4.30}$$

where it is noted that the dependence on t_o has vanished.

Equations (4.29) and (4.30) give us the true "matched filter," which is the optimum linear filter for signal-plus-white noise.

In this case, the frequency response function is proportional to the conjugate of the Fourier transform of the signal multiplied by the delay factor $e^{-j\omega t_o}$. In this case, we can easily determine the impulse response from (4.29) by noting that

$$
\begin{aligned}
h_{\text{opt}}(t) &= \frac{1}{2\pi}\int_{-\infty}^{\infty} d\omega H_{\text{opt}}(\omega)e^{j\omega t} \\
&= K'\,\frac{1}{2\pi}\int_{-\infty}^{\infty} d\omega S_{\text{i}}^{*}(\omega)e^{j\omega(t-t_{\text{o}})} \\
&= K'\left(\frac{1}{2\pi}\int_{-\infty}^{\infty} d\omega S_{\text{i}}(\omega)e^{-j\omega(t-t_{\text{o}})}\right)^{*} \\
&= K'[\text{s}_{\text{i}}(t_{\text{o}}-t)]^{*} = K'\text{s}_{\text{i}}(t_{\text{o}}-t)
\end{aligned}
\tag{4.31}
$$

Equation (4.31) tells us that the impulse response of the matched filter is proportional to a replica of the input signal waveform that is shifted by the amount t_{o} and inverted in time. To illustrate this, Figure 4.3a shows a signal waveform $\text{s}_{\text{i}}(t)$ and Figure 4.3b, c, and d show the matched filter impulse responses at three instants of time. It is clear from Figure 4.3b, c, and d that the matched filter is noncausal, because $h_{\text{opt}}(t)$

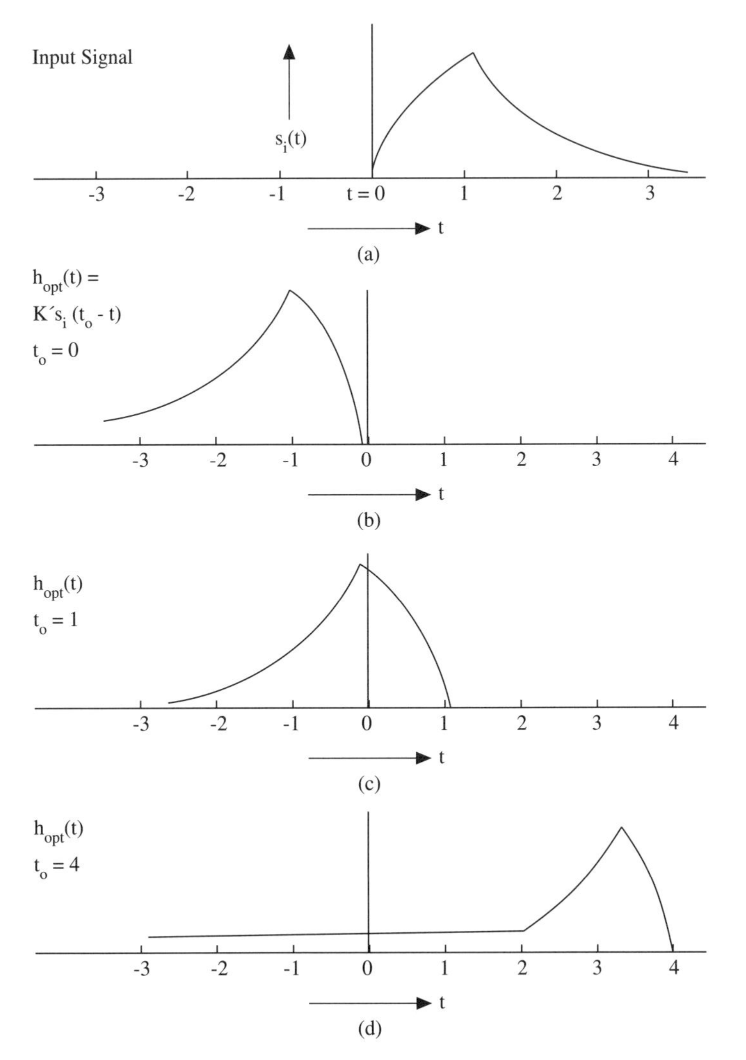

FIGURE 4.3
Matched filter impulse response.

$= 0$ for $t < 0$ while $s_i(t) = 0$ for $t < 0$. Hence, because causality is a requirement for a real-world linear filter, one cannot design a perfect matched filter in practice. However, Figure 4.3d illustrates the fact that it is possible to build a very good approximation to a matched filter provided the delay time t_o is sufficiently large to allow only a negligible nonzero value in the region $t < 0$. For example, a perfect rectangular pulse could be matched filtered causally provided t_o exceeds the pulse width. This would also be true for any pulse that decays to zero absolutely outside of a fixed time interval. This is illustrated in Figure 4.4a for a rectangular pulse and in Figure 4.4b for a pulse that is Gaussian between $t = 0$ and $t = \tau_p$ and zero for $t > \tau_p$ and $t < 0$.

We now return to (4.30) and note that we could express $S_i(\omega)$ as the inverse Fourier transform of $s_i(t)$, with the result (after changing the order of integration)

$$\frac{1}{2\pi}\int_{-\infty}^{\infty} d\omega |S_i(\omega)|^2 = \int_{-\infty}^{\infty} dt_1 \int_{-\infty}^{\infty} dt_2 s_i(t_1) s_i(t_2)\left(\frac{1}{2\pi}\int_{-\infty}^{\infty} d\omega e^{j\omega(t_1-t_2)}\right)$$

$$= \int_{-\infty}^{\infty} dt s_i^2(t) = E_s = \text{total signal energy} \tag{4.32}$$

where we have used the equivalent of (4.7) to obtain the result.

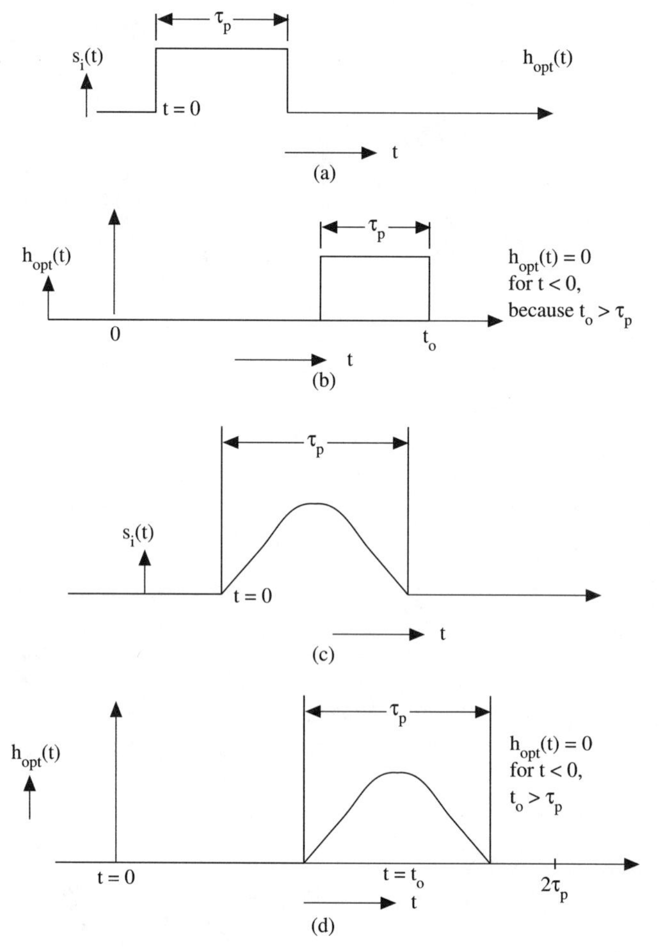

FIGURE 4.4
A matched filter that approximates causality.

Equation (4.32), known as Parseval's theorem, expresses the fact that we can determine the total signal *energy* by integrating signal *power* $s_i^2(t)$ over all time or alternatively by integrating the spectral distribution of power $|S_i(\omega)|^2$ over all frequency space.

Finally we use (4.32) in (4.30), with the result

$$\rho_o^{(o)} = \frac{E_s}{N_o} = \frac{\text{total input signal energy}}{\text{input noise power density}} \tag{4.33}$$

where

$$E_s = \int_{-\infty}^{\infty} dt s_i^2(t) = \frac{1}{2\pi}\int_{-\infty}^{\infty} d\omega |S_i(\omega)|^2 \tag{4.33a)'}$$

$$N_o = \frac{\sigma_{ni}^2}{B_N} \tag{4.33b)'}$$

where $\sigma_{ni}^2 = \langle n_i^2 \rangle$ = mean-square input noise and B_N = input noise bandwidth.

There are some interesting and important observations that can be made through matched filter theory, and some of these are quite pertinent to radar.

First, if the signal waveform occupies a finite time interval T_s, then from (4.33a)′, we can write (using the time integration)

$$E_s = \overline{s_i^2(t)} T_s \tag{4.34}$$

where $\overline{s_i^2(t)}$ is the time averaged signal power. We then define the input SNR as

$$\rho_i = \frac{\overline{s_i^2(t)}}{\sigma_{ni}^2} \tag{4.35}$$

It follows from (4.33), (4.34), and (4.35) that

$$\rho_o^{(o)} = \left[\frac{\overline{s_i^2(t)}}{\sigma_{ni}^2}\right] B_N T_s = \rho_i B_N T_s \tag{4.36}$$

Equation (4.36) tells us that the *improvement* in SNR due to matched filtering is

$$\frac{\rho_o^{(o)}}{\rho_i} = B_N T_s \tag{4.36)'}$$

i.e., SNR improvement is the "time bandwidth product," i.e., product of the input noise bandwidth and the signal duration.

In the sections to follow, we will discuss applications of the matched filter to radar problems.

4.2 NARROWBAND SIGNAL AND NOISE

A radar signal at the IF stage of a receiver has the form of $s_7(t; \boldsymbol{\Omega}_7)$ as in Eq. (3.25) in Section 3.7. To adapt $s_7(t; \boldsymbol{\Omega}_7)$ to this general discussion, we assume that it has the generic form

$$s_i(t) = a\cos(\omega_o t + \psi)p(t) \tag{4.37}$$

where $p(t)$ is assumed to vary slowly and where a and ψ are assumed constant over many cycles of $\cos(\omega_o t + \psi)$ (e.g., thousands to millions) and where ω_o is the frequency at that stage.

The noise waveform is modeled as a "narrowband" random signal, of the form

$$\begin{aligned} n_i(t) &= x_n(t)\cos\omega_o t + y_n(t)\sin\omega_o t \\ &= a_n(t)\cos[\omega_o t + \psi_n(t)] \end{aligned} \tag{4.38}$$

where

$$x_n(t) = a_n(t)\cos\psi_n(t)$$
$$y_n(t) = -a_n(t)\sin\psi_n(t)$$

or

$$a_n(t) = \sqrt{x_n^2(t) + y_n^2(t)}$$
$$\psi_n(t) = \tan^{-1}\left(\frac{-y_n(t)}{x_n(t)}\right)$$

and where $x_n(t)$, $y_n(t)$, $a_n(t)$, and $\psi_n(t)$ vary slowly compared with $\cos\omega_o t$ or $\sin\psi_o t$, i.e., they remain effectively constant over many cycles.

This type of noise waveform and the signal waveform of (4.37) (also "narrowband" but we ignore the slow time variation of a and ψ during the period required to implement processing) are standard for modeling radar signals at RF or IF and will be used in theoretical discussions from this point on. The "quadrature" components of the noise $x_n(t)$ and $y_n(t)$ (the terminology means they are "in quadrature" with each other, or equivalently 90° apart in phase) have the following properties.

Zero mean:

$$\langle x_n(t)\rangle = \langle y_n(t)\rangle = 0 \text{ for all } t \tag{4.39a}$$

Mean-square:

$$\begin{aligned} \overline{\langle n_i^2(t)\rangle} &= \langle x_n^2\rangle\,\overline{\cos^2(\omega_o t)} + \langle y_n^2(t)\rangle\,\overline{\sin^2(\omega_o t)} + 2\langle x_n(t)y_n(t)\rangle\,\overline{\cos\omega_o t\sin\omega_o t} \\ &= \left[\frac{\langle x_n^2(t)\rangle + \langle y_n^2(t)\rangle}{2}\right] + \left[\frac{\langle x_n^2(t)\rangle - \langle y_n^2(t)\rangle}{2}\right]\overline{\cos 2\omega_o t} \\ &\quad + \overline{\langle x_n(t)y_n(t)\rangle}\,\overline{\sin(2\omega_o t)} \end{aligned} \tag{4.39b}$$

The assumed properties of the mean squares of $x_n(t)$ and $y_n(t)$ are

$$\langle x_n^2(t)\rangle = \langle y_n^2(t)\rangle \tag{4.39c}$$

$$\langle x_n(t)y_n(t)\rangle = 0 \tag{4.39d}$$

Property (4.39c) eliminates the $\cos \omega_o t$ term of (4.39b) without time averaging and property (4.39d) eliminates the $\sin 2\omega_o t$ without time averaging. Also, time averaging will eliminate both of these terms without ensemble averaging, because $\overline{\cos 2\omega_o t}$ and $\overline{\sin 2\omega_o t}$ are both zero. Equations (4.39c, d) are consequences of the stationarity property assumed earlier, because they guarantee that time and ensemble averaging will give the same result. We should qualify that conclusion by noting that we have separated the time scales for time averaging into the short-time scale that applies to cosines or sines of $\omega_o t$ or $2\omega_o t$ and the long-time scale that applies to $x_n(t)$ and $y_n(t)$. That is quite legitimate because these time scales differ by enormous numbers of orders of magnitude, e.g., a factor of a million or more.

Properties (4.39c, d) are characteristic of narrowband *Gaussian* processes, but we do not need to assume Gaussian statistics for the discussion to follow, because the key ideas are qualitatively applicable regardless of the PDF.

A consequence of (4.38), (4.39a), and (4.39c, d) is

$$\langle n_i^2(t)\rangle = \frac{\langle x_n^2(t)\rangle + \langle y_n^2(t)\rangle}{2} = \frac{\langle a_n^2(t)\rangle}{2} = \langle x_n^2(t)\rangle = \langle y_n^2(t)\rangle \tag{4.39e}$$

This means that either quadrature component has a mean-square value equal to that of the noise waveform in $n_i(t)$ and it is one-half the mean-square of the amplitude $a_n(t)$.

A further property, following from (4.39e) and stationarity, is that

$$\langle x_n^2(t)\rangle = \langle y_n^2(t)\rangle = \langle x_n^2\rangle = \langle y_n^2\rangle = \frac{\langle a_n^2\rangle}{2} = \sigma_{ni}^2 \tag{4.39f}$$

i.e., all of these mean-square values are independent of time.

4.3 COHERENT FILTERING

Let us now reexamine the matched filter theory covered in Section 4.1 and try to apply it to narrowband signal-plus-noise as discussed in Section 4.2. This is necessary to apply the theory to a radar signal at the RF or IF stage of a receiver. Filtering at those stages is called "coherent filtering."

First, we will make a "white noise" assumption. We will also assume that the signal is an RF or IF rectangular pulse of arbitrary length T_s (Fig. 4.5). The input signal is

$$s_i(t)\begin{cases} = a\cos(\omega_o t + \psi) & \text{if } 0 \le t \le T_s \\ = 0 & \text{if } t < 0 \text{ or } t > T_s \end{cases} \tag{4.40}$$

The impulse response of a filter matched to this signal is

$$h_{opt}(t) = Ks_i(t_o - t) = Kap(t_o - t)\cos[\omega_o(t_o - t) + \psi] \tag{4.41}$$

where

$$\begin{aligned} p(t) &= u(t) - u(t - T_s) = \text{rectangular pulse envelope} \\ &= 1 \quad \text{if } 0 \le t \le T_s \\ &= 0 \quad \text{otherwise} \end{aligned}$$

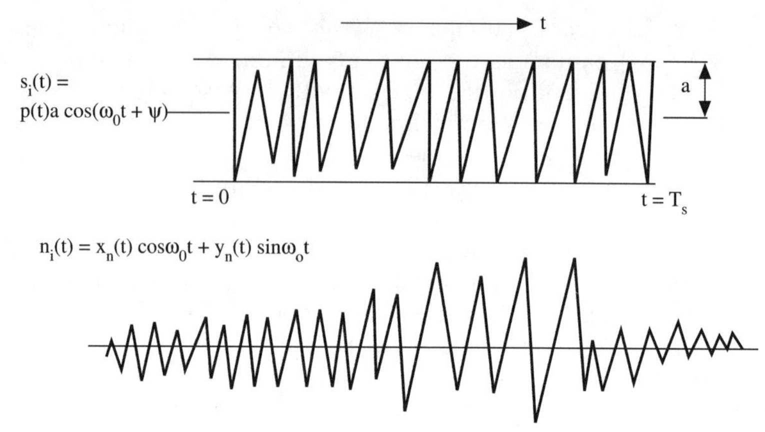

FIGURE 4.5
Radar pulse at RF or IF plus noise.

The matched filter *signal output* at an arbitrary instant of time t is [from (4.40) and (4.41)]

$$
\begin{aligned}
s_o(t) &= \int_{-\infty}^{\infty} d\tau h_{opt}(\tau) s_i(t-\tau) \\
&= Ka^2 \int_{-\infty}^{\infty} d\tau p(t_o-\tau)p(t-\tau)\cos[\omega_o(t-\tau)+\psi]\cos[\omega_o(t_o-\tau)+\psi] \\
&= \frac{Ka^2}{2}\int_{-\infty}^{\infty} d\tau p(t_o-\tau)p(t-\tau)\{\cos[\omega_o(t+t_o-2\tau)+2\psi]+\cos[\omega_o(t-t_o)]\} \\
&\cong \frac{Ka^2}{2}\cos[\omega_o(t-t_o)]\int_{-\infty}^{\infty} d\tau p(t_o-\tau)p(t-\tau)
\end{aligned} \tag{4.42}
$$

The final form of (4.42) is due to the fact that the waveform

$$\cos[\omega_o(t+t_o-2\tau)+2\psi]$$

undergoes many cycles during the pulse interval and effectively integrates to a negligible value. That assumption, when examined closely, may not be strictly valid at an IF of, say, 1 MHz and for a pulse length of the order of 1 μsec. But it is nearly always sufficiently accurate to be a safe assumption in radar system studies.

We can write the output signal $s_o(t)$ in the approximate form (from 4.42)

$$s_o(t) \approx \frac{Ka^2}{2}\cos(\omega_o\Delta t)\int_{-\infty}^{\infty} dt' p(t')p(t'+\Delta t) \tag{4.43}$$

where

$$\Delta t = t - t_o$$

The parameter Δt is the displacement of the time t from the instant $t = t_o$, the time instant at which the filter is optimized. The form of the output signal is

$$s_o(t) = A_o(\Delta t)\cos(\omega_o t + \psi + \Delta\psi) \tag{4.44}$$

which is an IF signal with a new amplitude level $A_o(\Delta t)$ and a phase shift $\Delta\psi$. The values of the output amplitude and a phase shift are

$$A_o(\Delta t) = \frac{Ka^2}{2}\int_{-\infty}^{\infty} dt' p(t')p(t' + \Delta t) \tag{4.44a$'$}$$

and

$$\Delta\psi = -\omega_o t_o - \psi \tag{4.44b$'$}$$

The amplitude is proportional to the input signal amplitude and depends on the time displacement Δt through the normalized autocorrelation function (NACF) of the modulation waveform given by

$$\Lambda(\Delta t) = \frac{\int_{-\infty}^{\infty} dt' p(t')p(t' + \Delta t)}{\int_{-\infty}^{\infty} dt' p^2(t')} \tag{4.45}$$

Using (4.45) in (4.44a)$'$, we can rewrite the latter as

$$A_o(\Delta t) = A_o(0)\Lambda(\Delta t) \tag{4.46}$$

where

$$A_o(0) = \frac{Ka^2}{2}\int_{-\infty}^{\infty} dt' p^2(t') = \text{signal output amplitude for zero time displacement}$$

and $\Lambda(\Delta t)$ = NACF of the amplitude modulation waveform $p(t)$, equal to unity when $\Delta t = 0$ and decreasing with increasing time displacement Δt.

The amplitude $A_o(0)$ is the amplitude of the output in the case where the filter is perfectly matched in time to the input signal, while the absolute value of $\Lambda(\Delta t)$ is a "degradation factor," which reduces the output signal amplitude as the time displacement Δt increases.

From (4.46), (4.44b)$'$, and (4.44), we have the final form of the output signal

$$s_o(t) = A_o(0)\Lambda(\Delta t)\cos(\omega_o \Delta t) \tag{4.47}$$

If the matching is perfect, then $\Delta t = 0$ and we have

$$s_o(t) = A_o(0) = \frac{Ka^2}{2}\int_{-\infty}^{\infty} dt' p^2(t') \tag{4.47$'$}$$

We now consider the noise output of the filter. For a narrowband noise centered at frequency ω_o, the spectrum is as shown in Figure 4.6.

The noise has passed through the RF and IF amplifier on its way to the point where we are applying matched filtering. Its spectrum has the shape of the IF passband. If that passband is approximately flat over the frequency region

$$\pm\omega_o - \pi B_N \leq \omega \leq \pm\omega_o + \pi B_N$$

as shown in Figure 4.6, then the mean noise output power, as obtained with the aid of (4.20), is

$$\begin{aligned}\langle n_o^2 \rangle &= \frac{1}{2\pi}\int_{-\infty}^{\infty} d\omega G_n(\omega)|H(\omega)|^2 \\ &= \frac{1}{2\pi}\left(\frac{N_o}{2}\right)\left\{\int_{-\infty}^{\infty} d\omega |H[\omega_o + (\omega - \omega_o)]|^2 \right. \\ &\quad \left. + \int_{-\infty}^{\infty} d\omega |H[-\omega_o + (\omega + \omega_o)]|^2 \right\}\end{aligned} \tag{4.48}$$

where it is assumed that no overlap occurs between the positive and negative frequency regions of the filter spectrum, as would always be the case for an IF that far exceeds the filter bandwidth. The final form of (4.48), then, is

$$\langle n_o^2 \rangle = N_o\left[\frac{1}{2\pi}\int_{-\infty}^{\infty} d\omega |H(\omega)|^2\right] \tag{4.49}$$

where $N_o = \sigma_{ni}^2/B_N$. In the case of the matched filter [Eq. (4.29)]

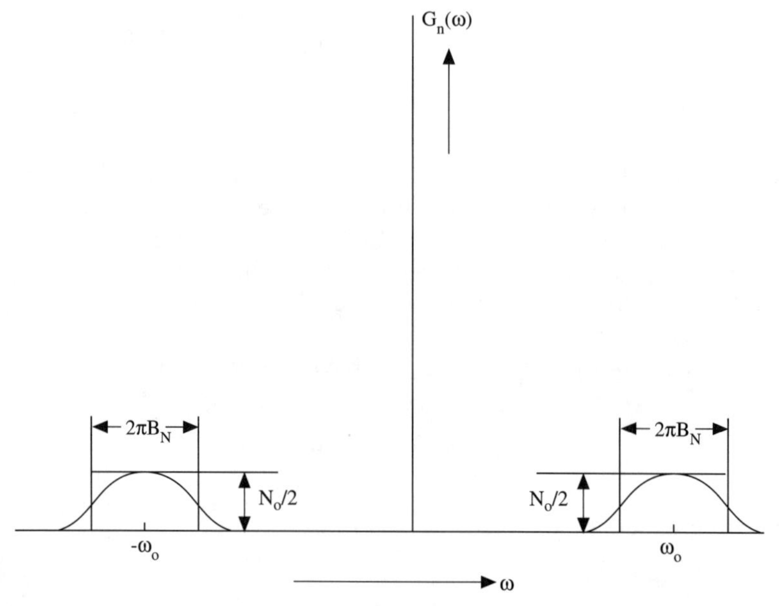

FIGURE 4.6
Narrowband noise spectrum—noise input to filter.

$$H_{\text{opt}}(\omega) = KS_{\text{i}}^{*}(\omega)e^{-j\omega t_{\text{o}}} \tag{4.50}$$

or even if the filter is not perfectly matched in time, so that its frequency response function is

$$H(\omega) = KS_{\text{i}}^{*}(\omega)e^{-j\omega t_{\text{o}}}e^{-j\omega\Delta t} \tag{4.51}$$

then from (4.49), (4.33a)′, and (4.37) with the pulse modulation function $p(t)$ included

$$\begin{aligned}\langle n_{\text{o}}^{2}\rangle &= K^{2}N_{\text{o}}\left(\frac{1}{2\pi}\int_{-\infty}^{\infty} d\omega |S_{\text{i}}(\omega)|^{2}\right)\\ &= K^{2}N_{\text{o}}\int_{-\infty}^{\infty} dt' s_{\text{i}}^{2}(t') = K^{2}N_{\text{o}}a^{2}\int_{-\infty}^{\infty} dt' p^{2}(t')\cos^{2}(\omega_{\text{o}}t' + \psi)\\ &= \frac{K^{2}N_{\text{o}}a^{2}}{2}\left[\int_{-\infty}^{\infty} dt' p^{2}(t') + \int_{-\infty}^{\infty} dt' p^{2}(t')\cos(2\omega_{\text{o}}t' + 2\psi)\right]\end{aligned} \tag{4.52}$$

The second term on the right-hand side (RHS) of (4.52) can be approximated as zero over a sufficiently large pulse length (i.e., $T_{\text{s}} >> 1/2\omega_{\text{o}}$), with the result

$$\langle n_{\text{o}}^{2}\rangle = \frac{K^{2}N_{\text{o}}a^{2}}{2}\int_{-\infty}^{\infty} dt' p^{2}(t') \tag{4.53}$$

We note that $\langle n_{\text{o}}^{2}\rangle$ is independent of Δt, i.e., the mismatch in time does not affect the output noise power.

To obtain the output SNR of this filter, whose amplitude waveform is $p(t)$, but which may be mismatched in time by an amount $\Delta t = t - t_{\text{o}}$, we write, with the aid of (4.47), (4.47)′, and (4.53)

$$\begin{aligned}\rho_{\text{o}}(\Delta t) &= \frac{[s_{\text{o}}(t)]^{2}}{\langle n_{\text{o}}^{2}\rangle} = \frac{[A_{\text{o}}(0)]^{2}[\Lambda(\Delta t)]^{2}\cos^{2}(\omega_{\text{o}}\Delta t)}{(K^{2}N_{\text{o}}a^{2}/2)\int_{-\infty}^{\infty} dt' p^{2}(t')}\\ &= \rho_{\text{o}}(0)[\Lambda(\Delta t)]^{2}\cos^{2}(\omega_{\text{o}}t)\end{aligned} \tag{4.54}$$

where

$$\begin{aligned}\rho_{\text{o}}(0) &= \text{output SNR for perfect matching in time } (\Delta t = 0)\\ &= \frac{(K^{2}a^{4}/4)\left[\int_{-\infty}^{\infty} pt' p^{2}(t')\right]^{2}}{(K^{2}a^{2}/2)N_{\text{o}}\left[\int_{-\infty}^{\infty} dt' p^{2}(t')\right]} = \frac{a^{2}}{2N_{\text{o}}}\int_{-\infty}^{\infty} dt' p^{2}(t') = \frac{a^{2}}{2\sigma_{\text{ni}}^{2}}B_{\text{N}}\int_{-\infty}^{\infty} dt' p^{2}(t')\end{aligned}$$

For a train of N_{p} rectangular pulses of duration τ_{p}, the most common type of radar signal, then

$$\rho_{\text{o}}(0) = \frac{a^{2}}{2\sigma_{\text{ni}}^{2}}N_{\text{p}}\tau_{\text{p}}B_{\text{N}} = \rho_{\text{i}}(B_{\text{N}})(N_{\text{p}}\tau_{\text{p}}) = \frac{E_{\text{s}}}{N_{\text{o}}} \tag{4.55}$$

where ρ_i is the input SNR, equal to $a^2/2\sigma_{ni}^2$, and $(B_N)(N_p\tau_p)$ is the "time-bandwidth product" $B_N T_s$ discussed generically in Section 4.1 and given as the ratio of output SNR to input SNR for a matched filter in Eq. (4.36)′.

At this point, let us return to the pulsed radar context and note that the SNR of the received signal at the input to the filter is [from (2.35)]

$$\rho_i = \frac{P_T G_{To} A_{eRo} \sigma_o f(\theta, \phi)}{(4\pi)^2 r^4 (kT_o B_N F_N)} \tag{4.56}$$

where B_R and B_N are equated and where the effects of angular offset due to antenna patterns and scattering patterns are lumped into the function $f(\theta, \phi)$. The output SNR for the filter perfectly matched in time, as obtained from (4.55) and (4.56), is

$$\rho_o = \frac{P_T G_{To} A_{eRo} \sigma_0 f(\theta, \phi) N_p \tau_p}{(4\pi)^2 r^4 (kT_o F_N)} \tag{4.57}$$

Equation (4.57) gives us the *maximum* possible SNR attainable through matched filtering of a pulsed radar signal, given that it were possible to match perfectly in time. It is a very useful tool in evaluating the best possible SNR performance based on absolutely ideal conditions, which are almost never quite attainable in practice. In the next few sections, the degradation effects of mismatch in the filter design due to ignorance of signal parameters will be discussed.

4.4 EFFECT OF TIME MISMATCH ON COHERENT "MATCHED" FILTER OUTPUT SNR: NONCOHERENT "MATCHED" FILTERING

It is somewhat of a contradiction in terms to refer to a "matched" filter that is mismatched in one of its signal parameters. But in radar, it is virtually impossible to know all the details of the return signal shape. However, to try to maximize SNR, the radar designer *attempts* to implement a matched filter, but must account for the ignorance of signal parameters in that design.

The first parameter about which the designer is generally ignorant is the signal delay time, an exact knowledge of which would require that he or she know the target range precisely. Before proceeding with a discussion of that, we return to (4.54) and use it to define the "degradation factor" as the ratio of the output SNR for time mismatch Δt to the output SNR for $\Delta t = 0$, given by

$$D(\Delta t) = \frac{\rho_o(\Delta t)}{\rho_o(0)} = D_{ncoh}(\Delta t) D_{coh}(\Delta\tau) \tag{4.58}$$

where

$$D_{coh}(\Delta t) = \cos^2(\omega_o \Delta t) = \frac{1}{2} + \frac{1}{2}\cos(2\omega_o \Delta t) \tag*{(4.58a)′}$$

assuming zero phase difference between signal and filter impulse response.

$$D_{\text{ncoh}}(\Delta t) = [\Lambda(\Delta t)]^2 = \left[\frac{\int_{-\infty}^{\infty} dt' p(t')p(t' + \Delta t)}{\int_{-\infty}^{\infty} dt' p^2(t')}\right]^2$$

$$= \text{square of NACF of pulse waveform} \qquad (4.58b)'$$

$D_{\text{coh}}(\Delta t)$ is the "coherent degradation factor," due to the fact that the "matching" of the filter includes matching of the IF waveform, i.e., the filtering is *coherent*.

The "noncoherent degradation factor," $D_{\text{ncoh}}(\Delta t)$, contains the effect of time mismatch in the pulse waveform, which is independent of the frequency ω_o and, for the simplest type of radar signal, can be separated out from the coherent degradation factor. Hence, those two effects can be studied separately.

It is obvious from (4.58a)′ that the degradation due to an attempt to match-filter coherently can be disastrous. If Δt has a value such that $2\omega_o\Delta t = (2n + 1)\pi$, $\cos(2\omega_o\Delta t) = -1$ and the output SNR is zero. Hence, it is very unsafe to use coherent matched filtering and expect to realize all of its SNR benefits unless it is possible to control the time mismatch very precisely. The obvious remedy is to use the *amplitude* if the filter output as the output parameter. That is known as "noncoherent" filtering, which leaves us with only the "noncoherent" degradation factor $D_{\text{ncoh}}(\Delta t)$ as given by (4.58b)′.

To implement noncoherent filtering, we can use "quadrature filters," as shown in Figure 4.7.

The "cosine" and "sine" filters in Figure 4.7 are both coherent "mismatched" filters that are matched in frequency and amplitude waveshape to the expected incoming signal but are 90° out of phase, or "in quadrature." Their impulse responses are

$$h_c(t) = Kp(t_o - t)\cos(\omega_o t) \qquad \text{for the cosine filter} \qquad (4.59a)$$

$$h_s(t) = Kp(t_o - t)\sin(\omega_o t) \qquad \text{for the sine filter} \qquad (4.59b)$$

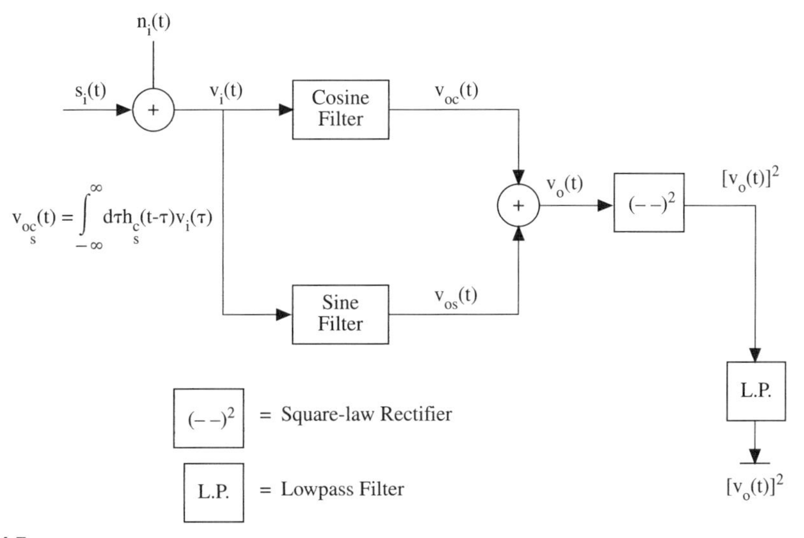

FIGURE 4.7
Quadrature filters.

The signal outputs of the two filters are [from (4.37) and (4.59a,b)]

$$
\begin{aligned}
s_{o\substack{c\\s}}(t) &= \int_{-\infty}^{\infty} d\tau h_{\substack{c\\s}}(t-\tau)s_i(\tau) \\
&= Ka \int_{-\infty}^{\infty} d\tau p(\tau - \Delta t)p(\tau) \begin{matrix}\cos\\ \sin\end{matrix}[\omega_o(t-\tau)] \cos(\omega_o\tau + \psi) \\
&= X_{o\substack{c\\s}}(\Delta t)\cos(\omega_o t) - Y_{o\substack{c\\s}}(\Delta t)\sin(\omega_o t) \qquad (4.60)
\end{aligned}
$$

where

$$
\begin{matrix} X_{oc}(\Delta t) \\ Y_{oc}(\Delta t) \end{matrix} = \frac{Ka}{2}\int_{-\infty}^{\infty} d\tau p(\tau)p(\tau-\Delta t)\begin{bmatrix} \cos\psi(1+\cos 2\omega_o\tau) - \sin\psi \sin 2\omega_o\tau \\ \sin\psi(1-\cos 2\omega_o\tau) - \cos\psi\sin 2\omega_o\tau \end{bmatrix}
$$

$$
\begin{matrix} X_{os}(\Delta t) \\ Y_{os}(\Delta t) \end{matrix} = \frac{Ka}{2}\int_{-\infty}^{\infty} d\tau p(\tau)p(\tau-\Delta t)\begin{bmatrix} \sin\psi(1-\cos 2\omega_o\tau) - \cos\psi \sin 2\omega_o\tau \\ -\cos\psi(1+\cos 2\omega_o\tau) + \sin\psi\sin 2\omega_o\tau \end{bmatrix}
$$

We can rewrite the output coefficients of (4.60) in the form [with the aid of (4.46)]

$$
\begin{matrix} X_{oc}(\Delta t) \\ Y_{oc}(\Delta t) \end{matrix} = A_o(0)\Lambda(\Delta t)\begin{bmatrix} \cos\psi \\ \sin\psi \end{bmatrix} \pm \frac{Ka}{2}\int_{-\infty}^{\infty} d\tau p(\tau)p(\tau-\Delta t)\begin{smallmatrix}\cos\\ \sin\end{smallmatrix}(2\omega_o\tau+\psi) \qquad (4.61a)
$$

$$
\begin{matrix} X_{os}(\Delta t) \\ Y_{os}(\Delta t) \end{matrix} = A_o(0)\Lambda(\Delta t)\begin{bmatrix} \sin\psi \\ -\cos\psi \end{bmatrix} \pm \frac{Ka}{2}\int_{-\infty}^{\infty} d\tau p(\tau)p(\tau-\Delta t)\begin{smallmatrix}\sin\\ \cos\end{smallmatrix}(2\omega_o\tau+\psi) \qquad (4.61b)
$$

The integrals in (4.61a,b) have oscillatory integrands that will result in negligibly small values of those integrals. The surviving terms, with the aid of (4.60), result in

$$
s_{oc}(t) = A_o(0)\Lambda(\Delta t)\cos(\omega_o t + \psi) \qquad (4.62a)
$$

$$
s_{os}(t) = A_o(0)\Lambda(\Delta t)\sin(\omega_o t + \psi) \qquad (4.62b)
$$

The signal portion of the sum $v_o(t) = v_{oc}(t) + v_{os}(t)$ shown in Figure 4.7, after square-law detection and lowpass filtering, is [with the aid of (4.58)]

$$
\overline{[s_{oc}(t) + s_{os}(t)]^2} = [A_o(0)\Lambda(\Delta t)]^2\left\{\frac{1}{2}[1+\overline{\cos(2\omega_o t + 2\psi)}] + \frac{1}{2}[1 - \overline{\cos(2\omega_o t + 2\psi)}] + \overline{\sin(2\omega_o t + 2\psi)}\right\} = [A_o(0)]^2 D_{ncoh}(\Delta t) \qquad (4.63)
$$

where the lowpass filtering action is depicted as a short-time averaging process, which filters out the components at the frequency $2\omega_o$.

It was established earlier that the noise power output of a coherent matched filter is independent of the time mismatch if such a mismatch exists [see Eqs. (4.48)–(4.53) and the discussion pertinent to these equations]. Since the time delay occurs in the *phase* of the signal, the same independence exists for any mismatch in phase regardless of the *cause* of that mismatch (i.e., whether it is caused by absence of knowledge of target range or ignorance of phase shifts that might occur in propagation or receiver circuitry). It follows that since the cosine and sine filters are both "matched" filters with mismatches in phase, their mean noise power outputs are the same. The square-law rectification operation yields, for signal-plus-noise outputs (where n_{oc} and n_{os} are cosine and sine filter noise outputs, respectively),

$$\begin{aligned}\langle v_o^2\rangle &= \langle (v_{oc} + v_{os})^2\rangle = \langle (s_{oc} + n_{oc} + s_{os} + n_{os})^2\rangle \\ &= (s_{oc} + s_{os})^2 + 2(s_{oc} + s_{os})(\langle n_{oc}\rangle + \langle n_{os}\rangle) \\ &\quad + \langle n_{oc}^2\rangle + \langle n_{os}^2\rangle + 2\langle n_{oc} n_{os}\rangle \end{aligned} \tag{4.64}$$

It was established earlier that the input noise has zero mean and is uncorrelated with the signal for a coherent filter in general [see Eqs. (4.16), (4.17a,b), (4.18a,b), and (4.19) and accompanying discussion] and a matched filter in particular. The output SNR of a linear rectification process (output is *amplitude*, not its square)

$$(\mathrm{SNR})_{\mathrm{out}} = \frac{[s_o(t)]^2}{\langle n_{oc}^2\rangle + \langle n_{os}^2\rangle} = \frac{[s_o(t)]^2}{2\langle n_{oc}^2\rangle} = \frac{1}{2}\rho_o \tag{4.65}$$

where ρ_o is the output of a *coherent* matched filter.

Equation (4.65) is obtained from (4.64) by recognizing that

$$\langle n_{oc}^2\rangle = \langle n_{os}^2\rangle \tag{4.66a}$$

as noted above and assuming that

$$\langle n_{oc} n_{os}\rangle = 0 \tag{4.66b}$$

which can be justified for narrowband noise by the argument

$$\begin{aligned}\langle n_{oc}(t) n_{os}(t)\rangle &= \int_{-\infty}^{\infty}\!\!\int d\tau_1 d\tau_2 \langle n_i(t-\tau_1) n_i(t-\tau_2)\rangle \\ &\quad \cdots \cos(\omega_o\tau_1)\sin(\omega_o\tau_2) p(\tau_1) p(\tau_2)\end{aligned} \tag{4.66c}$$

and assuming that the input noise ACF has the property

$$\langle n_i(t_1) n_i(t_2)\rangle = \frac{\sigma_{ni}^2}{B_N}\,\delta(t_1 - t_2) \tag{4.66d}$$

which results in (4.66c) taking the form

$$\langle n_{oc}(t) n_{os}(t)\rangle = \frac{\sigma_{ni}^2}{2B_N}\int_{-\infty}^{\infty} d\tau p^2(\tau)\sin(2\omega_o\tau_1) = 0 \tag{4.66e}$$

because the integral in (4.66e) is highly oscillatory and the function $\sin(2\omega_o\tau_1)$ integrated over many cycles with a relatively constant value of $p^2(\tau)$ yields a negligibly small integral for all practical purposes.

The property (4.66d) is tantamount to assuming noise that is "white" over the spectral band of $p(t)$. To justify it, we note (Wiener–Khintchine theorem) that the ACF and power spectrum of noise are Fourier transform pairs; for white noise $n(t)$

$$R_n(\tau) = \langle n(t)n(t+\tau)\rangle = \frac{1}{2\pi}\int_{-\infty}^{\infty} d\omega G_n(\omega)e^{j\omega\tau}$$

$$= \frac{\sigma_{ni}^2}{B_N}\left[\frac{1}{2\pi}\int_{-\infty}^{\infty} d\omega e^{j\omega\tau}\right] = \frac{\sigma_{ni}^2}{B_N}\delta(\tau) \tag{4.67}$$

[See Eqs. (4.5), (4.6), and (4.7) and associated discussion.]

The result (4.65) tells us that if a *linear rectifier* rather than a *quadratic rectifier* were used in the quadrature filter scheme illustrated in Figure 4.7 (equivalently if we were to take the *amplitude* of the matched filter as the output parameter rather than the complete IF signal), then we would lose exactly 3 dB of output SNR. Equivalently it would require a 3 dB increase in SNR at the filter input to achieve a specified output SNR with noncoherent rather than coherent matched filtering.

Since Figure 4.7 implies quadratic rectification rather than linear rectification, the result is slightly different from (4.65), but is in the same qualitative direction. As we will see in Chapter 5, when signal detection is covered, it requires a few more dB (not necessarily 3 dB, but usually somewhere in that neighborhood) to achieve a given level of detection performance with noncoherent filtering relative to that achievable with coherent filtering. That is the price that must be paid for ignorance of signal phase.

4.5 EFFECT OF RF PHASE MISMATCH ON COHERENT "MATCHED" FILTER OUTPUT SNR

The time delay and RF phase mismatches discussed in Section 4.4 constitute only one mechanism for generating a mismatch in the RF [and therefore IF, as explained in Section 3.7; see Eqs. (3.21) through (3.24)] phase ψ. The time delay appears in the IF signal as a part of the phase, and also appears in the modulation signal $p(t)$. But even if the time delay were exactly known, it would still be impossible in general to predict the RF phase of the received signal. To focus on this point, we note that the *expected* IF signal on which to base the matched filter design is of the form (given an exactly known frequency ω_o).

$$s_{exp}(t) = ap[t-(\tau_d)_{exp}]\cos\{\omega_o[t-(\tau_d)_{exp}]+\psi_{exp}\} \tag{4.68a}$$

whose subscript "exp" denotes "expected" value of a parameter.

The impulse response of a matched filter to "match" at $t = t_o$ is

$$h(t) = Ks_{exp}(t_o - t) = Kap(t_o - t - \tau_d)_{exp}$$

$$\cdots\cdots \cos\{\omega_o[t_o - t - (\tau_d)_{exp}] + \psi_{exp}\} \tag{4.68b}$$

The true signal is (where subscript "true" denotes true value of a parameter)

$$s_{true}(t) = ap[t - (\tau_d)_{true}] \cos\{\omega_o[t - (\tau_d)_{true}] + \psi_{true}\} \tag{4.68c}$$

The signal part of the filter output at time t_o is

$$\begin{aligned} s_o(t_o) &= \int_{-\infty}^{\infty} d\tau s_{true}(\tau) h(t - \tau) \\ &= Ka^2 \int_{-\infty}^{\infty} d\tau p[\tau - (\tau_d)_{true}] p[t_o - (t - \tau) - (\tau_d)_{exp}] \\ &\quad \cdots \cos\{\omega_o[\tau - (\tau_d)_{true}] + \psi_{true}\} \cos\{\omega_o[t_o - (t - \tau) - (\tau_d)_{exp} + \psi_{exp}]\} \\ &= Ka^2 \int_{-\infty}^{\infty} d\tau p[\tau - (\tau_d)_{true}] p[\tau - (\tau_d)_{exp} - \Delta t] \\ &\quad \cdots \frac{1}{2} \{\cos[\omega_o(\Delta\tau_d - \Delta t) - \Delta\psi] - \cos\{2\omega_o\tau - \omega_o[(\tau_d)_{true} + (\tau_d)_{exp}] \\ &\quad - \omega_o \Delta t + \psi_{true} + \psi_{exp})\} \end{aligned} \tag{4.69}$$

where $\Delta\psi = (\psi_{true} - \psi_{exp})$ and $\Delta\tau_d = (\tau_d)_{true} - (\tau_d)_{exp}$, $\Delta t = t - \tau$.

The second integrand in (4.69) is highly oscillatory (frequency $2\omega_o$) and hence is negligible in the integration. The result of the integration is

$$s_o(t) = \frac{Ka^2}{2} \cos[\omega_o(\Delta\tau_d - \Delta t) - \Delta\psi] \cdots \int_{-\infty}^{\infty} d\tau p[\tau - (\tau_d)_{true}]\, p[\tau - (\tau_d)_{exp} - \Delta t] \tag{4.70}$$

If the filter is matched at $t = t_o$, so that $\Delta t = 0$ and if $(\tau_d)_{exp} = (\tau_d)_{true}$ then (4.70) takes the form

$$s_o(t_o) = \left[\frac{Ka^2}{2} \int_{-\infty}^{\infty} d\tau p^2(\tau)\right] \cos(\Delta\psi) \tag{4.71}$$

From (4.71) we conclude that we can have a nearly perfect matched filter, matched in time delay and frequency, but if it is not matched in phase then the factor $\cos \Delta\psi$ appears on the signal output. The phase mismatch does not affect the output noise, hence the degradation factor on the SNR is

$$D(\Delta\psi) = \cos^2(\Delta\psi) = \frac{1}{2} + \frac{1}{2} \cos 2\Delta\psi \tag{4.72}$$

This is part of the "coherent" portion of the degradation factor $D_{coh}(\Delta t)$ defined in (4.58a)', denoting that the argument of $D_{coh}(\Delta t)$, $(\omega_o \Delta t)$, is a *part* of the mismatch in RF phase. To lump these two factors together, we should recognize that Δt is actually the mismatch in time delay (proportional to the mismatch in target range in a monostatic radar), given by

$$\Delta\tau_d = (\tau_d)_{true} - (\tau_d)_{exp} \tag{4.73a}$$

since the attempt to match the filter in time will be made at the time $(\tau_d)_{exp}$. The "phase mismatch" is

$$\Delta\psi = \psi_{true} - \psi_{exp} \tag{4.73b}$$

The total mismatch in (4.58) can be generalized to the form

$$D(\Delta\tau_d, \Delta\psi) = \frac{\rho_o(\Delta\tau_d, \Delta\psi)}{\rho_o(0)} = D_{ncoh}(\Delta\tau_d)D_{ncoh}(\omega_o\Delta\tau_d + \Delta\psi) \tag{4.74}$$

where

$$D_{ncoh}(\Delta\tau_d) = \frac{\displaystyle\int_{-\infty}^{\infty} dt'p(t')p(t' + \Delta\tau_d)}{\displaystyle\int_{-\infty}^{\infty} dt'p^2(t')} = \text{"noncoherent" degradation factor} \tag{4.74a}$$

and

$$\begin{aligned} D_{coh}(\omega_o\,\Delta\tau_d + \Delta\psi) &= \cos^2(\omega_o\Delta\tau_d + \Delta\psi) \\ &= \frac{1}{2} + \frac{1}{2}\cos(2\omega_o\Delta\tau_d + 2\Delta\psi) \\ &= \text{"coherent degradation factor"} \end{aligned} \tag{4.74b}$$

This generalization of (4.58) clearly separates the mechanisms for generating degradation factors in the matched filter output SNR. The first mechanism is uncertainty or "mismatch" in RF phase, which appears only in *coherent* filtering. The second mechanism is uncertainty in signal delay time, which appears in *both* coherent and noncoherent filtering. The "quadrature filter" technique discussed in Section 4.4 will eliminate the degradation in coherent filtering, which arises from both mechanisms, and substitute for it a 3 dB loss in output SNR, which is accepted as the price of uncertainty in RF phase and delay time. But it does *not* eliminate the degradation due to uncertainty in delay time that appears in the *noncoherent* filter output SNR, as given by (4.74a).

4.6 EFFECT OF FREQUENCY MISMATCH ON COHERENT OR NONCOHERENT "MATCHED" FILTER OUTPUT SNR

We have assumed throughout this discussion that the frequency of the received signal (and therefore the IF at the filter input) was known precisely enough to specify it in the design of the matched filter. In real radar applications, the Doppler shift of a signal that is received from a target is typically unknown. This is partially because we do not necessarily know how fast the target is moving. We also do not know the effect of extraneous signals from "clutter" sources, which might have Doppler shifts that will corrupt the return when such signals are superposed on the target return.

Suppose the radar designer *believes* that the return has a Doppler shift ω_d and *knows* that the transmitted RF signal frequency was ω_o. He or she *expects* to receive a signal with central frequency $\omega_{exp} = (\omega_o + \omega_d)$ and designs the matched filter accordingly. But due to effects unknown to the filter designer, the *true* received signal frequency is ω_{true}, which differs from ω_{exp} by an amount $\Delta\omega$. The same "frequency mismatch" will be experienced at IF, because the difference between the true and expected IF is [see Eq. (3.19) in Section 3.7]

$$\begin{aligned}(\omega_{IF})_{true} - (\omega_{IF})_{exp} &= (\omega_{true} - \omega_{LO}) - (\omega_{exp} - \omega_{LO}) \\ &= [\omega_{true} - (\omega_{exp} - \omega_{IF})] - [\omega_{exp} - (\omega_{exp} - \omega_{IF})] \\ &= \omega_{true} - \omega_{exp} = \Delta\omega \end{aligned} \tag{4.75}$$

Given the misconception that the actual frequency is ω_{exp}, the attempted "matched" filter is designed to have an impulse response

$$h(t) = p(t_o - t)\cos\{[(\omega_{IF})_{true} - \Delta\omega](t_o - t) + \psi]\} \tag{4.76a}$$

The input signal has the form

$$s_i(t) = p(t)\cos[(\omega_{IF})_{true}t + \psi] \tag{4.76b}$$

The signal output of the filter is, from (4.76a,b) (where we recall that $t - t_o = \Delta t$ = "time mismatch")

$$\begin{aligned} s_o(t) &= \int_{-\infty}^{\infty} d\tau s_i(\tau)h(t-\tau) \\ &= \int_{-\infty}^{\infty} d\tau p[t_o - (t-\tau)]\cos\{[(\omega_{IF})_{true} - \Delta\omega](t_o - t - \tau)] + \psi\} \\ &\quad \cdots p(\tau)\cos[(\omega_{IF})_{true}\tau + \psi] \\ &= \frac{1}{2}\int_{-\infty}^{\infty} d\tau p(\tau - \Delta t)p(\tau)\{\cos\{[(\omega_{IF})_{true} - \Delta\omega](\tau - \Delta t) \\ &\quad - (\omega_{IF})_{true}\tau\} + \cos[2(\omega_{IF})_{true}\tau - \Delta\omega\tau \\ &\quad - (\omega_{IF})_{true}\Delta t + \Delta\omega\Delta t + 2\psi]\} \end{aligned} \tag{4.76c}$$

The second term gives rise to an oscillatory integral that can be regarded as negligible for practical purposes [another way of saying this is that the filter passband rejects frequencies of the order of $2(\omega_{IF})_{true}$].

Neglecting that term we obtain

$$\begin{aligned} s_o(t) &= \frac{1}{2}\int_{-\infty}^{\infty} d\tau p(\tau)p(\tau - \Delta t)\cos[\Delta\omega\tau + (\omega_{IF})_{true}\Delta t - \Delta\omega\tau] \\ &= \frac{1}{2}\cos\{[(\omega_{IF})_{true} - \Delta\omega]\Delta t\}\int_{-\infty}^{\infty} d\tau p(\tau)p(\tau - \Delta t)\cos(\Delta\omega\tau) \\ &\quad - \frac{1}{2}\sin\{[(\omega_{IF})_{true} - \Delta\omega]\Delta t\}\int_{-\infty}^{\infty} d\tau p(\tau)p(\tau - \Delta t)\sin(\Delta\omega\tau) \\ &= A_c(\Delta\omega, \Delta t)\cos[(\omega_{IF})_{exp}\Delta t] \end{aligned}$$

$$- A_s(\Delta\omega, \Delta t) \sin[(\omega_{IF})_{exp}\Delta t]$$

$$= \sqrt{[A_c(\Delta\omega, \Delta t)]^2 + [A_s(\Delta\omega, \Delta t)]^2} \cos\left\{ [(\omega_{IF})_{exp}\Delta t] + \tan^{-1}\left[\frac{A_s(\Delta\omega, \Delta t)}{A_c(\Delta\omega, \Delta t)}\right]\right\} \tag{4.77}$$

where

$$\begin{matrix} A_c \\ A_s \end{matrix}(\Delta\omega, \Delta t) = \frac{1}{2}\int_{-\infty}^{\infty} d\tau p(\tau)p(t - \Delta t) \begin{matrix} \cos \\ \sin \end{matrix}(\Delta\omega\tau)$$

The noise output of the filter is independent of the frequency mismatch $\Delta\omega$. Since the noise is essentially spectrally flat over the filter's passband, the mismatch would have to be enormous before it could result in less noise being passed by the filter. The degradation factor on the output SNR due to *both* time and frequency mismatch is proportional to the square of the amplitude of (4.69) for coherent filtering and is 3 dB less for noncoherent filtering (see Section 4.4). The degradation factor is given by

$$D(\Delta\omega, \Delta t) = \left\{\frac{[A_c(\Delta\omega, \Delta t)]^2 + [A_s(\Delta\omega, \Delta t)]^2}{[A_c(0, 0)]^2 + [A_s(0, 0)]^2}\right\}$$

$$= \left[\frac{\int_{-\infty}^{\infty} d\tau p(\tau)p(\tau - \Delta t)\cos(\Delta\omega\tau)}{\int_{-\infty}^{\infty} d\tau p^2(\tau)}\right]^2 + \left[\frac{\int_{-\infty}^{\infty} d\tau p(\tau)p(\tau - \Delta t)\sin(\Delta\omega\tau)}{\int_{-\infty}^{\infty} d\tau p^2(\tau)}\right]^2 \tag{4.78}$$

4.7 THE AMBIGUITY FUNCTION

Since the concept of the ambiguity function is usually used in the context of study of range and Doppler resolution, it is not common to introduce it in connection with matched filters. But we have been dealing with the concept without using that nomenclature for the past few sections; hence it will be formally introduced here.

For a narrowband signal, it is convenient to use the "complex envelope" or "complex amplitude" representation $\hat{s}(t)$ and to include the central frequency in the signal waveform representation, i.e., to express the waveform as

$$s(t, \omega) = a(t)\cos[\omega t + \psi(t) + \psi_o] = \mathrm{Re}[\hat{s}(t)e^{j\omega t}] \tag{4.79}$$

where

$$\hat{s}(t) = \text{complex envelope of } s(t, \omega) = a(t)e^{j[\psi(t)+\psi_o]}$$

If we multiply $s(t, \omega)$ by $s(t + \Delta t, \omega + \Delta\omega)$ and time average the product, we obtain through (4.79)

$$\overline{s(t,\omega)s(t+\Delta\omega,\omega+\Delta\omega)}$$

$$= \overline{\frac{1}{2}[\hat{s}(t)e^{j\omega t} + \hat{s}^*(t)e^{-j\omega t}]\frac{1}{2}[\hat{s}(t+\Delta t)e^{j(\omega+\Delta\omega)(t+\Delta t)} + \hat{s}^*(t+\Delta r)e^{-j(\omega+\Delta\omega)(t+\Delta t)}]}$$

$$= \frac{1}{4}[\overline{\hat{s}(t)\hat{s}(t+\Delta t)e^{j(2\omega+\Delta\omega)t}}e^{j(\omega+\Delta\omega)\Delta t} + \overline{\hat{s}^*(t)\hat{s}(t+\Delta t)e^{-j(2\omega+\Delta\omega)t}}e^{-j(\omega+\Delta\omega)\Delta t}]$$

$$+ \frac{1}{4}[\overline{\hat{s}(t)\hat{s}^*(t+\Delta t)e^{-j\Delta\omega t}}e^{-j(\omega+\Delta\omega)\Delta t} + \overline{\hat{s}^*(t)\hat{s}(t+\Delta t)e^{j\Delta\omega t}}e^{j(\omega+\Delta\omega)\Delta t}]$$

$$= \frac{1}{2}\mathrm{Re}[\overline{\hat{s}(t)\hat{s}(t+\Delta t)e^{j(2\omega+\Delta\omega)t}}e^{j(\omega+\Delta\omega)\Delta t}] + \frac{1}{2}\mathrm{Re}[\overline{\hat{s}(t)\hat{s}^*(t+\Delta t)e^{-j\Delta\omega t}}e^{-j(\omega+\Delta\omega)\Delta t}] \quad (4.80)$$

It is assumed that

$$|\Delta\omega| << |\omega| \quad (4.81)$$

and noted that

$$\overline{\hat{s}(t)\hat{s}(t+\Delta t)e^{j(2\omega+\Delta\omega)t}} = \lim_{T\to\infty}\frac{1}{T}\int_{-T/2}^{T/2} dt\hat{s}(t)\hat{s}(t+\Delta t)e^{j(2\omega+\Delta\omega)t} \quad (4.81a)'$$

and

$$\overline{\hat{s}(t)\hat{s}^*(t+\Delta t)e^{-j\Delta\omega t}} = \lim_{T\to\infty}\frac{1}{T}\int_{-T/2}^{T/2} dt\hat{s}(t)\hat{s}^*(t+\Delta t)e^{-j\Delta\omega t} \quad (4.81b)'$$

If we assume that

$$|\hat{s}(t)| = 0 \qquad \text{for } |t| > |T| \quad (4.82)$$

as would be the case for any conceivable radar signal modulation envelope, then we can divide (4.81a,b)′ by $\frac{1}{2}\overline{|\hat{s}(t)|^2} = \frac{1}{2}\int_{-T/2}^{T/2} dt|\hat{s}(t)|^2$ before the limiting process is taken, resulting in the cancellation of $1/T$ in the fraction and the result

$$\Lambda(\Delta t,\Delta\omega) = \mathrm{Re}\left\{\left[\frac{\int_{-\infty}^{\infty} dt\hat{s}(t)\hat{s}(t+\Delta t)e^{j(2\omega+\Delta\omega)t}}{\int_{-\infty}^{\infty} dt|\hat{s}(t)|^2}\right]e^{j(\omega+\Delta\omega)\Delta t} + \left[\frac{\int_{-\infty}^{\infty} dt\hat{s}(t)\hat{s}^*(t+\Delta t)e^{-j\Delta\omega t}}{\int_{-\infty}^{\infty} dt|\hat{s}(t)|^2}\right]e^{-j(\omega+\Delta\omega)\Delta t}\right\} \quad (4.83)$$

where $\Lambda(\Delta t, \Delta\omega)$ is defined as the "ambiguity function" of the signal waveform $s(t, \omega)$ and is abbreviated as "AF."

The first term inside the "real part" bracket is negligibly small and can be set to zero because it involves a highly oscillatory integral, or equivalently a slowly varying function multiplied by a function varying at the frequency $2\omega + \Delta\omega$. That is the same familiar argument we have been using to eliminate that type of integral in the previous sections.

Having eliminated the oscillatory integral, we can rewrite (4.83) in the form

$$\Lambda(\Delta t, \Delta\omega) = \mathrm{Re}[\hat{\Lambda}(\Delta t, \Delta\omega)e^{-j(\omega+\Delta\omega)\Delta t}] \tag{4.84}$$

where $\hat{\Lambda}(\Delta t, \Delta\omega)$ is defined as the "complex ambiguity function" of $s(t, \omega)$ and is abbreviated as "CAF." It is the CAF that we will use in dealing with radar signals.

Actually the function $\Lambda(\Delta t, \Delta\omega)$ is a measure of the mismatches in time and frequency in matched filtering. It is the "mismatched filter" output (normalized) that we discussed in Sections 4.1 through 4.6, where the discrepancies between expected and true values of time delay and frequency are Δt and $\Delta\omega$, respectively. Its magnitude is a measure of the degradation in SNR that results from these mismatches.

Noting that $\hat{\Lambda}(\Delta t, \Delta\omega)$ is a complex quantity, we can express it in the form

$$\hat{\Lambda}(\Delta t, \Delta\omega) = |\hat{\Lambda}(\Delta t, \Delta\omega)|e^{j\psi_{AF}(\Delta t,\, \Delta\omega)} \tag{4.85}$$

$$|\hat{\Lambda}(\Delta t, \Delta\omega)| = \text{amplitude of CAF} = \left| \frac{\int_{-\infty}^{\infty} dt\hat{s}(t)\hat{s}^*(t + \Delta t)e^{-j\Delta\omega t}}{\int_{-\infty}^{\infty} dt|\hat{s}(t)|^2} \right|$$

$$\psi_{AF}(\Delta t, \Delta\omega) = \text{phase of CAF} = \tan^{-1}\left\{\frac{\mathrm{Im}[\hat{\Lambda}(\Delta t, \Delta\omega)]}{\mathrm{Re}[\hat{\Lambda}(\Delta t, \Delta\omega)]}\right\}$$

Using (4.84), Eq. (4.84) takes the form

$$\Lambda(\Delta t, \Delta\omega) = |\hat{\Lambda}(\Delta t, \Delta\omega)| \cos[(\omega + \Delta\omega)\Delta t - \psi_{AF}(\Delta t, \Delta\omega)] \tag{4.86}$$

i.e., $|\hat{\Lambda}(\Delta t, \Delta\omega)|$ is the amplitude of a signal at frequency $(\omega + \Delta\omega)$.

Based on the above analysis results, we can now use $|\hat{\Lambda}(\Delta t, \Delta\omega)|^2$ as the degradation factor in the output SNR of a matched filter possibly mismatched in time delay by Δt and frequency $\Delta\omega$ (RF phase mismatch is *not* included). From the analysis of the previous sections in Chapter 4, we can recast the "mismatched" filter output SNR results in terms of the square amplitude of the CAF, which contains the degradation factors $D_{\mathrm{ncoh}}(\Delta t)D_{\mathrm{coh}}(\Delta t)$ in (4.58) if $\Delta\omega$ is zero, or $D(\Delta\omega, \Delta t)$ in (4.78) if both Δt and $\Delta\omega$ are nonzero and not including the effect of RF phase mismatch $\Delta\psi$.

Our final result, then, is

$$\rho_o(\Delta t, \Delta\omega) = \rho_o^{(o)}|\hat{\Lambda}(\Delta t, \Delta\omega)|^2 \tag{4.87}$$

where

$\rho_o^{(o)}$ = output SNR of *perfectly* matched filter
Δt = time-delay mismatch
$\Delta\omega$ = frequency mismatch
$|\hat{\Lambda}(\Delta t, \Delta\omega)|^2$ = degradation factor due to mismatches = absolute square of CAF as given in (4.86)

The mismatch in RF phase $\Delta\psi$ is not explicitly indicated in (4.87).

The discussion to follow specializes the signal to a train of N_p rectangular RF pulses, a typical radar signal.

First, consider the single pulse case $N_p = 1$. From (4.85) with the aid of Figure 4.9, we have for this case

$$\hat{s}(t) = ap(t) = a[u(t + \tau_p/2) - u(t - \tau_p/2)] \tag{4.88a}$$

$$\hat{\Lambda}(\Delta t, \Delta\omega) = \frac{\int_{-\infty}^{\infty} dt p(t) p(t + \Delta t) e^{-j\Delta\omega t}}{\int_{-\infty}^{\infty} dt p^2(t)} \tag{4.88b}$$

where $p(t)$ is a rectangular pulse centered at $t = 0$ and of duration τ_p.

$$\hat{\Lambda}(0, \Delta\omega) = \frac{\int_{-\tau_p/2}^{\tau_p/2} dt e^{-j\Delta\omega t}}{\int_{-\tau_p/2}^{\tau_p/2} dt(1)} = \tau_p \frac{[(e^{j\Delta\omega\tau_p/2} - e^{-j\omega\tau_p/2}/(2j)\Delta\omega\tau_p/2)]}{\tau_p} = \operatorname{sinc}\left(\frac{\Delta\omega\tau_p}{2}\right) \tag{4.88c}$$

This is a (voltage, not power) degradation factor due to frequency mismatch $\Delta\omega$ where the filter is perfectly matched in time delay. It is also known as the (complex) "frequency ambiguity function" for a rectangular pulse. It is real, hence its phase $\psi_{AF}(0, \Delta\omega) = 0$ or 180°, depending on whether it is positive or negative; thus

$$\hat{\Lambda}(0, \Delta\omega) = |\hat{\Lambda}(0, \Delta\omega)| e^{j\psi_{AF}(0, \Delta\omega)} \tag{4.88d}$$

where

$$|\hat{\Lambda}(0, \Delta\omega)| = \left|\operatorname{sinc}\left(\frac{\Delta\omega\tau_p}{2}\right)\right| \qquad \text{(even function in } \Delta\omega\text{)}$$

$$\psi_{AF}(0, \Delta\omega) = \begin{cases} 0 & \text{if } \hat{A} > 0 \\ \pi & \text{if } \hat{A} < 0 \end{cases}$$

Sketches of $\hat{\Lambda}(0, \Delta\omega)$ and $|\hat{\Lambda}(0, \Delta\omega)|$ are shown in Figure 4.8a and b, respectively.

Now consider the "time ambiguity function" $\hat{\Lambda}(\Delta t, 0)$, which acts as the voltage degradation factor in the case where the filter is matched in frequency but not in time delay. It is given by

$$\hat{\Lambda}(\Delta t, 0) = \frac{\int_{-\infty}^{\infty} dt p(t) p(t + \Delta t)}{\int_{-\infty}^{\infty} dt p^2(t)} = \frac{1}{\tau_p} \int_{-\infty}^{\infty} dt p(t) p(t + \Delta t) \tag{4.88e}$$

and is recognizable as the normalized autocorrelation function (NACF) of the pulse amplitude waveform, defined earlier [Eq. (4.45)].

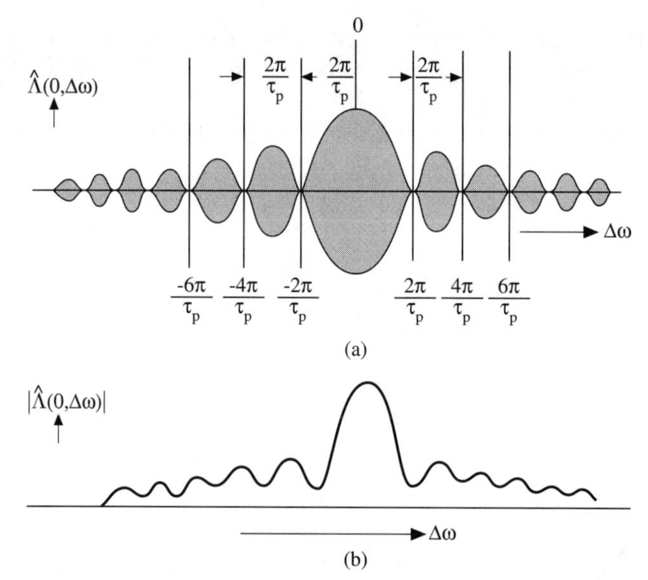

FIGURE 4.8
Frequency ambiguity function—rectangular pulse.

The result of the integration of (4.88f) is the "triangular function"

$$\hat{\Lambda}(\Delta t, 0) = \begin{cases} \left(1 - \dfrac{|\Delta t|}{\tau_p}\right), & \text{if } |\Delta t| \leq \tau_p \\ 0, & \text{if } |\Delta t| > \tau_p \end{cases} \tag{4.88f}$$

which is real and always positive and an even function in Δt.

The best approach to an explanation of (4.88f) is a graphic one, through Figure 4.9.

From Figure 4.9a and b, it is obvious that the product of $p(t)$ and $p(t + \Delta t)$ vanishes if $|\Delta t| > \tau_p$ because there is no overlap in that case. Hence $\hat{A}(\Delta t) = 0$ if $|\Delta t| > \tau_p$. From Figure 4.9a and c, we see that the degree of overlap increases with decreasing values of $|\Delta t|$ if $|\Delta t| < \tau_p$ and that there is complete overlap if $\Delta t = 0$. The end result of that reasoning is the function given by (4.88f).

Finally, if both Δt and $\Delta\omega$ are nonzero, we have

$$|\hat{\Lambda}(\Delta t, \Delta\omega)| = \frac{\displaystyle\int_{-\infty}^{\infty} dt p(t) p(t + \Delta t) e^{-j\Delta\omega\tau}}{\displaystyle\int_{-\infty}^{\infty} dt p^2(t)}$$

$$= \left[1 - \frac{|\Delta t|}{\tau_p}\right] \operatorname{sinc}\left[\frac{\Delta\omega\tau_p}{2}\left(1 - \frac{|\Delta t|}{\tau_p}\right)\right], \quad \text{if } |\Delta t| \leq \tau_p$$

$$= 0, \quad \text{otherwise} \tag{4.88g}$$

It is left as an exercise for the reader to derive (4.88g) from the rectangular pulse function $p(t)$.

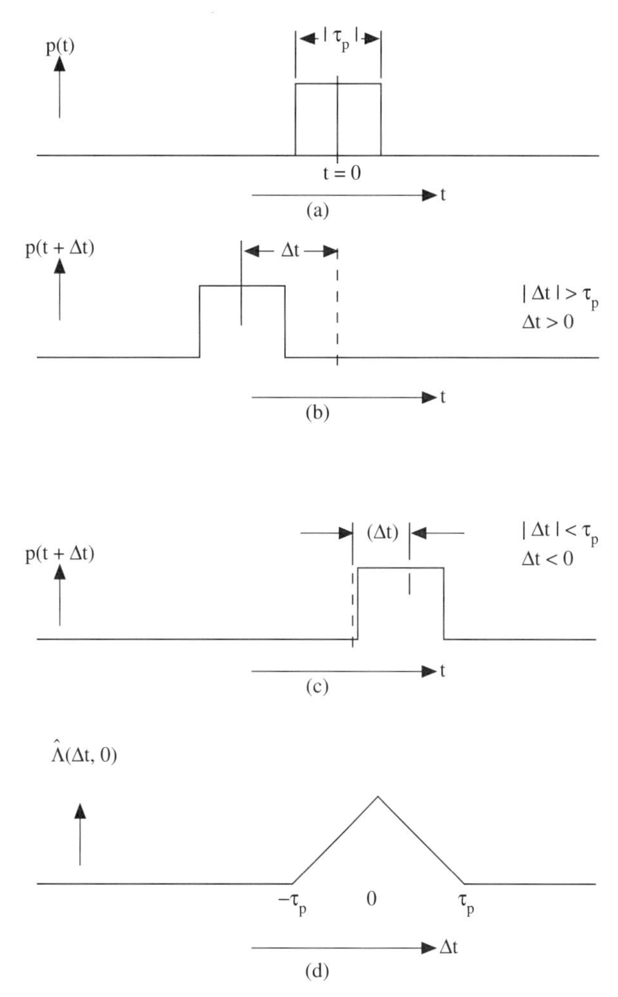

FIGURE 4.9
Time ambiguity function—rectangular pulse.

4.8 LOCATING A TARGET IN RANGE-DOPPLER SPACE

An interesting application of the ideas covered in the previous sections of Chapter 4 is that of achieving a specified minimum SNR (presumably for detection but possibly for another use of radar, e.g., measurement of a signal parameter or tracking) of a target whose range and line of sight velocity are a priori known only to a limited degree. Stated that way, this problem is specific to monostatic radar where range means path delay and line-of-sight velocity means Doppler.

Suppose the designer of a radar system begins with the knowledge that *if* the target of interest exists, it is within a range interval from R_1 to $R_1 + \delta R$ and within a Doppler interval from $V_c = -|\delta V/2|$ to $V_c = +|\delta V/2|$, as shown in Figure 4.10. The figure shows the "region of uncertainty" in the "range-Doppler plane."

The Doppler range uncertainty region should be specified as being bounded by $\pm$ the maximum possible relative speed between radar and target on either side, where $|\delta V/2|$ is that maximum speed and where $-|\delta V/2|$ is the minimum possible Doppler (calibrated in velocity units) for a target receding from the radar and $+|\delta V/2|$ is the maximum possible Doppler for a target approaching the radar. In many applications, approaching and receding targets are equally probable; hence that type

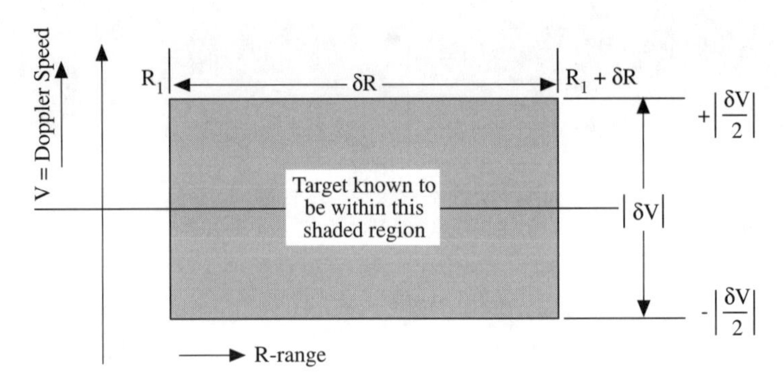

FIGURE 4.10
Range-Doppler uncertainty region.

of specification usually makes sense. In specifying range limits, of course, we account for the fact that range, by definition, *must* be positive.

We can immediately translate range R into the time delay $t = 2R/c$ and Doppler into frequency $\omega_d = 2\omega_o V/\lambda$ and thereby relate this discussion to the ambiguity function discussed in Section 4.7, which is usually expressed in terms of delay differences Δt and frequency differences $\Delta\omega$.

We first note that the dimensions of the range-Doppler uncertainty region calibrated in time delay and frequency are $\delta t = (2/c)\delta R$ and $\delta\omega = (2\omega_o/\lambda)\delta V$, respectively. These quantities δt and $\delta\omega$ are *not* the same as $\Delta t = (2/c)\Delta R$ and $\Delta\omega = (2\omega_o/\lambda)\Delta V$; the "mismatches" in the ambiguity function or degradation factors appear in Sections 4.4 through 4.7. The difference between the Δ and δ quantities will become apparent in the discussion to follow.

Suppose our radar designer needs a given minimum SNR to achieve the specified objective (e.g., target detection), that the true target has RCS σ_o, radar and signal parameters are P_T, f_o, G_{To}, A_{eRo}, τ_p, f_R, and receiver noise parameters are T_o, F_N, B_N, where the meanings of all of these parameters have been explained in Chapter 2, Eqs. (2.20)′ and (2.34). The transmitted signal is assumed to be a train of equally spaced rectangular pulses.

The first question the designer might ask is: "Suppose I construct a coherent matched filter to achieve the maximum SNR; what will that SNR be for a target with RCS = σ_o and a range R at the peak of both transmitting and receiving antenna beams, assuming no signal distortion and assuming that it is possible to have an exact knowledge of the RF phase of the received signal?" The answer, of course, is [see Eq. (2.38)]

$$\rho_o^{(o)} = \frac{E_s}{N_o} = \left[\frac{P_T G_{To} A_{eRo} \sigma_o d}{(4\pi)^2 R^4 (kT_o F_N)}\right] T_d \tag{4.89}$$

where T_d = total duration of pulse train and where d = pulse duty cycle = $f_r\tau_p = \tau_p/T_r$ = fraction of time that the pulses are "ON." The *true* SNR, due to an uncertainty Δt (*not* δt) in time delay and an uncertainty in Doppler shift $\Delta\omega$ (*not* $\delta\omega$), is

$$\rho_o = \rho_o^{(o)} |\hat{A}(\Delta t, \Delta\omega)|^2 \tag{4.90}$$

where $\rho_o^{(o)}$ is given by (4.89) and $\hat{A}(\Delta t, \Delta\omega)$ by (4.88g) in the general case where both range and Doppler are uncertain and in the special cases where *only* range or *only*

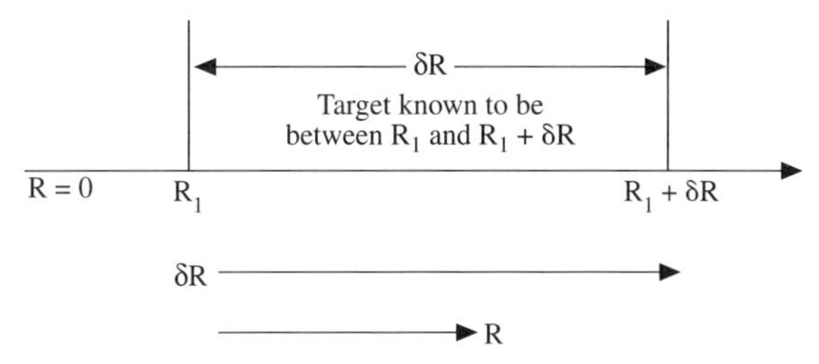

FIGURE 4.11
Range uncertainty region—Doppler known.

Doppler are uncertain and where $\hat{A}(\Delta t, 0)$ and $\hat{A}(0, \Delta\omega)$ are given by (4.88e) and (4.88d), respectively.

We begin the detailed discussion of this topic by assuming that the radar designer knows the Doppler precisely, i.e., he or she shows that the target is not moving and, therefore, $|\delta V| = 0$. This means that the only SNR degradation that can occur is due to absence of knowledge of range. The version of Figure 4.10 applicable to this case is Figure 4.11, showing a one-dimensional uncertainty region involving range only.

A filter that is matched in time to an instant $t_o = 2R_o/c$ will receive a signal from a target with a range near $R = R_o$ but will receive no signal from a target with a range far different from $R = R_o$. This matched filter is actually a "range gate," designed to respond only to targets within a small region around a given range. Figure 4.12 illustrates this by showing the response of such a range gate to a target at various ranges different from R_o. The amplitude response of the range gate is the ambiguity function (see Fig. 4.12).

$$\hat{A}\left[\frac{2\Delta R}{c}, 0\right] = \begin{cases} 1 - \left|\dfrac{2\Delta R}{c\tau_p}\right|, & \text{if } \Delta R \leq \dfrac{c\tau_p}{2} \\ 0, & \text{otherwise} \end{cases} \tag{4.91}$$

where $\Delta R = R - R_o$, and where R is the true target range and R_o the range to which the gate is set. If $R = R_o$, the peak signal strength is received and the output SNR of the gate is $\rho_o^{(o)}$ as given by (4.89). From (4.90) and (4.91) or graphically from Figure 4.12, we see that the actual SNR is reduced from $\rho_o^{(o)}$ in the amount 20 $\log_{10}|1 -$

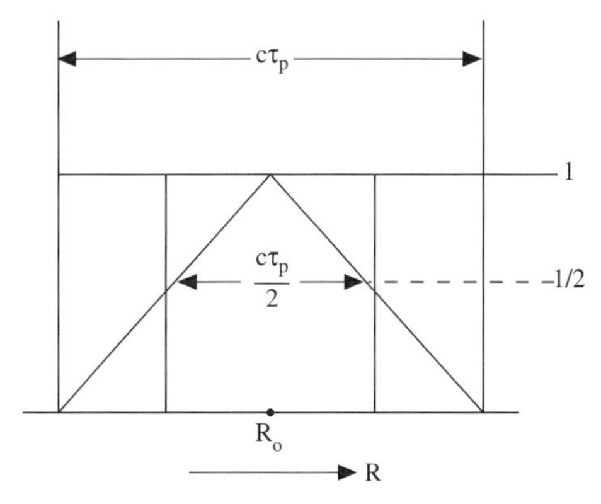

FIGURE 4.12
Amplitude response of range gate—known Doppler.

$(2/c\tau_p)([\frac{1}{2}]c\tau_p/2)|$ dB. If R differs from R_o by half a pulse length in range units, then the SNR is degraded by

$$2 \log_{10}\left|1 - \frac{2}{c\tau_p}\left(\frac{1}{2}\frac{c\tau_p}{2}\right)\right| = 20 \log_{10}\left|\frac{1}{2}\right| = -6 \text{ dB}$$

If the difference is one-quarter of a pulse length, then the degradation is

$$20 \log_{10}\left|1 - \frac{2}{c\tau_p}\left(\frac{1}{4}\frac{c\tau_p}{2}\right)\right| = 20 \log_{10}(3/4) = -2.5 \text{ dB}$$

A loss of 3 dB corresponds to $|R - R_o| = 0.3(c\tau_p/2)$ and a difference $|R - R_o| = 3/4(c\tau_p/2)$ results in a loss of $20 \log_{10}(1/4) = -12$ dB. To summarize the point, the discrepancy between gate setting and range cannot exceed the pulse duration calibrated in range units (neglecting range ambiguities due to multiple pulses, which will be discussed later but for the present will be ignored). If it does, then no signal will be received from that target. If is desired that no more than a 6 dB loss due to range mismatch will be tolerated, then the maximum allowable range gate width is *one* pulse length calibrated in range units. A 1-μsec pulse, for example, dictates a range-gate width no greater than 150 m, which means that the output SNR will be greater than or equal to $\rho_o^{(o)}/4$.

If a gate were designed to have a width of 300 m with a pulse length of 1 μsec, it would be possible for a target whose range differs from R_o by 150 m (1 pulse length) to show no output signal. The "safety factor" on the SNR in this case would have to be infinite. If we wanted a very large range gate, say 225 m, we would need a safety factor of 12 dB, i.e., we could assume an output SNR of at least $\rho_o^{(o)}/16$ if the target were defined as being "within the range gate." A general rule is to set the range gate width as a one pulse length in range units. Then if a target is "within the range gate," its return signal SNR equals or exceeds $\rho_o^{(o)}/16$, or equivalently, it suffers a maximum 6 dB loss from the ideal matched filter output SNR. To be "within" the range gate, its range cannot differ from the gate setting by more than half a pulse length.

The same ideas can be applied to Doppler uncertainty given a knowledge of range. Figure 4.13 shows the one-dimensional Doppler uncertainty region in the case where $|\delta R| = 0$. The matched filter response in this case is obtained from (4.88d) and if we "tune" the filter to the Doppler frequency 0, corresponding to no line-of-sight target motion (this means the return signal frequency is ω_o, the IF corresponding

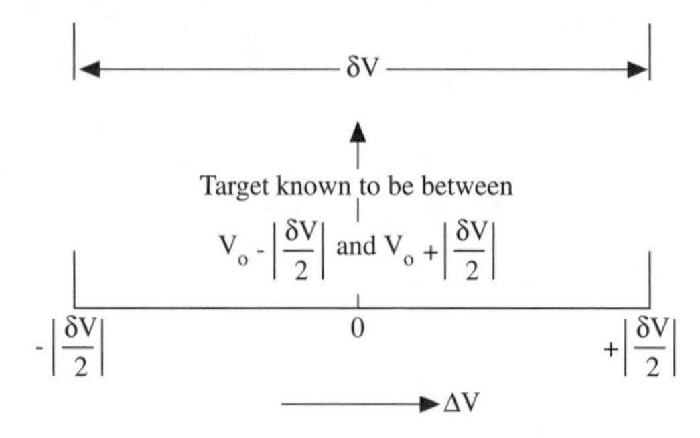

FIGURE 4.13
Doppler uncertainty region—range known.

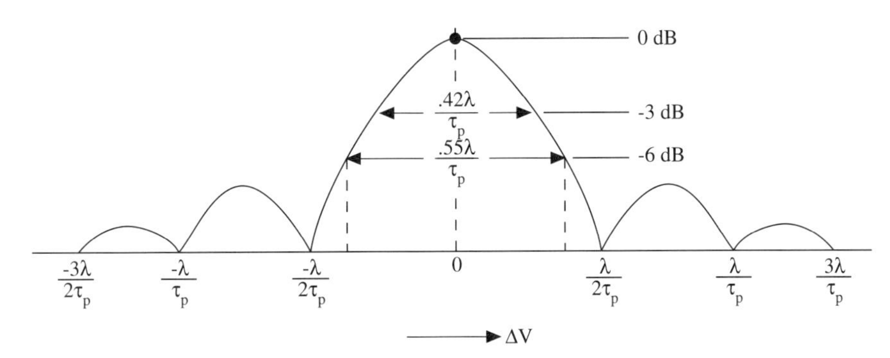

FIGURE 4.14
Amplitude response of Doppler gate—known range.

to the transmitted RF), then the displacement from ω_o in velocity units is given by $c\Delta\omega/2\omega_o$. Figure 4.14 shows the response of the filter tuned to central frequency ω_o to a target return displaced from zero LOS velocity by the amount $\Delta V = c\Delta\omega/2\omega_o$ with the range known absolutely.

The plot in Figure 4.14 is that of the amplitude of the ambiguity function $|\hat{A}(0, 2\omega_o\delta V/c)|$ as given by (4.88d) calibrated in velocity, which takes the form

$$\left|\hat{A}\left(0, \frac{2\omega_o}{c}\Delta V\right)\right| = \left|\mathrm{sinc}\left(\frac{\Delta V\tau_p\,\omega_o}{c}\right)\right| \tag{4.92}$$

REFERENCES

1. Raemer, H.R., "Statistical Communication Theory and Applications," Prentice-Hall, Englewood, NJ, 1969, Chapters 2, 3, and 6.
2. Davenport, W.B., and Root, W.L., "Theory of Random Signals and Noise." McGraw-Hill, New York, 1958.
3. DiFranco, J.V., and Rubin, W.L., "Radar Detection." Artech, Norwood, MA, 1980, Chapter 5.
4. Papoulis, A., "Probability, Random Variables and Stochastic Processes," 2nd ed. McGraw-Hill, New York, 1984.
5. Stark, H., and Woods, J.W., "Probability, Random Processes and Estimation Theory for Engineers." Prentice-Hall, Englewood Cliffs, NJ, 1986.

Chapter 5

Theory of Detection of Radar Signals in Additive Gaussian Noise

CONTENTS

We now specify that the noise additive to the received radar signal has Gaussian statistics. No specification of the statistics of the noise was made in Section 4. One reason for its specification as Gaussian in preparation for the discussion of detection of the signal in noise is ease of analysis. Fortunately, nature is kind enough to provide receiver noise that is approximately Gaussian much of the time. That is the good news. The bad news is that a significant amount of the noise that corrupts radar signals is *not* Gaussian.

Dismissing the bad news for the moment, let us focus on the good news. The basic thermal noise in a receiver is mostly Gaussian and much of the extraterrestrial radio noise that is naturally picked up by the radio receiver is also Gaussian. Hence if we know the performance characteristics of the radar in working against Gaussian noise, we know a great deal that is important in evaluating the radar.

The theory is highly tractable for Gaussian noise, so most of the "closed form" results available in the literature apply, strictly speaking, only to additive Gaussian noise.

A major source for some of the material shown here is "Radar Detection," by J.V. DiFranco and W.L. Rubin.[1] Other sources on radar detection,[2–10] elementary random function theory and detection theory in general will be useful to readers not thoroughly familiar with these topics or who wish to pursue the subject in greater detail.[11–15]

5.1 PROPERTIES OF A STATIONARY GAUSSIAN NOISE

Consider a noise waveform $n(t)$ that is a sample function of a stationary Gaussian random process. Its univariate probability density function (PDF) at time t is

$$p(n) = \frac{1}{\sqrt{2\pi\sigma_n^2}}\, e^{(n-\langle n\rangle)^2/2\sigma_n^2} \tag{5.1}$$

where $p(n)$ is independent of time and where

$\sigma_n^2 = \langle [n - \langle n\rangle]^2\rangle = \langle n^2\rangle - \langle n\rangle^2$ = mean-square deviation of noise, which is independent of time by virtue of the stationarity property
$\langle n(t)\rangle = \langle n\rangle$ = mean noise voltage, also independent of time

If the noise is *"narrowband,"* and *"zero mean,"* i.e., it can be expressed in the form (4.38) (repeated here for convenience)

$$n(t) = x_n(t)\cos\omega_o t + y_n(t)\sin\omega_o t \tag{5.2}$$

then (see Section 4.2) it has the properties (4.39a,b,c,d,e,f), also repeated below for convenience

$$\langle x_n(t)\rangle = \langle y_n(t)\rangle = 0 \tag{5.3a}$$

$$\langle x_n^2(t)\rangle = \langle x_n^2\rangle = \langle y_n^2(t)\rangle = \langle y_n^2\rangle = \langle n^2\rangle = \sigma_n^2 \tag{5.3b}$$

$$\langle x_n(t_1)y_n(t_2)\rangle = 0 \text{ for all } t_1, t_2 \tag{5.3c}$$

The joint (bivariate) PDF of x_n and y_n is

$$p(x_n, y_n) = \frac{1}{\sqrt{2\pi\langle x_n^2\rangle}} e^{-x_n^2/2\langle x_n^2\rangle} \frac{1}{\sqrt{2\pi\langle y_n^2\rangle}} e^{-y_n^2/2\langle y_n^2\rangle}$$

$$= \frac{1}{2\pi\sigma_n^2} e^{-(x_n^2+y_n^2)/2\sigma_n^2} \tag{5.4}$$

where (5.3a,b,c) must be invoked in deriving (5.4)from the general form of the bivariate Gaussian PDF, which involves the ACF of x_n and y_n, all of which are set equal to zero in (5.4). The form (5.4) expresses $p(x_n, y_n)$ as the product of the PDFs of x_n and y_n, i.e., assumes they are statistically independent. In the Gaussian case, (5.3c) implies statistical independence, but in general, the two variables being uncorrelated does not necessarily correspond one-to-one with statistical independence.

If (5.3a) does not apply, i.e., if $\langle x_n\rangle$ and $\langle y_n\rangle$ are nonzero, then the PDF of (5.4) can be generalized to the form

$$p(x, y) = \frac{1}{2\pi\sigma_n^2} e^{-[(x-\langle x_n\rangle)^2+(y-\langle y_n\rangle)^2]/2\sigma_n^2} \tag{5.5}$$

where $x = x_n + \langle x_n\rangle$, $y = y_n + \langle y_n\rangle$.

Equations (5.1) and (5.5) can be used in the case of a nonzero signal plus noise voltage, where $\langle n\rangle = 0$ in (5.1), as implied by (5.2) and (5.3a).

$$v(t) = s(t) + n(t) = x(t) \cos \omega_o t + y(t) \sin \omega_o t$$

$$= [x_s(t) + x_n(t)] \cos \omega_o t + [y_s(t) + y_n(t)] \sin \omega_o t \tag{5.6}$$

Dropping explicit indication of the time t in signal and noise, we note that (5.1), with the aid of (5.6), has the form

$$p(v/s) = \frac{1}{\sqrt{2\pi\sigma_n^2}} e^{-(v-s)^2/2\sigma_n^2} \tag{5.7a}$$

and (5.5) with the aid of (5.6) is

$$p(x, y/s) = \frac{1}{2\pi\sigma_n^2} e^{-[(x-x_s)^2+(y-y_s)^2]/2\sigma_n^2} \tag{5.7b}$$

where $/s$ indicates that these PDFs are conditional on the presence of a signal $s = x_s \cos \omega_o t + y_s \sin \omega_o t$.

The corresponding forms of (5.7a,b) with noise alone or zero signal are

$$p(v/n) = \frac{1}{\sqrt{2\pi\sigma_n^2}} e^{-v^2/2\sigma_n^2} \tag{5.8a}$$

$$p(x, y/n) = \frac{1}{2\pi\sigma_n^2} e^{-(x^2+y^2)/2\sigma_n^2} \tag{5.8b}$$

To show that (5.7b) implies (5.7a) for narrowband signal-plus-noise [also (5.8b) implies (5.8a) since (5.8a,b) are specializations of (5.7a,b)], we write

$$\begin{aligned} p(v/s) &= \int_{-\infty}^{\infty} dx \int_{-\infty}^{\infty} dy\, p(x, y/s)\delta[v - (x \cos \omega_o t + y \sin \omega_o t)] \\ &= \int_{-\infty}^{\infty} dx \int_{-\infty}^{\infty} dy\, p(x, y/s) \cdots \\ &\quad \frac{1}{2\pi}\int_{-\infty}^{\infty} d\xi\, e^{j\xi[v-(x \cos \omega_o t + y \sin \omega_o t)]} \\ &= \frac{1}{2\pi}\int_{-\infty}^{\infty} d\xi\, e^{j\xi v}\chi(\xi) \end{aligned} \tag{5.9}$$

where we have expressed the impulse function in the form (4.7) and changed the order of integration and where

$$\chi(\xi) = \int_{-\infty}^{\infty} dx \int_{-\infty}^{\infty} dy\, p(x, y/s) e^{-j\xi(x \cos \omega_o t + y \sin \omega_o t)} = \tag{5.9$'$}$$

Fourier transform of the PDF, known in general as the "characteristic function: or "CF" or "moment generating function" of the PDF $p(x,y)$.

In the particular case of the PDF (5.7b), (5.9)′ yields

$$\chi(\xi) = \chi_x(\xi)\chi_y(\xi) \tag{5.10}$$

where

$$\chi_x(\xi) = \frac{1}{\sqrt{2\pi\sigma_n^2}} \int_{-\infty}^{\infty} dx\, e^{-j\xi x \sin \omega_o t - [(x-x_s)^2/2\sigma_n^2]}$$

$$\chi_y(\xi) = \frac{1}{\sqrt{2\pi\sigma_n^2}} \int_{-\infty}^{\infty} dy\, e^{-j\xi y \sin \omega_o t - [(y-y_s)^2/2\sigma_n^2]}$$

By completing the square inside the Gaussian exponent, we obtain

$$\begin{aligned} \chi_x(\xi) &= \frac{e^{-j\xi x_s \cos \omega_o t}}{\sqrt{\pi}} \int_{-\infty}^{\infty} d\left(\frac{x}{\sqrt{2\sigma_n^2}}\right) e^{-1/2\sigma_n^2[(x-x_s)^2 + 2j\xi\sigma_n^2(x-x_s)\cos \omega_o t + (j\xi\sigma_n^2 \cos \omega_o t)^2]} \\ &\quad e^{+(j\xi\sigma_n^2 \cos \omega_o t)^2/2\sigma_n^2} \\ &\quad + e^{-j\xi x_s \cos \omega_o t} e^{(-\xi^2\sigma_n^2/2)\cos^2 \omega_o t} \end{aligned} \tag{5.11a}$$

and by the same procedure

$$\chi_y(\xi) = e^{-j\xi y_s \sin \omega_o t} e^{-\xi^2 \sigma_n^2 \sin^2 \omega_o t/2} \tag{5.11b}$$

From (5.10) and (5.11a, b)

$$\chi(\xi) = e^{-j\xi s} e^{-\xi^2 \sigma_n^2/2} \tag{5.11c}$$

where $s = x_s \cos w_o t + y_s \sin w_o t$.

Substitution of (5.11c) into (5.9) yields (again completing the square in the exponent)

$$\begin{aligned} p(v/s) &= \frac{1}{2\pi} \int_{-\infty}^{\infty} d\xi \, e^{j\xi(v-s)-(\xi^2\sigma_n^2/2)} \\ &= \frac{\sqrt{2}}{2\pi\sigma_n} \int_{-\infty}^{\infty} d\left(\frac{\sigma_n \xi}{\sqrt{2}}\right) e^{-[(\sigma_n\xi/\sqrt{2})^2 - 2j(\sigma_n\xi/\sqrt{2})(v-s)/2\sigma_n)\sqrt{2} + (j(v-s)/\sqrt{2}\sigma_n)^2]} \ldots \\ & e^{+(j(v-s)/\sqrt{2}\sigma_n)^2} = \frac{1}{\sqrt{2\pi\sigma_n^2}} e^{-(v-s)^2/2\sigma_n^2} \end{aligned} \tag{5.12}$$

which is the equivalent of (5.1) with $n = v$ and $\langle n \rangle = s$, i.e., the mean of the noise is the deterministic signal $s(t)$.

5.2 DETECTION PROBABILITY AND FALSE ALARM PROBABILITY

The detection problem is that of determination of the presence of a signal within the noise. The "detection probability" P_d is the conditional probability that, given that a signal *is* present, the signal-plus-noise falls within the range that will result in a "signal present" decision. The false alarm probability P_{fa} is the conditional probability that, given that *no* signal is present, the noise falls within a range that will result in a "signal present" decision, i.e., in the latter case, the detection criterion produces a false indication of signal presence, popularly known as a "false alarm."

Mathematically, these quantities are given by

$$P_d = \int_{v_T}^{\infty} dv \, p(v/s) \tag{5.13a}$$

$$P_{fa} = \int_{v_T}^{\infty} dv \, p(v/n) \tag{5.13b}$$

where v_T is a chosen "threshold" voltage level, such that, if $v(t)$ falls above that threshold, the decision will be "signal present" and if $v(t)$ falls below the threshold, the decision will be "noise alone," and where $p(v/s)$ and $p(v/n)$ are the conditional PDFs of v given the conditions "signal present" and "noise alone," respectively.

For Gaussian noise, we invoke (5.8a) for $p(v/n)$ and (5.12) for $p(v/s)$, and apply those expressions for $p(v/s)$ and $p(v/n)$ to (5.13a,b), with the results (see Fig.5.1)

$$P_{\mathrm{d}} = \int_{v_{\mathrm{T}}}^{\infty} dv \frac{1}{\sqrt{2\pi\sigma_{\mathrm{n}}^2}} e^{-(v-s)^2/2\sigma_{\mathrm{n}}^2} = \frac{1}{\sqrt{\pi}} \int_{v_{\mathrm{T}}/\sqrt{2}\sigma_{\mathrm{n}}}^{\infty} du' \; e^{-(u' \mp \sqrt{|R|})^2}$$

$$= \frac{1}{\sqrt{\pi}} \int_{(\hat{v}_{\mathrm{T}} \mp \sqrt{|R|}}^{\infty} du \; e^{-u^2} \tag{5.14a}$$

$$P_{\mathrm{fa}} = \int_{v_{\mathrm{T}}}^{\infty} \frac{dv \; e^{-v^2/2\sigma_{\mathrm{n}}^2}}{\sqrt{2\pi\sigma_{\mathrm{n}}^2}} = \frac{1}{\sqrt{\pi}} \int_{\hat{v}_{\mathrm{T}}}^{\infty} du \; e^{-u^2} \tag{5.14b}$$

where $\hat{v}_{\mathrm{T}} = v_{\mathrm{T}}/\sqrt{2}\sigma_{\mathrm{n}}$ = threshold voltage normalized to $\sqrt{2}$ times the root mean square (rms) noise level and $\sqrt{|R|} = |s|/\sqrt{2}\sigma_{\mathrm{n}} = 1/\sqrt{2}$ (voltage SNR).

We note that $u^2 = |R| = \frac{1}{2}(s/\sigma_{\mathrm{n}})^2 = \frac{1}{2}$ (power SNR).

One step further with (5.14a,b) leads us to

$$P_{\mathrm{d}} = \frac{1}{2}[1 - \mathrm{erf}(\hat{v}_{\mathrm{T}} - u)] \tag{5.15a}$$

$$P_{\mathrm{fa}} = \frac{1}{2}[1 - \mathrm{erf}(\hat{v}_{\mathrm{T}})] \tag{5.15b}$$

where $u = \pm\sqrt{|R|}$ and $\mathrm{erf}(x) = 2/\sqrt{\pi} \int_0^{\infty} dy \; e^{-y^2}$ = error function, which is tabulated in many sources (e.g., Abramowitz and Stegun,[16] pp. 310–311).

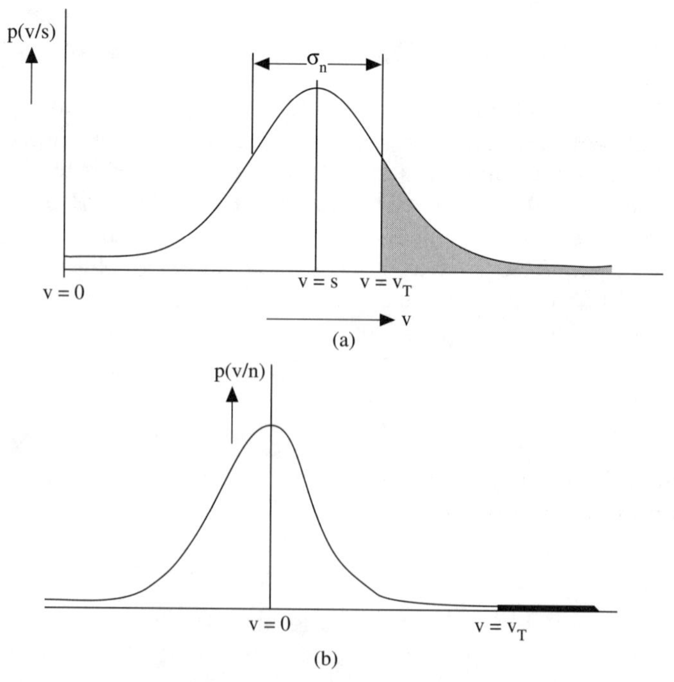

FIGURE 5.1
Illustration of detection and false alarm probabilities.

In Figure 5.3 we show graphically how P_d and P_{fa} both decrease with increasing threshold for a fixed SNR. The process under investigation is that of "coherent detection," in which the signal has undergone no nonlinear processing. First, in Figure 5.2, we show a rough plot of the error function.

From Figure 5.2, we see that the error function is an odd function of x and approaches +1 or −1 as $|x|$ becomes large in the positive or negative direction. This should give insight into the behavior of P_d and P_{fa} as given by (5.15a,b).

Figure 5.3 shows the effect of variation of the threshold on the detection probability for two signals, s_1 and s_2, where the SNR for s_1 is moderate (say about 3–6 dB) and that for s_2 is high (say about 10 dB). In Case (a), the threshold is set at $v_T=0$. The detection probability for both signals 1 and 2 is extremely high (near 100%) because both probability curves are high above the threshold. That is also true in Case (b), where the threshold is set at the rms noise level. But in Case (a), setting such a low threshold to ensure detection of both signals has a price. That price is the very high false alarm rate of 50%, i.e., a false alarm will occur on half the detection decisions. In Case (b), an attempt is made to reduce the false alarms by raising the threshold from 0 to σ_n. This reduces P_{fa} to about 25%, i.e., only a quarter of the output readings will be false alarms.

In Case (c), the threshold has been raised to s_1, which essentially eliminates the possibility of false alarms, or at least reduces P_{fa} to a negligibly small value. However the probability of detecting s_1 is now only 50%. In Case (c), the threshold has been raised even higher, i.e., to a value well above s_1 but well below s_2. This ensures that P_{fa} is even smaller than in Case (c) and preserves the condition that s_2 is detected on nearly every reading. However, it also reduces the probability of detecting s_1 to a very small value.

Figure 5.4 shows rough sketches of P_d and P_{fa} vs. normalized threshold $\hat{v}_T$, based on the behavior of the error function as illustrated in Figure 5.2.

From Figure 5.4a and b, we see that both P_{fa} and P_d asymptotically approach unity as the threshold $\hat{v}_T$ becomes more negative. As $\hat{v}_T$ becomes more positive, they both approach zero asymptotically. If $\hat{v}_T < \hat{v}_{T1}$, both P_d and P_{fa} are nearly unity, i.e., the chances of both detection and false alarm are very high. If $\hat{v}_T > \hat{v}_{T2}$, the chance of detection begins to decrease significantly while that of false alarm becomes very small. If $\hat{v}_T$ is between $\hat{v}_{T1}$ and $\hat{v}_{T2}$, significant changes can be made in both P_d and P_{fa} by changing the threshold. It is in that region that the tradeoff between high

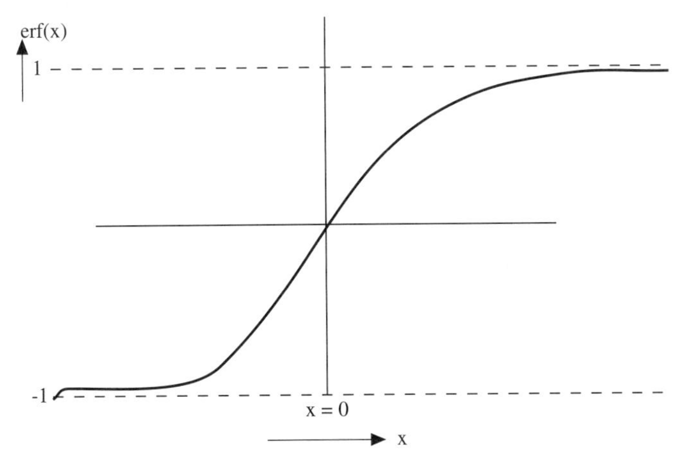

FIGURE 5.2
Error function.

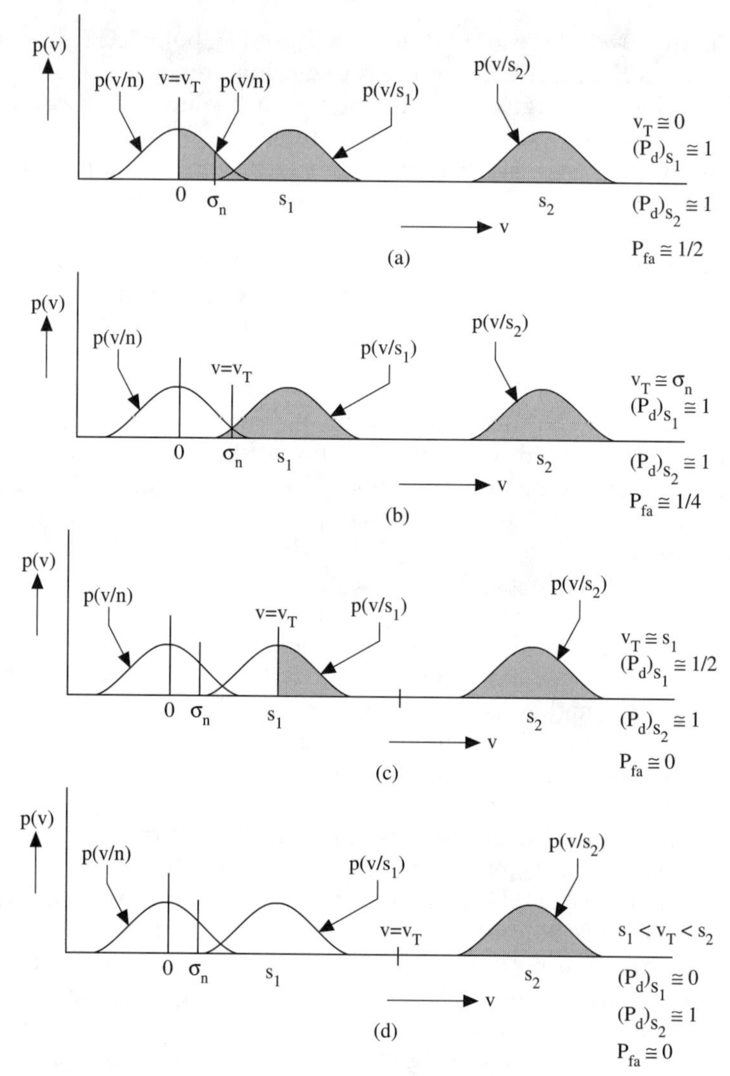

FIGURE 5.3
Effect of threshold on detection and false alarm probabilities—coherent detection.

threshold for minimizing false alarm and low threshold for maximizing detection probability becomes apparent.

5.3 COHERENT AND NONCOHERENT DETECTION

In Section 5.2, we discussed the basic principles of *coherent detection* of a signal in additive Gaussian noise. The fact that the noise is "narrowband" is irrelevant to the results for coherent detection. That kind of detection theory applies to radar signals at the RF or IF stage of a receiver, based on two assumptions, as follows:

1. The noise at the receiver input is Gaussian.
2. The heterodyning process behaves ideally (see Section 3.7) and the linearity of the signal-plus-noise is preserved.

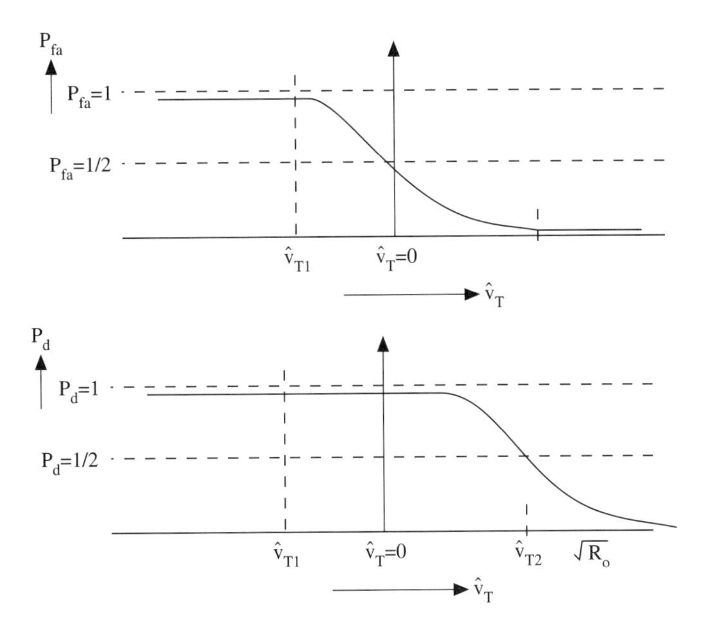

FIGURE 5.4
Coherent detection: P_d and P_{fa} vs. normalized threshold.

Based on Assumptions (1) and (2), we can conclude that the noise at the IF output stage is Gaussian and additive to the signal and that it has zero mean and is uncorrelated with the signal. The same conclusion applies to the noise at the output of a matched (or *attempted* matched, i.e., "mismatched") filter that might be used at the IF stage to increase the SNR (Sections 4.1 and 4.3) as much as possible before detection decisions are made.

This is because zero-mean Gaussian noise has the property that it remains zero-mean Gaussian after undergoing any *linear* process. Both the standard processes that occur in the receiver (Section 3.7) and the matched filter are linear, and hence noise that is zero-mean Gaussian at the receiver input remains zero-mean Gaussian after an attempt at matched filtering or any other kind of linear filtering.

A detection process that occurs at the IF stage or after an attempt at matched filtering or in general, "coherent filtering" as discussed in Section 4.3, is called "coherent detection" if the input to the detection process is still an IF signal. This means that the detection decision is made based on the value of the signal-plus noise before it has undergone any nonlinear operations. Referring to Figure 5.5, a value of the voltage waveform is sampled at the peak (hopefully) and that is the voltage value that is tested to determine whether it is above or below a threshold for the detection decision. That procedure is called "coherent detection."

To implement coherent detection, it must be possible to choose a sample of the signal-plus-noise at the coherent filter output that is somewhere near the *peak value* of that output. Noise will corrupt the waveform such that the choice of the peak value is not certain. It is also necessary to know the signal phase to know the location of the peak. This is exactly the same limitation as exists in realizing the maximum SNR at the output of a matched filter. A detection decision based on the wrong choice of sample point might yield disastrous results. For example, the first possible sample in Figure 5.5 is well above the threshold and the signal peak is *enhanced* by noise and not displaced. That sample will yield a high P_d. The second

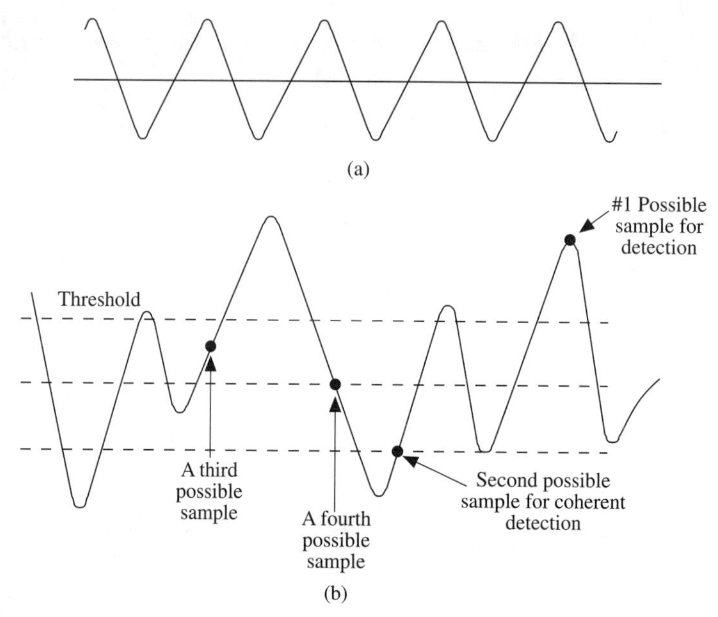

FIGURE 5.5
Coherent detection. (a) $s_0(t)$ = coherent filter output signal. (b) $v_0(t) = s_0(t) + n_0(t)$ = coherent filter output signal-plus-noise.

sample is below threshold because the noise has displaced the peak and the time instant where the peak *should* have occurred has a zero output. The third and fourth samples are also below threshold because the sample has been chosen at the wrong time due to ignorance of phase, although the intended signal peak has not been displaced by noise.

The conclusion from all this is that successful coherent detection is based on the assumption that phase is exactly known and its excursions can be controlled. Equivalently, it is based on the ability to match the IF signal to an IF reference signal with exactly the same phase $\pm\pi n$ and thereby maximize the SNR at the input to the detection decision process. The implication of our assumption is that the phase of the coherent filter output will be exactly known and therefore the benefits of coherent matched filtering can be maximally realized in optimizing detection. The detection probability for coherent detection is *always* increased by increasing SNR *if phase is known;* it always pays to use matched filtering or come as close as possible to it before initiating a detection process. Rough sketches of P_d vs. $u = \pm\sqrt{|R|}$ for coherent detection with fixed threshold levels are shown in Figure 5.6, to demonstrate the monotonic increase of P_d with SNR. These sketches are based on (5.15a) and carry the restriction that $\hat{v}_t > 0$ and $u > 0$.

Two generic observations can be made from the sketches in Figure 5.6. One is that the detection probability is 50% when u is exactly equal to the normalized threshold and the other is that P_d increases beyond 50% as u begins to exceed $\hat{v}_T$ in such a way that increases in SNR significantly increase the detection probability. When u becomes about three to five times the normalized threshold, then P_d levels off near unity and is not increased by much as u increases further. The message in this is that with sufficiently high SNR values, small increases in SNR may not increase detection probability enough to be worth the trouble. That conclusion must be qualified by the observation that in a high performance radar, a detection probability of 99.9% may be desired and may be attainable with an SNR of (say) 15 dB, where

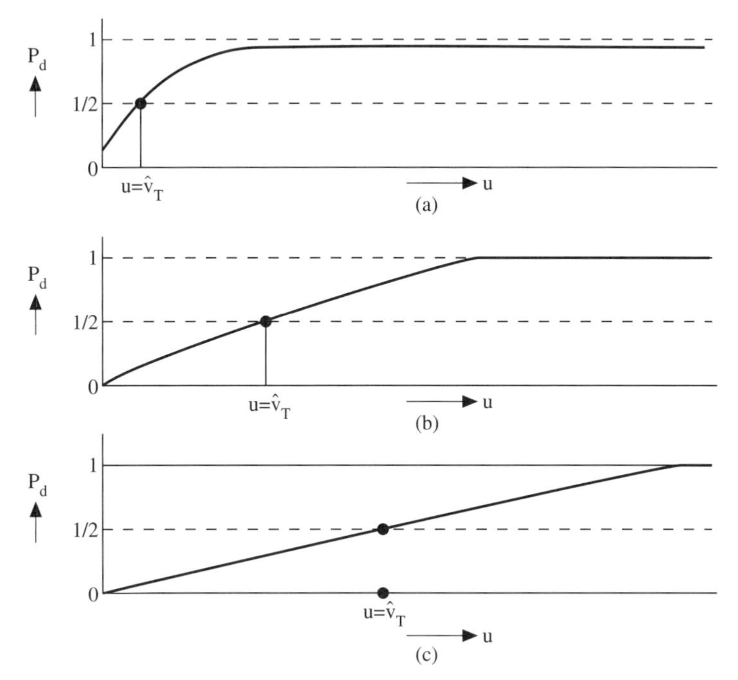

FIGURE 5.6
Coherent detection probability vs. SNR: $u > 0$, $\hat{v}_T > 0$.

14 dB may produce a detection probability of 95%. The extra dB *might* be worth the trouble. That is a strategic value judgment for the particular radar application.

Now suppose we have no hope of knowing the signal phase or do not want to take the trouble to attain phase information in the signal to be detected. In that case, we use the *amplitude* of the coherent filter output as the input to the detection process. Equivalently, we can refer back to Section 4.4 and observe that absence of phase knowledge in attempting matched filtering drove us to use "noncoherent matched filtering," or the "quadrature filtering" scheme illustrated in Figure 4.7 and explained in Eqs. (4.59a) through (4.65) in order to extract the signal amplitude at a cost of 3 dB SNR loss. The amplitude $|v_0|$ is always positive, of course, and the detection statistics are different. This detection methodology is called "noncoherent detection."

To examine noncoherent detection, we recall that the two quadrature components of the "narrowband" signal-plus-noise" at the coherent filter output have the joint PDF given by (5.7b) for nonzero signal and (5.8b) for zero signal. The PDF of the squared amplitude $|v|^2 = (x^2 + y^2)$ is obtained in the general form

$$
\begin{aligned}
p(|v|^2) &= \int_{-\infty}^{\infty} dx \int_{-\infty}^{\infty} dy\, p(x, y)\delta[|v|^2 - (x^2 + y^2)] \\
&= \frac{1}{2\pi}\int_{-\infty}^{\infty} d\xi\, e^{j\xi|v|^2}\int_{-\infty}^{\infty} dx \int_{-\infty}^{\infty} dy\, p(x, y)e^{-j\xi(x^2+y^2)} \\
&= \frac{1}{2\pi}\int_{-\infty}^{\infty} d\xi\, e^{j\xi|v|^2}\chi_{|v|^2}(\xi)
\end{aligned}
\tag{5.16}
$$

where $X_{|v|}{}^2 = \int_{-\infty}^{\infty} dx \int_{-\infty}^{\infty} dy\, p(x,y)\, e^{-j\xi(x^2+y^2)}$ = characteristic function for $|v|^2$.

For the noise-alone case with narrowband Gaussian noise, the PDF is given by (5.8b), from which it follows through (5.16) that

$$\chi_{|v|^2} = \chi_{x^2}\chi_{y^2} = \left(\frac{1}{\sqrt{2\pi\sigma_n^2}}\right)\left(\int_{-\infty}^{\infty} dx\; e^{-(x^2/2\sigma_n^2)-j\xi x^2}\right)^2$$

$$= \left(\frac{1}{\sqrt{1+j\xi 2\sigma_n^2}}\right)^2 = \frac{1}{1+j2\xi\sigma_n^2} \tag{5.17}$$

From (5.10) and (5.11a,b)

$$p(|v|^2) = \int_{-\infty}^{\infty} d\xi \, \frac{e^{j\xi|v|^2}}{1+j2\xi\sigma_n^2} \tag{5.18}$$

From a table of Fourier transforms we could obtain the inverse Fourier transform of $1/(1+j2\xi\sigma_n^2)$ and thereby obtain

$$p(|v|^2) = \frac{1}{2\sigma_n^2}\, e^{-v^2/2\,\sigma_n^2} \tag{5.19}$$

where we remove the absolute value sign from v to simplify the notation.

The characteristic function method used above is retained because it is useful in later developments involving pulse trains. But there is a much simpler way to obtain (5.19). From Figure 5.7, we note that the PDF of (5.8b) can be viewed as the PDF of v^2 by changing from rectangular coordinates (x, y) to cylindrical coordinates (v, ϕ). First we can always write

$$p(x, y)dx\, dy = p(v, \phi)dv\, d\phi \tag{5.20}$$

where $vdvd\phi$ is the area increment of the circular shell and therefore $p(v, \phi)dvd\phi$ is the probability that v and ϕ fall within the shell. Then $p(v, \phi)$ is defined as the joint PDF of v and ϕ. But $x = v\cos\phi$, $y = v\sin\phi$ and hence

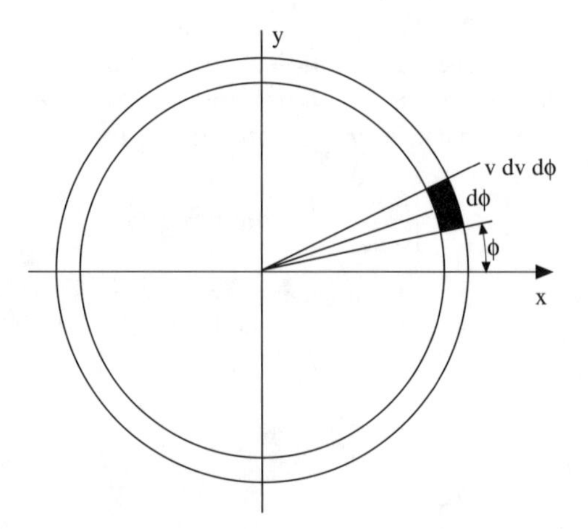

FIGURE 5.7
Area element for PDF of $v^2 = x^2 + y^2$.

$$p(x, y) = p(v \cos \phi, v \sin \phi) \tag{5.21}$$

Noting that $dx\, dy = vdvd\phi$ we can express (5.20) in the form [with the aid of (5.21)]

$$p(v, \phi)dv\, d\phi = p(v \cos \phi, v \sin \phi)\, vdvd\phi$$

or equivalently

$$p(v, \phi) = vp(v \cos \phi, v \sin \phi) \tag{5.22}$$

To obtain the amplitude PDF from the joint PDF of amplitude and phase, we integrate out the phase with the general result

$$p(v) = \int_0^{2\pi} d\phi\, p(v, \phi) = v \int_0^{2\pi} d\phi\, p(v \cos \phi, v \sin \phi) \tag{5.23}$$

For the specific PDF of (5.8b), (5.23) gives us

$$\begin{aligned} p(v/n) &= \frac{v}{2\pi\sigma_n^2} \int_0^{2\pi} d\phi\, e^{-[(v \cos \phi)^2 + (v \sin \phi)^2]/2\sigma_n^2} \\ &= \frac{v}{\sigma_n^2}\, e^{-v^2/2\sigma_n^2} \end{aligned} \tag{5.24}$$

Also, it is true in general that

$$p(v^2)d(v^2) = p(v)\, dv \tag{5.25}$$

or equivalently

$$p(v^2)2v\, dv = p(v)\, dv \tag{5.25$'$}$$

or

$$p(v^2) = \frac{p(v)}{2v} \tag{5.25$''$}$$

from which we obtain through (5.24)

$$p(v^2/n) = \frac{1}{2\sigma_n^2}\, e^{-v^2/2\sigma_n^2} \tag{5.26}$$

which is the same as (5.19), obtained by the characteristic function method.

Equation (5.24) is the *Rayleigh amplitude PDF,* which applies to the amplitude of a zero-mean narrowband Gaussian process. Equation (5.26) is the *exponential amplitude-square PDF,* which applies to the squared amplitude (or power) of such a process.

We can apply the same reasoning to the case of nonzero signal, i.e., the PDF given by (5.7b), substituted into (5.23), yields

$$\begin{aligned}
p(v/s) &= v\int_0^{2\pi} d\phi\, p(v\cos\phi, v\sin\phi) \\
&= \frac{v}{2\pi\sigma_n^2}\int_0^{2\pi} d\phi\, e^{-[(v\cos\phi - a_s\cos\phi_s)^2 + (v\sin\phi - a_s\sin\phi_s)^2]/2\sigma_n^2} \\
&= \frac{v}{2\pi\sigma_n^2}\int_0^{2\pi} d\phi\, e^{-[(v^2 + a_s^2 - 2va_s\cos(\phi-\phi_s)]/2\sigma_n^2} \\
&= \frac{ve^{-(v^2+a_s^2)/2\sigma_n^2}}{\sigma_n^2}\left[\frac{1}{2\pi}\int_0^{2\pi} d\phi'\, e^{(va_s/\sigma_n^2)\cos\phi'}\right] \\
&= \frac{ve^{-(v^2+a_s^2)/2\sigma_n^2}}{\sigma_n^2} I_0\left[\frac{va_s}{\sigma_n^2}\right]
\end{aligned} \tag{5.27}$$

where $x_s = a_s\cos\phi_s$, $y_s = a_s\sin\phi_s$, and where $I_o(x)$ is the zero-order modified Bessel function.

The PDF given by (5.27) is the *Rice PDF* (or *Rice–Nakagami PDF*), which applies to a narrowband Gaussian process with a *nonzero* mean.

From (5.24) and (5.27) we can obtain the false-alarm probability P_{fa} and the detection probability P_d, respectively, for noncoherent detection. These are

$$\begin{aligned}
P_{fa} &= \int_{v_T}^{\infty} dv\, p(v/n) = \int_{v_T}^{\infty} dv\, \frac{v}{\sigma_n^2} e^{-v^2/2\sigma_n^2} \\
&= \int_{v_T^2/2\sigma_n^2}^{\infty} dx\, e^{-x} = e^{-\hat{v}_T^2}
\end{aligned} \tag{5.28a}$$

where $\hat{v}_T = v_T/\sqrt{2}\sigma_n$ = normalized threshold.

$$\begin{aligned}
P_d &= \int_{v_T}^{\infty} dv\, p(v/s) = \int_{v_T}^{\infty} dv\, \frac{v}{\sigma_n^2} e^{-(v^2+a_s^2)/2\sigma_n^2} I_0\left[\frac{va_s}{\sigma_n^2}\right] \\
&= \int_{\hat{v}_T}^{\infty} dx\, xe^{-(x^2+\eta^2)/2} I_0(x\eta) = Q(\hat{v}_T, \eta)
\end{aligned} \tag{5.28b}$$

where $x = v/\sigma_n$, $\eta = a_s/\sigma_n$, $\hat{v}_T = v_T/\sigma_n = \sqrt{2}\hat{v}_T$, and $Q(x,y)$ = "Marcum Q-function" tabulated in classical papers by Marcum and Swerling.[17,18]

Sketches of $p(v/n)$ and $p(v/s)$ from (5.24) and (5.27), respectively, are shown in Figure 5.8a and b, respectively.

The peak of the Rayleigh curve occurs where

$$\frac{\partial p(v)}{\partial v} = \frac{1}{\sigma_n^2}\left[e^{-v^2/2\sigma_n^2} - \left(\frac{2v}{2\sigma_n^2}\right)ve^{-v^2/2\sigma_n^2}\right] = 0 \tag{5.29}$$

or equivalently $v = \sigma_n$.

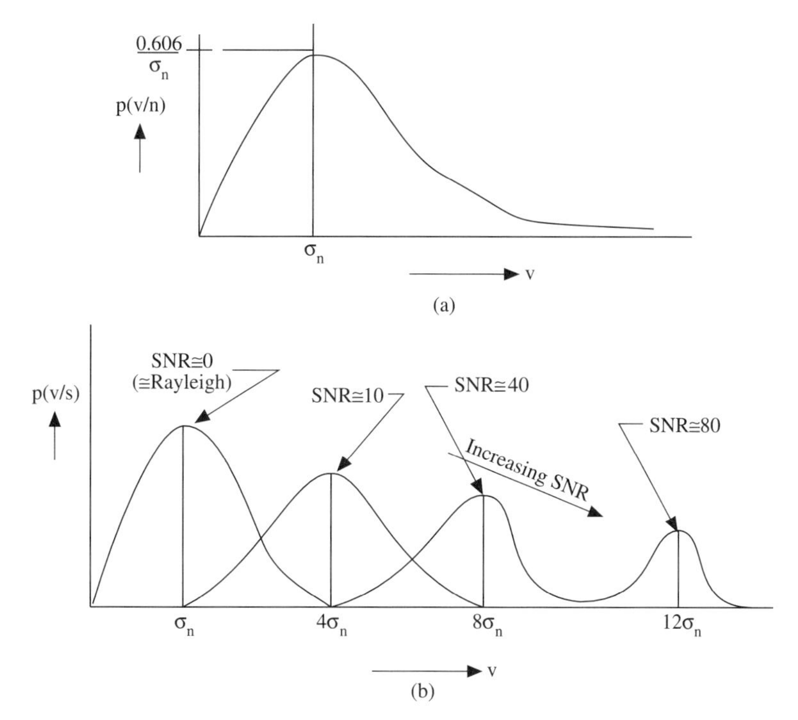

FIGURE 5.8
Sketches of $p(v/n)$ and $p(v/s)$—noncoherent detection.

At the peak, the value of $p(v/n)$ is $(1/\sigma_n)\, e^{-1/2} = 0.606/\sigma_n$.

As the SNR increases, the PDF for the nonzero signal case is seen to be centered at a value close to the value of the signal voltage and to become more narrow, approaching an impulse function as the SNR become infinite.

We can use Figure 5.8b redrawn as Figure 5.9 to illustrate the same point that was illustrated by Figures 5.3a,b,c,d, concerning the choice of threshold and the tradeoff between low false alarm rate and high detection probability.

Again, we consider the choice of threshold at $\hat{v}_{T1}$. The detection probabilities (areas under the PDFs of v given s_1, s_2, or s_3) appear to be (roughly) $(P_d)_{s1} =$ 85%, $(P_d)_{s2} =$ 100%, $(P_d)_{s3} =$ 100%, and the false alarm probability is about 30%. Choosing a larger $\hat{v}_T$, say at $\hat{v}_{T2}$, lowers P_{fa} to about 5–10%, but reduces $(P_d)_{s1}$ to

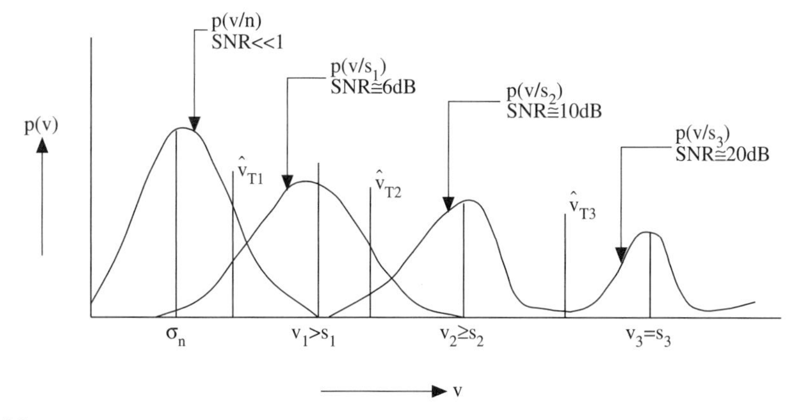

FIGURE 5.9
Effect of threshold choice in noncoherent detection.

about 30% and has little effect on $(P_d)_{s2}$ or $(P_d)_{s3}$. Choosing a still larger threshold, at $\hat{v}_{T3}$, virtually guarantees detection of s_3 but reduces the chance of detecting s_2 to about 5–10% and that of detecting s_1 to near zero. It also reduces the false alarm rate to nearly zero. The effects, as would be expected, are qualitatively the same as for coherent detection, but this kind of detection can have no negative signal outputs and negative thresholds would make no sense in this case. With coherent detection, a negative threshold is possible, but we would require that the detection law be based on the voltage being *below,* rather than *above* the threshold for a successful detection or a false alarm.

Let us examine the difference in the voltage waveforms for coherent vs. noncoherent detection, using Figure 5.10.

It is clear from the figure that the negative peaks of the IF waveform will all be below the threshold for coherent detection and most of those peaks will be above the threshold with noncoherent detection, while the positive peaks above threshold will be above threshold for both coherent and noncoherent detection. However, not all of the high peaks are due to signal; many of them may be due to noise. Thus false alarms would seem more likely with noncoherent detection and it *appears* that detection at any given time is more likely, i.e., at a given time sample, if a signal *is* present, it is more likely that the signal-plus-noise will be above threshold and therefore a successful detection will occur. However, that effect is offset by the fact that rectification lowers the SNR and hence there is more noise to corrupt the signal. The net effect is that noncoherent detection is inferior to coherent detection, as we will see later when we "get quantitative." Higher input SNR is required to obtain the same level of detection performance with noncoherent detection relative to coherent detection.

A sketch of the false alarm probability P_{fa}, as given by (5.28a), vs. normalized threshold $\hat{v}_T$, is shown in Figure 5.11, and compared with P_{fa} for coherent detection, as given by (5.15b) for positive threshold levels.

For low threshold levels, it is clear from Figure 5.11 that false alarm rates are higher for noncoherent detection but for higher threshold levels the difference becomes less pronounced. That is consistent with what we could infer from Figure 5.10, which would show far fewer peaks above threshold as the threshold is raised in either case, but not much difference if the threshold is sufficiently high.

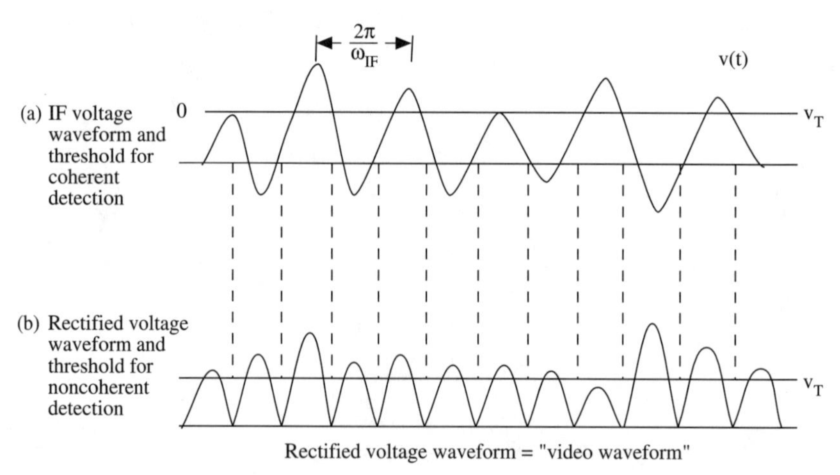

FIGURE 5.10
Voltage waveforms—coherent and noncoherent.

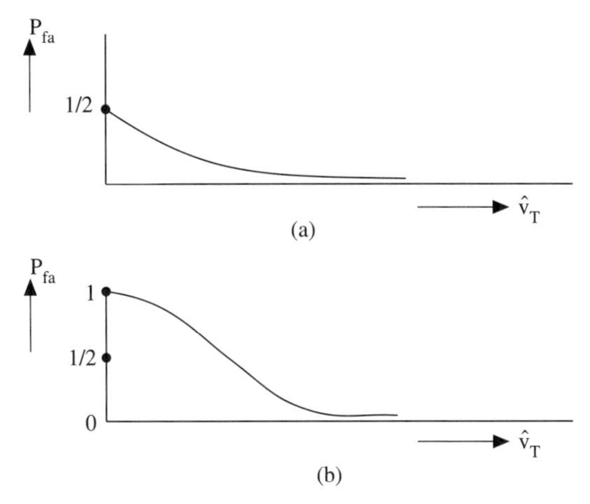

FIGURE 5.11
False alarm probability vs. normalized threshold. (a) Coherent detection. (b) Noncoherent detection.

Finally, Figures 5.12 and 5.13 are plots of P_d vs. SNR for different values of P_{fa}. These curves are from DiFranco and Rubin,[1] "Radar Detection," which has a large number of such curves covering a wide range of cases for radar detection. The curves in Figure 5.12 are for coherent detection and those in Figure 5.13 are for noncoherent detection of a single pulse. It is necessary to specify a "single pulse" in the case of noncoherent detection because integration of a pulse train can be either coherent or noncoherent. If it is noncoherent, then the detection performance is dependent on the number of pulses. If it is coherent, then the number of pulses is irrelevant and the same detection performance can be realized with a single pulse of energy E_s, or, say, 10 pulses each with energy $E_s/10$ or 1000 pulses each with energy $E_s/1000$. Only the total energy in the signal is important in coherent detection and hence it is not necessary to specify the number of pulses, only the total signal energy E_s relative to the noise power density, which gives the SNR at the matched filter output, whether the matching is to a single pulse or a train of pulses.

Figure 5.12 shows P_d (on probability paper) vs. the *peak* SNR R for an "exactly known signal." That phrase implies coherent matched filtering prior to the detection decision, given that coherent matched filtering requires exact knowledge of the IF signal waveshape including the phase. That process is what we have been calling "coherent detection." The *peak* SNR means the ratio of the squared amplitude to the mean-square noise at the matched filter output, i.e.,

$$R_o = \frac{a_{so}^2}{\sigma_{no}^2} = \frac{[s_o(0)]^2}{\langle n_o^2 \rangle} \tag{5.30}$$

where the subscript o implies output of the filter.

The more commonly used SNR is $a_{so}^2/2\sigma_{no}^2$, or half the SNR indicated in (5.30) and used in the DiFranco–Rubin curves.

Since P_{fa} depends on the threshold, and of course decreases monotonically with increasing threshold, a choice of P_{fa} to specify one of the curves in Figure 5.12 is tantamount to a choice of threshold, the higher curves corresponding to lower thresholds. It is evident that the difference in dB of SNR required for a given value of P_d does not change much for higher SNR but is very pronounced for lower SNR. It is also evident that changes in threshold (or equivalently changes in P_{fa}) do not

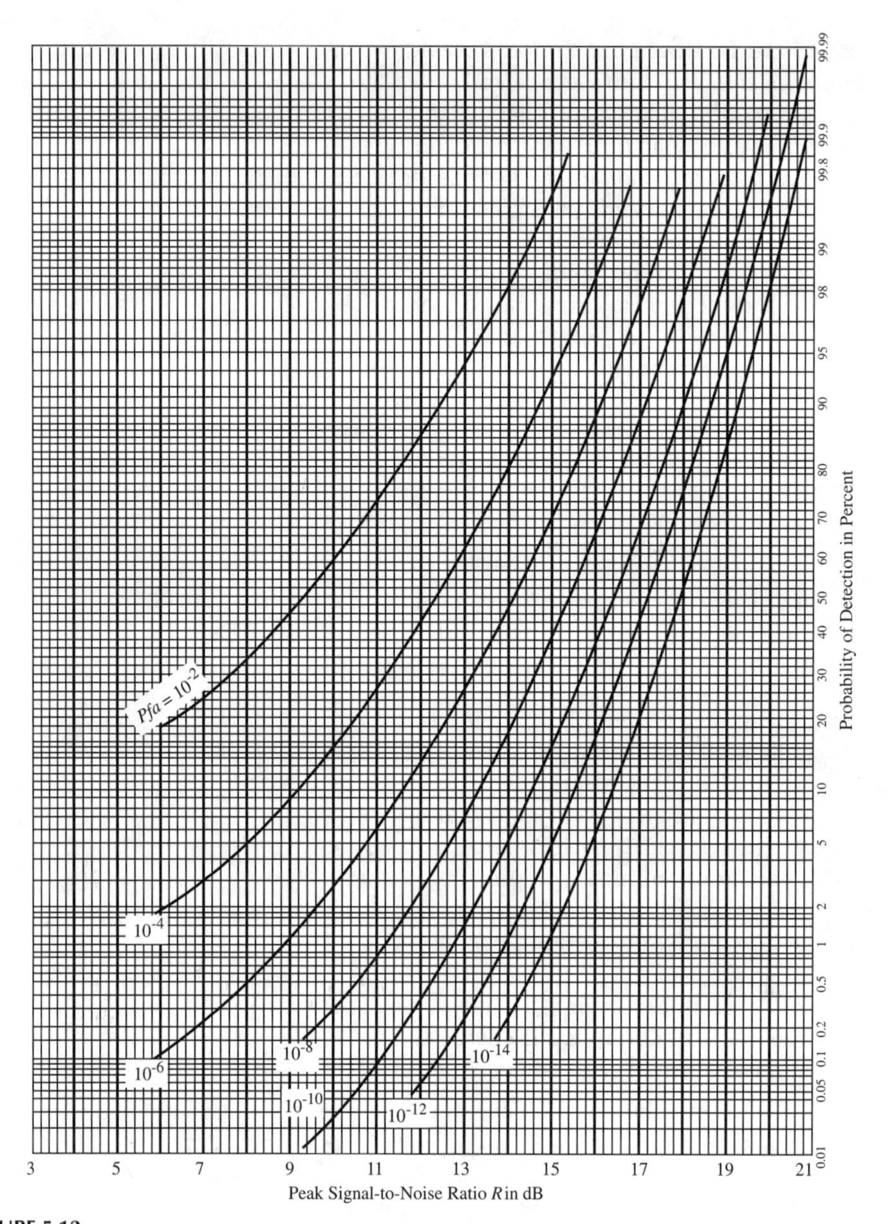

FIGURE 5.12
Coherent detection probabilities vs. SNR. From DiFranco and Rubin,[1] Figure 9.2.3 of "Radar Detection," by permission of the authors.

have a large effect on required SNR for a given detection probability at low P_{fa} (e.g., $P_{fa} = 10^{-14}$ to 10^{-8}) but have a much larger effect at high P_{fa} (e.g., 10^{-8} to 10^{-2}).

These curves are merely plots of P_d from (5.15a) where [from (5.15b)]

$$\hat{v}_T = \text{erf}^{-1}(1 - 2P_{fa}) \tag{5.31}$$

For example, if we specify $P_{fa} = 10^{-2}$, then $\hat{v}_T = \text{erf}^{-1}(0.98)$, which can be determined from a table of the error function.

Figure 5.13 has the same format and the plots are of P_d vs. R_o. However, it is indicated that signal phase is unknown. That dictates that the "matched" filtering

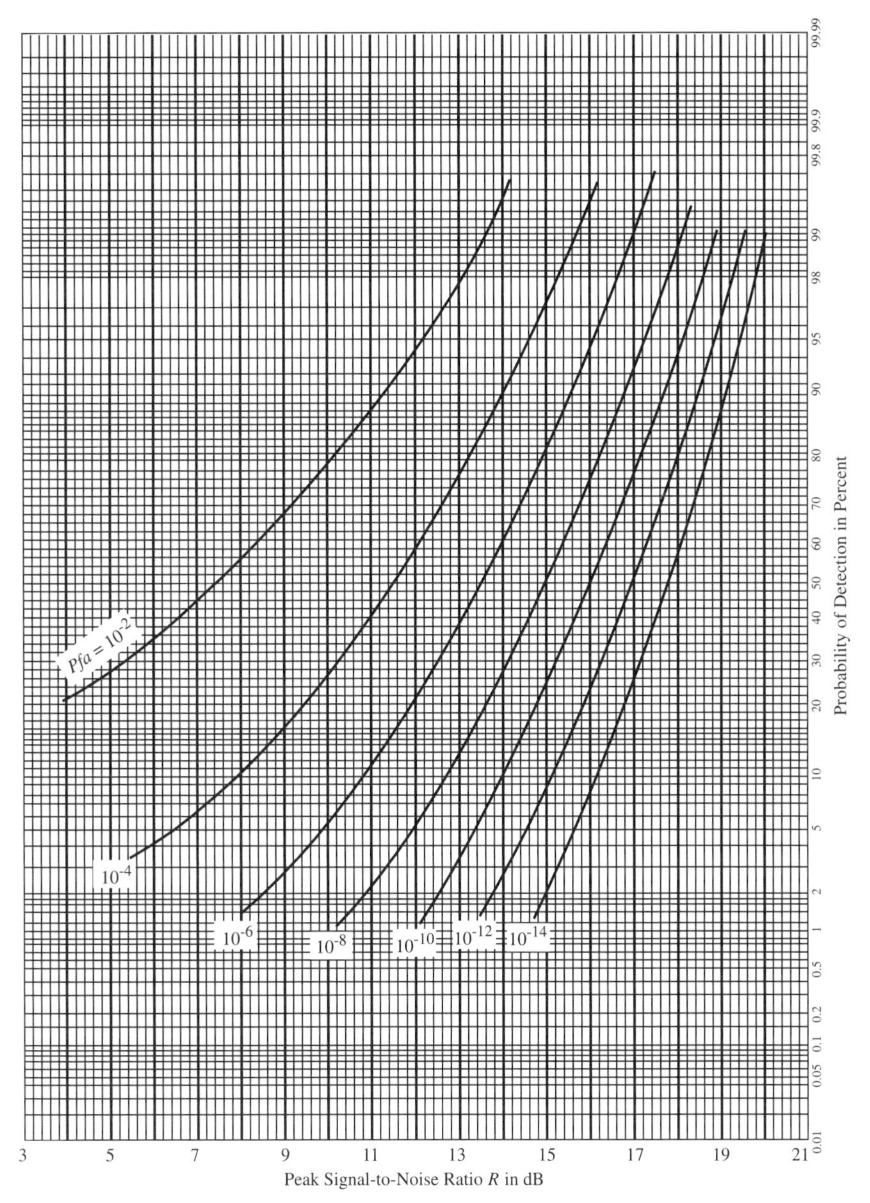

FIGURE 5.13
Noncoherent detection probabilities vs. SNR. From DiFranco and Rubin,[1] Figure 9.3.7 of "Radar Detection," by permission of the authors.

be noncoherent, i.e., the detection decision is made from the amplitude or envelope of the filter output, i.e., "noncoherent detection."

The curves in Figure 5.13 are plots of P_d from (5.28b), where [from (5.28a)], $\hat{v}_T$ is given by

$$\hat{v}_T = \sqrt{-\ln P_{fa}} \tag{5.32}$$

for example, if $P_{fa} = 10^{-2}$, then $\hat{v}_T = \sqrt{4.605} = 2.146$ and if $P_{fa} = 10^{-8}$, $\hat{v}_T = \sqrt{18.4} = 4.29$.

To compare coherent and noncoherent detection performance, we compare the SNR values required to obtain a given level of detection performance (say P_d = 95%, $P_{fa} = 10^{-6}$) in the two cases, using Figures 5.12 and 5.13. The required values of R_o are 16 and 16.6 dB for coherent and noncoherent cases, respectively. Thus for that specific performance level 0.6 dB more SNR is required to achieve the specified values of P_d and P_{fa}. Table 5.1 shows the same comparison for some other levels.

The overall conclusion from Table 5.1 is that the price to be paid for lack of a priori phase knowledge for detection probabilities of 90% or over (detection probabilities much lower than that are usually not very practical in radar, because detection of only 9 targets out of 10 is not considered good performance for a radar) is not very significant with single pulses, being of the order of about 1 dB or less of increase in required SNR with noncoherent detection. With a pulse train, the conclusion is still valid, *if* the quadrature filtering technique is used on the entire pulse train before the detection decision is made. That is conditional on the signal phase remaining stable throughout the pulse train, so that the sum of the squares of cosine and sine filter outputs is nearly that predicted by the theory in Sections 4.4 and 4.5. That procedure can be called "coherent pulse integration," implying that the pulses are added in voltage, not power, before taking the amplitude. Application of quadrature filtering to each individual pulse, then taking the output amplitude or amplitude-square (i.e., sum of squares of cosine and sine filter outputs) and adding pulse amplitudes *or* amplitude-square values, is called "noncoherent pulse integration." The latter is an entirely different procedure than the former, and results in a much larger differential SNR requirement for a given level of detection performance. This will be discussed later under "integration loss."

5.4 EFFECT OF SIGNAL FLUCTUATIONS

In all of the earlier discussions in this chapter, it has been assumed that the target signal amplitude is fixed throughout the duration of the detection process.

Actual target signal amplitudes *do* fluctuate. Although the signal amplitude a_s may be fixed during the detection process, it is not in general a priori known.

TABLE 5.1

Detection Threshold Comparisons for Nonfluctuating Signals

Specified Performance Level		Required Values of Peak SNR R in dB		Difference (Noncoherent R_o − Coherent R_o) in dB
P_d	P_{fa}	Coherent	Noncoherent	NCOH − COH
90%	10^{-14}	19.1	19.2	+0.1
	10^{-6}	15.4	16.0	+0.6
	10^{-2}	11.2	12.4	+1.2
95%	10^{-14}	19.4	19.6	+0.2
	10^{-6}	16.0	16.6	+0.6
	10^{-2}	12.0	13.2	+1.2
98%	10^{-14}	19.8	20.0	+0.2
	10^{-6}	16.6	17.0	+0.4
	10^{-2}	13.0	14.0	+1.0
99%	10^{-13}	20.0	20.2	+0.2
	10^{-6}	17.0	17.8	+0.8
	10^{-2}	13.4	14.4	+1.0

That does not affect the efficacy of matched filtering, because the amplitude is not one of the matching parameters. It *does,* however, affect the probability of detection, because its calculation requires integration over the PDF of the unknown signal amplitude. To analyze this effect, we return to (5.27) and rewrite it in the form

$$p[v/s(a_s)] = v\int_0^{2\pi} d\phi\, p(v\cos\phi, v\sin\phi) \tag{5.33}$$

$$= v\int_0^{2\pi} d\phi \int_0^{\infty} da_s\, p(a_s)p(v\cos\phi, v\sin\phi/a_s)$$

where the notation accounts explicitly for the presence of the random variable a_s, where the dependence of $p(v\cos\phi, v\sin\phi)$ on a_s is indicated notationally, and where $p(a_s)$ is the PDF of a_s.

Specializing (5.33) to the bivariate Gaussian function as in (5.27), but now integrating over both ϕ and the amplitude a_s, we have

$$p[v/s(a_s)] = \frac{ve^{-v^2/2\sigma_n^2}}{2\pi\sigma_n^2}\int_0^{\infty} da_s a_s \int_0^{2\pi} d\phi \cdots$$
$$e^{-[a_s^2-2va_s\cos(\phi-\phi_s)]/2\sigma_n^2}\, e^{-a_s^2/2\sigma_s^2} \tag{5.34}$$

We assume a Rayleigh fluctuation in the amplitude of the signal. That type of fluctuation amplitude PDF would arise from a target consisting of a large number of statistically independent scattering centers whose aggregate (due to the Central Limit Theorem, to be discussed in Section 6.1) would be well approximated by the bivariate Gaussian function. As we know from discussions of the noise PDF in Sections 5.1, 5.2, and 5.3, the amplitude of a zero-mean random process with a bivariate Gaussian distribution has a Rayleigh distribution [see Eq. (5.24)], i.e.

$$p(a_s) = \frac{a_s}{\sigma_s^2}\, e^{-a_s^2/2\sigma_s^2} \tag{5.35}$$

where $\sigma_s^2 = \frac{1}{2}\langle a_s^2\rangle$ and $p(a_s)$ peaks at $a_s = \sigma_s$.

To calculate $\langle a_s^2\rangle$ in terms of σ_s^2, we write

$$\langle a_s^2\rangle = \int_0^{\infty} da_s \frac{a_s}{\sigma_s^2} a_s^2\, e^{-a_s^2/2\sigma_s^2} = 2\sigma_s^2\int_0^{\infty} d\left(\frac{a_s^2}{2\sigma_s^2}\right)\frac{a_s^2}{2\sigma_s^2}\, e^{-a_s^2/2\sigma_s^2}$$

$$= 2\sigma_s^2\int_0^{\infty} dx\, xe^{-x} = 2\sigma_s^2 \tag{5.35a)$'$}$$

The peak of the Rayleigh PDF occurs where

$$\frac{\partial p(a_s)}{\partial a_s} = 0 = \frac{e^{-a_s^2/2\sigma_s^2}}{\sigma_s^2}\left[1 - \frac{2a_s^2}{2\sigma_s^2}\right], \text{ or } a_s = \sigma_s \tag{5.35b)$'$}$$

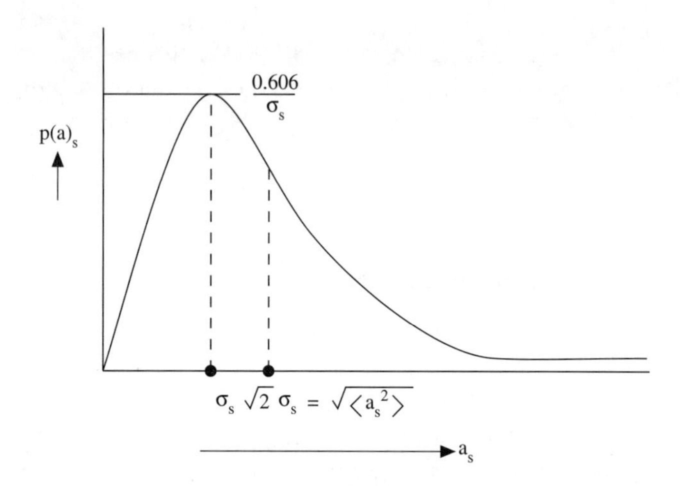

FIGURE 5.14
Rayleigh PDF.

where

$$p(\sigma_s) = \frac{1}{\sigma_s} e^{-1/2} = \frac{0.606}{\sigma_s} \qquad (5.35c)'$$

The Rayleigh function reaches its $1/e$ point relative to the peak at $a_s = (a_s)_{1/e}$, i.e.,

$$\frac{p[(a_s)_{1/e}]}{p(\sigma_s)} = \frac{1}{e} \qquad (5.35d)'$$

or

$$\frac{(a_s)^2_{1/e} - \sigma_s^2}{\sigma_s^2} = 2\left\{\ln\left[\frac{(a_s)_{1/e}}{\sigma_s}\right] + 1\right\}$$

The Rayleigh PDF was illustrated in Figure 5.8 in the context of receiver noise. It is illustrated in more detail above in Figure 5.14 in the present context.

Continuing with the calculation in (5.34) with the assumed PDF of (5.35), we obtain

$$p[v/s(a_s)] = \frac{ve^{-v^2/2\sigma_n^2}}{\sigma_n^2\sigma_s^2} \int_0^\infty da_s a_s e^{-(a_s^2/2)[(1/\sigma_s^2)+(1/\sigma_n^2)]} I_0\left(\frac{va_s}{\sigma_n^2}\right) \qquad (5.36)$$

From DiFranco and Rubin[1] [p. 311, Sec. 9.4, Eq. (9.4.11)], we have the integral on a_s in (5.36) expressed as a special case of the more general integral

$$L = \int_0^\infty dt\ t^{\mu-1} I_\nu(\alpha t) e^{-p^2t^2}$$

$$= \frac{\Gamma[(\mu + \nu/2)](\alpha/2p)^\nu}{2p^\mu\Gamma(\nu + 1)} e^{\alpha^2/4p^2}\ {}_1F_1\left[\frac{(\nu - \mu)}{2} + 1, \nu + 1; \frac{-\alpha^2}{4p^2}\right] \qquad (5.37)$$

where ${}_1F_1$ is the confluent hypergeometric function, represented by the power series [DiFranco and Rubin,[1] Eq. (9.4–13), p. 311]

$$ {}_1F_1(a, c, z) = 1 + \frac{az}{c} + \frac{a(a+1)z^2}{2!c(c+1)} + \cdots \tag{5.37a)'} $$

The integral in (5.36) has the specific forms of the parameters in (5.37) as follows:

$$ \mu = 2 $$

$$ \nu = 0 \qquad\qquad z = \frac{-\alpha^2}{4p^2} = \frac{-v^2\sigma_s^2}{2(\sigma_s^2 + \sigma_n^2)\sigma_n^2} $$

$$ \alpha = \frac{v}{\sigma_n^2} \qquad\qquad \frac{\mu + \nu}{2} = 1 $$

$$ p^2 = \frac{1}{2}\left[\frac{1}{\sigma_s^2} + \frac{1}{\sigma_n^2}\right] \qquad\qquad a = \frac{(\nu - \mu)}{2} + 1 = 0 $$

$$ c = \nu + 1 = 1 $$

From (5.37) and (5.37a)′ with these parameters we have

$$ \begin{aligned} L &= \frac{\Gamma(1)(\alpha/2p)^0}{2\{1/2\,[(1/\sigma_s^2) + (1/\sigma_n^2)]\}\Gamma(1)}\, e^{+v^2\sigma_s^2/2(\sigma_s^2+\sigma_n^2)\sigma_n^2}\, {}_1F_1\left[0, 1, \frac{-v^2\sigma_s^2}{2\sigma_n^2(\sigma_s^2 + \sigma_n^2)}\right] \\ &= \frac{\sigma_s^2\sigma_n^2}{\sigma_s^2 + \sigma_n^2}\, e^{v^2\sigma_s^2/2\sigma_n^2(\sigma_s^2+\sigma_n^2)}[1 + 0 + 0 + \cdots 0] \\ &= \frac{\sigma_s^2\sigma_n^2}{\sigma_s^2 + \sigma_n^2}\, e^{+v^2\sigma_s^2/2\sigma_n^2(\sigma_s^2+\sigma_n^2)} \end{aligned} \tag{5.37b)'} $$

From (5.36) and (5.37b)′ we obtain

$$ \begin{aligned} p[v/s(a_s)] &= \frac{v}{\sigma_s^2 + \sigma_n^2}\, e^{-(v^2/2\sigma_n^2)+[v^2\sigma_s^2/2\sigma_n^2(\sigma_s^2+\sigma_n^2)]} \\ &= \frac{v}{\sigma_s^2 + \sigma_n^2}\, e^{-v^2/2(\sigma_n^2+\sigma_s^2)} \end{aligned} \tag{5.37c)'} $$

The calculation of P_d now turns out to be much simpler than in the nonfluctuating signal case.

From (5.37c)′ and (5.13a)

$$ P_d = \int_{v_T}^{\infty} dv\, \frac{ve^{-v^2/2(\sigma_s^2+\sigma_n^2)}}{(\sigma_s^2 + \sigma_n^2)} = e^{-\hat{v}_T^2/(1+\rho_s)} \tag{5.38} $$

where $\hat{v}_T$ = normalized threshold = $v_T/\sqrt{2}\sigma_n$ and $\rho_s = \sigma_s = (\sigma_s/\sigma_n)^2$ = SNR as we now define it for a fluctuating signal.

A rough sketch of P_d and P_{fa} vs. $\hat{v}T$ is shown in Figure 5.15 for variation in ρ_s. P_d and P_{fa} are obtained from (5.38) and (5.28a), respectively.

Also from (5.38) and (5.28a),

$$\hat{v}_T = \sqrt{-\ln P_{fa}}$$

and hence

$$P_d = e^{\ln P_{fa}/(1+\rho_s)} = e^{-|\ln P_{fa}|/(1+\rho_s)} \tag{5.39}$$

which provides a direct functional relationship between P_d and P_{fa}, showing a monotonic increasing relationship between them, as expected.

The detection probabilities vs. SNR for this case are shown in Figure 5.16, taken from Figure 9.4.1 on p. 312 of DiFranco and Rubin.[1]

Another case of interest is that of a "one-dominant-plus-Rayleigh" fluctuating target signal, given by (see DiFranco and Rubin, Eq. (9.5-1), p. 313, where $A \rightarrow a_s$, $A_o \rightarrow \hat{\sigma}_s$)

$$p(a_s) = \frac{9a_s^3}{2\hat{\sigma}_s^4} e^{-3a_s^2/2\hat{\sigma}_s^2} \tag{5.40}$$

This distribution corresponds to a target with a large number of scattering centers of roughly comparable amplitude but with a single scattering center whose amplitude far exceeds the others and hence that dominates the scattered wave amplitude. If the amplitude of the dominant center were deterministic, that would lead to a Rice distribution, but since it is a random variable, the overall PDF is that given by (5.40) (without proof).

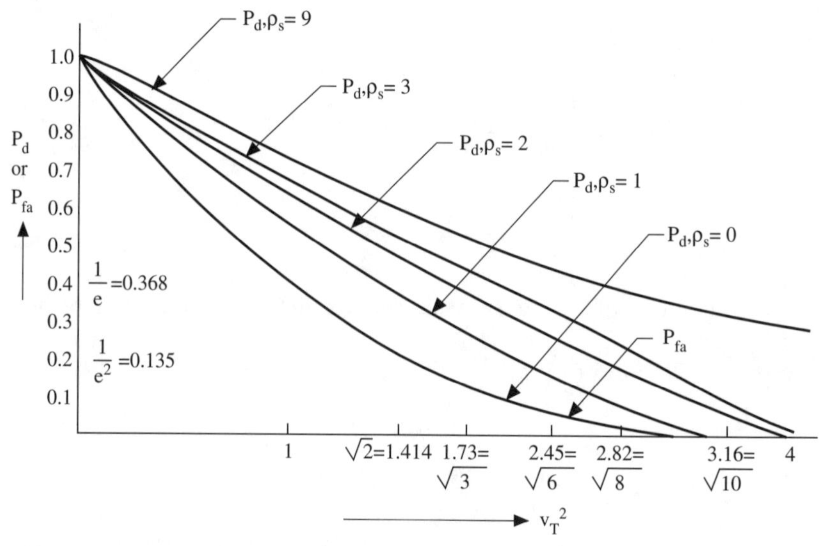

FIGURE 5.15
P_d and P_{fa} vs. $\bar{v}_T$—Rayleigh-fluctuating signal.

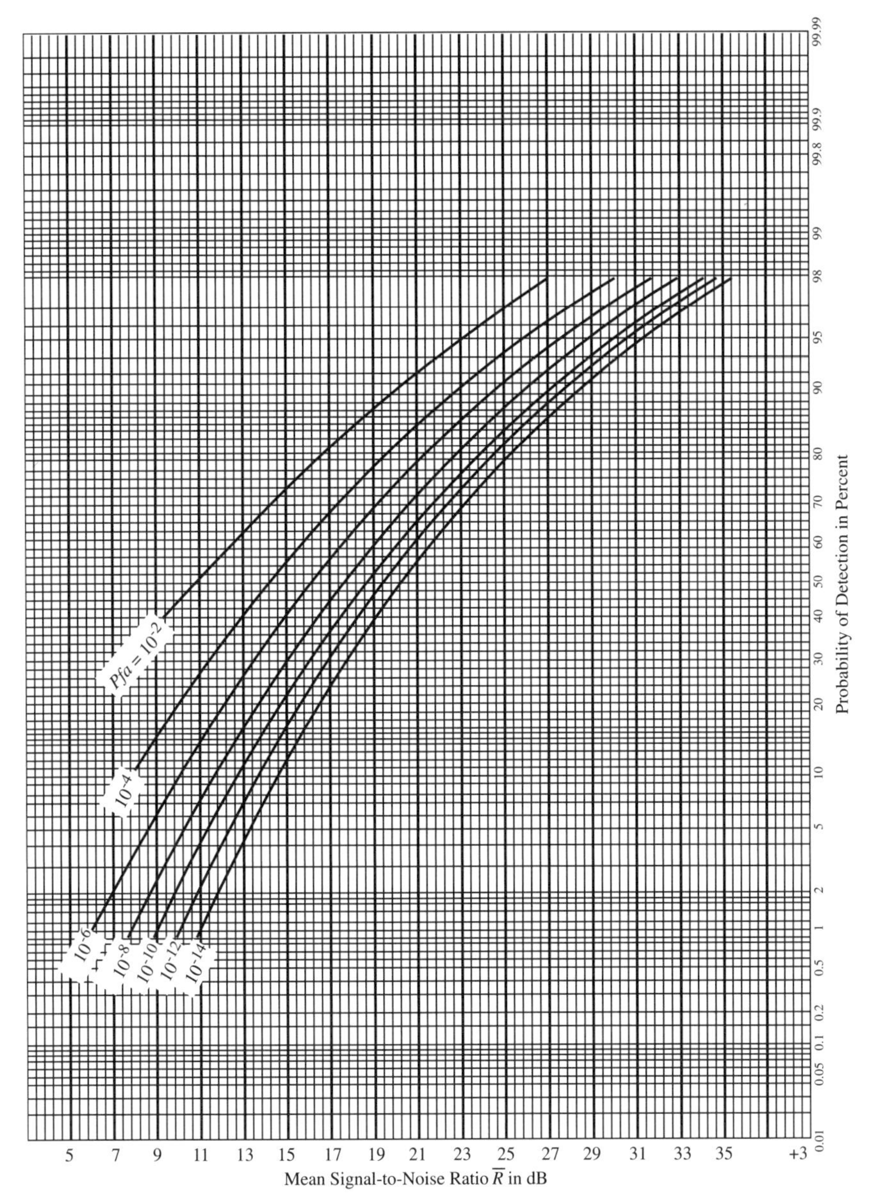

FIGURE 5.16
Optimum detection characteristic for a Rayleigh fluctuating target (unknown phase). From DiFranco and Rubin,[1] Figure 9.4.1 of "Radar Detection" by permission of the authors.

Substituting (5.40) rather than the Rayleigh PDF (5.35) into (5.33), we would obtain an analog of (5.36) of the form

$$p[v/s(a_s)] = \frac{9ve^{-v^2/2\sigma_n^2}}{2\sigma_n^2\hat{\sigma}_s^4}\int_0^\infty da_s\, a_s^3 e^{-(a_s^2/2)[(3/\hat{\sigma}_s^2)+(1/\sigma_n^2)]} I_0\left(\frac{va_s}{\sigma_n^2}\right) \tag{5.41}$$

From (5.37), (5.37a)′, and (5.41), this time with

$$\mu = 4$$

$$\nu = 0 \qquad z = \frac{-\alpha^2}{4p^2} = \frac{-v^2\hat{\sigma}_s^2}{2\sigma_n^2(\hat{\sigma}_s^2 + 3\sigma_n^2)}$$

$$\alpha = \frac{v}{\sigma_n^2} \qquad \frac{\mu + \nu}{2} = 2$$

$$p^2 = \frac{1}{2}\left(\frac{3}{\hat{\sigma}_s^2} + \frac{1}{\sigma_n^2}\right) \qquad a = \frac{(\nu - \mu)}{2} + 1 = -1$$

$$c = \nu + 1 = 1$$

From (5.37a)′ with these parameters

$${}_1F_1(a, c, z) = 1 + \frac{v^2\hat{\sigma}_s^2}{2\sigma_n^2(\hat{\sigma}_s^2 + 3\sigma_n^2)} + 0 + \cdots + 0 \cdots \tag{5.41a)'}$$

and from (5.37)

$$L = \frac{\Gamma(2)(\alpha/2p)^0 e^{v^2\hat{\sigma}_s^2/2\sigma_n^2(\hat{\sigma}_s^2+3\sigma_n^2)}}{2\Gamma(1)\left\{\frac{1}{2}\left[(3/\hat{\sigma}_s^2) + (1/\sigma_n^2)\right]\right\}^2}\left[1 + \frac{v^2\hat{\sigma}_s^2}{2\sigma_n^2(\hat{\sigma}_s^2 + 3\sigma_n^2)}\right] \tag{5.41b)'}$$

and finally from (5.33), (5.41), and (5.41a,b)′

$$\begin{aligned} p[v/s(a_s)] &= \frac{9L}{2\sigma_n^2\hat{\sigma}_s^4} = \frac{9v\sigma_n^2}{(\hat{\sigma}_s^2 + 3\sigma_n^2)^2}\, e^{-3v^2/2(3\sigma_n^2+\hat{\sigma}_s^2)} \cdots \\ &\left[1 + \frac{\hat{\sigma}_s^2 v^2}{2\sigma_n^2(\hat{\sigma}_s^2 + 3\sigma_n^2)}\right] \\ &= \frac{9v}{\sigma_n^2(3 + \hat{\rho}_s)^2}\, e^{-3v^2/2\sigma_n^2(3+\hat{\rho}_s)}\left[1 + \frac{\hat{\rho}_s v^2}{2\sigma_n^2(3 + \hat{\rho}_s)}\right] \end{aligned} \tag{5.41c)'}$$

where $\hat{\rho} = \hat{\sigma}_s^2/\sigma_n^2$.

From (5.41c)′ and (5.13a)

$$\begin{aligned} P_d &= \int_{v_T}^{\infty} dv\, p[v/s(a_s)] = \frac{9(2)}{3\sigma_n^2(3 + \hat{\rho}_s)}\int_{v=v_T}^{\infty} d\left(\frac{\sqrt{3}v}{\sqrt{2\sigma_n^2(3 + \hat{\rho}_s)}}\right)\cdots \\ &\left(\frac{\sqrt{3}v}{\sqrt{2\sigma_n^2(3 + \hat{\rho}_s)}}\right) e^{-(\sqrt{3}v/\sqrt{2\sigma_n^2(3+\hat{\rho}_s)})^2}\left[1 + \frac{\hat{\rho}_s}{3}\left(\frac{\sqrt{3}v}{\sqrt{2\sigma_n^2(3 + \hat{\rho}_s)}}\right)^2\right] \\ &= \frac{3}{(3 + \hat{\rho}_s)}\int_{(\sqrt{3}v_T/\sqrt{3+\hat{\rho}_s})^2}^{\infty} d(u^2)e^{-u^2}\left(1 + \frac{\hat{\rho}_s}{3}u^2\right) = \left(\frac{3}{3 + \hat{\rho}_s}\right)\int_{[3\hat{v}_T^2/(3+\hat{\rho}_s)]}^{\infty} dx\, e^{-x}\left(1 + \frac{\hat{\rho}_s}{3}x\right) \end{aligned} \tag{5.42}$$

Finally

$$P_d = e^{-3\hat{\upsilon}_T^2/(3+\hat{\rho}_s)}\left(\frac{3}{3+\hat{\rho}_s}\right)\left\{1 + \frac{\hat{\rho}_s}{3} + \frac{\hat{\upsilon}_T^2}{[1+(3/\hat{\rho}_s)]}\right\} \tag{5.43}$$

Noting that (from (5.40))

$$\begin{aligned} \langle a_s^2 \rangle &= \frac{9}{2\hat{\sigma}_s^4}\int_0^\infty da_s\, a_s^5 e^{-3a_s^2/2\sigma_s^2} \\ &= \frac{4\hat{\sigma}_s^2}{3}\int_0^\infty d\left(\sqrt{\frac{3}{2}}\frac{a_s}{\hat{\sigma}_s}\right)\left(\sqrt{\frac{3}{2}}\frac{a_s}{\hat{\sigma}_s}\right)^5 e^{-(\sqrt{3/2}(a_s/\hat{\sigma}_s))^2} \\ &= \frac{4\hat{\sigma}_s^2}{3}\int_{-\infty}^0 dx\, x^5 e^{x^2} = \frac{2\hat{\sigma}_s^2}{3}\int_{-\infty}^0 dy\, y^2 e^y = \frac{4\hat{\sigma}_s^2}{3} \end{aligned} \tag{5.44}$$

In terms of $\rho_s = \langle a_s^2 \rangle/\sigma_n^2 = \frac{4}{3}\hat{\rho}_s$, we have $\hat{\rho}_s = \frac{3}{4}\rho_s$ and [from (5.43) and (5.44)]

$$P_d = \frac{e^{-\hat{\upsilon}_T^2/2(1+(\rho_s/4))}}{(1+(\rho_s/4))}\left\{1 + \frac{\rho_s}{4} + \frac{\hat{\upsilon}_T^2}{(1+(4/\rho_s))}\right\} \tag{5.45}$$

Equation (5.45) is equivalent to Eq. (9.5.9-a) on p. 315 of DiFranco and Rubin,[1] where $\bar{\mathcal{R}} \rightarrow \rho_s$ and $\ell n\, P_{fa} \rightarrow \hat{\upsilon}_T^2$, consistent with (5.28a).

Figure 5.17 shows detection probabilities for the case of one-dominant-plus Rayleigh fluctuations. This is Figure 9.5.1 on p. 314 of DiFranco and Rubin.[1] The values of P_d are obtained from (5.45) with $\hat{\upsilon}_t = \sqrt{|\ell n\, P_{fa}|}$ based on (5.28a).

A comparison between Rayleigh and one-dominant-plus Rayleigh signal fluctuation results for P_d vs ρ_s can be made by setting the minimum acceptable P_d and maximum acceptable P_{fa} values and determining the minimum values of ρ_s required to attain these specifications. Further comparisons of these two fluctuation cases with the case of the nonfluctuating signal are shown in Table 5.2. Also shown is the comparison with the case of noncoherent detection with no signal fluctuations. The format for Table 5.2 is the same as that for Table 5.1.

It is evident from Table 5.2 that very significant differences exist between the SNR requirements for various performance levels due to target signal fluctuations. The most striking differences are for the Rayleigh case compared with the nonfluctuating signal case. For very high performance levels, e.g., 99% P_d and $P_{fa} = 10^{-14}$, it requires 17 dB more SNR to achieve the specified performance level for Rayleigh fluctuations. For comparatively low performance levels, e.g., P_d = 90% and $P_{fa} = 10^{-2}$, only about a 7 dB difference is observed. In general, it appears to require between about 8 and 15 dB more SNR to achieve detection probabilities over 90% and false alarm rates of about 10^{-6}. The differences are significantly lower for one-dominant-plus Rayleigh fluctuations, varying from about 3 to 5 dB for 90% detection to 7 to 9 dB for 99%. In all cases, the differences are greater for higher performance levels.

To explain the results for the Rayleigh case, we need only look at the possibility of very large amplitude swings when the signal is present due to its fluctuations. Some of those amplitude swings interfere *destructively* with noise swings and produce lower total signal-plus-noise amplitude than would otherwise be present. If the

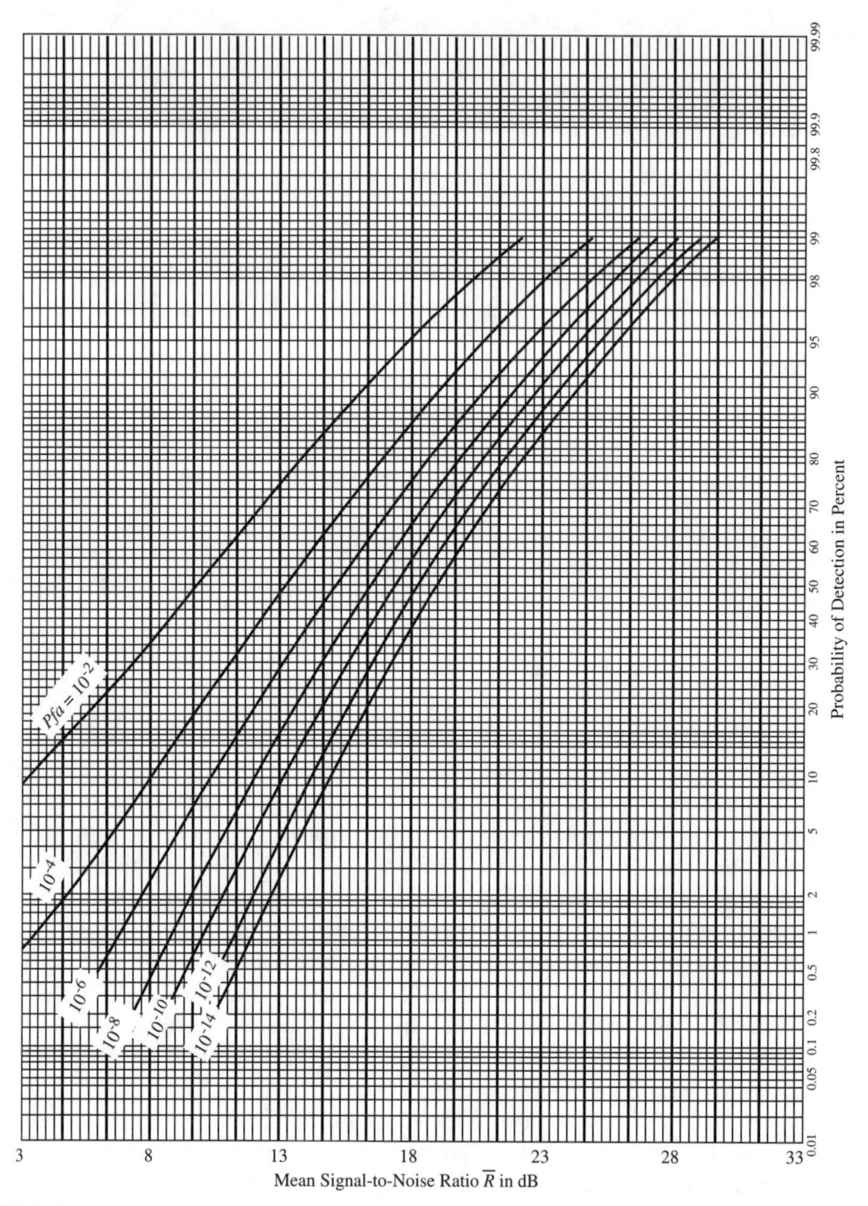

FIGURE 5.17
Detection characteristic for a one-dominant-plus Rayleigh fluctuating target. From DiFranco and Rubin,[1] Figure 9.5.1 of "Radar Detection," by permission of the authors.

threshold is set very high to attain low false alarm rates, then the signal amplitude swings are more likely to drive the total amplitude below threshold.

In the one-dominant-plus Rayleigh case, the same phenomenon occurs, of course, and qualitatively the trends are the same. However, in that case the one-dominant scattering center produces most of the fluctuations and the net result is that a smaller fraction of the signal energy is involved in the process of destructive interference that drives the total amplitude below the threshold; hence the process tends to be more dominated by noise than by signal fluctuations and the differences with the nonfluctuating case are not as pronounced.

TABLE 5.2

Detection Threshold Comparisons for Fluctuating Signals

Specified Performance Level		Required Values of Peak SNR in dB—Rounded off: Noncoherent Detection—Single Pulse			Difference in dB		
P_d	P_{fa}	(1) No Fluctuations	(2) Rayleigh	(3) One-Dominant-Plus Rayleigh	(2) − (1)	(3) − (1)	(3) − (2)
90%	10^{-14}	19.2	28	24	+9	+5	−4
	10^{-6}	16.0	24	20	+8	+4	−4
	10^{-2}	12.4	19	15	+7	+3	−4
95%	10^{-14}	19.6	31	26	+11	+6	−5
	10^{-6}	16.6	27	21	+10	+4	−6
	10^{-2}	13.2	23	17	+10	+4	−6
98%	10^{-14}	20.0	35	28	+15	+8	−7
	10^{-6}	17.0	31	24	+14	+7	−7
	10^{-2}	14.0	26	20	+12	+6	−6
99%	10^{-14}	20.2	37	29	+17	+9	−8
	10^{-6}	17.8	34	26	+16	+8	−8
	10^{-2}	14.4	29	21	+14	+7	−8

5.5 OPTIMUM DETECTION STRATEGY

We have already learned (in Chapter 4) that an attempt at matched filtering is an "optimum" strategy for maximizing SNR. We also learned (in this chapter) that in all cases studied so far, detection performance improves with increasing SNR. Based on that, we might well conclude that an optimum detection strategy is to matched-filter the incoming signal, then make the detection decision. That is indeed the case, but we can learn more about optimization of the detection strategy by asking the question, "Given a set of N sample values of the signal-plus-noise waveform," what is the strategy that maximizes a particular measure of detection performance (other than SNR) based on the use of those N sample points?" to formulate that problem, we first define a measure of detection performance. In general we will choose the "measure of quality" Q, where

$$Q = K_A P_d + K_B(1 - P_{fa}) \tag{5.46}$$

where K_A and K_B are weighing factors whose assignment is based on a value judgment as to which type of favorable event is most desirable, a correct decision if a signal *is* present or a correct decision if *no* signal is present. In the former case K_A is chosen to exceed K_B and in the latter case, K_B exceeds K_A. If $K_A = P_S$, $K_B = P_n$, then Q is the probability of a favorable decision (i.e., the probability that the signal is present times the conditional probability that the decision made is "signal present" plus the probability that no signal is present times the probability that the decision is "noise alone," where P_s and P_n are, respectively, the a priori probabilities of signal presence and noise alone). For that assignment of K_A and K_B, the strategy of *maximizing* Q is called the "ideal observer" criterion and Q is the "probability of success."

Equivalent to (5.46) is

$$C = 1 - Q = K'_A(1 - P_d) + K'_B P_{fa} \tag{5.47}$$

where C is now a measure of degradation or absence of quality. The strategy of *minimizing* C is now optimal, where C is known as "cost," "risk," or "loss" and the criterion is variously called "minimum average cost" or "minimum average risk" or "minimum average loss." The main point of (5.47) is that $(1 - P_{\mathrm{d}})$ and P_{fa} are the probabilities of two *unfavorable* decisions, or *errors*, P_{fa} being that of a false alarm and $(1 - P_{\mathrm{d}})$ being that of a "false rest" or "false dismissal," where a target signal is actually present but the detection strategy missed it. If K_{A}' and K_{B}' are P_{s} and P_{n}, respectively, then C is the "probability of error." If other weightings are put on K_{A}' and K_{B}', their choice indicates a value judgment on the relative importance of avoiding false rests and false alarms.

Whichever of these criteria is to be used, we begin with a set of sample values of the IF signal-plus-noise waveform $v(t)$ at times $(t_1, t_2, \ldots, t_N)$ as shown in Figure 5.18. These sample values are $(v_1, v_2, \ldots, v_N)$, where $v_k = v(t_k)$.

We now denote the vector of N sample points by $\mathbf{v} = (v_1, v_2, \ldots, v_N)$. The joint multivariate PDF of $\mathbf{v}$ given noise alone is $p(\mathbf{v}/n)$ and the joint multivariate PDF of $\mathbf{v}$ given a signal $s(t)$ with a set of possibly random parameters $(\Omega_2, \Omega_2, \ldots, \Omega_M)$ characterizing the signal waveform is $p[\mathbf{v}/s(\Omega)]$, where $\mathbf{\Omega}$ is a vector representing the M signal parameters. The joint multivariate PDF of the vector $\mathbf{\Omega}$ is $p(\mathbf{\Omega})$.

Based on these definitions, the detection probability P_{d} and false alarm probability P_{fa} are

$$P_{\mathrm{d}} = \int \underset{R_{\mathrm{s}}}{\cdots} \int d^N\mathbf{v} \underset{\mathbf{\Omega}\text{-space}}{\int \cdots \int} d^M\mathbf{\Omega} p(\mathbf{\Omega}) p[\mathbf{v}/s(\mathbf{\Omega})] \tag{5.48a}$$

$$P_{\mathrm{fa}} = \int \underset{R_{\mathrm{s}}}{\cdots} \int d^N\mathbf{v} p(\mathbf{v}/n) \tag{5.48b}$$

where R_{s} is a region in the N-dimensional hyperspace of the variables $(v_1, v_2, \ldots, v_N)$. If $\mathbf{v}$ is within that region, then the decision will be "signal present." This is illustrated in Figure 5.19 for the case $N = 3$.

In (5.48a,b), the PDF $p[\mathbf{v}/s(\mathbf{\Omega})]$ is averaged over the distribution of the signal parameters $(\Omega_1, \Omega_2, \ldots, \Omega_M)$. The designation "$\mathbf{\Omega}$-space" is the M-dimensional hyperspace of these parameters, and $d^N\mathbf{v}$ and $d^M\mathbf{\Omega}$ are, respectively, the incremental elements $(dv_1\, dv_2 \ldots dv_N)$ and $(d\Omega_1 d\Omega_2 \ldots d\Omega_M)$.

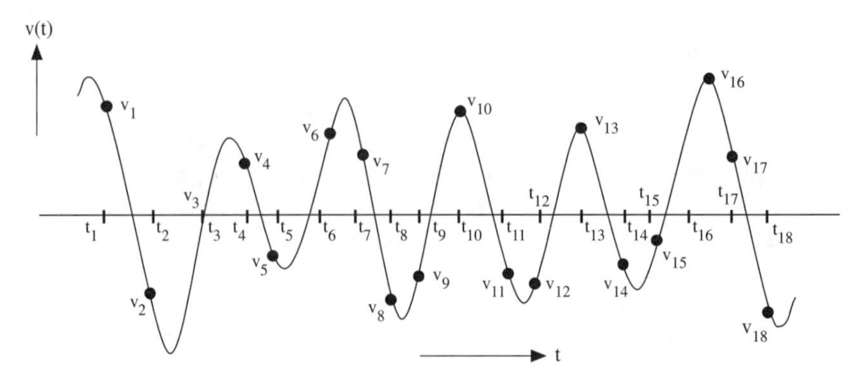

FIGURE 5.18
IF sample points for optimum detection.

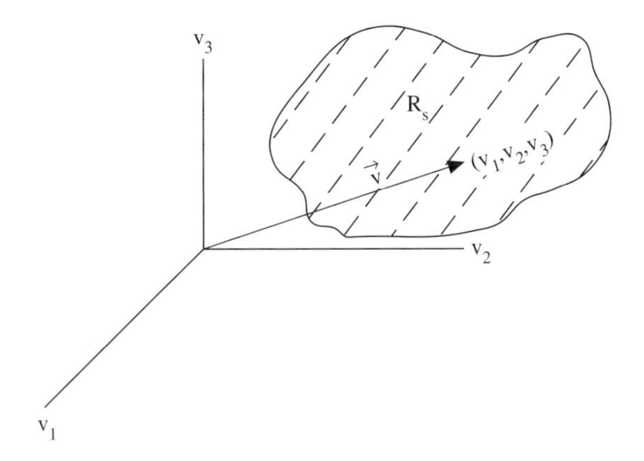

FIGURE 5.19
Hyperspace region R_s for positive decision. R_n = space other than R_s. If (v_1, v_2, v_3) is in R_n, *then the decision is "noise alone." If* (v_1, v_2, v_3) is in R_s, *then the decision is "signal present."*

We now rewrite (5.48a) in the form

$$P_{\mathrm{d}} = \int \underset{R_{\mathrm{s}}}{\cdots} \int d^N \mathbf{v} p(\mathbf{v}/n) \overline{L}(\mathbf{v}) \tag{5.49}$$

where

$$\overline{L}(\mathbf{v}) = \underset{\Omega\text{-space}}{\int \cdots \int} d^M \boldsymbol{\Omega} p(\boldsymbol{\Omega}) L(\mathbf{v}/\boldsymbol{\Omega}) \tag{5.49$'$}$$

and where

$$\begin{aligned} L(\mathbf{v}/\boldsymbol{\Omega}) &= p[\mathbf{v}/s(\boldsymbol{\Omega})]/p(\mathbf{v}/n) \\ &= \text{"likelihood ratio" for the set of signal parameter values } \boldsymbol{\Omega} \\ &= \Omega_1, \Omega_2, \ldots, \Omega_M \end{aligned}$$

and

$$\overline{L}(\mathbf{v}) = \text{likelihood ratio averaged over all parameter values } \Omega_1, \Omega_2 \ldots, \Omega_M$$

Application of (5.46), (5.49), and (5.49)′ to determination of Q leads to

$$\begin{aligned} Q &= K_{\mathrm{A}} \int \underset{R_{\mathrm{s}}}{\cdots} \int d^M \mathbf{v} p(\mathbf{v}/n) \overline{L}(\mathbf{v}) \\ &\quad + K_{\mathrm{B}} = \left[1 - \int \underset{R_{\mathrm{s}}}{\cdots} \int d^N \mathbf{v} p(\mathbf{v}/n)\right] \\ &= K_{\mathrm{B}} + K_{\mathrm{A}} \int \underset{R_{\mathrm{s}}}{\cdots} \int d^N \mathbf{v} p(\mathbf{v}/n) \left[\overline{L}(\mathbf{v}) - \frac{K_{\mathrm{B}}}{K_{\mathrm{A}}}\right] \end{aligned} \tag{5.50}$$

The maximum possible value of Q will be attained if the region R_s is chosen such that the integrand $p(\mathbf{v}/n)\,[\bar{L}(\mathbf{v} - (K_B/K_A)]$ is always positive at every point within R_s.

The obvious optimal strategy following from (5.50) is to (1) compute $\bar{L}(\mathbf{v})$, the likelihood ratio averaged over all unknown signal parameters, and (2) set a threshold $\bar{L}_T = (K_B/K_A)$ on $\bar{L}(\mathbf{v})$: (3) let the decision rule for this be: If $\bar{L}(\mathbf{v}) \geq \bar{L}_T$, decide that a signal is present, and if $\bar{L}(\mathbf{v}) < \bar{L}_T$, decide that the voltage waveform $v(t)$ contains only noise.

We now specialize the discussion to the case of a completely deterministic signal in Gaussian noise. In this case, the PDF of the unknown signal parameters $\mathbf{\Omega}$ is an impulse function centered at known and fixed values of these parameters Ω_{k0}, $k = 1, 2, \ldots, M$, i.e.

$$p(\mathbf{\Omega}) = \delta(\mathbf{\Omega}_1 - \mathbf{\Omega}_{10})\delta(\mathbf{\Omega}_2 - \mathbf{\Omega}_{20}) \cdots \delta(\mathbf{\Omega}_M - \mathbf{\Omega}_{M0}) \tag{5.51}$$

where $\delta(x)$ = unit impulse function.

Then, from (5.49)′ and (5.51)

$$\bar{L}(\mathbf{v}) = L(\mathbf{v}/\mathbf{\Omega}_0) \tag{5.52}$$

where $\mathbf{\Omega}_0 = (\Omega_{01}, \Omega_{02}, \ldots, \Omega_{0M})$

The multivariate noise PDFs with and without signal presence are

$$p[\mathbf{v}/s(\mathbf{\Omega}_0)] = \frac{1}{2\pi\sigma_n^2}\, e^{-(1/2\sigma_n^2)\sum_{k=-N}^{N}(v_k - s_k)^2} \tag{5.53a}$$

where it has been assumed that the $(2N + 1)$ noise samples are statistically independent and where $s_k = s(t_k)$, and

$$p(\mathbf{v}/n) = \frac{1}{2\pi\sigma_n^2}\, e^{-(1/2\sigma_n^2)\sum_{k=-N}^{N} v_k^2} \tag{5.53b}$$

The likelihood ratio, obtained by taking the ratio of (5.53a) to (5.53b), is

$$\begin{aligned}\bar{L}(\mathbf{v}) &= e^{-(1/2\sigma_n^2)\sum_{k=-N}^{N}(v_k^2 - 2v_k s_k + s_k^2 - v_k^2)} \\ &= [e^{-(1/2\sigma_n^2)\sum_{k=-N}^{N} s_k^2}][e^{(1/\sigma_n^2)\sum_{k=-N}^{N} s_k v_k}] = K e^{(1/\sigma_n^2)\sum_{k=-N}^{N} s_k v_k]}\end{aligned} \tag{5.54}$$

The factor on the left is independent of the voltage samples and hence just a constant *in this case* (it will be important in the next case to be considered, so it will be retained).

Since the factor on the right is a monotonically increasing function of its exponent, the strategy is to take the log* of $L(\mathbf{v})$ as given in (5.54) and set a threshold on it, i.e.,

$$\log \bar{L}(\mathbf{v}) = \log K + \frac{1}{\sigma_n^2}\sum_{k=1}^{N} s_k v_k \tag{5.54$'$}$$

* Equation (5.54)′ implies that the "log" is the *natural* log; however, the base of the logarithm affects only a scale factor and hence is unimportant in the end result.

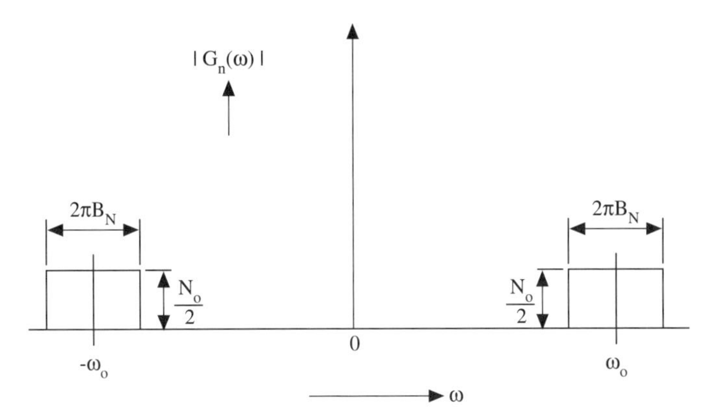

FIGURE 5.20
Spectrum of bandlimited white IF noise.

Then, if

$$\sum_{k=-N}^{N} s_k v_k \geq \overline{\overline{L}}_{\mathrm{T}}, \qquad \text{call "signal present"}$$

(where $\overline{\overline{L}}_{\mathrm{T}} = \sigma_{\mathrm{n}}^2 \log(\overline{L}_{\mathrm{T}}/K) =$ threshold level) and if

$$\sum_{k=-N}^{N} s_k v_k < \overline{\overline{L}}_{\mathrm{T}}, \qquad \text{call "noise alone"} \tag{5.55}$$

The processor output $\Sigma_{k=-N}^{N} s_k\, v_k$ is observed to be a "digital cross-correlator," in which a replica of the expected signal waveform is constructed and its sampled version is used as a reference signal to be multiplied by the incoming voltage waveform and the product $v_k s_k$ summed over all samples. Referring to Chapter 4, we see that this is nothing more or less than the equivalent of a digitized version of a matched filter. In fact, we will show that it is equivalent to a matched filter (analog version) provided we are dealing with bandlimited "white" noise, which has a flat spectrum over a given frequency interval and is zero outside that interval (Fig. 5.20). That is a very good approximation of the noise at the IF stage of a radio receiver if we account for the fact that there are bandlimited white noise spectra in both positive and negative frequency regions. The autocorrelation function (ACF) of this noise is the Fourier transform of the noise spectrum

$$\begin{aligned}
R_{\mathrm{n}}(\tau) &= \frac{N_0}{2}\left[\frac{1}{2\pi}\int_{\omega_0-\pi B_{\mathrm{N}}}^{\omega_0+\pi B_{\mathrm{N}}} d\omega e^{j\omega\tau} + \frac{1}{2\pi}\int_{-\omega_0-\pi B_{\mathrm{N}}}^{-\omega_0+\pi B_{\mathrm{N}}} d\omega e^{j\omega\tau}\right] \\
&= \langle n(t)n(t+\tau)\rangle \\
&= \frac{\sigma_n^2}{4\pi B_{\mathrm{N}}}\left[\frac{e^{j\omega_0\tau}(e^{j\pi B_{\mathrm{N}}\tau} - e^{-j\pi B_{\mathrm{N}}\tau})}{j\tau} + \frac{e^{-j\omega_0\tau}(e^{j\pi B_{\mathrm{N}}\tau} - e^{-j\pi B_{\mathrm{N}}\tau})}{j\tau}\right] \\
&= \sigma_n^2\left[\frac{e^{j\omega_0\tau} + e^{-j\omega_0\tau}}{2}\right]\frac{\sin(\pi B_{\mathrm{N}} t)}{\pi B_{\mathrm{N}}\tau} \\
&= \sigma_{\mathrm{n}}^2 \cos(\omega_{\mathrm{o}}\tau)\,\mathrm{sinc}(\pi B_{\mathrm{N}}\tau)
\end{aligned} \tag{5.56}$$

This ACF is shown in Figure 5.21.

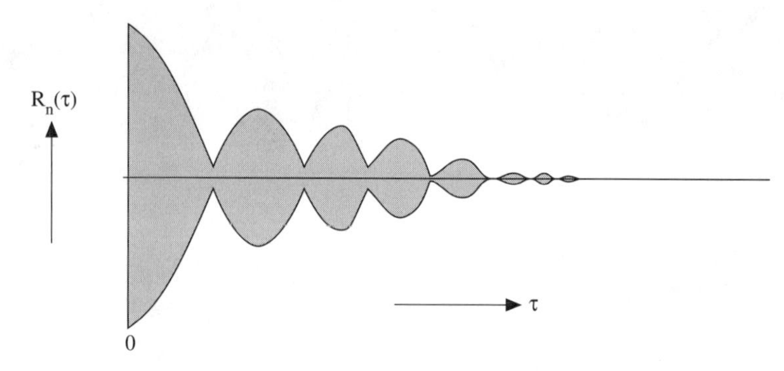

FIGURE 5.21
ACF of bandlimited white IF noise.

The nulls of the amplitude of $R_n(\tau)$ [which is $|R_n(\tau)| = \sigma_n^2|\text{sinc}\,(\pi B_N\,\tau)|$] are at $\pi B_N\tau = n\pi$, where n is any integer other than zero. Thus if $\tau = 1/B_N$, then the ACF is zero, and hence noise samples are separated by intervals $1/B_N$. The signal-plus-noise voltage waveform $v(t) = s(t) + n(t)$ has a bandwidth limited by that of the noise, since the latter always has a bandwidth larger than (or at least as large as) that of the signal. According to the sampling theorem, all of the information about the waveshape $v(t)$ can be recovered from samples taken at intervals $1/B_N$, where B_N is the total (two-sided) bandwidth of $v(t)$. We can sample it at those intervals and represent it as

$$v(t) = \sum_{k=-N}^{N} v_K \,\text{sinc}\left[\pi B_N\left(t - \frac{k}{B_N}\right)\right] \tag{5.57a}$$

The signal portion of $v(t)$ can also be represented this way, i.e.,

$$s(t) = \sum_{k=-N}^{N} s_k \,\text{sinc}\left[\pi B_N\left(t - \frac{k}{B_N}\right)\right] \tag{5.57b}$$

since $s(t)$ has a bandwidth less than or equal to B_N.

Using (5.57a,b) in representing the integrals $\int_{-T/2}^{T/2} dt\; s(t)\,v(t)$ or $\int_{-T/2}^{T/2} dt\; s^2(t)$, or $\int_{-T/2}^{T/2} dt\; v^2(t)$, we have [where the sampling interval is $1/B_N$ and therefore the number of samples minus one $= 2N = B_N T$, T being the duration of $v(t)$,

$$\int_{T/2}^{T/2} dt \begin{bmatrix} v(t) & s(t) \\ s(t) & s(t) \\ v(t) & v(t) \end{bmatrix}$$

$$= \sum_{l,m=-N}^{N} \begin{bmatrix} v_l & s_m \\ s_l & s_m \\ v_l & v_m \end{bmatrix} \int_{-N/B_N}^{N\backslash B_N} dt\;\text{sinc}\left[\pi B_N\left(t - \frac{l}{B_N}\right)\right] \text{sinc}\left[\pi B_N\left(t - \frac{m}{B_N}\right)\right] \tag{5.58}$$

$$= 2\sum_{l=-N}^{N} \begin{bmatrix} v_l s_l \\ s_l^2 \\ v_l^2 \end{bmatrix} \int_0^{\infty} dt\;\text{sinc}^2\left[\pi B_N\left(t - \frac{l}{B_N}\right)\right] = \frac{1}{B_N}\sum_{l=-N}^{N} \begin{bmatrix} v_l s_l \\ s_l^2 \\ v_l^2 \end{bmatrix}$$

(from Spiegel,[19] #15.36, p. 96).

Using (5.58) to determine the likelihood ratio for a set of signal parameters $\boldsymbol{\Omega}_1, \Omega_2, \ldots, \Omega_M$, we have

$$L(\mathbf{v}/\boldsymbol{\Omega}) = [e^{-(B_N/2\pi\sigma_n^2)\int_0^T dt s^2(t)}][e^{(B_N/\pi\sigma_n^2)\int_0^T dt\, s(t)v(t)}] \tag{5.59}$$

where we have (for notational convenience) shifted the time axis by $T/2$ and thereby set the limits on the integrals at $t = 0$ and $t = T$.

By setting the sampling interval at $1/B_N$, we have obtained assurance that the assumption of statistical independence of the noise samples is valid. That conclusion follows from (5.56) and the fact that the general multivariate Gaussian PDF for a set of samples $v_1, v_2, \ldots, v_N$ contains an exponential factor of the form

$$F = e^{-C\sum_{j,k=1}^{N} R_{jk}(v_j - \langle v_j \rangle)(v_k - \langle v_k \rangle)} \tag{5.60}$$

where R_{jk} is the cross-correlation function (CCF) of the samples v_j and v_k.

From (5.56), we know that $R_n(\tau)$ has nulls at intervals of $\tau = 1/B_N$, if v_j and v_k are separated by an interval $\tau = 1/B_N$, then $R_{jk} = R_n\,(1/B_N)$, which we know to be zero. It follows that the factor F in (5.60) has the value unity, since $R_{jk} = 0$ in every term of the exponent. This leads to the conclusion that the samples are statistically independent *because* they are mutually uncorrelated. That is true in general only for *Gaussian* random processes and not necessarily for processes with non-Gaussian statistics.

We now return to (5.59) and represent $s(t)$ and $v(t)$ as narrowband signals of the form $x(t)\cos\omega_o t + y(t)\sin\omega_o t$, with the results [where $u(t)$ and $v(t)$ are generic narrowband signal waveforms]

$$\begin{aligned}
\int_0^T dt u(t) w(t) &= \int_0^T dt x_u(t) x_w(t) \cos^2 \omega_o t \\
&\quad + \int_0^T dt\, y_u(t) y_w(t) \sin^2 \omega_o t + \int_0^T dt[x_u(t) y_w(t) \\
&\quad + x_w(t) y_u(t)] \sin \omega_o t \cos \omega_o t \\
&= \frac{1}{2}\left\{\int_0^T dt[x_u(t)x_w(t) + y_u(t)y_w(t)]\right\} \\
&\quad + \frac{1}{2}\left\{\int_0^T dt[x_u(t)x_w(t) - y_u(t)y_w(t)] \cos 2\,\omega_o t\right\} \\
&\quad + \frac{1}{2}\left\{\int_0^T dt[x_u(t)y_w(t) + x_w(t)y_u(t)] \sin 2\,\omega_o t\right\}
\end{aligned} \tag{5.61}$$

Since it is always true in this application that

$$T >> \frac{1}{2\omega_o} \tag{5.62}$$

it follows from (5.61) with the aid of (5.62) that the integral can be approximated by

$$\int_0^T dt u(t) w(t) = \frac{1}{2} \int_0^T dt a_u(t) a_w(t) \cos[\psi_u(t) - \psi_w(t)] \tag{5.63}$$

where $[a_u(t), \psi_w(t)]$ and $[a_w(t), \psi_w(t)]$ are the slowly varying amplitude and phase pair of $u(t)$ and $w(t)$, respectively. Letting $u(t)$ and $w(t)$ both represent $s(t)$, (5.63) gives us

$$\int_0^T dt s^2(t) = \frac{1}{2} \int_0^T dt a_s^2(t) \tag{5.64a}$$

and if $u(t)$ and $w(t)$ represent $s(t)$ and $v(t)$, respectively, we obtain

$$\int_0^T dt s(t) v(t) = \frac{1}{2} \int_0^T dt a_s(t) a_v(t) \cos[\psi_s(t) - \psi_v(t)] \tag{5.64b}$$

Substitution of (5.64a,b) into (5.59) yields

$$L(\mathbf{v}/\boldsymbol{\Omega}) = \left(e^{(-B_N/4\sigma_n^2)\int_0^T dt a_s^2(t)}\right)\left(e^{(B_N 2\sigma_n^2)\int_0^T dt\, a_v(t) a_s(t) \cos[\psi_s(t) - \psi_v(t)]}\right) \tag{5.65}$$

We now make the further assumption that both amplitude and phase of the signal waveform $s(t)$ remain approximately constant over the interval 0–T, i.e.,

$$a_s(t) = a_s = \sqrt{x_s^2 + y_s^2}$$
$$\psi_s(t) = \psi_s = \tan^{-1}\left(\frac{-y_s}{x_s}\right), \qquad 0 \le t \le T \tag{5.66}$$

Equation (5.66) would be particularly well-suited to the case of a single rectangular pulse. Assuming such a signal waveform for the moment, (5.65) aided by (5.66) yields

$$L(\mathbf{v}/\boldsymbol{\Omega}) = e^{-a_s^2(B_N T/4\sigma_n^2)} e^{2a_s(B_N T/4\sigma_n^2)(x_{vo}\cos\psi_s + y_{vo}\sin\psi_s)} \tag{5.67}$$

where

$$x_{vo} = \frac{1}{T}\int_0^T dt x_v(t) \tag{5.67a)'}$$

$$y_{vo} = \frac{1}{T}\int_0^T dt y_v(t) \tag{5.67b)'}$$

and where the integrals in (5.67a,b)′ are time averages of the quadrature components of $v(t)$, $x_v(t)$ and $y_v(t)$, over the interval from 0 to T.

An alternative way to write (5.67) is

$$L(\mathbf{v}/\boldsymbol{\Omega}) = e^{-a_s^2(B_N T/4\sigma_n^2)} e^{2a_s(B_N T/4\sigma_n^2) a_{vo}\cos(\psi_s + \psi_{vo})} \tag{5.68}$$

where

$$a_{vo} = \sqrt{x_{vo}^2 + y_{vo}^2}$$

$$\psi_{vo} = \tan^{-1}\left(\frac{-y_{vo}}{x_{vo}}\right)$$

We now consider specific cases of interest in radar applications.

Case 1: All signal waveform parameters known; this is the case considered earlier [Eqs. (5.51) through (5.55)]. In this case, we can choose the signal parameters as follows:

$$\Omega_1 = a_s; \qquad \Omega_{10} = a_{so}$$

$$\Omega_2 = \psi_s; \qquad \Omega_{20} = \psi_{20}$$

From Eq. (5.52) and Eqs. (5.54), (5.54)′, and (5.55) applied with (5.68)

$$\overline{L}(\mathbf{v}) = L(\mathbf{v}/\mathbf{\Omega}) = L(\mathbf{v}/a_{so}, \psi_{so}) = e^{-a_{so}^2(B_N T/4\sigma_n^2)} \ldots$$

$$e^{2a_{so}(B_N T/4\sigma_n^2)a_{vo}\cos(\psi_{so}+\psi_{vo})} \tag{5.69}$$

and the decision rule is

If a_{vo} cos $(\psi_{so} + \psi_{vo})$ exceeds threshold; call "signal present"

If a_{vo} cos $(\psi_{so} + \psi_{vo})$ is less than threshold; call "noise alone"

Case 2: Amplitude of signal a_s is known and equal to a_{so} (i.e., no amplitude fluctuations), but phase is completely unknown, hence its a priori PDF is

$$p(\Omega_2) = p(\psi_s) = \frac{1}{2\pi}; \qquad 0 \leq \psi_s \leq 2\pi \tag{5.69$'$}$$

From (5.68), (5.69), and (5.49)′

$$\overline{L}(\mathbf{v}) = \left[\frac{1}{2\pi}\int_0^{2\pi} d\psi_s e^{2a_{so}(B_N T/4\sigma_n^2)a_{vo}\cos(\psi_s+\psi_{vo})}\right] e^{-a_{so}^2(B_N T/4\sigma_n^2)}$$

$$= e^{-a_{so}^2(B_N T/4\sigma_n^2)} I_o[2a_{so}(B_N T/4\sigma_n^2)a_{vo}] \tag{5.70}$$

(from Spiegel,[19] #9.6.16, p. 376).

The modified Bessel function $I_o(x)$ is a monotonic increasing function of x, and since x is a positive number in this case, the implication of (5.70) is the decision rule

$$\text{If } a_{vo} \geq \text{threshold}, \qquad \text{call "signal present"}$$

$$\text{If } a_{vo} < \text{threshold}, \qquad \text{call "noise alone"} \tag{5.71}$$

The decision rule (5.71) dictates that the *amplitude* of the matched filter output is the quantity on which the threshold should be set. That implies that the quadrature filtering operation, as described in Section 4.4, should be applied to the incoming waveform $v(t)$ and the decision threshold based on its output, which is proportional to the amplitude of the output of a coherent matched filter.

In Section 4.4, this method was discussed as a way to deal with a priori ignorance of phase in maximizing SNR. We have now learned that it is the theoretical optimum strategy from the viewpoint of maximizing detection probability, i.e., there is no other technique for compensating for phase ignorance that will result in better detection performance.

Case 3: Amplitude *and* phase both unknown. Equation (5.69)′ applies to phase. If we assume a Rayleigh amplitude PDF as given by (5.35), Eqs. (5.49)′ and (5.68) result in

$$\overline{L}(\mathbf{v}) = \frac{1}{2\pi}\int_0^{2\pi} d\psi_s \int_0^{\infty} da_s \frac{a_s}{\sigma_s^2} e^{-(a_s^2/2)[(1/\sigma_s^2)+(B_N T/2\sigma_n^2)]} \ldots$$

$$e^{2a_s(B_N T/4\sigma_n^2)a_{vo}\cos(\psi_s+\psi_{vo})} = \int_0^{\infty} \frac{da_s a_s}{\sigma_s^2} e^{-(a_s^2/2)[(B_N T/2\sigma_n^2)+(1/\sigma_s^2)]}$$

$$= \cdots I_o\left(\frac{B_N T}{2\sigma_n^2} a_{vo} a_s\right) \qquad (5.72)$$

If we integrate (5.72) first on ψ_s and then on a_s, we will obtain an integral on a_s of the same form as (5.37). With $\nu = 0$, $\mu = 2$, $\alpha = B_N T/2\sigma_n^2$ and $p^2 = \frac{1}{2}[(1/\sigma_s^2) + (B_N T/2\sigma_n^2)]$. The result, equivalent to (5.37b)′ with v replaced by $(B_N T\, a_{vo}/2)$ and σ_n^2 replaced by $(2\sigma_n^2/B_N T)$, is

$$\overline{L}(\mathbf{v}) = \frac{1}{[(1/\sigma_s^2) + (B_N T/2\sigma_n^2)]} e^{a_{vo}^2\sigma_s^2(B_N T/2)^2/2[\sigma_s^2+(2\sigma_n^2/B_N T)]} \qquad (5.72a)'$$

Again, since the positive exponential is a monotonic increasing function of its argument, the optimal decision rule should be based on the square of the filter output amplitude a_{vo}^2, or equivalently on the output amplitude a_{vo} itself.

The same line of reasoning with the one-dominant-plus Rayleigh PDF as given by (5.40) leads to the end-result [Eq. (5.41b)′ with σ_n^2 replaced by $2\sigma_n^2/B_N T$ and v by $(B_N T\, a_{vo}/2)$]

$$\overline{L}(\mathbf{v}) = 1\Big/\left[\frac{3}{\sigma_s^2} + \frac{B_N T}{2\sigma_n^2}\right]\exp\left[a_{vo}^2\sigma_s^2\left(\frac{B_N T}{6}\right)^2\Big/2\left(\frac{\sigma_s^2}{3} + \frac{2\sigma_n^2}{B_N T}\right)\right]\cdots$$

$$\left\{1 + \frac{a_{vo}^2(B_N T/2)^2\sigma_s^2}{4\sigma_n^2[\sigma_s^2 + (6\sigma_n^2/B_N T)]}\right\} \qquad (5.72b)'$$

Taking the logarithms of (5.72a,b)′ and consolidating the factors not dependent on a_{vo}^2, the quantity to be used for the decision rule for optimum detection would be

$$V = K_1 + K_2 a_{vo}^2, \qquad \text{for Rayleigh} \qquad (5.73a)$$

$$K_1' + K_2' a_{vo}^2 + \ln[1 + K_3' a_{vo}^2], \qquad \text{for one-dominant-plus Rayleigh} \qquad (5.73b)$$

In (5.73a,b), the constants K_1, K_2, ... are quantities dependent on the ratio of σ_s^2 to σ_n^2, and on the product $B_N T$. Unless one is interested in the exact value of the threshold, the interpretation of (5.73a) is that optimum detection would be achieved by computing a_{vo} or a_{vo}^2 and setting a threshold on it in the Rayleigh case.

That of (5.73b) is that optimum detection with one-dominant-plus Rayleigh would be achieved by computing the quantity

$$a_{vo}^2\left[1 + \frac{1}{K_2' a_{vo}^2}\ln(1 + K_3' a_{vo}^2)\right]$$

and setting a threshold on it. However, this computation would generally be regarded as too complicated for a practical system. If the second term is small compared with the first, then computation of a_{vo}^2 would suffice. Otherwise, a suitable approximation would be necessary, depending on the values of K_2' and K_3'. All of this is not really necessary if one focuses on the fact that V in (5.72b)′ is a monotonic increasing function of a_{vo}^2 and that buried somewhere in that expression is a threshold value on that quantity that would optimize detection. Again, unless the precise threshold value is important to us, that conclusion would suffice. Since the threshold is usually chosen empirically based on strategic value judgments that are not purely technical, we will end the discussion here and merely conclude that a detection of a single pulse with total a priori ignorance of phase would be optimized by computing the amplitude a_{vo} or squared amplitude a_{vo}^2 and basing the decision rule on a_{vo} or a_{vo}^2 being above a chosen threshold. This would be true whether the signal amplitude is fixed or fluctuating, except that the "optimum" thresholds would be dependent on the ratio of mean-square fluctuation to mean-square noise. Moreover, for the "single observation" case (a single-pulse matched-filter output in the radar case), the choice of a_{vo} or a_{vo}^2 give exactly equivalent results because the PDF of a_{vo} and that of a_{vo}^2 are related through

$$p(a_{vo}^2)\,d(a_{vo}^2) = p(a_{vo})\,da_{vo} \tag{5.74}$$

or equivalently

$$p(a_{vo}^2) = \frac{p(a_{vo})}{2a_{vo}} \tag*{(5.74)′}$$

Then in general [from (5.74)]

$$P_d = \int_{v_T}^{\infty} da_{vo} p(a_{vo}/s) = \int_{v_T^2}^{\infty} d(a_{vo}^2) p(a_{vo}^2/s) \tag{5.75a}$$

$$P_{fa} = \int_{v_T}^{\infty} da_{vo} p(a_{vo}/n) = \int_{v_T^2}^{\infty} d(a_{vo}^2) p(a_{vo}^2/n) \tag{5.75b}$$

From (5.75a,b), it follows that the only difference between use of a_{vo} and a_{vo}^2 for a single observation is a change in the threshold from v_T to v_T^2.

5.6 OPTIMUM DETECTION OF PULSE TRAINS

We now proceed to a case of great importance in radar, that of detection of a pulse train. The first point, which was already made in Section 5.5, is that the signal waveform is totally irrelevant to optimization of detection if the signal parameters are completely known. [That case was discussed in Section 5.5, leading to Eqs. (5.51), (5.52) and (5.55).] In that case, coherent detection is applied and the optimal processing scheme is a digital cross-correlator, shown in Section 5.5 to be equivalent to an analog cross-correlator in the case of bandlimited white noise and hence the procedure is equivalent to that of matched filtering if the detection process is based on a set of sample IF voltage values.

Even if phase is unknown, a pulse train can be treated in the same way as a single pulse if the quadrature filtering technique is used to eliminate the effect of unknown phase, as the optimization theory covered under "Case 2" [discussion surrounding Eqs. (5.69) and (5.70)] dictates for optimal detection in this case. This is still true in the fluctuating signal case [Case 3, Eqs. (5.72) through (5.73) and associated discussions].

The point of departure is the case where the decision is made to take the amplitude or amplitude-squared of *each pulse* and base the detection decision on a sum of amplitudes or amplitude-squared values over the entire pulse train. In that case, this process of "pulse integration" results in a loss, i.e., a higher individual pulse SNR will be required to achieve the same level of detection performance. The comparison may be with "coherent integration" of the pulses, i.e., the reference signal for matched filtering or (equivalently) cross-correlation is a train of IF pulses synchronized with the incoming pulse train in frequency, phase, and PRF. If phase is unknown, then the pulses may still be integrated coherently, but this time with two reference signals in phase quadrature with each other. If noncoherent integration is to be used, then the incoming pulses are first rectified, either linearly or quadratically, and the reference signal for cross-correlation is a train of video pulses. The first two alternatives are shown in Figure 5.22.

To analyze the results shown in Figure 5.22, we model the incoming and reference signals $s_i(t)$ and $s_R(t - \Delta t)$ as follows:

$$s_i(t) = a_i \cos(\omega_o t + \psi) \sum_{k=0}^{N_p - 1} p(t - kT_r) \tag{5.76a}$$

$$s_R(t - \Delta t) = a_R \cos[\omega_o(t - \Delta t) + \psi] \sum_{l=0}^{N_p - 1} p(t - \Delta t - lT_r) \tag{5.76b}$$

where a_i and a_R are signal amplitudes, ψ is the phase of either signal (assumed matched), $p(t)$ is the pulse amplitude of either signal (again assumed matched), and T_r is the PRF of either signal (also matched).

The cross-correlator output in this case is [using the cosine product formula $\cos A \cos B = \frac{1}{2}[\cos(A + B) + \cos(A - B)]$ and using only the difference term, based on the fact that $\tau_p >> 1/\omega_o$ and hence the integral of the sum term over the pulses [whose integrand is $\ldots \cos(2\omega_o t + \cdots)$ is negligible]

$$\begin{aligned} s_o(\Delta t) &= \int_{-\infty}^{\infty} dt\, s_i(t) s_R(t - \Delta t) \\ &= \frac{a_i a_R}{2} \cos(\omega_o \Delta t) s_{oA}^{(N_p)}(\Delta t) \end{aligned} \tag{5.77}$$

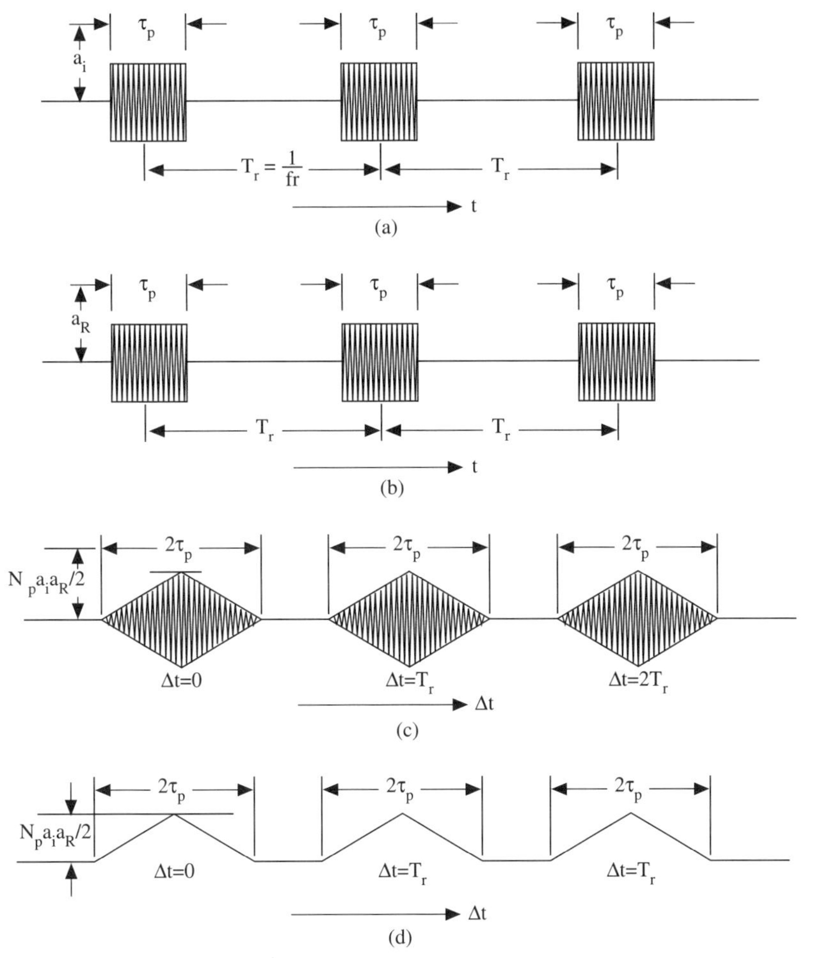

FIGURE 5.22
Coherent pulse integration. (a) Incoming pulse train (IF pulses). (b) References pulse train—"coherent pulse integration." (c) Coherent pulse integration signal output (IF signal). (d) Coherent pulse integration with quadrature filtering—output is amplitude.

where

$$s_{oA}^{(N_p)}(\Delta t) = \sum_{l,m=0}^{N_p-1} \int_{-\infty}^{\infty} dt\ p(t - lT_r)p(t - mT_r - \Delta t)$$

$$= \text{cross-correlation function (CCF) of the pulse train amplitude modulation function } p(t) \qquad (5.77a)'$$

Let us rewrite (5.77a)′ in the form

$$s_{oA}^{(N_p)}(\Delta t) = \sum_{l=0}^{N_p-1} \left\{ \sum_{m=0}^{N_p-1} \int_{-\infty}^{\infty} dt'\ p(t')p[t' - [(m - l)T_r + \Delta t]] \right\} \qquad (5.77b)'$$

Consider the simple cases where $N_p = 2$ and $N_p = 3$, and call the internal summation $S_{il}(\Delta t)$:

$N_p = 2$:

$$S_{il}(\Delta t) = \int_{-\infty}^{\infty} dt'\ p(t')p(t' + lT_r - \Delta t) + \int_{-\infty}^{\infty} dt'\ p(t')p[t' + (l-1)T_r - \Delta t] \tag{5.78a}$$

and

$$\begin{aligned} s_{oA}^{(2)}(\Delta t) &= \int_{-\infty}^{\infty} dt'\ p(t')p(t' - \Delta t) + \int_{-\infty}^{\infty} dt'\ p(t')p(t' - T_r - \Delta t) \\ &\quad + \int_{-\infty}^{\infty} dt'\ p(t')p(t' + T_r - \Delta t) + \int_{-\infty}^{\infty} dt'\ p(t')p(t' - \Delta t) \\ &= 2\int_{-\infty}^{\infty} dt'\ p(t')p(t' - \Delta t) + \left\{\int_{-\infty}^{\infty} dt'\ p(t')p[t' - (\Delta t + T_r)]\right. \qquad (5.78a)' \\ &\quad \left. + \int_{-\infty}^{\infty} dt'\ p(t')p[t' - (\Delta t - T_r)]\right\} \end{aligned}$$

$N_p = 3$:

$$S_{il}(\Delta t) = \int_{-\infty}^{\infty} dt'\ p(t')p(t' + lT_r - \Delta t) + \int_{-\infty}^{\infty} dt'\ p(t')p[t' + (l-1)T_r - \Delta t] + \int_{-\infty}^{\infty} dt'\ p(t')p[t' + (l-2)T_r - \Delta t] \tag{5.78b}$$

and (dropping the limits on the integrals for convenience)

$$\begin{aligned} s_{oA}^{(3)}(\Delta t) &= \int dt'\ p(t')p(t' - \Delta t) + \int dt'\ p(t')p(t' - T_r - \Delta \mathrm{t}) \\ &\quad + \int dt'\ p(t')p(t' - 2T_r - \Delta t) + \int dt'\ p(t' + T_r - \Delta t) \\ &\quad + \int dt'\ p(t')p(t' - \Delta t) + \int dt'\ p(t')p(t' - T_r - \Delta t) \\ &\quad + \int dt'\ p(t')p(t' + 2T_r - \Delta t) + \int dt'\ p(t')p(t' + T_r - \Delta t) \\ &\quad + \int dt'\ p(t')p(t' - \Delta t) \end{aligned} \qquad (5.78b)'$$

$$= 3\int dt'\, p(t')p(t' - \Delta t) + 2\left\{\int dt'\, p(t')p[t' - (\Delta t - T_r)]\right.$$

$$+ \int dt'\, p(t')p[t' - (\Delta t + T_r)] + \int dt'\, p(t')p[t' - (\Delta t - 2T_r)]$$

$$\left. + \int dt'\, p(t')p[t' - (\Delta t + 2T_r)]\right\}$$

In general, if this process were continued indefinitely, the result, for a large value of N_p, would be

$$\begin{aligned} s_{oA}^{(N_p)}(\Delta t) = {} & N_p s_{op}(\Delta t) + (N_p - 1)[s_{op}(\Delta t - T_r) \\ & + s_{op}(\Delta t + T_r)] + (N_p - 2)[s_{op}(\Delta t - 2T_r) \\ & + s_{op}(\Delta t + 2T_r)] + \cdots \\ & + 3\{s_{op}[\Delta t - (N_p - 3)T_r] + s_{op}[\Delta t + (N_p - 3)T_r]\} \\ & + 2\{s_{op}[\Delta t - (N_p - 2)T_r] + s_{op}[\Delta t + (N_p - 2)T_r]\} \\ & + \{s_{op}[\Delta t - (N_p - 1)T_r] + s_{op}[\Delta t + (N_p - 1)T_r]\} \end{aligned} \tag{5.79}$$

where in general

$$s_{op}(T) = \int_{-\infty}^{\infty} dt'\, p(t')p(t' - T) = \text{ACF of the pulse amplitude function } p(t) \qquad (5.79)'$$

Figure 5.22 is an illustration of a case where (a) the pulses shown are within the interior region of the pulse train and (b) there are a very large number of pulses. Equivalently, the figure shows the case of an infinite pulse train, except for the indication that the amplitude of the integrated output at $\Delta t = 0$ is equal to N_p, the number of pulses, which is corroborated by (5.79). As long as N_p far exceeds (say) 10, the first 10 terms of (5.79) will have amplitudes of approximately N_p (since $N_p - 1 \simeq N_p$, $N_p - 2 \simeq N_p, \ldots$) and the integrated output looks approximately like Figure 5.22c at IF and like Figure 5.22d after rectification. Both (5.79) and Figure 5.22 demonstrate that the peaks of the output occur when $\Delta t = \pm kT_r$, where k is an integer and that the output is zero where Δt is between $kT_r + \tau_p$ and $[(k + 1)\, T_r - \tau_p]$, i.e., when the pulses in the integrand of (5.77b)′ do not overlap.

Equation (5.79) will be useful later when ambiguity functions and range ambiguities are discussed in the context of range resolution considerations. For the moment, let us pass to the case of "noncoherent pulse integration," illustrated in Figure 5.23.

In this case, the input to the process (Fig. 5.23a) has already been rectified and is a train of video pulses. So is the reference pulse train (Fig. 5.23b) for what will now be called "video cross-correlation," equivalent to "noncoherent pulse integration" in the case of a pulse train.

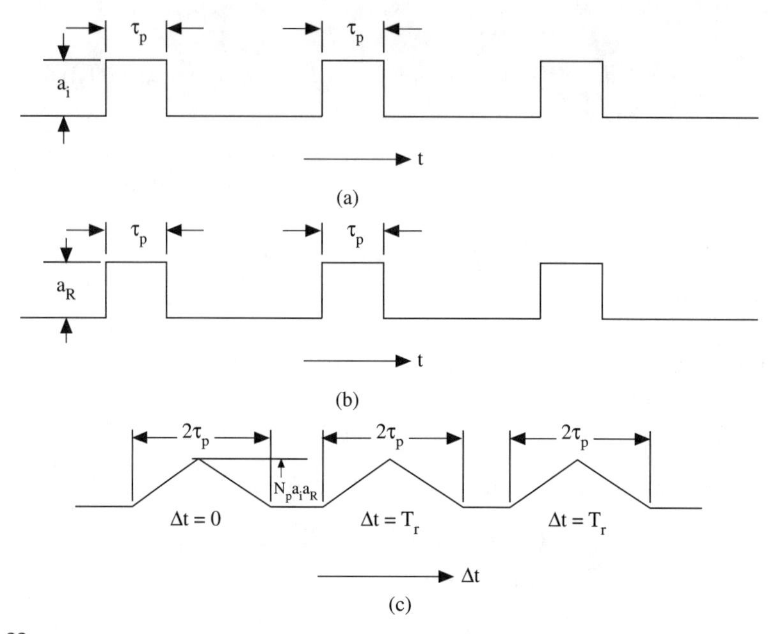

FIGURE 5.23
Noncoherent pulse integration. (a) Incoming pulse train—video pulses. (b) Reference pulse train—video pulses. (c) Output.

In this case, the portion of the analysis above and beyond Eq. (5.77) is still applicable and the conclusions still apply. The process can now be described by writing

$$s_o(\Delta t) = \int_{-\infty}^{\infty} dt\ s_i(t)s_R(t - \Delta t) = a_i a_R s_{oA}^{(N_p)}(\Delta t) \tag{5.80}$$

where

$$s_{oA}^{(N_p)}(\Delta t) = \sum_{l,m=0}^{N_p-1} \int_{-\infty}^{\infty} dt\ p(t - lT_r)p(t - mT_r - \Delta t) \tag{5.80$'$}$$

where (5.80) differs from (5.77) only because of the absence of the factor $(a_i\ a_R/2)$ $\cos(\omega_o\ \Delta t)$ and (5.80)′ is identical to (5.77a)′. Equations (5.77b)′, (5.78a, b), (5.78a,b)′, (5.79), and (5.79)′ would then follow (5.80)′ with no changes.

We now return to detection theory and ask the "optimal" detection question in a different way. Suppose we wanted to base the detection decision on a number of video samples rather than IF samples. The assumption behind this new view of optimum detection is that coherent cross-correlation is infeasible because we do not know the IF signal phase. In this case, the signal-plus-noise samples $v(t_1), v(t_2), \ldots, v(t_N)$ are $(a_1, a_2, \ldots, a_N)$, where a_k is the amplitude of the video waveform at $t = t_k$. The PDF of the amplitude is now Ricean because the amplitude of signal-plus-noise has a nonzero mean. Then the likelihood ratio (assuming again that the noise samples are uncorrelated) is given by the product of likelihood ratios such as that of (5.70) for individual pulses.

Thus, for the case where all parameters except phase are a priori known

$$\overline{L}(\mathbf{v}) = \prod_{k=1}^{N_p} \overline{L}_k(\mathbf{v}) \tag{5.81}$$

where

$$\overline{L}_k(\mathbf{v}) = e^{-a_{sok}^2(B_N T/4\sigma_n^2)} I_o\left[a_{sok}\left(\frac{B_N T}{2\sigma_n^2}\right)a_{vok}\right] \tag{5.81$'$}$$

and where a_{sok} and a_{vok} are the amplitudes of the pure signal and signal-plus-noise waveforms, respectively.

In the case of known (except phase) parameters, the optimum strategy is now to compute the product

$$V_o' = \prod_{k=1}^{N_p} I_o(\beta_k a_{vok}) \tag{5.82}$$

where $\beta_K = (B_N T/2\sigma_n^2)\, a_{sok}$ and put a threshold on it for the detection decision.

For rectangular pulses, $a_{sok} = a_{so}$ = constant for all pulses; hence $\beta_k = \beta = (B_N T/2\sigma_n^2) a_{so}$.

If $\beta a_{vok} << 1$, then we can invoke the approximation to the modified Bessel function that holds for small arguments (first two terms of the power series).

If $\beta a_{vok} >> 1$, we can use the asymptotic form for large arguments. These two extreme approximations are [from (5.82) and (24.35) on p. 138 of Spiegel[19]]

$$V_o' = \prod_{k=1}^{N_p}\left[1 + \frac{(\beta a_{vok})^2}{4}\right] = 1 + \frac{\beta^2}{4}\sum_{k=1}^{N_p} a_{vok}^2 + \cdots O(a_{vok}^4)$$

$$\simeq 1 + \frac{\beta^2}{4}\sum_{k=1}^{N_p} a_{vok}^2 \tag{5.82a$'$}$$

in the case where $\beta a_{vo} << 1$ (small SNR, since βa_{vo} is proportional to the square of the SNR).

For $\beta a_{vo} >> 1$ (large SNR) [from (5.82) and (24.107) on p. 143 of Spiegel[19]]

$$V_o' = \prod_{k=1}^{N_p} \frac{e^{\beta a_{vok}}}{\sqrt{2\pi\beta a_{vok}}} = \frac{e^{\beta \sum_{k=1}^{N_p} a_{vok}}}{\prod_{k=1}^{N_p}\sqrt{2\pi\beta a_{vok}}} \tag{5.82b$'$}$$

Taking the log of V_o' results in

$$\log V_o' = \beta \sum_{k=1}^{N_p} a_{vok} - \frac{1}{2}\sum_{k=1}^{N_p} \log(2\pi\beta a_{vok}) \tag{5.82c$'$}$$

For sufficiently large values of N_p, the first term overwhelms the second and (5.82c)′ becomes

$$\log V_o' = \beta \sum_{k=1}^{N_p} a_{vok} \tag{5.82d)$'$}$$

Equations (5.82a)′ and (5.82d)′, respectively, tell us that under certain conditions (low and high SNR), optimal detection of a set of video pulses is achieved by summing the outputs of a video cross-correlator over all the pulses and setting a threshold on the sum for the detection decision. The difference in these two extreme cases is that we are told to use quadratic rectification on each pulse in the low SNR case and linear rectification on each pulse in the high SNR case.

Strictly speaking, if one really wanted to "optimize" the detection process, one would compute the likelihood ratio as given by (5.81)′, or a quantity monotonically increasing with it, and base the threshold setting on that quantity. That procedure is virtually never applied rigorously (to the author's knowledge). A procedure dictated by intuition is to compute the sum $\sum_{k=1}^{N_p} a_{vok}$ or the sum $\sum_{k=1}^{N_p} a_{vok}{}^2$ and base the decision process on that sum. Optimization theory tell us that one of those procedures is truly optimum for low SNR and the other is truly optimum for high SNR. Moreover, studies of required SNR for given levels of detection performance using the square-law form $\sum_{k=1}^{N_p} a_{vok}{}^2$ indicate that the required SNR values for reasonable radar performance levels (e.g., $P_d \geq 85\%$, $P_{fa} \leq 10^{-3}$) are always much greater than unity and hence optimization theory tends toward the high SNR case, where we *should* be using linear rectification and computing $\sum_{k=1}^{N_p} a_{vok}$. Results obtained for both cases show very little difference in the two types of rectification. Calculations of P_d and P_{fa} are more tractable in the quadratic case than in the linear case, and hence those are the results used in the detection theory expounded in DiFranco and Rubin and the P_d results we are using in this book.

The most mathematically efficient way to calculate $p(\mathbf{v}/s)$ and $p(\mathbf{v}/n)$ for a pulse train is to use the "characteristic function" concept introduced in Section 5.1. This is the procedure used in DiFranco and Rubin.[20] The assumption behind the analysis to follow is that the detection decision will be based on the sum of squared-amplitudes of video-cross-correlator outputs and that these outputs are mutually statistically independent. The quantity V_o, the processor output, is (where K is an arbitrary proportionality constant)

$$V_o = K \sum_{k=1}^{N_p} a_{vok}^2 \tag{5.83}$$

where a_{vok} is the amplitude of the kth pulse.

The PDF of V_o is the product of PFDs for the individual pulses, whose noise components are statistically independent, i.e.,

$$p(V_o/s) = \int_0^\infty \cdots \int da_{vo1} \cdots da_{voN_p} \prod_{k=1}^{N_p} p[a_{vok}/(s_{ok} + n_{ok})] \cdots \delta\left(V_o - K \sum_{k=1}^{N_p} a_{vok}^2\right) \tag{5.84a}$$

$$p(V_o/n) = \int_0^\infty \cdots \int da_{vo1} \cdots da_{voN_p} \prod_{k=1}^{N_p} p(a_{vok}/n_{ok}) \cdots$$
$$\delta\left(V_o - K \sum_{k=1}^{N_p} a_{vok}^2\right) \tag{5.84b}$$

where $p[a_{vok}/(s_{ok} + n_{ok})]$ is the conditional PDF of a_{vok} given a signal output s_{ok} and a noise output n_{ok} for the kth pulse [hence the amplitude square output for the kth pulse is the amplitude square of $(s_{ok} + n_{ok})$] and $p(a_{vok}/n_{ok})$ is the conditional PDF of a_{vok} given noise output n_{ok} only (hence the amplitude square output for the kth pulse is a $a_{nok}{}^2$, where a_{nok} is the noise amplitude).

The interpretation of (5.84a,b) is as follows: The PDF of V_o is the joint PDF of the pulse outputs integrated over all points where $V_o = K\sum_{k=1}^{N_p} a_{vok}{}^2$. Points where that condition does not hold give a zero contribution to the integral.

Given the definition of the unit impulse in Eq. (4.7), repeated below for convenience

$$\delta(x) = \frac{1}{2\pi}\int_{-\infty}^{\infty} d\xi\, e^{\pm j\xi x} \tag{5.85}$$

we can rewrite (5.84a,b) with the aid of (5.85) and change the order of integration, with the results

$$p(V_o/s) = \frac{1}{2\pi}\int_{-\infty}^{\infty} d\xi\, e^{-j\xi V_o}\chi(\xi/s) \tag{5.86a}$$

$$p(V_o/n) = \frac{1}{2\pi}\int_{-\infty}^{\infty} d\xi\, e^{-j\xi V_o}\chi(\xi/n) \tag{5.86b}$$

where $\chi(\xi/s)$ and $\chi(\xi/n)$ are the respective characteristic functions for the "signal present" and "noise alone" cases, where those characteristic functions are given by

$$\chi(\xi/s) = \prod_{k=1}^{N_p} \chi_k(\xi/s_{ok}) \tag{5.86a$'$}$$

$$\chi(\xi/n) = \prod_{k=1}^{N_p} \chi_k(\xi/n_{ok}) \tag{5.86b$'$}$$

and where $\chi_k(\xi/s_{ok})$ and $\chi_k(\xi/n_{ok})$ are the characteristic functions for the kth pulse in the signal present and noise alone cases, respectively, given by

$$\chi_k(\xi/s_{ok}) = \int_0^\infty da_{vok}\, p(a_{vok}/s_{ok})e^{j\xi K a_{vok}^2} \tag{5.86a$''$}$$

$$\chi_k(\xi/n_{ok}) = \int_0^\infty da_{vok}\, p(a_{vok}/n_{ok})e^{j\xi K a_{vok}^2} \tag{5.86b$''$}$$

To evaluate (5.86a,b)″, we need a specific PDF of a_{vok} for each case. For a nonfluctuating signal, $p(a_{vok}/s_{ok})$ is the Rice PDF as given by (5.27). In the present context, that PDF is

$$p(a_{vok}) = \frac{a_{vok}}{\sigma_{nok}^2} e^{-(a_{vok}^2+a_{sok}^2)/2\sigma_{nok}^2} I_o\left[\frac{a_{vok}a_{sok}}{\sigma_{nok}^2}\right] \tag{5.87a}$$

For the noise alone case, the PDF is Rayleigh, as given by (5.24), which in the present context is

$$p(a_{vok}) = \frac{a_{vok}}{\sigma_{nok}^2} e^{-a_{vok}^2/2\sigma_{nok}^2} \tag{5.87b}$$

where in both cases $\sigma_{nok}^2 = \langle n_{ok}^2 \rangle$.

Substituting (5.87a,b) into (5.86a,b)″, we obtain for the signal-present case

$$\chi_k(\xi/s_{ok}) = \int_0^\infty da_{vok} \frac{a_{vok}}{\sigma_{nok}^2} e^{-a_{sok}^2/2\sigma_{nok}^2} e^{-a_{vok}^2(1-j\xi)/2\sigma_{nok}^2} \cdots$$

$$I_o\left[\sqrt{2}\left(\frac{a_{vok}}{\sqrt{2}\sigma_{nok}}\right)\left(\frac{a_{sok}}{\sigma_{nok}}\right)\right] \tag{5.88a}$$

if K in (5.86a,b)″ is chosen as $1/2\sigma_{nok}^2$. For the noise-alone case

$$\chi_k(\xi/n_{ok}) = \int_0^\infty da_{vok} \frac{a_{vok}}{\sigma_{nok}^2} e^{-a_{vok}^2(1-j\xi)/2\sigma_{nok}^2} \tag{5.88b}$$

Equations (5.88a,b) can be rewritten in the forms (where $u_k = a_{vok}^2/2\sigma_{nok}^2$ and $\rho_{ok} = a_{sok}^2/\sigma_{nok}^2$ = peak SNR)

$$\chi_k(\xi/s_{ok}) = e^{-\rho_{ok}/2} \int_0^\infty du_k\, e^{-u_k} I_o(\sqrt{2u_k\rho_{ok}}) e^{ju_k\xi} \tag{5.88a$'$}$$

$$\chi_k(\xi/n_{ok}) = \int_0^\infty du_k\, e^{-u_k} e^{j\xi u_k} \tag{5.88b$'$}$$

From Fourier transform tables (e.g., Campbell and Foster Tables,[21] cited by DiFranco and Rubin)

$$\chi_k(\xi/s_{ok}) = \frac{e^{-\rho_{ok}/2}}{1-j\xi} e^{j\rho_{ok}\xi/2(1-j\xi)} \tag{5.88a$''$}$$

$$\chi_k(\xi/n_{ok}) = \frac{1}{1-j\xi} \tag{5.88b$''$}$$

The results (5.88a,b)″ could also be obtained through (5.37) and (5.37a)′ applied directly to the forms (5.88a,b), where the integrand is complex and where

$$\mu = 2 \qquad a = \left(\frac{\nu - \mu}{2}\right) + 1 = 0$$

$$\nu = 0 \qquad b = \nu + 1 = 1$$

$$\alpha = \frac{a_{sok}}{\sigma_{nok}^2} \qquad c = \frac{-\alpha^2}{4p^2} = \frac{-\rho_{ok}}{2(1 - j\xi)}$$

$$p^2 = \frac{1}{2\sigma_{nok}^2}(1 - j\xi)$$

Noting that the noise $\sigma_{nok}{}^2 = \sigma_{no}{}^2$ is the same for all pulses, it follows through (5.88a,b)″ that the final forms of the characteristic functions given by (5.86a,b)′ are

$$\chi(\xi/s) = \frac{1}{(1 - j\xi)^{N_p}} e^{-1/2(\sum_{k=1}^{N_p} \rho_{ok})} e^{j\xi/2(1-j\xi)(\sum_{k=1}^{N_p} \rho_{ok})}$$

$$= \frac{1}{(1 - j\xi)^{N_p}} e^{-N_p\bar{\rho}_o/2} e^{j\xi N_p \bar{\rho}_o/2(1-j\xi)} \tag{5.89a}$$

where $\bar{\rho}_o = (1/N_p)\sum_{k=1}^{N_p} \rho_{ok}$ = average (over all pulses) of the peak value of the SNR for a single pulse

$$\chi(\xi/n) = \frac{1}{(1 - j\xi)^{N_p}} \tag{5.89b}$$

From p. 650.0 of Campbell and Foster,[21] and (5.86a,b), we obtain

$$p(V_o/s) = e^{-N_p\bar{\rho}_o/2} \int_{-\infty}^{\infty} d\xi \frac{1}{(1 - j\xi)^{N_p}} e^{-j\xi\{V_o - [N_p\bar{\rho}_o\xi/2(1-j\xi)]\}}$$

$$= \left(\frac{2V_o}{N_p\bar{\rho}_o}\right)^{(N_p-1)/2} e^{-V_o-(N_p\bar{\rho}_o/2)} I_{N_p-1}(\sqrt{2N_p\bar{\rho}_o V_o}) \tag{5.90a}$$

where $I_{N_p-1}(x)$ is the modified Bessel function of order $(N_p - 1)$.

$$p(V_o/n) = \int_{-\infty}^{\infty} d\xi \frac{1}{(1 - j\xi)^{N_p}} e^{-j\xi V_o} \tag{5.90b}$$

$$= \lim_{\bar{\rho}_o \to 0} \left\{ \left(\frac{2V_o}{N_p\bar{\rho}_o}\right)^{(N_p-1)/2} e \frac{-V_o(2N_p\bar{\rho}_o V_o)^{(N_p-1)/2}}{2^{N_p-1}(N_p - 1)!} \left[1 + \frac{(2N_p\bar{\rho}_o V_o)}{2^2(N_p)} + \frac{(2N_p\bar{\rho}_o V_o)^2}{2.2^4(N_p)(N_p + 1)} + \cdots \right]\right\}$$

$$= \frac{(2V_o)^{N_p-1} e^{-V_o}}{2^{N_p-1}(N_p - 1)!} = \frac{V_o^{N_p-1} e^{-V_o}}{(N_p - 1)!}, \qquad V_o \geq 0$$

$$= 0, \qquad V_o < 0$$

The use of the power series representation of the modified Bessel function for small arguments in (5.90b) is necessitated by the fact that (5.90a) is an indeterminate form for $\bar{\rho}_o = 0$ and the method used cancels out $\bar{\rho}_o$ as $\bar{\rho}_o$ approaches zero.

The detection and false alarm probability for a train of N_p pulses are obtained by integrating $p(V_o/s)$ and $p(V_o/n)$, respectively, from the threshold value V_{oT} to infinity. Remember that this is the case of *noncoherent integration* of the pulse train and that our objective is to obtain P_d and P_{fa} for this more realistic situation and compare the results with those for coherent pulse integration.

The detection and false alarm probabilities are [from (5.90a,b)]

$$P_d = e^{N_p\bar{\rho}_o/2}\int_{V_{oT}}^{\infty} dV_o\left(\frac{2V_o}{N_p\bar{\rho}_o}\right)^{(N_p-1)/2} I_{N_p-1}(\sqrt{2N_p\bar{\rho}_o V_o})e^{-V_o} \tag{5.91a}$$

(equivalent to Eq. (10.4–27) on p. 348 of DiFranco and Rubin[1]) and

$$P_{fa} = \frac{1}{(N_p-1)!}\int_{V_{oT}}^{\infty} dV_o\, V_o^{N_p-1}e^{-V_o} = \frac{(-1)^{N_p-1}}{(N_p-1)!}\int_{-\infty}^{\infty} d(e^x)x^{N_p-1}$$

$$= \frac{1}{(N_p-1)!}e^{-V_{oT}}(V_{oT})^{N_p-1}\left[1 + \frac{(N_p-1)}{V_{oT}} + \frac{(N_p-1)(N_p-2)}{V_{oT}^2} + \cdots\right] \tag{5.91b}$$

(through integration by parts), where V_{oT} is the threshold [already normalized through choice of K in (5.83) as the reciprocal of noise level and therefore a dimensionless quantity]. Equation (5.91b) is equivalent to Eq. (10.4–20) on p. 347 of DiFranco and Rubin.

The curve families shown in Figure 5.24a–j are plots of P_d vs. SNR for fixed values of P_{fa} obtained from DiFranco and Rubin,[1] Figures 10.4.1 through 10.4.10, pp. 349–358. These curves were obtained from numerical computations on (5.91a) and (5.91b), and apply to the case under discussion here, that of noncoherent pulse integration using quadratic rectification of each pulse, where the target signal does *not* fluctuate.

In Figure 5.24a–j, the parameter R_p is $\bar{\rho}_o$ in our notation. It is the *peak* SNR of a single pulse, averaged over all pulses. The number of pulses, N_p in our notation, is designated as N by DiFranco and Rubin. Also, the false alarm rate P_{fa} is given on these curves by

$$P_{fa} = \frac{0.693}{n'}$$

where $n' = 10, 10^3, 10^6, 10^8$, or 10^{10}, unlike the DiFranco and Rubin curves used previously (Figs. 5.12, 5.13, 5.16, and 5.17), where P_{fa} is always 10^{-m}, where m is a positive integer. For purposes of comparison, the following table gives the correspondence.

n'	$P_{fa} = 0.693/n'$	Closest Value of 10^{-m}
10	$6.93(10^{-2})$	$\approx 10^{-1}$
10^2	$6.93(10^{-3})$	$\approx 10^{-2}$
10^3	$6.93(10^{-4})$	$\approx 10^{-3}$
10^4	$6.93(10^{-5})$	$\approx 10^{-4}$
10^5	$6.93(10^{-6})$	$\approx 10^{-5}$
10^6	$6.93(10^{-7})$	$\approx 10^{-6}$
10^7	$6.93(10^{-8})$	$\approx 10^{-7}$
10^8	$6.93(10^{-9})$	$\approx 10^{-8}$
10^9	$6.93(10^{-10})$	$\approx 10^{-9}$
10^{10}	$6.93(10^{-11})$	$\approx 10^{-10}$

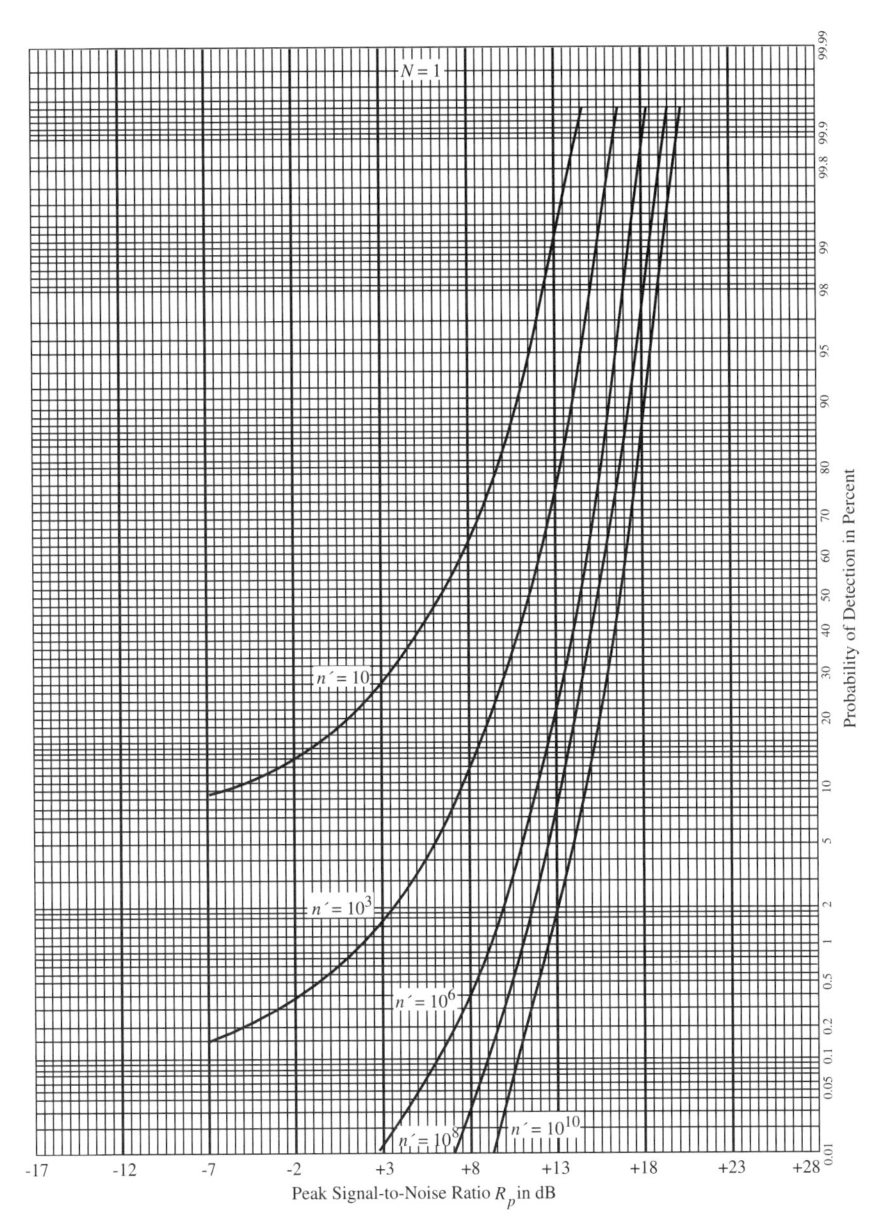

FIGURE 5.24a
Probability of detecting a nonfluctuating target (square-law detector), N eq 1. [N = number of pulses incoherently integrated: $P_{fa} = 0.693/n'$.] From DiFranco and Rubin,[1] Figure 10.4.1 of "Radar Detection," by permission of the authors.

5.7 PULSE INTEGRATION LOSS

When pulses are integrated noncoherently, there is a definite loss relative to the case where they are integrated coherently. In the latter case, pulse integration is part of the process of coherent matched filtering and the minimum required SNR for a given level of detection performance is decreased by N_p relative to the case of a single pulse. That is because the matched filter output SNR depends only on *total signal energy,* independently of how the power is distributed in time.

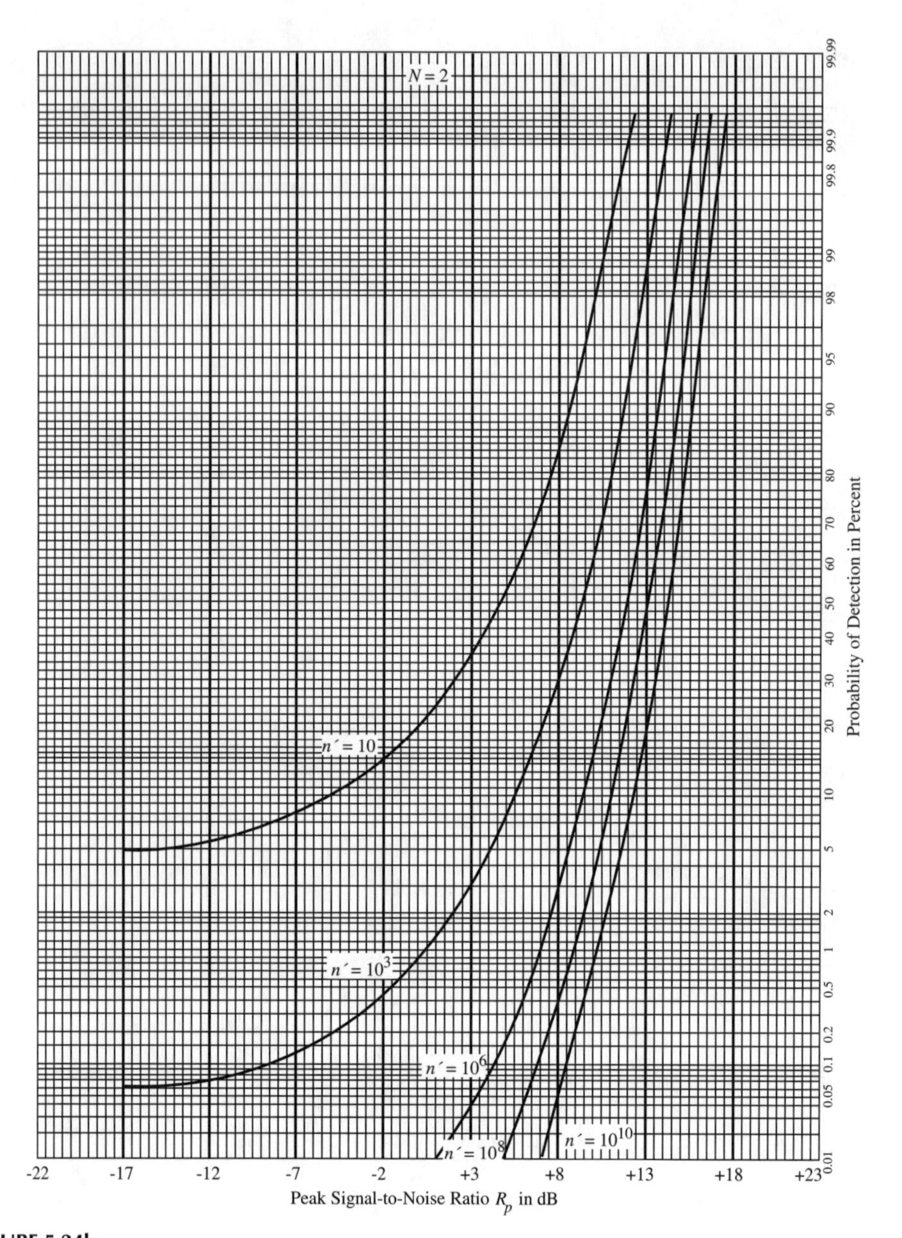

FIGURE 5.24b
Probability of detecting a nonfluctuating target (square-law detector), $N = 2$. [N = number of pulses incoherently integrated; $P_{fa} = 0.693/n'$.] From DiFranco and Rubin,[1] Figure 10.4.2 of "Radar Detection," by permission of the authors.

To illustrate the phenomenon of "integration loss" due to *noncoherent* integration of pulses, let us examine Figure 5.24a–j and determine the required SNR for four performance levels for various numbers of pulses, from $N_p = 1$ to $N_p = 3000$. That information, obtained from the curves, is given in Table 5.3. We will first define integration loss. It is the ratio of the required *individual* pulse SNR for a given performance level with *noncoherent* integration relative to that required with coherent integration. The latter is the ratio

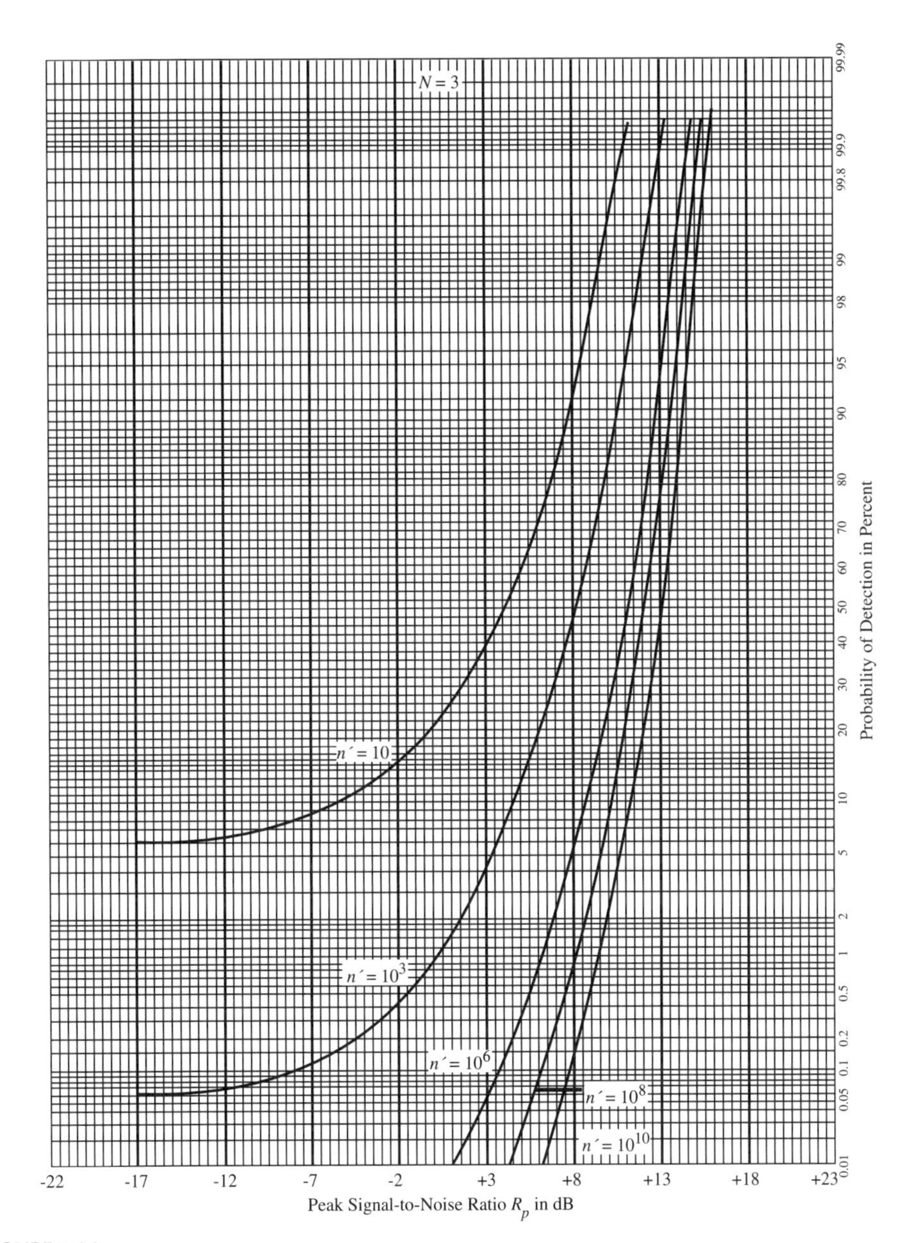

FIGURE 5.24c
Probability of detecting a nonfluctuating target (square-law detector), $N = 3$. [N = number of pulses incoherently integrated; $P_{fa} = 0.693/n'$.] From DiFranco and Rubin,[1] Figure 10.4.3 of "Radar Detection," by permission of the authors.

$$(\bar{\rho}_o)_1^{(N_p-\mathrm{coh})} = \frac{\bar{\rho}_o^{(\mathrm{coh})}}{N_p} \tag{5.92}$$

where $(\bar{\rho}_o)_1^{(N_p-\mathrm{coh})}$ is the required single pulse SNR with the specified P_d and P_{fa} values if a train of N_p pulses is first integrated *coherently* and then *noncoherent* detection is performed, and $(\rho_o)^{(\mathrm{coh})}$ is the required SNR for the same performance level if only one of those pulses is *noncoherently detected* after *coherent matched filtering.*

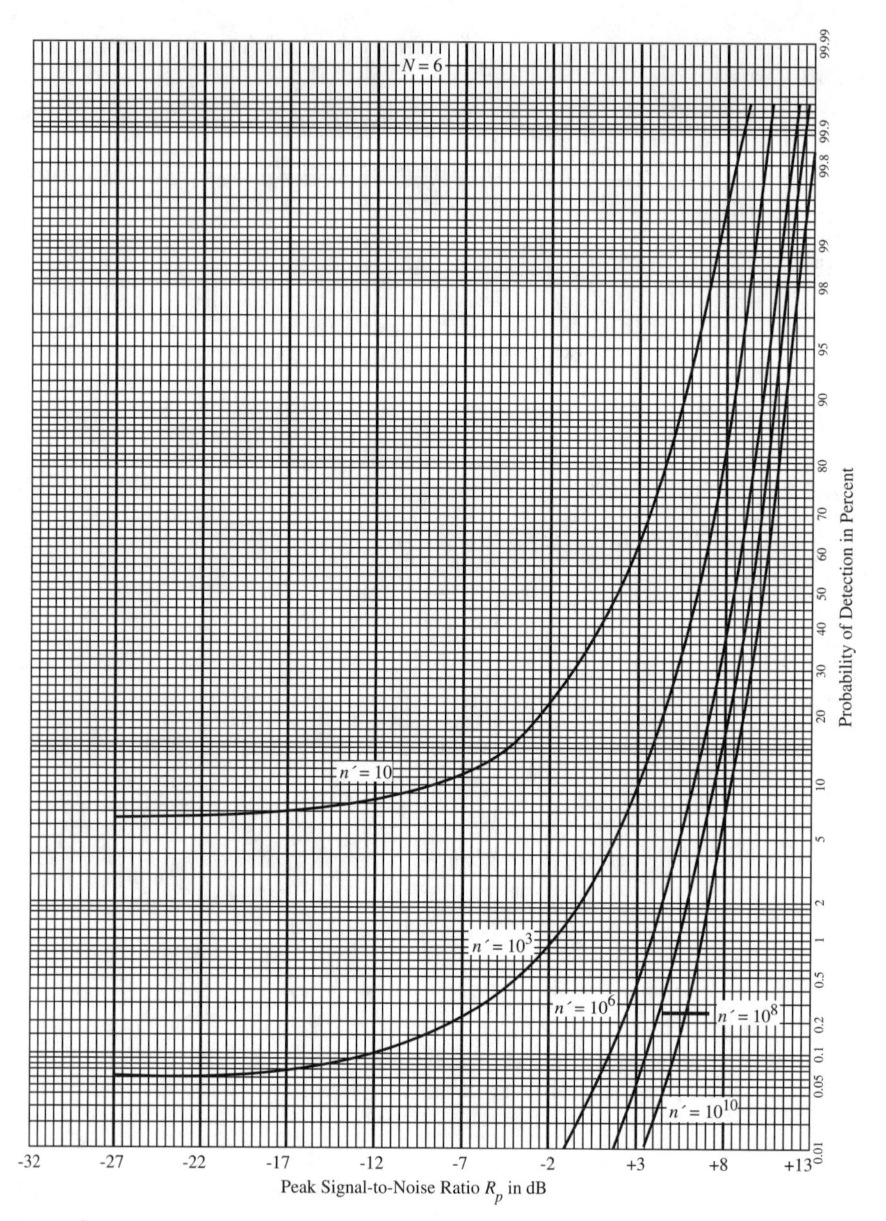

FIGURE 5.24d
Probability of detecting a nonfluctuating target (square-law detector), $N = 6$. [N = number of pulses incoherently integrated; $P_{fa} = 0.693/n'$.] From DiFranco and Rubin,[1] Figure 10.4.4 of "Radar Detection," by permission of the authors.

The integration loss, defined in dB, is [with the aid of (5.92)]

$$L_{dB} = 10\log_{10}\left[\frac{(\bar{\rho}_o)_1^{(N_p-\text{ncoh})}}{(\bar{\rho}_o)_1^{(N_p-\text{coh})}}\right] = 10\log_{10}\left[\frac{N_p(\bar{\rho}_o)^{(N_p-\text{ncoh})}}{(\rho_o)_{N_p}^{(\text{coh})}}\right] \tag{5.93}$$

where $(\bar{\rho}_o)_1^{(N_p-\text{ncoh})}$ is the required single pulse SNR with noncoherent integration of N_p pulses.

Since more SNR is always required for a given performance level with noncoherent integration than with coherent integration, the numerator of (5.93)

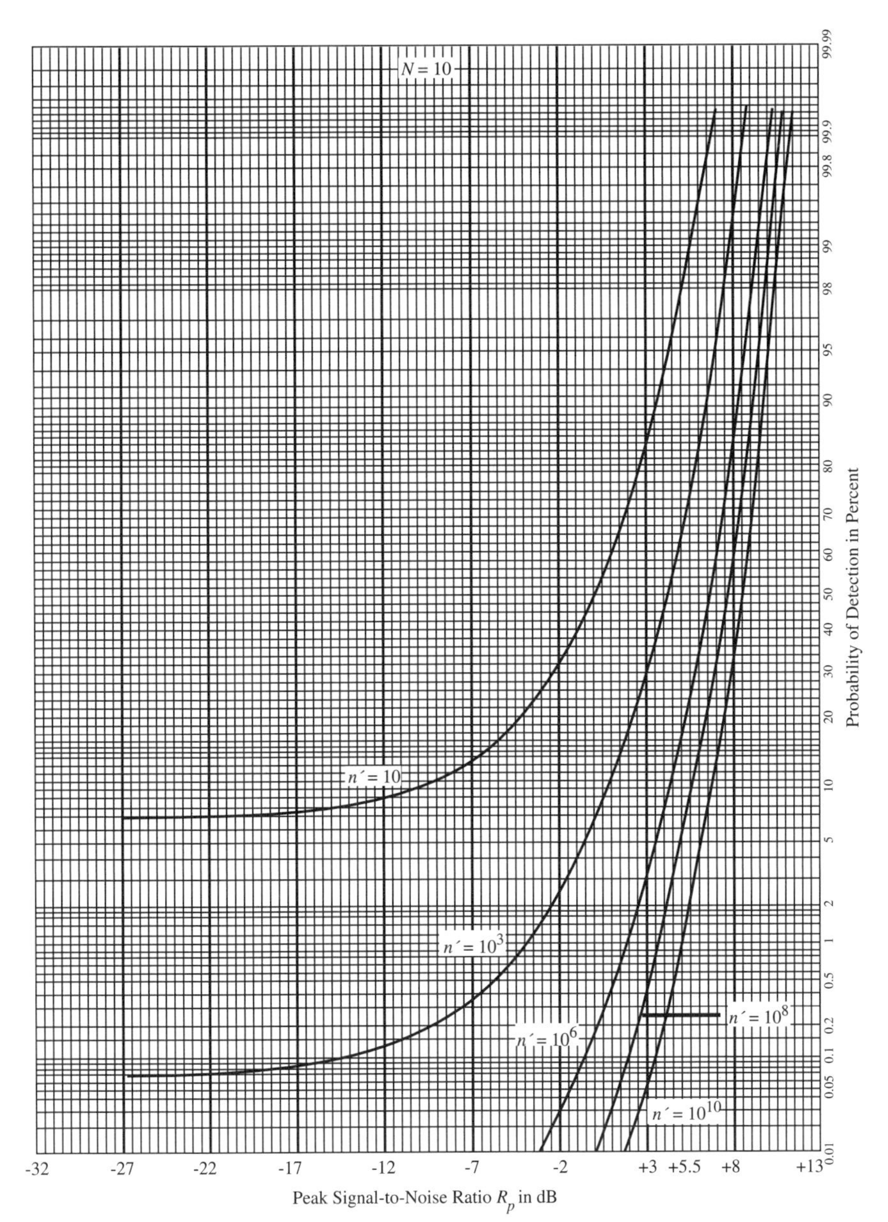

FIGURE 5.24e
Probability of detecting a nonfluctuating target (square-law detector), $N = 10$. [N = number of pulses incoherently integrated; $P_{fa} = 0.693/n'$.] From DiFranco and Rubin,[1] Figure 10.4.5 of "Radar Detection," by permission of the authors.

always exceeds the denominator and hence L_{dB} is always positive, always representing a *loss*, never a gain.

Table 5.3 shows the comparative values of required SNR for four performance levels with varying numbers of pulses noncoherently integrated. It also shows the integration loss as given by (5.93) in each case. This information is from the curves of Figure 5.24a–j.

The "collapsing loss," defined in DiFranco and Rubin[1] [Eq. (12.26) on p. 466] as the increase in SNR required for a specified detection probability due to integration of noise-alone pulses along with signal-plus-noise pulses, is not included in Table

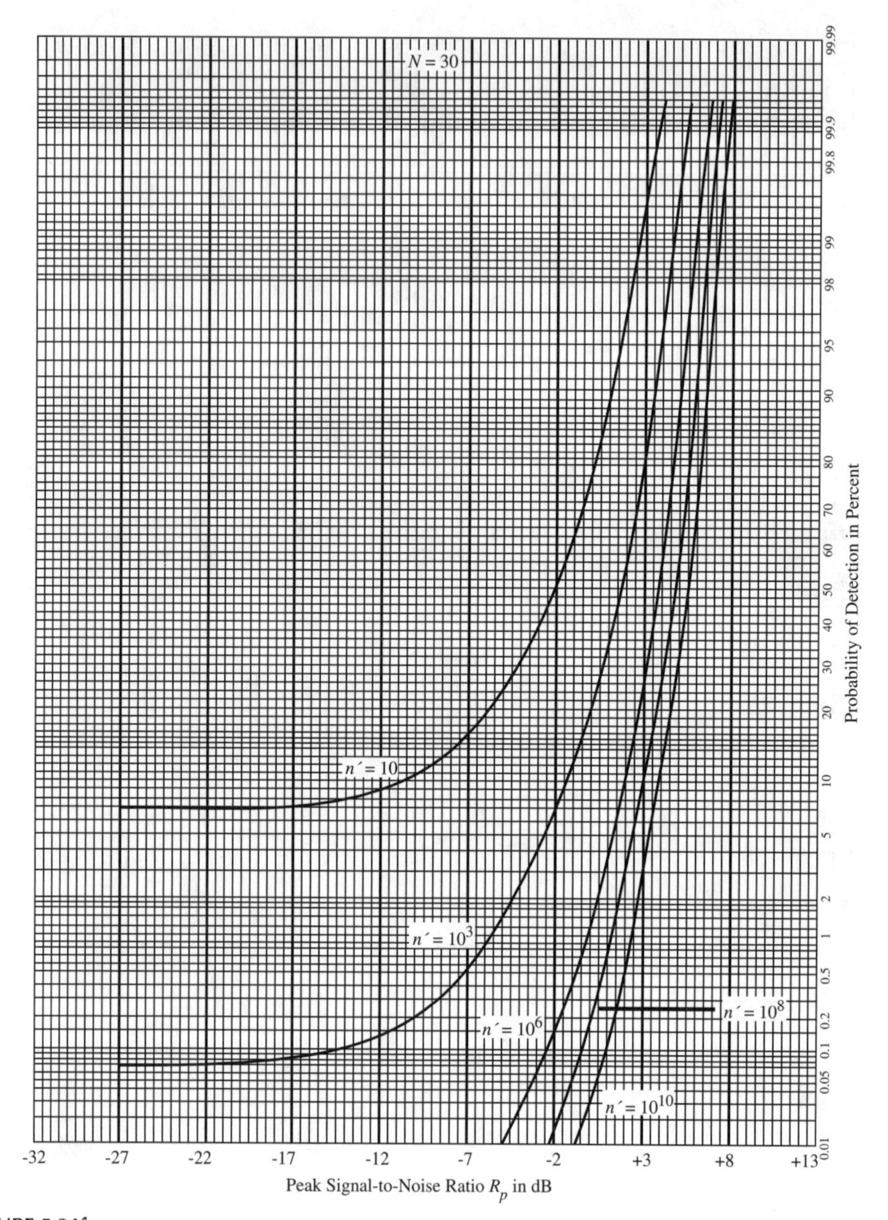

FIGURE 5.24f
Probability of detecting a nonfluctuating target (square-law detector), $N = 30$. [N = number of pulses incoherently integrated; $P_{fa} = 0.693/n'$.] From DiFranco and Rubin,[1] Figure 10.4.6 of "Radar Detection," by permission of the authors.

5.3. That loss occurs, for example, when a radar return contains information on target range, elevation angle, and azimuth angles but the output is "collapsed" into a two-dimensional display, e.g., range and azimuth as in a plan-position indicator (PPI).

The PPI display shows the distribution of the radar return power in the range-azimuth plane. This is a *two*-dimensional display, but the original information was the distribution of return power in the *three*-dimensional range-azimuth-elevation volume (i.e., in spherical coordinates). To "collapse" the output from three to two dimensions in this case, pulses within a given range gate are integrated as the antenna beam scans over a span of elevation angles while it is approximately fixed

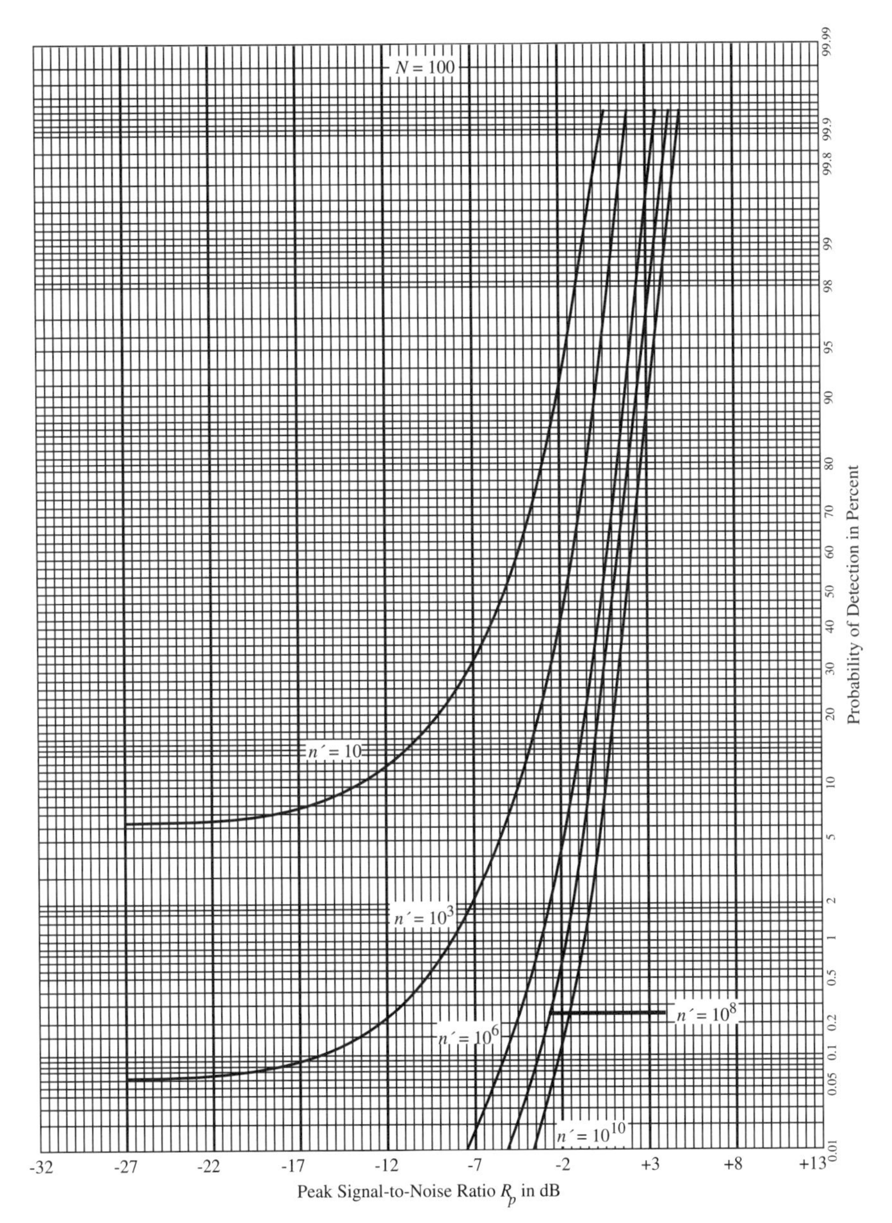

FIGURE 5.24g
Probability of detecting a nonfluctuating target (square-law detector), $N = 100$. [N = number of pulses incoherently integrated; $P_{fa} = 0.693/n'$.] From DiFranco and Rubin,[1] Figure 10.4.7 of "Radar Detection," by permission of the authors.

at a particular azimuth angle. Thus, for a given range and azimuth, the returns at all elevation angles covered by the scan are integrated. The only pulses containing signal energy are those occurring while the antenna scans in elevation past the target. The remaining returns are noise-alone pulses.

Coverage of collapsing loss in standard radar texts can be found, for example, DiFranco and Rubin[1] (pp. 464–467), Skolnick[4] (pp. 59–60), Rohan[7] (pp. 77–78), and Blake[8] (pp. 376–378).

From Table 5.3 it is evident that integration loss becomes substantial as the number of pulses approaches numbers such as 100 to 3000, although it is very small

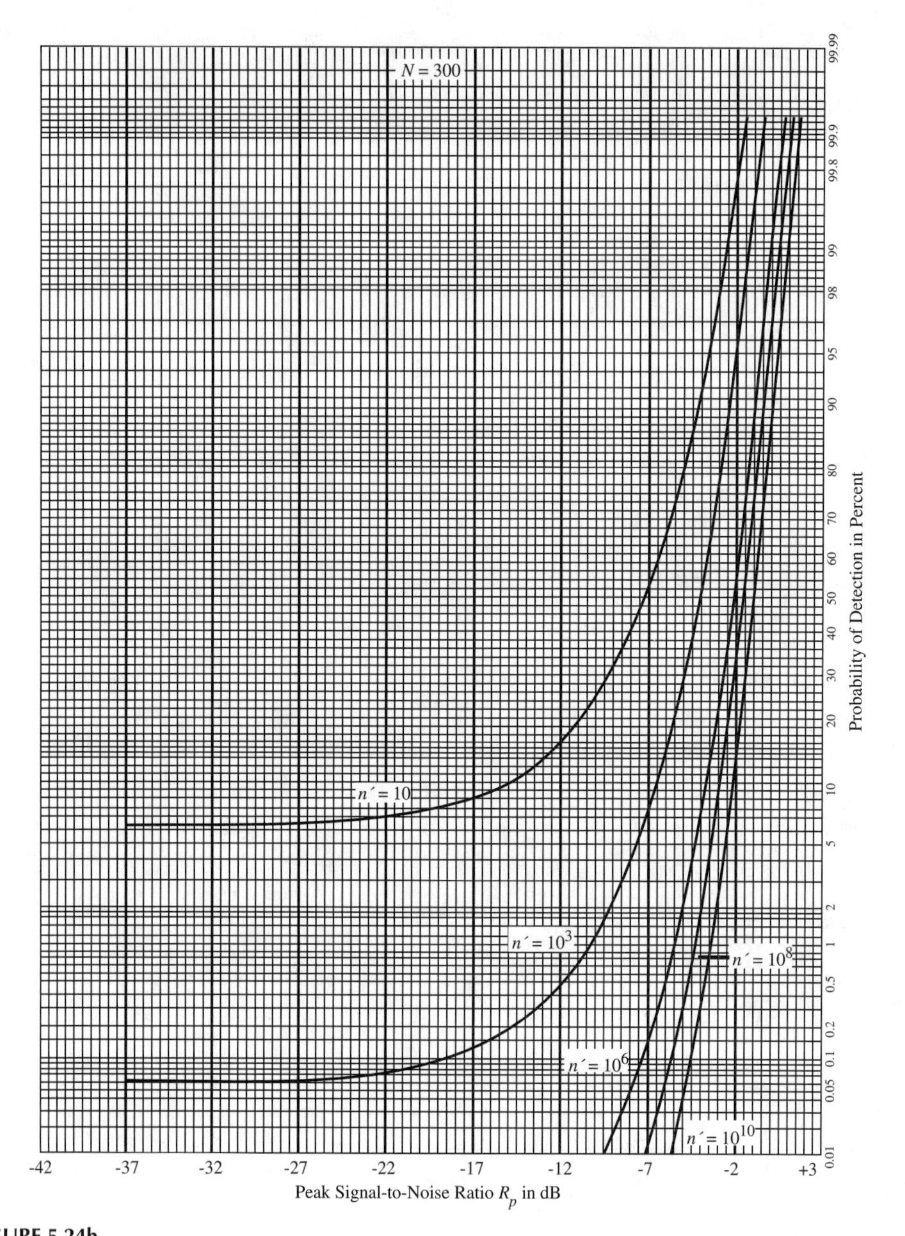

FIGURE 5.24h
Probability of detecting a nonfluctuating target (square-law detector), $N = 300$. [N = number of pulses incoherently integrated; $P_{fa} = 0.693/n'$.] From DiFranco and Rubin,[1] Figure 10.4.8 of "Radar Detection," by permission of the authors.

for values of N_p below and including 6. As N_p becomes large, further increases vary as ($10 \log_{10} \sqrt{N_p} - 5.5$). It will be left as an exercise to use Figure 5.24a–j and 5.12 to plot curves of L_{dB} vs. N_p for a few fixed values of P_d and P_{fa} (similar to Figs. 10.5.1 and 10.5.2 of DiFranco and Rubin[1]) and to obtain the general rule given by DiFranco and Rubin[1] (p. 390) by graphic means:

$$L_{dB} = 10 \log_{10}\sqrt{N_p} - 5.5 \tag{5.94}$$

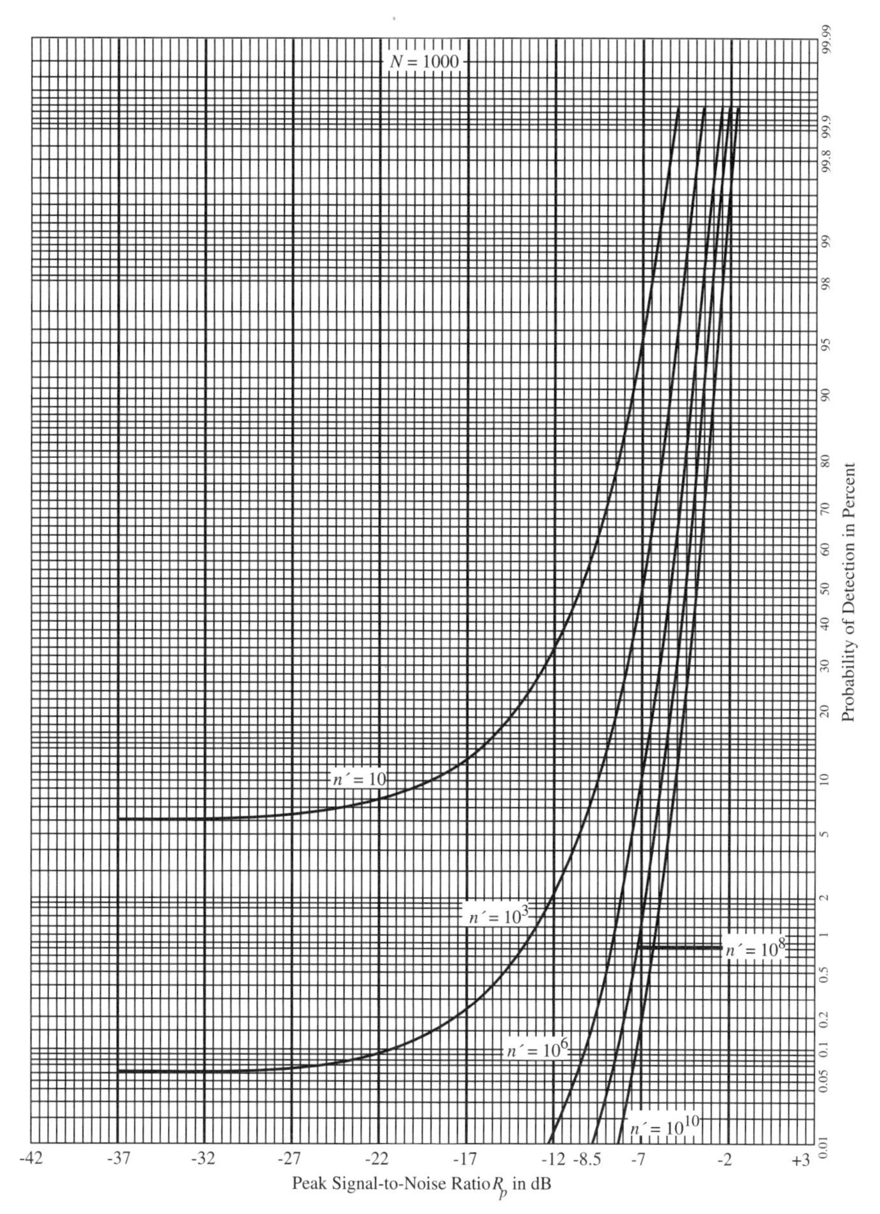

FIGURE 5.24i
Probability of detecting a nonfluctuating target (square-law detector), $N = 1000$. [N = number of pulses incoherently integrated; $P_{fa} = 0.693/n'$.] From DiFranco and Rubin,[1] Figure 10.4.9 of "Radar Detection," by permission of the authors.

The effect of collapsing loss on these results is shown in DiFranco and Rubin[1] [Section 12.7 (E), pp. 464–467]. It is evident from (5.86a,b)′ that the characteristic function for a sum of $(l + m)$ quadratically detected pulses, where l of those pulses contain signal-plus-noise and m of them contain noise alone, is given by

$$\chi(\xi/l + m, m) = \prod_{k=1}^{l} \chi_k(\xi/s_{ok}) \prod_{j=1}^{m} \chi_j(\xi/n_{oj}) \qquad (5.94a)'$$

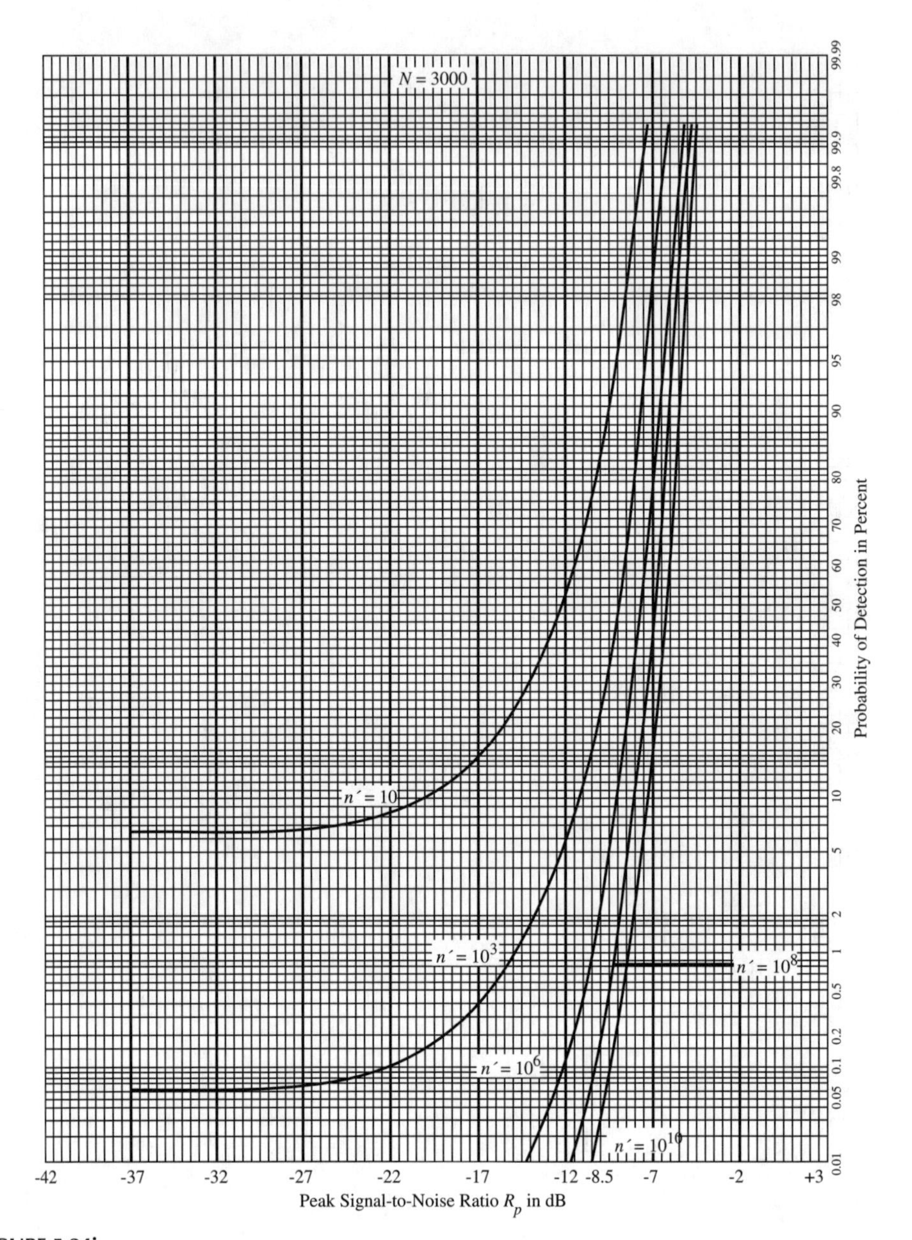

FIGURE 5.24j
Probability of detecting a nonfluctuating target (square-law detector), $N = 3000$. [N = number of pulses incoherently integrated; $P_{fa} = 0.693/n'$.] From DiFranco and Rubin,[1] Figure 10.4.10 of "Radar Detection," by permission of the authors.

where the SNR for each of the signal-plus-noise pulses is ρ_o.

In DiFranco and Rubin,[1] beginning with Eq. (12.7-3) on p. 465, which is the equivalent of (5.94a)′, it is shown [Eq. (12.7-5) on p. 466] that $\chi(\xi/l + m, m)$ is the characteristic function for a sum of $(l + m)$ quadratically detected pulses containing signal-plus noise, but with SNR for each pulse reduced to ρ_o/r_c, where r_c is the "collapsing ratio," given by

$$r_c = \frac{l + m}{l} \tag{5.94b$'$}$$

TABLE 5.3

Integration Loss vs. Number of Pulses[a]

N_P	A	B	L_{dB}	A	B	L_{dB}	A	B	L_{dB}	A	B	L_{dB}
1	18.5	18.5	0	17.5	17.5	0	15.5	15.5	0	13.5	13.5	0
2	16	15.5	0.5	15	14.5	0.5	14	12.5	1.5	11	10.5	0.5
3	15	13.7	1.3	14	12.7	1.3	12	10.7	1.3	10	8.7	1.3
6	12	10.7	1.3	11.5	9.7	1.8	10	7.7	2.3	7	5.7	1.3
10	11.5	8.5	3.0	9.5	7.5	2.0	8.5	5.5	3.0	5.5	3.5	2.0
30	7.5	3.7	3.8	5.5	2.7	2.8	4.5	0.7	3.8	3	−1.3	4.3
100	3.5	−1.5	5.0	3	−2.5	5.5	2	−4.5	6.5	0	−6.5	6.5
300	0	−6.3	6.3	−0.5	−7.3	6.8	−1	−9.3	8.3	−3	−11.3	8.3
1000	−2.5	−11.5	9.0	−3	−12.5	9.5	−4.5	−14.5	10.0	−6	−16.5	10.5
3000	−5.5	−16.3	10.8	−6	−17.3	11.3	−6.5	−19.3	12.8	−8	−21.3	13.3
	$P_d = 99\%$			$P_d = 95\%$			$P_d = 90\%$			$P_d = 85\%$		
	$P_{fa} = 6.939(10^{-10})$			$P_{fa} = 6.93\ (10^{-8})$			$P_{fa} = 6.93\ (10^{-6})$			$P_{fa} = 6.93\ (10^{-3})$		

[a] $A = (\bar{\rho}_o)_{req}^{(ncoh)}$ = required single-pulse SNR-dB, noncoherent integration
$B = (\bar{\rho}_o)_{req}^{(coh)}$ = required single-pulse SNR-dB, coherent integration
$L_{dB} = A - B$ = integration loss
N_p = Number of pulses integrated

This leads to the conclusion that the detection probability for this integration of l signal-plus-noise pulses with SNR $= \rho_o$ and m noise-alone pulses is the same as that for $(l + m)$ signal-plus-noise pulses with SNR $= \rho_o/r_c$.

The collapsing loss can be defined as the ratio of the integration loss for $(l + m)$ pulses to that for integration of l pulses. That definition, as given in Eq. (2.52) on p. 59 of DiFranco and Rubin[1] is

$$L_c(m, l) = L_i(m + l)/L_i(l) \tag{5.94c$'$}$$

where $L_c(m, l)$ is the collapsing loss with integration of l signal-plus-noise pulses, $L_i(m + l)$ is the integration loss with l signal-plus-noise pulses and m noise-alone pulses, and $L_i(l)$ is the integration loss with l signal-plus-noise pulses.

Another definition, that given in DiFranco and Rubin,[1] is

$$L_c(m + l) = (\rho_{req})_{m+l}/(\rho_{req})_l \tag{5.94d$'$}$$

where $(\rho_{req})_n$ is the required SNR for a specified detection probability with integration of n pulses.

5.8 EFFECT OF TARGET SIGNAL FLUCTUATION ON DETECTION OF PULSE TRAINS: SWERLING MODELS

Target signal fluctuations can occur over a wide range of possible time scales. The important parameter in deciding how to account for these fluctuations is the ratio of the pulse repetition period (PRP) to the correlation time of the signal amplitude fluctuations.

Let the PRP be denoted by *Tr* and the correlation time of fluctuations by τ_C. The following conditions might prevail, as illustrated in Figure 5.25: It is assumed that the return pulses have constant amplitude in the absence of fluctuations and that the ACF of the signal amplitude is exponential.

Figure 5.25a shows the constant-amplitude video pulse train. Figures 5.25b and c show the rapid and unpredictable fluctuations of the pulse amplitudes in the case where the PRP *Tr* far exceeds the correlation time τ_{Cf} of the fluctuations (Fig. 5.25b) and the autocorrelation function (ACF) in this case (Fig. 5.25c), where the latter falls off to negligible values at values of τ well below the PRP, indicating that the fluctuations are uncorrelated from pulse-to-pulse. Equivalently, we can say that each pulse amplitude is independent of the previous or subsequent pulse amplitudes.

Figures 5.25d and e show the opposite extreme case, where the correlation time far exceeds the PRP. In this case, Figure 5.25e shows that the pulse amplitudes are correlated over the duration of the pulse train, and Figure 5.25d shows that the pulses (only one possibility) may be slowly decreasing in amplitude, but do so in a smooth and predictable way, i.e., one could infer the amplitude of a given pulse to some degree by knowing the trend for previous pulses. In the extreme, the entire pulse train has the same amplitude, but it may differ from its value without fluctuations.

If one prefers dealing with frequency rather than time, the following equivalent views are possible.

1. Case $Tr \gg \tau_{cf}$ is equivalent to ($fr \ll B_f$) where B_f is the spectral bandwidth of the fluctuating amplitude, which is the reciprocal of the correlation time, and *fr* is the PRF, the reciprocal of *Tr*.
2. Case $Tr \ll \tau_{cf}$ equivalent to $fr \gg B_f$.

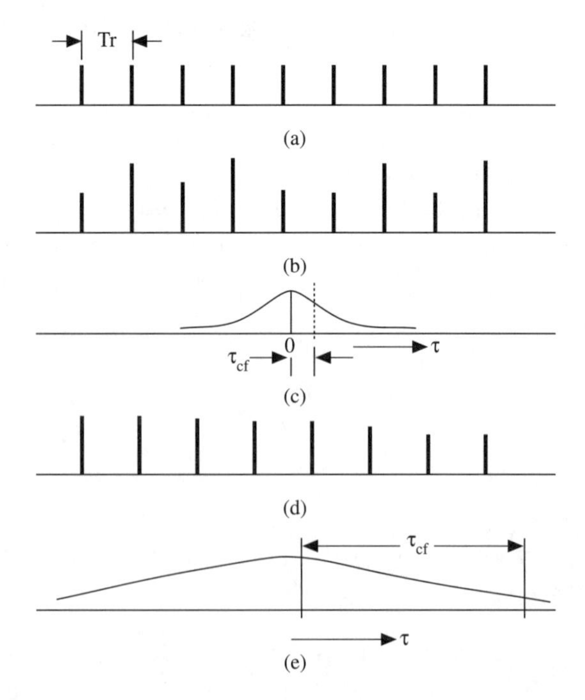

FIGURE 5.25
Pulse amplitude fluctuations. (a) Video pulse train without fluctuations. (b) Fluctuating signal; $Tr >> \tau_{cf}$. (c) ACF for case $Tr >> \tau_{cf}$. (d) Fluctuating signal; $Tr << \tau_{cf}$. (e) ACF for case $Tr << \tau_{cf}$.

There is an infinite range of intermediate cases between the two extremes illustrated in the figure, but those two cases are the ones that are easily calculated and hence the analytical results available focus on those cases. This is not a severe limitation in radar system design. Adequate rough estimates can be made from these cases that can help to guide the choice of lower and upper limits on design parameters.

The models for these two extreme cases are called the "Swerling models," after the co-author of the classical report on the subject by Marcum and Swerling.[17,18] The following Swerling models are cited by DiFranco and Rubin[1] (Chapter 11, pp. 373–445).

Swerling 1: Rayleigh amplitude fluctuations. This is the case where the fluctuations are slow compared to the PRF, i.e., the case $Tr \ll \tau_{cf}$ or $fr \gg B_f$ as illustrated in Figure 5.25d and e. The analytical model in this case assumes that the amplitude of the entire pulse train is a single random variable with Rayleigh statistics.

Swerling 2: Rayleigh amplitude fluctuations. In this model, the amplitude of each pulse is a random variable and the pulse amplitudes are statistically independent. This is the case where the fluctuations are rapid compared with the PRF, i.e., the case $Tr \gg \tau_{cf}$ or $fr \ll B_f$ as illustrated in Figure 5.25b and c.

Swerling 3: Same as Swerling 1 except that the amplitude fluctuation statistics are one-dominant-plus Rayleigh.

Swerling 4: Same as Swerling 2 with one-dominant-plus Rayleigh statistics.

In formulating the Swerling models the initial phases of individual pulses are assumed to be statistically independent and uniformly distributed.

Effects of antenna scanning: In the usual radar parlance, Swerling 1 or 3 is called "scan-to-scan" fluctuations and Swerling 2 or 4 is "pulse-to-pulse" fluctuations. This is because in a typical search radar (for example), the antenna is scanning across the target within a finite time. While the peak of the antenna beam is scanning past the target, tens, hundreds, or thousands of pulses are being transmitted. Approximating the antenna as a flat beam of width $\Delta\phi$, and the scan rate as $\dot{\phi}$, the period during which the antenna beam is "on target" is (Fig. 5.26)

$$\Delta t = \frac{\Delta\phi}{\dot{\phi}} \tag{5.95}$$

Figure 5.26 shows a typical search radar situation in which the radar is scanning continuously in azimuth over 360°. A linear scanning pattern gives the azimuthal beam angle as a function of time, i.e.,

$$\phi(t) = \phi_0 + \dot{\phi}(t - t_0) \tag{5.96}$$

where $\dot{\phi}$ is the fixed scanning rate in degrees (or radians if preferred) per second and ϕ_0 is the azimuth at time t_0.

The return pulse train, as shown in Figure 5.26c, traces out the shape of the antenna beam as it scans past the target. The antenna pattern is pictured as having a strong main lobe and a set of much weaker sidelobes.

In Figure 5.26d we show an "equivalent" pulse train that is approximated as flat over a portion of the main lobe. The purpose of that is to enable us to define an "equivalent duration" of the pulse train (since there is always *some* pulse energy, even far out in the sidelobes, except, of course, at the nulls) and to approximate the pulse train as having fixed amplitude over that duration, which corresponds to the antenna beamwidth $\Delta\phi$, related to Δt through (5.95). (All of this ignores target signal fluctuations, of course.)

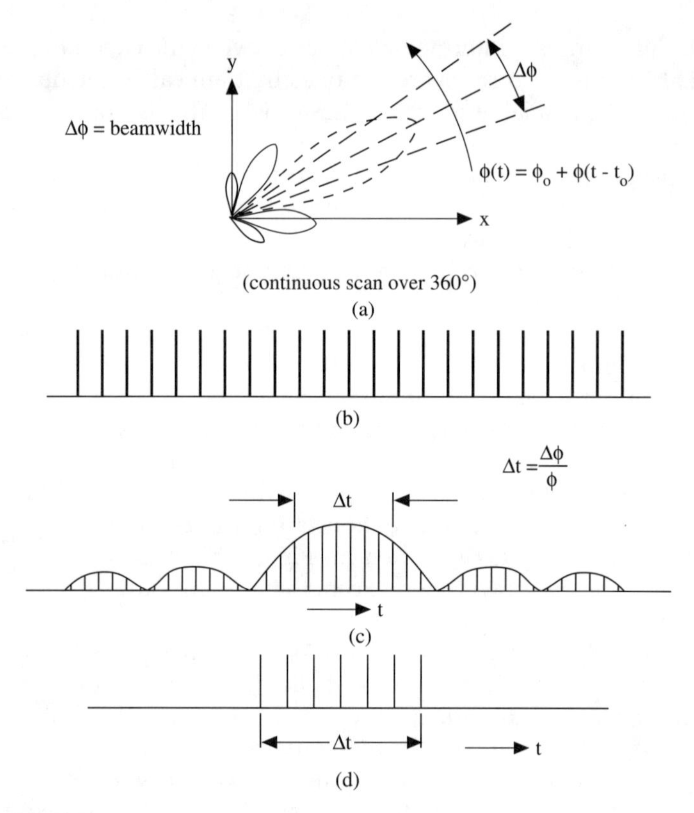

FIGURE 5.26
Typical search radar situation. (a) Azumuthally scanning antenna. (b) Transmitted pulse train. (c) Received pulse train. (d) Equivalent received pulse train.

Returning to the Swerling models, Swerling 1 or 3 is called a "scan-to-scan" model because the period, Δt, during which the beam is "on target," is small compared with the correlation time τ_{cf}, so the amplitude is approximately constant over Δt and we can regard the pulse amplitude as a single random variable over each scan. The detection decision is then based on the sum of amplitude-squared values over Δt. In this case, Tr, which is much smaller than Δt, must also be small compared to τ_{cf}.

If Δt is large compared to τ_{cf}, *and* $Tr \gg \tau_{cf}$, then we have the "pulse-to-pulse" case (Swerling 2 or 4), where pulse amplitudes are mutually independent random variables.

In the context of scanning antennas, the conditions for Swerling 1 or 3 are [from (5.95)]

$$\Delta\phi << \dot{\phi}\tau_{cf} = \frac{\dot{\phi}}{B_f} \tag{5.97a}$$

where $\Delta\phi$ is the beamwidth, $\dot{\phi}$ the scan rate, τ_{cf} the fluctuation correlation time, and B_f the fluctuation bandwidth.

The conditions for Swerling 2 or 4 are

$$\Delta\phi >> \dot{\phi}\tau_{cf} \qquad and \qquad Tr >> \tau_{cf}$$

or equivalently

$$\Delta\phi >> \frac{\dot{\phi}}{B_f} \qquad and \qquad fr << B_f \tag{5.97b}$$

To treat the pulse-train detection problem for the Swerling 1 or 3 case, we return to (5.90a) and allow $\bar{\rho}_o$ to be a random variable with Rayleigh (Swerling 1) or one-dominant-plus Rayleigh (Swerling 3) statistics. We must then average (5.90a) over the distribution of $\bar{\rho}_o$. Before doing this, we note that the PDF of $\bar{\rho}_o$ is that of the amplitude-square, or power, and hence, whatever the amplitude PDF, the PDF of amplitude-square is related to it through the expression

$$p(a_s^2)d(a_s^2) = p(a_s)\, da_s \tag{5.98}$$

or equivalently

$$p(a_s^2) = \frac{p(a_s)}{2a_s} \tag{5.98$'$}$$

For the Rayleigh PDF [from (5.35) with the aid of (5.35a)′]

$$p(a_s) = \frac{2a_s}{\langle a_s^2 \rangle} e^{-a_s^2/\langle a_s^2 \rangle} \tag{5.99a}$$

and hence

$$p(a_s^2) = \frac{1}{\langle a_s^2 \rangle} e^{-a_s^2/\langle a_s^2 \rangle} \tag{5.99b}$$

which is the exponential distribution. Since

$$\bar{\rho}_o = \frac{a_s^2}{\sigma_n^2} \qquad \text{and} \qquad \langle \bar{\rho}_o \rangle = \frac{\langle a_s^2 \rangle}{\sigma_n^2} \tag{5.99c}$$

then

$$p(\bar{\rho}_o) = p(a_s^2)\frac{d(a_s^2)}{d(\bar{\rho}_o)} = \sigma_n^2 p(a_s^2) = \frac{1}{\langle \bar{\rho}_o \rangle} e^{-\bar{\rho}_o/\langle \bar{\rho}_o \rangle} \tag{5.99d}$$

Through steps analogous to (5.99a) to (5.99d), we obtain for the one-dominant-plus Rayleigh PDF [from (5.40) with the aid of (5.44)]

$$p(a_s) = \frac{8a_s^3}{\langle a_s^2 \rangle^2} e^{-2a_s^2/\langle a_s^2 \rangle} \tag{5.100a}$$

$$p(a_s^2) = p(a_s)\frac{d(a_s)}{d(a_s^2)} = \frac{4a_s^2}{\langle a_s^2 \rangle^2} e^{-2a_s^2/\langle a_s^2 \rangle} \tag{5.100b}$$

From (5.99c) it then follows that

$$p(\bar{\rho}_o) = p(a_s^2)\frac{d(a_s^2)}{d(\bar{\rho}_o)} = \sigma_n^2 p(a_s^2) = \frac{4\bar{\rho}_o}{\langle\bar{\rho}_o\rangle^2}\, e^{-2\bar{\rho}_o/\langle\bar{\rho}_o\rangle} \tag{5.100c}$$

For the Swerling 1 or 3 case, where a_s is a single random variable over the entire pulse train, Eq. (5.90a) averaged over the random variable $\bar{\rho}_o$ gives us

$$p(V_o/s) = \int_0^\infty d\bar{\rho}_o\, p(\bar{\rho}_o)\left(\frac{2V_o}{N_p\bar{\rho}_o}\right)^{(N_p-1)/2} e^{-V_o-(N_p\bar{\rho}_o/2)} I_{N_p-1}(\sqrt{2N_p\bar{\rho}_o V_o}) \tag{5.101}$$

For Swerling 1 and 3, respectively, we substitute (5.99c) and (5.100c), respectively, into (5.101) with the results

$$[p(V_o/s)]_{\text{Swerling 1}} = \left(\frac{2V_o}{N_p}\right)^{(N_p-1)/2} \frac{e^{-V_o}}{\langle\bar{\rho}_o\rangle}$$

$$\times \int_0^\infty d\bar{\rho}_o\, \bar{\rho}_o^{[(1-N_p)/2]} e^{-[(N_p/2)+(1/\langle\bar{\rho}_o\rangle)]\bar{\rho}_o} I_{N_p-1}(\sqrt{2N_p V_o \bar{\rho}_o}) \tag{5.102a}$$

$$[p(V_o/s)]_{\text{Swerling 3}} = 4\left(\frac{2V_o}{N_p}\right)^{(N_p-1)/2} \frac{e^{-V_o}}{\langle\bar{\rho}_o\rangle^2}$$

$$\times \int_0^\infty d\bar{\rho}_o\, \bar{\rho}_o^{[(3-N_p)/2]-[(N_p/2)+(2/\langle\bar{\rho}_o\rangle)]\bar{\rho}_o} I_{N_p-1}(\sqrt{2N_p V_o \bar{\rho}_o}) \tag{5.102b}$$

If (5.102a,b) are considered integrals over the variable $\sqrt{\bar{\rho}_o}$, they have the form amenable to the use of (5.37), with $\nu = 2N_p - 1$ in general, $\mu = (3 - N_p)/2$, $p^2 = (N_p/2) + (1/\langle\bar{\rho}_o\rangle)$ in (5.102a), and $\mu = (5 - N_p)/2$, $p^2 = (N_p/2 + (2/\langle\bar{\rho}_o\rangle)$ in (5.102b).

The detection probability

$$P_d = \int_{V_{oT}}^\infty dV_o\, p(V_o/s) \tag{5.103}$$

can be determined in both cases, i.e., by specification of the integrand in (5.103) from (5.102a) or (5.102b).

For Swerling 2 or 4, the mathematical problem of calculating P_d is quite different. In those cases, the averaging over amplitude is done for each pulse, since the signal amplitudes are statistically independent from pulse-to-pulse.

To determine the PDF of the sum of square amplitudes, we return to (5.37c)′ for Swerling 2 and (5.41c)′ for Swerling 4, from which we obtain the PDF of the individual pulse voltage with the signal amplitude already averaged out, assuming a Rayleigh PDF in (5.37c)′ and a one-dominant-plus Rayleigh in (5.41c)′. To obtain the PDF of the sum of amplitude square in the Swerling 2 (Rayleigh) case, we first determine the PDF of v^2 for each pulse. From (5.37c)′, where v_k = voltage of the kth pulse

$$p[v_k^2/s(a_s)] = \frac{p[v_k/s(a_s)]}{2v_k} = \frac{1}{2(\sigma_s^2 + \sigma_n^2)} e^{-v_k^2/2(\sigma_s^2+\sigma_n^2)} \tag{5.104}$$

The characteristic function for the kth pulse squared voltage normalized to $2\sigma_n^2[u_k = (v_k^2/2\sigma_n^2)]$ is

$$\chi_k[\xi/s(a_s)] = \int_0^\infty du_k\, e^{-[u_k/(1+\rho_s)]+j\xi u_k(1+\rho_s)} \tag{5.105}$$

$$= \frac{1}{1 - j\xi(1+\rho_s)} \quad \text{where} \quad \rho_s = \frac{\sigma_s^2}{\sigma_n^2}$$

The overall characteristic function is

$$\chi[\xi/s(a_s)] = \prod_{k=1}^{N_p} \frac{1}{[1 - j\xi(1+\rho_s)]^k} = \frac{1}{[1 - j\xi(1+\rho_s)]^{N_p}} \tag{5.106}$$

The PDF of the sum $V_o = \Sigma_{k=1}^{N_p} (v_k^2/2\sigma_n^2)$ is

$$p[V_o/s(a_s)] = \int_{-\infty}^{\infty} d\xi\, e^{-j\xi V} \frac{1}{[1 - j\xi(1+\rho_s)]^{N_p}}$$

$$= \frac{1}{(N_p - 1)!(1+\rho_s)^{N_p}} V_o^{N_p-1} e^{-V_o/(1+\rho_s)} \tag{5.107}$$

(from Campbell and Foster[21] tables or other Fourier transform tables).

The detection probability is

$$P_d = \frac{1}{(N_p - 1)!} \int_{V_{oT}}^{\infty} d\left(\frac{V_o}{1+\rho_s}\right)\left(\frac{V_o}{1+\rho_s}\right)^{N_p-1} e^{-(V_o/(1+\rho_s))}$$

$$= 1 - I\left[\frac{V_{oT}}{\sqrt{N_p}(1+\rho_s)}, N_p - 1\right] \tag{5.108}$$

where $I(a,b)$ = incomplete gamma function

$$= \int_0^{a\sqrt{1+b}} dx \frac{e^{-x}x^2}{b!} \tag{5.109}$$

A similar procedure will produce a more complicated result for the Swerling 4 model. The results for Swerling 1 and 3, as obtained through (5.102a,b) and (5.103), are given in DiFranco and Rubin,[1] as are the results for Swerling 2 [equivalent to (5.107) and (5.108)] and for Swerling 4, obtained by the procedure analogous to Eqs. (5.104) through (5.108) for a one-dominant-plus Rayleigh PDF. Some of these results are also given in Eaves and Reedy,[3] Meyer and Mayer,[2] and in various other references previously cited in this chapter.

5.9 SUMMARY OF RADAR DETECTION HIGHLIGHTS

5.9.1 Optimization of Detection: Summary

The assumption made in our detection theory discussions is that the radar target signal is corrupted by bandlimited white Gaussian noise. The theory shows that the optimal processing scheme is to compute the "likelihood ratio," given in general by

$$\overline{L}(\mathbf{v}) = \int_{\boldsymbol{\beta}\text{-space}} \cdots \int d^M \boldsymbol{\beta} L[\mathbf{v}/s(\boldsymbol{\beta})]p(\boldsymbol{\beta}) \tag{5.110}$$

where $\boldsymbol{\beta} = (\beta_1, \ldots, \beta_M)$ is a set of target signal parameters that may be partially deterministic but are usually partially random, where $p(\boldsymbol{\beta})$ is the joint PDF of the parameters $\boldsymbol{\beta}$, $\mathbf{v} = (v_1, v_2, \ldots, v_N)$ = vector of sample values, and

$$L[\mathbf{v}/s(\boldsymbol{\beta})] = \frac{p[v/s(\boldsymbol{\beta})]}{p(\mathbf{v}/n)} =$$

likelihood ratio for a *fixed* set of parameters $\boldsymbol{\beta}$, where $p[\mathbf{v}/s(\boldsymbol{\beta})]$ = joint conditional PDF of $(v_1, v_2, \ldots, v_N)$ if a target signal with parameter values $\boldsymbol{\beta}$ is present and $p(\mathbf{v}/n)$ is that PDF with noise alone.

For the assumptions made, namely statistically independent narrowband signal-plus-noise samples with bandlimited white additive Gaussian noise, it was shown that for a signal waveform whose parameters $\boldsymbol{\beta}$ (amplitude, phase, waveshape, etc.) are a priori known

$$L[\mathbf{v}/s(\boldsymbol{\beta})] = e^{-(B_N a_s^2/2\sigma_n^2)\int_{-\infty}^{\infty} dt\, \cos^2(\omega_o t + \psi_s)p^2(t)} \cdots$$
$$e^{(B_N a_s/\sigma_n^2)\int_{-\infty}^{\infty} dt\, v(t)\cos(\omega_o t + \psi_s)p(t)} \tag{5.111}$$

where the signal waveshape is assumed to be

$$s(t) = a_s \cos(\omega_o t + \psi_s)p(t) \tag{5.112}$$

and $p(t)$ is the amplitude modulation waveform (e.g., pulse or train of pulses).

Equation (5.111) is the basis of the conclusion that the optimum processing scheme is coherent cross-correlation of the incoming waveform $v(t)$ with a replica of the expected signal waveform.

Under the usual assumption that $p(t)$ is nearly constant over many cycles of the IF ω_o, (5.111) reduces to

$$L(\mathbf{v}/a_s, \psi_s) = e^{-(B_N a_s^2/4\sigma_n^2)T_{\text{eff}}} \cdots \tag{5.113}$$
$$e^{(B_N a_s/\sigma_n^2)V_o \cos[\psi_s - \tan^{-1}(V_{os}/V_{oc})]}$$

where the parameters β_1, β_2 are now specified as signal amplitude a_s and phase ψ_s, where T_{eff} = "effective signal duration" =

$$\int_{-\infty}^{\infty} dt\, p^2(t)$$

and where

$$V_{oc} = \int_0^T dt\, v(t)p(t) \cos \omega_o t = \text{“cosine filter” output}$$

$$V_{os} = \int_0^T dt\, v(t)p(t) \sin \omega_o t = \text{“sine filter” output}$$

$$V_o = \sqrt{V_{oc}^2 + V_{os}^2}$$

If the signal phase ψ_s is a priori *unknown,* but the signal amplitude a_{so} is a priori *known,* then the averaging (5.113) over the random phase distribution of ψ_s results in

$$L[\mathbf{v}/s(a_s)] = e^{-(B_N a_s^2/4\sigma_n^2)T_{eff}} I_o\left[\frac{B_N a_s}{\sigma_n^2} V_o\right] \tag{5.114}$$

Equation (5.114) is a basis for use of quadrature filtering in the case of unknown phase, or “noncoherent detection.” Since $I_o(x)$ is monotonically increasing with x, it tells us that we should compute and set a threshold on a quantity proportional to V_o, or V_o^2, the latter being the sum of squares of cosine and sine filter outputs and the former being its square root, or equivalently the amplitude of the output of a coherent cross-correlator.

5.9.2 Coherent Detection: Summary

For coherent detection, the detection and false alarm probabilities are (with no signal amplitude fluctuations) [Eqs. (5. 15a,b)]

$$P_d = \frac{1}{2}[1 - \text{erf}(\hat{v}_T - u)] \tag{5.115a}$$

$$P_{fa} = \frac{1}{2}[1 - \text{erf}(\hat{v}_T)] \tag{5.115b}$$

where

$u = \hat{v}/\sqrt{2}\sigma_n = (1/\sqrt{2})\sqrt{\rho};\ \rho = v^2/\sigma_n^2 =$ peak SNR-power units
$v_T = vT/\sqrt{2}\sigma_n =$ normalized threshold voltage
$v_T =$ actual signal-plus-noise voltage
$\sigma_n =$ rms noise in volts

With coherent detection, it is assumed that all signal parameters including phase are exactly known, so (5.115a,b) apply equally well to single pulses or pulse trains, provided the parameters are exactly known and remain stable throughout the signal duration. Therefore fluctuations in a_s are *not* included in the theory.

We can solve (5.115b) for $\hat{v}_T$ as a function of P_{fa} and substitute it into (5.115a), with the result

$$P_d = \frac{1}{2}\{1 - \text{erf}[\text{erf}^{-1}(1 - 2P_{fa}) - u]\} \tag{5.116}$$

which gives us P_d for any given value of P_{fa}. That is the basis of Figure 5.12, from which we can find the required SNR to attain a given detection probability for a specified false alarm rate.

Optimization of coherent detection could be accomplished through computation of the output of a coherent cross-correlator as in (5.111) and the setting of a threshold $\hat{v}_T$ on that quantity. In that case

$$\hat{v}_T = \frac{v_{oT}}{\sqrt{2}\sigma_{no}} \tag{5.117a}$$

$$u = \frac{v_o}{\sqrt{2}\sigma_{no}} = \frac{1}{\sqrt{2}}\rho_o^{(o)} \tag{5.117b}$$

where

v_o = output voltage of coherent cross-correlator or equivalently a matched filter
σ_{no} = rms noise output of a matched filter
v_{oT} = threshold on matched filter output
$\rho_o^{(o)}$ = SNR of coherent matched filter output = E_s/N_o

5.9.3 Noncoherent Detection, Single Pulse, No Amplitude Fluctuations: Summary

In this case [see (5.28a,b)]

$$P_d = Q(\hat{v}_T, \eta) \tag{5.118a}$$

$$P_{fa} = e^{-\hat{v}_T^2} \tag{5.118b}$$

where $\hat{v}_T = v_T/\sigma_n\sqrt{2}$, $\eta = \sqrt{2}(a_s/\sigma_n) = \sqrt{2}\sqrt{\rho}$, ρ = peak SNR = a_s^2/σ_n^2, and where

$$Q(v_T, \eta) = \text{"Marcum } Q\text{-function"}$$

$$= \int_{\hat{v}_T}^{\infty} dx\, x e^{-(x^2+\eta^2)/2} I_o(x\eta) \tag{5.119}$$

Solving (5.118b) for $\hat{v}_T$ in terms of P_{fa}, we can rewrite (5.118a) in the form [counterpart of (5.116)]

$$P_d = Q(\sqrt{-\ln P_{fa}}, \eta) \tag{5.120}$$

The curves of Figures 5.13 are based on 5.120.

As in the coherent case, optimization theory instructs us to matched-filter the signal-plus-noise before making the detection decision but deviates from the coherent case in adding a further instruction to use the *amplitude* of the matched filter output as the basis for the decision, or equivalently to use the "quadrature filtering" technique. If those instructions are followed, then

$$\eta = \sqrt{2}\,\sqrt{\rho_o^{(0)}},\ \rho_o^{(0)} = \text{output SNR of matched filter} \tag{5.121a}$$

$$\sigma_n = \sigma_{no} = \text{rms noise output of matched filter} \tag{5.121b}$$

$$v_T = v_{To} = \text{threshold on matched filter output voltage} \tag{5.121c}$$

5.9.4 Noncoherent Detection, Single Pulse, with Amplitude Fluctuations: Summary

In this case, it is noted that the probability for noncoherent detection of a single pulse, as given by (5.118a), is proportional to a_s, the signal amplitude. If a_s is a random variable, then η is also a random variable and the detection probability must include the process of averaging over the distribution of η. Equation (5.118a) would then be revised and would take the form

$$P_d = \int_0^\infty d\eta\, p(\eta) Q(\overline{v}_T, \eta) \tag{5.122}$$

where $p(\eta)$ is the PDF of η.

Two PDFs were covered, the Rayleigh PDF and the one-dominant-plus Rayleigh. In terms of the variable η in (5.118a), these are:

Rayleigh [Eq. (5.35)]

$$p(\eta) = \frac{2\eta}{\langle \eta^2 \rangle} e^{-\eta^2/\langle \eta^2 \rangle} \tag{5.123}$$

In terms of the RCS σ, proportional to the square of η, (5.123) becomes the exponential distribution [Eq. (5.99b), where a_s^2 is proportional to σ].

$$p(\sigma) = \frac{1}{\langle \sigma \rangle} e^{-\sigma/\langle \sigma \rangle} \quad \text{or} \quad \frac{1}{\langle \overline{\rho}_o \rangle} e^{-\overline{\rho}_o/\langle \overline{\rho}_o \rangle} \tag{5.124}$$

One-dominant-plus Rayleigh [Eqs. (5.40) and (5.44)]

$$p(\eta) = \frac{8\eta^2}{\langle \eta^2 \rangle^2} e^{-2\eta^2/\langle \eta^2 \rangle} \tag{5.125}$$

The corresponding form of the PDF of σ is [Eq. (5.100b), where a_s^2 is proportional to σ]

$$p(\sigma) = \frac{4\sigma}{\langle \sigma \rangle^2} e^{-2\sigma/\langle \sigma \rangle} \tag{5.126}$$

Detection probabilities for a single pulse with distributions given by (5.123) or (5.125) are computed by substituting (5.123) or (5.125) into (5.122). The results are: Rayleigh case [Eq. (5.38)]:

$$P_d = e^{-\hat{v}_T^2/(1+\rho_s)} \tag{5.127}$$

where $\rho_s = \langle a_s^2 \rangle / 2\sigma_n^2$ and $\hat{v}_T = \sqrt{-\ln P_{fa}}$. The detection curves of Figure 5.16 are based on 5.127. One-dominant-plus Rayleigh case [Eq. (5.45)]:

$$P_d = \frac{e^{-\hat{v}_T^2/[1+(\rho_s/4)]}}{[1 + (\rho_s/4)]} \left[1 + \frac{\rho_s}{4} + \frac{\hat{v}_T}{1 + (4/\rho_s)} \right] \tag{5.128}$$

where $\rho_s = \langle a_s^2 \rangle / 2\sigma_n^2$ and $\hat{v}_T = \sqrt{-\ln P_{fa}}$

The detection curves of Figure 5.17 are based on (5.128).

5.9.5 Noncoherent Detection of Pulse Trains, No Amplitude Fluctuations: Summary

Under the assumption of statistical independence of pulses (nearly always valid with radar pulses), a near-optimal detection strategy against bandlimited white Gaussian noise is to set the decision threshold on the sum

$$V_o = \sum_{l=1}^{N_p} K_l a_{vol}^2,\ N_p = \text{number of pulses} \tag{5.129}$$

where a_{vol} is the matched filter output amplitude for the lth pulse and K_l is a proportionality constant to allow for the possibility that signal amplitudes are different for each pulse. In practice one generally assumes K_l = constant K and simply sums the squared amplitudes, i.e.,

$$V_o = K \sum_{l=1}^{N_p} a_{vol}^2 \text{ [see Eq. (5.83)]} \tag{5.129$'$}$$

or the absolute amplitudes. The theory [Eqs. (5.82a)′ and (5.82b)′] tells us that the sum of amplitude-squares (quadratic rectification of each pulse) is optimal for low SNR and the sum of amplitudes (linear rectification of each pulse) is optimal for high SNR. But results exist that show very little difference between the two in practice and hence the quadratic rectification option is chosen for the analysis because it is more easily tractable.

The detection and false alarm probabilities P_d and P_{fa} are (Eqs. 5.91a,b) [where K in (5.129)′ is chosen as $1/2\sigma_n^2$]

$$P_d = e^{-N_p\bar{\rho}_o/2} \int_{V_{oT}}^{\infty} dV_o \left(\frac{2V_o}{N_p\bar{\rho}_o} \right)^{(N_p-1)/2} I_{N_p-1}(\sqrt{2N_p\bar{\rho}_o V_o}) e^{-V_o} \tag{5.130a}$$

(evaluated numerically)

$$P_{fa} = \frac{1}{(N_p - 1)!} e^{-V_{oT}} (V_{oT})^{N_p - 1} \left[1 + \frac{(N_p - 1)}{V_{oT}} + \frac{(N_p - 1)(N_p - 2)}{V_{oT}^2} + \cdots \right] \tag{5.130b}$$

where the threshold V_{oT} is already normalized to $2\sigma_{no}^2$ (σ_{no}^2 being the matched filter output rms noise for a single pulse-same for all pulses) and $\bar{\rho}_o$ being the single pulse SNR time averaged over the pulse train (in case the signal amplitude varies deterministically over the pulse train, which it may well do because of antenna scanning). This "averaging" is *not* over an ensemble of random variables; it is strictly deterministic time averaging.

The detection curves of Figure 5.24a–j are based on Eqs. (5.130a,b). These curves demonstrate the phenomenon of "*integration loss,*" defined by [see Eqs. (5.92) and (5.93)]

$$L_{dB} = 10 \log_{10}\left(\frac{N_p \bar{\rho}_o^{(N_p - \text{ncoh})}}{(\bar{\rho}_o)^{(\text{coh})}} \right) \tag{5.131}$$

where $\bar{\rho}_o^{(N_p - \text{ncoh})}$ is the required single pulse SNR with *noncoherent* integration of N_p pulses, $(\bar{\rho}_o)^{(\text{coh})}/N_p$ is the required single-pulse SNR with *coherent* integration of N_p pulses (followed by quadrature filtering on the pulse train output), and "required" means required for a specified performance level, e.g., $P_d = 95\%$, $P_{fa} = 10^{-4}$.

As a general rule, it is found that for sufficiently large values of N_p [Eq. (5.94)]

$$L_{dB} = 10 \log_{10}\sqrt{N_p} - 5.5 \tag{5.132}$$

Equation (5.132) is useful for making rough estimates of integration loss.

5.9.6 Noncoherent Detection of Pulse Trains with Amplitude Fluctuations: Swerling Models: Summary

In this case there are two limiting situations that are analytically tractable.

1. The amplitude fluctuations are slow compared with the duration of the pulse train. Equivalently the correlation time of the fluctuations is very large compared with the duration of the pulse train, *or* also equivalently, the fluctuation bandwidth is small compared with the bandwidth of the pulse train. This assumption allows the amplitude of the pulse train to be treated as a *single* random variable. The averaging over signal amplitude is done on the entire pulse train. This limiting case is known as Swerling 1 if the amplitude statistics are Rayleigh and Swerling 3 if they are one-dominant-plus Rayleigh.
2. The amplitude fluctuations are statistically independent from pulse-to-pulse. Equivalently the correlation time of the amplitude fluctuations is small compared with the interpulse period *or,* also equivalently, the fluctuation bandwidth is large compared with the PRF. This assumption allows the amplitudes of the individual pulses to be treated as mutually statistically independent random variables. The averaging over amplitude is done for each individual pulse *before* determining the PDF of the sum of squared amplitudes, whereas in the other limiting case (1), the PDF of the sum of amplitude-squares is first calculated as if the amplitude were deterministic, then the average of that PDF is taken over the fluctuation PDF of the amplitude. The limiting case 2 is known as Swerling 2 if the amplitude statistics are Rayleigh and as Swerling 4 if they are one-dominant-plus Rayleigh.

The Swerling 1 or 3 cases are often called "scan-to-scan" fluctuations because, with a scanning antenna, a given number of pulses are integrated while the antenna scans past the target, and it is assumed that the fluctuating amplitude remains constant during the period while the beam is "on target." By the time the next scan comes around, the amplitude has fluctuated to a different value that remains fixed during the "pass" over the target. The assumption justifying use of Swerling 1 or 3 is that the amplitude correlation time, while large compared to the "on target" period of the scanning beam, is small compared to the scanning period (reciprocal of scanning rate), so the amplitude fluctuations can be considered as statistically independent from scan-to-scan.

The Swerling 2 or 4 cases are often called "pulse-to-pulse" fluctuations, because the amplitudes are statistically independent from pulse-to-pulse.

For Swerling 1 or 3, the detection probability is obtained by extending the calculation in (5.130a), including the average over the distribution of amplitude inside the integrand, i.e., the PDF of V_o is

$$[p(V_o/s)]_{\text{fixed } a_s} = e^{-N_p\bar{\rho}_o/2}\left(\frac{2V_o}{N_p\bar{\rho}_o}\right)^{(N_p-1)/2} I_{N_p-1}(\sqrt{2N_p\bar{\rho}_o V_o})e^{-V_o} \tag{5.133a}$$

and the amplitude-averaged form is [Eq. (5.101)]

$$p(V_o/s) = \int_0^\infty da_s\, p(a_s)[p(V_o/s)]_{\text{fixed } a_s} \tag{5.133b}$$

Then

$$P_d = \int_{V_{oT}}^\infty dV_o\, p(V_o/s) \tag{5.133c}$$

For Swerling 1 [Eq. (5.102a)]

$$P_d = \int_{V_{oT}}^\infty dV_o \left(\frac{2V_o}{N_p}\right)^{(N_p-1)/2} \frac{e^{-V_o}}{\langle\bar{\rho}_o\rangle} \times \int_0^\infty d\bar{\rho}_o\, \bar{\rho}^{[(1-N_p)/2]} e^{-[(N_p/2)+(1/\langle\bar{\rho}_o\rangle)]\bar{\rho}_o} I_{N_p-1}(\sqrt{2N_p V_o\bar{\rho}_o}) \tag{5.133d}$$

and for Swerling 3 [Eq. (5.102b)]

$$P_d = \int_{V_{oT}}^\infty dV_o \left(\frac{2V_o}{N_p}\right)^{(N_p-1)/2} \frac{e^{-V_o}}{\langle\bar{\rho}_o\rangle^2} \times \int_0^\infty d\bar{\rho}_o\, \bar{\rho}_o^{[(3-N_p)/2]} e^{-[(N_p/2)+(2/\langle\bar{\rho}_o\rangle)]\bar{\rho}_o} I_{N_p-1}(\sqrt{2N_p V_o\bar{\rho}_o}) \tag{5.133e}$$

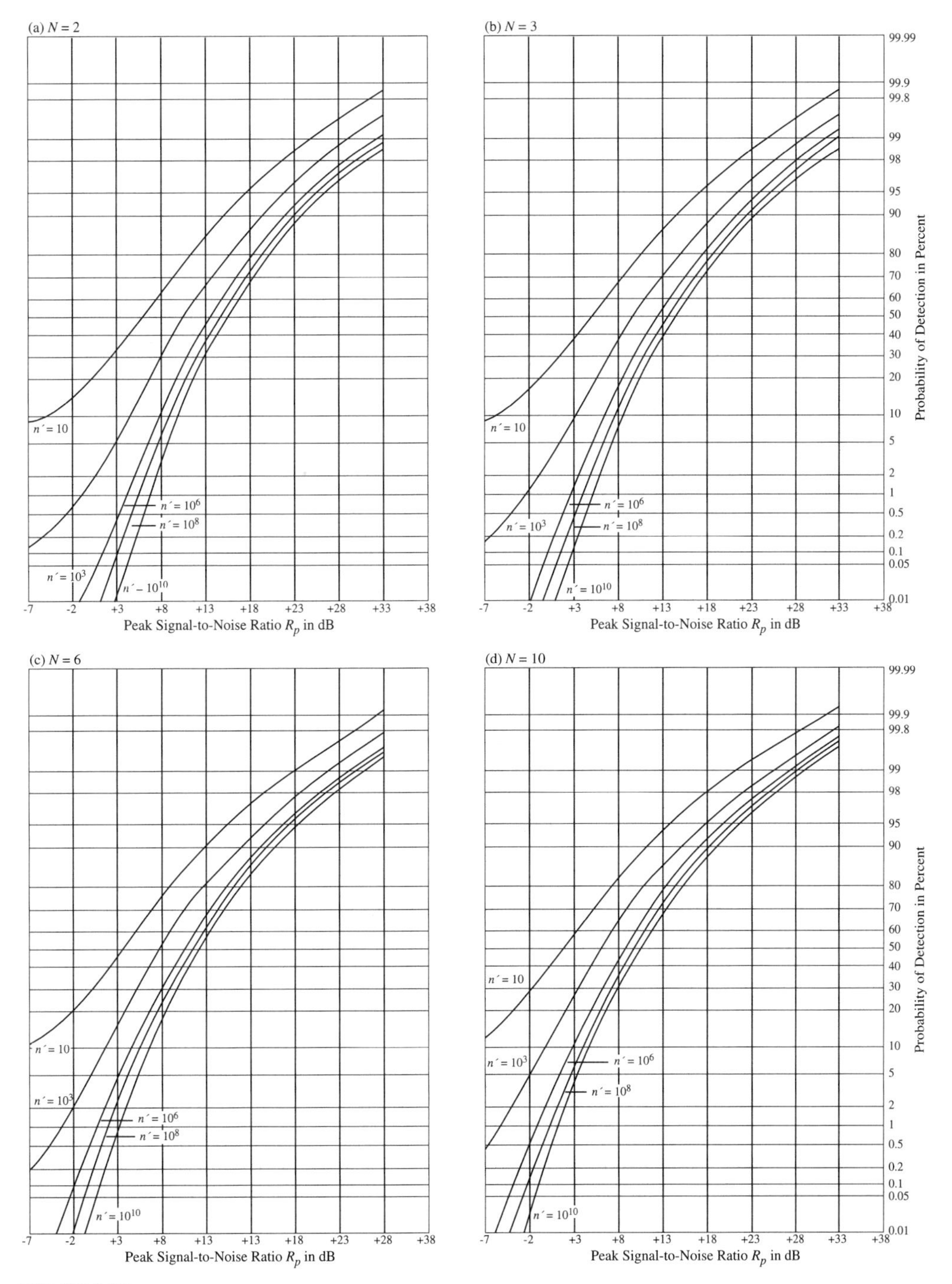

FIGURE 5.27
Detection curves for Swerling 1. From DiFranco and Rubin,[1] "Radar Detection," by permission of the authors.

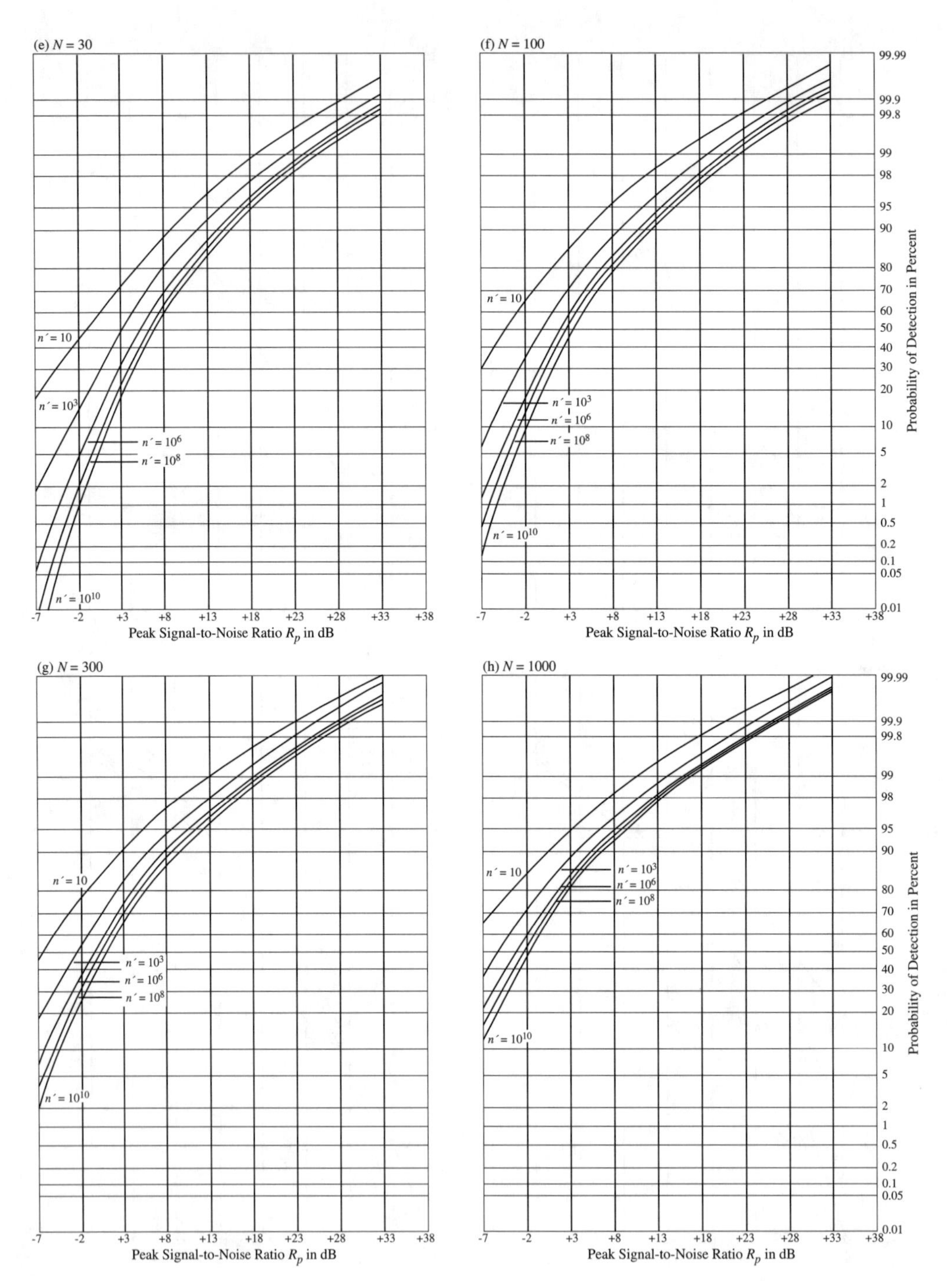

FIGURE 5.27
Continued.

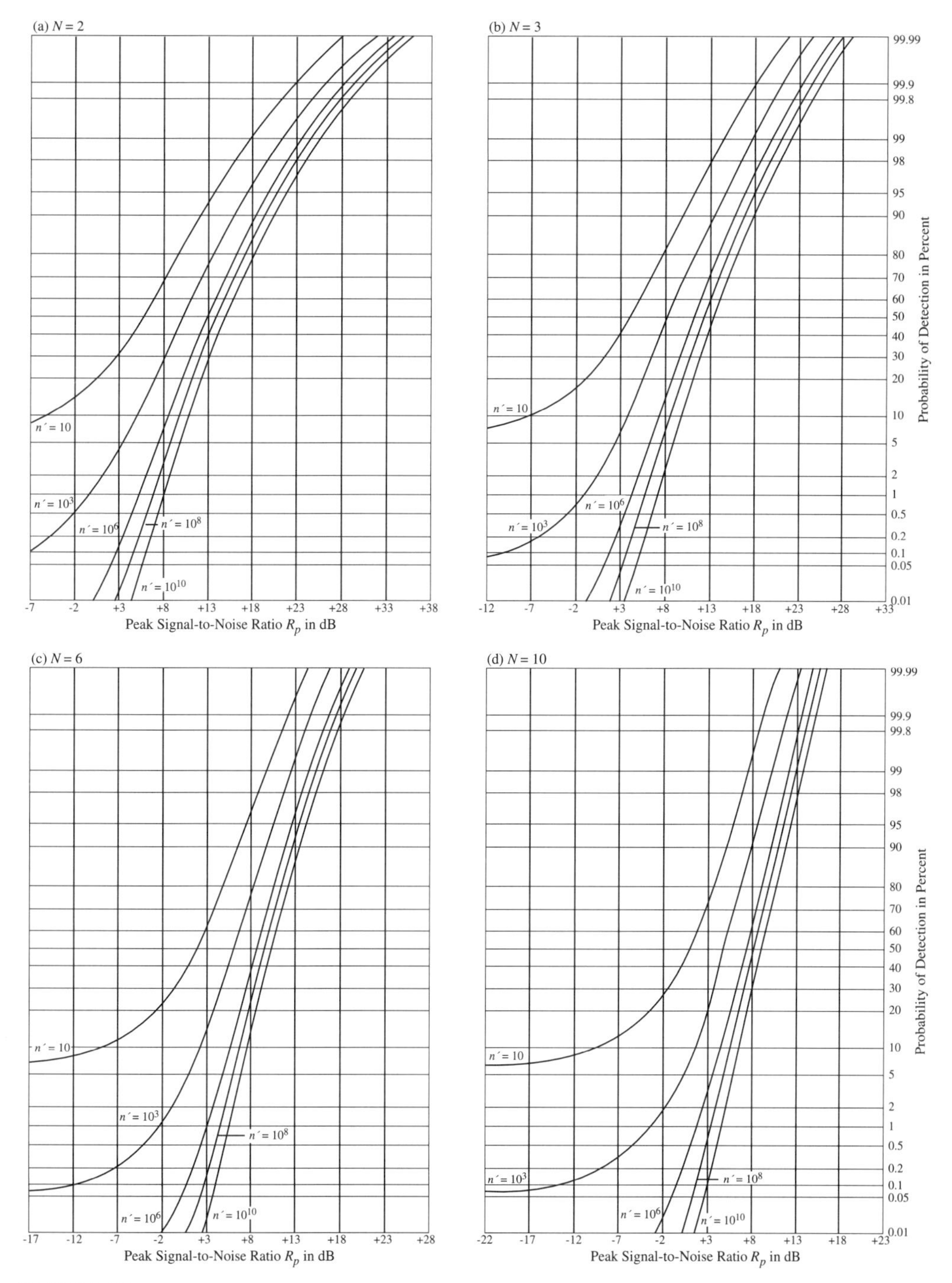

FIGURE 5.28
Detection curves for Swerling 2. From DiFranco and Rubin,[1] "Radar Detection," by permission of the authors.

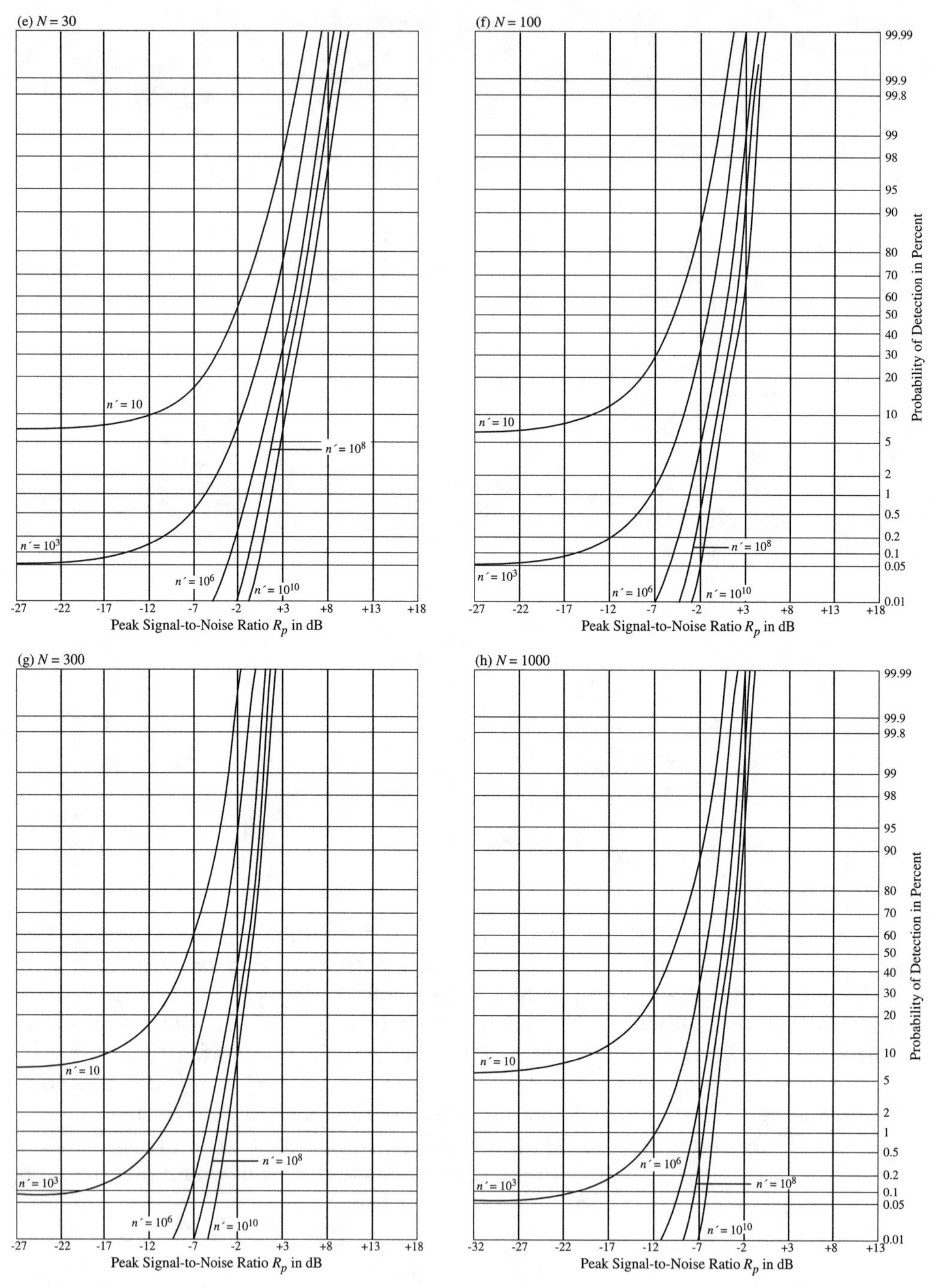

FIGURE 5.28
Continued.

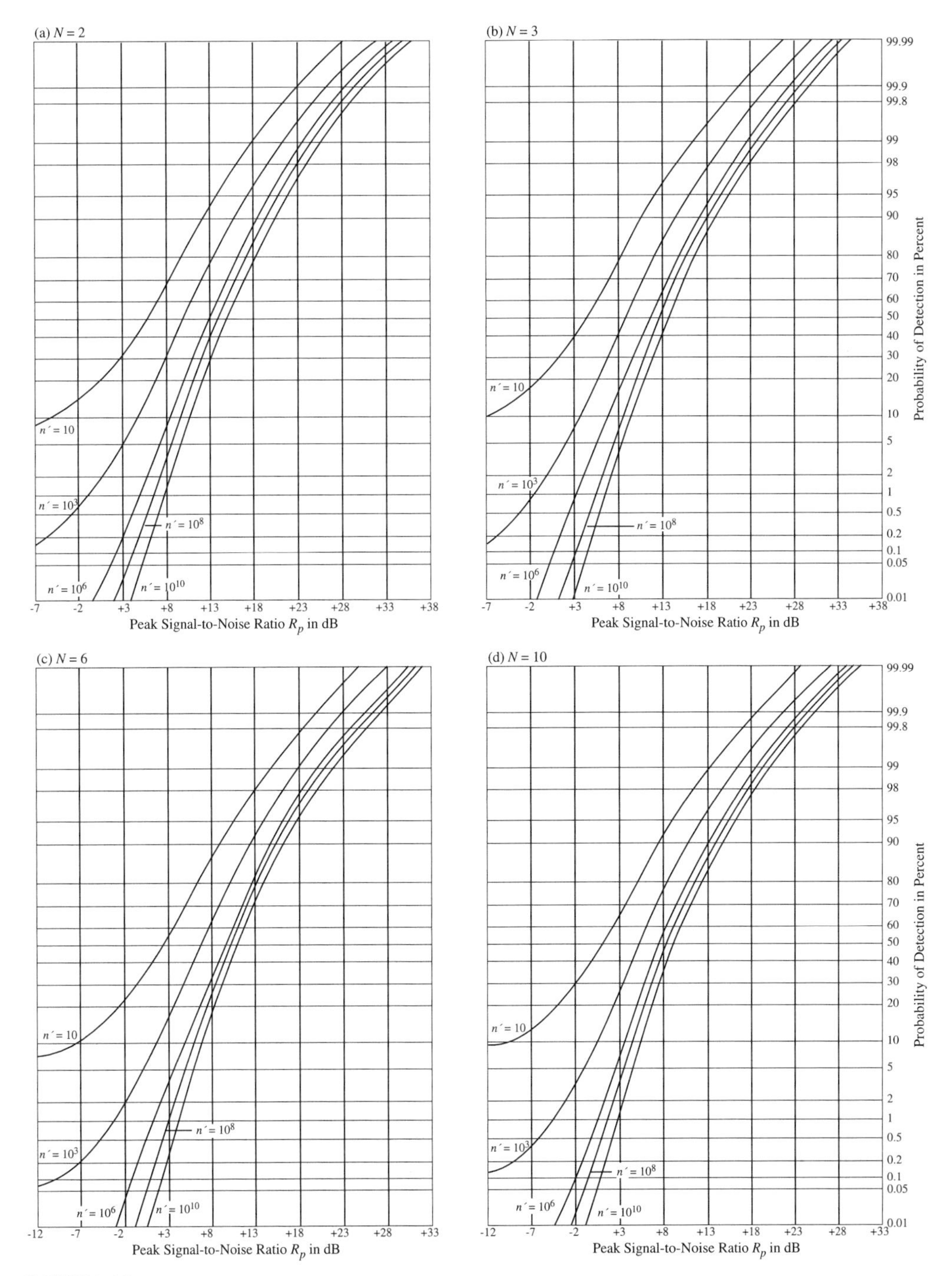

FIGURE 5.29
Detection curves for Swerling 3. From DiFranco and Rubin,[1] "Radar Detection," by permission of the authors.

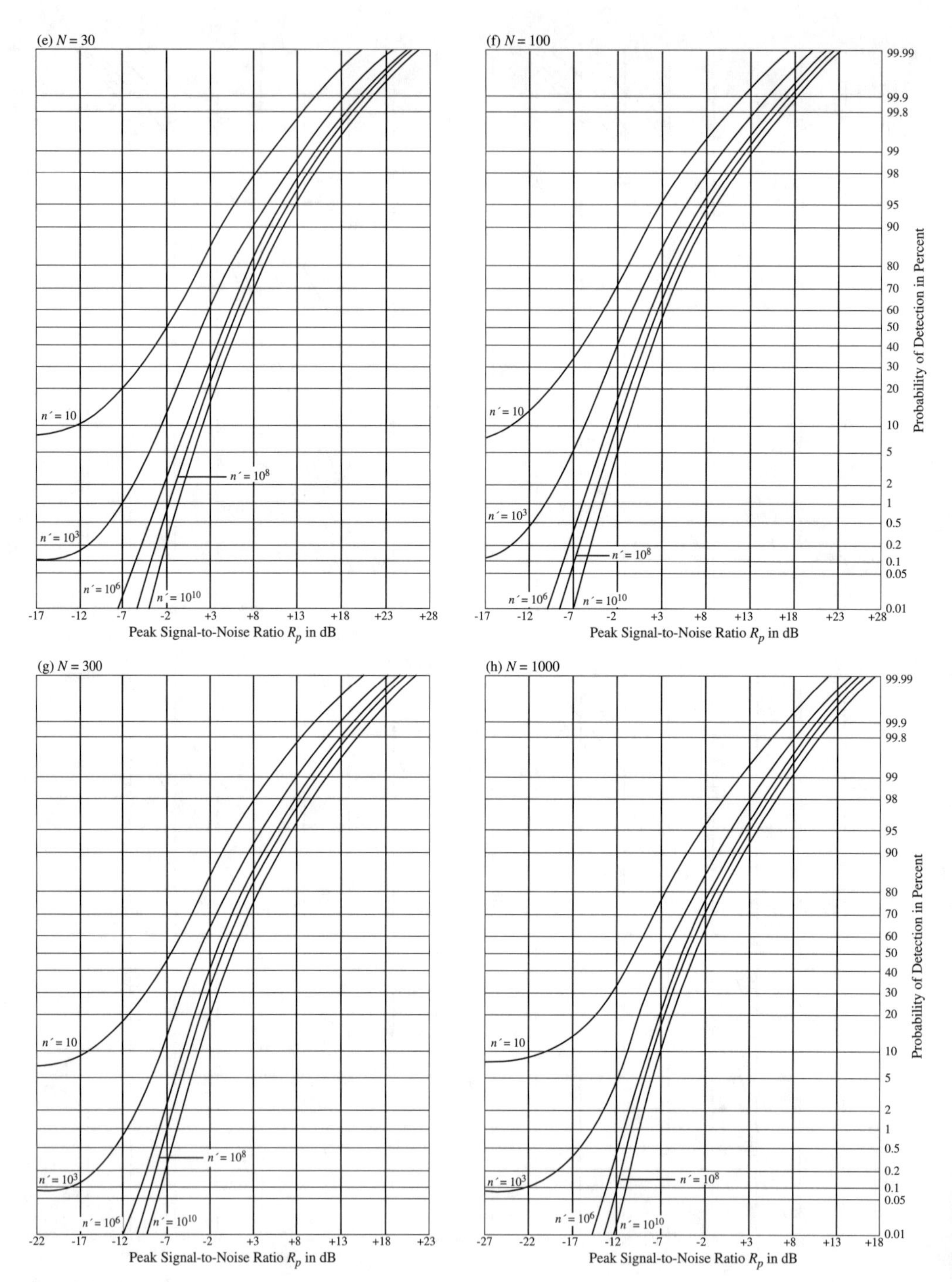

FIGURE 5.29
Continued.

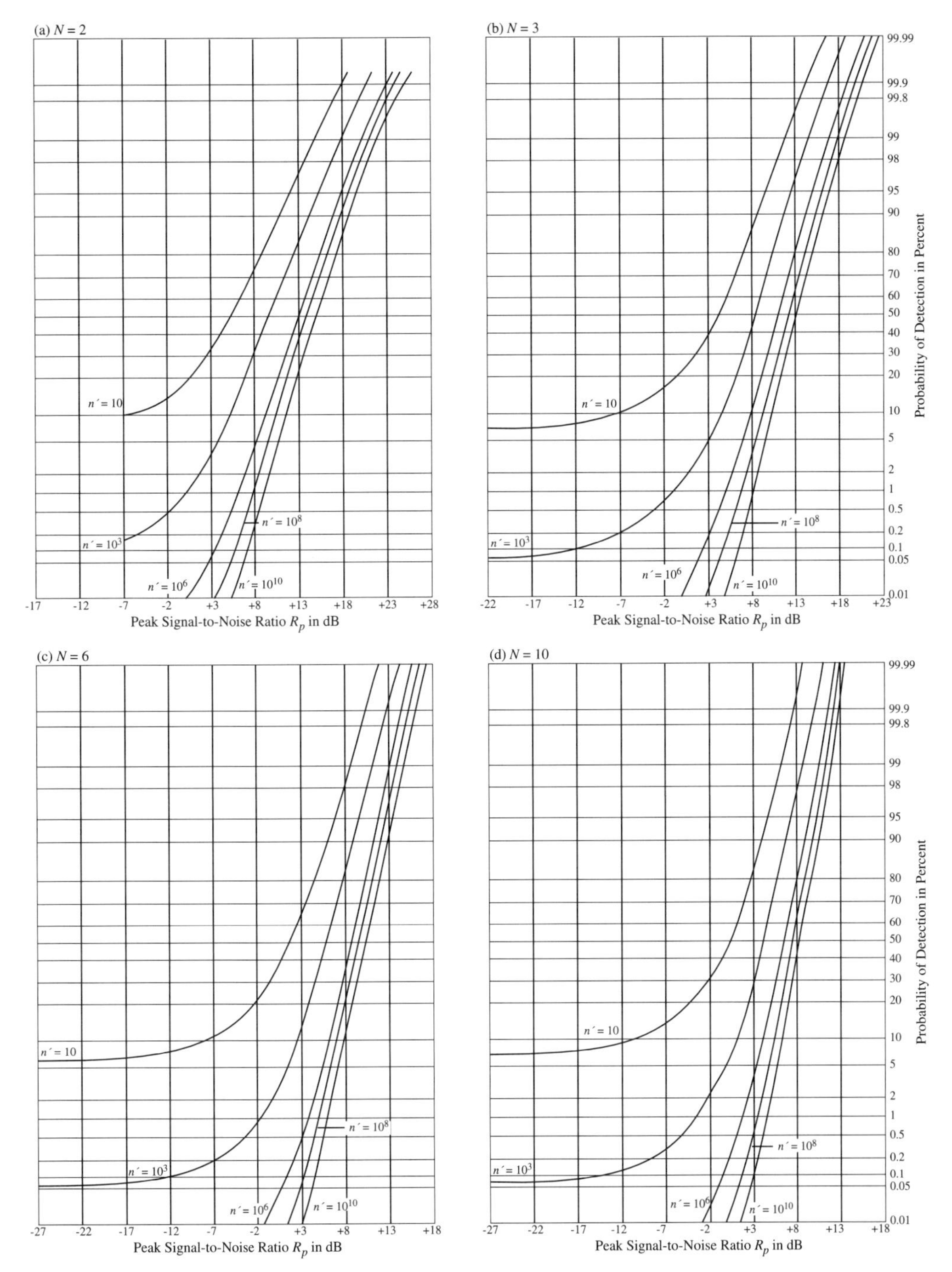

FIGURE 5.30
Detection curves for Swerling 4. From DiFranco and Rubin,[1] "Radar Detection," by permission of the authors.

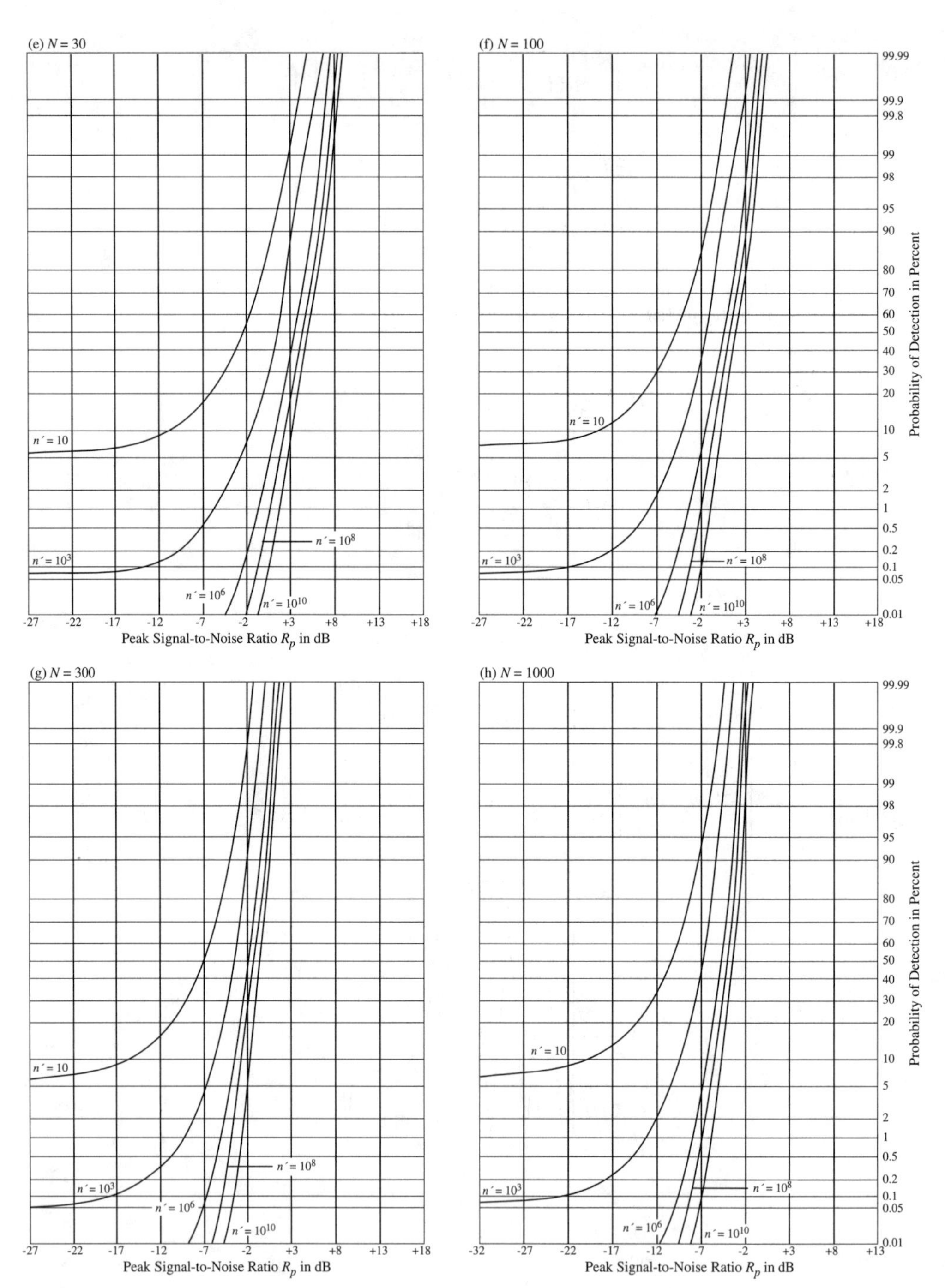

FIGURE 5.30
Continued.

The Swerling 2 or 4 results are obtained by first using the single pulse PDF as in (5.37c)′ or (5.41c)′, i.e., for the kth pulse with Rayleigh statistics

$$p[v_k/s(a_{sk})] = \frac{v_k}{\sigma_{sk}^2 + \sigma_n^2}\, e^{-v_k^2/2(\sigma_n^2+\sigma_{sk}^2)} \tag{5.134a}$$

where $\sigma_{sk}^2 = \langle a_{sk}^2\rangle/2$, and with one-dominant-plus Rayleigh statistics,

$$p[v_k/s(a_{sk})] = \frac{9v_k}{(\hat{\sigma}_s^2 + 3\sigma_n^2)^2}\, e^{-v_k^2/2(3\sigma_n^2+\hat{\sigma}_{sk}^2)} \tag{5.134b}$$

where $\hat{\sigma}_{sk}^2 = \frac{3}{4}\langle a_{sk}^2\rangle$.

The next step is to calculate the PDF of the sum $V_o = K\Sigma_{k=1}^{N_p} v_k^2$ using (5.134a,b) for the single pulse PDF in the Swerling 2 and 4 cases, respectively. The averaging over amplitude has already been done in obtaining (5.134a,b) so the calculation to be performed is

$$P_d = \int_{V_{oT}}^{\infty} dV_o\, p(V_o/s) \tag{5.135}$$

where $V_o = K\Sigma_{k=1}^{N_p} v_k^2$.

The PDF of V_o for Swerling 2 is (Eq.5.107)

$$p(V_o/s) = \frac{1}{(N_p - 1)!(1 + \rho_s)^{N_p}}\, V_o^{N_p-1} e^{-V_o/(1+\rho_s)} \tag{5.136}$$

The integration in (5.135) using the PDF in (5.136) gives [see Eqs. (5.108), (5.109)] for the Swerling 2 case

$$P_d = 1 - I\left[\frac{V_{oT}}{\sqrt{N_p}(1 + \rho_s)}, N_p - 1\right] \tag{5.137}$$

where $I(a,b)$ = incomplete Gamma function =

$$\int_0^{a\sqrt{1+b}} dx\, \frac{e^{-x}x^2}{b!}$$

The counterpart of (5.137) for Swerling 4 is obtained numerically.

The detection curves on the pages immediately following are those for the Swerling models 1, 2, 3, and 4, obtained from the results (5.133d) for Swerling 1, (5.133e) for Swerling 3, (5.137) for Swerling 2, and its counterpart for Swerling 4.

In all cases, the false alarm rates corresponding to the threshold level V_{oT} are obtained from (5.130b), which is obviously unaffected by signal fluctuations because it involves only noise.

The integration loss can be determined graphically from the curves for given numbers of pulses from its definition (5.131).

REFERENCES

1. DiFranco, J.V., and Rubin, W.L., "Radar Detection." Artech, Norwood, MA, 1980.
2. Meyer, D.P., and Mayor, H.A., "Radar Target Detection." Academic Press, New York, 1973.
3. Eaves, J.L., and Reedy, E.K. (eds.), "Principles of Modern Radar." Van Nostrand Reinhold, New York, 1987, Chapter 9, by J.D. Echard.
4. Skolnick, M.I., "Introduction to Radar Systems," 2nd ed. McGraw-Hill, New York, 1980. Chapter 10.
5. Barton, D.K., "Modern Radar System Analysis." Artech, Norwood, MA, 1988, Chapter 2.
6. Raemer, H.R., "Statistical Communication Theory and Applications." Prentice-Hall, Englewood Cliffs, NJ, 1969, Chapters 4 and 7.
7. Rohan, P., "Surveillance Radar Performance Prediction." Peter Peregrinus Ltd, London (UK), 1983. Chapter 3.
8. Blake, L.V., "Radar Range-Performance Analysis." Artech, Norwood, MA, 1986, Chapter 2 and Appendix 2A, by L.W. Brooks.
9. Nathanson, F.E., Reilly, J.P., and Cohen, M.N., "Radar Design Principles," 2nd ed. McGraw-Hill, New York, 1991, Chapter 3, by F.E. Nathanson and J.P. Reilly.
10. Levanon, N., "Radar Principles." Wiley, New York, 1988, Chapter 3.
11. X.Y. Hou, Norinaga, N., and Namekawa, T. "Direct Evaluation of Radar Detection Probabilities." IEEE Trans. AES-23 (4), 418–424, 1987.
12. Schnidman, D.A., "The Calculation of the Probability of Detection and the Generalized Marcum Q-Function." IEEE Trans. IT-35(2), 389–400, 1989.
13. Schnidman, D.A. "Radar Detection Probabilities and Their Calculation." IEEE Trans. AES-31(3), 928–950, 1955.
14. Davenport, W.B., and Root, W.L., "Theory of Random Signals and Noise." McGraw-Hill, New York, 1958.
15. Papoulis, A., "Probability, Random Variables and Stochastic Processes," 2nd ed. McGraw-Hill, New York, 1984.
16. Abramowitz, M., and Stegun, I.A., "Handbook of Mathematical Functions." Dover, New York, 1965.
17. Marcum, J.L., and Swerling, P., "A Statistical Theory of Target Detection by Pulsed Radar." IRE Trans. Prof. Group on Information Theory, IT-6(2), 59–267, 1960.
18. Swerling, P., "More on Detection of Fluctuating Targets." IRE Trans. IT-11(3), 459–460, 1965.
19. Spiegel, M.R., "Mathematical Handbook" (Schaum's Outline Series in Mathematics), McGraw-Hill, New York, 1968.
20. DiFranco, J.V., and Rubin, W.L., "Radar Detection." Artech, Norwood, MA, 1980, Section 3.12, pp. 94–96 and Chapters 9, 10, 11.
21. Campbell, G.A., and Foster, R.M., "Fourier Integrals." Van Nostrand, New York, 1948.

Chapter 6

MODELS FOR TARGET SIGNAL FLUCTUATION STATISTICS

CONTENTS

In Section 5.8 we discussed the effects of target signal fluctuations on detection probabilities. However, in that discussion, the emphasis was on the detection probabilities and the effects of antenna scanning. No mention was made of the mechanisms by which target fluctuations are generated, and no mention was made of amplitude fluctuation statistics other than Rayleigh and one-dominant-plus Rayleigh. Coverage of some of those topics is the purpose of this chapter.

6.1 THE CENTRAL LIMIT THEOREM

The central limit theorem of probability theory is the basis for the Gaussian statistics observed in many random processes encountered in radar. It states in essence that the sum of a large number of statistically independent random variables has a PDF approaching Gaussian as the number of variables becomes arbitrarily large. The beauty of the theorem is that it appears to work very well even if the number of variables summed is not enormously large, e.g., the sum of as few as 10 truly statistically independent random variables has a PDF closely approximated by a Gaussian function *regardless* of the PDFs of the individual variables.

Very few sources on radar contain a derivation of the central limit theorem.[1,2] Since it has consequences important in radar analysis, a derivation following the lines of that in DiFranco and Rubin[2] (Section 3.13, pp. 96–98) is presented below. It is not absolutely rigorous from a mathematician's viewpoint, but is presented in that source as a "plausibility argument."

The characteristic function of a Gaussian PDF is

$$\begin{aligned}\chi(\xi) &= \frac{1}{\sqrt{2\pi\sigma^2}}\int_{-\infty}^{\infty} dv\ e^{j\xi v}e^{-(1/2\sigma^2)(v-\langle v\rangle)^2} \\ &= \frac{e^{j\xi\langle v\rangle}}{\sqrt{2\pi\sigma^2}}\int_{-\infty}^{\infty} dv\ e^{-(1/2\sigma^2)[(v-\langle v\rangle)^2-2j\xi\sigma^2(v-\langle v\rangle)+(j\xi\sigma^2)^2]}\ \ldots \\ & e^{(j\xi\sigma^2)^2/2\sigma^2} = e^{j\xi\langle v\rangle}e^{-\xi^2\sigma^2/2}\end{aligned} \tag{6.1}$$

(through completion of the square in the exponent) where $\sigma^2 = \langle (v - \langle v\rangle)^2\rangle$ = variance of v.

The derivation is directed toward demonstrating that the PDF of a sum of statistically independent random variables has a characteristic function of the form of (6.1).

Consider N variables $(v_1, v_2, \ldots, v_N)$ that are mutually statistically independent. The sum of these variables is denoted by w and given by

$$w = \sum_{k=1}^{N} v_k \tag{6.2}$$

We define a "standardized sum variable" with zero-mean and unit variance, as follows:

$$W_s = \frac{w - \langle w\rangle}{\sqrt{\langle (w - \langle w\rangle)^2\rangle}} = \frac{\sum_{k=1}^{N}(v_k - \langle v_k\rangle)}{\sigma_w} \tag{6.3}$$

where

$$\langle w\rangle = \sum_{k=1}^{N}\langle v_k\rangle = \text{mean of } w$$

or equivalently

$$\langle W_s\rangle = 0$$

and the variance of W is

$$\langle [W_s - \langle W_s \rangle]^2 \rangle = \langle W_s^2 \rangle = \left\langle \left[\left(\frac{w - \langle w \rangle}{\sigma_w} \right) \right]^2 \right\rangle = 1$$

The characteristic function of W_s is

$$\begin{aligned} \chi(\xi/W_s) &= \int \cdots \int dv_1 \cdots dv_N e^{j\xi W_s} \prod_{k=1}^{N} p(v_k) \\ &= \prod_{k=1}^{N} \int_{-\infty}^{\infty} dv_k \, p(v_k) e^{[j\xi(v_k - \langle v_k \rangle)]/\sigma_w} \\ &= e^{-j\xi \Sigma_k \langle v_k \rangle / \sigma_w} \prod_{k=1}^{N} \int_{-\infty}^{\infty} dv_k \, p(v_k) e^{j\xi v_k/\sigma_w} \\ &= e^{-j\xi \langle w \rangle / \sigma_w} \left[\chi\left(\frac{\xi}{\sigma_w} \middle/ v \right) \right]^N \end{aligned} \tag{6.4}$$

on the assumption that all the variates v_k have the same PDF (and therefore the same characteristic function $\chi[(\xi/\sigma_w)/v]$).

We now assume that $\chi[(\xi/\sigma_w)/v]$ can be expanded in a power series, i.e.,

$$\begin{aligned} \chi\left(\frac{\xi}{\sigma_w} \middle/ v \right) &= \int_{-\infty}^{\infty} dv \, p(v) e^{j(\xi/\sigma_w)v} \\ &= 1 + \frac{j\xi \langle v \rangle}{\sigma_w} - \frac{\xi^2 \langle v^2 \rangle}{2!\sigma_w^2} - \frac{j\xi^3 \langle v^3 \rangle}{3!\sigma_w^3} + \frac{\xi^4 \langle v^4 \rangle}{4!\sigma_w^4} + \cdots \end{aligned} \tag{6.5}$$

To simplify the discussion, zero mean is assumed and the variance is denoted by σ_v; thus (6.5) takes the form

$$\chi\left(\frac{\xi}{\sigma_w} \middle/ v \right) = 1 - \frac{\xi^2 \sigma_v^2}{2\sigma_w^2} - \frac{j\xi^3 M_3}{6\sigma_w^3} + \frac{\xi^4 M_4}{24\sigma_w^4} + \cdots \tag{6.6}$$

where M_k is the kth moment of v, equal to $\langle v^k \rangle$.

Substituting (6.6) into (6.4), we obtain

$$\chi(\xi/W_s) = \left(1 - \frac{\xi^2 \sigma_v^2}{2\sigma_w^2} - \frac{j\xi^3 M_3}{6\sigma_w^3} + \frac{\xi^4 M_4}{24\sigma_w^4} + \cdots \right)^N \tag{6.7}$$

where the assumption $\langle v_k \rangle = 0$ has removed the factor $e^{-j\xi \langle w \rangle / \sigma_w}$ in (6.4).

We now note that (due to the assumption that the mean of each variate is zero and the variances of the variates are all the same and equal to σ_v^2) Eq. (6.2) gives us the result

$$\sigma_w^2 = \langle w^2 \rangle = \sum_{k=1}^{N} \langle v^2 \rangle + \sum_{\substack{j,k=1\\ j\neq k}}^{N}{}' \langle v_j v_k \rangle = \sum_{k=1}^{N} \langle v^2 \rangle = N\sigma_v^2 \tag{6.8}$$

(since the variables are uncorrelated, the cross terms in the second sum all vanish).

Since (6.8) implies that $\sigma_v = \sigma_w/\sqrt{N}$, it follows from (6.7) that

$$\chi(\xi/W_s) = \left(1 - \frac{\xi^2}{2N} - \frac{j\xi^3 M_3}{6N^{3/2}\sigma_v^3} + \frac{\xi^4 M_4}{24N^2\sigma_v^4} \cdots\right)^N \tag{6.9}$$

As N becomes very large, the third term and beyond approach zero more rapidly than the second and $\chi(\xi/W_s)$ approaches the limit

$$[\chi(\xi/W_s)]_{N\to\infty} = \lim_{N\to\infty}\left(1 - \frac{\xi^2}{2N}\right)^N = e^{-\xi^2/2} \tag{6.10}$$

The PDF corresponding to (6.10) is

$$\begin{aligned} p(W_s) &= \frac{1}{2\pi}\int_{-\infty}^{\infty} d\xi\, e^{-(\xi^2/2)-j\xi W_s} = \frac{1}{2\pi}\int_{-\infty}^{\infty} d\xi\, e^{-1/2[\xi^2+(2jW_s)\xi]} \\ &= \frac{1}{2\pi}\int_{-\infty}^{\infty} d\xi\, e^{-1/2[\xi^2+2(jW_s)\xi+(jW_s)^2]} e^{+(jW_s)^2/2} \\ &= \frac{1}{\sqrt{2\pi}} e^{-W_s^2/2} \end{aligned} \tag{6.11}$$

(which is the Gaussian PDF with unit variance and zero mean).

The argument above contains assumptions about zero mean for each variate and the same PDF for each variate. However, it can be generalized to include variations in PDF and nonzero means, but at a considerable cost in mathematical complexity. From a general philosophical viewpoint, all we need from this demonstration is the general idea that a sum of statistically independent random variables has a PDF close to Gaussian if there are enough of them. That is used in qualitative arguments that support Gaussian statistics for a wide variety of target or clutter situations.

6.2 RUDIMENTARY TARGET SCATTERING CONCEPTS

Consider a typical radar target, e.g., an aircraft. Over much of the microwave spectrum, particularly the high end of it, where wavelengths are of the order of 30 cm or less (from L band through the millimeter bands) the target spans many wavelengths. Scattering theory in its very general form shows the electric or magnetic fields of a wave scattered from a target as a complicated vector integral over the currents induced on the surface of the target by the wave from the radar transmitter that impinges on the target. This integral contains the relative phases of the induced currents at every point on the surface of the target. Also, if the target spans many wavelengths in the direction of the radar-target line-of-sight, then each incremental point will have a different time delay e^{-2jkr}, where $k = 2\pi/\lambda$. This is illustrated in Figure 6.1.

If the target subtends a sufficiently small angle at the radar site, then from the target's viewpoint, the rays impinging on any two points, say Point 1 and Point 2 illustrated in Figure 6.1, appear to come from the same direction although there is really a small change in direction between the two rays. This is another way of saying that an observer at the radar would see the target as a point. The waves between target and radar (both incident and reflected) would appear as plane waves propagating between target and radar. Their sphericity would be negligible if the lateral extent of the target (i.e., normal to the line-of-sight) were small compared to the range.

The sum of received signal voltages from two points on the target (for monostatic radar), labeled Point 1 and Point 2, would have the form (neglecting doppler differences)

$$s(t) = a_1 \cos\{\omega_o[t - (2r_1/c)] + \psi_1\} + a_2 \cos\{\omega_o[t - (2r_2/c)] + \psi_2\} \tag{6.12}$$

where a_k and ψ_k are the amplitude and phase values that result from scattering from Point k and propagation mechanisms along the paths. The path delays $2\omega_o r_1/c$ and $2\omega_o r_2/c$ are specific contributions to the effective phases of the return from #1 and #2. These are the contributions that will be the focus of attention in what follows.

It is customary to use complex envelope concepts in dealing with radar waves, hence (6.12) can be recast in the form (where the number of points has been increased to N rather than 2)

$$\begin{aligned} s(t) &= \sum_{k=1}^{N} a_k \cos\left(\omega_o t + \psi_k - \frac{2\omega_o r_k}{c}\right) \\ &= \mathrm{Re}\left(\sum_{k=1}^{N} a_k e^{j(\omega_o t + \psi_k - (2\omega_o r_o/c) - (2\omega_o/c)\Delta r_k)}\right) \\ &= \mathrm{Re}\left(e^{j\{\omega_o[t-(2r_o/c)]\}} \sum_{k=1}^{N} \hat{s}_k e^{-j4\pi\Delta r_k/\lambda}\right) \end{aligned} \tag{6.13}$$

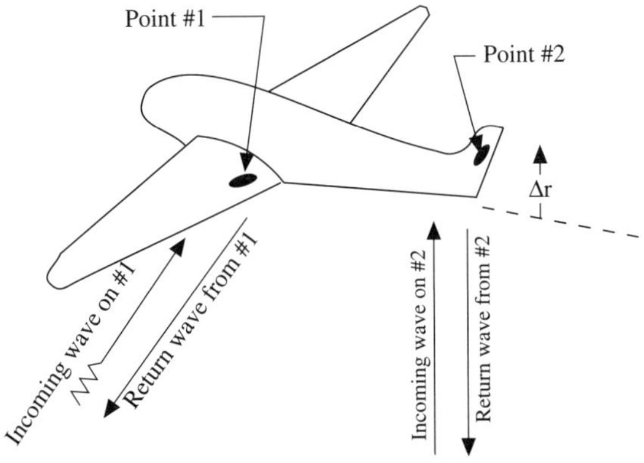

FIGURE 6.1
Combining returns from different points on target.

where $r_k = r_o + \Delta r_k$ and where $\hat{\hat{s}}_k$ is the "complex envelope" of the signal waveform $a_k \cos\{\omega_o[t - (2r_o/c)]\}$, given by

$$\hat{\hat{s}}_k = a_k e^{j\psi_k} \tag{6.13a)'}$$

and $2\omega_o r_k/c = 4\pi r_k/\lambda$ is the two-way *electrical* path delay of Point k, i.e., it appears as a contribution to the phase in radians and is proportional to the ratio of the range to the wavelength.

We have separated r_k into r_0, the range to a central point on the target, designated as its "centroid," and Δr_k, the deviation from the centroid. This separation reflects the fact that it is the delay *difference* between prominent scattering points on the target and not the absolute delays that influence the return.

The quantity we will deal with in analysis will be the complex envelope of the total return from the scattering centers on the target, as it appears in (6.13), i.e.,

$$\hat{\hat{s}} = \sum_{k=1}^{N} \hat{\hat{s}}_k e^{-j4\pi\Delta r_k/\lambda} \tag{6.13b)'}$$

where $\hat{\hat{s}}_k$ is proportional to the square-root of the RCS of the kth scattering center, or

$$\hat{\hat{s}}_k = C\hat{s}_k, \qquad \hat{\hat{s}} = C\hat{s} \tag{6.13c)'}$$

where $\hat{s}_k = \sqrt{\sigma_k}$, σ_k being the RCS of the kth point in square meters and C being a proportionality constant.

Finally, we have the effective complex envelope, given by

$$\hat{s} = \sum_{k=1}^{N} \sqrt{\sigma_k} e^{-j4\pi\Delta r_k/\lambda} \tag{6.13d)'}$$

Figure 6.2 illustrates linear distributions of six scattering centers on a target as they relate to aspect angle.

In Figure 6.2a, the linear distribution of six scattering centers is along the line-of-sight. In Figure 6.2b, the distribution of these same six centers has been rotated through an angle θ with respect to the line-of-sight. To demonstrate the effect of this rotation in a mathematically simple way, we assume that there are N centers, that they are equally spaced ($\Delta r_k = \Delta r$), and all have the same RCS ($\sigma_k = \sigma$). From (6.13d)′ we have for the case shown

$$\hat{s} = \sqrt{\sigma} \sum_{k=1}^{N} e^{-j(4\pi\Delta r/\lambda)k} = \sqrt{\sigma} \frac{(1 - e^{-j4\pi N\Delta r/\lambda})}{(1 - e^{-j4\pi\Delta r/\lambda})} \tag{6.14a}$$

Whose magnitude-square (effective RCS of the aggregate of scattering centers, called σ_{eff}) is

$$\sigma_{\text{eff}} = |\hat{s}|^2 = \sigma \left| \sin\left(\frac{2\pi N\Delta r}{\lambda}\right) \Big/ \sin\left(\frac{2\pi\Delta r}{\lambda}\right) \right| \tag{6.14b}$$

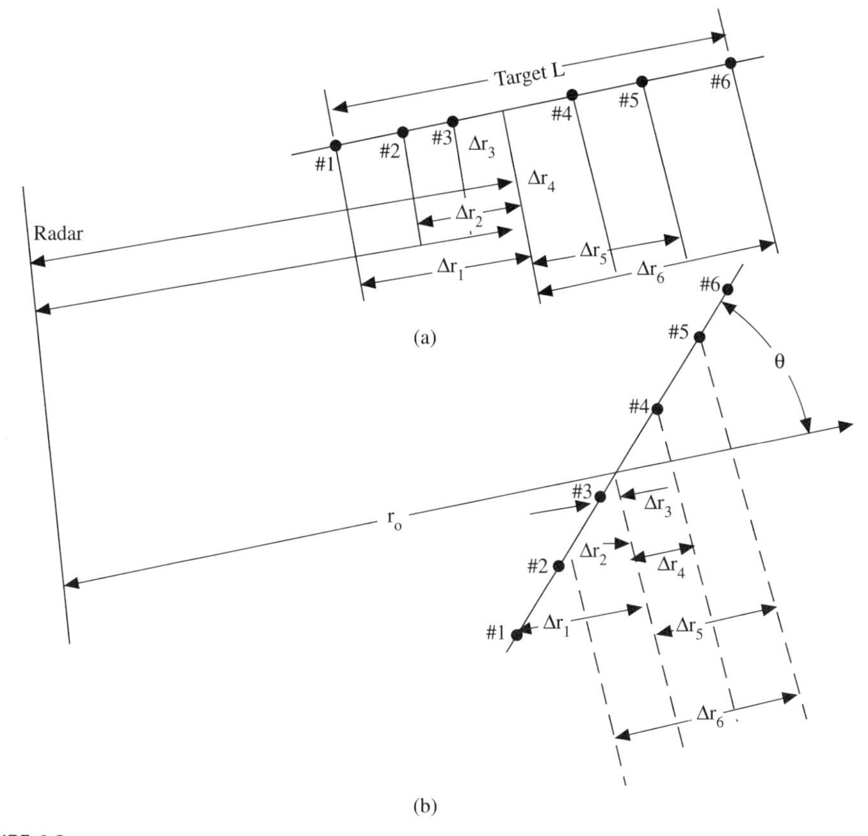

FIGURE 6.2
Linear distributions of six scattering centers on a target—relation to aspect angle.

For the case shown in Figure 6.2b, where the line of scattering centers is rotated through an angle θ, and where the spacing interval Δr has the same meaning as in Figure 6.2a,

$$\sigma_{\text{eff}}(\theta) = |\hat{s}(\theta)|^2 = \sigma(\theta)\left|\sin\left(\frac{2\pi N \Delta r \cos\theta}{\lambda}\right)\Big/\sin\left(\frac{2\pi\Delta r\cos\theta}{\lambda}\right)\right|^2 \quad (6.14c)$$

where the possible dependence of the RCS on θ is explicitly indicated in the notation. If the line is normal to the line-of-sight, $\theta = 90°$ and (6.14c) is

$$\sigma_{\text{eff}}\left(\frac{\pi}{2}\right) = \left|\hat{s}\left(\frac{\pi}{2}\right)\right|^2 = \sigma\left(\frac{\pi}{2}\right) \quad (6.14d)$$

In this highly oversimplified illustrative example, we see that the aggregate of scattering centers can have a strongly fluctuating RCS due to natural oscillatory dynamics of the aircraft in its rolling, yawing, and pitching motions. This can cause a strong RCS fluctuation through peaks and nulls while the aircraft is in what seems to be straight and level flight. If it is maneuvering, then the fluctuations should be even stronger.

To illustrate how strong an effect fluctuations in the angle θ can have on the effective RCS, see Figure 6.3a and b where $\sigma_{eff}(\theta)$ is sketched as a function of cos θ for four scattering centers. In these cases σ(θ) is assumed independent of θ, but in reality the RCS of an individual scattering center can be a strong function of the aspect angle θ. For example, as the center rotates, it may face the radar at a certain angle with concave surfaces where it appears as a corner reflecter. At other angles, it may face the radar as a large flat plane at a wide angle off the normal. In the former case, it would return a great deal of power and in the latter case, almost no power.

For the moment we will ignore the variation of σ(θ) with angle and concentrate on the variation of $\sigma_{eff}(\theta)$ with cos θ as shown in Figure 6.3a for the case of four scattering centers. It is seen that as cos θ varies from 0 to a value $\lambda/8\Delta r$, the effective RCS changes from its peak value of four times the RCS of an individual scattering center to zero, then goes through another fluctuation to a value lower than 4σ, then goes through another one and one-half lobes before it repeats itself. For high frequency radars, where λ is very small, the sensitivity to θ can be enormous. For example, suppose the separation between scattering centers is 3 m. For X-band radar, where λ = 3 cm, cos θ changes from 0 to 1/8 as $\sigma_{eff}(\theta)$ fluctuates between its

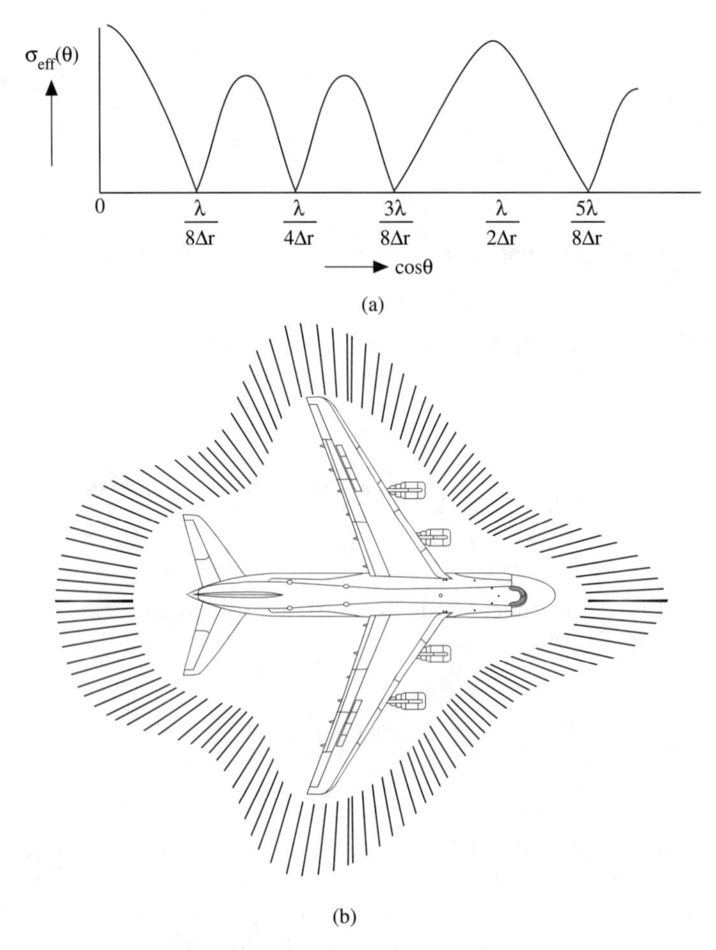

FIGURE 6.3
RCS fluctuations due to linear distribution of scattering centers—angle dependence. (a) $\sigma_{eff}(\theta)$ vs. cos θ. (b) Typical RCS pattern—polar plot.

peak value and its first null. As θ changes from 90° to 80°, cos θ increases from 0 to 0.174. The arguments of the numerator and denominator of $\sigma_{\text{eff}}(\theta)$ are 1.392π and 34.8π, respectively, leading to a rather wild fluctuation in RCS for an angular excursion of 10°.

Such effects (*including* the effect of angular dependence of the RCS of individual scattering centers) lead to typical polar plots for aircraft or missile targets, such as that in the rough sketch in Figure 6.3b. These polar plots show the RCS fluctuations of RCS as the aspect angle of a typical radar target varies over 360°. Such plots can be found in many well-known radar textbooks (e.g., Eaves and Reedy,[3] Skolnick,[4] and DiFranco and Rubin[5]).

There is a great deal more that could be said about the physical mechanisms that cause RCS fluctuations. But since the focus of this section is on the *statistics* of the fluctuations, we will now return to the role of the central limit theorem in generating fluctuations that have Rayleigh or one-dominant-plus Rayleigh amplitude statistics, or other statistics that have a Rayleigh-like constituent.

The generic target model based on "scattering centers," as given by (6.13) or (6.13a,b)′, is useful in the study of amplitude statistics. In the general form, we can write the complex envelope of the return signal voltage (normalized to the square root of the return power for unit RCS) as

$$\hat{s} = \sum_{k=1}^{N} \sqrt{\sigma_k}\, e^{-j\Delta\psi_k} \tag{6.15}$$

where $\Delta\psi_k = (4\pi\Delta r_k/\lambda) - \psi_k$ and where ψ_k is the phase shift due to scattering and propagation effects for the kth scattering center.

If $\sqrt{\sigma_k}$ does not vary widely from center-to-center, and can be considered as a random variable *from the viewpoint of the radar observer,* then $\hat{s}$ can be viewed as a sum of statistically independent random variables with unknown amplitude statistics and uniform phase distributions. The rationale for that is the enormous *ignorance* of the radar observer (nothing pejorative intended) about the physical mechanisms acting on the scattering from any individual center. From his or her viewpoint, there is nothing predictable about either the amplitude or the phase of any of these individual returns, because such predictability would require a precise knowledge (within fractions of centimeters for the higher frequencies such as X-band or above) of the distance-from-radar of each center and a precise knowledge (within a very few degrees) of the orientation of each center. As a matter of fact, even the issue of what portions of the target constitute "scattering centers" is never precisely known. Moreover, even if the target geometry *were* precisely known, the available theory, as complicated as it sometimes becomes, is not adequate to produce a precise knowledge of the return from a small portion of the target. (As a final point, if it were possible to characterize the target return precisely, one would not need a radar.)

For all these reasons, one would have to regard the returns from the "scattering centers" as random variables in the sense that they are unpredictable. That applies especially to the phases, for which a uniform PDF is the only one that makes sense [because a phase angle of (say) 10° is just as likely as (say) 37°, because the factor $\Delta r/\lambda$ in the phase angle can change the phase factor so radically as aspect angle varies]. Also it is not feasible to assess interdependence between center returns (although surely *some* dependence exists) so they must be regarded as statistically independent.

The central limit theorem, then, tells us that $\hat{s}$ as given by (6.15), being a sum of statistically independent random variables, ought to be well approximated as having Gaussian statistics. Since $\hat{s}$ is really a narrowband Gaussian random variable, then its *amplitude* will have a Rayleigh PDF if it has zero mean and a Rice PDF is it has a nonzero *deterministic* mean. The mean itself (which generally arises from a steady specular reflection from a segment of the target that behaves like an infinite flat plane reflector) can be expected to vary slowly over a typical detection interval and is not predictable because it is generally orientation-dependent. Hence the mean must *also* be regarded as a random variable. The one-dominant-plus Rayleigh distribution is an approximation to a case where there is a single dominant scatterer that *may* have a random fluctuation but that remains constant over the time-scale of interest, plus a sum of statistically independent zero mean random scatterers having RCS values of roughly the same order of magnitude. That is only one variation of the possible scenarios. Another is the presence of more than one (say two, three, or even more) scatterers whose RCS values are all high compared with the background sum of (roughly) equal-amplitude random scatterers but that may be different from each other. We *could* call that, e.g., "2 (or 3)-dominant-plus Rayleigh." The way to deal with a vast variety of these possibilities is to use a chi-square distribution, discussed in the next section.[6–9]

6.3 CHI-SQUARE TARGET MODELS

The chi-square (χ^2) PDF is a PDF of *power, not* voltage, although, of course, each such PDF has a corresponding voltage PDF through the transformation

$$p(\chi) = 2\chi p(\chi^2) \tag{6.16}$$

which was invoked many times in Chapter 5. To write the general χ^2 PDF of degree $2k$, we will use the RCS σ as the power variable. This χ^2 PDF with $\chi^2 = \sigma$ is

$$p(\sigma; k) = \frac{k}{\Gamma(k)\langle\sigma\rangle}\left(\frac{k\sigma}{\langle\sigma\rangle}\right)^{k-1} e^{-(k\sigma/\langle\sigma\rangle)} \tag{6.17}$$

where the Gamma function $\Gamma(k) = (k-1)!$ if k is an integer. The corresponding PDF of $\sqrt{\sigma} = \chi$, from the transformation (6.16), is

$$p(\chi; k) = \frac{2k}{\Gamma(k)\langle\chi^2\rangle}\left(\frac{k}{\langle\chi^2\rangle}\right)^{k-1} \chi^{2k-1} e^{-k\chi^2/\langle\chi^2\rangle} \tag{6.18}$$

We note that for $k = 1$

$$p(\sigma; 1) = \frac{1}{\langle\sigma\rangle} e^{-\sigma/\langle\sigma\rangle} \tag{6.19a}$$

which is the exponential PDF [from Eq. (5.35), with the aid of (5.35a)′ and (6.16)] and

$$p(\chi; 1) = \frac{2\chi}{\langle\chi^2\rangle} e^{-\chi^2/\langle\chi^2\rangle} \tag{6.19b}$$

which is the Rayleigh PDF [Eqs. (5.35), (5.35a)′]

For $k = 2$

$$p(\sigma; 2) = \frac{4\sigma}{\langle\sigma\rangle^2} e^{-2\sigma/\langle\sigma\rangle} \tag{6.20a}$$

and

$$p(\chi; 2) = \frac{8\chi^3}{\langle\chi^2\rangle^2} e^{-2\chi^2/\langle\chi^2\rangle} \tag{6.20b}$$

where (6.20b) is the one-dominant-plus Rayleigh PDF for amplitude χ and (6.20a) is the corresponding form for the power or RCS. These are obtained from Eqs. (5.40) with the aid of (5.44) and (6.16). Some of the χ^2 PDFs of degree higher than 2 were used by Weinstock[10] in modeling of some standard target shapes, e.g., cylinders and prolate spheroids, where a χ^2 PDF with $k=4$ was used for a stabilized cylinder and $k=0.6$ to 1.4 for a randomly oriented cylinder (Meyer and Mayer,[11] 1973, p. 64). The case $k=4$ yields, from (6.17) and (6.18)

$$p(\sigma; 4) = \frac{256\sigma^3}{3!\langle\sigma\rangle^4} e^{-4\sigma/\langle\sigma\rangle} \tag{6.21a}$$

$$p(\chi; 4) = \frac{512\chi^7}{3!\langle\chi^2\rangle^4} e^{-4\chi^2/\langle\chi^2\rangle} \tag{6.21b}$$

In Figure 6.4, some rough sketches of the χ^2 PDFs of RCS for a few values

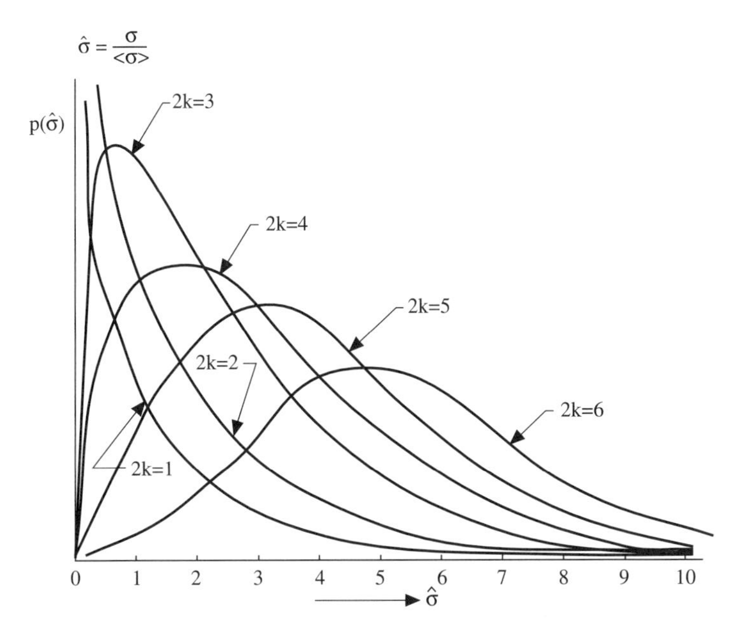

FIGURE 6.4
Chi-square PDFs vs. $\sigma/\langle\sigma\rangle$. From Meyer and Mayer[11] (Figure 3.18a on p. 65) by permission of Academic Press and the authors.

of $2k$ are shown. These are from Figure 3.18a on p. 65 of Meyer and Mayer[11] and are plots of Eq. (6.17) in terms of the normalized RCS $\sigma/\langle\sigma\rangle = \hat{\sigma}$, i.e.,

$$p(\hat{\sigma}; k) = \frac{k}{\Gamma(k)} (k\hat{\sigma})^{k-1} e^{-k\hat{\sigma}} \tag{6.22}$$

It is evident from Figure 6.4 that higher values of k produce PDFs that have longer "tails" than the Rayleigh or other low-degree chi-square PDFs. That is the feature often observed in experimentally determined target amplitude or amplitude-squared PDFs that leads people to seek PDFs other than Rayleigh. Since the chi-square can produce as long a tail as desired for large enough choice of k, it is one of those classes of PDF that can be used to fit to actual data when the measured PDFs are found to have longer tails than can be explained with a Rayleigh PDF.

The methods of finding detection probabilities analytically, expounded in Chapter 5, can be used with the general chi-square PDF of (6.17) or (6.18). Results exist for higher degree chi-square PDFs but will not be presented here. Reference sources on those and other PDFs will appear at the end of this chapter.[6–9,12,13]

It should be remarked again that the *exact* "one-dominant-plus Rayleigh" PDF that applies to a single dominant deterministic scatterer plus a background of independent random scatterers of smaller RCS is the Rice PDF, given by Eq. (5.27) in general. In the present context, it is

$$p(\sigma) = \frac{(1+s)}{\langle\sigma\rangle} e^{-[s+(\sigma/\langle\sigma\rangle)(1+s)]} I_o\left(2\sqrt{\frac{\sigma}{\langle\sigma\rangle} s(1+s)}\right) \tag{6.23}$$

where s = ratio of RCS of dominant scatterer to that of the Rayleigh-distributed sum of small scatterers. Equation (6.23) is easily deduced from the general Rice PDF for amplitude as given in (5.27), with the aid of (6.16), which converts (5.27) into a PDF for amplitude-squared.

The form given in (6.23) is that given by Skolnick[4] [p. 50, Eq. (2.41)]. Alternative forms are given in other sources, e.g., Meyer and Mayer,[11] whose rendition of the Rice PDF is [p. 56, Eq. (3.91)]

$$p(\sigma) = \frac{1}{\langle\sigma_R\rangle} e^{-[s+(\sigma/\langle\sigma_R\rangle)]} I_o\left(2\sqrt{\frac{s\sigma}{\langle\sigma_R\rangle}}\right) \tag{6.23$'$}$$

where $\langle\sigma_R\rangle$ is the RCS of the Rayleigh aggregate.

The forms (6.23) and (6.23)′ can be reconciled by recognizing that

$$\langle\sigma\rangle = \sigma_T + \langle\sigma_R\rangle = \langle\sigma_R\rangle(1+s) \tag{6.23$''$}$$

where σ_T = RCS of the large dominant scatterer.

We note that (6.23)′ and (5.27) would be equivalent if (5.27) were converted to the PDF of $[v/\sqrt{2}]^2$, where $\sigma = [v/\sqrt{2}]^2$, $\langle\sigma_R\rangle = \sigma_n^2$ and $s = a_s^2/2\sigma_n^2$.

6.4 LOGNORMAL PDFs

The lognormal PDF of amplitude or RCS is the basis of another class of target model.[14–18] Like the higher-degree χ^2, it also has longer tails than the Rayleigh. It has the form (using a_s for amplitude and σ for amplitude-squared)

$$p(\sigma) = \frac{1e^{-[[\ln(\sigma)-\ln(\sigma_m)]^2/2\rho_s^2]}}{\sigma\rho_s\sqrt{2\pi}} \tag{6.24a}$$

where ρ_s = standard deviation of ln $[\sigma/\sigma_m]$. =

$$\sqrt{\left\langle\left[\ln\left(\frac{\sigma}{\sigma_m}\right) - \left\langle\ln\left(\frac{\sigma}{\sigma_m}\right)\right\rangle\right]^2\right\rangle}$$

σ_m = median value of RCS, and from (6.16) where $\chi = a_s$

$$p(a_s) = \frac{1}{a_s\rho_s}\sqrt{\frac{2}{\pi}}\, e^{-[(2\ln a_s - \ln\sigma_m)^2/2\rho_s^2]} \tag{6.24b}$$

The lognormal PDF should arise from targets whose scattering exhibits some multiplication of random components in conjunction with addition of random components. When there are many paths between radar and target (i.e., multipath or multiple scattering) it would seem that the amplitude of the signal would involve both additive and multiplicative effects. Insofar as the log of a product is a sum, it is not out of the question that the *amplitude* of the signal, having been obtained from a nonlinear operation on the sum of return signal *voltages,* might contain a dominant component that is essentially a product of independent random variables. In such a case the central limit theorem would indicate that the log of the amplitude would have a Gaussian (equivalently "normal") distribution.

It has been observed that returns from targets with a few very large scatterers tend toward a lognormal PDF, which is characterized by a longer tail than would be observed with a Rayleigh PDF, i.e., an emphasis on the larger amplitudes.

The above qualitative remarks notwithstanding, there is no *precise* physical theory (to the author's knowledge) that would predict a lognormal PDF for a specific target.

To point up the fact that the lognormal PDF is truly the normal (i.e., Gaussian) PDF of the log of σ, we note that it is required for any PDF $p(\sigma)$ that

$$\int_0^\infty d\sigma\, p(\sigma) = 1$$

For the lognormal PDF (6.24a)

$$\begin{aligned}
\int_0^\infty d\sigma\, p(\sigma) &= \frac{1}{\rho_s\sqrt{2\pi}}\int_{\sigma=0}^{\sigma=\infty}\left(\frac{d\sigma}{\sigma}\right)e^{-(1/2\rho_s^2)[\ln(\sigma/\sigma_m)]^2} \\
&= \frac{1}{\rho_s\sqrt{2\pi}}\int_{\ln(\sigma/\sigma_m)=\ln(0)=-\infty}^{\ln(\sigma/\sigma_m)=\ln(\infty)=\infty} d\left(\ln\left(\frac{\sigma}{\sigma_m}\right)\right) e^{-(1/2\rho_s^2)[\ln(\sigma/\sigma_m)]^2} \\
&= \frac{1}{\sqrt{\pi}}\int_{-\infty}^{\infty} d\left(\frac{x}{\rho_s\sqrt{2}}\right)e^{-(x/\rho_s\sqrt{2})^2} = 1
\end{aligned} \tag{6.24c}$$

The mean RCS with the lognormal distribution is given by

$$
\begin{aligned}
\langle\sigma\rangle &= \int_0^\infty d\sigma\, \sigma p(\sigma) = \sigma_m \int_0^\infty d\left(\frac{\sigma}{\sigma_m}\right) \frac{1}{\rho_s\sqrt{2\pi}} e^{-[\ln(\sigma/\sigma_m)]^2/2\rho_s^2} \\
&= \frac{\sigma_m}{\rho_s\sqrt{2\pi}} \int_0^\infty dx\, e^{-(\ln x)^2/2\rho_s^2} = \frac{\sigma_m}{\rho_s\sqrt{2\pi}} \int_{-\infty}^\infty dy\, e^y e^{-y^2/2\rho_s^2} \\
&= \frac{\sigma_m}{\sqrt{\pi}} \int_{-\infty}^\infty dy'\, e^{-(y'^2 - 2(\rho_s^2/2)y')} = \frac{\sigma_m}{\sqrt{\pi}} \sqrt{\pi}\, e^{+\rho_s^2/2}
\end{aligned} \tag{6.24d}
$$

or equivalently

$$
\frac{\langle\sigma\rangle}{\sigma_m} = \text{ratio of mean to median RCS} = e^{\rho_s^2/2} \tag{6.24d)'}
$$

If $\ln \sigma/\sigma_m$ has a large standard deviation, i.e., if ρ_s is very large, (6.24d)′ tells us that the mean-to-median ratio can be very large. For example if $\rho_s = 10$, $\langle\sigma\rangle/\sigma_m$ would be about about $5.2(10^{21})$. Even if ρ_s is as small as 4, $\langle\sigma\rangle/\sigma_m$ would be about 3000 and if $\rho_s = 2$, it would be about 74. This means that a lognormal target with scattering centers that have a large fluctuation from their mean values can have a very large mean RCS even if its median value is only moderately high, i.e., large RCS parts of the target are emphasized.

6.5 THE WEIBULL PDF

The Weibull PDFs for amplitude a_s and RCS σ are [19–24]

$$
p(\sigma) = \lambda\beta\sigma^{\beta-1}e^{-\lambda\sigma^\beta} \tag{6.25a}
$$

$$
p(a_s) = 2a_s\lambda\beta a_s^{2\beta-2}e^{-\lambda a_s^{2\beta}} \tag{6.25b}
$$

where β and λ are shape and magnitude parameters, respectively, and are both positive numbers.

If $\beta = 1$ and $\lambda = 1/\langle\sigma\rangle$ or $1/\langle a_s^2\rangle$, Eq. (6.25a) yields the exponential PDF

$$
p(\sigma) = \frac{1}{\langle\sigma\rangle} e^{-\sigma/\langle\sigma\rangle} \tag{6.26a}
$$

and Eq. (6.25b) results in

$$
p(a_s) = \frac{2a_s}{\langle a_s^2\rangle} e^{-a_s^2/\langle a_s^2\rangle} \tag{6.26b}
$$

which is, of course, the Rayleigh PDF.

Beyond this case, the Weibull PDF does not become any of the standard cases we have encountered before for any given values of λ and β. It should also be stated that no physical target models give rise naturally to a Weibull distribution.

6.6 THE *K*-DISTRIBUTION

In recent years, there has been considerable interest in the radar community in the "*K*-distribution." The *K*-amplitude PDF is given by[25–27]

$$P(a_s; \nu) = \frac{2b}{\Gamma(\nu + 1)} \left(\frac{ba_s}{2}\right)^{\nu+1} K_\nu(ba_s) \tag{6.27a}$$

where a_x is the amplitude, ν is a number exceeding -1, $\Gamma(x)$ is the gamma function, $K_\nu(x)$ is the modified Bessel function of the second kind of order ν, and b is a real number.

The PDF of the squared amplitude $\sigma = a_s^2$ corresponding to the *K*-distribution is [from (6.16) and (6.27a)]

$$P(\sigma; \nu) = \frac{P(a_s; \nu)}{2a_s} = \frac{b^{\nu+2}\sigma^{\nu/2}}{2^{\nu+1}\Gamma(\nu + 1)} K_\nu(b\sqrt{\sigma}) \tag{6.27b}$$

The mean values of a_s and σ are [from (6.27a) and (6.27b), respectively]

$$\begin{aligned} \langle a_s \rangle &= \frac{b^{\nu+2}}{\Gamma(\nu + 1)2^\nu} \int_0^\infty da_s a_s^{\nu+2} K_\nu(ba_s) \\ &= \frac{1}{2^\nu \Gamma(\nu + 1)b} \int_0^\infty dx\, x^{\nu+2} K_\nu(x) \end{aligned} \tag{6.28a}$$

and

$$\begin{aligned} \langle \sigma \rangle &= \frac{b^{\nu+2}}{2^{\nu+1}\Gamma(\nu + 1)} \int_0^\infty d\sigma\, \sigma^{(\nu+1)/2} K_\nu(b\sqrt{\sigma}) \\ &= \frac{1}{2^\nu \Gamma(\nu + 1)b^2} \int_0^\infty dx\, x^{\nu+3} K_\nu(x) \end{aligned} \tag{6.28b}$$

The *K*-distribution contains most of the features of the distributions we have already discussed in this chapter and degenerates to some of them under certain conditions. It is a "long-tailed distribution"[26] and the length of the tail increases as the quantity $(1 + \nu)$ decreases.

This distribution was proposed in Jakeman and Pusey[25] as a means of modeling sea echo that does not fit a Rayleigh distribution. In recent years, it has been extended to include some classes of land clutter and polarimetric returns from targets.

The quantity b in (6.27a) can be related to the mean-square of a_s or the mean of σ through the expression

$$b = 2\sqrt{\frac{(\nu + 1)}{\langle a_s^2 \rangle}} = 2\sqrt{\frac{(\nu + 1)}{\langle \sigma \rangle}} \tag{6.29}$$

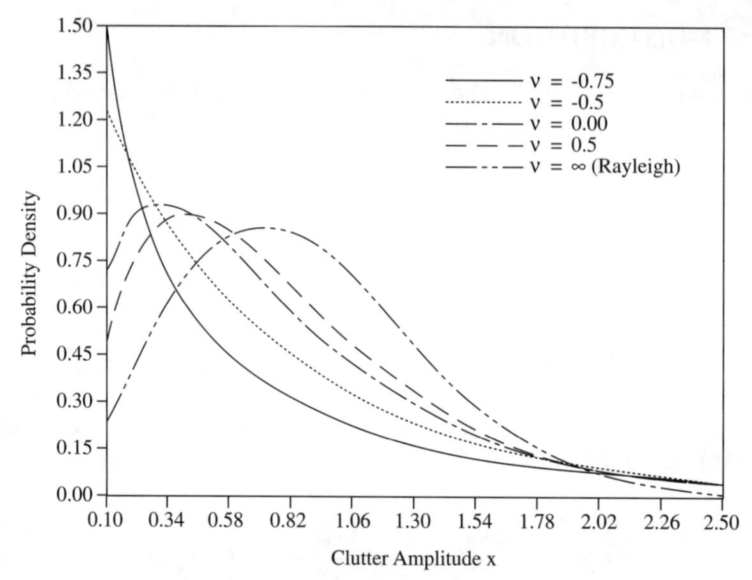

FIGURE 6.5
The *K*-distribution. From Raghavan,[27] IEEE Trans. AES 27, 238, 1991, by permission of IEEE, © 1991 IEEE.

A set of plots of $p(a_s; \nu)$ for $\nu = -0.75, -0.5, 0, 0.5, 2, 4, 6$, and ∞, taken from a paper by Raghavan,[27] is shown in Figure 6.5. From these plots it is evident that for negative values of ν, the PDF decreases rapidly with the amplitude a_s for $a_s >$ 0.1 and shows no peak within that range. For $\nu = 0$, it increases with a_s for small values, reaches a peak value of about 90% at $a_s \approx 0.34$, and then decays with further increases in a_s. The same general behavior is observed for $\nu > 0$, but as ν is increased from 0 to 0.5, the location of the peak increases to about 0.46 and its ampliutude decreases to about 0.85. Further increases in ν through $\nu = 2$, 4, and 6 do not change the peak's amplitude or location very much. The former is at about 85% and the latter at $a_s \approx 50$ to 60. For an infinite value of ν, the peak is at $a_s \approx 0.75$ and still has an amplitude near 85%. It can be shown[26] that the *K*-distribution approaches Rayleigh as ν approaches infinity and approaches Weibull as ν approaches -0.5.

REFERENCES

1. Lawson, J.L., and Uhlenbeck, G.E., "Threshold Signals" (MIT Rad. Lab Series, Vol. 24). McGraw-Hill, New York, 1950, Sections 3.5, 3.6.
2. DiFranco J.V., and Rubin, W.L., "Radar Detection." Artech, Norwood, MA, 1980, Section 3.13.
3. Eaves, J.L., and Reedy, E.K. (eds.), "Principles of Modern Radar." Van Nostrand Reinhold, New York, 1987, p. 344.
4. Skolnick, M.I., "Introduction to Radar Systems," 2nd ed. McGraw-Hill, New York, 1980, pp. 39, 40.
5. DiFranco, J.V., and Rubin, W.L., "Radar Detection." Artech, Norwood, MA, 1980, pp. 26, 27, 28.
6. Skolnick, M.I., "Introduction to Radar Systems," 2nd ed. McGraw-Hill, New York, 1980, pp. 49–51 [e.g., Eq. (2.40), p. 49].
7. Blake, L.V., "Radar Range Performance Analysis." Artech, Norwood, MA, 1986, pp. 49, 70, 73 [e.g., Eq. (2.40)], 117 [e.g., Eq. (3.37)].
8. Levanon, N., "Radar Principles." Wiley, New York, 1988, p. 32 [e.g., Eqs. (2.31–2.33)].
9. Mitchell, R.L., and Walker, J.F., "Recursive Methods for Computing Detection Probabilities." IEEE Trans. AES-7(4), 671–676, 1971.
10. Weinstock, W., "Target Cross-Section Models for Radar Systems Analysis." Ph.d. Thesis, University of Pennsylvania, 1964.
11. Meyer, D.P., and Mayer, H.A., "Radar Target Detection." Academic Press, New York 1973.

12. Eaves, J.L., and Reedy, E.K. (eds.), "Principles of Modern Radar." Van Nostrand Reinhold, New York, 1987, Section 11.3, "Target Fluctuation Models," pp. 343–348 of Chapter 11, by C.R. Barrett, Jr., pp. 343–367.
13. Nathanson, F.E., Reilly, J.P., and Cohen, M.N., "Radar Design Principles," 2nd ed. McGraw-Hill, New York, 1991, pp. 167–169 and p. 175 (e.g., Table 5.6 on pp. 167, 168), in Chapter 5, by Nathanson and Reilly, pp. 147–214.
14. Blake, L.V., "Radar Range Performance Analysis." Artech, Norwood, MA, 1986, pp. 319, 320 [e.g., Eq. (7.38) on p. 319); also pp. 77, 78 [e.g., Eq. (2.52) on p. 77)].
15. Skolnick, M.I., "Introduction to Radar Systems," 2nd ed. McGraw-Hill, New York, 1980, p. 51 [e.g., Eq. (2.42)]; also p. 479 [e.g., Eqs. (13.11a, b)].
16. Levanon, N., "Radar Principles." Wiley, New York, 1988, pp. 34, 35 [e.g., Eq. (2.43) on p. 34 and (2.44)–(2.47) on p. 35].
17. Heidbreder, G.R., and Mitchell, R.L., "Detection Probabilities for Lognormally Distributed Signals." IEEE Trans. AES-23(3), 5–13, 1967.
18. Schnidman, D.A., "Calculation of probability of detection for lognormal target fluctuations," IEEE Trans. AES-27, 1, January 1991, pp. 172–174.
19. Eaves, J.L., and Reedy, E.K., "Principles of Modern Radar." Van Nostrand Reinhold, New York, 1987, Section 10.2.2, pp. 292–294 (e.g., Figure 10.6-c on p. 293).
20. Skolnick, M.I. "Introduction to Radar Systems," 2nd ed. McGraw-Hill, New York, 1980, p. 480 [e.g., Eq. (13.12)].
21. Blake, L.V., "Radar Range Performance Analysis." Artech, Norwood, MA, 1980, p. 326 [e.g., Eq. (7.47)].
22. Levanon, N. "Radar Principles." Wiley, New York, 1988, pp. 33–34 [e.g., Eqs. (2.34)–(2.42)].
23. Nathanson, F.E., Reilly, J.P., and Cohen, M.N., "Radar Design Principles," 2nd ed. McGraw-Hill, New York, 1991, pp. 300, 301 [e.g., Eqs. (7.9)–(7.11)]; also pp. 324–336, 315, 324, 336–338.
24. Scheher, "Radar Detection in Weibull Clutter." IEEE Trans. AES-12, 736–743, 1976.
25. Jakeman, E., and Pusey, P.N., "A Model for Non-Rayleigh Sea Echo." IEEE Trans. AP-24(6), 806–814, 1976.
26. Fante, R.L., "Detection of Multi-scatter Targets in K-Distributed Clutter." IEEE Trans. AP-32(12), 1358–1363, 1984.
27. Raghavan, R.S., "A Method of Estimating Parameters of K-Distributed Clutter." IEEE Trans. AES-27(2), 238–246, 1991.

Chapter 7

Transmitted Signal Waveform Design; Range and Doppler Resolution

CONTENTS

In this chapter, we will discuss the topic of design of the transmitted radar signal. The objective of the discussion is to impart an understanding of the tradeoffs between range and doppler resolution and between resolution in those parameters and signal-to-noise ratio (or equivalently target detectability). In the design of a transmitted signal, there is often a necessary compromise between achievement of a given level of resolution and that of a specified level of detection performance, given a maximum available transmitted signal energy. The parameter variations that affect those compromises will be the subject of this chapter.*

7.1 THE RANGE-DOPPLER AMBIGUITY FUNCTION

Let us assume a transmitted radar signal of the usual form

$$s_T(t, \omega) = a(t)\cos[\omega t + \psi(t)] \tag{7.1}$$

Target 1 at range r and with doppler $-\omega_d$ (monostatic radar), assuming no distortion in transmission or scattering, will return a signal of the form (see Fig. 7.1)

$$\begin{aligned} s_{R1}(t, \omega) &= Ks_T\left(t - \frac{2r}{c}, \omega - \omega_d\right) \\ &= Ka\left(t - \frac{2r}{c}\right)\cos\left[(\omega - \omega_d)\left(t - \frac{2r}{c}\right) + \psi\left(t - \frac{2r}{c}\right)\right] \end{aligned} \tag{7.2}$$

where K is a constant scale factor.

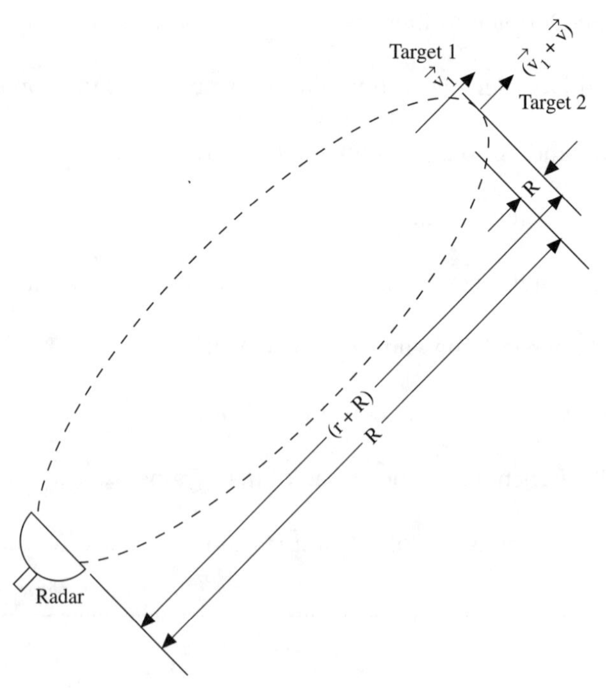

FIGURE 7.1
Return from two targets slightly displaced in range and doppler.

* The material in this chapter is developed from first principles and hence it is self-contained. For the reader's benefit, the reference list at the end of the chapter indicates the location of coverage of these topics in a few established texts on radar.[1–7]

A second target with range slightly displaced from r (by an amount R) and slightly displaced in doppler from ω_d (by an amount Ω) returns a signal of the form

$$s_{R2}(t, \omega) = s_T\left[t - \frac{2}{c}(r + R), \omega - (\omega_d + \Omega)\right]$$
$$= Ka\left[t - \frac{2}{c}(r + R)\right]\cos\left\{[\omega - (\omega_d + \Omega)]\left[t - \frac{2}{c}(r + R)\right]\right.$$
$$\left. + \psi\left[t - \frac{2}{c}(r + R)\right]\right\} \tag{7.3}$$

where it is assumed that there is no significant change in the coefficient K for the two targets.

With respect to the last assumption, the more basic assumption behind it is that both targets return the same energy, or equivalently

$$E_s = \int_{-\infty}^{\infty} dt\, s_{R1}^2(t, \omega) = \int_{-\infty}^{\infty} dt\, s_{R2}^2(t, \omega) \tag{7.4}$$

In the interest of enhanced calculation efficiency, we will use complex envelope representations of the two signals, i.e.,

$$s_{Rk}(t, \omega) = \mathrm{Re}[\hat{s}_{Rk}(t, \omega)e^{j\omega t}] \tag{7.5}$$

where, from (7.2) and (7.3)

$$\hat{s}_{R1}(t, \omega) = Ka\left(t - \frac{2r}{c}\right)e^{j[-\omega_d t - (\omega - \omega_d)(2r/c) + \psi[t - (2r/c)]]}$$

$$\hat{s}_{R2}(t, \omega) = Ka\left[t - \frac{2}{c}(r + R)\right]e^{j[-(\omega_d + \Omega)t}$$
$$-[\omega - (\omega_d + \Omega)]\frac{2}{c}(r + R) + \psi\left[t - \frac{2}{c}(r + R)\right]\Big\}$$

Now let us calculate the time average of the mean-square difference between the two return signals, using the complex envelope representations as in (7.5), and normalize that time average to the total signal energy as given by (7.4). The normalized mean-square difference (numerator of the expression indicated above) is

$$\overline{[\Delta s]^2} = \int_{-\infty}^{\infty} dt[s_{R1}(t, \omega) - s_{R2}(t, \omega)]^2 \tag{7.6a}$$

We note that (from 7.5))

$$s_{R1}(t, \omega) - s_{R2}(t, \omega) = \mathrm{Re}\{[\hat{s}_{R1}(t, \omega) - \hat{s}_{R2}(t, \omega)]e^{j\omega t}\}$$
$$= \frac{1}{2}\{[\hat{s}_{R1}(t, \omega) - \hat{s}_{R2}(t, \omega)]e^{j\omega t} + [\hat{s}_{R1}^*(t, \omega) - \hat{s}_{R2}^*(t, \omega)]e^{-j\omega t}\} \tag{7.6b}$$

From which it follows that (dropping the arguments in the functions)

$$[s_{R1}(t,\omega) - s_{R2}(t,\omega)]^2 = \frac{1}{4}[(\hat{s}_{R1} - \hat{s}_{R2})^2 e^{2j\omega t} + (\hat{s}^*_{R1} - \hat{s}^*_{R2})^2 e^{-2j\omega t}$$

$$+ (\hat{s}_{R1} - \hat{s}_{R2})(\hat{s}^*_{R1} - \hat{s}^*_{R2}) + (\hat{s}^*_{R1} - \hat{s}^*_{R2})(\hat{s}_{R1} - \hat{s}_{R2})]$$

$$= \frac{1}{2}\,\mathrm{Re}[(\hat{s}_{R1} - \hat{s}_{R2})^2 e^{2j\omega t} + (\hat{s}_{R1} - \hat{s}_{R2})(\hat{s}^*_{R1} - \hat{s}^*_{R2})] \qquad (7.6c)$$

The first term, when integrated over time from $-\infty$ to $+\infty$, will be negligibly small, based on the usual assumption that $(\hat{s}_{R1} - \hat{s}_2)$ is slowly varying compared with $e^{2j\omega t}$. The final result is

$$\overline{[\Delta s]^2} = \frac{1}{2}\,\mathrm{Re}\left(\int_{-\infty}^{\infty} dt\ |\hat{s}_{R1} - \hat{s}_{R2}|^2\right) \qquad (7.6d)$$

$$= \frac{1}{2}\int_{-\infty}^{\infty} dt\ |\hat{s}_{R1} - \hat{s}_{R2}|^2 \qquad \text{(since the integrand is already real)}$$

where

$$|\hat{s}_{R1} - \hat{s}_{R2}|^2 = |\hat{s}_{R1}(t,\omega)|^2 + |\hat{s}_{R2}(t,\omega)|^2 - 2\,\mathrm{Re}[\hat{s}_{R1}(t,\omega)\hat{s}^*_{R2}(t,\omega)]$$

The total signal energy for either signal, from (7.4), (7.5) and the counterpart of (7.6c), is

$$E_s = \int_{-\infty}^{\infty} dt\ s^2_{R1}(t,\omega) = \frac{1}{4}\int_{-\infty}^{\infty} dt\ [\hat{s}^2_{R1} e^{2j\omega t} + (\hat{s}^*_{R1})^2 e^{-2j\omega t}$$

$$+ \hat{s}_{R1}\hat{s}^*_{R1} + \hat{s}^*_{R1}\hat{s}_{R1}] = \frac{1}{2}\,\mathrm{Re}\left(\int_{-\infty}^{\infty} dt\ (\hat{s}^2_{R1} e^{2j\omega t} + |\hat{s}_{R1}|^2)\right)$$

$$= \frac{1}{2}\int_{-\infty}^{\infty} dt\ |\hat{s}_{R1}(t,\omega)|^2 \qquad (7.6e)$$

The normalized mean-square difference is given by

$$D = \frac{\int_{-\infty}^{\infty} dt[s_{R1}(t,\omega) - s_{R2}(t,\omega)]^2}{\int_{-\infty}^{\infty} dt\ s^2_{R1}(t,\omega)} \qquad (7.7)$$

which, with the aid of (7.6d) and (7.6e), becomes

$$D = \frac{\frac{1}{2}\left\{\int_{-\infty}^{\infty} dt\ |\hat{s}_{R1}(t,\omega)|^2 + \int_{-\infty}^{\infty} dt\ |\hat{s}_{R2}(t,\omega)|^2 - 2\,\mathrm{Re}\left(\int_{-\infty}^{\infty} dt\ \hat{s}_{R1}(t,\omega)\hat{s}^*_{R2}(t,\omega)\right)\right\}}{\frac{1}{2}\int_{-\infty}^{\infty} dt\ |\hat{s}_{R1}(t,\omega)|^2}$$

$$(7.7)'$$

Because of (7.4), the first two integrals in the numerator are equal and are both equal to the denominator integral; hence it follows that

$$D = 2\left\{1 - \mathrm{Re}\left[\frac{\int_{-\infty}^{\infty} dt\ \hat{s}_{R1}(t,\omega)\hat{s}^{*}_{R2}(t,\omega)}{\int_{-\infty}^{\infty} dt\ |\hat{s}_{R1}(t,\omega)|^2}\right]\right\} \tag{7.7a)''}$$

Had we written D without the benefit of the complex envelope representation, directly from (7.7), it would have been [again with the aid of (7.4)]

$$D = 2\left\{1 - \left[\frac{\int_{-\infty}^{\infty} dt\ s_{R1}(t,\omega)s_{R2}(t,\omega)}{\int_{-\infty}^{\infty} dt\ s^{2}_{R1}(t,\omega)}\right]\right\} \tag{7.7b)''}$$

The square-bracketed quantity in (7.7b)″ is defined as the "ambiguity function" (AF) of the signal waveform $s_{R1}(t,\omega)$ and that in (7.7a)″ is proportional to the "complex ambiguity function" (CAF) of the *complex envelope* of $s_{R1}(t,\omega)$, or the "complex ambiguity function" of $s_{R1}(t,\omega)$. When written in the general form as given by (7.7b);", this function appears as the normalized *cross*-correlation function of $s_{R1}(t,\omega)$ and $s_{R2}(t,\omega)$. But since $s_{R2}(t,\omega)$ is the *same* function as $s_{R1}(t,\omega)$ but displaced in the amounts $2R/c$ and Ω in time and frequency, respectively, the AF is really more like an *auto*correlation function of the transmitted waveform.

Finally, from (7.2), (7.3), and (7.7a,b)″, we obtain

$$\begin{aligned} \Lambda(\tau,\Omega) &= \text{ambiguity function (AF)} \\ \text{of } s_T(t,\omega) &= \frac{\int_{-\infty}^{\infty} dt\ s_T(t,\omega)s_T(t-\tau,\omega-\Omega)}{\int_{-\infty}^{\infty} dt\ s^{2}_{T}(t,\omega)} \\ &= \mathrm{Re}[\hat{\Lambda}(\tau,\Omega)e^{+j(\omega-\Omega)\tau}] \end{aligned} \tag{7.8}$$

where $\tau = 2R/c$ and $\hat{\Lambda}(\tau,\Omega)$ = the CAF of $\hat{s}_T(t,\omega)$ [usually known as the complex ambiguity function of the waveform $s_T(t,\omega)$ itself] given by

$$\hat{\Lambda}(\tau,\Omega) = \frac{\int_{-\infty}^{\infty} dt\ \hat{s}_T(t,\omega)\hat{s}^{*}_{T}(t-\tau,\omega-\Omega)e^{j\Omega t}}{\int_{-\infty}^{\infty} dt\ |\hat{s}_T(t,\omega)|^2} \tag{7.8)'}$$

The phase factor $e^{+j(\omega-\Omega)\tau}$ inside the "real part" symbol in (7.8) is stripped-off in the definition of the complex ambiguity function. To understand the rationale for this, we note that [from (7.2) and (7.3)] the numerator of $\hat{\Lambda}(\tau,\Omega)$ is

$$\text{Num}[\hat{\Lambda}(\tau, \Omega)] = \int_{-\infty}^{\infty} dt\ \hat{s}_{R1}(t, \omega)\hat{s}^*_{R2}(t, \omega) \tag{7.9a}$$

$$= \int_{-\infty}^{\infty} dt\ a_T(t)a_T(t - \tau)e^{j[\psi_T(t)-\psi_T(t-\tau)]}e^{j\omega t}e^{-j(\omega-\Omega)(t-\tau)}$$

$$= \left[\int_{-\infty}^{\infty} dt\ a_T(t)a_T(t - \tau)e^{j[\psi_T(t)-\psi_T(t-\tau)]}e^{j\Omega t}\right]e^{+j(\omega-\Omega)\tau}$$

and that

$$\hat{s}_T(t, \omega) = a_T(t)e^{j\psi_T(t)}$$
$$\hat{s}_T(t - \tau, \omega) = a_T(t - \tau)e^{j\psi_T(t-\tau)} \tag{7.9b}$$

resulting in the form given by (7.8)′.

It is customary in radar studies to use the form given in (7.8)′, relying on the *complex* ambiguity function to provide information about the signal waveform. Stripping off the factor $e^{j(\omega-\Omega)\tau}$ does not affect the *amplitude* of $\hat{\Lambda}(\tau, \Omega)$, from which the important information is extracted.

There is an alternative definition of the CAF in terms of $S(\omega)$, the Fourier transform of the signal waveform $s(t, \omega)$. To discuss this, we let ω be the central frequency ω_0, which has been understood throughout the previous discussion. Calling it ω_0 prevents confusion with ω, the frequency variable. Returning to (7.1), we rewrite it in terms of its Fourier transform. $S_T(\omega,\omega_o)$, i.e.,

$$s_T(t; \omega_o) = \frac{1}{2\pi}\int_{-\infty}^{\infty} d\omega\ S_T(\omega, \omega_o)e^{j\omega t} \tag{7.10a}$$

Then if $t \rightarrow t - \tau$, $\omega_o \rightarrow \omega_o - \Omega$

$$s_T(t - \tau; \omega_o - \Omega) = \frac{1}{2\pi}\int_{-\infty}^{\infty} d\omega\ S_T(\omega; \omega_o - \Omega)e^{j\omega(t-\tau)} \tag{7.10b}$$

Using (7.10a,b) in the numerator of (7.8), and changing the order of integration, we have

$$\text{Num}[\Lambda(\tau, \Omega)] = \frac{1}{2\pi}\int_{-\infty}^{\infty} d\omega_1 \int_{-\infty}^{\infty} d\omega_2 \cdots$$
$$S_T(\omega; \omega_o)S_T(\omega; \omega_o - \Omega)e^{-j\omega_2\tau} \cdots \tag{7.11a}$$
$$\left[\frac{1}{2\pi}\int_{-\infty}^{\infty} dt\ e^{j(\omega_1+\omega_2)t}\right]$$

Since the bracketed quantity in (7.11a) is the impulse function $\delta(\omega_1 + \omega_2)$ and $S_T(-\omega; \omega_o) = S_T^*(\omega; \omega_o)$, it follows that

$$\text{Num}[\Lambda(\tau, \Omega)] = \frac{1}{2\pi}\int_{-\infty}^{\infty} d\omega\ S_T(\omega, \omega_o)S_T^*(\omega; \omega_o - \Omega)e^{j\omega\tau} \tag{7.11b}$$

The denominator of (7.8) is the same as (7.11b) for $\Omega = 0$, $\tau = 0$, or

$$\text{Denom}[\Lambda(\tau, \Omega)] = \frac{1}{2\pi} \int_{-\infty}^{\infty} d\omega \; |S_T(\omega, \omega_o)|^2 \tag{7.11c}$$

From (7.11b,c), we can express the ambiguity function in the form

$$\Lambda(\tau, \Omega) = \frac{\displaystyle\int_{-\infty}^{\infty} d\omega \; S_T(\omega, \omega_o) S_T^*(\omega, \omega_o - \Omega) e^{j\omega\tau}}{\displaystyle\int_{-\infty}^{\infty} d\omega \; |S_T(\omega, \omega_o)|^2} \tag{7.12}$$

The forms (7.8) [or (7.8)′] or (7.12) can be used interchangeably to represent the ambiguity function, depending on whether the time waveform $s_T(t,\omega_o)$ or the frequency spectrum or Fourier transform $S_T(\omega,\omega_o)$ is more readily available for the transmitted signal.

7.2 RANGE AND DOPPLER RESOLUTION VIA THE AMBIGUITY FUNCTION

It is easily recognized from the form (7.8) that the AF is identical to the "degradation factor" (Chapter 4) due to mismatch in time-delay or Doppler in the output of a filter that is supposed to be matched to the incoming signal in both range and Doppler, under the assumption of no distortion of the transmitted signal waveshape (aside from these mismatches, of course) on its way from the transmitter to target and back to the receiver. In previous discussions, we used the AF only to study the degradation in SNR (or equivalently the degradation in target detectability) without reference to other radar performance factors. In this section the AF will be used to study the issue of resolution in range and Doppler.

To relate the discussion to resolution, we return to (7.7a,b)″, which, with the aid of (7.8), can be rewritten in the form

$$D(\tau, \Omega) = 2[1 - \Lambda(\tau, \Omega)] \tag{7.13}$$

We recall that $D(\tau,\Omega)$ is the normalized mean-square difference between the two returns differing only in time-delay and Doppler, in the amounts τ and Ω.

If $D(\tau,\Omega)$ changes very rapidly as τ and/or Ω departs from zero by a small amount, good resolution in τ and/or Ω is implied. If τ or Ω must deviate from zero by a large amount before a perceptible difference in $D(\tau,\Omega)$ occurs, that would indicate poor resolution. Since $D(\tau,\Omega)$ is proportional to 1 minus the ambiguity function, then the sensitivity of the latter to small variations in τ and/or Ω can also be used as a measure of resolution.

The situation is illustrated in Figure 7.2 for the case $\Omega = 0$ (range resolution only). A similar situation holds for the case $\tau = 0$ (Doppler resolution only).

Figure 7.2a shows a case of somewhat poor resolution in τ, which is proportional to range. The mean-square difference $D(\tau,0)$ starts at zero for $|\tau| = 0$ and increases slowly as $|\tau|$ increases, until it asymptotically approaches 2 as $|\tau|$ becomes very large. The ambiguity function must be unity for $|\tau| = 0$ and decays to a negligibly small value. The "width" of the ambiguity function [e.g., the separation between

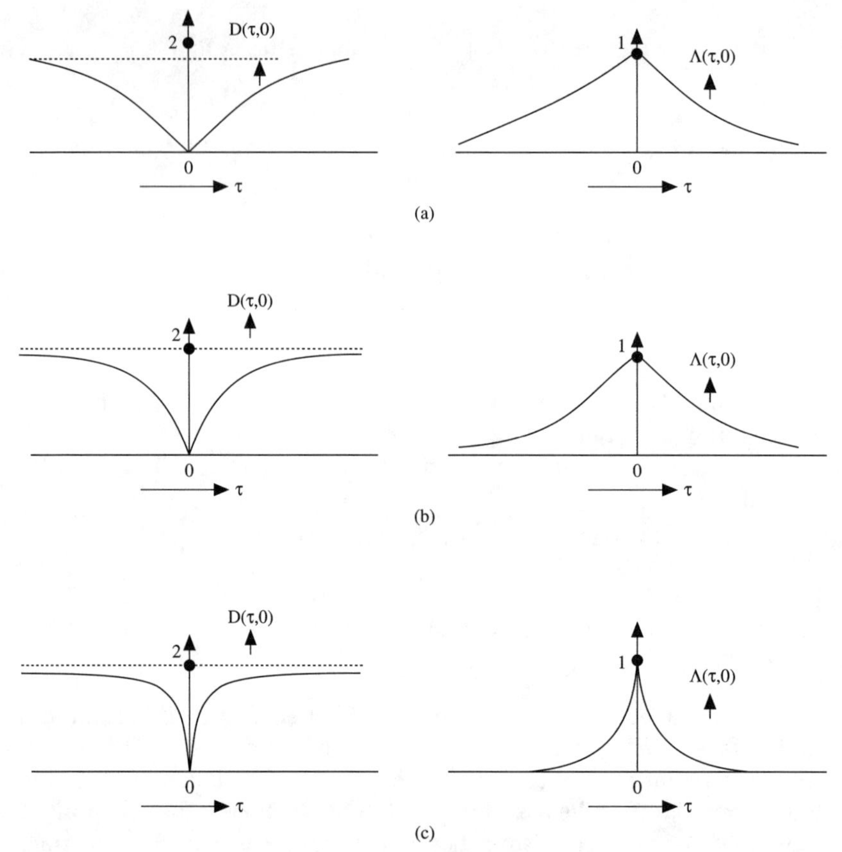

FIGURE 7.2
Mean-square difference, ambiguity function, and range resolution.

the two points on either side of $|\tau| = 0$ where $|\wedge(\tau,0)| = 1/2]$ acts as a measure of range resolution when it is calibrated in range units.

Figure 7.2b and c shows intermediate and good resolution, respectively. The qualitative behavior is the same as that of Case (a) but the "widths" of the ambiguity function decrease from (a) to (b) and then to (c), showing improved range resolution from (a) to (c).

To study range and Doppler resolution, we define the "range ambiguity function" (RAF) and "Doppler ambiguity function" (DAF) as follows: From (7.8) and (7.8)′, the range ambiguity function is given by

$$\Lambda(\tau, 0) = \frac{\displaystyle\int_{-\infty}^{\infty} dt\; s_T(t, \omega_o) s_T(t - \tau, \omega_o)}{\displaystyle\int_{-\infty}^{\infty} dt\; s_T^2(t, \omega_o)} \tag{7.14}$$

$$= \mathrm{Re}[\hat{\Lambda}(\tau, 0) e^{j\omega_o \tau}]$$

where [noting that $\hat{s}\,(t, \omega_o)$ is not a function of ω_o and therefore removing ω_o from its argument]

$$\hat{\Lambda}(\tau, 0) = \frac{\displaystyle\int_{-\infty}^{\infty} dt\ \hat{s}_T(t)\hat{s}_T^*(t - \tau)}{\displaystyle\int_{-\infty}^{\infty} dt\ |\hat{s}_T(t)|^2} \tag{7.14$'$}$$

and the Doppler ambiguity function by

$$\begin{aligned}\Lambda(0, \Omega) &= \frac{\displaystyle\int_{-\infty}^{\infty} dt\ s_T(t, \omega_o)s_T(t, \omega_o - \Omega)}{\displaystyle\int_{-\infty}^{\infty} dt\ s_T^2(t, \omega_o)} \\ &= \mathrm{Re}[\hat{\Lambda}(0, \Omega)]\end{aligned} \tag{7.15}$$

where

$$\hat{\Lambda}(0, \Omega) = \frac{\displaystyle\int_{-\infty}^{\infty} dt\ |\hat{s}(t)|^2 e^{j\Omega t}}{\displaystyle\int_{-\infty}^{\infty} dt\ |\hat{s}_T(t)|^2} \tag{7.15$'$}$$

From (7.12), using the Fourier transform of the signal waveform rather than the waveform itself, we obtain for the RAF

$$\Lambda(\tau, 0) = \frac{\displaystyle\int_{-\infty}^{\infty} d\omega\ |S_T(\omega, \omega_o)|^2 e^{j\omega\tau}}{\displaystyle\int_{-\infty}^{\infty} d\omega\ |S_T(\omega, \omega_o)|^2} \tag{7.16}$$

and for the DAF

$$\Lambda(0, \Omega) = \frac{\displaystyle\int_{-\infty}^{\infty} d\omega\ S_T(\omega, \omega_o)S_T^*(\omega, \omega_o - \Omega)}{\displaystyle\int_{-\infty}^{\infty} d\omega\ |S_T(\omega, \omega_o)|^2} \tag{7.17}$$

The quantity given by (7.14) is easily recognized as a normalized autocorrelation function (NACF) of $s_T(t, \omega_o)$ and that given by (7.14)′ as the complex NACF applicable to the complex envelope of a narrowband signal waveform. The quantity given by (7.15) is a normalized ACF in frequency space, i.e., the normalized integral of the product of the waveform and a frequency-shifted but otherwise undistorted version of itself.

Let us consolidate the above results in one place. In general, we have for the range-Doppler AF [from (7.8), (7.8)′ and (7.12)]

$$\Lambda(\tau, \Omega) = \frac{\int_{-\infty}^{\infty} dt\ s_T(t, \omega_o) s_T(t - \tau, \omega_o - \Omega)}{\int_{-\infty}^{\infty} dt\ s_T^2(t, \omega_o)} = \frac{\int_{-\infty}^{\infty} d\omega\ S_T(\omega, \omega_o) S_T^*(\omega, \omega_o - \Omega)^{j\omega\tau}}{\int_{-\infty}^{\infty} d\omega\ |S_T(\omega, \omega_o)|^2}$$

$$= \mathrm{Re}[\hat{\Lambda}(\tau, \Omega) e^{j(\omega - \Omega)\tau}] \tag{7.18}$$

where

$$\hat{\Lambda}(\tau, \Omega) = \frac{\int_{-\infty}^{\infty} dt\ \hat{s}_T(t) \hat{s}_T^*(t - \tau) e^{j\Omega t}}{\int_{-\infty}^{\infty} dt\ |\hat{s}_T(t)|^2}$$

from which the range-ambiguity function is

$$\Lambda(\tau, 0) = \frac{\int_{-\infty}^{\infty} dt\ s_T(t, \omega_o) s_T(t - \tau, \omega_o)}{\int_{-\infty}^{\infty} dt\ s_T^2(t, \omega_o)} = \frac{\int_{-\infty}^{\infty} d\omega\ |S_T(\omega, \omega_o)|^2 e^{j\omega\tau}}{\int_{-\infty}^{\infty} d\omega\ |S_T(\omega, \omega_o)|^2}$$

$$= \mathrm{Re}[\hat{\Lambda}(\tau, 0) e^{j\omega_o\tau}] \tag{7.18$'$}$$

where

$$\hat{\Lambda}(\tau, 0) = \frac{\int_{-\infty}^{\infty} dt\ \hat{s}_T(t)\ \ \hat{s}_T^*(t - \tau)}{\int_{-\infty}^{\infty} dt\ |\hat{s}_T(t)|^2}$$

and the Doppler ambiguity function is

$$\Lambda(0, \Omega) = \frac{\int_{-\infty}^{\infty} dt\ s_T(t, \omega_o) s_T(t, \omega_o - \Omega)}{\int_{-\infty}^{\infty} dt\ s_T^2(t, \omega_o)} = \frac{\int_{-\infty}^{\infty} d\omega\ S_T(\omega, \omega_o) S_T^*(\omega, \omega_o - \Omega)}{\int_{-\infty}^{\infty} d\omega\ |S_T(\omega, \omega_o)|^2}$$

$$= \mathrm{Re}[\hat{\Lambda}(0, \Omega)] \tag{7.18$''$}$$

where

$$\hat{\Lambda}(0, \Omega) = \frac{\int_{-\infty}^{\infty} dt |\ \hat{s}_T(t)|^2 e^{j\Omega t}}{\int_{-\infty}^{\infty} dt\ |\hat{s}_T(t)|^2}$$

To study range and/or Doppler resolution using ambiguity functions, we must first define those properties. Range resolution is defined in terms of the "minimum resolvable range difference." This, in turn, can be defined as the smallest difference in range between two otherwise identical target returns where the sum of both returns is discernible as coming from the two different targets at two different ranges. Below that value of range difference, the superposition of the two returns would appear to come from a single target at a range that is somewhere near the average of the two ranges. An analogous definition could be devised for Doppler resolution by simply replacing "range" with "doppler."

Strictly speaking, there is no absolute limit of resolution without consideration of noise. The ambiguity function formalism does not involve noise, but can still be very useful in generating guidelines for determination of the limits of range and doppler resolution. P.M. Woodward, in his classical work on the subject,[1] defined three quantities for this purpose, as follows:

$$\begin{aligned}\text{TRC} &= \text{"time resolution constant"}\\ &= K_o \int_{-\infty}^{\infty} d\tau \, |\Lambda(\tau, 0)|^2 = \frac{1}{\text{FSP}}\end{aligned} \tag{7.19a}$$

where

$$\begin{aligned}\text{FSP} &= \text{"frequency span"}\\ \text{FRC} &= \text{"frequency resolution constant"}\\ &= \frac{K_o}{2\pi} \int_{-\infty}^{\infty} d\Omega \, |\Lambda(0, \Omega)|^2 = \frac{1}{\text{TSP}}\end{aligned} \tag{7.19b}$$

where

$$\text{TSP} = \text{"time span"}$$

$$\text{AA} = \frac{K_o}{2\pi} \int_{-\infty}^{\infty} d\tau \int_{-\infty}^{\infty} d\omega \, |\Lambda(\tau, \Omega)|^2 = \text{"area of ambiguity"} \tag{7.19c}$$

where $K_o = 1$ for a baseband signal and $K_o = 2$ for a narrowband signal. Woodward[1] [Eq. (5), p. 117 and (15), p. 119] defines TRC and FRC, which he calls T and F, respectively, in terms of $\hat{\Lambda}$ rather than Λ. Our definitions in (7.19a,b,c)′ are consistent with those of Woodward.

To understand how these quantities can give us useful insights in studying radar signals, we will use a single rectangular pulse as an example. Before doing this, we will express them in terms of the complex AF $\hat{\Lambda}(\tau,\Omega)$. From (7.8), we can write

$$\begin{aligned}|\Lambda(\tau, \Omega)|^2 &= \left|\frac{1}{2} [\hat{\Lambda}(\tau, \Omega)e^{-j(\omega-\Omega)\tau} + \hat{\Lambda}^*(\tau, \Omega)e^{j(\omega-\Omega)\tau}]\right|^2\\ &= \frac{1}{4} \{2|\hat{\Lambda}(\tau, \Omega)|^2 + 2[|\hat{\Lambda}(\tau, \Omega)|^2 \cos[2(\omega - \Omega)\tau + \Psi)]\}\\ &= \frac{1}{2} |\hat{\Lambda}(\tau, \Omega)|^2\{1 + \cos[2(\omega - \Omega)\tau + 2\Psi]\}\end{aligned} \tag{7.19d}$$

where Ψ is the phase of the AF. Hence it follows that we can alternatively define the above quantities as

$$\text{TRC} = \int_{-\infty}^{\infty} d\tau \, |\hat{\Lambda}(\tau, 0)|^2 = \frac{1}{\text{FSP}} \tag{7.19a$'$}$$

$$\text{FRC} = \frac{1}{2\pi} \int_{-\infty}^{\infty} d\Omega \, |\hat{\Lambda}(0, \Omega)|^2 = \frac{1}{\text{TSP}} \tag{7.19b$'$}$$

$$\text{AA} = \frac{1}{2\pi} \int_{-\infty}^{\infty} d\tau \int_{-\infty}^{\infty} d\Omega \, |\hat{\Lambda}(\tau, \Omega)|^2 \tag{7.19c$'$}$$

since the cosine term, being an even function in τ and Ω, integrates to zero in all three of these expressions.

Now consider the example of the rectangular RF or IF pulse of duration τ_P centered at $t = 0$ with frequency ω_o. For this case (Fig. 7.3)

$$\hat{s}_T(t) = u\left(t + \frac{\tau_P}{2}\right) - u\left(t - \frac{\tau_P}{2}\right) \tag{7.20a}$$

and the FT of $s_T(t, \omega_o)$ is

$$\begin{aligned} S_T(\omega, \omega_o) &= \int_{-\tau_P/2}^{\tau_P/2} dt \cos(\omega_o t + \psi) e^{-j\omega t} \\ &= \frac{e^{j\psi}}{2} \int_{-\tau_P/2}^{\tau_P/2} dt \, e^{-j(\omega - \omega_o)t} + \frac{e^{-j\psi}}{2} \int_{-\tau_P/2}^{\tau_P/2} dt \, e^{-j(\omega + \omega_o)t} \\ &= e^{j\psi} \operatorname{sinc}\left[\frac{(\omega - \omega_o)\tau_P}{2}\right] + e^{-j\psi} \operatorname{sinc}\left[\frac{(\omega + \omega_o)\tau_P}{2}\right] \end{aligned} \tag{7.20b}$$

$t = \frac{-\tau_p}{2}$ $\frac{2\pi}{\omega_o}$ $t = \frac{\tau_p}{2}$

1

0

-1

$t = 0$

(a)

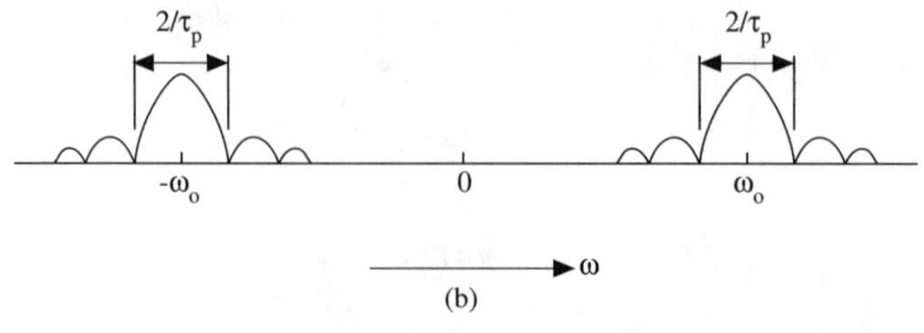

(b)

FIGURE 7.3
Rectangular IF pulse and its Fourier transform (spectrum). (a) Pulse waveform. (b) Fourier transform.

The -3dB bandwidth (separation in Hertz frequency between -3 dB points) of this waveform is determined by the condition $|\text{sinc}\,(2\pi B\tau_p/(2)2)| = 1/\sqrt{2}$, from which it follows (approximately) that

$$\text{sinc}\left(\frac{\pi B\tau_p}{2}\right) \simeq 1 - \frac{(\pi B\tau_p)^2}{3!} = \frac{1}{\sqrt{2}} \tag{7.20c}$$

or $(\pi B\tau_p/2) = \sqrt{6}\,[1 - (1/\sqrt{2})]$, from which it follows that

$$B = \frac{0.844}{\tau_p} \simeq \frac{1}{\tau_p} \tag{7.20d}$$

From (7.20a) and (7.18)

$$\hat{\Lambda}(t, \Omega) = \int_{-\infty}^{\infty} dt\, e^{j\Omega t}\left[u\left(t + \frac{\tau_p}{2}\right) - u\left(t - \frac{\tau_p}{2}\right)\right]\left\{u\left[t + \left(\frac{\tau_p}{2} - \tau\right)\right] - u\left[t - \left(\frac{\tau_p}{2} + \tau\right)\right]\right\}$$

$$= \int_{-\infty}^{\infty} dt\, e^{j\Omega t}[f_1(t, \tau_p, \tau) - f_2(t, \tau_p, \tau) - f_3(t, \tau_p, \tau) + f_4(t, \tau_p, \tau)] \tag{7.21a}$$

where

$$f_1(t, \tau_p, \tau) = u\left(t + \frac{\tau_p}{2}\right) \quad \text{if } \frac{-\tau_p}{2} \geq \frac{-\tau_p}{2} + \tau \quad (\tau \leq 0)$$

$$u\left[t + \left(\frac{\tau_p}{2} - \tau\right)\right] \quad \text{if } \frac{-\tau_p}{2} \leq \frac{-\tau_p}{2} + \tau \quad (\tau \geq 0)$$

$$f_2(t, \tau_p, \tau) = u\left(t - \frac{\tau_p}{2}\right) \quad \text{if } \frac{\tau_p}{2} \geq \frac{-\tau_p}{2} + \tau \quad (\tau \leq \tau_p)$$

$$u\left[t + \left(\frac{\tau_p}{2} - \tau\right)\right] \quad \text{if } \frac{\tau_p}{2} \leq \frac{-\tau_p}{2} + \tau \quad (\tau \geq \tau_p)$$

$$f_3(t, \tau_p, \tau) = u\left(t + \frac{\tau_p}{2}\right) \quad \text{if } \frac{-\tau_p}{2} \geq \frac{\tau_p}{2} + \tau \quad (\tau \leq -\tau_p)$$

$$u\left[t - \left(\frac{\tau_p}{2} + \tau\right)\right] \quad \text{if } \frac{-\tau_p}{2} \leq \frac{\tau_p}{2} + \tau \quad (\tau \geq -\tau_p)$$

$$f_4(t, \tau_p, \tau) = u\left(t - \frac{\tau_p}{2}\right) \quad \text{if } \frac{\tau_p}{2} \geq \frac{\tau_p}{2} + \tau \quad (\tau \leq 0)$$

$$u\left[t - \left(\frac{\tau_p}{2} + \tau\right)\right] \quad \text{if } \frac{\tau_p}{2} \leq \frac{\tau_p}{2} + \tau \quad (\tau \geq 0)$$

If $\tau \leq -\tau_p$, then $\tau \leq 0$, and

$$f_1 - f_2 - f_3 + f_4 = u\left(t + \frac{\tau_p}{2}\right) - u\left(t - \frac{\tau_p}{2}\right) - u\left(t + \frac{\tau_p}{2}\right) + u\left(t - \frac{\tau_p}{2}\right) = 0$$

If $\tau \geq \tau_p$, then $\tau \geq 0$, $\tau \geq -\tau_p$, and

$$f_1 - f_2 - f_3 + f_4 = u\left[t + \left(\frac{\tau_p}{2} - \tau\right)\right] - u\left[t + \left(\frac{\tau_p}{2} - \tau\right)\right] - u\left[t - \left(\frac{\tau_p}{2} + \tau\right)\right] + u\left[t - \left(\frac{\tau_p}{2} + \tau\right)\right] = 0$$

If $-\tau_p \leq \tau \leq 0$, then $\tau \leq \tau_p$ and

$$f_1 - f_2 - f_3 + f_4 = u\left(t + \frac{\tau_p}{2}\right) - u\left(t - \frac{\tau_p}{2}\right) - u\left[t - \left(\frac{\tau_p}{2} + \tau\right)\right] + u\left(t - \frac{\tau_p}{2}\right) = u\left(t + \frac{\tau_p}{2}\right) - u\left[t - \left(\tau + \frac{\tau_p}{2}\right)\right]$$

If $0 \leq \tau \leq \tau_p$, then $\tau \geq -\tau_p$ and

$$f_1 - f_2 - f_3 + f_4 = u\left[t + \left(\frac{\tau_p}{2} - \tau\right)\right] - u\left(t - \frac{\tau_p}{2}\right) - u\left[t - \left(\frac{\tau_p}{2} + \tau\right)\right] + u\left[t - \left(\frac{\tau_p}{2} + \tau\right)\right] = u\left[t - \left(\tau - \frac{\tau_p}{2}\right)\right] - u\left(t - \frac{\tau_p}{2}\right)$$

and hence

$$\begin{aligned}
\hat{\Lambda}(\tau, \Omega) &= 0 \text{ if } |\tau| \geq \tau_p \\
&= \frac{1}{\tau_p}\int_{-\tau_p/2}^{\tau+\tau_p/2} dt\, e^{j\Omega t} = -e^{-j\Omega\tau_p/2} + e^{j\Omega[\tau+(\tau_p/2)]} \qquad \text{if } -\tau_p \leq \tau \leq 0 \\
&= \frac{1}{\tau_p}\int_{\tau-\tau_p/2}^{\tau_p/2} dt\, e^{j\Omega t} = -e^{j\Omega[\tau-(\tau_p/2)]} + e^{j\Omega\tau_p/2} \qquad \text{if } 0 \leq \tau \leq \tau_p
\end{aligned} \tag{7.21b}$$

We can express (7.21b) in the form

$$\hat{\Lambda}(\tau, \Omega) = u(\tau_p - |\tau|)e^{j\Omega\tau/2}\left\{\frac{e^{j\Omega\tau_p/2}[1 - (|\tau|/\tau_p)] - e^{-j\Omega\tau_p/2}[1 - (|\tau|/\tau_p)]}{j\Omega\tau_p}\right\}$$

$$= e^{j(\Omega\tau/2)}u(\tau_p - |\tau|)\left(1 - \frac{|\tau|}{\tau_p}\right)\text{sinc}\left[\frac{\Omega\tau_p}{2}\left(1 - \frac{|\tau|}{\tau_p}\right)\right] \quad (7.22)$$

We note that (7.22) is identical to (4.88g), which we encountered in Chapter 4 in connection with the ambiguity function as a measure of degradation in the output of a "mismatched" filter, i.e., an attempted matched filter that has mismatches in range and doppler. The mismatches were denoted by Δt and $\Delta\omega$ in (4.88g) and by τ and Ω in (7.22). Otherwise the two expressions are the same. That is hardly a surprise, because this is the same ambiguity function we studied earlier, but for a different purpose.

The specializations of (7.22) for range-alone and Doppler-alone cases are

$$\hat{\Lambda}(\tau, 0) = u(\tau_p - |\tau|)\left(1 - \frac{|\tau|}{\tau_p}\right) \quad (7.22a)'$$

and

$$\hat{\Lambda}(0, \Omega) = \text{sinc}\left(\frac{\Omega\tau_p}{2}\right) \quad (7.22b)'$$

Equations (7.22a)′ and (7.22b)′ are, respectively, the familiar "triangular function" we have encountered before and a function of Ω that is identical to the Fourier transform of the rectangular pulse amplitude function. These are shown in Figure 7.4.

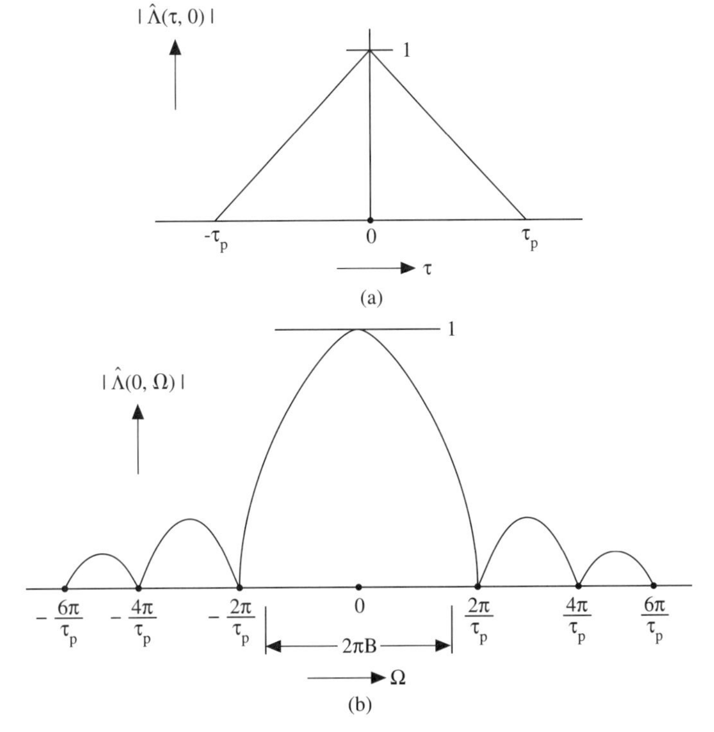

FIGURE 7.4
Ambiguity functions for rectangular pulses. (a) Range ambiguity function. (b) Doppler ambiguity function.

The calibration of Figure 7.4b (nulls at $\pm 2\pi n/\tau_p$; $n = 1,2,3,\ldots$) follows from (7.22b)′, since $\mathrm{sinc}(\Omega\tau_p/2) = 0$ when $\Omega\tau_p/2 = n\pi$, $n = \pm 1, \pm 2, \pm 3, \ldots$, etc.

Now we will focus on the determination of the quantities in (7.19a,b,c)′, the time resolution constant TRC, frequency resolution constant FRC, the reciprocals of TRC and FRC, known as the frequency span FSP, and the time span TSP, respectively, and finally the area of ambiguity AA, all for the rectangular pulse.

First, the time resolution constant [from (7.19a)′, (7.22a)′, and (7.20d)] is

$$\mathrm{TRC} = \int_{-\tau_p}^{\tau_p} d\tau \left(1 - \frac{|\tau|}{\tau_p}\right)^2 = 2\tau_p \int_0^1 dx(1-x)^2 = \frac{2}{3}\tau_p = \frac{0.563}{B} \tag{7.23a}$$

The frequency span [from (7.23a)] is

$$\mathrm{FSP} = 1/\mathrm{TRC} = \frac{3}{2\tau_p} = 1.776B \tag{7.23b}$$

The frequency resolution constant [from (7.19b)′, (7.22b)′, and (7.20d)] is

$$\mathrm{FRC} = \frac{1}{2\pi}\int_{-\infty}^{\infty} d\Omega\, \mathrm{sinc}^2\left(\frac{\Omega\tau_p}{2}\right) = \frac{4}{2\pi\tau_p}\int_0^{\infty} dx\, \mathrm{sinc}^2(x) = \frac{1}{\tau_p} = 1.185B \tag{7.23c}$$

(from Spiegel,[8] #15.36, P. 96). The time-span [from (7.23c)] is

$$\mathrm{TSP} = 1/\mathrm{FRC} = \tau_p = \frac{0.844}{B} \tag{7.23d}$$

Finally, the area of ambiguity [from (7.22) and (7.19c)′] is

$$\begin{aligned}
\mathrm{AA} &= \frac{1}{2\pi}\int_{-\tau_p}^{\tau_p} d\tau\left(1 - \frac{|\tau|}{\tau_p}\right)^2 \int_{-\infty}^{\infty} d\Omega\, \mathrm{sinc}^2\left[\frac{\Omega\tau_p}{2}\left(1 - \frac{|\tau|}{\tau_p}\right)\right] \\
&= \frac{2\tau_p}{2\pi}\left(\frac{2}{\tau_p}\right)\int_0^1 dx(1-x)^2 \int_{-\infty}^{\infty} dy\, \mathrm{sinc}^2[y(1-x)] \\
&= \frac{2}{\pi}\int_0^1 dx \frac{(1-x)^2}{(1-x)}\int_{-\infty}^{\infty} dz\, \mathrm{sinc}^2(z) = \left(\frac{2}{\pi}\right)\left(\frac{1}{2}\right)\pi = 1
\end{aligned} \tag{7.23e}$$

The results that emerge from this exercise are as follows:

$$\mathrm{TRC} = \frac{2}{3}\tau_p = \frac{0.563}{B}, \qquad \mathrm{FSP} = \frac{1.5}{\tau_p} = 1.776B \tag{7.24a}$$

$$\mathrm{FRC} = \frac{1}{\tau_p} = 1.185B, \qquad \mathrm{TSP} = \tau_p = \frac{0.844}{B} \tag{7.24b}$$

$$(\mathrm{TRC})(\mathrm{FRC}) = \frac{1}{(\mathrm{FSP})(\mathrm{TSP})} = \frac{2}{3} \tag{7.24c}$$

$$\mathrm{AA} = 1 \tag{7.24d}$$

Before discussing the meanings of the results (7.24a,b,c,d), let us consider another pulse shape, the Gaussian pulse with the same total energy as the rectangular pulse of duration τ_p (Fig. 7.5)

The complex envelope is

$$\hat{s}_T(t) = e^{-1/2(t/\Delta t)^2} \tag{7.25a}$$

and the FT of $s_T(t, \omega_o)$ is

$$\begin{aligned}
S_T(\omega, \omega_o) &= \int_{-\infty}^{\infty} dt\, e^{-1/2(t/\Delta t)^2} \cos(\omega_o t + \psi) e^{-j\omega t} \\
&= \frac{e^{j\psi}}{2} \int_{-\infty}^{\infty} dt\, e^{-j(\omega-\omega_o)t-(t^2/2\Delta t^2)} + \frac{e^{-j\psi}}{2} \int_{-\infty}^{\infty} dt\, e^{-j(\omega+\omega_o)t-(t^2/2\Delta t^2)} \\
&= \frac{e^{j\psi}}{2} \int_{-\infty}^{\infty} dt\, e^{(-1/2(\Delta t)^2)\{t^2+2[j(\omega-\omega_o)(\Delta t)^2]t\}} + \frac{e^{-j\psi}}{2} \int_{-\infty}^{\infty} dt\, e^{(-1/2(\Delta t)^2)\{t^2+2[j(\omega+\omega_o)(\Delta t)^2]t\}} \\
&= \Delta t\sqrt{2\pi}\left\{\frac{e^{j\psi}}{2} e^{-[(\omega-\omega_o)\Delta t]^2/2} + \frac{e^{-j\psi}}{2} e^{-[(\omega+\omega_o)\Delta t]^2/2}\right\}
\end{aligned} \tag{7.25b}$$

To determine Δt in terms of τ_p, the duration of a rectangular pulse of the same energy, we equate the energies of the two pulses, i.e.,

$$\begin{aligned}
\int_{-\infty}^{\infty} dt\, |\hat{s}_T(t)|^2_{\text{rect}} &= \int_{-\tau_p/2}^{\tau_p/2} dt(1) = \tau_p = \int_{-\infty}^{\infty} dt\, |\hat{s}_T(t)|^2_{\text{Gauss}} \\
&= \int_{-\infty}^{\infty} dt\, e^{-t^2/(\Delta t)^2} = \Delta t\sqrt{\pi}
\end{aligned} \tag{7.25c}$$

We can rewrite (7.25a,b) with the aid of (7.25c), with the results

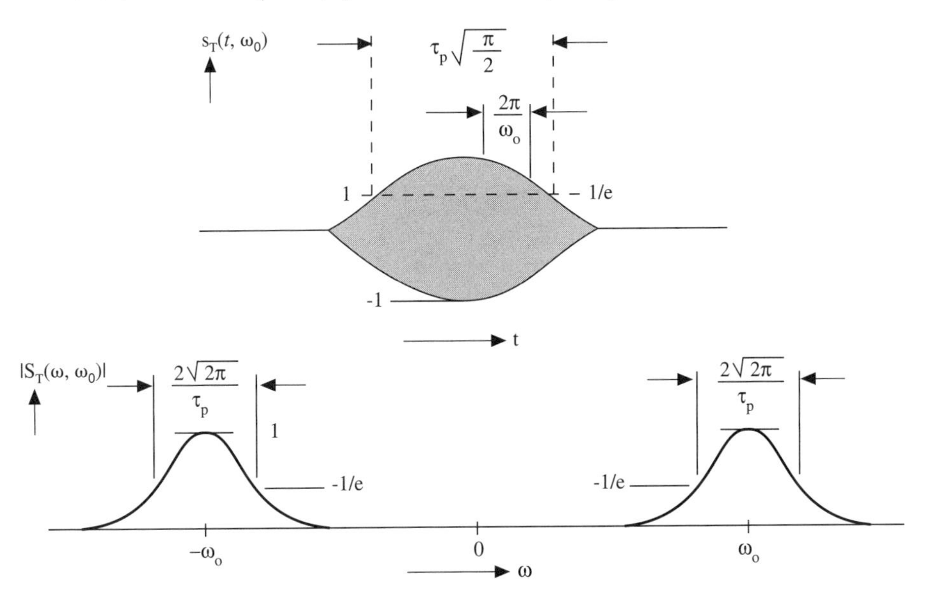

FIGURE 7.5
Gaussian IF pulse and its Fourier transform (spectrum).

$$\hat{s}_T(t) = e^{-\pi/2(t/\tau_p)^2} \tag{7.25a)'$$

$$S_T(\omega, \omega_o) = \tau_p\sqrt{2}\left\{\frac{e^{j\psi}}{2}\, e^{-[(\omega-\omega_o)\tau_p]^2/2\pi} + \frac{e^{-j\psi}}{2}\, e^{-[(\omega+\omega_o)\tau_p]^2/2\pi}\right\} \tag{7.25b)'$$

The complex ambiguity function for this case [from (7.18)] is

$$\hat{\Lambda}(\tau, \Omega) = \frac{\displaystyle\int_{-\infty}^{\infty} dt\; e^{j\Omega t}\, e^{-(\pi/2)(t/\tau_p)^2-(\pi/2)[(t-\tau)/\tau_p]^2}}{\displaystyle\int_{-\infty}^{\infty} dt\; e^{-2(\pi/2)(t/\tau_p)^2}} \tag{7.26a}$$

The numerator of $\hat{\Lambda}(\tau, \Omega)$ can be determined from the usual completion of the square in the exponential, i.e.

$$\text{Num}[\hat{\Lambda}(\tau, \Omega)] = e^{-\pi\tau^2/2\tau_p^2}\int_{-\infty}^{\infty} dt\; e^{-(\pi/\tau_p^2)\{t^2-2[1/2[\tau+(j\tau_p^2\Omega/\pi)]]t\}}$$

$$= \left(\frac{\tau_p}{\sqrt{\pi}}\right)\sqrt{\pi}\; e^{-\pi\tau^2/2\tau_p^2}e^{+(\pi/4\tau_p^2)[\tau+(j\tau_p^2\Omega/\pi)]^2} \tag{7.26b}$$

$$\text{Num}[\hat{\Lambda}(\tau, \Omega)] = \tau_p e^{-(\pi\tau^2/2\tau_p^2)+(\pi\tau^2/4\tau_p^2)-(\tau_p^2\Omega^2/4\pi)}\; e^{j\Omega\tau}$$

$$= \tau_p e^{-\pi/4[(\tau/\tau_p)^2+(\tau_p\Omega/\pi)^2]}\; e^{j\Omega\tau}$$

The denominator, obtained by setting τ and Ω to zero in (7.26b) is

$$\text{Denom}[\hat{\Lambda}(\tau, \Omega)] = \tau_p \tag{7.26c}$$

Finally, the CAF is

$$\hat{\Lambda}(\tau, \Omega) = e^{(-\pi/4)[(\tau/\tau_p)^2+(\tau_p\Omega/\pi)^2]}e^{j\Omega\tau} \tag{7.26d}$$

The range-only and Doppler-only forms of $|\hat{\Lambda}(\tau, \Omega)|$ are separable in this case, i.e.

$$|\hat{\Lambda}(\tau, \Omega)| = |\hat{\Lambda}(\tau, 0)||\hat{\Lambda}(0, \Omega)| \tag{7.26e}$$

but if we maintain the complex forms, the separability is lost because of the factor $e^{j\Omega\tau}$. The range-and-Doppler ambiguity functions are

$$|\hat{\Lambda}(\tau, \Omega)| = e^{-(\pi/4)(\tau/\tau_p)^2} \tag{7.26f}$$

$$|\hat{\Lambda}(0, \Omega)| = e^{-(\pi/4)(\tau_p\Omega/\pi)^2} \tag{7.26g}$$

If we define the bandwidth of the Gaussian pulse as the spacing between -3 dB points, then from (7.25b)′, the condition $e^{-1/2\pi\,(W\tau_p/2)^2} = 1/\sqrt{2}$ (where $W = 2\pi B$ and B is the -3 dB bandwidth in Hertz) gives rise to the relationship between B and τ_p.

$$B = \frac{\left(\sqrt{\frac{\ln 2}{\pi}}\right)}{\tau_p} = \frac{0.4697}{\tau_p} \simeq \frac{1}{2\tau_p} \tag{7.26h}$$

If we wanted to express Ω as $2\pi F$ and express the Doppler AF in terms of bandwidth in Hertz rather than pulse width, then we could rewrite (7.26g) in the form

$$\hat{\Lambda}(0, 2\pi F) = e^{-(\pi/4)(0.8825)(F/B)^2} \tag{7.26i}$$

The analogs of (7.23a,b,c,d,e) are [from (7.26b,h) and (7.19a)′]

$$\text{TRC} = \int_{-\infty}^{\infty} d\tau\, e^{-2(\pi/4)(\tau/\tau_p)^2} = \tau_p\sqrt{\frac{2}{\pi}}\int_{-\infty}^{\infty} dx\, e^{-x^2} = \sqrt{2}\tau_p = 1.414\tau_p$$
$$= \frac{0.664}{B} \tag{7.27a}$$

[and with the aid of (7.26h)]

$$\text{FSP} = \frac{1}{\text{TRC}} = \frac{1}{\sqrt{2}\tau_p} = \frac{0.707}{\tau_p} = B\sqrt{\frac{\pi}{2\ln 2}} = 1.5B \tag{7.27b}$$

[from (7.26c) and (7.19b)′]

$$\text{FRC} = \frac{1}{2\pi}\int_{-\infty}^{\infty} d\Omega\, e^{-2(\Omega\tau_p)^2/4\pi} = \frac{\sqrt{2\pi}}{2\pi\tau_p}\int_{-\infty}^{\infty} dx\, e^{-x^2} \tag{7.27c}$$
$$= \frac{1}{\sqrt{2}\tau_p} = \frac{0.707}{\tau_p} = 1.5B$$

[and with the aid of (7.26h)]

$$\text{TSP} = \frac{1}{\text{FRC}} = \sqrt{2}\tau_p = 1.414\tau_p = \frac{0.664}{B} \tag{7.27d}$$

[from (7.26a) and (7.19c)′ and (7.27b,d)].

$$\text{AA} = \frac{1}{2\pi}\int_{-\infty}^{\infty} d\tau \int_{-\infty}^{\infty} d\Omega\, e^{(-\pi/2)[(\tau/\tau_p)^2 + (\Omega\tau_p/\pi)^2]}$$
$$= \left(\int_{-\infty}^{\infty} d\tau\, e^{-(\pi/2)(\tau/\tau_p)^2}\right)\left(\frac{1}{2\pi}\int_{-\infty}^{\infty} d\Omega\, e^{-(\Omega\tau_p)^2/2\pi}\right) \tag{7.27e}$$
$$= (\text{TRC})(\text{FRC}) = \left(\frac{1}{\sqrt{2}\tau_p}\right)(\sqrt{2}\tau_p) = 1$$

The summary of results for this case, analogous to Eqs. (7.24a,b,c,d) for the rectangular case, is

$$\text{TRC} = \sqrt{2}\tau_p = \frac{0.664}{B}, \qquad \text{FSP} = \frac{1}{\sqrt{2}\tau_p} = 1.5B \tag{7.28a}$$

$$\text{FRC} = \frac{1}{\sqrt{2}\tau_p} = 1.5B, \qquad \text{TSP} = \sqrt{2}\tau_p = \frac{0.664}{B} \tag{7.28b}$$

$$(\text{TRC})(\text{FRC}) = \frac{1}{(\text{FSP})(\text{TSP})} = 1 \tag{7.28c}$$

$$\text{AA} = 1 \tag{7.28d}$$

One observation that we can make by comparison of these two cases is that the area of ambiguity AA is unity in both cases. It can be shown that this is true for *any* pulse shape. To demonstrate this, we invoke (7.18) and (7.19c)′, leading to

$$\text{AA} = \left[\frac{1}{2\pi}\int_{-\infty}^{\infty} d\tau \int_{-\infty}^{\infty} d\Omega \iint_{-\infty}^{\infty} dt_1\, dt_2\, \hat{s}_T\,(t_1)\hat{s}_T^*(t_1 - \tau) \cdots\right.$$

$$\left.\hat{s}_T^*(t_2)\hat{s}_T(t_2 - \tau)e^{j\Omega(t_1 - t_2)}\right] / \ldots$$

$$\left[\iint_{-\infty}^{\infty} dt_1\, dt_2 |\hat{s}_T(t_1)|^2 |\hat{s}_T(t_2)|^2\right] \tag{7.29a}$$

whose numerator is (after changing the order of integration)

$$\text{Num (AA)} = \int_{-\infty}^{\infty} d\tau \int_{-\infty}^{\infty} dt_1\, \hat{s}_T(t_1)\hat{s}_T^*(t_1 - \tau) \int_{-\infty}^{\infty} dt_2\, \hat{s}_T^*(t_2)\hat{s}_T(t_2 - \tau) \cdots$$

$$\left(\frac{1}{2\pi}\int_{-\infty}^{\infty} d\Omega\, e^{j\Omega(t_1 - t_2)}\right) \tag{7.29b}$$

Noting that the parenthesized integral in (7.29b) is the unit impulse function $\delta(t_1 - t_2)$, we obtain

$$\text{Num (AA)} = \int_{-\infty}^{\infty} dt_1\, |\hat{s}_T(t_1)|^2 \int_{-\infty}^{\infty} d\tau\, |\hat{s}_T(t_1 - \tau)|^2$$

$$= \left[\int_{-\infty}^{\infty} dt\, |\hat{s}_T(t)|^2\right]^2 \tag{7.29c}$$

which is the same as the denominator in (7.29a), leading to the result

$$\text{AA} = 1 \tag{7.29d}$$

regardless of the shape of the waveform.

We can perform a similar calculation for TRC, using the frequency form of (7.18)′, and the fact that $\frac{1}{2\pi}\int_{-\infty}^{\infty} d\tau\, e^{j(\omega_1 - \omega_2)\tau} = \delta(\omega_1 - \omega_2)$

$$\text{TRC} = \frac{\displaystyle\int_{-\infty}^{\infty} d\tau \iint_{-\infty}^{\infty} d\omega_1\, d\omega_2\, |S_T(\omega_1;\, \omega_o)|^2 |S_T(\omega_2;\, \omega_o)|^2 e^{j(\omega_1-\omega_2)\tau}}{\left[\displaystyle\int_{-\infty}^{\infty} d\omega\, |S_T(\omega;\, \omega_o)|^2\right]^2}$$

$$= \frac{2\pi \displaystyle\int_{-\infty}^{\infty} d\omega\, |S_T(\omega;\, \omega_o)|^4}{\left[\displaystyle\int d\omega\, |S_T(\omega;\, \omega_o)|^2\right]^2} \tag{7.29e}$$

For FRC, we can use the CAF form in (7.18)″ with the result

$$\text{FRC} = \frac{\dfrac{1}{2\pi}\displaystyle\int_{-\infty}^{\infty} d\Omega \iint_{-\infty}^{\infty} dt_1\, dt_2\, |\hat{s}_T(t_1)|^2 |\hat{s}_T(t_2)|^2 e^{j\Omega(t_1-t_2)}}{\left[\displaystyle\int_{-\infty}^{\infty} dt\, |\hat{s}_T(t)|^2\right]^2}$$

$$= \int_{-\infty}^{\infty} dt\, |\hat{s}_T(t)|^4 \Big/ \left[\int_{-\infty}^{\infty} dt\, |\hat{s}_T(t)|^2\right]^2 \tag{7.29f}$$

If the signal spectrum were perfectly flat and had a Hertz bandwidth B, it is evident from (7.29e) that TRC would be $2\pi(2\pi B)/(2\pi B)^2 = 1/B$. If the signal waveform were perfectly flat and had duration τ_p (rectangular pulse), (7.29f) tells us that FRC would be $\tau_p/(\tau_p)^2 = 1/\tau_p$. Although these two cases are incompatible, the fact remains that for *any* pulse shape, it is intuitively evident that

$$\tau_p = \frac{\kappa}{B} \tag{7.29g}$$

where κ is a positive number usually between about 0.4 and 2.0 (i.e., "of the order of 1"), that

$$\text{TRC} = \frac{1}{\text{FSP}} = \frac{a}{B} \tag{7.29h}$$

and that

$$\text{FRC} = \frac{1}{\text{TSP}} = \frac{b}{\tau_p} \tag{7.29i}$$

where a and b are both positive numbers "of the order of 1." The product of TRC and FRC, reciprocal to the product of FSP and TSP is [from (7.29g,h,i)]

$$(\text{TRC})(\text{FRC}) = \frac{1}{(\text{FSP})(\text{TSP})} = \frac{ab}{\kappa} = d \tag{7.29j}$$

where d is again a positive number somewhere near 1. In the case of the rectangular and Gaussian pulses with the same total energy, $d = 0.667$ and 1, respectively, as shown by (7.24c) and (7.28c), respectively, whereas AA is 1 in both cases. Both the product in (7.29j) and the area of ambiguity AA are illustrative of the tradeoff between range and Doppler resolution (i.e., good resolution in one implies poor resolution in the other and vice versa).

The tradeoff is related to the fact that a wide signal spectrum implies a short-duration signal and a narrow signal spectrum implies a long-duration signal. The TRC is close to the reciprocal of the bandwidth and the FRC is close to the reciprocal of the signal duration. The frequency span, as shown by taking the reciprocal of (7.29e), is *not* identical to the bandwidth (except in the case of a flat spectrum), but is a measure of the frequency space occupied by the signal spectrum, which in most cases is very close to the bandwidth.

The time span, the reciprocal of (7.29f), is a measure of how much time is actually occupied by the signal, only equal to the pulse duration in the case of a rectangular pulse, but usually very close to the pulse duration regardless of the pulse shape. When the frequency span and time span are multiplied together, their product is always close to unity, as is the product of duration and bandwidth. The distinction between these two products is not terribly important for simple pulse shapes, but becomes quite important for cases where the bandwidth is much greater than the reciprocal pulse duration, as in the case of frequency modulated or "chirp" pulses.

To illustrate the points discussed above, we show "ambiguity diagrams" for rectangular and Gaussian pulses in Figures 7.6 and 7.7, respectively. These are three-dimensional sketches of the amplitude of the ambiguity function in the two-dimensional space of τ and Ω.*

To understand these diagrams, we begin with the rectangular pulse case (Fig. 7.6) which is based on Eq. (7.22), whose absolute value is

$$|\hat{\Lambda}(\tau, \Omega)| = u(\tau_p - |\tau|)\left(1 - \frac{|\tau|}{\tau_p}\right)\left|\operatorname{sinc}\left[\frac{\Omega\tau_p}{2}\left(1 - \frac{|\tau|}{\tau_p}\right)\right]\right| \tag{7.30}$$

It is evident from (7.30) that the positions of the nulls are dependent on the value of $|\tau|$. The nulls occur when $\Omega = \Omega_n$, where

$$\frac{\Omega_n\tau_p}{2}\left[1 - \frac{|\tau|}{\tau_p}\right] = n\pi; \qquad n = \pm 1, +2, +3 \cdots \tag{7.31a}$$

or equivalently

$$\Omega_n = \frac{2n\pi}{[1 - (|\tau|/\tau_p)]\tau_p}$$

Thus, for example

$$\begin{aligned}\Omega_n &= \frac{2\pi n}{\tau_p} \quad \text{if } |\tau| = 0; \qquad \Omega_n = \frac{20n\pi}{\tau_p} \quad \text{if } |\tau| = 0.9\tau_p \\ &= \frac{4\pi n}{\tau_p} \quad \text{if } |\tau| = \frac{\tau_p}{2} \qquad\quad = \infty \text{ if } |\tau| = \tau_p\end{aligned} \tag{7.31b}$$

* Ambiguity diagrams are useful graphic tools for study of radar signal resolution and such diagrams for various classes of radar signal waveforms can be found in many radar texts.[9,10]

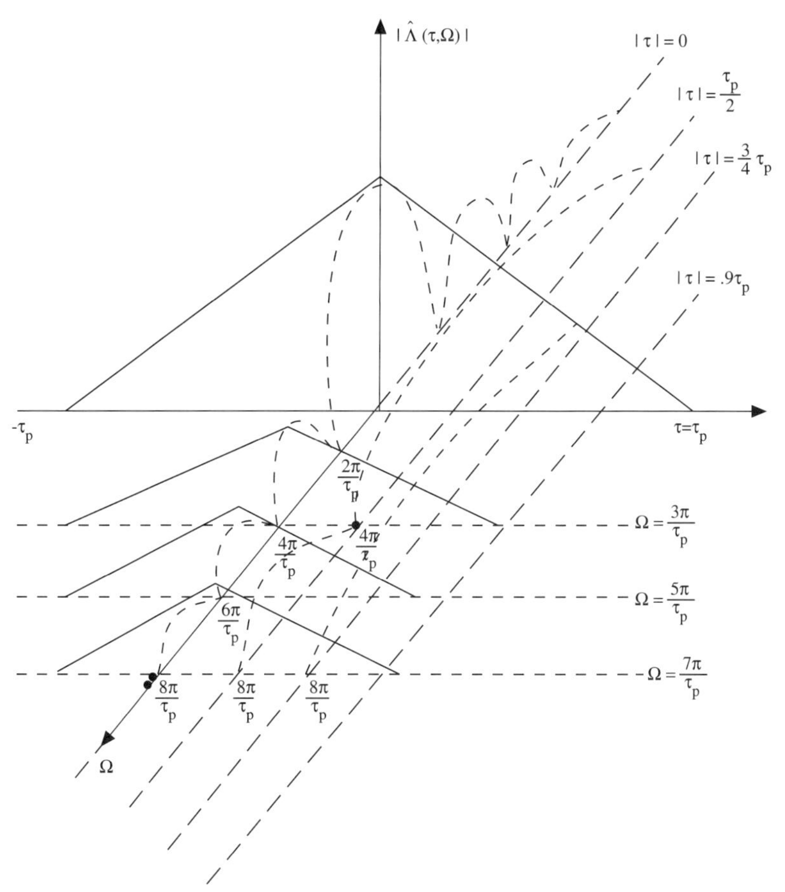

FIGURE 7.6
Ambiguity diagram for rectangular pulse.

$$= \frac{8\pi n}{\tau_p} \qquad \text{if } |\tau| = \frac{3}{4}\tau_p$$

The implication of (7.31b) is that the width of the AF in Doppler space becomes very large as $|\tau|$ approaches τ_p, and becomes infinite as $|\tau|$ becomes equal to τ_p. Figure 7.6 illustrates this. Also, if Ω falls on one of the sidelobe peaks in Doppler space, then $(\Omega\tau_p/2)[1 - (|\tau|/\tau_p)] = [n + (1/2)]\,\pi$ and the sine function takes the value $1/[n + (1/2)]\,\pi$. The projections on the planes $|\Omega| = 3\pi/\tau_p$, $|\Omega| = 5\pi/\tau_p$, $|\Omega| = 7\pi/\tau_p$, etc. are reduced in amplitude in amounts 1/3, 1/5, 1/7 etc. Thus the amplitude of the AF is reduced as the Doppler increases for fixed time delay $|\tau|$ and the width increases as $|\tau|$ increases for fixed Doppler.

Figure 7.7 shows an ambiguity diagram for the Gaussian pulse, based on Eq. (7.26d), whose amplitude is

$$|\hat{\Lambda}(\tau, \Omega)| = e^{(-\pi/4)[(\tau/\tau_p)^2 + (\tau_p\Omega/\pi)^2]} \tag{7.32}$$

Figure 7.7a shows a three-dimensional sketch of the AF. Figure 7.7b shows a two-dimensional projection on the τ–Ω plane. The ellipses on (b) show the contours of constant $|\hat{\lambda}(\tau,\Omega)$ for the case of a wide pulse and that of a narrow pulse. It is a consequence of the result that the area of ambiguity is always unity (Eq. 7.28d) and that an attempt to enhance range resolution by using a narrow pulse (large bandwidth), which will produce a small value of TRC, must be achieved at the cost of reduced Doppler resolution (high value of FRC). The wide-pulse case shows that an attempt to enhance Doppler resolution by using a wide pulse will produce a low value of FRC but a high value of TRC, thereby degrading range resolution. These contour plots of the ambiguity function serve to quantify that tradeoff by showing graphically how high a price must be paid in reduced range resolution for an attempt to enhance Doppler resolution and vice versa. The key to the tradeoff is the result that the area of ambiguity AA is *always* constant regardless of the shape of the waveform [Eq.(7.29d)] and an attempt to reduce its width in one direction (τ or Ω) is always accompanied by a widening in the other direction (Ω or τ). To summarize the point, high resolution can be achieved in *either* range *or* doppler but not both simultaneously. The ultimate limit of good Doppler resolution is a CW signal, which would appear on Figure 7.7b as a horizontal straight line of infinite extent. The ultimate limit of good range resolution is a pulse that is an impulse function, which would appear as a vertical straight line of infinite extent.

To get the contours illustrated in Figure 7.7b, we will specify the contour $|\hat{\Lambda}(\tau, \Omega| = 1/\sqrt{2}$ ($-$ 3 dB points), resulting in the ellipse

$$\left(\frac{\tau}{a}\right)^2 + \left(\frac{\Omega}{b}\right)^2 = 1 \tag{7.33}$$

where

$$a = \text{dimension in } \tau = \tau_p\sqrt{\frac{2}{\pi}\ln 2} = 0.664\tau_p$$

$$b = \text{dimension in } \Omega = \frac{\sqrt{2\pi \ln 2}}{\tau_p} = \frac{2.087}{\tau_p} \tag{7.33a)$'$}$$

and where a and b can be expressed in terms of bandwidth through (7.26h) ($\tau_p = 1/2B$ for Gaussian pulse) or in terms of TRC or FRC through (7.27a,c)

$$a = \frac{0.332}{B} = \frac{1}{2}\,(\text{TRC})$$

$$b = 4.174B = 2.78\,(\text{FRC}) \tag{7.33b)$'$}$$

To get into the radar context, it would be helpful to calibrate a and b in terms of range and Doppler. That could be done by simply expressing the pulse length τ_p in terms of its range equivalent Δr_p, i.e.

$$\tau_p = \frac{2}{c}\Delta r_p = (6.7)(10^{-9})\Delta r_p \tag{7.33c)$'$}$$

or expressing B in terms of its doppler speed equivalent ΔV

$$B = \frac{2f_0\Delta V}{c} = \frac{2\Delta V}{\lambda} \tag{7.33d)$'$}$$

Substituting (7.33c)′ into the form (7.33a)′ would give us a calibration in range and substituting (7.33d)′ into (7.33b)′ a calibration in Doppler.

Let us now consider a signal pulse with frequency modulation. If the FM is linear, this is usually called a "chirp" pulse. It is used in an attempt to enhance range resolution with a pulse much longer than would ordinarily be employed in a high-range resolution radar. The ambiguity function is an excellent tool for the study of chirp pulses. It shows how one can make use of a large bandwidth in a way that enhances range resolution without too large a compromise in Doppler resolution.

The linear FM signal has the form

$$s_T(t, \omega_o) = p(t) \cos[(\omega_o + \dot{\omega}_M t)t + \psi] \tag{7.34a}$$

where $\dot{\omega}_M$ = rate of change of transmitted frequency in radians per (second)^2 and $p(t)$ is the pulse amplitude function. The complex envelope of $s_T(t, \omega_o)$ is

$$\hat{s}_T(t) = p(t)e^{j\psi}e^{j\dot{\omega}_M t^2} \tag{7.34b}$$

The CAF is [from (7.8)′]

$$\hat{\Lambda}(\tau, \Omega) = \frac{\displaystyle\int_{-\infty}^{\infty} dt\ p(t)p(t-\tau)e^{j\dot{\omega}_M t^2}e^{-j\dot{\omega}_M (t-\tau)^2}e^{j\Omega t}}{\displaystyle\int_{-\infty}^{\infty} dt\ p^2(t)} \tag{7.34c}$$

Chirping can be done with any pulse amplitude function $p(t)$, but the easiest case to analyze is the Gaussian pulse, where [from (7.25a)′]

$$p(t) = e^{(-\pi/2)(t/\tau_p)^2} \tag{7.34d}$$

Using (7.34d), Eq. (7.34c) takes the form

$$\hat{\Lambda}(\tau, \Omega) = \frac{\displaystyle\int_{-\infty}^{\infty} dt\ e^{(-\pi/2)(t/\tau_p)^2-(\pi/2)[(t-\tau)/\tau_p]^2+j\dot{\omega}_M t^2-j\dot{\omega}_M(t-\tau)^2+j\Omega t}}{\displaystyle\int_{-\infty}^{\infty} dt\ e^{-\pi(t/\tau_p)^2}} \tag{7.34e}$$

The denominator has the value τ_p as in (7.26c), being unaffected by the FM process. Then, if $\hat{\tau} = \tau/\tau_p$,

$$\hat{\Lambda}(\tau, \Omega) = \int_{-\infty}^{\infty} dt'\ e^{-(\pi/2)[t'^2+(t'-\hat{\tau})]^2} \ldots$$

$$e^{j\dot{\omega}_M \tau_p^2[2\hat{\tau}t'-\hat{\tau}^2]+j\Omega\tau_p t'} \tag{7.34f}$$

The t'^2 terms cancel each other out in the imaginary part of the exponent and we are left with an expression not greatly different from that arising in the non-chirped Gaussian pulse case, i.e.

$$\hat{\Lambda}(\tau, \Omega) = e^{-(\pi/2)(\tau/\tau_p)^2-j\dot{\omega}_M\tau_p^2(\tau/\tau_p)^2} \ldots$$

$$\int_{-\infty}^{\infty} \mathrm{d}t'\ \mathrm{e}^{-\pi\{t'^2-2((\tau/2\tau_p)+j[(\dot{\omega}_M\tau_p^2\hat{\tau}/\pi)+(\Omega\tau_p/2\pi)])t'\}} \tag{7.34g}$$

Since we are interested only in the *amplitude* $|\hat{\Lambda}(\tau, \Omega)|$, we now complete the square in the exponent in the usual way and obtain the result

$$
\begin{aligned}
|\hat{\Lambda}(\tau, \Omega)| &= \left| e^{-(\pi/2)(\tau/\tau_p)^2} e^{+\pi((\tau/2\tau_p)+(j\tau_p/2\pi)(2\dot{\omega}_M\tau+\Omega))^2} \right| \\
&= e^{-(\pi/2)(\tau/\tau_p)^2+(\pi/4)(\tau/\tau_p)^2-\pi\{(\Omega\tau_p/2\pi)^2+(\tau_p/2\pi)^2[(2\dot{\omega}_M\tau)^2+4\dot{\omega}_M\tau\Omega]\}} \\
&= e^{-(\pi/4)[(\tau/\tau_p)^2+(\Omega\tau_p/\pi)^2+(2\dot{\omega}_M\tau_p/\pi)^2\tau^2+(4\dot{\omega}_M\tau_p^2/\pi)(\tau\Omega)]} \\
&= e^{-(\pi/4)\{[1+(2\dot{\omega}_M\tau_p^2/\pi)^2](\tau/\tau_p)^2+(\Omega\tau_p/\pi)^2+(4\dot{\omega}_M\tau_p^2/\pi)(\tau\Omega)\}}
\end{aligned}
\tag{7.34h}
$$

The exponent is clearly an ellipse in the (τ, Ω) plane, but it is rotated with respect to the τ and Ω axes. The new axes are designated as $(\tau'\text{–}\Omega')$ and are rotated through an angle ϕ' with respect to the $(\tau\text{–}\Omega)$ axes. Thus, comparing with the form in (7.26b), we can write

$$
\begin{aligned}
\left(\frac{\tau'}{\tau_p}\right)^2 + \left(\frac{\tau_p\Omega'}{\pi}\right)^2 &= \left[1 + \left(\frac{2\dot{\omega}_M\tau_p^2}{\pi}\right)^2\right]\left(\frac{\tau}{\tau_p}\right)^2 \\
&\quad + \left(\frac{\Omega\tau_p}{\pi}\right)^2 + 4\dot{\omega}_M\tau_p^2\left(\frac{\Omega\tau_p}{\pi}\right)\left(\frac{\tau}{\tau_p}\right)
\end{aligned}
\tag{7.34i}
$$

where

$$
\begin{aligned}
\tau' &= \tau\cos\phi' + \Omega\sin\phi' \\
\Omega' &= -\tau\sin\phi' + \Omega\cos\phi'
\end{aligned}
$$

It is left as an exercise to find the angle ϕ' corresponding to given values of $\dot{\omega}_M$. It suffices for the moment to say that the rotation of the elliptical contours obtained from (7.33) and shown in Figure 7.7 increases monotonically with the degree of frequency modulation, i.e., with the value of $\dot{\omega}_M$. It is obvious from (7.34h) that the original form (7.32) is recovered when $\dot{\omega}_M = 0$. Sketches showing the effect of chirping on the AF are shown in Figure 7.8.

To calculate the various quantities related to resolution as in (7.28a,b,c,d) we form the following manipulations:

$$
\begin{aligned}
\text{TRC} &= \int_{-\infty}^{\infty} d\tau\, |\hat{\Lambda}(\tau, 0)|^2 = \int_{-\infty}^{\infty} d\tau\, e^{-(\pi/2)[1+(2\dot{\omega}_M\tau_p^2/\pi)^2](\tau/\tau_p)^2} \\
&= \frac{\sqrt{2}\tau_p}{\sqrt{1+(2\dot{\omega}_M\tau_p^2/\pi)^2}}
\end{aligned}
\tag{7.35a}
$$

$$
\text{FSP} = \frac{1}{\text{TRC}} = \frac{\sqrt{1+(2\dot{\omega}_M\tau_p^2/\pi)^2}}{\sqrt{2}\tau_p}
\tag{7.35b}
$$

$$
\text{FRC} = \frac{1}{2\pi}\int_{-\infty}^{\infty} d\Omega\, |\hat{\Lambda}(0, \Omega)|^2 = \frac{1}{2\pi}\int_{-\infty}^{\infty} d\Omega\, e^{-(\pi/2)(\Omega\tau_p/\pi)^2} = \frac{1}{\sqrt{2}\tau_p}
\tag{7.35c}
$$

$$
\text{TSP} = \frac{1}{\text{FRC}} = \sqrt{2}\tau_p
\tag{7.35d}
$$

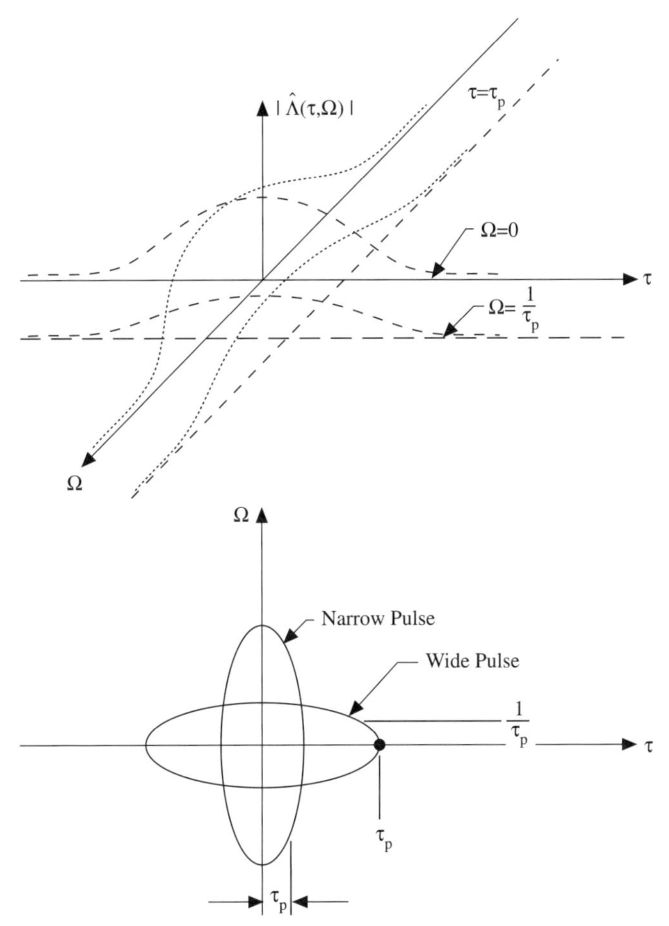

FIGURE 7.7
Ambiguity function for Caussian pulse. (a) 3 dimensional ambiguity function. (b) 2-dimensional projection of (τ-Ω) plane-contours of constant $|\hat{\Lambda}(\tau,\Omega)|$.

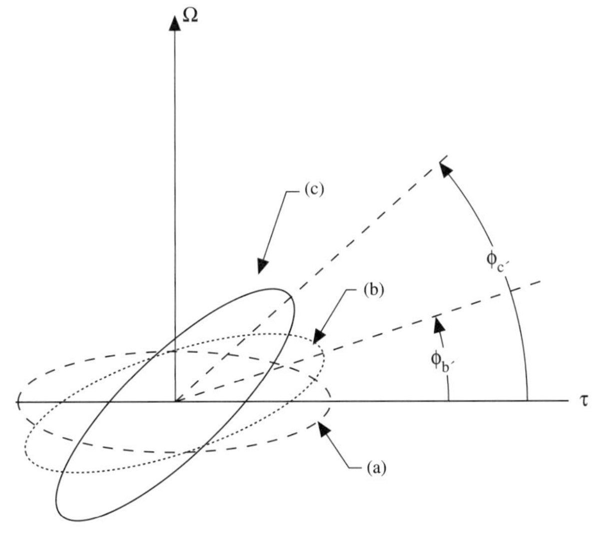

FIGURE 7.8
Effect of FM (chirp) technique on ambiguity function with Gaussian envelope. (a) $\dot{\omega}_M = 0$ (no frequency modulation) – – –. (b) $(2\dot{\omega}_M\tau_p^2/\pi)^2 << 1$, ····. (c) $(2\dot{\omega}_M\tau_p^2/\pi)^2 = 1$, ——.

$$\text{AA} = \frac{1}{2}\int_{-\infty}^{\infty} d\hat{\tau} \int_{-\infty}^{\infty} d\hat{\Omega}\, e^{-(\pi/2)\{[1+(2\dot{\omega}_M\tau_P^2/\pi)^2]\hat{\tau}^2+(\hat{\Omega})^2+2\dot{\omega}_M\tau_P^2\hat{\tau}\hat{\Omega}\}}$$

$$= \frac{1}{2}\int_{-\infty}^{\infty} d\hat{\Omega}\, e^{-\pi\hat{\Omega}^2/2} I(\hat{\Omega}) \tag{7.35e}$$

where, if

$$\gamma = \left[\frac{\dot{\omega}_M\tau_P^2\hat{\Omega}}{1 + (2\dot{\omega}_M\tau_{P/\pi}^2)^2}\right]$$

$$I(\hat{\Omega}) = \int_{-\infty}^{\infty} d\tau\, e^{-(\pi/2)[1+(2\dot{\omega}_M\tau_P^2/\pi)^2][\hat{\tau}^2+2\gamma\hat{\tau}]}$$

$$= \sqrt{\frac{2}{\pi}\frac{1}{[1 + (2\dot{\omega}_M\tau_P^2/\pi)^2]}}\sqrt{\pi}\, e^{+(\pi/2)[1+(2\dot{\omega}_M\tau_P^2/\pi)^2]\gamma^2}$$

where the usual device of completing the square in the exponent has been employed to evaluate the integral. Carrying out the integral over $\hat{\Omega}$, we obtain

$$\text{AA} = \frac{1}{\sqrt{2}}\frac{1}{\sqrt{1 + (2\dot{\omega}_M\tau_P^2/\pi)^2}}$$

$$\times \int_{-\infty}^{\infty} d\hat{\Omega}\, e^{-(\pi/2)\hat{\Omega}^2\{[1+(2\dot{\omega}_M\tau_P^2/\pi)^2-(2\dot{\omega}_M\tau_P^2/\pi)^2]/1+(2\dot{\omega}_M\tau_P^2/\pi)^2\}} = 1 \tag{7.35f}$$

which merely serves to confirm that the general result (7.29d) holds and thereby acts as a check on the correctness of (7.34h).

To interpret the results (7.35a,b,c,d), in terms of bandwidth, we calculate the Fourier transform of the chirp pulse waveform and find the −3 dB points on the spectral curve. From (7.34a) and (7.34d)

$$S_T(\omega, \omega_o) = \frac{1}{2}\int_{-\infty}^{\infty} dt\, e^{-j\omega t} e^{-(\pi/2)(t/\tau_P)^2}[e^{j\psi}\, e^{j(\omega_o+\dot{\omega}_M t)t}$$

$$+ e^{-j\psi}\, e^{-j(\omega_o+\dot{\omega}_M t)t}] = \frac{1}{2}\hat{S}_T(\omega - \omega_o, 0) \tag{7.36}$$

$$+ \frac{1}{2}\hat{S}_T(\omega + \omega_o, 0)$$

where (since we are only interested in the *magnitude* of $\hat{S}_T$)

$$|\hat{S}_T(\omega, 0)| = \left|\int_{-\infty}^{\infty} dt\, e^{-j\omega t} e^{-(\pi/2)t^2[(1/\tau_P^2)\pm j(2\dot{\omega}_M/\pi)]}\right| =$$

(again using the device of completing the square in the exponent)

$$|\hat{S}_T(\omega, 0)| = K\left|\int_{-\infty}^{\infty} dt'\, e^{-(\pi/2)[1\pm j(2\dot{\omega}_M\tau_p^2/\pi)](t'^2+2\alpha t')}\right|$$

$$= \left|K' e^{(\pi/2)[1\pm j\,(2\dot{\omega}_M\tau_p^2/\pi)]\alpha^2}\right| \qquad (7.36a)'$$

where K and K' are scale factors irrelevant to the desired end result (since we are interested only in the variation of $|\hat{S}_T|$ with ω, not its absolute magnitude), and where

$$\alpha = \frac{j(\omega\tau_p/\pi)}{1 \mp j(2\dot{\omega}_M\tau_p^2/\pi)}$$

Ignoring scale factors, we have

$$|\hat{S}_T(\omega, 0)| = e^{-\,(\omega\tau_p)^2/\{2\pi[1+(2\dot{\omega}_M\tau_p^2/\pi)^2]\}} \qquad (7.36b)'$$

Finally, the bandwidth B is determined from the expression

$$e^{-\{[2\pi(B/2)\tau_p]^2/2\pi[1+(2\dot{\omega}_M\tau_p^2/\pi)^2]\}} = \frac{1}{\sqrt{2}} \qquad (7.36c)'$$

from which it follows that

$$B = (B)_{GCP} = \frac{1}{\tau_p}\sqrt{\frac{\ln 2}{\pi}}\,\sqrt{1 + (2\dot{\omega}_M\tau_p^2/\pi)^2} \qquad (7.36d)'$$

Since B for the "unchirped" Gaussian pulse is

$$(B) = \frac{1}{\tau_p}\sqrt{\frac{\ln 2}{\pi}} \qquad (7.36e)'$$

Thus the ratio of bandwidths (chirped relative to unchirped) is

$$\frac{B_{GCP}}{B} = \sqrt{1 + \left(\frac{2\dot{\omega}_M\tau_p^2}{\pi}\right)^2} \qquad (7.36f)'$$

Equation (7.36f)′ tells us that the bandwidth is increased by chirping the pulse and that the degree of enhancement increases with the frequency sweep rate $\dot{\omega}_M$. However, (7.36f)′ is inadequate to quantitatively describe the effect because it contains τ_p, the pulse duration, and seems to show an *increase* in B_{GCP} with increasing τ_p, violating our intuitive ideas about an inverse relationship between bandwidth and pulse duration. To try to understand what is happening, we refer to Figure 7.9, which shows the chirp pulse whose frequency variation during the pulse period τ_P, in Hertz, is Δf and hence

$$\dot{\omega}_M = 2\pi \dot{f}_M = \frac{2\pi\Delta f}{\tau_P} \tag{7.36g)'}$$

We now rewrite (7.36d)′, using (7.36g)′ and square both sides, with the result

$$(B_{GCP}\tau_P)^2 = \frac{\ln 2}{\pi}[1 + (4\Delta f \tau_P)^2] \tag{7.36h)'}$$

with a fixed value of Δf, we can now solve (7.36h)′ for τ_P with the result

$$\tau_P = \frac{(1/B_{GCP})\sqrt{\ln 2/\pi}}{\sqrt{1 - (16\ln 2/\pi)(\Delta f/B_{GCP})^2}} \tag{7.36i)'}$$

valid only if

$$\left|\frac{\Delta f}{B_{GCP}}\right| < \frac{1}{4}\sqrt{\frac{\pi}{\ln 2}} = 0.532 \tag{7.36j)'}$$

Equation (7.36j)′ tells us that the total range of frequency excursion during a pulse cannot exceed about half the bandwidth.

Also, rewriting (7.36f)′ with the aid of (7.36g)′ and (7.36e)′, we obtain

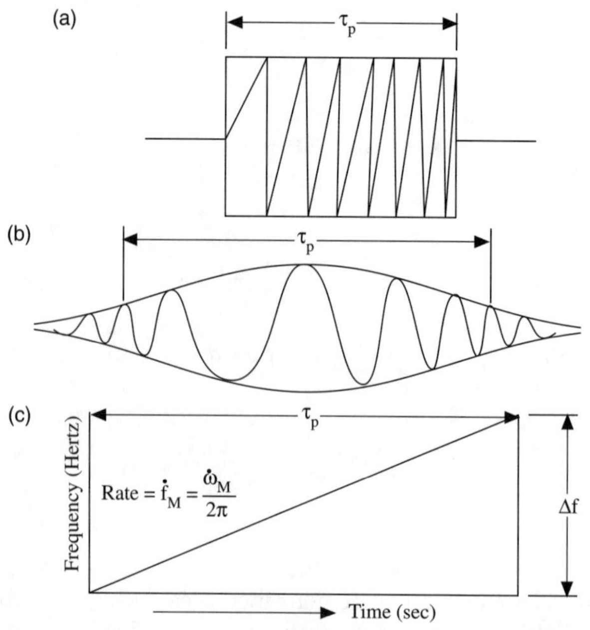

FIGURE 7.9
Linear FM (chirp) pulse. (a) Rectangular pulse envelope. (b) Gaussian pulse envelope. (c) Variation of frequency with time.

$$\frac{B_{\text{GCP}}}{B} = \sqrt{1 + \frac{16\ln 2}{\pi}\left(\frac{\Delta f}{B}\right)^2} = \sqrt{1 + 3.53\left(\frac{\Delta f}{B}\right)^2} \tag{7.36k)'$$

which shows that the enhancement in bandwidth due to chirping depends positively on the ratio of the frequency excursion to the bandwidth of the original (unchirped) pulse.

From (7.36d)′ and (7.35a,b,c,d,e) with the aid of (7.36g,i)′, we can consolidate the results for the chirp pulse with Gaussian envelope as we did in (7.28a,b,c,d) for the "nonchirped" pulse with Gaussian envelope (where B is now the bandwidth of the chirped pulse).

$$\text{TRC} = \frac{\sqrt{2}\tau_p}{\sqrt{1 + 16(\Delta f\tau_p)^2}} = \frac{1}{B}\sqrt{\frac{2\ln 2}{\pi}} \tag{7.37a}$$

$$\text{FSP} = \frac{1}{\text{TRC}} = \frac{\sqrt{1 + 16(\Delta f\tau_p)^2}}{\sqrt{2}\tau_p} = B\sqrt{\frac{\pi}{2\ln 2}}$$

$$\text{FRC} = \frac{1}{\sqrt{2}\tau_p} = \frac{B}{\sqrt{2}}\sqrt{\frac{\pi}{\ln 2}}\sqrt{1 + \left(\frac{16\ln 2}{\pi}\right)\left(\frac{\Delta f}{B}\right)^2} \tag{7.37b}$$

$$\text{TSP} = \frac{1}{\text{FRC}} = \frac{\sqrt{2\ln 2/\pi}}{B\sqrt{1 + (16\ln 2/\pi)(\Delta f/B)^2}}$$

$$(\text{TRC})(\text{FRC}) = \frac{1}{(\text{FSP})(\text{TSP})} = \frac{1}{\sqrt{1 + 16(\Delta f\tau_p)^2}} \tag{7.37c}$$

$$\text{AA} = 1 \tag{7.37d}$$

Comparing (7.37a,b,c,d) with (7.28a,b,c,d), we can express the ratios of the quantities related to resolution as follows (for fixed value of τ_p)

$$\frac{(\text{TRC})_{\text{chirp}}}{(\text{TRC})_{\text{nonchirp}}} = \frac{1}{\sqrt{1 + 16(\Delta f\tau_p)^2}} \tag{7.38a}$$

$$\frac{(\text{FSP})_{\text{chirp}}}{(\text{FSP})_{\text{nonchirp}}} = \sqrt{1 + 16(\Delta f\tau_p)^2} \tag{7.38b}$$

$$\frac{(\text{FRC})_{\text{chirp}}}{(\text{FRC})_{\text{nonchirp}}} = 1 \tag{7.38c}$$

$$\frac{(\text{TSP})_{\text{chirp}}}{(\text{TSP})_{\text{nonchirp}}} = 1 \tag{7.38d}$$

$$\frac{[(\text{TSP})(\text{FSP})]_{\text{chirp}}}{[(\text{TSP})(\text{FSP})]_{\text{nonchirp}}} = \sqrt{1 + 16(\Delta f\tau_p)^2} \tag{7.38e}$$

From (7.36f)′, we have for the ratio of bandwidths in the two cases

$$\frac{(B)_{\text{chirp}}}{(B)_{\text{nonchirp}}} = \sqrt{1 + 16(\Delta f\tau_p)^2} \tag{7.39a}$$

Equation (7.39a) also gives the ratios of the "time-bandwidth product" $B\tau_p$ in the two cases. Since that product is unity in the nonchirp case,

$$\frac{(B\tau_p)_{\text{chirp}}}{(B\tau_p)_{\text{nonchirp}}} = \sqrt{1 + 16(\Delta f \tau_p)^2} = \frac{(B)_{\text{chirp}}}{(B)_{\text{nonchirp}}} \tag{7.39b}$$

Again, this ratio is the same as the ratios of the frequency spans, the inverse ratios of the minimum resolvable time delay (range) difference as measured through the TRC and the ratios of the product of time span and frequency span.

The results all point to the enhancement of frequency span or of bandwidth (not exactly the same thing in general but in this case they are close) and the resulting enhancement of time delay (range) resolution, due to the linear FM or "chirp" technique, without a corresponding change in time span (again not exactly the same entity as duration as usually defined but very close in this particular case) or in the ability to resolve in Doppler frequency space. The "bottom line" of this is that a long pulse can be used without the consequent penalty in range resolution and retaining the Doppler resolution inherent in the use of a longer pulse.

The increased range resolution is attained by using a large bandwidth in order to chirp the pulse.

This technique is one of a class of techniques known as "pulse compression." The ratio of bandwidths given in (7.39a) is related to the "compression ratio." The compression ratio will be discussed later in general in Section 7.9, but at this point, it can be loosely defined as the ratio of range resolution constant without chirping relative to that *with* chirping, or the time span–frequency span product for a pulse (equal to 1 for the unchirped pulse and exceeding 1 for the chirped pulse), or the time–bandwidth product. All of these quantities are identical for the chirp pulse with Gaussian envelope. The larger the compression ratio, the greater the enhancement of range resolution due to the technique employed.

7.3 RESOLUTION AND AMBIGUITIES FOR PULSE TRAINS

It is important in radar application to extend the concepts covered in Section 7.2 to pulse trains. We will confine coverage to trains of identical pulses.

Before discussing ambiguity functions for pulse trains, we will first discuss the concept of bandwidth for pulse trains, in order to enable us to interpret the results in terms of bandwidth.

A train of N_p identical and uniformly spaced pulses has the form

$$s_{N_p}(t) = \sum_{k=0}^{N_p-1} s_1(t - kTr) \tag{7.40}$$

where Tr is the PRP or reciprocal PRF and $s_1(t)$ is the waveform for an individual pulse.

The Fourier transform of $s(t)$ is

$$\begin{aligned} S_{N_p}(\omega) &= \int_{-\infty}^{\infty} dt\, s_{N_p}(t)e^{-j\omega t} \\ &= \sum_{k=0}^{N_p-1} \int_{-\infty}^{\infty} dt\, s_1(t - kTr)e^{-j\omega(t-kTr)}e^{j\omega kTr} \\ &= \left[\sum_{k=0}^{N_p-1} e^{j\omega kTr}\right] \int_{-\infty}^{\infty} dt'\, s_1(t')e^{-j\omega t'} \end{aligned} \tag{7.41}$$

Thus the Fourier transform of the pulse train is that of the individual pulse multiplied by the factor in brackets. The latter, through the truncated geometric series

$$\sum_{n=0}^{N-1} x^n = \frac{1 - x^N}{1 - x} \tag{7.42}$$

is given by

$$\sum_{k=0}^{N_p-1} e^{j\omega kTr} = \left[\sum_{n=0}^{N_p-1} (e^{j\omega Tr})^n\right]$$

$$= e^{j\omega Tr}\left[\frac{1 - e^{j\omega N_p Tr}}{1 - e^{j\omega Tr}}\right] = e^{j\omega Tr + (j\omega N_p Tr/2) - (j\omega Tr/2)} \ldots$$

$$\frac{\sin(\omega N_p Tr/2)}{\sin(\omega Tr/2)} \tag{7.43}$$

The magnitude $|S_{N_p}(\omega)|$ is given by

$$|S_{N_p}(\omega)| = |S_1(\omega)|\left|\frac{\sin(N_p\omega Tr/2)}{\sin(\omega Tr/2)}\right| \tag{7.44}$$

where $|S_1(\omega)|$ is the spectrum of a single pulse and $|S_{N_p}(\omega)|$ is that of the entire pulse train.

Figure 7.10 shows the spectrum of a train of N_p identical and uniformly spaced "ultrashort" pulses, i.e., pulses with the property that $(fr\ \tau_p) << N_p$. The pulses are at video, not IF, i.e., the central frequency is zero, not $\pm\omega_o$. These rough plots are calibrated in Hertz, not radians/second.

To obtain the results needed for the figure, we note that the nulls of numerator and denominator of the factor in (7.44) occur at $N_p\omega Tr/2 = 2\pi N_p f/2fr = n\pi$ and

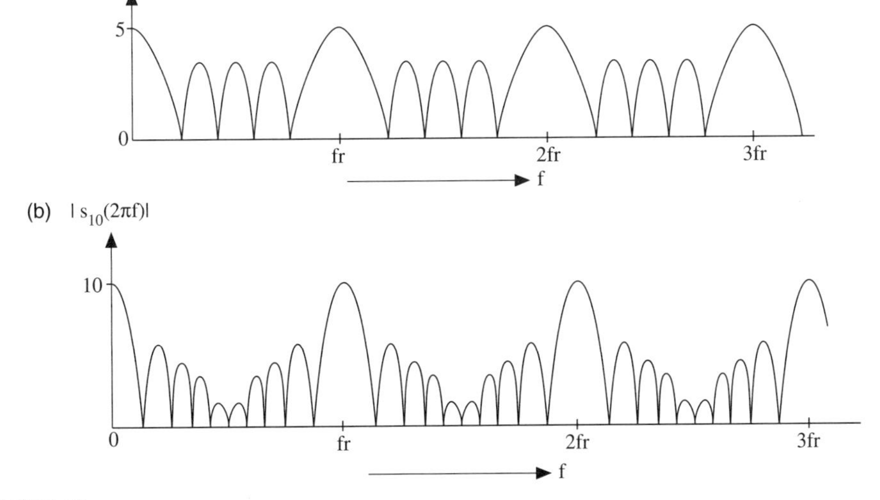

FIGURE 7.10
Spectrum of train of identical uniformly spaced ultrashort pulses: $f > 0$, $(fr\ \tau_p) << N_p$. (a) $N_p = 5$. (b) $N_p = 10$.

$\omega Tr/2 = 2\pi f/2fr = n\pi$, respectively, where n is any integer other than zero. Thus in Hertz, nulls of the numerator occur at $f = \pm fr/N_p, \pm 2fr/N_p, \pm 3fr/N_p, \ldots \pm nfr/N_p$ and those of the demoninator at $f = \pm fr, \pm 2fr, \pm 3fr, \ldots, \pm nfr$. At $f = 0$, nulls of both numerator and denominator occur, resulting in an indeterminate form. To evaluate it, we need only return to the original expression that generated the factor in (7.44), i.e., rewrite (7.44) in the form

$$|S_{N_p}(\omega)| = |S_1(\omega)| \left| \sum_{n=0}^{N_p-1} e^{j\omega Trn} \right| \tag{7.44a)'}$$

If $\omega Tr = 2\pi m$, where m is any integer, then

$$\left| \sum_{n=0}^{N_p-1} e_j^{2\pi mn} \right| = \sum_{n=0}^{N_p-1} (1) = N_p \tag{7.44b)'}$$

where $m = 0, \pm 1, \pm 2, \ldots.$

The factor attains its maximum value, N_p, when all the terms in the summation are in-phase and hence all phase factors have weight unity, leading to the result (7.44b)′. That event occurs wherever

$$f = mfr, \qquad m = 0, \pm 1, \pm 2, \ldots \tag{7.44c)'}$$

Thus the factor is N_p at all denominator nulls, i.e., it repeats itself at integral multiples of the PRF. The resulting behavior is that shown in Figure 7.10a for $N_p = 5$ and Figure 7.10b for $N_p = 10$. In both cases the plots are for $f \geq 0$ only, since the behavior for $f \leq 0$ is the mirror image of that for $f \geq 0$.

By specifying the pulses as "ultrashort," we are simply saying that the pulse duty cycle is very small compared with the number of pulses, and hence on the time-scale of the interpulse period, we can approximate them as impulse functions; hence on Figure 7.10, there is no perceptable decay in amplitude over many times the PRF. In Figure 7.11, we show a case where the duty cycle is 1/5 and there are a very large number of pulses (specifically $N_p = 100$)

If the pulses are rectangular, the spectrum of a single pulse is a sinc function $\text{sinc}(\pi f \tau_p)$, whose first null occurs at $f = 1/\tau_p$. This first null of the pulse-train factor in (7.44) is at frN_p. If the bandwidth of a single pulse is approximately $1/\tau_p$ and that of the pulse train is roughly half the spacing between first nulls on either side of the main peak, then the ratio of the bandwidth of the pulse train to that of a single pulse is roughly

$$\frac{B_{N_p}}{B_1} = \frac{fr\tau_p}{N_p} = \frac{\tau_p}{(N_p Tr)} \tag{7.45}$$

where B_{N_p} and B_1 are bandwidths of pulse-train and individual pulse, respectively. The reduction in bandwidth is approximately the ratio of duty cycle to the number of pulses. In the example shown in Figure 7.11 the bandwidth is reduced by about 1/500 by using 100 pulses with a 1/5 duty cycle. The expression on the extreme right′ in (7.45) shows the ratio of bandwidths to be equivalent to the ratio of the duration of one pulse to the total amount of time that the pulses are "on," i.e., the

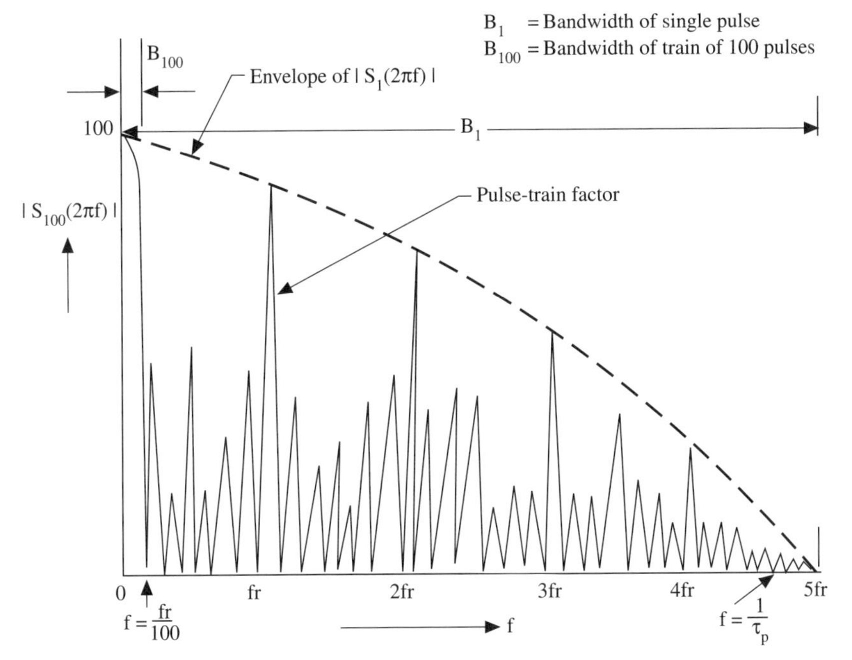

FIGURE 7.11
Spectrum of train of identical rectangular pulses; $fr\ \tau_p = 0.2$, $N_p = 100$.

total time space occupied by the pulse train. That interpretation will be used in discussing resolution and ambiguity functions for pulse trains.

We now return to the ambiguity function. The CAF for the pulse train of the form (7.40) has a numerator [from (7.18)] of the form

$$\text{Num}[\hat{\Lambda}_{N_p}(\tau, \Omega)] = \sum_{l,m=0}^{N_p-1} \int_{-\infty}^{\infty} dt\ \hat{s}_1(t - lTr)\hat{s}_1^*(t - mTr - \tau)e^{j\Omega t} \tag{7.46a}$$

and the denominator

$$\text{Denom}\ [\hat{\Lambda}_{N_p}(\tau, \Omega)] = \sum_{l,m=0}^{N_p-1} \int_{-\infty}^{\infty} dt\ \hat{s}_1(t - lTr)\hat{s}^*(t - mTr) \tag{7.46b}$$

By changing the variable of integration from t to $(t - lTr)$ in (7.46a,b), we arrive at the forms

$$\text{Num}[\hat{\Lambda}_{N_p}(\tau, \Omega)] = \sum_{l,m=1}^{N_p} e^{j\Omega lTr} \int_{-\infty}^{\infty} dt'\ \hat{s}_1(t')\hat{s}_1^*\{t' - [\tau + (m - l)Tr]\}e^{j\Omega t'} \tag{7.46c}$$

$$\text{Denom}[\hat{\Lambda}_{N_p}(\tau, \Omega)] = \sum_{l,m=1}^{N_p} \int_{-\infty}^{\infty} dt'\ \hat{s}_1(t')\hat{s}_1^*[t' - (m - l)Tr] \tag{7.46d}$$

The only conditions under which the factors in the integrand of the denominator overlap significantly are when $(m - l)Tr = 0$, resulting in

$$\text{Denom}[\hat{\Lambda}_{N_p}(\tau, \Omega)] = \sum_{m=1}^{N_p} \int_{-\infty}^{\infty} dt' |\hat{s}(t')|^2 = N_p \int_{-\infty}^{\infty} dt' |\hat{s}(t')|^2 \tag{7.46e}$$

Combining (7.46e) and (7.46c), we obtain

$$\hat{\Lambda}_{N_p}(\tau, \Omega) = \frac{1}{N_p} \sum_{l,m=1}^{N_p} e^{j\Omega l Tr} \hat{\Lambda}_1[\tau + (m - l)Tr, \Omega] \tag{7.46f}$$

where

$$\hat{\Lambda}_1(\tau, \Omega) = \text{CAF of individual pulse}$$

$$= \frac{\int_{-\infty}^{\infty} dt\, \hat{s}_1(t)\hat{s}_1^*(t - \tau)e^{j\Omega t}}{\int_{-\infty}^{\infty} dt |\hat{s}_1(t)|^2}$$

Equation (7.46f) can be written in a more illuminating form, accounting for the fact that all terms in the ℓ summation for which $\ell = m - k$ are the same for any fixed value of k, i.e.,

$$\begin{aligned}
\hat{\Lambda}_{N_p}(\tau, \Omega) = \sum_{m=1}^{N_p} \{&[\text{Terms}(l - m = 0)] + [(\text{Terms } l - m = 1) + (\text{Terms } l - m = -1)] \\
&+ [(\text{Terms } l - m = 2) + (\text{Terms } l - m = -2)] \\
&+ [(\text{Terms } l - m = 3) + (\text{Terms } l - m = -3)] + \cdots \\
&+ [(\text{Terms } l - m = N_p - 2) + [\text{Terms } l - m = -(N_p - 2)]] \\
&+ [(\text{Terms } l - m = N_p - 1) + [\text{Terms } l - m = -(N_p - 1)]] \\
&+ [(\text{Terms } l - m = -N_p) + (\text{Terms } l - m = -\ N_p)]\}
\end{aligned} \tag{7.46g}$$

or

$$\begin{aligned}
\hat{\Lambda}_{N_p}(\tau, \Omega) = \frac{1}{N_p} \sum_{m=1}^{N_p} e^{j\Omega m Tr} \{&N_p \hat{\Lambda}_1(\tau, \Omega) \\
&+ (N_p - 1)[\hat{\Lambda}_1(\tau, + Tr, \Omega)e^{-j\Omega Tr} + \hat{\Lambda}_1(\tau - Tr, \Omega)e^{j\Omega Tr}] \\
&+ (N_p - 2)[\hat{\Lambda}_1(\tau + 2Tr, \Omega)e^{-j2\Omega Tr} + \hat{\Lambda}_1(\tau - 2Tr, \Omega)e^{j2\Omega Tr}] \\
&+ (N_p - 3)[\hat{\Lambda}_1(\tau + 3Tr, \Omega)e^{-j3\Omega Tr} + \hat{\Lambda}_1(\tau - 3Tr, \Omega)e^{j3\Omega Tr}] \\
&+ \cdots \\
&+ (3)[\hat{\Lambda}_1[\tau + (N_p - 3)Tr, \Omega]e^{-j(N_p - 3)\Omega Tr} \\
&+ \hat{\Lambda}_1[\tau - (N_p - 3)Tr, \Omega]e^{j(N_p - 3)\Omega Tr}] \\
&+ (2)[\hat{\Lambda}_1[\tau + (N_p - 2)Tr, \Omega]e^{-j(N_p - 2)\Omega Tr} \\
&+ \hat{\Lambda}_1[\tau - (N_p - 2)Tr, \Omega]e^{j(N_p - 2)\Omega Tr}] \\
&+ (1)[\hat{\Lambda}_1[\tau + (N_p - 1)Tr, \Omega]e^{-j(N_p - 1)\Omega Tr} \\
&+ \hat{\Lambda}_1[\tau - (N_p - 1)Tr, \Omega]e^{j(N_p - 1)\Omega Tr}]\}
\end{aligned} \tag{7.46h}$$

To interpret (7.46h), let us consider a simple case where $N_p = 5$, illustrated along the delay (range) axis in Figure 7.12 for rectangular pulses.

From (7.46h), we have for this case (it is clear from Figure 7.12 that there is no overlapping of terms, given that $\tau_p >> Tr$, as is usually the case for radar pulses)

$$
\begin{aligned}
|\hat{\Lambda}_5(\tau, \Omega| = \frac{1}{5}\{&5|\hat{\Lambda}_1(\tau, \Omega)| \\
&+ 4[|\hat{\Lambda}_1(\tau + Tr, \Omega)| + |\hat{\Lambda}_1(\tau - Tr, \Omega)|] \\
&+ 3[|\hat{\Lambda}_1(\tau + 2Tr, \Omega)| + |\hat{\Lambda}_1(\tau - 2Tr, \Omega)|] \\
&+ 2[|\hat{\Lambda}_1(\tau + 3Tr, \Omega)| + |\hat{\Lambda}_1(t - 3Tr, \Omega)|] \\
&+ 1[|\hat{\Lambda}_1(\tau + 4Tr, \Omega)| + |\hat{\Lambda}_1(\tau - 4Tr, \Omega)|]\}
\end{aligned}
\tag{7.46i}
$$

The decreasing amplitudes of the AF peaks as $|\tau|$ increases are due to the fact that there are fewer overlapping pulses in the product that forms the AF as we proceed from $\tau = 0$ to $\tau = \pm Tr$. In (7.46h), dissecting the summations that make up the first few terms, we note that the first term consists of the aggregate of all terms for which $l = m$. Those are the terms for which all N_p pulses of $s(t,\omega_o)$ overlap with all N_p pulses of $s(t - \tau,\omega_o)$, τ being zero. If $l = m - 1$, only $(N_p - 1)$ pulses overlap, because one pulse in the delayed train $s(t - \tau,\omega_o)$ is nonoverlapping and $s(t - \tau,\omega_o)$ is to the left of $s(t,\omega_o)$. If $l = m + 1$, then the nonoverlapping pulse in $s(t - \tau,\omega_o)$ is to the right of $s(t,\omega_o)$. In either case, there are only $(N_p - 1)$ overlapping pulses. One can continue increasing $|\tau|$ to $2Tr$, in which case $(N_p - 2)$ pulses overlap, to $3Tr$, in which case $(N_p - 3)$ pulses overlap, etc., until one reaches $(N_p - 1)Tr$, in which the

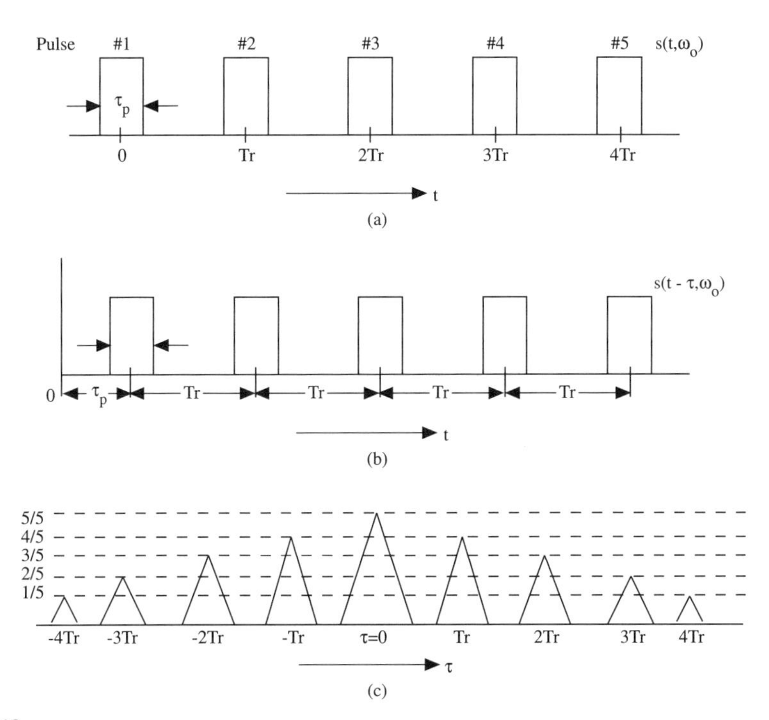

FIGURE 7.12
Ambiguity function for train of five rectangular pulses. (a) Pulse train. (b) Delayed pulse train. (c) Ambiguity function amplitude.

number of overlapping pulses is $N_p - (N_p - 1)$. It is graphically evident from Figure 7.12 that beyond $(N_p - 1)Tr$, there is no overlap and the AF becomes zero.

For a much larger train of pulses, say $N_p = 1000$, then the decay in amplitude of the peaks of the AF is much smaller, even imperceptable. For example, if $N_p = 1000$, the first set of "sidelobe" peaks on either side of the peak at $\tau = 0$ has amplitude 0.999, whereas the amplitude of the peak at $\tau = 0$ is 1. The second set of peaks has amplitude 0.998, the third has 0.997, etc. It is clear from this line of reasoning that the peaks that occur at multiples of the PRP Tr would have equal amplitude for an infinite pulse train. This leads to the subject of range ambiguities.

7.4 RANGE AMBIGUITIES

Figure 7.13 shows return pulse trains for targets at different ranges as they might appear on a time base calibrated in range units. Figure 7.13a shows the transmitted train of 10 pulses, separated by a range interval $c/2fr$. The return from a target at range r_1, which is less than $c/2fr$, is shown in (b). In (c), the return from a target at a range r_2, which exceeds $c/2fr$ but is less than $2(c/2fr)$ and $3(c/2fr)$ is shown. The returns in (c) and (d) are "multiple time-around echoes."

It is obvious from comparisons of (b), (c), and (d) that

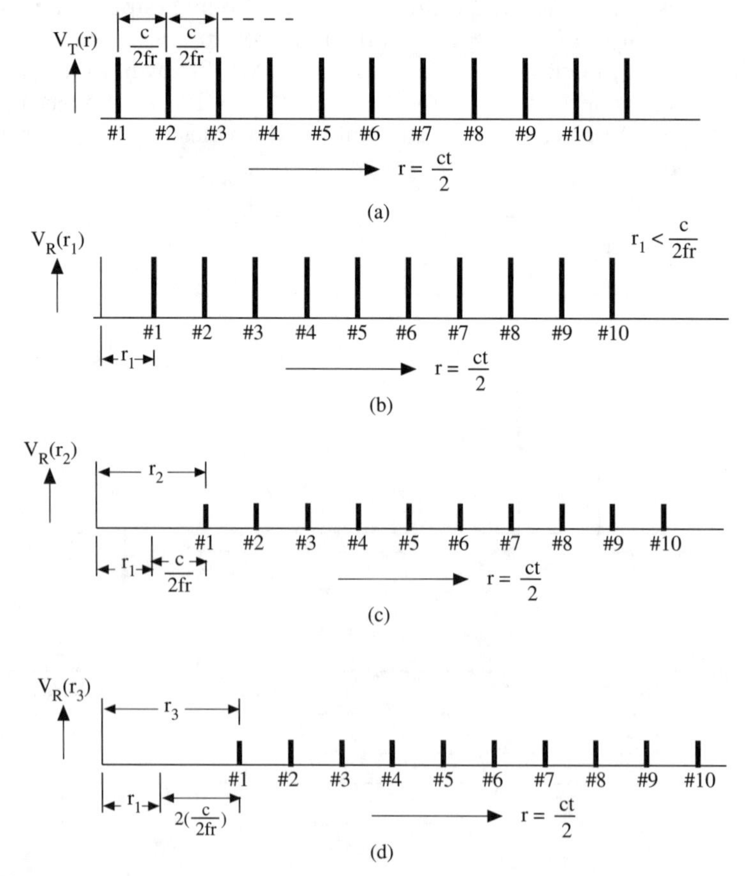

FIGURE 7.13
Return pulse trains from targets at various ranges. (a) Transmitted pulse train. (b) Returns from range r_1. (c) Returns from range $r_2 = r_1 + (c/2fr)$. (d) Returns from range $r_3 = r_1 + 2(c/2fr)$.

1. Pulse k from the target at range r_2 is in the same location on the time delay (range) scale as pulse $(k + 1)$ from the target at range r_1.
2. Pulse k from the range r_3 is in the same location on the scale as pulse $(k + 1)$ from range r_2 and pulse $(k + 2)$ from range r_1.
3. If we had another target at range $r_N = r_1 + (N - 1)(c/2fr)$, then its kth pulse would be in the same location on the scale as pulse $(k + 1)$ from range $r_N - c/2fr$, pulse $(k + 2)$ from range $r_N - 2(c/2fr)$, pulse $(k + 3)$ from range $r_N - 3(c/2fr)$, etc.

The phenomenon illustrated in Figure 7.13 is that of "ambiguous ranges" or "range ambiguities." In a certain sense, any range r could be regarded as ambiguous, since a return pulse from a target at that range has the same location on the time delay scale as return pulses from all ranges that *exceed* r by multiples of $c/2fr$. However, as the range *increases* the return power decreases in general, unless, of course, the RCS of the distant target is orders of magnitude larger than that of the nearby target. But if we are confining attention to targets of roughly comparable RCS, and if we can define a "maximum range" r_{max} as that at which the return exceeds the "minimum detectable signal" (the usual definition of maximum range), then range ambiguities are of no concern if $r_{max} < c/2fr$, as in Figure 7.13b. That is because targets at ranges beyond r_{max} are below the receiver noise level.

The range ambiguities of concern, then, are those wherein targets at a range *lower* than that of a given target have the same location on the scale as those of the target of interest. This means that a range gate set for the correct range of a given target will capture returns not only from that target but also returns from targets at ranges lower than that of the subject target by multiples of $c/2fr$. The latter returns are generally stronger than those from the target of interest because they are closer to the radar than the target and undergo a $1/r^4$ loss.

For example, the target at range $r_2 = r_1 + (c/2fr)$ in Figure 7.13c is ambiguous in the sense that a target at range r_1 appears in a gate set to r_2. That target can be said to have *one* ambiguous range. A target at $r_3 = r_1 + 2(c/2fr)$ has *two* range ambiguities, one at $r_1 + (c/2fr)$, the other at r_1.

The general ground rule, evident from Figure 7.13 and the accompanying text, is

$$\text{If } r_{max} < \frac{c}{2fr} \text{ or equivalently } fr < \frac{c}{2r_{max}} \qquad \textit{no} \text{ range ambiguities} \tag{7.47a}$$

$$\text{If } \frac{c}{2fr} < r_{max} < 2\left(\frac{c}{2fr}\right), \text{ or } \frac{c}{2r_{max}} < fr < 2\left(\frac{c}{2r_{max}}\right) \qquad \textit{one} \text{ range ambiguity} \tag{7.47b}$$

$$\text{If } 2\left(\frac{c}{2r_{max}}\right) < fr < 3\left(\frac{c}{2r_{max}}\right), \qquad \textit{two} \text{ range ambiguities} \tag{7.47c}$$

$$\vdots$$

$$\text{If } (N - 1)\frac{c}{2r_{max}} < fr < N\frac{c}{2r_{max}}, \qquad (N - 1) \text{ range ambiguities} \tag{7.47d}$$

Based on (7.47a,b,c,d), we define the "maximum unambiguous range" as

$$r_{\text{unamb}} = \frac{c}{2fr} \tag{7.48}$$

If the maximum range r_{max} is below r_{unamb}, the range ambiguities present no problem. If that is an important consideration in the design of a radar system (as it well might be, because it obviously leads to erroneous range indications) then the PRF is limited by (7.47a), i.e., the maximum allowable PRF to avoid range ambiguities is

$$(fr)_{\substack{\text{max}\\ \text{unamb}}} = \frac{c}{2r_{\text{max}}} \tag{7.49}$$

If it is necessary to increase the PRF beyond $(fr)_{\text{max unamb}}$, then one can use one of the various available methods of combatting range ambiguities, such as that of variable PRF or the "Chinese remainder theorem" (e.g., Skolnick,[7] pp. 19.12 to 19.17).

It is notable that for a truly "infinite" pulse train, range ambiguities would seem unavoidable because there is no pulse numbering possible. But since "infinity" is only an abstraction, we can always number the return pulses, and the *first* return pulse tells us the true range for a given target. But in a practical radar, the pulse train is range-gated, resulting in a summation over many pulses, and hence a processing scheme to combat the effect of range ambiguities must account for that.

Before discussing the issue of *combatting* range ambiguities, we return to the ambiguity function of a pulse train as shown in Figure 7.12. We should remember that the AF illustrated in that figure is really the output of a "mismatched" filter, where the variable τ, if calibrated in range units $c\tau/2$, would be the mismatch in range. Equivalently the mismatched filter is a range gate, in the sense that it produces peaks near the range value to which it is set and no outputs beyond a pulse length on either side of those peaks. The near-periodicity of the range-gate output is an indication of the range ambiguities we have been discussing above. If there were an enormous number of pulses, i.e., approximating an infinite number for practical purposes, then the peaks would be essentially identical and the AF would be truly periodic. With a smaller number of pulses, the decay in the AF is substantial within a small period of time and there is less chance of confusing the secondary peaks with the primary peak, since they can be distinguished on the basis of amplitude. Various levels of range ambiguity are illustrated in Figure 7.14, which shows sketches of the AF for different values of PRF for fixed duration of the entire pulse train. We note that

$$N_{\text{p}} = frT_{\text{d}} \tag{7.50}$$

where T_d is the duration of the pulse train. The sketches are calibrated in Δr, the difference between range gate setting and true target range.

The number of range-ambiguous peaks of nearly equal amplitude is very large within the interval $cT_{\text{d}}/2$ in Case (a), where $fr >> 1/T_{\text{d}}$. In Cases (b) and (c), as the PRF is decreased substantially, the spacing between the ambiguous peaks and the reduction in amplitude within the duration of the pulse train is more pronounced, thus mitigating the effect of range ambiguities.

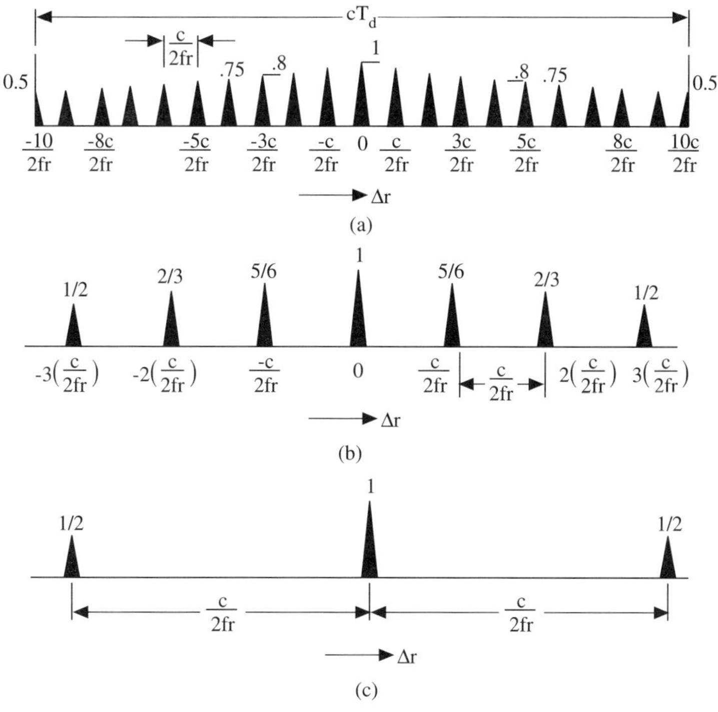

FIGURE 7.14
Cuts of range-ambiguity function vs. PRF for fixed T_d. (a) $N_p = frT_d = 20$; $fr >> 1/T_d$. (b) $N_p = frT_d = 6$. (c) $N_p = frT_d = 2$.

7.5 DOPPLER AMBIGUITIES

The key to understanding of Doppler ambiguities is (7.44), depicted graphically in Figure 7.10 for $N_p = 5$ and 10. Figure 7.10 shows that the factor on the signal spectrum

$$F(f) = \left|\frac{\sin(\pi N_p(f/fr))}{\sin(\pi(f/fr))}\right| \tag{7.51}$$

written in terms of Hertz frequency f, repeats itself at integral multiples of the PRF for a fixed total number of pulses N_p. It is also noted that the ambiguity function for a pulse train, as given by (7.46h), contains the factor $\sum_{m=1}^{N_p} e^{j\Omega Tr}$, which, by the same reasoning as used to obtain (7.44) from (7.41) through (7.42) and (7.43), gives rise to an amplitude factor

$$F(\Omega) = \left|\frac{\sin(\Omega N_p/2fr)}{\sin(\Omega/2fr)}\right| \tag{7.52}$$

on the ambiguity function.

Comparison of (7.52) and (7.51) indicates that the same factor that appears in the frequency spectrum of a pulse train in terms of radian frequency ω also appears in the ambiguity function, where the radian frequency variable Ω replaces ω. Spectral plots like those in Figures 7.10 and 7.11 could be duplicated for the ambiguity function where the argument is Δf, the frequency mismatch in the output of an attempted matched filter, expressed in Hertz. Both of these figures show peaks at integral multiples of the PRF. Figure 7.11 accounts for the spectrum of the pulse shape as well as the factor arising from repeated pulses.

Let us now consider the AF as the output of a Doppler filter with no range mismatch. Frequency is calibrated in doppler speed units, i.e.,

$$\Delta f = \frac{\Omega}{2\pi} = \frac{2\Delta V}{\lambda_0} \tag{7.53}$$

The factor $F(\Omega)$ in (7.52), with the aid of (7.53), is given by

$$F(\Delta V) = \left|\frac{\sin(2\pi\Delta V N_p/fr\lambda_0)}{\sin(2\pi\Delta V/fr\lambda_0)}\right| \tag{7.54}$$

and λ_o is the wavelength corresponding to the central radar frequency f_0.

Figure 7.15 shows the function $F(\Delta V)$ sketched as a function of ΔV for fixed wavelength and for three values of fr as it relates to the maximum expected Doppler

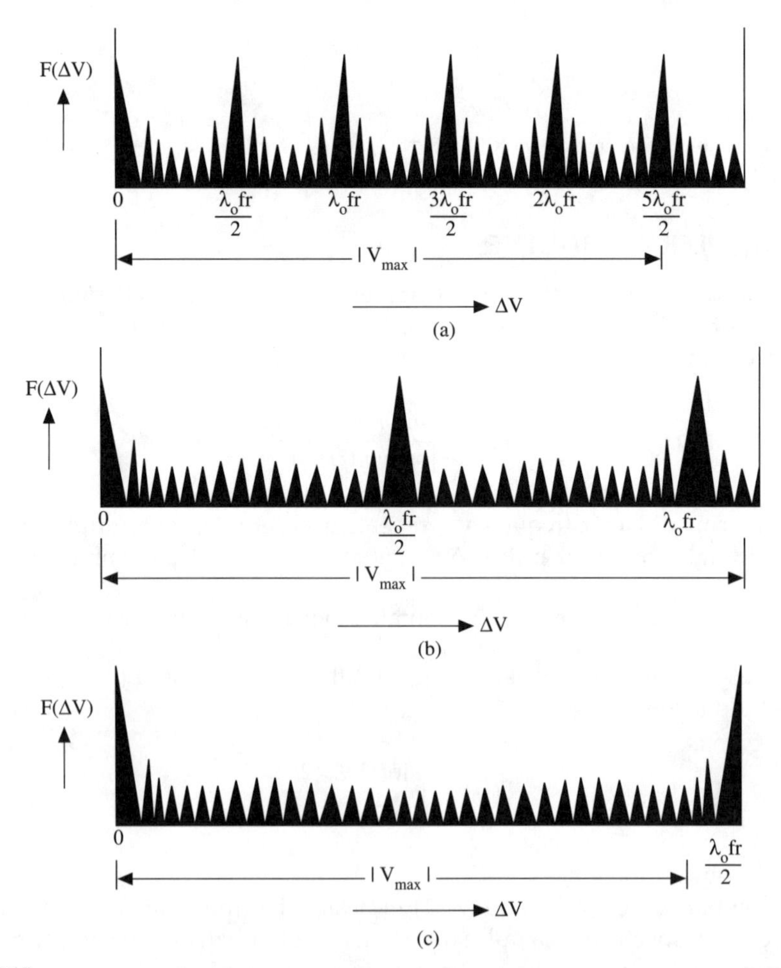

FIGURE 7.15
Factor on AF due to multiple pulses vs. ΔV. (a) $fr = 0.4|V_{max}|/\lambda_0$. (b) $fr = |V_{max}|/\lambda_0$. (c) $fr = 2|V_{max}|/\lambda_0$.

speed $2|V_{max}|$ (shown as $|V_{max}|$ in Figure 7.15 because the negative ΔV region is not shown.)

It is implied in Figure 7.15 that there are a large number of pulses, i.e.,

$$N_p = frT_d >> 1 \quad (7.55)$$

For fixed wavelength, the spacing between peaks (which represent ambiguous Doppler speeds) increases with PRF. Hence, as seen from comparison of Cases (a), (b), and (c), the number of unwanted Doppler peaks (not including the peak at $\Delta V = 0$, which is the peak used to determine the Doppler speed), within the interval from 0 to $|V_{max}|$ decreases from 5 in Case (a) through 2 in Case (b) and finally to none in Case (c). The implication is that if the maximum Doppler speed to be expected (based on a priori information about expected targets in the radar scenario) does not exceed the parameter $\lambda_0 fr/2$, then doppler ambiguities will be no problem. If $|V_{max}|$ exceeds $\lambda_0 fr$ but does not exceed $3\lambda_0 fr/2$, as in Case (b), then there are four unwanted doppler ambiguities (two for ΔV positive and another two for ΔV negative, not shown in the figure). If $|V_{max}|$ exceeds $\lambda_0 fr/2$ as in Case (c), we have eight Doppler ambiguities, four for $\Delta V > 0$ and four for $\Delta V < 0$. The criterion for avoidance of Doppler ambiguities, then, is

$$fr > \frac{2}{\lambda_0}|V_{max}| \quad (7.56)$$

which means, of course, that we must choose the PRF at a high enough value, in accordance with (7.56), to avoid doppler ambiguities. However, in Section 7.4, criteria are given for minimizing *range* ambiguities by keeping the PRF *low.* That implies a tradeoff between high PRF for Doppler ambiguity avoidance and low PRF for range ambiguity avoidance. This tradeoff will be discussed in Section 7.6. Before closing this section, we present the Doppler analogs of (7.47a,b,c,d) and (7.48), as follows:

$$\text{If } fr > \frac{2|V_{max}|}{\lambda_0}, \qquad \textit{no} \text{ Doppler ambiguities} \quad (7.57a)$$

$$\text{If } \frac{1}{2}\left(\frac{2|V_{max}|}{\lambda_0}\right) < fr < \left(\frac{2|V_{max}|}{\lambda_0}\right); \qquad \textit{two} \text{ Doppler ambiguities} \quad (7.57b)$$

$$\text{If } \frac{1}{3}\left(\frac{2|V_{max}}{\lambda_0}\right) < fr < \frac{1}{2}\left(\frac{2|V_{max}|}{\lambda_0}\right); \qquad \textit{four} \text{ Doppler ambiguities} \quad (7.57c)$$

$$\text{If } \frac{1}{N}\left(\frac{2|V_{max}|}{\lambda_0}\right) < fr < \frac{1}{(N-1)}\left(\frac{2|V_{max}|}{\lambda_0}\right); \quad (7.57d)$$

$2(N - 1)$ Doppler ambiguities

Based on (7.57a,b,c,d), we define the "maximum unambiguous Doppler velocity" as

$$\Delta V_{\text{unamb}} = \frac{\lambda_0 fr}{2} \tag{7.58}$$

7.6 TRADEOFF BETWEEN RANGE AND DOPPLER AMBIGUITIES

The tradeoff between range and Doppler ambiguity avoidance is best investigated through the range–Doppler ambiguity function for a train of uniformly spaced identical pulses. From (7.46h) calibrated in range and Doppler velocity units and (7.54), which details the factor due to multiple pulses, we can write (assuming no pulse overlaps)

$$\begin{aligned}
|\hat{\Lambda}_{N_p}(\Delta r, \Delta V)| = F(\Delta V)\Bigg\{ & |\hat{\Lambda}_1(\Delta r, \Delta V)| \\
& + \xi_1\left[\left|\hat{\Lambda}_1\left(\Delta r + \frac{c}{2fr}, \Delta V\right)\right| + \left|\hat{\Lambda}_1\left(\Delta r - \frac{c}{2fr}, \Delta V\right)\right|\right] \\
& + \xi_2\left[\left|\hat{\Lambda}_1\left(\Delta r + \frac{c}{fr}, \Delta V\right)\right| + \left|\hat{\Lambda}_1\left(\Delta r - \frac{c}{fr}, \Delta V\right)\right|\right] + \cdots \\
& + \xi_k\left[\left|\hat{\Lambda}_1\left(\Delta r + \frac{kc}{2fr}, \Delta V\right)\right| + \left|\hat{\Lambda}_1\left(\Delta r - \frac{kc}{2fr}, \Delta V\right)\right|\right] + \cdots \Bigg\}
\end{aligned} \tag{7.59}$$

where

$\xi_k = \dfrac{(N_p - k)}{N_p}$ = amplitude factor on kth range peak on either side of main range peak

$|\hat{\Lambda}_1(\Delta r, \Delta V)|$ = amplitude of CAF of single pulse in range–Doppler units

$$= \frac{\int_{-\infty}^{\infty} dt\ \hat{s}_1(t)\hat{s}_1^*[t - (2/c)\Delta r]\ e^{j4\pi\Delta Vt/\lambda_0}}{\int_{-\infty}^{\infty} dt\ |\hat{s}_1(t)|^2}$$

$$\text{F}(\Delta V) = \text{factor due to multiple pulses} = \left|\sin\left(\frac{2\pi\Delta V}{fr\lambda_0}N_{\text{p}}\right)\Big/\sin\left(\frac{2\pi\Delta V}{fr\lambda_0}\right)\right|$$

Figure 7.16 shows a three-dimensional sketch of the AF given by (7.59) for a "bed of spikes," the case where each individual pulse is very narrow relative to the PRF and thus can be approximated as an impulse. The condition required is equivalent to

$$\frac{\tau_{\text{p}}}{Tr} = \tau_{\text{p}} fr << 1 \tag{7.60}$$

Since $\tau_{\text{p}}/Tr = \tau_{\text{p}} fr$ is the pulse duty cycle, (7.60) says in effect that the duty cycle is a very small fraction of unity, or that the pulses are "on" for only a tiny fraction of the available time.

Figure 7.17 shows the positions of the peaks for the "bed of spikes" on the plane of Δr and ΔV. Using this type of pulse waveform removes the effect of the

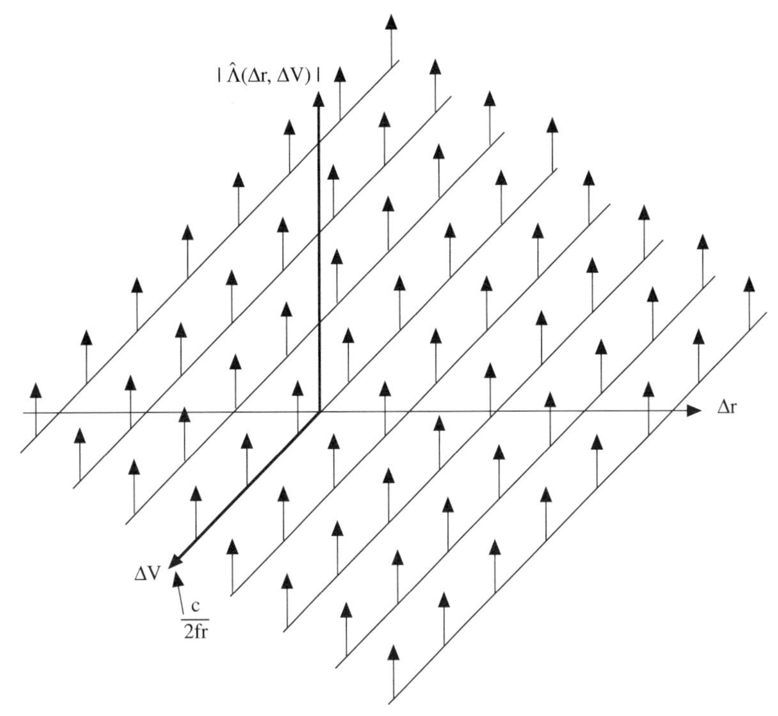

FIGURE 7.16
Graphical depiction of "bed of spikes" ambiguity function. Assumptions: Uniformly spaced identical pulse train—duty cycle << 1.

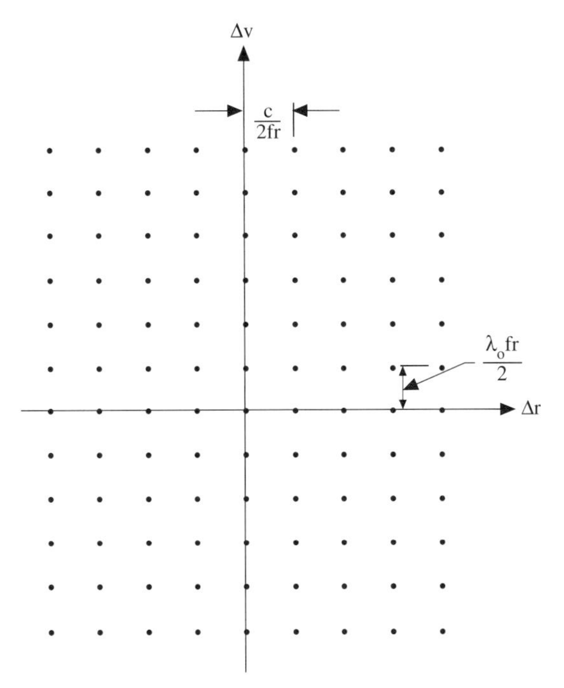

FIGURE 7.17
Positions of ambiguity peaks—"bed of spikes." Assumptions: Train of uniformly spaced identical pulses with duty cycle << 1.

individual pulse spectrum, which is now flat from $-\infty$, to $+\infty$, and thereby simplifies the discussion (since the individual pulse spectrum is an extraneous issue in discussing range and Doppler ambiguities).

It is qualitatively evident from Figure 7.17 that range ambiguities can be reduced by decreasing the PRF and Doppler ambiguities can be reduced by increasing the PRF. The quantitative extent of this tradeoff is best investigated by problem exercizes for realistic radar scenarios. The problem assignment associated with this chapter will contain some problems involving the tradeoff. Some insight into the tradeoff is gained by noting that the spacing between ambiguity peaks in range is $c/2fr$ and that between Doppler ambiguity peaks is $\lambda_0 fr/2$. Then

$$P_{\mathrm{rd}} = \Delta_{\substack{\mathrm{amb}\\ \mathrm{range}}} \Delta_{\substack{\mathrm{amb}\\ \mathrm{Doppler}}} = \left(\frac{c}{2fr}\right)\left(\frac{\lambda_0 fr}{2}\right) = \frac{c\lambda_0}{4} \tag{7.61}$$

where

$\Delta_{\mathrm{amb\ range}}$ = separation between range ambiguity peaks
$\Delta_{\mathrm{amb\ Doppler}}$ = separation between Doppler ambiguity peaks
P_{rd} = product of separations between ambiguity peaks in range and Doppler.

The obvious inference from (7.61) is that any attempt to increase $\Delta_{\mathrm{amb\ range}}$ by decreasing fr will be offset by a decrease in $\Delta_{\mathrm{amb\ doppler}}$ and vice versa, which is the tradeoff we have been referring to.

7.7 RESOLUTION CONSIDERATIONS WITH PULSE TRAINS

It was pointed out in Section 7.2 that the TRC and FRC for a signal waveform can be obtained directly from Eqs. (7.29e) and (7.29f), respectively. For a uniform pulse train with no pulse overlap, Eq. (7.29f) gives us

$$(\mathrm{FRC})_{N_{\mathrm{p}}} = \frac{\displaystyle\int_{-\infty}^{\infty} dt\ |\hat{s}_{\mathrm{T}}(t)|^4}{\left[\displaystyle\int_{-\infty}^{\infty} dt\ |\hat{s}_{\mathrm{T}}(t)|^2\right]^2} = \frac{\displaystyle\sum_{k=0}^{N_{\mathrm{p}}-1}\int_{-\infty}^{\infty} dt\ p^4(t)}{\left[\displaystyle\sum_{k=0}^{N_{\mathrm{p}}-1}\int_{-\infty}^{\infty} dt\ p^2(t)\right]^2}$$

$$= \frac{N_{\mathrm{p}}}{N_{\mathrm{p}}^2}\left\{\frac{\displaystyle\int_{-\infty}^{\infty} dt\ p^4(t)}{\left[\displaystyle\int_{-\infty}^{\infty} dt\ p^2(r)\right]^2}\right\} = \frac{1}{N_{\mathrm{p}}}(\mathrm{FRC})_1 \tag{7.62a}$$

where $(\mathrm{FRC})_1$ and $(\mathrm{FRC})_{N_{\mathrm{p}}}$ are the frequency resolution constants for one of the pulses and the entire pulse train, respectively. The time span TSP, the reciprocal of FRC, is

$$(\mathrm{TSP})_{N_{\mathrm{p}}} = N_{\mathrm{p}}\,(\mathrm{TSP})_1 \tag{7.62b}$$

where subscripts 1 and N_{p} refer to a single pulse and the entire train, respectively.

Equation (7.62b) is a somewhat obvious result, or could be considered a validation of the assertion that the reciprocal of FRC really *is* the total amount of time occupied by the signal. The amount of time occupied by a train of N_p identical pulses is obviously N_p times that occupied by a single pulse.

Invoking (7.29e) and (7.44) to obtain the time-resolution constant gives us (with appropriate and obvious notational modifications)

$$(\mathrm{TRC})_{N_p} = \frac{2\pi \int_{-\infty}^{\infty} d\omega \, |S_{N_p}(\omega)|^4}{\left[\int_{-\infty}^{\infty} d\omega \, |S_{N_p}(\omega)|^2\right]^2}$$

$$= \frac{2\pi \int_{-\infty}^{\infty} d\omega \, |S_1(\omega)|^4 [\sin(N_p \omega Tr/2)/\sin(\omega Tr/2)]^4}{\left[\int_{-\infty}^{\infty} d\omega \, |S_1(\omega)|^2 \{[\sin(N_p \omega Tr/2)/\sin(\omega Tr/2)]^2\}\right]^2} \qquad (7.62c)$$

Equation (7.62c) would appear difficult to evaluate in the general case, but in the typical case of low duty cycle, where the pulse spectrum is extremely wide compared with that of the multiple-pulse factor, the function $|S_1(\omega)|^4$ and $|S_1(\omega)|^2$ in the numerator and denominator, respectively, can be approximated as flat over the region where the ratio of sine functions is large and these factors can be removed from the integrals and will then appear as the ratio $|S_1(\omega)|^4/(|S_1(\omega)|^2)^2 = 1$. In that case, which should be a good approximation for most standard radar applications, we have

$$(\mathrm{TRC})_{N_p} = \frac{2\pi \int_{-\infty}^{\infty} d\omega \, [\sin(N_p \omega Tr/2)/\sin(\omega Tr/2)]^4}{\left[\int_{-\infty}^{\infty} d\omega \, \{[\sin(N_p \omega Tr/2)/\sin\,(\omega Tr/2)]\}^2\right]^2}$$

$$= \frac{\frac{\pi}{fr} \int_{-\infty}^{\infty} dx \, [\sin(N_p x)/\sin \mathrm{x}]^4}{\left\{\int_{-\infty}^{\infty} dx \, [\sin(N_p x)/\sin x]^2\right\}^2} \qquad (7.62d)$$

If the integrals in (7.62d) were tabulated, we would be able to obtain the answer in one step. However, that is not the case. We can get an approximate answer by approximating $\sin(N_p x)/\sin x$ as a sinc function (which is not a bad approximation for very small x-values), i.e.,

$$\frac{\sin(N_p x)}{\sin x} \simeq \frac{\sin(N_p x)}{x} = N_p \,\mathrm{sinc}(N_p x) \qquad (7.62e)$$

In this case (from 15.36 on p. 96 of Spiegel[8])

$$\left\{\int_{-\infty}^{\infty} dx \left[\left(\frac{\sin(N_p x)}{\sin x}\right)^2\right]\right\}^2 = \left\{N_p \int_{-\infty}^{\infty} d(N_p x)[\text{sinc}(N_p x)]^2\right\}^2 = (N_p \pi)^2 \quad (7.62f)$$

and (from 15.59 on p. 97 of Spiegel[8])

$$\int_{-\infty}^{\infty} dx \left(\frac{\sin(N_p x)}{\sin x}\right)^4 \simeq N_p^4 \int_{-\infty}^{\infty} dx \ \text{sinc}^4(N_p x) = N_p^3 \frac{2\pi}{3} \quad (7.62g)$$

From (7.62d,f,g), it follows that (approximately)

$$(\text{TRC})_{N_p} \simeq \frac{\pi}{fr} \frac{N_p^3(2\pi/3)}{N_p^2 \pi^2} = \frac{2N_p}{3fr} \quad (7.62h)$$

The frequency span, the reciprocal of $(\text{TRC})_{N_p}$ is

$$(\text{FSP})_{N_p} \simeq \frac{3}{2} \frac{fr}{N_p} \quad (7.62i)$$

The result (7.62i) can be viewed alongside the result (7.45), which tells us that the bandwidth of a pulse train is approximately fr/N_p when "bandwidth" is defined as the separation between -3 dB points on the spectral curve. The frequency span FSP is very close to the bandwidth, and for a spectral function that is nearly flat over a frequency region and falls off quite rapidly beyond that region, the frequency span *is* the bandwidth. Also the result (7.62h) confirms, with the aid of (7.45), that the smallest resolvable range difference for the pulse train is roughly the reciprocal bandwidth and hence that a pulse train degrades range resolution relative to a single pulse in the amount of its decrease in bandwidth relative to the single pulse. The result (7.62a) tells us that the pulse train has better Doppler resolution than the single pulse in the amount of the relative increase in time occupied by the signal. Both of these results confirm the general principle that range resolution can generally be improved by increasing the bandwidth and Doppler resolution by increasing the amount of time occupied by signal energy.

7.8 AMBIGUITY FUNCTION OF UNORTHODOX PULSE TRAINS

To calculate AFs of unorthodox types of pulse trains, designed to improve resolution or alleviate the effects of ambiguities, we first note some simple relationships that we have encountered before and used so far only for conventional pulse trains. Using these relationships will simplify evaluation of *un*conventional pulse trains.

First, note the CAF of a pulse $\hat{s}_1(t)$ at $t = 0$ (Fig. 7.18) and that of the same pulse at $t = T_{12}$, where the pulse shape is arbitrary. Then add a third pulse shaped the same as the first two at $t = T_{13}$.

The pulses are identically shaped, hence

$$\hat{s}_2(t) = \hat{s}_1(t - T_{12}) \quad (7.63a)$$

$$\hat{s}_3(t) = \hat{s}_1(t - T_{13}) \quad (7.63b)$$

The denominators of the CAFs of all three pulses are

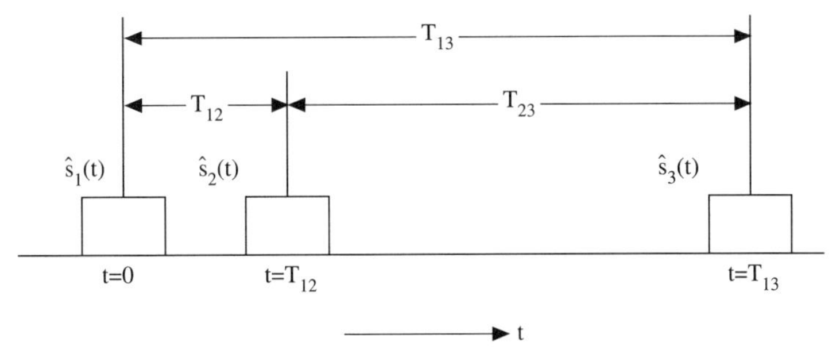

FIGURE 7.18
Example—three identical pulses with irregular spacing.

$$\text{Denom}[\hat{\Lambda}_{\substack{1\\2\\3}}(\tau,\Omega)] = \int_{-\infty}^{\infty} dt\,|\hat{s}_1(t)|^2 = \int_{-\infty}^{\infty} dt\,|\hat{s}_2(t)|^2 = \int_{-\infty}^{\infty} dt\,|\hat{s}_3(t)|^2 \quad (7.64a)$$

The numerators of $\hat{\Lambda}_1$, $\hat{\Lambda}_2$, and $\hat{\Lambda}_3$ are

$$\text{Num}[\hat{\Lambda}_1(\tau,\Omega)] = \int_{-\infty}^{\infty} dt\,\hat{s}_1(t)\hat{s}_1^*(t-\tau)e^{j\Omega t} \quad (7.64b)$$

$$\begin{aligned}\text{Num}[\hat{\Lambda}_2(t,\Omega)] &= \int_{-\infty}^{\infty} dt\,\hat{s}_2(t)\hat{s}_2^*(t-\tau)e^{j\Omega t}\\ &= \int_{-\infty}^{\infty} dt\,\hat{s}_1(t-T_{12})\hat{s}_1^*(t-T_{12}-\tau)e^{j\Omega t} \quad (7.64c)\\ &= \int_{-\infty}^{\infty} d(t-T_{12})\hat{s}_1(t-T_{12})\hat{s}_1^*[(t-T_{12})-\tau]e^{j\Omega(t-T_{12})}e^{j\Omega T_{12}}\\ &= e^{j\Omega T_{12}}\int_{-\infty}^{\infty} dt'\,\hat{s}_1(t')\hat{s}_1^*(t'-\tau)e^{j\Omega t'} = e^{j\Omega T_{12}}\text{Num}[\hat{\Lambda}_1(\tau,\Omega)]\end{aligned}$$

It follows from (7.64c) and (7.63a) that

$$\hat{\Lambda}_2(\tau,\Omega) = e^{j\Omega T_{12}}\hat{\Lambda}_1(\tau,\Omega) \quad (7.64d)$$

The steps (7.64c) and (7.64d) can also be repeated for $\hat{s}_3(t)$ with the aid of (7.63b) and (7.64b), with the result

$$\hat{\Lambda}_3(\tau,\Omega) = e^{j\Omega T_{13}}\hat{\Lambda}_1(\tau,\Omega) = e^{j\Omega T_{23}}\hat{\Lambda}_2(\tau,\Omega) \quad (7.64e)$$

When a signal consists of a superposition of two or more identical pulses with arbitrary spacing between them, the CAF contains cross-terms between the pulses. Consider as an illustrative example the signal with complex amplitude

$$\begin{aligned}\hat{s}(t) &= \hat{s}_1(t) + \hat{s}_2(t) + \hat{s}_3(t)\\ &= \hat{s}_1(t) + \hat{s}_1(t-T_{12}) + \hat{s}_1(t-T_{13}) \quad (7.65a)\end{aligned}$$

as shown in Figure 7.18.

The denominator of the AF is (assuming *no overlap* between pulses) and with the aid of (7.64a))

$$\text{Denom}[\hat{\Lambda}(\tau, \Omega)] = \int_{-\infty}^{\infty} dt\, |\hat{s}_1(t)|^2 + \int_{-\infty}^{\infty} dt\, |\hat{s}_2(t)|^2 + \int_{-\infty}^{\infty} dt\, |\hat{s}_3(t)|^2$$

$$= 3 \int_{-\infty}^{\infty} dt\, |\hat{s}_1(t)|^2 \tag{7.65b}$$

The numerator (which *does* contain the effects of overlap for certain values of τ) is

$$\begin{aligned}\text{Num}[\hat{\Lambda}(\tau, \Omega)] = B(\tau, \Omega) = \int_{-\infty}^{\infty} dt\, e^{j\Omega t}\{&\hat{s}_1(t)\hat{s}_1^*(t - \tau) \\ &+ \hat{s}_2(t)\hat{s}_2^*(t - \tau) + \hat{s}_3(t)\hat{s}_3^*(t - \tau) + [\hat{s}_1(t)\hat{s}_2^*(t - \tau) \\ &+ s_2(t)\hat{s}_1^*(t - \tau)] + [\hat{s}_1(t)\hat{s}_3^*(t - \tau) + \hat{s}_3(t)\hat{s}_1^*(t - \tau)] \\ &+ [\hat{s}_2(t)\hat{s}_3^*(t - \tau) + \hat{s}_3(t)\hat{s}_2^*(t - \tau)]\}\end{aligned} \tag{7.65c}$$

Using the notation

$$A_{lm}(\tau, \Omega) = \int_{-\infty}^{\infty} dt\, \hat{s}_l(t)\hat{s}_m^*(t - \tau)e^{j\Omega t} \tag{7.65d}$$

we can rewrite (7.65c) in the form

$$\begin{aligned}B(\tau, \Omega) = {} & [A_{11}(\tau, \Omega) + A_{22}(\tau, \Omega) + A_{33}(\tau, \Omega)] \\ & + \{[A_{12}(\tau, \Omega) + A_{21}(\tau, \Omega)] \\ & + [A_{13}(\tau, \Omega) + A_{31}(\tau, \Omega)] \\ & + [A_{23}(\tau, \Omega) + A_{32}(\tau, \Omega)]\}\end{aligned} \tag{7.65e}$$

From (7.64e), the top line of (7.65e) (term involving the *same* index for each signal in that term) is

$$B_{mm}(\tau, \Omega) = A_{11}(\tau, \Omega)(1 + e^{j\Omega T_{12}} + e^{j\Omega T_{13}}) \tag{7.65f}$$

The generic-cross term A_{lm} (τ, Ω) is

$$\begin{aligned}A_{lm}(\tau, \Omega) &= \int_{-\infty}^{\infty} dt\, \hat{s}_l(t)\hat{s}_m^*(t - \tau)e^{j\Omega t} \\ &= \int_{-\infty}^{\infty} dt\, \hat{s}_l(t)\hat{s}_l^*(t - T_{lm} - \tau)e^{j\Omega t} \\ &= \int_{-\infty}^{\infty} dt\, \hat{s}_l(t)\hat{s}_l^*[t - (\tau + T_{lm})]e^{j\Omega t}\end{aligned} \tag{7.65g}$$

$$= A_{ll}(\tau + T_{lm}, \Omega) = e^{j\Omega T_{1l}} A_{11}(\tau + T_{lm}, \Omega)$$

where $T_{lm} = T_{1m} - T_{1sL}$.

Also for this same l and m

$$A_{ml}(\tau, \Omega) = \int_{-\infty}^{\infty} dt\, \hat{s}_m(t)\hat{s}_m^*(t - T_{ml} - \tau)e^{j\Omega t}$$

$$= \int_{-\infty}^{\infty} dt\, \hat{s}_m(t)\hat{s}_m^*[t - (\tau - T_{lm})]e^{j\Omega t} = e^{j\Omega T_{1m}} A_{11}(\tau - T_{lm}, \Omega) \quad (7.65h)$$

From (7.65a,. . .,h), (7.64d,e), and (7.64a)

$$\begin{aligned}\hat{\Lambda}(\tau, \Omega) = \frac{1}{3}\{&\hat{\Lambda}_1(\tau, \Omega)[1 + e^{j\Omega T_{12}} + e^{j\Omega T_{13}}] \\ &+ [\hat{\Lambda}_1(\tau + T_{12}, \Omega) + \hat{\Lambda}_1(\tau - T_{12}, \Omega)e^{j\Omega T_{12}}] \\ &+ [\hat{\Lambda}_1(\tau + T_{13}, \Omega) + \hat{\Lambda}_1(\tau - T_{13}, \Omega)e^{j\Omega T_{13}}] \\ &+ e^{j\Omega T_{12}}[\hat{\Lambda}_1(\tau + T_{23}, \Omega) + \hat{\Lambda}_1(\tau - T_{23}, \Omega)e^{j\Omega T_{23}}]\}\end{aligned} \quad (7.66)$$

The range-only CAF is

$$\begin{aligned}\hat{\Lambda}(\tau, 0) = \Lambda_1(\tau, 0) &+ \frac{1}{3}\{[\hat{\Lambda}_1(\tau + T_{12}, 0) + \hat{\Lambda}_1(\tau - T_{12}, 0)] \\ &+ [\hat{\Lambda}_1(\tau + T_{13}, 0) + \hat{\Lambda}_1(\tau - T_{13}, 0)] \\ &+ [\hat{\Lambda}_1(\tau + T_{23}, 0) + \hat{\Lambda}_1(\tau - T_{23}, 0)]\}\end{aligned} \quad (7.67)$$

The doppler-only CAF is

$$\begin{aligned}\hat{\Lambda}(0, \Omega) = \frac{\hat{\Lambda}_1(0, \Omega)}{3}&(1 + e^{j\Omega T_{12}} + e^{j\Omega T_{13}}) \\ &+ \frac{1}{3}\{[\hat{\Lambda}_1(T_{12}, \Omega) + \hat{\Lambda}_1^*(-T_{12}, \Omega)e^{j\Omega T_{12}}] \\ &+ [\hat{\Lambda}_1(T_{13}, \Omega) + \hat{\Lambda}_1^*(-T_{13}, \Omega)\, e^{j\Omega T_{13}}] \\ &+ [\hat{\Lambda}_1(T_{23}, \Omega) + \hat{\Lambda}_1^*(-T_{23}, \Omega)e^{j\Omega T_{23}}]\}\end{aligned} \quad (7.68)$$

The range-only AF is sketched in Figure 7.19 for rectangular pulses that are widely separated, i.e.

$$T_{lm} >> \tau_p \quad (7.69)$$

where τ_p is the pulse duration.

If the assumption (7.69) holds, then the cross-terms in (7.68) are negligible because $\hat{\Lambda}(T_{lm},\Omega)$ is zero for all practical purposes if $T_{lm} >> \tau_p$. Then

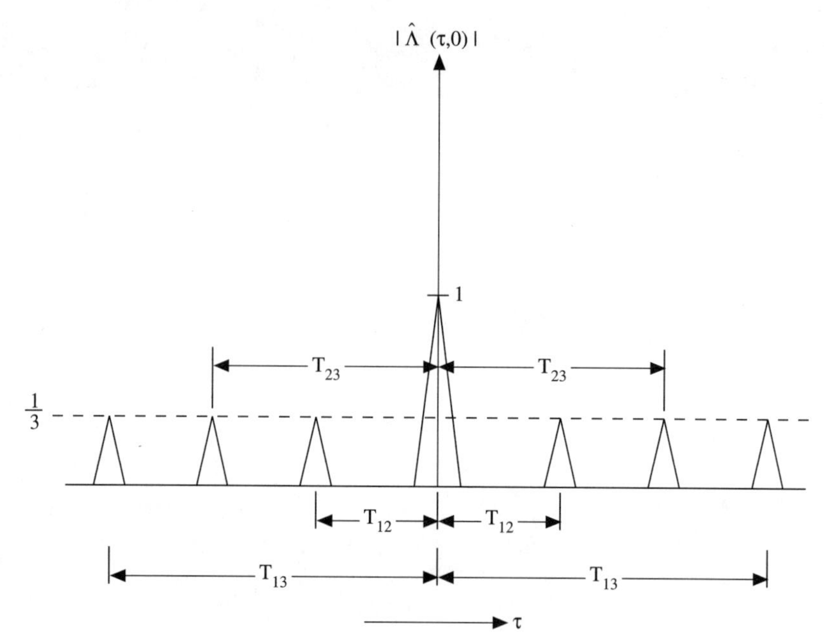

FIGURE 7.19
Range-only AF for train of three identical widely separated pulses.

$$\hat{\Lambda}(0,\Omega) = \frac{\hat{\Lambda}_1(0,\Omega)}{3}\left(1 + e^{j\Omega T_{12}} + e^{j\Omega T_{13}}\right) \tag{7.70}$$

and

$$\begin{aligned} |\hat{\Lambda}(0,\Omega)| &= \frac{1}{3}|\hat{\Lambda}_1(0,\Omega)|\sqrt{[(1+\cos\Omega T_{12}+\cos\Omega T_{13})^2 + (\sin\Omega T_{12}+\sin\Omega T_{13})^2]} \\ &= \frac{1}{3}|\hat{\Lambda}_1(0,\Omega)|\sqrt{\{3+2(\cos\Omega T_{12}+\cos\Omega T_{13}+\cos\Omega T_{23})\}} \\ &= \frac{1}{\sqrt{3}}|\hat{\Lambda}_1(0,\Omega)|\sqrt{1+\frac{2}{3}(\cos\Omega T_{12}+\cos\Omega T_{13}+\cos\Omega T_{23})} \end{aligned} \tag{7.71}$$

The lesson to be learned from Figure 7.19 is that variation of the PRF to avoid coincidence between the peaks of the cross-terms will alleviate the range-ambiguity problem by producing only *isolated* cross-term peaks, each of which has an amplitude of $1/N_p$ relative that of the main peak at $\tau = 0$. ($N_p = 3$ in this simple case but the result is easily generalized to arbitrary N_p.)

If the PRF is fixed, the cross-term peaks will accumulate at the multiples of the PRF and produce the result shown in Figure 7.14, where the amplitudes of the ambiguity peaks are comparable to those of the main peak. The AF for the variable PRF case, which is illustrated by Figure 7.19 for the simple case of three pulses, more closely resembles the "thumbtack" ambiguity function most *desirable* for avoidance of range ambiguities as the number of pulses increases, since the amplitudes of the cross-term peaks relative to the main peak amplitude are $1/N_p$. The fixed PRF case, as illustrated in Figure 7.14, gives rise to a "bed of spikes" or "bed of nails" AF, which is most *undesirable* for avoidance of range ambiguities.[9,10]

The limitation on this effect is that a number of pulses *so* large that some of the cross-term peaks will overlap and therefore accumulate to large values will begin to aggravate the range-ambiguity problem. Also multiple PRFs that are integral multiples of each other might produce undesirable cross-term peaks.

7.9 MORE ON TECHNIQUES OF PULSE COMPRESSION

In Sections 7.1 through 7.8 we alluded to pulse compression both directly and indirectly. In the present section, some techniques that can be used to reduce the effective duration of a radar pulse and thereby improve the range resolution will be discussed. In particular, the discussion of linear FM ("chirp") techniques will be extended and there will be some discussion of phase coding techniques.[11–17]

The standard practice is to refer to a pulse as having an "uncompressed pulse duration" τ_u and a "compressed pulse duration" τ_c and to refer to the ratio of uncompressed-to-compressed pulse duration as the "compression ratio, i.e.,

$$r_c = \frac{\tau_u}{\tau_c}$$

where τ_u, τ_c, and r_c are, respectively, the uncompressed pulse duration, compressed pulse duration, and compression ratio.

The compressed pulse duration is a measure of the range resolution attainable with the pulse; hence the larger the compression ratio, the greater the reduction in the minimum resolvable range difference attainable with the particular pulse compression technique under investigation.

The key operational advantage of pulse compression is that it allows a long pulse (with its accompanying high Doppler resolution) to be used without the commensurate penalty in loss of range resolution. This was discussed earlier, in Section 7.2, using linear frequency modulation of the pulse as a specific example of a pulse compression technique.

To begin the discussion, we refer back to Section 7.2, where we discussed the effect of "chirping" or linear FM on a pulse with Gaussian amplitude shape. Now we focus on the chirp pulse with rectangular amplitude shape. Consider the waveform as given by (7.34a), where $p(t)$ is the amplitude function, in this case a rectangular pulse of width τ_p centered at $t = 0$. The Fourier transform of this waveform is

$$\begin{aligned} S_T(\omega) &= \frac{1}{2} e^{j\psi} \int_{-\tau_p/2}^{\tau_p/2} dt\, e^{-j[(\omega-\omega_o)t - \dot{\omega}_M t^2]} \\ &\quad + \frac{1}{2} e^{-j\psi} \int_{-\tau_p/2}^{\tau_p/2} dt\, e^{-j[(\omega+\omega_o)t + \dot{\omega}_M t^2]} \\ &= \frac{1}{2} e^{j\psi} \hat{S}(\omega - \omega_o) + \frac{1}{2} e^{-j\psi} \hat{S}(\omega + \omega_o) \end{aligned} \tag{7.72}$$

where $\hat{S}(\omega)$ is the Fourier transform for the baseband signal $p(t)\cos(\dot{\omega}_M t^2)$, given by

$$\hat{S}(\omega) = \int_{-\tau_P/2}^{\tau_P/2} dt\, e^{-j\omega t + j\dot{\omega}_M t^2} \tag{7.72$'$}$$

Assuming no overlap between $\hat{S}(\omega - \omega_o)$ and $\hat{S}(\omega + \omega_o)$, it is sufficient to deal with $\hat{S}(\omega)$ in the analysis.

Completion of the square in the exponent inside the integral in (7.72)′ leads to

$$\hat{S}(\omega) = e^{\pm j\omega^2/4\dot{\omega}_M} \int_{-\tau_P/2}^{\tau_P/2} dt\, e^{\pm j\dot{\omega}_M [t \mp (\omega/2\dot{\omega}_M)]^2} \tag{7.73a}$$

or

$$\hat{S}(\omega) = e^{\mp j\omega^2/4\dot{\omega}_M} \left[\int_{-1/2[\tau_P \pm \omega/\dot{\omega}_M]}^{1/2[\tau_P \mp (\omega/\dot{\omega}_M)]} dt'\, \cos(\dot{\omega}_M t'^2) \pm j \int_{-1/2[\tau_P \pm \omega/\dot{\omega}_M]}^{1/2[\tau_P \mp (\omega/\dot{\omega}_M)]} dt'\, \sin(\dot{\omega}_M t'^2) \right] \tag{7.73b}$$

The integrals are recognizable as Fresnel integrals, i.e.

$$C(X) = \int_0^x dt\, \cos\left(\frac{\pi t^2}{2}\right) \tag{7.74a}$$

$$S(X) = \int_0^x dt\, \sin\left(\frac{\pi t^2}{2}\right) \tag{7.74b}$$

The amplitude spectrum $|\hat{S}(\omega)|$, based on (7.74a,b), is given by

$$|\hat{S}(\omega)| = \sqrt{[C(X_1) - C(X_2)]^2 + [S(X_1) - S(X_2)]^2} \tag{7.75}$$

where

$$X_1 = \sqrt{\frac{\dot{\omega}_M}{2\pi}} \left[\tau_P \mp \frac{\omega}{\dot{\omega}_M} \right]$$

$$X_2 = \sqrt{\frac{\dot{\omega}_M}{2\pi}} \left[\tau_P \pm \frac{\omega}{\dot{\omega}_M} \right]$$

A plot of (7.75) as a function of frequency $\omega/2\pi$ for various values of $\dot{f}_M = \dot{\omega}_M/2\pi$ is shown in Figure 7.20.

It is evident from Figure 7.20 that the spectrum of a chirp pulse with a rectangular envelope is nearly flat over a band covering the frequency swing of the modulation and falls off rapidly beyond that region. The spectrum of the rectangular pulse of the same duration without FM is also shown for comparison purposes. The ratio of the bandwidths of these two pulses (FM relative to non-FM) is shown to always exceed unity. That quantity is a measure of the compression ratio.

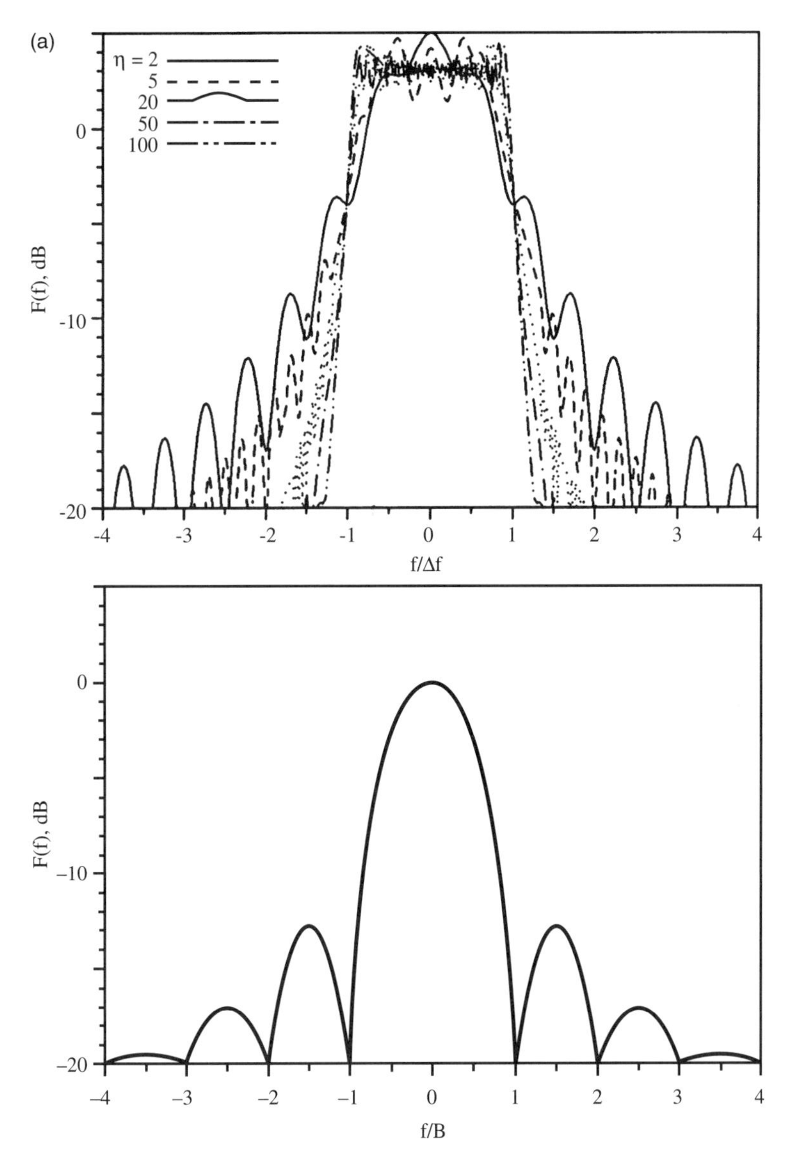

FIGURE 7.20
Spectrum of chirp pulse with rectangular envelope. (a) Chirp pulse spectrum for various compression ratios. (b) Unchirped pulse spectrum.

Now consider the complex ambiguity function of the rectangular envelope chirp pulse, given by (7.34c), where $p(t)$ is the rectangular pulse waveform. Integration of (7.34c) is no more difficult if $|\dot{\omega}_M| > 0$ than if $|\dot{\omega}_M| = 0$, because the quadratic terms in the exponent inside the integrand cancel each other out leaving only terms linear in t. The result of the integration is essentially the same as (7.22) where Ω is replaced by $(\Omega + 2\dot{\omega}_M\tau)$, i.e.,

$$|\hat{\Lambda}(\tau, \Omega)| = u(\tau_p - |\tau|)\left(1 - \frac{|\tau|}{\tau_p}\right) \operatorname{sinc}\left[\frac{(\Omega + 2\dot{\omega}_M\tau)\tau_p}{2}\left(1 - \frac{|\tau|}{\tau_p}\right)\right] \tag{7.76}$$

Consider the case where $\Omega=0$, where (7.76) becomes the output of a filter in the receiver, which is perfectly matched in frequency but mismatched in time delay in the amount τ; thus (where $\tau' = \tau/\tau_p$)

$$|\hat{\Lambda}(\tau, 0)| = |u[\tau_p(1 - |\tau'|)](1 - |\tau'|)\,\text{sinc}[\pi r_c' \tau'(1 - |\tau'|)]| \qquad (7.76)'$$

where $r_c' = \dot{\omega}_M \tau_p^2/\pi = (\Delta f\, \tau_p)$, Δf being the frequency excursion.

We see from (7.76)′ that $|\hat{\Lambda}(\tau,0)|$ is the product of $|\hat{\Lambda}(\tau,0)|$ for the pure rectangular pulse [as given by (7.22a)′ and illustrated in Fig. 7.4a] and $|\hat{\Lambda}(0, 2\dot{\omega}_M \tau[1 - (|\tau|/\tau_p)]|$ [as given by (7.22b)′ where $\Omega \rightarrow 2\dot{\omega}_M \tau[1 - (|\tau|/\tau_p)]$.

In Section 7.2, we discussed the meaning of the dimensionless ratio $\dot{\omega}_M \tau_p^2$ in the context of the chirp pulse with Gaussian envelope and arrived at Eqs. (7.36f)′ and (7.39a), which related that quantity to the ratio of the bandwidth of the chirped pulse to that of the unchirped pulse, where the quantity itself, expressed through (7.36g)′, is

$$\dot{\omega}_M \tau_p^2 = 2\pi \Delta f \tau_p = \pi\left(\frac{2\Delta f}{B}\right) = \pi r_c' \qquad (7.77a)$$

where Δf is the frequency excusion in Hertz due to the modulation and B is the bandwidth of the unchirped pulse in Hertz. The compression ratio, alluded to at the end of Section 7.2, could be defined for the case of the chirp pulse with rectangular envelope as the ratio $(2\Delta f/B)$, which is the reason it is named r_c'. The reason for the prime on r_c' is that the basic definition of compression ratio, as given at the beginning of Section 7.9 and named r_c, is the ratio of the true ("uncompressed") pulse duration τ_u, which we know to be τ_p or $(1/B)$, to the "compressed" pulse duration τ_c, which we have not yet defined precisely. Hence we do not yet know whether the definition as twice the ratio of frequency excursion-to-uncompressed pulse bandwidth is equivalent to the more fundamental definition as the ratio of uncompressed-to-compressed pulse duration. For r_c and r_c' to be equivalent, τ_c must be defined as $(1/2\Delta f)$. To investigate this, we note that nulls of the sinc function in (7.76)′ occur inside the triangular function envelope (those *outside* that envelope are irrelevant) whenever $|n|$ is any positive integer. The quadratic equation arising from that condition and the other required condition (i.e., that $|\tau'| \leq 1$ and that the sinc function nulls do not coincide with those of its argument, which occurs if $|n| = 0$, or with those of the triangular function, which occurs if $|\tau'|=1$) has the solution

$$|\tau'| = \frac{1}{2}\left[1 \pm \sqrt{1 - \frac{4|n|}{r_c'}}\right], \qquad |n| = 1, 2, 3, \ldots \qquad (7.77b)$$

Since $|\tau'|$ must be real, another required condition is

$$|n| \leq r_c'/4 \qquad (7.77c)$$

implying that the number of nulls within the envelope is limited by the size of r_c'. If $r_c' = 4$, there is only a single null at $|\tau| = 0.5\tau_p$. If $r_c' > 4$, there are at least two nulls, at $|\tau| = 0.5\tau_p\,[1 \pm \sqrt{1 - (4/r_c')}]$. If $r_c' > 8$, two more nulls arise, at $|\tau| = 0.5\tau_p[1 \pm \sqrt{1 - (8/r_c')}]$. As r_c' increases more nulls appear inside the envelope, which means that the main lobe of the sinc function becomes increasingly narrow and more sidelobes appear inside the triangular envelope. The situation is illustrated in Figure 7.21, where rough sketches of $|\Lambda(\tau,0)|$ are shown for a few values of r_c'.

The first null always occurs at $\tau = 0.5\tau_p[1 - \sqrt{1 - 4/r_c'}]$. If $r_c' \gg 4$, this can be approximated by

$$\tau_1 = 0.5\tau_p[1 - (1 - 2r_c')] = \frac{\tau_p}{r_c'} \qquad (7.77d)$$

If the "compressed pulse duration" τ_c is defined as the spacing between $\tau = 0$ and the first null at $\tau = \tau_1$, then the compression ratio as defined at the beginning of Section 7.9 is

$$r_c = \frac{\tau_u}{\tau_c} = \frac{\tau_p}{\tau_c} = \frac{2}{[1 - \sqrt{1 - (4/r_c')}]} \quad \text{if } r_c' \geq 4 \qquad (7.77e)$$

and

$$r_c = r_c' = \frac{2\Delta f}{B} \quad \text{if } r_c' >> 4 \qquad (7.77f)$$

If $r_c' < 4$, there are no nulls within the triangular envelope, and the "compressed pulse length" must be defined through the triangular envelope itself, weighted by the main lobe of the sinc function.

To summarize, we observe that the width of a pulse can be reduced substantially by the chirp technique applied to a rectangular pulse, as was already shown for the Gaussian pulse in Section 7.2. In that discussion, it was shown through Figure

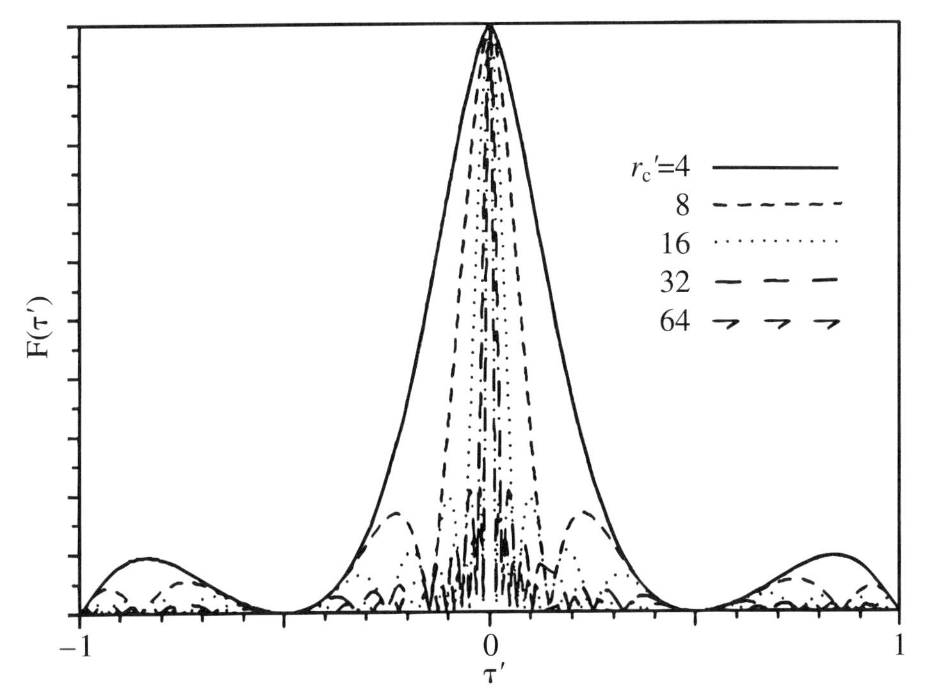

FIGURE 7.21
Time ambiguity functions for chirp pulse with rectangular envelope.

7.8 how the axis of the ambiguity diagram is increasingly rotated in the (τ–Ω) plane as the rate of frequency variation is increased, thereby decreasing the width in the τ-direction of the contour plot of $|\hat{\Lambda}(\tau,\Omega)|$ at $\Omega = 0$. The same phenomenon occurs with the rectangular pulse.

Pulse compression can also be achieved by means other than the chirp technique. Another widely used class of techniques is binary phase ("biphase") coding, where the pulse is constructed from a sequence of N bits each of duration τ_p/N. The amplitude of each bit is the same as that of the original pulse; hence the pulse energy is conserved. The modulation by these short binary pulses, whose phase is either zero or 180°, increases the bandwidth and thereby improves the time-delay resolution of the pulse. A great deal of research has been done on various phase shift coding schemes.[18–21] A very prominent set of biphase codes is the Barker codes,[22–25] which have the property that the sidelobe levels never exceed $1/N$.

In addition to biphase sequence codes, there are also "polyphase" codes, in which the phase of each pulse is varied over more than two states.[26–28] For example, "Frank polyphase codes," which are of length N^2, where N is an integer, can be used for discrete approximations to the linear chirp waveform.[26,29–31]

The CAF of a biphase sequence, such as a Barker code, is given by generalizing (7.46f) to allow for the phase shift on each bit pulse, all other properties of that pulse being the same for all of them. The generalization is that $\hat{\Lambda}_1(\tau,\Omega)$ is modified to reflect the fact that the cross-ambiguity function for different pairs of bits are different. We rewrite (7.46f) in the form

$$\hat{\Lambda}(\tau,\Omega) = \frac{1}{N}\left\{\sum_{l=1}^{N} e^{j\Omega l\tau_p/N}\hat{\Lambda}_{ll}(\tau,\Omega) + \sum_{\substack{l,m=1\\ l\neq 1}}^{N}{}' e^{j\Omega l\tau_p/N}\hat{\Lambda}_{lm}\left[\tau + (l-m)\frac{\tau_p}{N}, \Omega\right]\right\} \tag{7.78a}$$

where

$$\hat{\Lambda}_{lm}(\tau,\Omega) = \frac{\int_{-\infty}^{\infty} dt\ \hat{s}_l(t)\hat{s}_m^*(t-\tau)e^{j\Omega t}}{\int_{-\infty}^{\infty} dt\ |\hat{s}_l(t)|^2}$$

and where $|\hat{s}_l t)|^2 = |\hat{s}_m t)|^2$.

For a biphase sequence, $\hat{\Lambda}_{ll}(\tau,\Omega)$ is the same for all l but $\hat{\Lambda}_{lm}(\tau,\Omega)$ for $l \geq m$ is the same as $\hat{\Lambda}_{ll}(\tau,\Omega)$ if the two bits are in-phase and $-\hat{\Lambda}_{ll}(\tau,\Omega)$ if they are 180° out of phase. Consider the example of a five-bit Barker code of the form + + + − +. From (7.78a)

$$\begin{aligned}|\hat{\Lambda}(\tau,0)| = \Big|\frac{1}{5}\{&\hat{\Lambda}_1(\tau,0)[1+1+1+1+1] \\ &+ \hat{\Lambda}_{11}(\tau+\tau_p/5,0)[1+1-1-1] + \hat{\Lambda}_{11}(\tau-\tau_p/5)[1+1-1-1] \\ &+ \hat{\Lambda}_{11}(\tau+2\tau_p/5,0)[1-1+1] + \hat{\Lambda}_{11}(\tau-2\tau_p/5,0)[1-1+1] \\ &+ \hat{\Lambda}_{11}(\tau+3\tau_p/5,0)[-1+1] + \hat{\Lambda}_{11}(\tau-3\tau_p/5,0)[-1+1] \\ &+ \hat{\Lambda}_{11}(\tau+4\tau_p/5,0)[+1] + \hat{\Lambda}_{11}(\tau-4\tau_p/5,0)[+1]\}\Big|\end{aligned} \tag{7.78b}$$

$$= |\hat{\Lambda}_{11}(\tau, 0)| + \frac{1}{5}[|\hat{\Lambda}_{11}(\tau - 0.4\tau_p)| + |\hat{\Lambda}_{11}(\tau + 0.4\tau_p)|]$$

$$+ \frac{1}{5}[|\hat{\Lambda}_{11}(\tau - 0.8\tau_p)| + |\hat{\Lambda}_{11}(\tau + 0.8\tau_p)|]$$

A sketch-plot of (7.78b) is shown in Figure 7.22a.

By the same reasoning, we can determine the CAF for the 13-bit Barker code $+ + + + + - - + + - + - +$ (the longest known Barker code). There are 12 cross-terms involving bits separated by 1, i.e., 12, 23, 34, etc., of which 6 either do not involve bits 6, 7, 10, or 12 *or* involve two of those bits and are therefore positive, and 6 *do* involve one of those bits and a positive bit and are therefore negative. Adding all of these bits as the coefficients of the cross-CAF's $\hat{\Lambda}_{11}(\tau \pm \tau_p/13)$ with their corresponding signs produces a coefficient of zero for those terms and cancels them out.

Extending that line of reasoning to all terms, noting (as shown in Fig. 7.22b) that any pair involving only one of the bits 6, 7, 10, or 12 is negative and all others, including those involving two of those bits, are positive, we can conclude that the coefficients on the terms $\hat{\Lambda}_{11}[\tau \pm (2n + 1)\tau_p/13)]$ are all zero and those on the terms $\hat{\Lambda}_{11}(\tau \pm 2n\tau_p/13)$ are +1 or −1, resulting in a main lobe of unit amplitude and nonoverlapping sidelobes each of amplitude 1/13, i.e.,

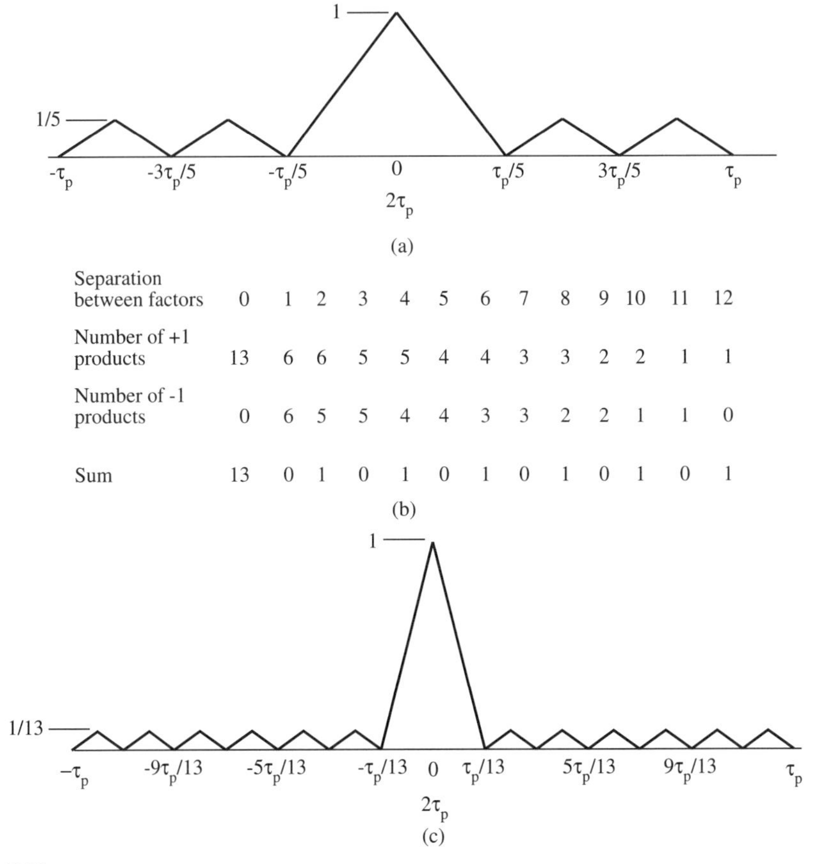

Separation between factors	0	1	2	3	4	5	6	7	8	9	10	11	12
Number of +1 products	13	6	6	5	5	4	4	3	3	2	2	1	1
Number of -1 products	0	6	5	5	4	4	3	3	2	2	1	1	0
Sum	13	0	1	0	1	0	1	0	1	0	1	0	1

FIGURE 7.22
Illustration of Barker codes. (a) 5-bit Barker code. (b) Logic behind 13-bit Barker code. (c) 13-bit Barker code.

$$
\begin{aligned}
|\hat{\Lambda}(\tau, \Omega)| = |\hat{\Lambda}_{11}(\tau, 0)| &+ \frac{1}{13}[\hat{\Lambda}_{11}(\tau - 2\tau_p/13, 0) + \hat{\Lambda}_{11}(\tau + 2\tau_p/13, 0)] \\
&+ \frac{1}{13}[\hat{\Lambda}_{11}(\tau - 4\tau_p/13, 0) + \hat{\Lambda}_{11}(\tau + 4\tau_p/13, 0) \\
&+ \frac{1}{13}[\hat{\Lambda}_{11}(\tau - 6\tau_p/13, 0) + \hat{\Lambda}_{11}(\tau + 6\tau_p/13, 0) + \cdots \\
&+ \frac{1}{13}[\hat{\Lambda}_{11}(\tau - 12\tau_p/13, 0) + \hat{\Lambda}_{11}(\tau + 12\tau_p/13, 0)]
\end{aligned}
\tag{7.78c}
$$

Equation (7.78c) is shown in Figure 7.22c.

For the Barker codes, because of the low sidelobes, the width of the ambiguity function is roughly that of the main lobe, implying that the compressed pulse width is roughly τ_p/N and hence the compression ratio is about N.

The topics of biphase and polyphase coding sequences and variable frequency sequences[32–34] to achieve pulse compression are extensive. These topics will not be discussed further here. They are covered in more detail in some of the references cited in this chapter.

REFERENCES

1. Woodward, P.M., "Probability and Information Theory." McGraw-Hill, New York, 1953.
2. Skolnick, M.I., "Introduction to Radar Systems," 2nd ed. McGraw-Hill, New York, 1980, Chapter 11, Sections 11.4, 11.5, pp. 411–434.
3. Barton, D.K., "Modern Radar System Analysis," Artech, Norwood, MA, 1988, Chapter 5, Sections 5.1 and 5.2, pp. 209–230.
4. Levanon, N., "Radar Principles." Wiley, New York, 1988, Chapters 6, 7, and 8.
5. Carpentier, M.H., "Principles of Modern Radar Systems." Artech, Norwood, MA, 1988, Chapter 3, Sections 3.6–3.13, pp. 81–116.
6. Nathanson, F.E., Reilly, J.P., and Cohen, M.N., "Radar Design Principles," 2nd ed. McGraw-Hill, New York, 1991, Chapters 8 and 11, by F.E. Nathanson.
7. Skolnick, M.I., "Radar Handbook," 1st ed McGraw-Hill, New York, 1970 or 2nd ed. McGraw-Hill, New York, 1990.
8. Spiegel, M., "Mathematical Handbook" (Schaum's Outline on Mathematics). Dover, New York, 1968.
9. Skolnick, M.I., "Introduction to Radar Systems," 2nd ed. McGraw-Hill, New York, 1980, pp. 418, 419.
10. Levanon, N., "Radar Principles." Wiley, New York, 1988, p. 143.
11. Eaves, J.L., and Reedy, E.K., "Principles of Modern Radar." Van Nostrand Reinhold, New York, 1987, Chapter 15, by M.N. Cohen.
12. Skolnick, M.I., "Introduction to Radar Systems," 2nd ed. McGraw-Hill, New York, 1980, Section 11.5, pp. 420–434.
13. Barton, D.K., "Modern Radar System Analysis." Artech, Norwood, MA, 1988, Section 5.2, pp. 220–231.
14. Carpentier, M.H., "Principles of Modern Radar Systems." Artech, Norwood, MA, 1988, Section 4.5, pp. 134–151.
15. Ulaby, F.T., Moore, R.K., Fung. A.K., "Microwave Remote Sensing," Vol. II. Addison Wesley, Reading, MA; 1982. Sections 7-5.2, 7-5.3, and 7-5.4, pp. 512–530.
16. Cooke, C.E., and M. Bernfeld, "Radar Signals." Academic Press, New York, 1967.
17. Nathanson, F.E., Reilly, J.P., and Cohen, M.N. "Radar Design Principles," 2nd. ed. McGraw-Hill, New York, 1991, Chapter 12, by M.N. Cohen and F.E. Nathanson, and Chapter 13, by F.E. Nathanson and M.N. Cohen.
18. Nathanson, F.E., Reilly, J.P., and Cohen, M.N. "Radar Design Principles," 2nd. ed. McGraw-Hill, New York, 1991, in particular pp. 533–582.
19. Ulaby, F.T., Moore, R.K., and Fung, A.K., "Microwave Remote Sensing," Vol. II. Addison Wesley, Reading, MA, 1982, Section 7-5.4, pp. 525–530.

20. Skolnick, M.I., "Introduction to Radar Systems," 2nd ed. McGraw-Hill, New York, 1980, pp. 428–431.
21. Levanon, N., "Radar Principles." Wiley, New York, 1988, pp. 152–163.
22. Barker, R.H., "Group Synchronizing of Binary Digital Systems." In "Communication Theory" (W. Jackson, ed.). Academic Press, New York, 1953, pp. 273–287.
23. Nathanson, F.E., Reilly, J.P., and Cohen, M.N., "Radar Design Principles," 2nd ed. McGraw-Hill, New York, 1991, in particular pp. 538–539.
24. Ulaby, F.T., Moore, R.K., and Fung, A.K., "Microwave Remote Sensing," Vol. II. Addison Wesley, Reading, MA, 1982, in particular pp. 528–530.
25. Eaves, J.L., and Reedy, E.K., "Principles of Modern Radar." Van Nostrand Reinhol, New York Section 15.3.1, pp. 480–483; by M.N. Cohen.
26. Nathanson, F.E., Reilly, J.P., and Cohen, M.N., "Radar Design Principles," 2nd ed. McGraw-Hill, New York, 1991, Section 12.5, pp. 559–565.
27. Lewis, B., and Kretchmer, F., "Linear Frequency Modulation Derived Polyphase Pulse Compression Codes." IEEE Trans. AES-18 (5), 637–641, 1982.
28. Siraswamy, R., "Multiphase Complementary Codes." IEEE Trans. Inform. Thry. IT-24(5), 546–552, 1978.
29. Frank, R.L., "Polyphase Codes with Good Nonperiodic Correlation Properties," IEEE Trans. Inform. Thry. IT-9(1), 43–45, 1963.
30. Levanon, N., "Radar Principles." Wiley, New York, 1988, pp. 153–159.
31. Nathanson, F.E., Reilly, J.P., and Cohen, M.N., "Radar Design Principles," 2nd ed. McGraw-Hill, New York, 1991, in particular pp. 559–561.
32. Eaves, J.L., and Reedy, E.K., "Principles of Modern Radar." Van Nostrand Reinhold, New York, 1987, Secs. 15.2.2, 15.2.3, pp. 475–478, by M.N. Cohen.
33. Levanon, N., "Radar Principles." Wiley, New York, 1988, pp. 145–152.
34. Nathanson, F.E., Reilly, J.P., and Cohen, M.N., "Radar Design Principles," 2nd ed. McGraw-Hill, New York, 1991, in particular Sections 13.6–13.13, pp. 599–634.

Chapter 8

Spatially Extended Targets and Clutter

CONTENTS

Previously we did not consider radar targets that are extended in space. All targets considered subtended a negligibly small angle at the radar receiver (and transmitter as well in the bistatic case) and could therefore be viewed as "point" targets.

Radar problems become more interesting when they involve targets that are extended in range and angle with respect to the radar receiver. In this chapter, such targets will be discussed. First, aggregates of point targets distributed over a set of ranges and angles will be considered, followed by consideration of radar returns from a continuous distribution of reflectivity. The latter begins with surface distributions as encountered in land and water surface clutter and then follows with volume distributions as encountered in precipitation clutter.

A few words are in order concerning terminology. The word "target" usually refers to an object that the radar wants to see clearly in order to detect it or extract information from it. The word "clutter" refers to *unwanted* returns from features within the field of view of the radar. In any radar scenario, as long as the antenna beam has a finite beamwidth (i.e., other than a "pencil beam" that sees objects only in a single direction; only an idealization that is impossible to realize in practice) and unless it has a range gate of zero width (again impossible to realize in practice because it requires infinite bandwidth), the radar will usually see unwanted reflecting objects. Hence clutter is a pervasive cause of corruption of radar signals.* Today's high-powered radars are essentially clutter-limited for most targets within specified range limits. They can be designed to overcome noise at enormous ranges but very often encounter clutter environments that overwhelm the noise.

One more point on terminology. It can be said that "one-person's clutter is another person's desired target." This means that any object or set of objects or continuum of matter that has nonzero reflectivity can be seen by a radar and there is nothing in the physics of radar scattering that can distinguish desired targets from undesired targets (clutter). Hence if we are dealing with a terrain mapping or imaging radar, then return from ground terrain is the desired target. Also, for a weather radar, returns from rain and falling snow are the desired targets. Hence it is obvious that these classes of return would not be called "clutter" in those classes of radar. But in most radars we deal with, the desired target is a man-made object such as an aircraft and all other returns including ground-scattering and rain-scattering *and* including *other* aircraft are designated as clutter.

8.1 RETURN FROM AN AGGREGATE OF POINT TARGETS

Consider (Fig. 8.1) an aggregate of point targets at various ranges and angles with respect to the radar receiver. We confine attention for the moment to a monostatic radar. For the seven point targets shown in the figure, it is evident that numbers 3, 4, and 5 are within the mainlobe of the antenna *and* within the range gate, and number 6 is within the range gate but outside the mainbeam, being near the peak of the first sidelobe above the mainbeam. Number 7 is near the peak of the first sidelobe below the mainbeam but is outside the range gate and hence not visible. Number 2 is within the mainbeam but also outside the range gate and therefore is also not visible. Number 1 is both outside the mainbeam and outside the range gate. The returns from these point targets would be weighted by the beam pattern and the range gate shape function. The latter has the triangular-function shape in the case of rectangular pulses and the Gaussian shape for Gaussian-shaped pulses. In either case, or for any other reasonable pulse shape, it can be regarded as negligibly small or approximately zero for ranges outside a given span around the range to which it is set. Hence certain targets can validly be designated as "outside the range gate." In the case of the antenna pattern, the situation is somewhat more "fuzzy." The conventional procedure is to regard a return as essentially zero if the target is a certain number of beamwidths away from the peak of the mainbeam. But clutter

* Location of material on clutter in standard radar texts is indicated in the reference list for Chapter 8.[1–6]

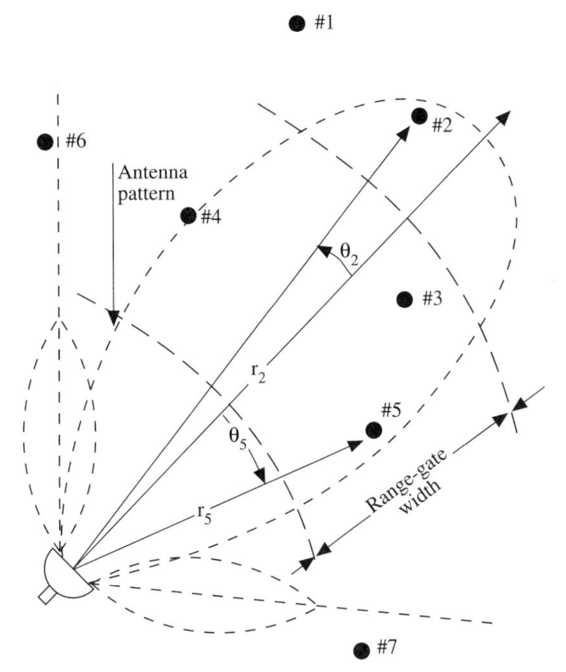

FIGURE 8.1
Aggregates of point targets.

targets that fall near the peak of the first or second sidelobe, if there are enough of them or if their reflectivity is high, can sometimes return enough energy to be significant. Therefore such clutter sources cannot necessarily be ignored. A case in point is shown in Figure 8.2, where the antenna beam is pointed horizontally. The desired target, an aircraft, and a patch of ground that may have a high reflectivity, are both within the range gate. The target is within the mainbeam of the antenna and the patch of ground within the range gate is within the first sidelobe. Since the patch of ground is an extended target, its total RCS may be high enough to return power comparable to that of the aircraft.

Now let us consider an aggregate of N point targets. The kth target is at range r_k and direction angles relative to the radar (θ_k,ϕ_k) (see Fig. 8.3, illustrating this for three point targets). The radar-based coordinate system is designated as (x,y,z) and the coordinates (r_k,θ_k,ϕ_k) are the standard spherical coordinates for the kth target.

From the monostatic radar equation (2.16), the power return from target k is

$$P_{RK} = P_{RO}\frac{g^2(\theta_k,\ \phi_k)\sigma_{kO}g_k^2(\theta_k',\ \phi_k')}{r_k^4} \tag{8.1}$$

where

$P_{RO} = \dfrac{P_T G_O^2 \lambda^2}{(4\pi)^3}$

$g(\theta_k,\ \phi_k)$ = one-way power pattern of antenna (same for transmission and reception) in direction of k

σ_{k0} = peak RCS of k (for backscattering)

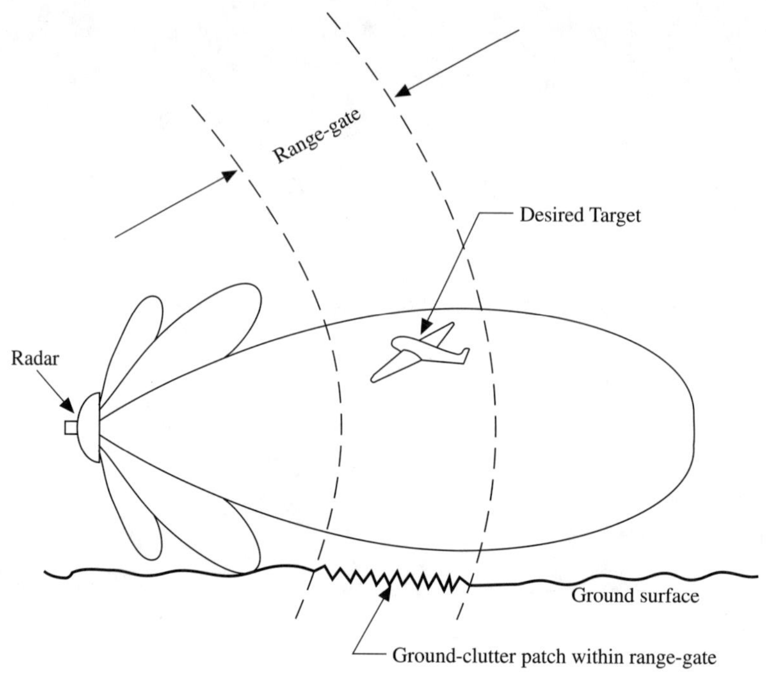

FIGURE 8.2
Ground clutter return from sidelobes.

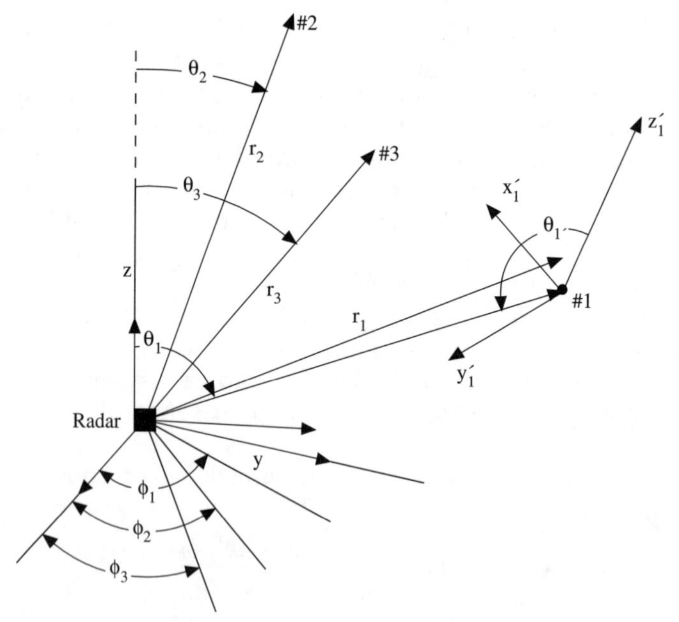

FIGURE 8.3
Coordinate geometry for three point targets.

$g_k(\theta'_k, \phi'_k)$ = power pattern of backscattering from k in terms of the direction angles (θ'_k, ϕ'_k), the spherical angles of the radar with respect to the target in a coordinate system (x'_k, y'_k, x'_k) whose origin is on target k (shown in Fig. 8.3 for target 1 only) and whose spherical coordinates are $(r'_k, \theta'_k, \phi'_k)$. Note that the target is assumed to be a "reciprocal scatterer" whose angular directivity pattern is the same for receptivity to the incident wave and distribution of power in the scattered wave.

The summation of returns from all N of the point targets, in principle, takes place "coherently," i.e., it is a summation of the "complex voltages" from all of the targets. The complex voltage corresponding to the return from the kth target has an amplitude proportional to the square root of (8.1) and a phase shift that is a sum of the path delay due to the target range r_k and the aggregate of phase shifts due to the antenna pattern and the directivity pattern for the scatterer. We can write the complex voltage for the kth target from (8.1) as follows:

$$V_{Rk} = \frac{C\sqrt{P_{RO}}\sqrt{\sigma_{ko}}}{r_k^2} f^2(\theta_k, \phi_k) f_k^2(\theta_k', \phi_k') e^{-2j\omega r_k/c} \tag{8.2}$$

where C is a possibly complex scale factor associated with conversion from power to voltage in the receiver, $e^{-2j\omega r_k/c}$ is the factor accounting for two-way path delay and $f(\theta_k, \phi_k) = \sqrt{g(\theta_k, \phi_k)}\, e^{j\psi_a(\theta_k, \phi_k)}$ = one-way complex voltage pattern for antenna in direction of k, whose phase is $\psi_a(\theta_k, \phi_k)$ and $f_k(\theta_k', \phi_k') = \sqrt{g_k(\theta_k', \phi_k')}\, e^{j\psi_b(\theta_k', \phi_k')}$ = complex voltage directivity pattern for scattering from target k, whose phase is ψ_b (θ_k', ϕ_k').

Let us suppress the angular dependence of f_k (θ_k', ϕ_k') and take the view that the target RCS σ_k is a number with a fixed value for the specific geometry of the radar and the kth target, depending on the angular orientation of k relative to the radar and generally being different for different targets. We can then separate the angle-dependent and range-dependent portions of each return and write the coherent summation of complex voltage from all N targets in the form

$$V_{RN} = C_o \sum_{k=1}^{N} a_k h(\theta_k, \phi_k) \frac{e^{-2j\omega r_k/c}}{r_k^2} \tag{8.3}$$

where

C_o = factor that is constant and same for all targets = $C\sqrt{P_{RO}}$

$\dfrac{e^{-2j\omega r_k/c}}{r_k^2}$ = factor dependent on r_k

a_k = factor that is dependent on k but approximately independent of r_k, θ_k, ϕ_k (i.e., its explicit dependence on those coordinates is suppressed) = $\sqrt{\sigma_k}\sqrt{g_k}$

$h(\theta_k,\phi_k)$ = factor dependent on angles (θ_k, ϕ_k) = $\sqrt{g(\theta_k, \phi_k)}\, e^{j\psi_a(\theta_k, \phi_k)}$

The weighting due to the antenna pattern is contained in $h(\theta_k, \phi_k)$ but so far we have not included a weighting due to the range gating operation. Let us define such a weighting factor by

$m(r)$ = range-gating function that weights the voltage returns according to the chosen pulse shape and range-gating method (8.4)

Inserting $m(r)$ into (8.3), we obtain the final form of V_{RN}

$$V_{RN} = C_o \sum_{k=1}^{N} a_k h(\theta_k, \phi_k) \left[\frac{m(r_k)}{r_k^2} e^{-2j\omega r_k/c} \right] \tag{8.5}$$

The received power from the aggregate of N targets is [from (8.5) with the aid of (8.3)]

$$P_{RN} = |C_o|^2 \sum_{\mu,\nu=1}^{N} a_\mu a_\nu^* h(\theta_\mu, \phi_\mu) h^*(\theta_\nu, \phi_\nu) \cdots \frac{m(r_\mu)m(r_\nu)}{r_\mu^2 r_\nu^2} e^{-(j2\omega/c)(r_\mu - r_\nu)} \tag{8.6}$$

In general a_k has some randomness in it due to fluctuations in both amplitude and phase. We can express a_k in the form

$$a_k = \langle a_k \rangle + \Delta a_k \tag{8.7}$$

where

$\langle a_k \rangle$ = quantity proportional to mean value of $\sqrt{\sigma_k}$ (ensemble mean)
Δa_k = deviation from ensemble mean

The mean voltage return is

$$\langle V_{RN} \rangle = C \sum_{k=1}^{N} \langle a_k \rangle h(\theta_k, \phi_k) \left[\frac{m(r_k)}{r_k^2} e^{-2j\omega r_k/c} \right] \tag{8.8a}$$

and using (8.7) in (8.6), the mean power return is obtained, i.e.

$$\langle P_{RN} \rangle = |C_o|^2 \left\{ \sum_{\mu=1}^{N} \langle |a_\mu|^2 \rangle \frac{|h(\theta_\mu, \phi_\mu)|^2}{r_\mu^4} |m(r_\mu)|^2 + \sum_{\substack{\mu,\nu=1 \\ \mu \neq \nu}}^{N}{}' \langle a_\mu a_\nu^* \rangle \frac{h(\theta_\mu, \phi_\mu) h^*(\theta_\nu, \phi_\nu)}{r_\mu^2 r_\nu^2} m(r_\mu) m^*(r_\nu) \cdots e^{-j2\omega(r_\mu - r_\nu)/c} \right\} \tag{8.8b}$$

where (if $\langle a_\mu \rangle$ = mean value of a_μ, Δa_μ = deviation from mean),

$$\langle |a_\mu|^2 \rangle = |\langle a_\mu \rangle|^2 + \langle |\Delta a_\mu|^2 \rangle \tag{8.8c}$$

$$\langle a_\mu a_\nu^* \rangle = \langle a_\mu \rangle \langle a_\nu^* \rangle + \langle \Delta a_\mu \Delta a_\nu^* \rangle \tag{8.8d}$$

8.2 RETURN FROM CONTINUOUS DISTRIBUTIONS OF REFLECTION

Let us concentrate on the first term of (8.8b), which is the noncoherent term that characterizes most clutter. If we assume that the return power is given by that first term alone,* we have

* An assumption buried in the neglect of the second term is that the returns from any two different targets are mutually uncorrelated. This is not necessarily true for all clutter environments.

$$\langle P_{RN}\rangle = |C_o|^2 \sum_{\mu=1}^{N} \sigma_\mu \frac{|h(\theta_\mu, \phi_\mu)|^2}{r_\mu^4} |m(r_\mu)|^2 \tag{8.9}$$

where the factor $|g_k|^2$ has been set equal to 1 and the averaging sign has been omitted from σ_μ.

Equation (8.9) holds for the noncoherent summation of returns from N discrete point targets. To generalize it to a continuous distribution of reflecting material, we can devise a definition of the "RCS per unit volume" $\sigma_v(r,\theta,\phi)$ and integrate over all angles θ,ϕ weighted by the antenna pattern and integrate over all r with the weighting of the range-gating function $m(r)$. The generalization of (8.9) to volume clutter is

$$\langle P_{RN}\rangle = |C_o|^2 \int_0^\infty dr\, r^2 \int_0^\pi d\theta \sin\theta \int_0^{2\pi} d\phi \cdots$$

$$\sigma_v(r, \theta, \phi) \frac{|h(\theta, \phi)|^2 |m(r)|^2}{r^4} \tag{8.10}$$

where $\sigma_v(e,\theta,\phi)$ is the RCS per unit volume (in reciprocal meters) at the position (r,θ,ϕ).

Equation (8.10) is used to determine the power return from precipitation. For clutter from a surface, e.g., ground clutter or sea clutter, we define the "normalized RCS" (often called the NRCS and notated here as $\sigma^{(o)}$)* as the RCS per unit surface area, which is a dimensionless quantity and is usually given in dB. The generalization of (8.10) for this case is

$$\langle P_{RN}\rangle = |C_o|^2 \iint_S dx'\, dy' \cdots$$

$$\sigma^{(o)}\left\{ r(x', y'), \theta(x', y'), \phi(x', y')|h[\theta(x', y'), \phi(x', y')]|^2 \cdots\right.$$

$$\left. - \frac{|m[r(x', y')]|^2}{|[r(x', y')]|^4} \right\} \tag{8.11}$$

where $dx'\, dy'$ is an area increment of the surface of the clutter area S' (which may be, for example, the earth's surface) and where the notation indicates that each point (r,θ,ϕ) intersects with the clutter surface and therefore $r(x',y')$ is the value of r at the intersection point and $\theta(x',y')$ and $\phi(x',y')$ have analogous meanings.

One more case is the line distribution of reflectivity, for which the analog of (8.11) is

$$\langle P_{RN}\rangle = |C_o|^2 \int_{s_A'}^{s_B'} ds' \cdots$$

$$\sigma_L\left(r(s'), \theta(s'), \phi(s')|h[\theta(s'), \phi(s')]|^2 \cdots \frac{|m[r(s')]|^2}{[r(s')]^4} \right) \tag{8.12}$$

where s' is a coordinate that follows the distribution of reflectivity, the coordinates

* In some radar literature the NRCS is denoted by σ_o.

r,θ,ϕ of the line element ds' are indicated by the notation $r(s')$, $\theta(s')$, $\phi(s')$, and the element extends from s'_A to s'_B. The symbol σ_L denotes the RCS per unit length (in meters) at the line element location.

8.3 GEOMETRY OF SURFACE CLUTTER

In radar engineering practice it is often necessary to estimate the clutter levels likely to be encountered in a given scenario. To that end, it is important to have simplified expressions that can yield quick estimates that are useful for setting design objectives. However, these estimates do not account for many of the complexities of electromagnetic scattering from terrain features, in particular, the variability of the results due to changes in moisture content, wind speeds, etc.

In spite of such limitations, these simplified expressions are still the best tools available for determining clutter parameters needed quickly, accurate to within a few dB.

The first simplified expressions to be discussed are those obtainable from specializations of (8.11) for clutter return from the surface of the earth.

To use (8.11) in that context, we refer to Figure 8.4. Before explaining the diagrams in the figure, we note that in Figure 8.4 we are using the "flat earth" approximation. This implies that the curvature of the earth is not a significant factor in determination of clutter power. For ranges of less than a few miles (say 10 miles or less) this approximation usually gives negligible error. For ranges beyond 50 miles, particularly over 100 miles, it is sometimes very important to account for earth curvature.

Definitions of angles are important in the discussion to follow. The "depression angle," denoted by β in Figure 8.4a, is the angle between the direction of the incident wave and the horizontal in the vertical plane. The "grazing" angle α is the angle between the plane of the ground and the direction of the source of the incident wave as seen from the illuminated point, also measured in the vertical plane. The "vertical plane" referred to, which happens to be the x–z plane in the chosen geometry, is the "plane of incidence," defined as the plane containing the vector in the direction of the incident wave and the unit vector normal to the plane from which the reflection is occurring. In this case the "normal unit vector" $\hat{n}$ points vertically upward and the direction of the incident wave source from the reflection point is indicated by the unit vector $\hat{v}$. In Figure 8.4a, we define the incident wave graphically as in the direction of the peak of the radar antenna beam, but the definitions of α and β could also be defined for any illuminated point on the earth's surface wherever it appeared within the antenna beam. In that sense there is an infinity of values of α and β, but the ones we specify are those corresponding to the peak of the antenna beam direction.

The other angle to be defined is θ_i, the "angle of incidence," the angle between the vector $\hat{v}$ in the direction of the source of the incident wave at the reflection point and the surface normal at that point. That angle is the complement of the grazing angle α. For the flat earth approximation, $\alpha = \beta$, $\theta_i = (\pi/2) - \beta$. But that is not true if earth curvature is accounted for, as is shown in Figure 8.5, where it is evident that $\alpha < \beta$ and $\theta_i > [(\pi/2) - \beta]$.

The case depicted in Figure 8.4 is that for a *small* grazing angle or equivalently a *large* angle of incidence, where Δr, the width of the radar pulse or the range gate (which are roughly equivalent in this context) is large enough to ensure that the patch of ground illuminated by the antenna beam (shown as having beamwidth $\Delta\theta$ in the elevation plane) lies entirely within the range gate. This is the case of "pulse-

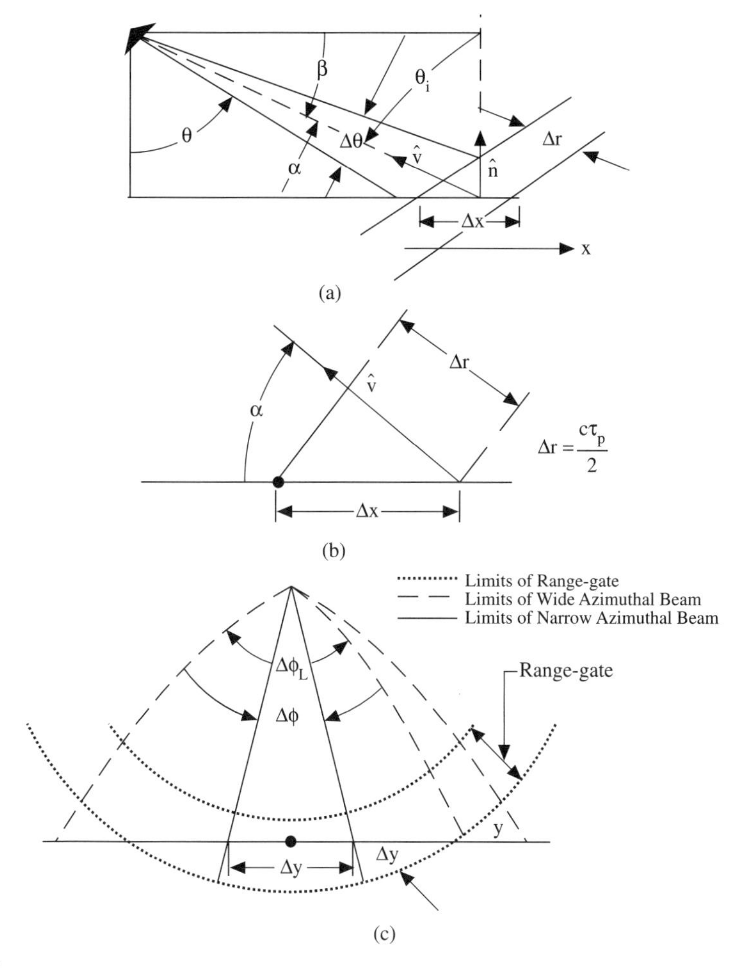

FIGURE 8.4
Ground clutter geometry—airborne radar-pulse-limited. Case of pulse (or range gate)-limited clutter illumination in $+x$ direction. Flat earth approximation. Narrow beam. β = depression angle, α = grazing angle, θ_i = angle of incidence, $\hat{n}$ = normal unit vector at reflection point, $\hat{v}$ = unit vector in direction of source of incident wave at reflection point. (a) x-z plane view. (b) Clutter patch: x-z plane view. (c) y-z plane (slanted) view.

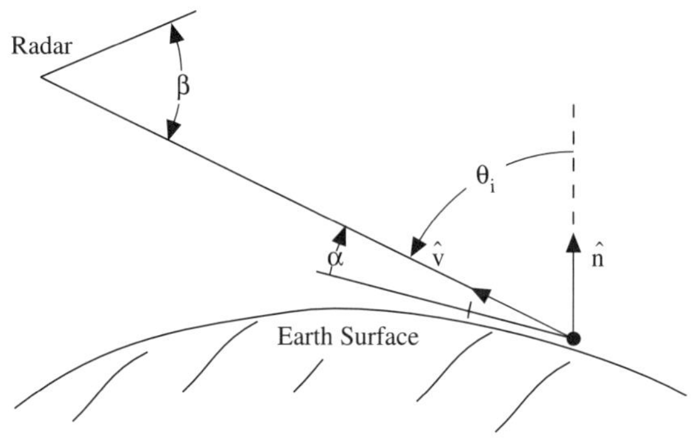

FIGURE 8.5
Ground-clutter geometry with earth curvature.

width-limited" earth-surface clutter, which is not affected by the *elevation* beamwidth $\Delta\theta$, but *is* affected by the *azimuthal* beamwidth $\Delta\phi$, as shown in Figure 8.4c. Figure 8.4c shows a slanted view of the illuminating beam parallel to the y–z plane and shows the entire clutter patch width Δy in that plane to be within the range gate as long as the azimuthal beamwidth is sufficiently small to preclude the sphericity of the range gate from eliminating some of the ground clutter within the beam. A *large* beamwidth $\Delta\phi_L$ is shown with the dashed radial lines in Figure 8.4c. That demonstrates the possibility that the clutter return would be limited by the range gate in the azimuthal direction if the beam were extremely wide in that direction. In the usual case the azimuthal beam is small and includes all the clutter within the range gate (solid radial lines on Figure 8.4c).

To determine the clutter power for the pulse-limited case shown in Figure 8.4, we specialize (8.11) to the case where $\sigma^{(o)}$ is constant throughout the region illuminated by the beam and/or pulse. That is the only way to get handy formulas involving a simple dependence on beamwidth and pulse-width. That specialization of (8.11) is supplemented by the assumption that we can approximate the value of r as constant over the illuminated region and that the approximate value of r is r_o, that corresponding to the center of the antenna beam. We assume that (x_o, y_o) is the center of the ground clutter patch, its x and y widths are Δx and Δy, respectively, and that the beam pattern can be separated into two factors, one for elevation θ and one for azimuth ϕ, given by $h_\theta(\theta)$ and $h_\phi(\phi)$, respectively. These assumptions applied to (8.11) in the case depicted in Figure 8.4, with $\Delta\phi$ small enough to contain the entire clutter patch within the pulse width, result in a specialized version of (8.11), to be derived below.

The constraint that exists is that (r,θ,ϕ) cannot be specified independently in this case. The integration over θ is constrained by the relationships in the x–z plane.

$$x = r\sin\theta = r_o\sin\theta = r_o\cos\alpha \tag{8.13a}$$

$$dx = r_o\cos\theta\, d\theta \tag{8.13b}$$

Thus the integration on θ is replaced by the integration on x over the extent of the patch. The integration on ϕ is replaced by an integration on y. The resulting form of (8.11) is

$$\langle P_{RN}\rangle = \frac{|C_o|^2\sigma^{(o)}}{r_o^4}\int_{x_o-(\Delta x/2)}^{x_o+(\Delta x/2)} dx \int_{y_o-(\Delta y/2)}^{y_o+(\Delta y/2)} dy \left|h_\theta\left[\sin^{-1}\left(\frac{x_o}{r_o}\right)\right]\right|^2 \cdots$$

$$\left|h\phi\left(\frac{y_o}{r_o}\right)\right|^2 |m(r_o)|^2 \tag{8.14}$$

where it is evident from the geometry of Figure 8.4c that

$$dy = r_o\, d\phi \tag{8.15a}$$

and hence

$$\Delta y = r_o\Delta\phi \tag{8.15b}$$

Finally, since $|m(r_o)|^2 = 1$, the result of the integrations in (8.14) for sufficiently small $\Delta\phi$ and Δr is

$$\langle P_{RN} \rangle = \frac{|C_o|^2 \sigma^{(o)}}{r_o^4} \Delta x \Delta y \tag{8.16a}$$

where (see Fig. 8.4b)

$$\Delta x = \frac{\Delta r}{\cos \alpha} \tag{8.16b}$$

and

$$\Delta y = r_o \Delta \phi \tag{8.16c}$$

The final form of (8.16a), with the aid of (8.16b,c) is

$$\langle P_{RN} \rangle = \frac{|C_o|^2 \sigma^{(o)}}{r_o^3} \left(\frac{c\tau_p}{2} \right) \Delta \phi \sec \alpha \tag{8.17}$$

Equation (8.17) shows a significant departure from the usual radar equation. For the case of low grazing angle, where the return is pulse-width limited, it is shown to be directly proportional to the pulse-width and to the azimuthal beamwidth, directly proportional to the secant of the grazing angle or the cosecant of the incidence angle (i.e., monotonically *increasing* with the grazing angle α) and *above all*, varying as the inverse *third* power of range rather than the usual fourth power as for a point target.

We will now consider another extreme case, that shown in Figure 8.6. This is the case of large grazing angle or equivalently small angle of incidence, where the beam is pointed nearly vertically downward. In that case, as is evident from the figure, the ground patch is entirely within the range gate but is limited by the elevation beamwidth. In this case, (8.16a) still applies with the assumptions made earlier (flat earth, $r = r_o$ everywhere on illuminated patch, etc.) but

$$\Delta x = \frac{r_o \Delta \theta}{\sin \alpha} = r_o \Delta \theta \csc \alpha \tag{8.18a}$$

$$\Delta y = r_o \Delta \phi \quad \text{[same as (8.16c)]} \tag{8.18b}$$

From (8.16a) and (8.18a,b)

$$\langle P_{RN} \rangle = \frac{|C_o|^2 \sigma^{(o)}}{r_o^2} (\Delta \theta \Delta \phi) \csc \alpha \tag{8.19}$$

Note that the return power in this case is independent of the pulse-width, varies linearly with the product of the elevation and azimuthal beamwidth, monotonically *decreases* with grazing angle (varying as it does with the cosecant of α, which decreases with increasing α), and varies as the *inverse-square* of range rather than the inverse fourth power as for a point target.

To summarize these results, we have the following three cases.

A. Point target—RCS $= \sigma$
Target at peak of beam-range $= r_o$

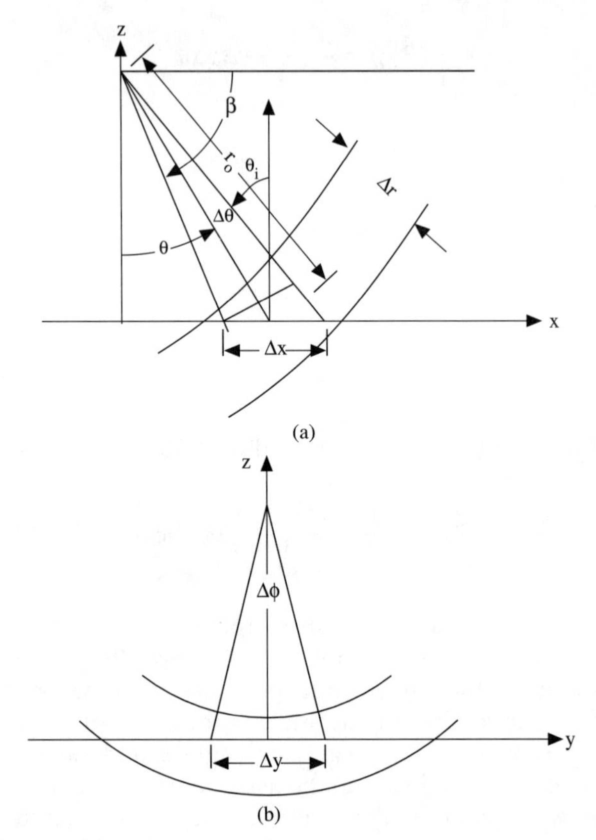

FIGURE 8.6
Ground clutter geometry-airborne radar-beam-limited. (a) x-z plane view. (b) y-z plane view.

$$P_R = \frac{P_T G_o^2 \lambda^2 \sigma}{(4\pi)^3 r_o^4} \tag{8.20a}$$

B. Surface patch NRCS $= \sigma^{(o)}$
Patch at peak of beam; *small* grazing angle α; patch small enough to appear as a point to radar; range to patch center $= r_o$

$$P_R \simeq \frac{P_T G_o^2 \lambda^2 \sigma^{(o)}}{(4\pi)^3 r_o^3} \Delta\phi \left(\frac{c\tau_p}{2}\right) \sec\alpha \tag{8.20b}$$

C. Same as B but α *large*

$$P_R \simeq \frac{P_T G_o^2 \lambda^2 \sigma^{(o)}}{(4\pi)^3 r_o^2} \Delta\theta\Delta\phi \csc\alpha$$

$$\simeq \frac{P_T G_o \lambda^2 \sigma^{(o)}}{(4\pi)^2 r_o^2} \csc\alpha \tag{8.20c}$$

The second form of (8.20c) is due to the relationship $G_o = 4\pi/\Delta\Omega$ [see Eq. (2.29), Chapter 2] where $\Delta\Omega$ is the solid angle of coverage of the beam, well approximated by the product of elevation and azimuth beamwidth, i.e., given

$$G_o = \frac{4\pi}{\Delta\Omega} \tag{8.21a}$$

where $\Delta\Omega$ = solid angle of coverage, and

$$\Delta\Omega = \Delta\theta\Delta\phi \tag{8.21b}$$

we can obtain the second form of (8.20c) from the first form.

8.4 GEOMETRY OF VOLUME CLUTTER

There are so many possible configurations of volume clutter that the reasonable way to obtain a simple formula for it is to consider the reflecting material to occupy the range-angle space occupied by the beam and range gate and to assume σ_V to be constant over that space. From (8.10) with $\sigma_V(r,\theta,\phi) = \sigma_V$ = constant and assuming the volume illuminated to be small enough to consider r equal to r_o, the range to the center of that volume.

With those assumptions, we can express (8.10) in the form

$$\langle P_{RN} \rangle = \frac{|C_o|^2}{r_o^4} \sigma_V \Delta V \tag{8.22a}$$

where ΔV is the volume occupied by the clutter-producing material. If the beam and range gate are entirely occupied by this material, then, with the aid of (8.21b)

$$\Delta V = (r_o^2 \Delta\Omega)(\Delta r) = r_o^2 \left(\frac{c\tau_P}{2}\right)\Delta\Omega = \left(\frac{c\tau_P}{2}\right)\Delta\theta\Delta\phi \tag{8.22b}$$

where $\Delta\Omega$ is the solid angle of coverage, and where

$$\Delta\Omega = \int_0^{\pi} d\theta \sin\theta \int_0^{2\pi} d\phi |h(\theta, \phi)|^2$$

$$\Delta r = \frac{c\tau_P}{2} = \int_{-\infty}^{\infty} dr |m(r)|^2$$

Combining (8.22a) and (8.22b), we obtain

$$\langle P_{RN} \rangle = \frac{P_T G_o^2 \lambda^2}{(4\pi)^3 r_o^2} \sigma_V \left(\frac{c\tau_P}{2}\right)\Delta\theta\Delta\phi \tag{8.22c}$$

According to (8.22c), volume clutter return varies as the inverse square of range and is proportional to pulse-width and the product of elevation and azimuthal beamwidths (Fig. 8.7).

8.5 SIGNAL-TO-CLUTTER RATIO AND SIGNAL-TO-INTERFERENCE RATIO

The signal-to-clutter ratio, or SCR, is the ratio of the return from a point target at the same range as the center of the clutter patch and also at the peak of the antenna beam. We will confine attention in what follows to surface clutter as discussed in Section 8.3 but the reader could produce analogous results for volume clutter based on Section 8.4.

From (8.20a,b,c), the SCR is *for small grazing angle*

$$\mathrm{SCR} = \left(\frac{P_{\mathrm{Rt}}}{P_{\mathrm{Rc}}}\right)_{\text{small } \alpha} = \frac{P_{\mathrm{T}} G_{\mathrm{o}}^2 \lambda^2 \sigma_{\mathrm{t}}/(4\pi)^3 r_{\mathrm{o}}^4}{[P_{\mathrm{T}} G_{\mathrm{o}}^2 \lambda^2 \sigma^{(\mathrm{o})}/(4\pi)^3 r_{\mathrm{o}}^3] \Delta\phi (c\tau_{\mathrm{p}}/2) \sec \alpha}$$

$$= \frac{\sigma_{\mathrm{t}} \cos \alpha}{\sigma^{(\mathrm{o})} r_{\mathrm{o}} \Delta\phi (c\tau_{\mathrm{p}}/2)} \tag{8.23a}$$

where P_{Rt} and P_{Rc} are power returns from target and clutter, respectively. *For large grazing angle*

$$\mathrm{SCR} = \left(\frac{P_{\mathrm{Rt}}}{P_{\mathrm{Rc}}}\right)_{\text{large } \alpha} = \frac{\sigma_{\mathrm{t}} \sin \alpha}{\sigma^{(\mathrm{o})} r_{\mathrm{o}}^2 (\Delta\theta \Delta\phi)} \tag{8.23b}$$

A major point about both (8.23a) and (8.23b) (and any intermediate case not covered by the simple analysis leading to these limiting results) is that the SCR cannot be influenced by increasing the transmitted power. That is quite obvious, of course (without *any* analysis) because the increased return power resulting from that measure affects target and clutter returns equally. The other major point is that the inverse dependence on range is weaker than that of the SNR (which of course is

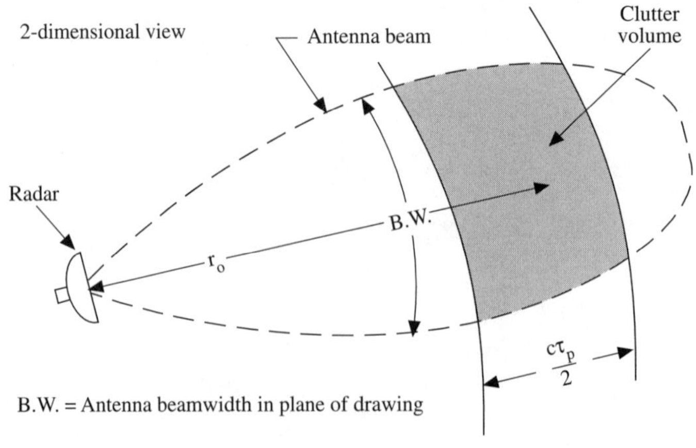

FIGURE 8.7
Volume clutter geometry.

$1/r_o^4$). Since the range dependence is still an inverse one, however, it is still possible to define a "maximum range" r_{cmax} for a given SCR, or equivalently, for a minimum value of target signal power that will exceed the clutter. That range is given by solving (8.23a) or (8.23b) for r_o with fixed SCR

$$r_{cmax} = \frac{\sigma_t \cos\alpha}{\sigma^{(o)}\Delta\phi(c\tau_p/2)(SCR)_{min}} \tag{8.24a}$$

if α is *small* (e.g., $\alpha \leq 30°$) and

$$r_{cmax} = \sqrt{\frac{\sigma_t \sin\alpha}{\sigma^{(o)}(\Delta\theta\Delta\phi)(SCR)_{min}}} \tag{8.24b}$$

if α is *large*, (e.g., $\alpha \geq 50°$).

If α is in the intermediate region (e.g., $30° < \alpha < 50°$), then the dependence of SCR on r_o is somewhere between $1/r$ and $1/r^2$ and the dependence of r_{cmax} on $\sigma_t/\sigma^{(o)}$, $(SCR)_{min}$, and other variables on the right-handside (8.24a,b) is somewhere between linear and square root.

The only *controllable* variables in (8.23a,b) or (8.24a,b) are the pulse duration τ_p and the beamwidths $\Delta\theta$ and $\Delta\phi$, so these relationships present possible tradeoff problems in deciding on those parameters in design. For example, if a long pulse-length or a large azimuthal beamwidth were desirable goals for a particular design specification, e.g., good detection performance in noise, but low grazing angle clutter was a particularly important problem, then (8.24a) would present us with a reason for reducing pulse length or azimuthal beamwidth to improve the range against clutter. That is one of an infinity of possible tradeoffs in radar system design.

For completely "clutter-limited" radar scenarios, i.e., those for which there is sufficient power to overcome noise in all cases of interest, Eqs. (8.23a,b) or (8.24a,b) can be used to guide design against clutter. However, in many systems the range of scenarios to cover may be so broad as to include regions of noise-limited *and* clutter-limited operation. In this case, a better indicator than the SCR is the signal-to-interference ratio, or SIR, where "interference" means clutter power plus noise power. This parameter is defined as

$$SIR = \frac{P_{Rt}}{P_{Rc} + P_n} \tag{8.25a}$$

where P_{Rt} and P_{Rc} are target and clutter returns, respectively, and P_n is noise power.

We can write (8.25a) in one of two more convenient forms, as follows:

$$SIR = \frac{SNR}{1 + CNR} = \frac{SNR}{1 + (SNR/SCR)} \tag{8.25b}$$

where CNR = clutter-to-noise ratio = P_{Rc}/P_n and SNR = signal-to-noise ratio = P_{Rt}/P_n. This form would be most convenient in noise-limited operation, where SNR $<<$ SCR and hence SIR $\simeq$ SNR.

Another possible form of (8.25a) is

$$\mathrm{SIR} = \frac{\mathrm{SCR}}{1 + (1/\mathrm{CNR})} = \frac{\mathrm{SCR}}{1 + (\mathrm{SCR}/\mathrm{SNR})} \tag{8.25c}$$

which would be most convenient in clutter-limited regimes, where SCR << SNR and hence SIR ≃ SCR.

The range dependence of SIR in the low grazing angle case [from (8.23a) and (8.25c)] is

$$(\mathrm{SIR})_{\alpha\text{-small}} = \frac{\sigma_t \cos\alpha/\sigma^{(o)} r_o \Delta\phi (c\tau_p/2)}{1 + [(4\pi)^3 r_o^3 (kT_o B_R F_N)\cos\alpha / P_T G_o^2 \lambda^2 \sigma^{(o)} \Delta\phi (c\tau_p/2)]} \tag{8.26a}$$

and for the high grazing angle case [from (8.20c), (8.23b), and (8.25c)]

$$(\mathrm{SIR})_{\alpha\text{-large}} = \frac{[\sigma_t/\sigma_{(o)} \sin\alpha (\Delta\theta\Delta\phi) r_o^2]}{1 + [(4\pi)^3 r_o^2 \sin\alpha (kT_o B_R F_N)/P_T G_o^2 \lambda^2 \sigma^{(o)} \Delta\theta\Delta\phi]} \tag{8.26b}$$

It will be left as an exercise to determine r_{cmax} for a specified value of $(\mathrm{SIR})_{min}$, which is somewhat more complicated than the analogous procedure for SCR, which led to (8.24a,b). Also, these procedures could be done for volume clutter, using (8.22c).

8.6 STATISTICS OF CLUTTER AMPLITUDE

Clutter signals, like those from targets, are aggregates of returns from individual scatterers and hence usually contain enough "randomness" to be considered as sample functions of random processes. Much of the apparent randomness is due to ignorance, i.e., the inability to model a clutter target exactly because it is impossible to know all the details of its motion. Since the same is true of target signals, it is not surprising that the statistics of clutter returns have much in common with those of the aggregate of "scattering centers" that constitutes a desired target.

The major difference between clutter and standard point targets like aircraft is that clutter occupies a much larger region in space, i.e. (all of or) a significant fraction of the antenna beam and/or the range gate. It also generally occupies a larger region in Doppler space, since an aggregate of clutter sources can have a wide range of velocities. The latter feature affects its spectrum but does not significantly affect its amplitude PDF.

The reader is referred to Chapter 6, where the statistics of the amplitude fluctuations of signals from targets was discussed in some detail. Because of the central limit theorem, both target and clutter signals often exhibit Rayleigh amplitude statistics, both being essentially aggregates of returns from large numbers of point scatterers that are statistically independent. At other times, when there is one large dominant scatterer superposed on a background of smaller independent scatterers, the statistics are Ricean, which can be approximated by the particular chi-square PDF called "one-dominant-plus Rayleigh." In still other cases, the amplitude PDFs are observed to have longer tails than would be predicted through Rayleigh or Ricean PDFs so they are modeled with higher order chi-square, lognormal, Weibull, or *K*-distributions. All of that, discussed in Chapter 6 for targets, also applies to clutter. A good introduction to amplitude statistics of clutter would result from a review of Chapter 6. All of the PDFs discussed there apply to clutter.

One might ask the question, "If target and clutter returns have the same amplitude fluctuation statistics and the same mean power values in a particular radar scenario, how is it possible to detect the target in clutter?" The answer is that if the mean power in target and clutter returns is roughly comparable, then the more similar the clutter and target fluctuation PDFs, the more difficult the detection problem. For that reason, one must look for another discriminant that can distinguish typical clutter signals from typical target signals. Two that come to mind are Doppler spectra and polarization characteristics. These properties of clutter will be discussed in a later section of Chapter 8.

Some sample experimental curves are presented below, showing amplitude PDFs of clutter. These are taken from standard references.[1,3,7,8] First, in Figure 8.8, we show three plots of $W(P)dP$, the probability that the received clutter power is within a small incremental region between P and $P + dP$ (proportional to the PDF of power or RCS) vs. $P/\langle P\rangle$, the received power normalized to its mean value. The experimental results are the histograms and the continuous curves are Ricean PDFs fitted as closely as possible to the experimental results. These results, for heavily wooded and rocky terrain samples, are from M.W. Long, "Radar Reflectivity of Land and Sea,"[7] a well-known reference on radar returns from terrain. The original reference from which these results were taken was D.E. Kerr, "Propagation of Short Radio Waves,"[8] which was Volume 13 of the MIT Radiation Laboratory Series and is a classical (and still very useful) reference on radar wave propagation and scattering.

It is intuitively evident that the return from terrain should be approximatable by a Ricean PDF, because such returns often have a significantly large specular component that would constitute a mean voltage value, plus noncoherent returns from aggregates of smaller independent scatterers. As the central limit theorem tells us, this should produce a Ricean PDF.

In Figure 8.9 (also from Long[7]) some cumulative probability distributions of return power in dB are shown. The quantity plotted (if RCS rather than power is used as the argument, these two quantities being mutually proportional) is

$$P_{\mathrm{cum}}(\sigma_o) = \int_0^{\sigma_o} d\sigma\, p(\sigma) \tag{8.27}$$

i.e., the probability that σ does not exceed σ_0 which obviously increases with increasing σ_0 and plots as nearly a straight line against σ_0 or power in dB. These experimental plots are compared with lognormal and Rayleigh distributions for sea clutter returns in two different radar bands and for two different polarizations. The agreement appears excellent, especially in the case of L-band returns with HH polarization.

Another experimental result compared with a lognormal PDF is shown in Figure 8.10. Again it is a cumulative probability plot from (8.27) but with the power scale reversed in direction, thus explaining the negative rather than positive slope as in Figure 8.9. The case shown in Figure 8.10 is for trees at 95 GHz with horizontal polarization. It is taken from Eaves and Reedy, "Principles of Modern Radar."[3]

Another sea clutter result, Figure 8.11, from M. Skolnick, "Introduction to Radar Systems,"[1] shows cumulative PDFs vs. clutter RCS in dB at X-band for two sea states. This time, the probability that σ *exceeds* σ_0 is plotted vs. σ_0, producing a negative slope where σ_0 increases to the right. The equation is

$$\overline{P}_{\mathrm{cum}}(\sigma_o) = 1 - P_{\mathrm{cum}}(\sigma_o) = 1 - \int_0^{\sigma_0} d\sigma\, p(\sigma) = \int_{\sigma_0}^{\infty} d\sigma\, p(\sigma) \tag{8.28}$$

where $p_{\mathrm{cum}}\,(\sigma_0)$ is the quantity in (8.27). In Figure 8.11, we see that the comparison

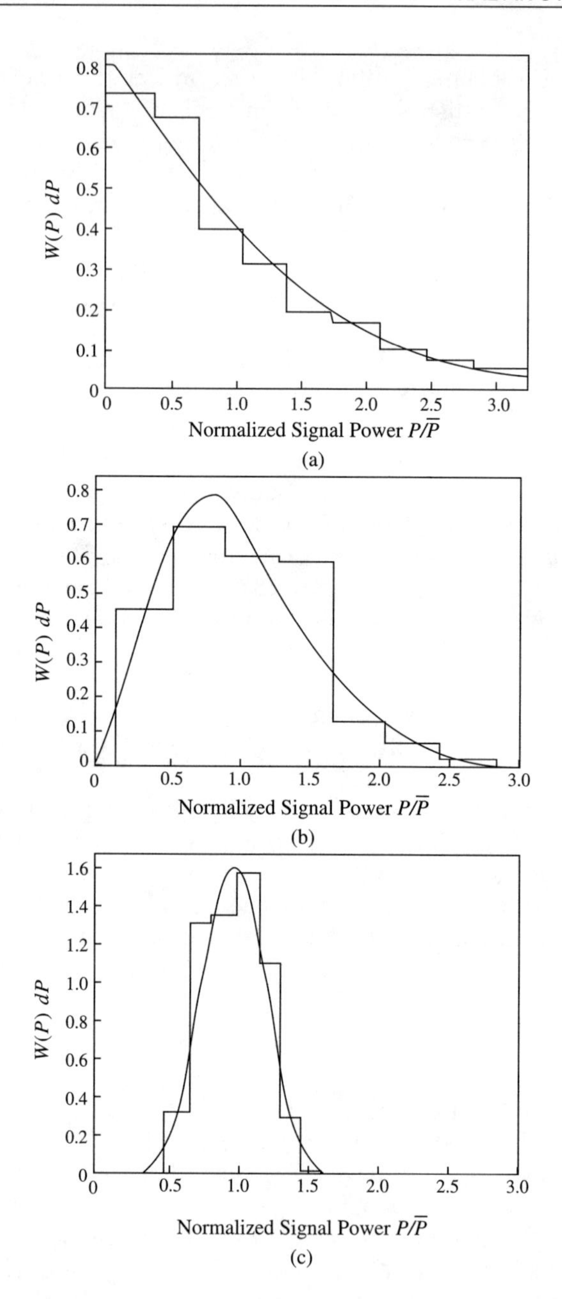

FIGURE 8.8
Several probability distributions for ground clutter at 9.2 cm. Experimental data are shown by histograms. The continuous curves are Ricean distributions. (a) Heavily wooded terrain, wind speed 25 mph, $m^2 = 0.8$. (b) Heavily wooded terrain, wind speed 10 mph, $m^2 = 5.2$. (c) Rocky terrain, wind speed 10 mph, $m^2 = 30$.From Figure 5.3 on p. 150 of M.W. Long, "Radar Reflectivity of Land and Sea." Artech, Norwood, MA, 1983.[7] Original source: D.E. Kerr, "Propagation of Short Radio Waves." McGraw-Hill, New York, 1951; by permission of McGraw-Hill.[8]

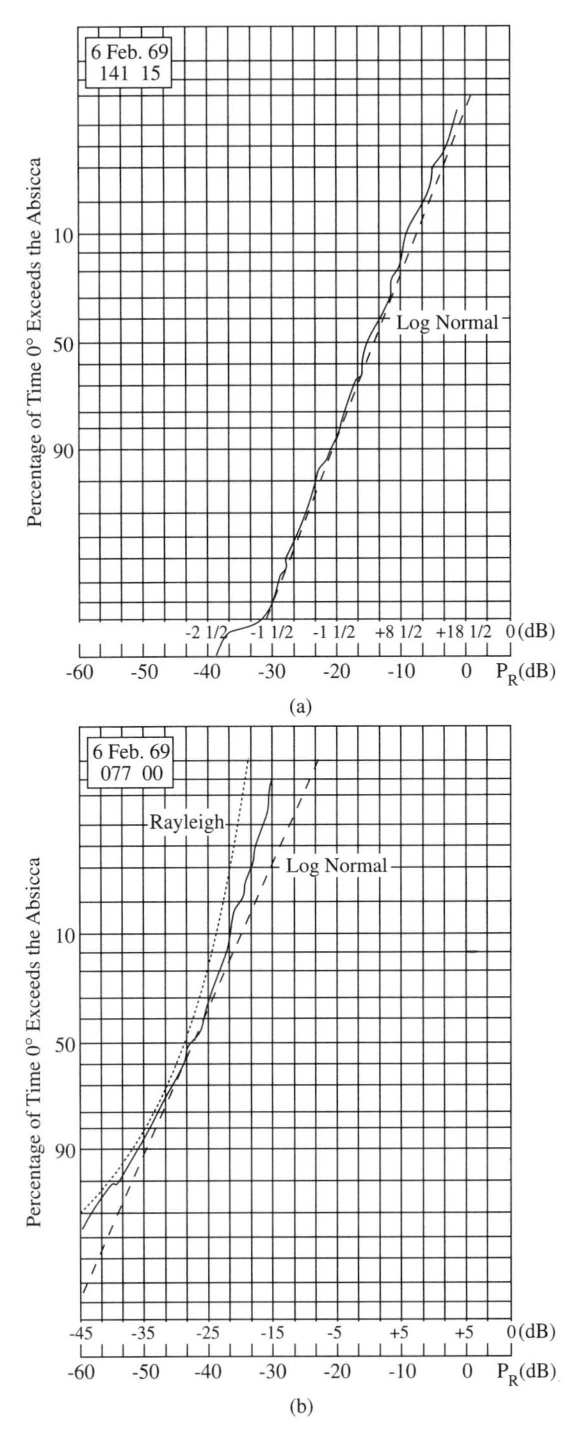

FIGURE 8.9
Samples of cumulative probability distributions for echo power: (a) is for L band and HH polarization; (b) is for P band and VV polarization. From Figure 5.22 on p. 187 of M.W. Long, "Radar Reflectivity of Land and Sea." Artech, Norwood, MA, 1983.[7] Original source: Daley, J.C., Davis, W.T., Duncan, J.R., and Laing, M. B. "NRL Terrain Clutter Study, Phase II." NRL Report 6749, October 21, 1968.

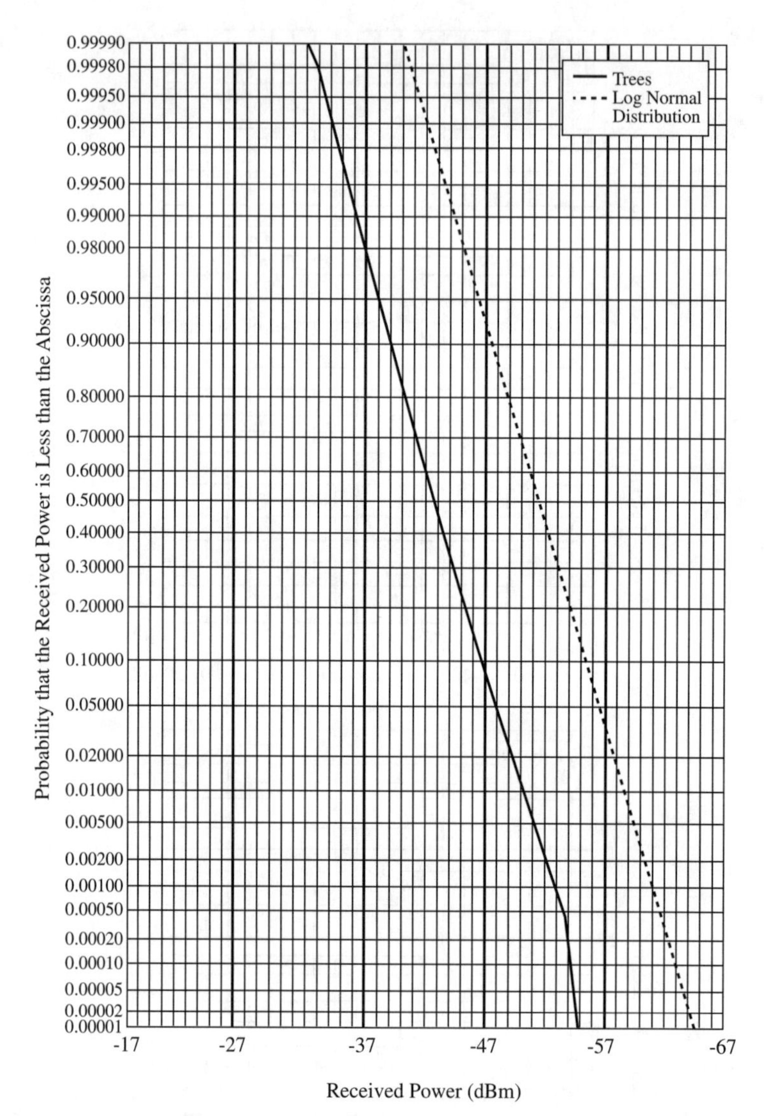

FIGURE 8.10
Cumulative probability distribution of the received power from deciduous trees: 95 GHz, horizontal polarization, and 4.1° depression angle. From Figure 10.28 on p. 317 of Eaves and Reedy, "Principles of Modern Radar." Van Nostrand, 1987.[3] Original source: Currie, N.C., Dyer, F.B., and Hayes, R.D., "Radar Land Clutter Measurements at 9.5, 16, 35 and 95 GHz." Technical Report No. 3 on Contract DAA25-73-0256, Georgia Institute of Technology, Atlanta, March 1975.

with the lognormal distribution is considerably better than that with the Rayleigh distribution for both sea states shown. The larger sea state indicates a rougher sea. It is evident that as the sea state increases, the agreement with lognormal improves and that with Rayleigh worsens. That is expected because a larger sea state should produce more high amplitude fluctuations due to the presence of more facets of sea surface that are normal to the radar line-of-sight.

Many more examples could be shown, but a general conclusion is that some forms of clutter do not show statistics that can be explained with a Rayleigh PDF and that clutter returns with large amplitude fluctuations are better explained with lognormal or possibly Weibull or higher order chi-squared PDFs, and those with large means are better explained with Ricean PDFs.

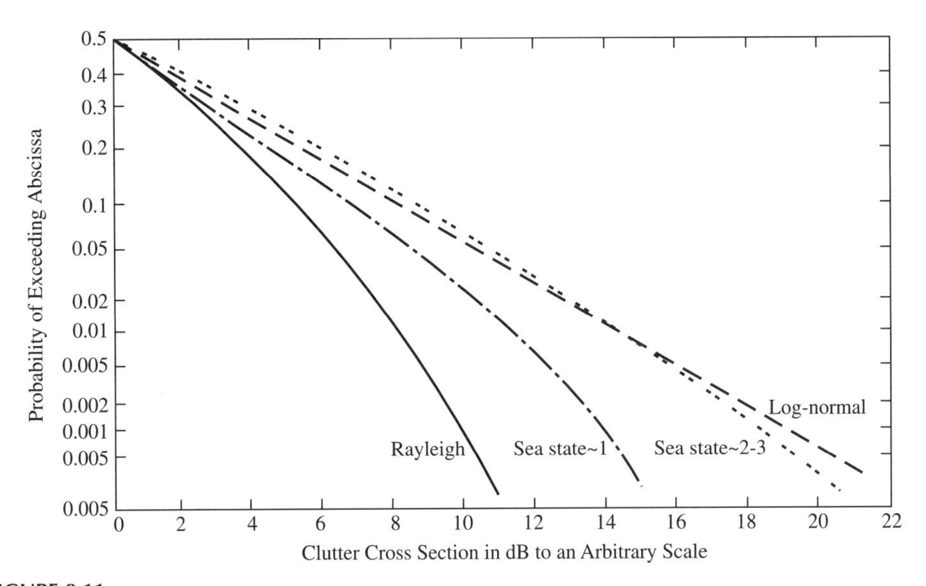

FIGURE 8.11
Experimental statistics of vertical polarization, X-band sea clutter for two sea states. Pulse width = 20 nsec; beamwidth = 0.5°, grazing angle = 4.7°, range = 2 nmi. The Rayleigh distribution is shown for comparison. Dashed curve shows attempt to fit the sea state 2–3 data to lognormal distribution. From Figure 13.5 on p. 479 of Skolnick, "Introduction to Radar Systems," (2nd ed.) McGraw-Hill, New York, 1982.[1] Original source: Trunk, G.V., and George, S.F., "Detection of Targets in Non-Gaussian Sea Clutter." IEEE Trans. Aerospace Electronic Syst. AES-8, 195–204, 1972, by permission of IEEE: © IEEE 1972.

8.7 MEAN RCS VALUES FOR CLUTTER SOURCES

The subject of radar scattering from terrain of various kinds (not to mention scattering from discrete objects that sometimes constitute desired targets and sometimes clutter) has occupied the attention of hundreds of investigators for over 50 years. A few important general references on this vast subject area, usually called "microwave remote sensing," are included in the reference list.[7–12] In the present section, a few brief highlights of direct use to radar engineers will be presented.

The attributes of clutter sources that are of greatest importance to a radar engineer planning or using a particular radar system in a given environment are

1. mean RCS or reflectivity
2. amplitude PDF
3. Doppler spectrum of clutter return
4. *spatial properties*
   a. spatial distribution of clutter sources
   b. spatial correlation properties of clutter amplitude
5. *temporal properties*
   a. time variation of clutter amplitude
   b. temporal correlation properties of clutter amplitude
6. polarization properties of clutter return.

Items 1 through 6 listed above are not necessarily mutually independent, but it is important to the radar designer to be able to separate them and conceive of them as if they *were* independent.

In Section 8.6, Item 2 in the above list was discussed. In the present section, Item 1, that of the mean RCS values to be expected from various clutter sources, will be discussed.

To discuss RCS, we first classify typical clutter sources:

1. *Surface clutter*
   a. sea surface (or freshwater surface) return
   b. land surface return—with little vegetation
   c. land surface return—vegetation covered (forested or agricultural terrain)
   d. land surface return—snow covered
2. *Volume clutter*
   a. rain return
   b. falling snow return
   c. chaff-cloud return
   d. return from large flocks of birds or insects
3. *Discrete clutter*
   a. return from aircraft
   b. return from land vehicles—cars or trucks
   c. return from ships
   d. return from man-made structures, e.g., buildings or towers
   e. individual trees

As has already been indicated, for surface clutter, we require a mean value of $\sigma^{(o)}$, the RCS per unit area, a pure number measured in dB. For volume clutter, we require σ_v (or η in standard notation), the volume reflectivity or RCS per unit volume, a number with dimensions of reciprocal distance, measured in reciprocal meters. For discrete clutter, we require the mean RCS of the object itself, a quantity measured in square meters or in dB relative to 1 m^2.

It is obvious that the clutter sources listed as "volume clutter" in our list are composed of aggregates of discrete scatterers, namely single raindrops (approximately spherical water drops) for Item a, snowflakes for Item b, chaff particles for Item c, and individual birds or insects for Item d. However, it is impractical to view these clutter sources as discrete scatterers except in theoretical analysis designed to determine the RCS of the entire assembly from the RCS of the individual object.

The resolution of the radar signal in range and angle is limited to theoretical values comparable to a radar wavelength. In an X-band radar, for example, where wavelength is about 3 cm, it would be possible *in principle* to resolve objects down to 3 cm. But other limitations, e.g., minimum range-gate width of (say) 50 m, minimum beamwidth of (say) 1° would not enable us to resolve a rain volume to within less than a few hundreds or thousands of raindrops, each of which has a diameter of the order of millimeters. Thus, for practical purpose, a rainstorm region is a volume scatterer. The same applies to the other volume sources mentioned.

Also, there is considerable volume scattering involved in scattering from these sources listed as "surface clutter." However, there is very little penetration of radar waves into the sea, somewhat more into freshwater, and still more into dry land terrain. For dry land terrain, the wave undergoes attenuation and phase shifts as it penetrates the medium, but its bulk scattering can still be viewed as coming from the surface on which the wave impinges. Its RCS, of course, is affected by the volume scattering phenomena that occur along its path, but it is still "surface scattering."

The theoretical approaches to scattering from the clutter sources listed above are legion. Many of these approaches require very complicated electromagnetic theory and at best produce results that would change radically with small modifications in geometry. It is sufficient to say that these approaches provide useful guidelines when they are tempered with intuitive considerations. For example, high frequency approximation techniques such as geometric theory of diffraction (GTD) or physical theory of diffraction (PTD) or standard geometric optics or physical optics approaches can be interpreted in terms of intuitive ideas about scattering from aggregates of points, with consideration of the differences in delay between these points. These approaches view an object or a patch of terrain as an aggregate of "scattering centers" whose effects are superposed. Numerical approaches such as method of moments often require enormous expenditures of CPU time* and the interpretation of the results is sometimes difficult. In the last analysis, only experimental confirmation of the RCS values obtained by numerical approaches (or *any* theoretical approaches) can establish confidence in their validity. Unless a scatterer is an infinite flat plane boundary between two homogeneous media, a set of layers of such media, a homogeneous sphere or layered sphere, an infinite homogeneous cylinder, or at worst a homogeneous spheroid, it cannot be analyzed *exactly*. Hence theoretical approaches are based on approximations of real-world scatterers as simple objects or on perturbation theory of one or two orders from those cases.

From the radar engineer's viewpoint, it is necessary to have an estimate of RCS only within a few dB. Clutter results are so dependent on moisture, temperature, and other meteorological parameters that they cannot possibly fit all conditions to be encountered within any given scenario. Hence excessive precision in specification of RCS is a form of self-delusion.

With this preamble, we will now introduce the concept of the "scattering matrix," which constitutes all of the information attainable about a source that can be viewed as a "point-target" from the radar's frame of reference. This is also called the "polarization matrix."

Referring to Figure 8.12, we have a wave that appears as a plane wave impinging on an object. Its electric field is in the plane of the paper in Figure 8.12a and normal to the plane of the paper in Figure 8.12b. The direction of propagation of the incident wave is denoted by the unit "wave vector" $\hat{k}_{inc}$ and that of the wave by *its* unit wave vector $\hat{k}_{sc}$. The "polarization" of each wave is specified in terms of the direction of its *electric* field vector $\mathbf{E}_{inc}$ or $\mathbf{E}_{sc}$, respectively. If the electric field vector $\mathbf{E}_{inc}$ is in the plane of the paper, that is designated as "vertical polarization." If it is normal to the plane of the paper, that is "horizontal polarization." The directions of magnetic field vectors **H** are normal to their corresponding **E** vectors, but we do not use **H** in specifying polarization, so it is irrelevant to this discussion.

The general*† bistatic case is illustrated in Figure 8.12. If the incident wave is vertically polarized, then we define (somewhat arbitrarily) the direction of vertical polarization for the scattered wave as that of the angle θ_{sc}, the spherical polar angle of the direction in which $\hat{k}_{sc}$ points. Thus the component of $\mathbf{E}_{sc}$ in the θ_{sc} direction is called the "vertically polarized scattered wave field." The component of $\mathbf{E}_{sc}$ in the ϕ_{sc} direction (normal to the θ_{sc} direction) is called the "horizontally polarized scattered wave field." These fields are denoted by $(E_{sc})_V$ and $(E_{sc})_H$, respectively, while the incident wave field components parallel and perpendicular to the plane of the paper are denoted by $(E_{inc})_V$ and $(E_{inc})_H$, respectively. The scattering matrix $[S]$ for this general case is defined by the expression

* These remarks should be tempered by the observation that available computer power is increasing very rapidly as of this writing; for example, MOM computations on scattering objects several wavelengths in dimension can now be done in much shorter CPU times than they could 5 years ago and faster algorithms are being developed continuously.

*† Actually the case depicted in Figure 8.12 is not the general case; it is the in-plane bistatic case, where the scattered wave vector is in the same plane as the incident wave vector. In general, the scattered wave vector can be "out-of-plane," meaning *not* in the same plane as the incident wave vector.

$$\begin{bmatrix}(E_{sc})_V\\(E_{sc})_H\end{bmatrix} = \begin{bmatrix}S_{VV} & S_{VH}\\S_{HV} & S_{HH}\end{bmatrix}\begin{bmatrix}(E_{inc})_V\\(E_{inc})_H\end{bmatrix} \tag{8.29}$$

where

$$S_{VV} = \left[\frac{(E_{sc})_V}{(E_{inc})_V}\right]_{(E_{inc})_H=0}, \qquad S_{VH} = \left[\frac{(E_{sc})_V}{(E_{inc})_H}\right]_{(E_{inc})_V=0}$$

$$S_{HV} = \left[\frac{(E_{sc})_H}{(E_{inc})_V}\right]_{(E_{inc})_H=0}, \qquad S_{HH} = \left[\frac{(E_{sc})_H}{(E_{inc})_H}\right]_{(E_{inc})_V=0}$$

For the monostatic geometry characteristic of most radar applications, the view is that of Figure 8.13, where the clutter target is a horizontal patch of ground.

The designation of "vertical" or "horizontal" polarization in Figure 8.13 is that used in the radio community, where "vertical polarization" refers to the electric field being in a vertical plane relative to the earth and "horizontal polarization" means the **E** field is parallel to the plane of the earth. The backscattered wave, in general, has a "copolarized" component, colloqually known as "copol," which is polarized parallel to the incident wave field and a "cross-polarized" component, known as "crosspol," with polarization perpendicular to that of the incident wave. In (8.29), the copol components are S_{VV} and S_{HH} and the crosspol components are S_{VH} and S_{HV}. The physics of the scattering process usually favors copol return over crosspol return because vertically polarized incident waves tend to excite vertical

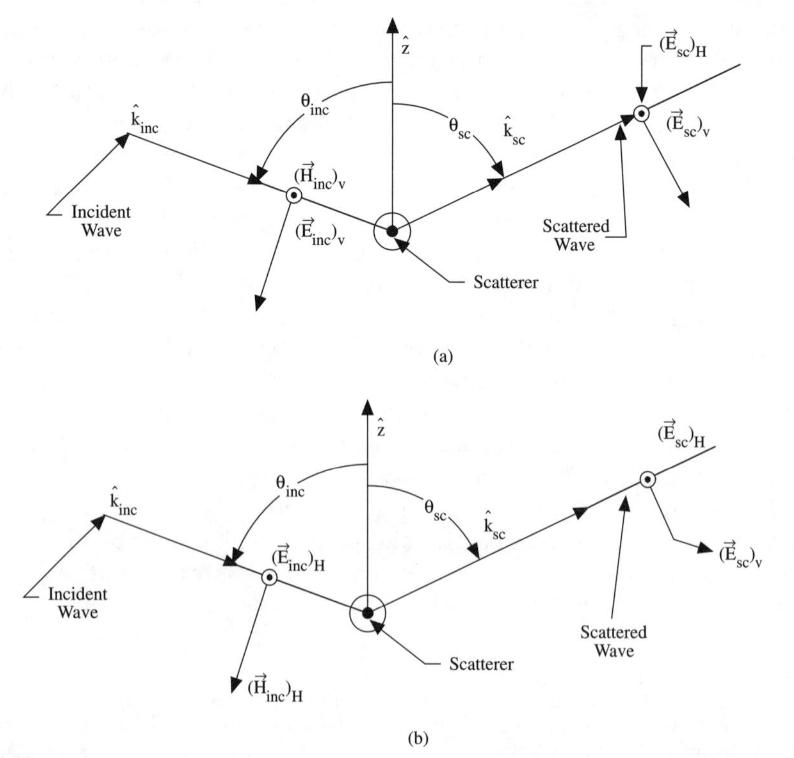

FIGURE 8.12
Geometry of "scattering matrix" or "polarization matrix." (a) Vertically polarized incident wave. (b) Horizontally polarized incident wave.

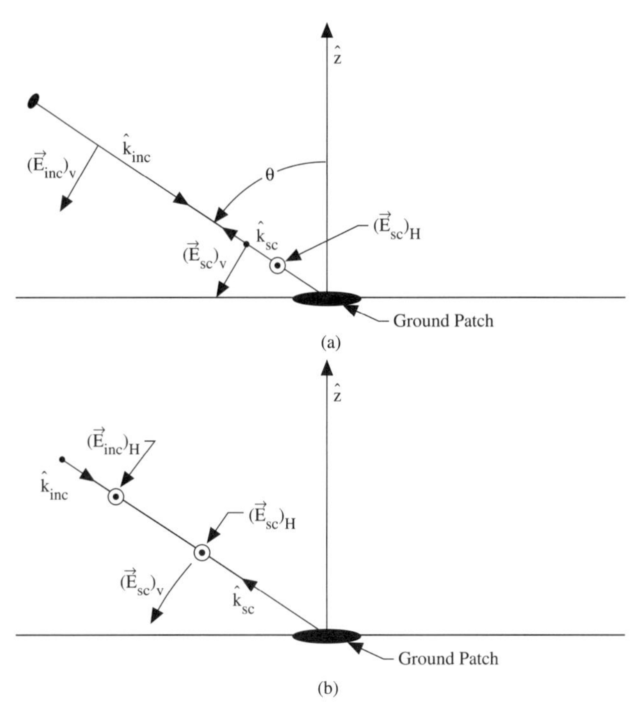

FIGURE 8.13
Scattering matrix-monostatic radar. (a) Vertical polarization. (b) Horizontal polarization.

currents on the object, which in turn generate vertically polarized backscattered waves. The analogous phenomena occur for horizontally polarized incident waves. Hence the scattering matrix $[S]$ is usually *almost* diagonal, e.g., for a perfectly flat and smooth ground patch, we would have

$$[S] \approx \begin{bmatrix} S_{VV} & 0 \\ 0 & S_{HH} \end{bmatrix} \tag{8.30}$$

Because of surface tilt and randomness in the scattering medium, a more usual situation is

$$[S] = \begin{bmatrix} S_{VV} & S_{VH} \\ S_{HV} & S_{HH} \end{bmatrix} \tag{8.31}$$

where $|S_{VHHV}| << |S_{VV}|$ or $|S_{HH}|$ i.e., *some* crosspol scattering occurs but it is much smaller than the copol return.

A most useful tool for the radar engineer is the matrix involving the RCS values for copol and crosspol scattering, which we can develop from $[S]$ as in (8.29) through the definition of the RCS in terms of electric fields of incident and scattered waves. For any given polarizations of the two waves, we define the RCS as

$$\sigma_{AB} = (\mathrm{RCS})_{AB} = \lim \left[4\pi r^2 \left| \frac{(E_{sc})_{AB}}{(E_{inc})_{AB}} \right|^2 \right] \tag{8.32}$$

where A and B denote V or H as appropriate and r is the range.

A matrix we can formulate from the definition (8.32) will be denoted by $S_{\sqrt{\sigma}}$ and is given by

$$S_{\sqrt{\sigma}} = \begin{bmatrix} \sqrt{\sigma_{VV}}e^{j\psi_{VV}} & \sqrt{\sigma_{VH}}e^{j\psi_{VH}} \\ \sqrt{\sigma_{HV}}e^{j\psi_{HV}} & \sqrt{\sigma_{HH}}e^{j\psi_{HH}} \end{bmatrix} \tag{8.33}$$

where, if the limiting process in (8.32) is replaced by the assumption that r is large enough to be considered as infinite (from the viewpoint that the scattered spherical wave looks like a plane wave at a local observation point), then the element $(S_{\sqrt{\sigma}})_{AB}$ and (S_{AB}) of (8.29) are related by the expression

$$(S_{\sqrt{\sigma}})_{AB} = \sqrt{4\pi r^2 |S_{AB}|^2} \tag{8.34}$$

and where ψ_{AB} is the phase associated with the complex matrix element S_{AB}.

The radar engineer is often interested only in the amplitude and not in the phase of the scattered wave, except for the delay factor e^{-2jkr}, which is suppressed in the matrix $S_{\sqrt{\sigma}}$. Hence one can define the "RCS matrix" as

$$S_{\sigma} = \begin{bmatrix} \sigma_{VV} & \sigma_{VH} \\ \sigma_{HV} & \sigma_{HH} \end{bmatrix} \tag{8.35}$$

which is the matrix of the RCS values for the different polarizations, all of whose elements are, of course, real and positive. For surface or volume scattering (8.33) or (8.35) could be redefined for the NRCS $\sigma_{AB}{}^{(o)}$ or the volume reflectivity η_{AB}.

In radar engineering practice (as differentiated from the science of microwave remote sensing) it is important to have simple and convenient clutter models that can be used to estimate clutter RCS values for a wide range of cases that might be encountered. One such model for surface terrain clutter, often cited in the radar literature and usually called the "constant gamma model" if $N = 1$, is

$$\sigma^{(o)}(\alpha) = \gamma \sin^N \alpha = \gamma \cos^N \theta_i \tag{8.36}$$

where θ_i, the angle of incidence, is obviously the complement of the grazing angle α, where γ is the value of $\sigma^{(o)}$ at normal incidence, is constant, and is of course the maximum value of $\sigma^{(o)}$, and where N is a number near (but not necessarily exactly equal to) + 1. Levanon[5] (pp. 81–82) cites a general model for surface terrain that is a generalization of (8.36) and covers both land and sea terrain. That model is

$$\begin{aligned} \sigma^{(o)}(\alpha) &= C\frac{\sin^N \alpha}{\cos^M \alpha}; \qquad 0^\circ \leq \alpha \leq \alpha_G \\ &= \sigma^{(o)}(\alpha_G); \qquad \alpha_G \leq \alpha \leq 90^\circ \end{aligned} \tag{8.37}$$

where

$$-1 \leq N \leq 3, \qquad 0 \leq M \leq 1 \text{ for land}$$

$$1 \leq N \leq 2, \qquad 1 \leq M \leq 5 \text{ for sea}$$

and where the usual assumed values of C and α_G are $C = 0.01$ and $\alpha_G \leq 85^\circ$.

The behavior of $\sigma^{(o)}$ vs. grazing angle as observed experimentally can usually be fitted to a function of the form (8.37). Plotted against grazing angle, it begins with a very small value of 0°, increases rapidly with α for small α, increases somewhat less rapidly between $\alpha \simeq 10°$ and $\alpha \approx 70°$, then eventually reaches a plateau near 86° and remains relatively fixed up to normal incidence ($\alpha = 90°$). Since one often sees plots of $\sigma^{(o)}$ vs. θ_i, the angle of incidence, which slopes downward rather than upward, such a plot is shown in Figure 8.14b, along with the plot of $\sigma^{(o)}(\alpha)$ vs. α in Figure 8.14a. These sample plots are in dB, so they would be fitted with the function

$$[\sigma^{(o)}(\alpha)]_{dB} = 10 \log_{10} C + 10N \log_{10}|\sin \alpha| - 10M \log_{10}|\cos \alpha| \tag{8.38a}$$

for Figure 8.14a and

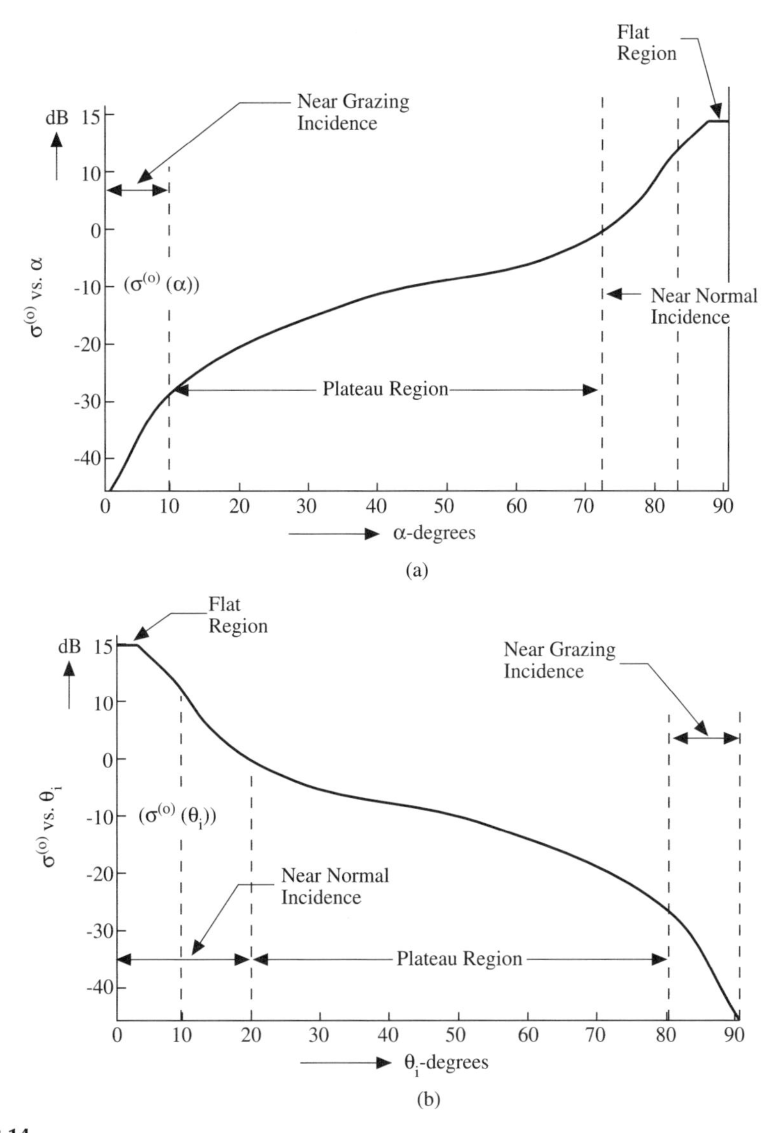

FIGURE 8.14
Typical variation of NRCS vs. α or θ_i. (a) NRCS vs. grazing angle. (b) NRCS vs. angleof incidence. Part (a) similar to Figure 4.9 on p. 79 of N. Levanon, "Radar Principles." 1988, by permission of John Wiley & Sons, Inc.

$$[\sigma^{(o)}(\theta_i)]_{dB} = 10 \log_{10} C + 10N \log_{10}|\cos \theta_i| - 10M \log_{10}|\sin \theta_i| \tag{8.38b}$$

for Figure 8.14b (close approximation to Levanon's[5] Figure 4.9 on p. 79).

The division of the behavior of $\sigma^{(o)}$ vs. α or θ_i into four regions adds another region to the three usually defined. The largest return is at normal incidence where a great deal of specular return is present. That return remains nearly constant for a few degrees from normal incidence. If the surface is relatively smooth, there is a rather abrupt transition as the specular reflection recedes with increasing angle of incidence. If the surface has considerable roughness, then there is a large amount of diffuse reflection along with the specular return (diffuse reflection is by definition nearly uniform over all angles and is caused by a high degree of roughness) resulting in a slow gradual decrease in $\sigma^{(o)}$ with further increases in θ_i until the plateau region is reached (between $\theta_i \simeq 20°$ and $\theta_i \approx 80°$). Finally in the region near grazing incidence the NRCS begins to decrease rapidly with θ_i, nearing zero at grazing incidence. In that region, the return is dominated by destructive interference between direct and ground-reflected ray paths, which results in negligible returns at grazing incidence.

Most experimental or validated theoretical plots of $\sigma^{(o)}$ vs. α or θ_i look something like Figure 8.14a if plotted against α or Figure 8.14b if plotted against θ_i. A few examples are shown in Figure 8.15 (from Ulaby et al.,[10] "Microwave Remote Sensing," vol. II, Fig. 11–44, p. 879) for dry snow-covered ground at various frequencies, Figure 8.16 (Ulaby et al.,[10] Vol. II, Fig. 11–38, p. 871) for corn canopy at 9 GHz for different soil moisture contents, Figure 8.17 (From Eaves and Reedy,[3] Fig.10-15,

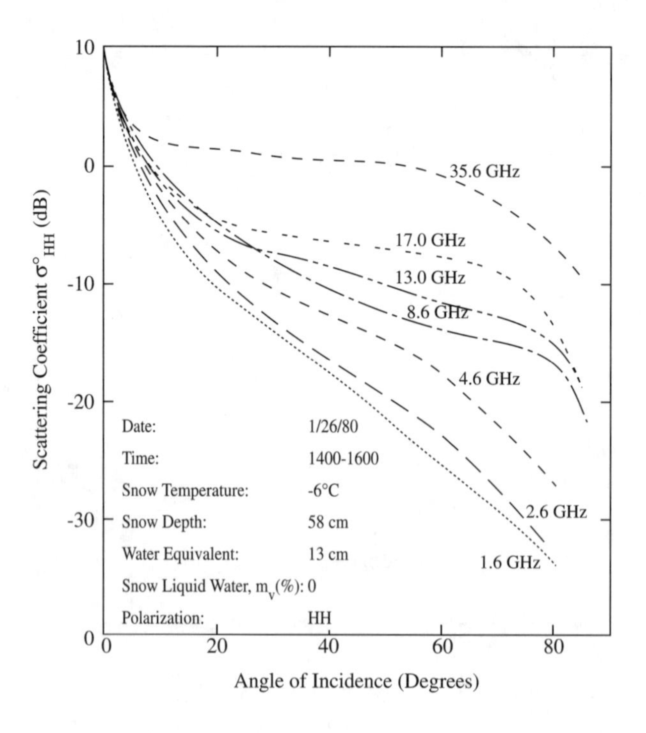

FIGURE 8.15
Angular patterns of measured backscattering coefficent of dry snow. From Figure 11.44 on p. 879 of Ulaby et al., "Microwave Remote Sensing," Vol. II. Addison and Wesley, Reading, MA, 1982.[10] Original source: Stiles, W.H., Ulaby, F.T., Fung, A.K., and Aslam, A., "Radar Spectral Observations of Snow," 1981 IEEE International Geoscience and Remote Sensing Symposium (IGARSS'81), Digest, Washington, D.C., pp. 654–668, by permission of IEEE © IEEE 1981.

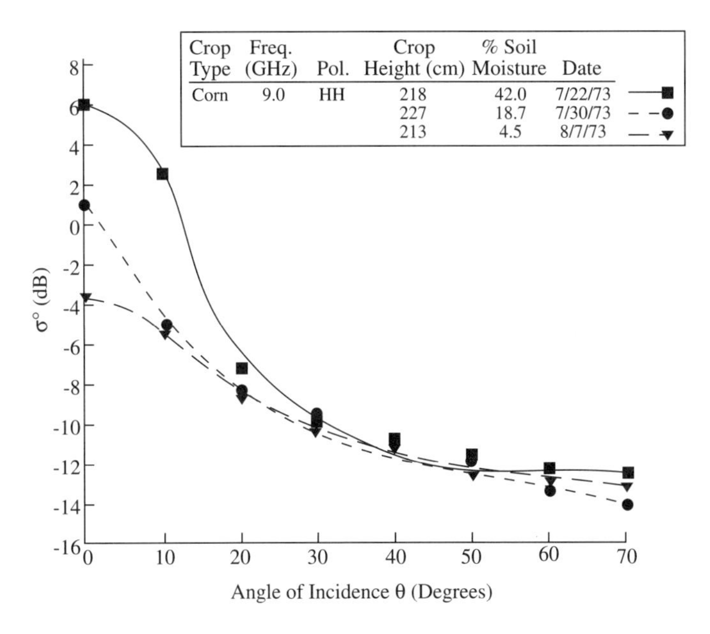

FIGURE 8.16
Measured angular pattern of the backscattering coefficient σ^0 at 9 GHz for a corn canopy under three different soil–moisture conditions. From Figure 11.38 on p. 871 of Ulaby et al. "Microwave Remote Sensing," Vol. II. Addison and Wesley, Reading, MA, 1982.[10] Original source: Ulaby, F. T., Bush, T.F., and Batlivala, P.P., "Radar response to Vegetation II: 8–18GHz Band." IEEE Trans. Ant. Prop. AP-23, 608–618, 1975, by permission of IEEE: © IEEE 1975.

p.305 for grass crops at X-band, and Figure 8.18 (from Eaves and Reedy,[3] Figures 10-32, 10-33 p. 325) for sea clutter. Note that in the last two figures $\sigma^{(o)}$ is plotted against depression angle. That is because one is never certain what the angle of incidence or grazing angle is but is quite sure of the depression angle because it is under the control of the observer. The tacit assumption is that grazing and depression angles are the same. Figure 8.19 shows why this may be far from true for hilly ground terrain or a rough sea surface. Also note that in some of these figures, the polarization conditions are specified, i.e., $\sigma^{(o)}$ is specified as $\sigma^{(o)}_{VV}$, $\sigma^{(o)}_{HH}$, etc. This serves to remind us that the polarization should be specified for RCS, since RCS may be radically different for different polarizations. In the general model (8.37) the polarization is not explicitly indicated, but it should be noted that for a given clutter source, values of C, N, M, and α_G will, in general, be different for VV, HH, VH, and HV (usually the same for VH and HV).

For volume clutter, empirical models have been generated from years of research experience on scattering from rain, falling snow, etc. A commonly used model [Barton,[4] p. 133, Eq. (3.5.12)] is

$$\sigma_v = 5.7 \times 10^{-14} R^{1.6}/\lambda^4 \tag{8.39}$$

applicable for $\lambda \geq 2$ cm (i.e., $f \leq 15$ GHz) where σ_V is the RCS per unit volume or "volume reflectivity" in m^{-1}, R is the rainfall rate in mm/hr, and λ is the wavelength in meters.

For dry snow [Barton,[4] p. 133, Eq. (3.5.13)]

$$\sigma_v = 1.2 \times 10^{-13} R_s^2/\lambda^4 \tag{8.40}$$

where R_s is a snowfall rate related to water content of the snow.

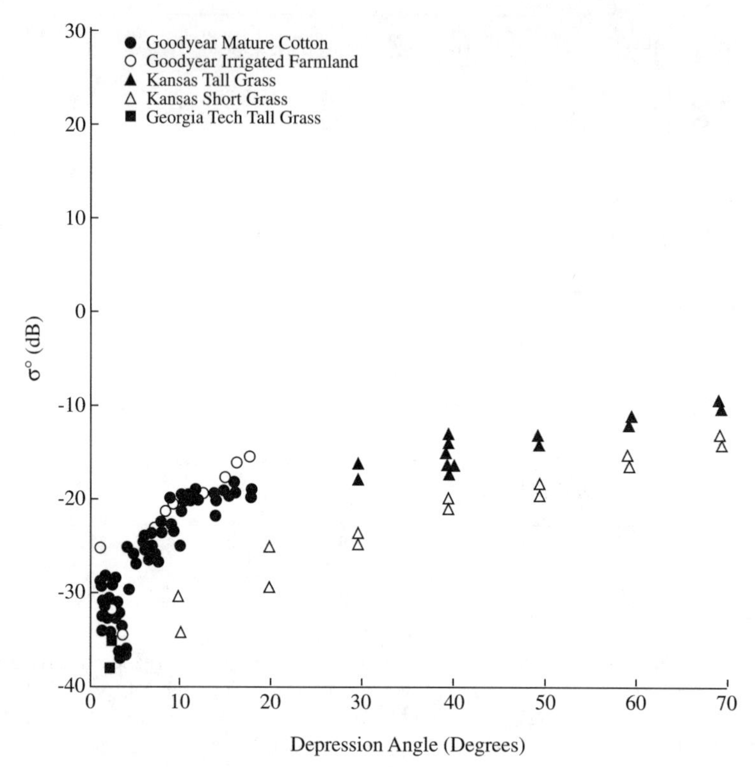

FIGURE 8.17
Radar reflectivity of grass/crops at X-band. From Figure 10.15 on p. 305 of Eaves and Reedy, "Principles of Modern Radar." Van Nostrand, Reinhold, New York, 1987.[3] Original sources: "Radar Terrain Return Study." Goodyear Aircraft Corp., Final Report on Contract NOAS-59-6186-C, GERA-463, Pheonix, September 1959. Stiles, W.H., Ulaby, F.T., and Wilson, E., "Backscatter Response of Roads and Roadside Surfaces." Sandia Report No. SAND78-7069, University of Kansas Center for Research, Lawrence, March 1969. Currie, N.C., Dyer, F.B., and Hayes, R.D., "Radar Land Clutter Measurements at 9.5, 16, 35 and 97 GHz." Technical Report No. 3, Contract DAA25-73-0256, Georgia Institute of Technology, Atlanta, March 1975.

The $1/\lambda^4$ dependence in (8.39) and (8.40) arises from the fact that raindrops or snowdrops are very small compared to wavelength at frequencies above 15 GHz, and the more complicated Mie scattering theory holds for spheres with radii comparable to wavelength. Resonances occur at certain sphere radii and the scattering becomes a much more complicated function of λ.

A more general model for rain that contains (8.39) as a special case is given in Eaves and Reedy[3] (p. 335)

$$\sigma_V = AR^B \text{m}^2/\text{m}^3 \tag{8.41}$$

where A and B are given for various frequencies. From Table 10-9 in Eaves and Reedy[3] (p. 338), A has values 1.3×10^{-8}, 1.2×10^{-6}, 4.2×10^{-5}, and 1.5×10^{-5}, at $f = 9.4$, 35, 70, and 95 GHz, respectively, and B is 1.6, 1.6, 1.1, and 1.0 at these respective frequencies.

Skolnick[1] [p. 501, Eq. (13.23)] presents the same rain model as Barton in a slightly different form, i.e.,

$$\sigma_V = 7f^4R^{1.6} \times 10^{-12} \text{ m}^2/\text{m}^3 \tag{8.42}$$

(where f is in Gigahertz) which is the same as (8.39) and close (but not equal) to

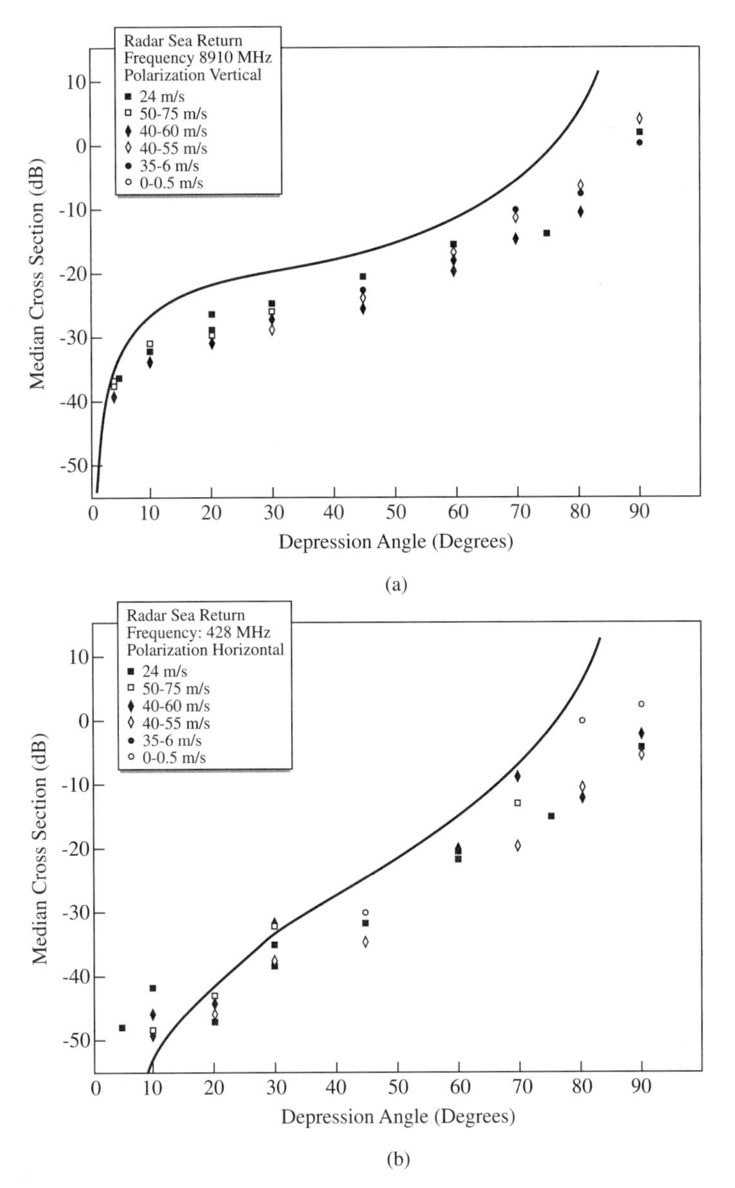

FIGURE 8.18
(a) The variation of median σ_{vv} with grazing angle and wind speed. X-band. (b) The variation of median σ_{HH} with grazing angle and wind speed, X-band. From Figure 10.33 on p. 325 of Eaves and Reedy, "Principles of Modern Radar." Van Nostrand, Reinhold, New York, 1987.[3] Original source: Guinard, N.W., and Daley, J.C., "An Experimental Study of a Sea Clutter Model." Proc. IEEE, 58, 435–550, 1970; by permission of IEEE: © IEEE 1970.

(8.41) for $f = 9.4$ GHz (the only frequency cited in Eaves and Reedy[3] that falls within the condition $f \leq 15$ GHz).

The main point to be gleaned from (8.39), (8.41), or (8.42) is that rain return varies roughly as the fourth power of frequency, between the 1.0 and 1.6 power of rainfall rate and that its reflectivity could be roughly estimated from one of these models for frequencies below about 15 GHz.

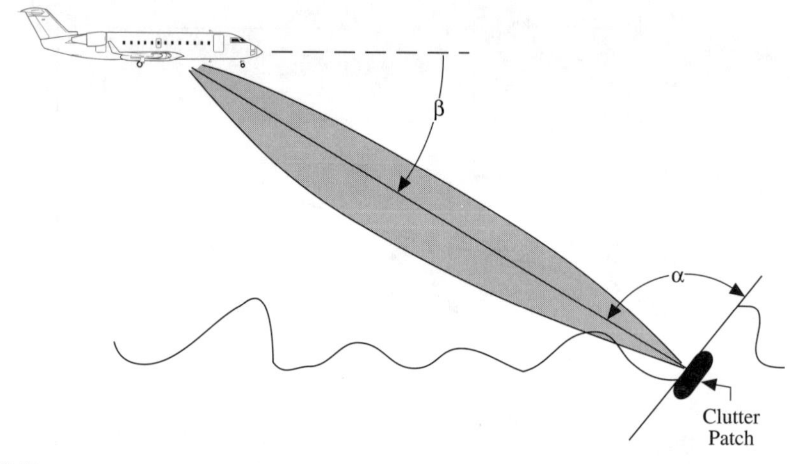

FIGURE 8.19
Depression angle and Grazing angle. Rough ground or sea surface: β = depression angle, about 40°; α = grazing angle, about 90°; θ_i = incidence angle, about 0°.

8.8 DOPPLER SPECTRUM OF CLUTTER

Doppler is one of the major discriminants used to detect targets in clutter. The discrimination is based on the fact that *most* targets have different Doppler spectra than *most* clutter returns.

Two rather classical empirical models for describing the spectral distribution of clutter are the Gaussian model (Skolnick,[1] pp. 131–132)

$$W(f) = W(f_o)e^{-2.77\left(\frac{f - f_o}{\Delta f_c}\right)^2} \tag{8.43a}$$

and a model that seems to fit some data better than the Gaussian model, having the form (Eaves and Reedy,[3] p. 318)

$$W(f) = \frac{W(f_o)}{1 + |2(f - f_o)/\Delta f_c|^N} \tag{8.43b}$$

In (8.43a,b),

N = positive real number
$W(f)$ = power density (power per unit frequency in watts/Hertz) at frequency f
f_o = central frequency of Doppler spectrum
Δf_c = Doppler bandwidth = spectral spacing between half-power (−3 dB) points on either side of f_o

Although it is important to have simple models like (8.43a,b) to estimate the Doppler characteristics of the clutter, those models are useful only if we know the "Doppler spread" or "Doppler bandwidth" Δf_c in (8.43a) and (8.43b), the exponent N in (8.43b) in order to determine the roll-off characteristics of the spectrum, and finally the power density at the central frequency f_o, given by $W(f_o)$ in (8.43a,b). The last of these would be obtained from a knowledge of the RCS of the particular class of clutter of concern. To get some idea of expected clutter spread, let us examine a few simple clutter geometries, illustrated in Figure 8.20 for ground or sea clutter in the presence of a desired target, i.e., an aircraft.

The geometries illustrated in Figure 8.20 are two-dimensional, although clutter also exists outside the plane of the paper. Two-dimensional illustrations are qualitatively valid, however, because the contributions normal to the target line-of-sight would have the same Doppler behavior in the qualitative sense, differing slightly in the Doppler magnitudes.

In all three geometries illustrated the target is within the range gate and within the mainlobe of the antenna beam and there is a land-or-water clutter contribution also within the mainbeam and the range gate. In Figure 8.20a we have a ground-based radar that, being stationary, has no velocity with respect to the earth. The target has a velocity vector $\mathbf{v}_T$. The illuminated clutter region, if it is land, can be considered stationary. Thus the target return has a Doppler of magnitude $|2/\lambda \mathbf{v}_T \cdot \hat{r}_{RT}|$, where $\hat{r}_{RT}$ is the unit vector from radar to target and λ is the wavelength. Since

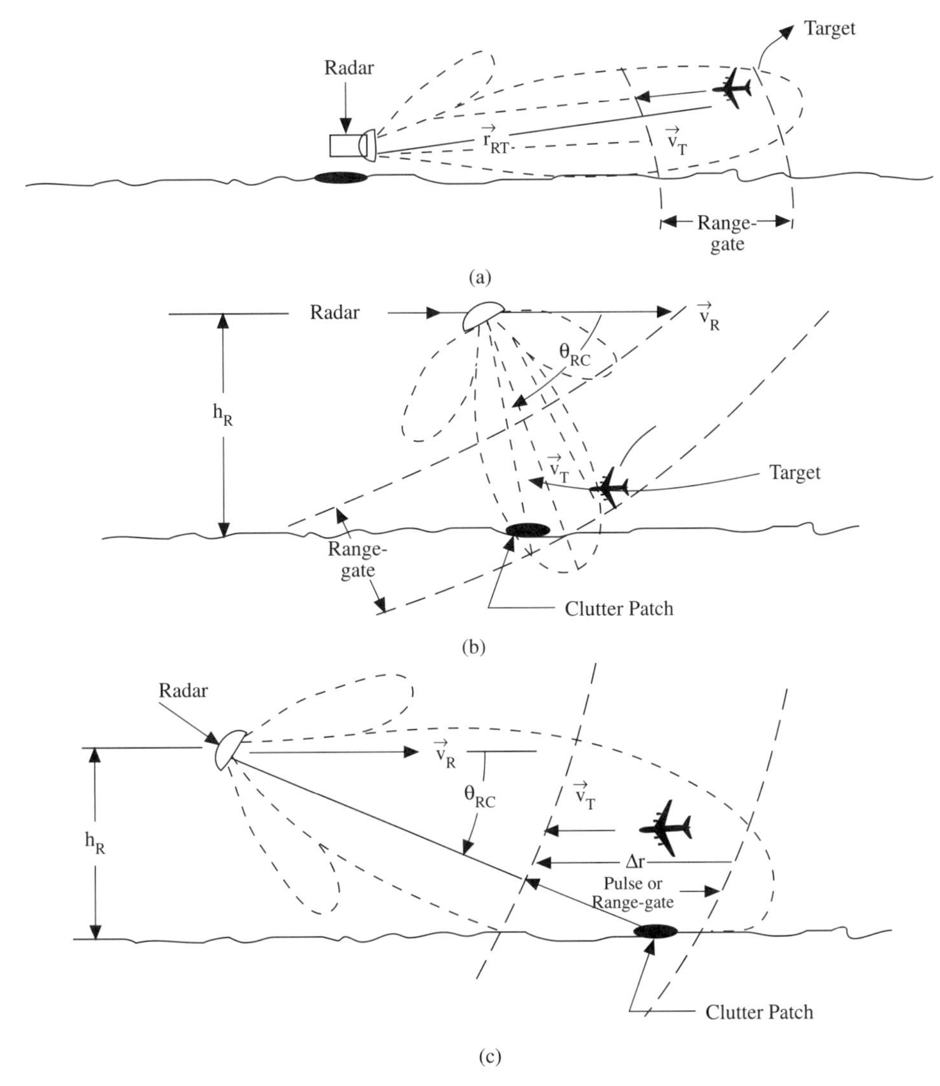

FIGURE 8.20
Illustrative clutter geometries. (a) Ground-based radar-land or sea clutter. (b) Airborne radar-land or sea clutter-high depression angles. (c) Airborne radar-land or sea clutter-low-to-moderate depression angles.

the radar and clutter are both stationary, the Doppler of the clutter return has magnitude zero. Thus the clutter and target spectra for a stationary ground-based radar are very narrow peaks, nearly "spikes" or impulse functions, at frequencies 0 and $|(2v_T/\lambda)\cos\theta_{RT}|$, respectively (where θ_{RT} is the angle between $\hat{r}_{RT}$ and $\mathbf{v}_T$). This is illustrated in Figure 8.21a. The angle θ_{RT} is changing rapidly with time, but the illustration does not account for that, because it is a "snapshot" at a specific time instant.

In the case of the clutter Doppler there is a variation in the value of θ_{RC} (the angle between the radar velocity vector and the line-of-sight between radar and clutter patch) for clutter patches throughout the area of intersection between the beam and the ground (since the clutter is beam limited in this case). If the beamwidth is $\Delta\theta$, then $\cos\theta_{RC}$ varies from $\sqrt{(r_{RCL}{}^2 - h_R{}^2)}/r_{RCL}$ to $\sqrt{r_{RCH}{}^2 - h_R{}^2)}/r_{RCH}$, where h_R is the radar altitude (see the diagram below) and r_{RCL} and r_{RCH} are the shortest and longest distances from radar to clutter patches within the beam, respectively. The Doppler for a given clutter patch at distance r_{RC} is $|(2/\lambda)|\mathbf{v}_R|\cos\theta_{RC}|$ (assuming the clutter is stationary), where θ_{RC} = angle between velocity vector and line-of-sight to clutter patch, and hence the doppler spread is between $|(2/\lambda)|\mathbf{v}_R|\cos\beta_{min}|$ and

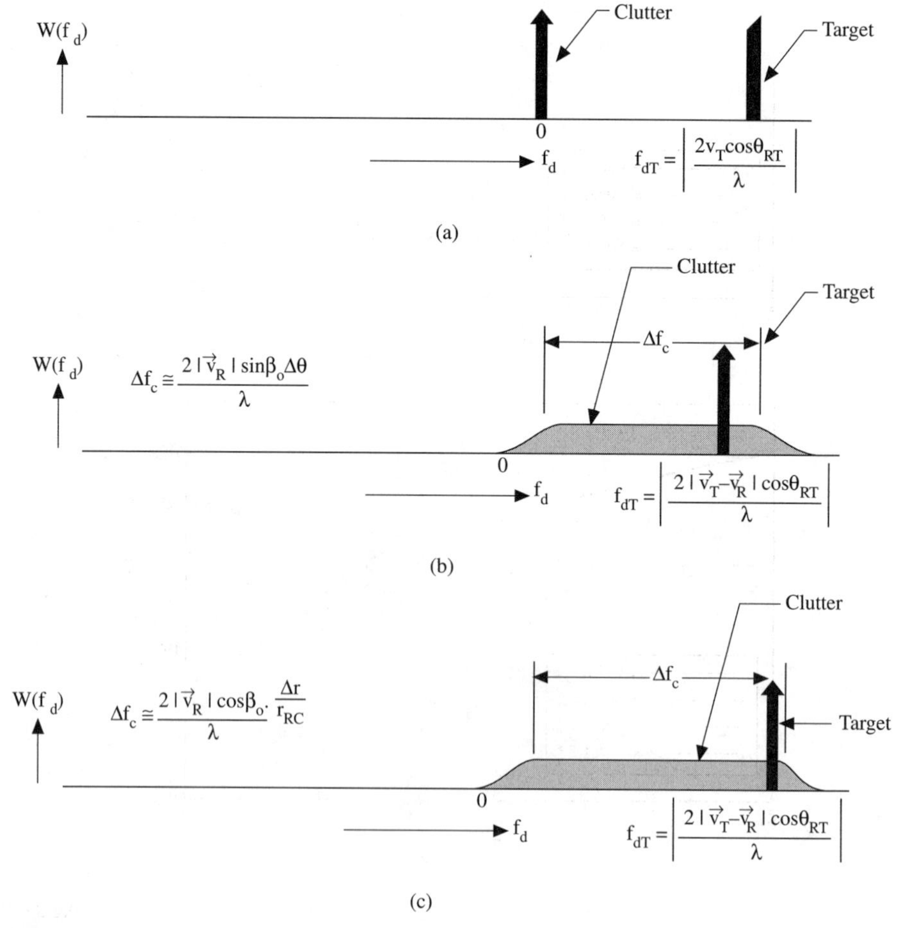

FIGURE 8.21
Illustrations of Doppler spectra $W(f_d)$. (a) Stationary ground-based radar, stationary land clutter, target is flying aircraft [Fig. 8.20(a)]. (b) Clutter and target same as (a), but airborne radar, high depression angle [Fig. 8.20(b)]. (c) Clutter and target same as (a) but airborne radar, low depression angle [Fig. 8.20(c)].

$|(2\lambda)|\mathbf{v}_R \cos \beta_{max}|$ where $\beta = \theta_{RC}$ (since $\mathbf{v}_R$ is horizontal) and β_{min} and β_{max} are minimum and maximum depression angles, respectively, of clutter patches within the beam. Since $\beta_{min} = \beta_{max} - \Delta\theta$, it follows that the Doppler spread for Case b (beam-limited ground clutter) is

$$\Delta f_c = \frac{2|\mathbf{v}_R|}{\lambda}\left[\cos\left(\beta_0 - \frac{\Delta\theta}{2}\right) - \cos\left(\beta_0 + \frac{\Delta\theta}{2}\right)\right] \tag{8.44a}$$

where β_0 is the depression angle at the center of the patch.

For small beamwidth, where $|\Delta\theta| << \pi/2$, $\cos \Delta\theta \simeq 1$, $\sin \Delta\theta \simeq \Delta\theta$, and (8.44a) can be approximated by

$$\Delta f_c \approx \frac{2|\mathbf{v}_R|}{\lambda} \sin \beta_o \Delta\theta \tag{8.44b}$$

Equation (8.44b) tells us that the Doppler spread for beam-limited ground clutter is roughly proportional to elevation beamwidth and to the sine of the depression angle at the center of the patch. This means that the spread is largest for normal incidence and decreases by about 0.707 for 45° depression angle.

Next we consider the case illustrated in Figure 8.20c, where the depression angle is small and the clutter is pulse-limited.

In that case, Eq. (8.44a) is changed to (where r_{RO} = distance to pulse center)

$$\Delta f_c = \frac{2|\mathbf{v}_R|\sqrt{r_{RO}^2 - h_R^2}}{\lambda}\left|\frac{1}{r_{RO} - (\Delta r/2)} - \frac{1}{r_{RO} + (\Delta r/2)}\right| \tag{8.45a}$$

because the outer limits of the clutter spectrum are defined by the limits of the pulse or (equivalently) range gate, which has a width Δr. For $\Delta r << 2r_{RO}$, virtually always the case for this geometry, the approximation to (8.45a) is

$$\Delta f_c = \frac{2|\mathbf{v}_R|\sqrt{r_{RO}^2 - h_R^2}}{\lambda \quad r_{RO}}\left[\left(1 + \frac{\Delta r}{2r_{RO}}\right) - \left(1 - \frac{\Delta r}{2r_{RO}}\right)\right] \simeq \frac{2|\mathbf{v}_R|}{\lambda r_{RO}} \cos \beta_o \Delta r \tag{8.45b}$$

For this case the Doppler spread is roughly proportional to the cosine of the depression angle and inversely proportional to the slant-range r_{RO}. For a long-enough pulse, Doppler spread can be quite substantial. In this case, it decreases with increasing depression angle, but since the pulse-limited case applies to small depression angles, the $\cos \beta_o$ dependence does not depart very much from unity.

Since the target Doppler spectrum is a spike at $|(2/\lambda)|\mathbf{v}_R - \mathbf{c}_T| \cos \theta_{RT}|$ in the cases b and c in Figure 8.21, the spectra of target and clutter for those cases show the spike for the target and a continuous spectrum for the clutter. Whether the target spike falls within the clutter spectrum and exactly *where* it falls within the spectrum depends on the target's velocity relative to that of the radar, but in general discrimination between target and clutter on the basis of differences in the Doppler spectrum depends very critically on the geometric variables, such as target and radar speeds and directions of motion and the size of the ground clutter patch illuminated by the radar.

Qualitative arguments like those above could be extended to sea clutter and precipitation clutter. Sea clutter may contain a Doppler component due to wind-induced motion of the water, implying that downwind and cross-wind clutter spectra have different central frequencies, reflecting the differences in line-of-sight velocity components. In addition to that, the individual reflecting facets of a rough sea surface will have different Dopplers, some of them ascending and some descending. The overall effect of that is a still broader Doppler spread than in the case of most ground clutter. There are some ground clutter sources that may exhibit wind-dependent Doppler, e.g., forested terrain, where leaves of trees move in response to wind currents.

Rain clutter may also be wind-dependent. The entire rainstorm area being illuminated by the radar has a mean horizontal velocity that determines the center of its Doppler spectrum if the motion has components along the radar-line-of-sight. Also, inhomogeneities in the rain intensity in different regions of the rainfall volume occur due to different drop-size distributions in those regions. Resulting spatial inhomogeneities occur in the velocities of raindrops of different sizes in response to wind currents, both the average wind currents and the statistical fluctuations due to wind gusts. All of this serves to spread out the Doppler spectrum.

8.9 SIDELOBE CLUTTER AND EFFECTS OF RANGE AND DOPPLER AMBIGUITIES

Sidelobe clutter was alluded to in Section 8.1 and is illustrated in Figure 8.2. In that figure, a situation is shown in which there is a prominent clutter patch within the first sidelobe and also within the range gate. In that illustration, where a beam is directed horizontally, there is no significant clutter return from the mainlobe, hence the only clutter is from the sidelobe. Had the range gate been set to a much higher value, there would have been no clutter return at all.

Embellishment of an illustration like that of Figure 8.2 will serve to demonstrate the role of range ambiguities in distinguishing clutter from target returns when some of the sidelobe clutter coincides with the range ambiguities. Consider the cases shown in Figure 8.22. Again the antenna is pointed horizontally and it has elevation sidelobe coverage in the direction of the ground. The three cases a, b, and c are labeled (a) low PRF, (b) medium PRF, and (c) high PRF, respectively. The target is within a range gate of width Δr at the maximum range of interest and is within the mainbeam in all three cases. The pulse repetition interval in range units is $c/2fr$, where fr is the PRF. In Case a, that interval is so long that no ambiguous ranges exist, because the first such range to the left of the target is well behind the radar and all of the others would be far beyond the target where presumably all extraneous targets and clutter sources would be well "out of range," since the $1/r^4$ loss would drive their returns well below the receiver noise level. In Case b, the PRF is approximately doubled relative to Case a, which is enough to create two ambiguous range intervals within the radar's field-of-view. Whereas only clutter patch 1, being within the range gate and within the mainlobe, would be seen in Case a, clutter patches 2 and 3 would also add to the total clutter in Case b, the first of these being in the mainlobe and the second in the first sidelobe.

Finally in Case c, the large increase in PRF has resulted in an additional three clutter patches, all of which can contribute to the total clutter power because they are either somewhere in the mainbeam or in the first or second sidelobe.

Although it is difficult to be precisely quantitative about the situations depicted in Figure 8.22, it is possible to do a very rough estimate of the effect of PRF on the signal-to-clutter ratio. In Case a, the SCR would be given by a small

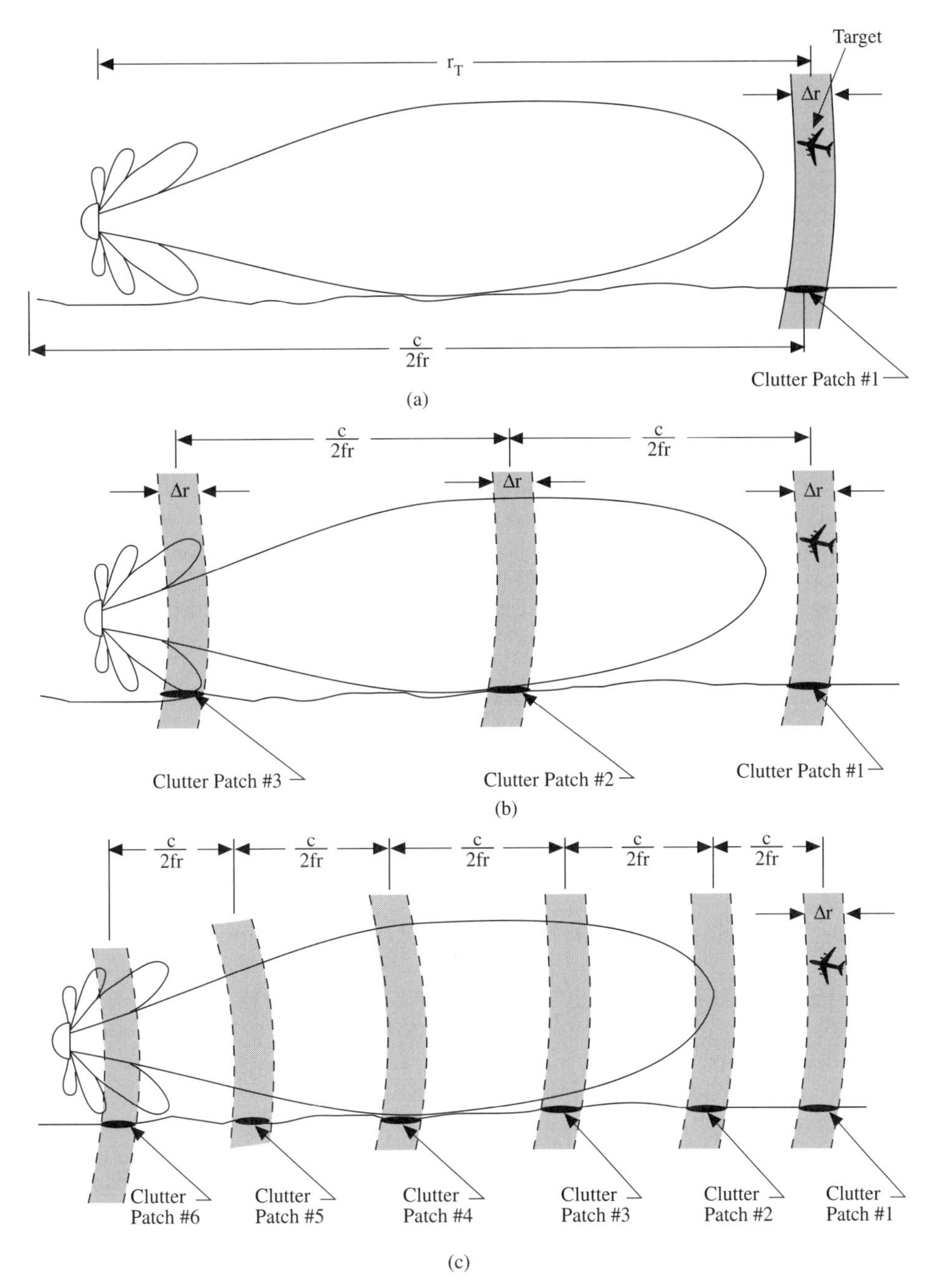

FIGURE 8.22
Sidelobe clutter and range ambiguities. (a) Low PRF. (b) Medium PRF. (c) High PRF.

modification of (8.23a), because, although the low grazing angle geometry basic to (8.23a) holds for the clutter, and the latter is pulse-limited in the elevation plane as is the case for (8.23a), the clutter patch is further down on the main beam than is the target and hence the clutter loses some power and gives a small enhancement of the SCR. Defining that "beam-loss" in absolute power units as L_b, the modified version of (8.23a) for our Case a would be

$$\mathrm{SCR} = \frac{\sigma_t L_{b_1} \cos\alpha}{\sigma^{(o)} r_T \Delta\phi (c\tau_p/2)} \tag{8.46a}$$

In Case b, the two additional contributions to the clutter power in the range gate set at range r_T due to ambiguities at $r_T - (c/2fr)$ and $r_T - (2c/2fr)$ (clutter patches 2 and 3) could be calculated by returning to Eq. (8.20b) and adding the power contributions from all three clutter patches, as follows:

$$P_{RC} = \frac{P_T G_o^2 \lambda^2 \sigma^{(o)} \Delta\phi}{(4\pi)^3} \left(\frac{c\tau_p}{2}\right) \left(\frac{\sec\alpha_1}{L_{b_1} r_1^3} + \frac{\sec\alpha_2}{L_{b_2} r_2^3} + \frac{\sec\alpha_3}{L_{b_3} r_3^3}\right) \tag{8.46b}$$

Equation (8.46b) is based on the assumptions that the NRCS $\sigma^{(0)}$ is uniform throughout the ground area containing the patches shown and that the ambiguities have nearly the same amplitude as the main range gate output ("infinite" pulse train). Subscripts 1, 2, and 3 on L_b, α, and r of course refer to clutter patches 1, 2, and 3, respectively.

Extrapolating from (8.46b) to Case c, invoking (8.46a) and noting that the range r_k is $[r_T - (k - 1)(c/2fr)]$ for clutter patch k, the degradation factors on SCR for the cases b and c are obtained from

$$(\mathrm{SCR})_{\text{case a}} = \frac{\sigma_t L_{b_1} \cos\alpha_1}{\sigma^{(o)} r_T \Delta\phi (c\tau_p/2)} \tag{8.47a}$$

$$\frac{(\mathrm{SCR})_{\text{case b}}}{(\mathrm{SCR})_{\text{case a}}} = 3 \Bigg/ \left[1 + \left(\frac{\cos\alpha_1}{\cos\alpha_2}\right)\left(\frac{L_{b_1}}{L_{b_2}}\right) \Bigg/ \left[1 - \frac{c}{2f_r r_T}\right]^3 + \left(\frac{\cos\alpha_1}{\cos\alpha_3}\right)\left(\frac{L_{b_1}}{L_{b_3}}\right) \Bigg/ \left[1 - \frac{c}{f_r r_T}\right]^3\right] \tag{8.47b}$$

$$\frac{(\mathrm{SCR})_{\text{case c}}}{(\mathrm{SCR})_{\text{case a}}} = 6 \Bigg/ \left\{1 + \sum_{k=1}^{5} \left(\frac{\cos\alpha_1}{\cos\alpha_{1+k}}\right)\left(\frac{L_{b_1}}{L_{b,1+k}}\right) \Bigg/ \left[1 - k\left(\frac{c}{2f_r r_T}\right)\right]^3\right\} \tag{8.47c}$$

The contribution of each successive clutter patch to the denominator of the degradation factor on the SCR in (8.47b) or (8.47c) serves to decrease the SCR by amounts which may be significant. As the ranges of the patches decrease (e.g., patches 3 in b and patches 5 and 6 in c), the factors $1/[1 - k[c/2fr]]^3$ in (8.47b) become large enough to introduce very large clutter contributions from patches at ambiguous ranges far removed from the site of the target. This is offset to some extent by the reduction in the factor $[L_{b_1}/L_{b,1+k}]$ as k increases, but not enough to offset the effect of the $1/r^3$ dependence, which accentuates the effect of the near-in clutter sources at ambiguous ranges. It is also offset by the enhancement factors 3 and 6 on (8.47b) and (8.47c), due to the fact that the target signal power is also seen in the ambiguous range gates, but again the effect of near-in ambiguous range clutter returns may overwhelm those enhancements.

Similar effects would occur with other clutter geometries. For example, in Figure 8.23, a rain clutter situation is depicted in which return from the rainfall volume occurs in sidelobes and at ambiguous ranges, although the target return itself is not affected by the rain.

The shadowed regions marked clutter volume 1, 2, 3, 4, and 5 are regions of rain clutter that would be visible to the radar. Clutter volumes 1 and 2 are in the range gate occupied by the target and within the first two sidelobes of the antenna beam, and hence would be visible at any PRF. But clutter volumes 3, 4, and 5 are

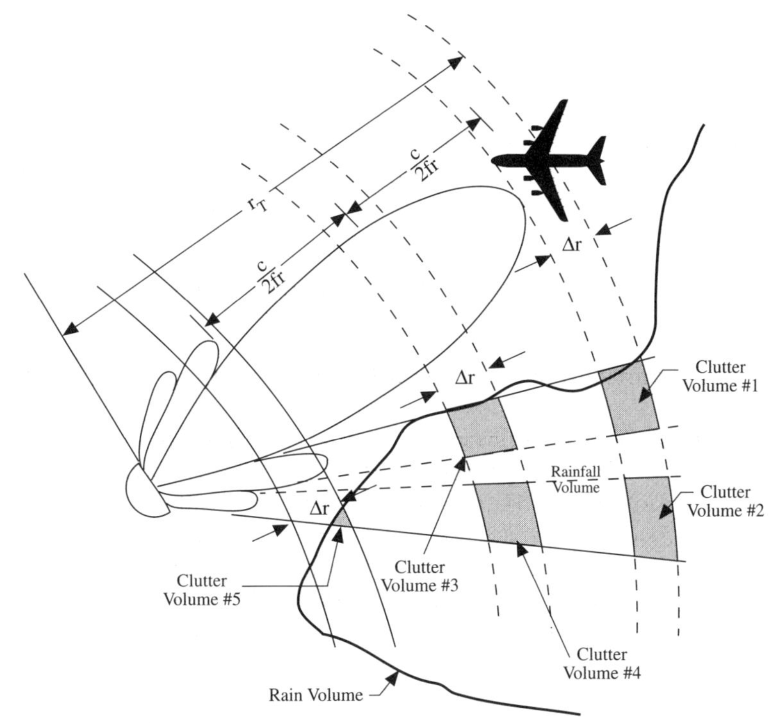

FIGURE 8.23
A rain clutter scenario involving range ambiguities.

in the sidelobes and within the second and third ambiguous range regions and hence would be seen only at a PRF large enough to produce these ambiguities. It is also evident that further increases in PRF resulting in more ambiguous ranges might produce more clutter volumes and hence decrease the SCR.

There is another point not explicitly mentioned in the above discussion, which has been focused on the issue of SCR reduction due to clutter from sidelobes and/or ambiguous ranges, thus affecting adversely the target detection probability. Another issue is that of increased false alarms due to clutter misinterpreted as target signals. If the range gates had been set at different points in Figures 8.22 and 8.23, far from the target range, clutter from ambiguous ranges in addition to the true range would have appeared, thus driving returns in a range gate *not* containing a target above the detection threshold and thereby generating a false alarm (or even if the desired target *is* present at some other range, resulting in an erroneous target range indication). In CFAR operation, where the threshold setting is based on a particular expected noise level, the superposition of all of this extraneous clutter on top of the noise can increase the false alarm rate beyond acceptable limits. Avoiding that by raising the threshold might reduce the detection probability to values below the specified level. Hence the thrust of design strategies in radars that must deal with situations like those depicted here is to *eliminate* the clutter returns or reduce them as much as possible.

A "first-order" measure to reduce the effects of sidelobe (and *some* mainlobe) clutter is to reduce or eliminate the range ambiguities by decreasing the PRF. The limit on that strategy is that a lower PRF enhances the *Doppler* ambiguities. To demonstrate that effect, consider again the situation illustrated in Figure 8.22. But this time we focus on the Doppler spectrum of the returns rather than their distribution in range. To that end, we use Figure 8.24a, b, and c for low, medium, and high PRF

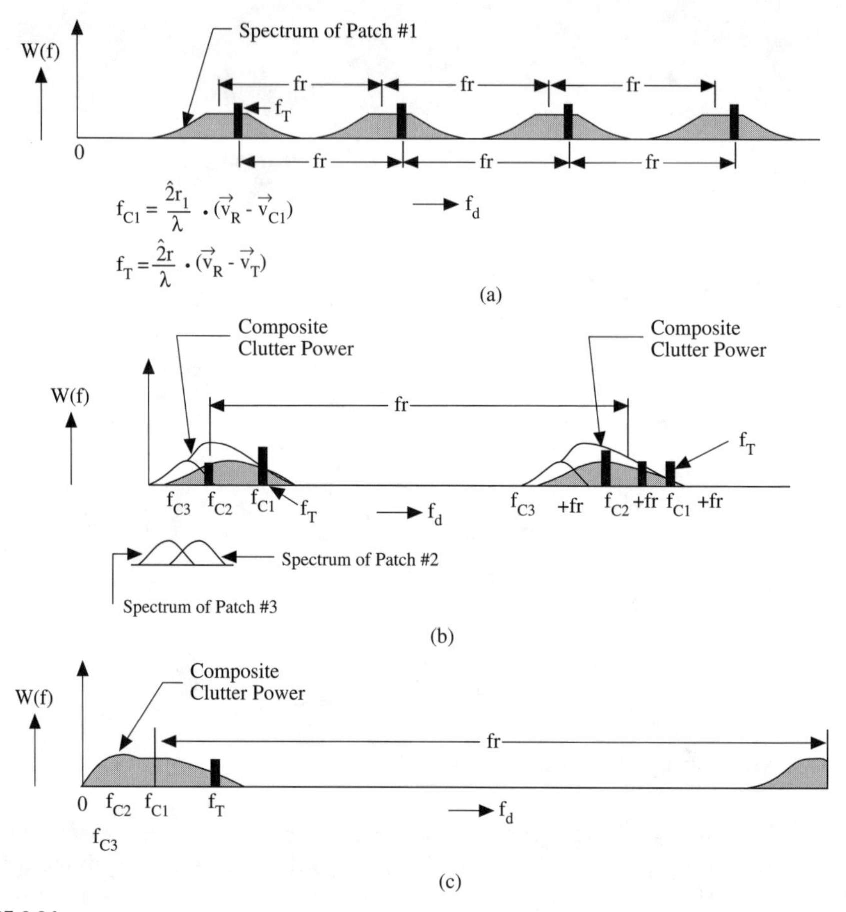

FIGURE 8.24
Doppler spectrum of target and clutter returns. (a) Low PRF (case in Figure 8.22a). (b) Medium PRF (case in Figure 8.22b). (c) High PRF (case in Figure 8.22c).

cases, respectively. These cases are the same as those in Figures 8.22a, b, and c, respectively. In each case, the radar is assumed to have a horizontal motion to the right, the target is assumed to be approaching, and the clutter patches are approximately stationary with (possibly) small random motions about their fixed positions, *or* in the case of a body of water a fixed horizontal speed with (possibly) small vertical fluctuations in position. In the first case (land terrain), the Doppler spectrum for each patch has its center frequency at $[(2/\lambda)v_R \cos \beta]$, where v_R is the radar speed and β is the instantaneous depression angle to the patch. In the latter case (water terrain), the spectrum for the patch has its center frequency determined by the *relative* velocity, i.e, it is given by $(2/\lambda)\hat{r} \cdot (\mathbf{v}_R - \mathbf{v}_C)$, where $\hat{r}$ is the instantaneous unit vector from the radar to the clutter patch and $(\mathbf{v}_R - \mathbf{v}_C)$ is the vector difference between radar and clutter patch velocities. The spectrum of each patch has some spread due to the variation of the depression angle (or equivalently the instantaneous line-of-sight relative velocity component) throughout the illuminated area, as discussed in Section 8.8. The target Doppler has a smaller spread than the clutter, as illustrated in Figure 8.21.

In Figure 8.24a, we show the spectral behavior for the scenario depicted in Figure 8.22a, where the target return spectrum has a narrow peak at $f_T = (2\hat{r}/\lambda) \cdot (\mathbf{v}_R - \mathbf{v}_T)$, where $\mathbf{v}_T$ is the target velocity vector. There is one clutter patch picked up in the target's range gate, called clutter patch 1 in Figure 8.22. Its central frequency is $f_o = (2/\lambda)\,\hat{r}_1 \cdot (\mathbf{v}_R - \mathbf{v}_{C1})$, where $\hat{r}_1$ is the instantaneous unit vector from radar to patch and $\mathbf{v}_{C1}$ is the velocity vector of the clutter patch. The patch's spectrum, being much wider than that of the target, is shown with its upper limit coinciding with the target spectrum. The implication is that the target Doppler is less than half the Doppler spread of the clutter patch and hence the target return *is* affected by the clutter, but in this case not much by the ambiguities. That is only because the ambiguities are still far enough apart to preclude the possibility that the target spectrum overlaps the second ambiguous Doppler spectrum for the clutter patch.

In Cases b and c, Figure 8.24 shows the effect of the additional clutter patches from the sidelobes that are in the ambiguous range regions. The major effect is to produce clutter spectra that peak at lower Dopplers than patch 1 because of the larger depression angles that reduce the line-of-sight components of relative velocity. The net effect of that is that the composite of contributions from these other patches generates a Doppler spectrum that may be more skewed toward the lower end and widens the overall spectrum.

Figure 8.24 does not fully illustrate the effect of interest. To do this, we consider Figure 8.25, in which we show a case where the PRF is less than the clutter bandwidth B_c.

In the case shown on the figure, the composite target-plus-clutter Doppler spectrum repeats itself at intervals of *fr*, but the clutter spectra at adjacent multiples of *fr* overlap and there is continuous coverage of Doppler space by clutter sources. Moreover, the small peaks that indicate the target or one of its Doppler ambiguities may be in regions of intense Doppler. This is a case where the SCR is low throughout the band due to Doppler ambiguities.

Operating at such a low PRF to avoid *range* ambiguities will make it exceedingly difficult to discriminate between target and clutter on the basis of Doppler. Obviously, there is a significant tradeoff problem in choosing the PRF, depending on the target clutter environment and other system parameters.

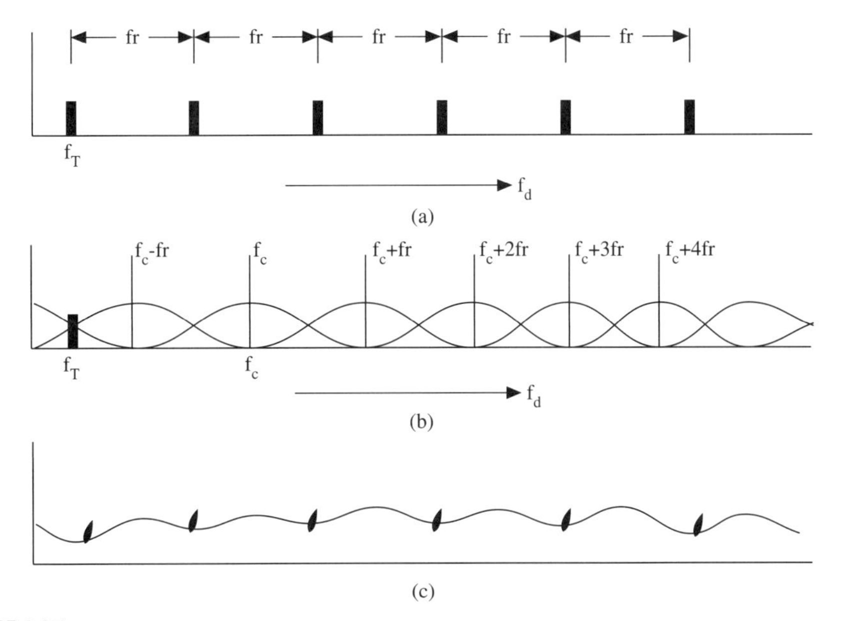

FIGURE 8.25
Doppler spectra. (a) Target spectrum—no clutter—Target central doppler = f_T. (b) True and ambiguous clutter bands—true clutter spectrum centered at f_c. (c) Composite target + clutter spectra.

REFERENCES

1. Skolnick, M.I., "Introduction to Radar Systems," 2nd ed. McGraw-Hill, New York, 1980, Chapter 13.
2. Blake, L.V., "Radar-Range Performance Analysis." Artech, Norwood, MA, 1986, Chapter 7.
3. Eaves, J.L., and Reedy, E.K. (eds.), "Principles of Modern Radar," Van Nostrand Reinhold, New York, 1987, Chapter 10, by N.C. Currie.
4. Barton, D.K., "Modern Radar System Analysis." Artech, Norwood, MA 1988, Sections 1.5, 3.6.
5. Levanon, N., "Radar Principles." Wiley, New York, 1988, Chapter 4, pp. 78–98.
6. Nathanson, F.E., Reilly, J.P., and Cohen, M.N., "Radar Design Principles," 2nd ed. McGraw-Hill, New York, 1991, Chapters 6 and 7, by F.E. Nathanson and J.P. Reilly.
7. M.W. Long, "Radar Reflectivity of Land and Sea." Artech, Norwood, MA, 1983.
8. Kerr, D.E. (ed.), "Propagation of Short Radio Waves." McGraw-Hill, New York, 1951.
9. Beckmann, P., and Spizzichino, A., "The Scattering of Electromagnetic Waves from Rough Surfaces." Artech, Norwood, MA, 1987.
10. Ulaby, F.T., Moore, R.K., and Fung, A.K., "Microwave Remote Sensing—Active and Passive," Vol. I, 1981, Vol. II, 1982, Addison-Wesley, Reading, MA; Vol. III, 1986, Artech, Norwood, MA.
11. Ulaby, F.T., and Dobson, M.C., "Radar Scattering Statistics for Terrain." Artech, Norwood, MA, 1989.
12. Tsang, L., Kong, J.A., and Shin, R.T., "Theory of Microwave Remote Sensing." John Wiley & Sons, New York, 1985.

Chapter 9

Methods of Combatting Clutter

CONTENTS

The methods of combatting clutter all depend very sensitively on the specifics of radar-clutter geometry, but the most common methods choose some property that is usually significantly different for typical target and clutter returns. That property is used as the discriminant between these two classes of returns. The most popular such discriminant is Doppler speed. We will focus attention on Doppler-speed-dependent techniques for suppression of clutter.[1–4]

9.1 INTRODUCTORY REMARKS ON MTI RADAR

The moving target indicator, or MTI,[1,2,5,6] is a device that suppresses returns from stationary targets and thereby (hopefully) pulls out the desired return from a moving target without the background return from stationary clutter that would usually be present. The motion, of course, must have a component along the target-radar line-of-sight, because a nonzero Doppler appears only if such a component is present. Any cross-line-of-sight differential velocity components between target and clutter cannot be distinguished by Doppler measurement.

9.2 DELAY-LINE CANCELLERS: BLIND SPEEDS

The most elementary type of MTI is the two-pulse delay-line canceller,[7] illustrated in Figure 9.1. This can also be called the single delay-line canceller, because it requires a single delay line to implement it, as shown in Figure 9.1.

This device is based on the assumption that bipolar video amplitudes of moving targets will change from pulse-to-pulse and those from clutter, assumed stationary, will not. Hence subtracting two adjacent bipolar video pulses should remove stationary clutter. The first IF pulse voltage signal is

$$v_1^{(IF)}(t) = p(t - t_0)\cos[(\omega_o + \omega_d)(t - t_0) + \psi] \tag{9.1a}$$

The second IF pulse signal is

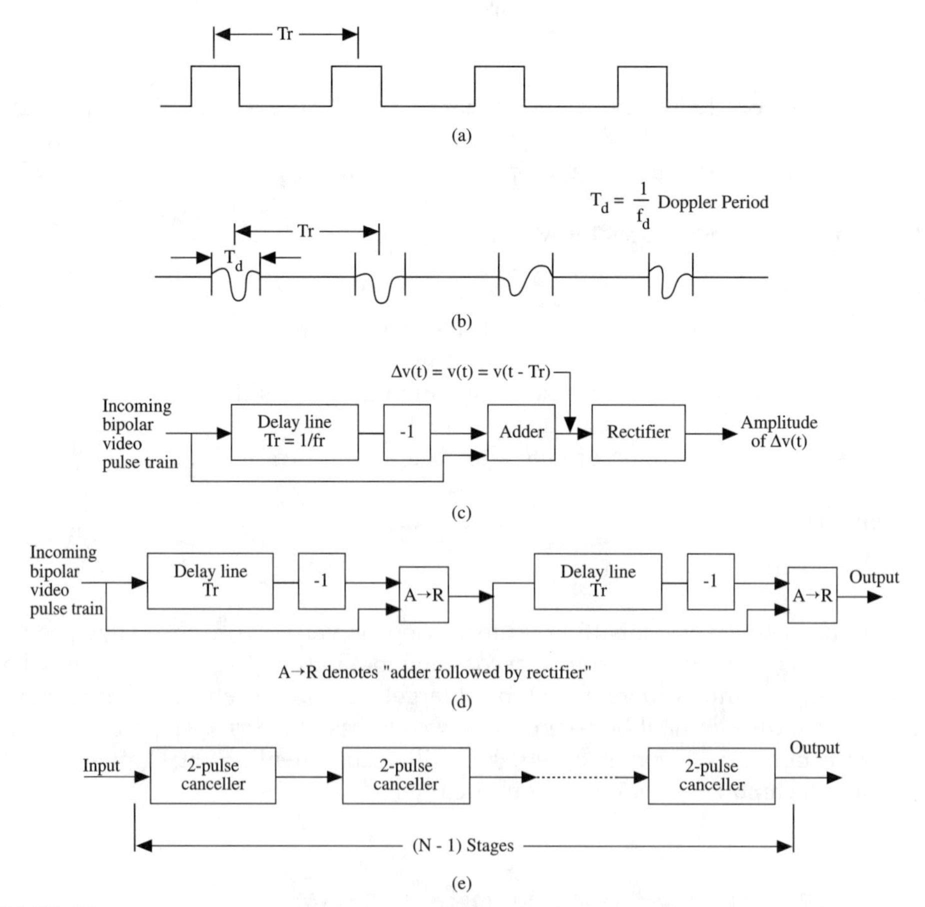

FIGURE 9.1
Delay-line cancellers. (a) Video pulse train from mixed targets. (b) Bipolar video pulse train from moving targets. (c) Two-pulse canceller. (d) Three-pulse canceller (two-two pulse cancellers in cascade). (e) (N-1)-pulse canceller ((N-1)-two pulse cancellers in cascade).

$$v_2^{(IF)}(t) = p[t - (t_0 + Tr)]\cos\{(\omega_o + \omega_d)[t - (t_0 + Tr) + \psi]\} \tag{9.1b}$$

where $Tr = 1/fr =$ PRP. The "bipolar video" signals resulting from down-conversion of the two signals $v_1^{(IF)}(t)$ and $v_2^{(IF)}(t)$ to baseband (where t_o and ψ can be set arbitrarily and hence are set at zero) are

$$v_1^{(BPV)}(t) = p(t)\cos(\omega_d t) \tag{9.2a}$$

$$v_2^{(BPV)}(t) = p(t - Tr)\cos[\omega_d(t - Tr)] \tag{9.2b}$$

For a standard pulse train,

$$p(t - nTr) = a[u(t - nTr) - u(t - (n+1)Tr)]; \qquad n = 0, 1 \tag{9.2c}$$

The video difference signal is

$$\begin{aligned}\Delta v(t) &= v_1^{(BPV)}(t) - v_2^{(BPV)}(t) \\ &= a\{\cos(\omega_d t) - \cos[\omega_d(t - Tr)]\} \\ &= a\{\cos(\omega_d t)[1 - \cos(\omega_d Tr)] - \sin(\omega_d t)\sin(\omega_d Tr)\}\end{aligned} \tag{9.3}$$

which is a sinusoidal waveform at the Doppler frequency. Its amplitude is

$$\begin{aligned}|\Delta v| &= a\sqrt{[1 - \cos(\omega_d Tr)]^2 + [\sin(\omega_d Tr)]^2} \\ &= a\sqrt{1 - 2\cos(\omega_d Tr) + 1} = a\sqrt{2}\,\sqrt{2}\,\sqrt{\frac{1 - \cos(\omega_d Tr)}{2}} \\ &= 2a\sqrt{\sin^2\left(\frac{\omega_d Tr}{2}\right)} = 2a\left|\sin\left(\frac{\omega_d Tr}{2}\right)\right| = 2a\left|\sin\left(\frac{\pi f_d}{fr}\right)\right|\end{aligned} \tag{9.4}$$

which vanishes for $f_d = 0$ (stationary target).

The output $|\Delta v|$ is shown in Figure 9.2. It has nulls at $f_d = 0, \pm fr, \pm 2fr, \pm 3fr, \ldots$, i.e., at a Doppler of zero and at all integral multiples of the PRF. The rejection of stationary targets (hopefully clutter and not desired targets) is the purpose of the device, but unfortunately the nulls at multiples of the PRF must also be present. These are called "blind speeds," because they represent line-of-sight velocities at which a desired target may be moving and, if so, would not be seen at the canceller output. It is clear that increasing the PRF would decrease the number of blind speeds within a given Doppler region, but would have the effect of increasing ambiguous ranges. Deferring a discussion of that familiar tradeoff in the MTI context, we now consider attempts to broaden the "notch" at zero Doppler and thereby enchance the rejection of stationary clutter. This can be done by adding another stage of delay line cancellation. This amounts to taking the difference between two adjacent pulse differences, i.e., referring to (9.2a,b,c),

$$\Delta^2 v = [v_1^{(BPV)}(t) - v_2^{(BPV)}(t)] - [v_2^{(BPV)}(t) - v_3^{(BPV)}(t)] \tag{9.5}$$

where (see 9.1a,b)

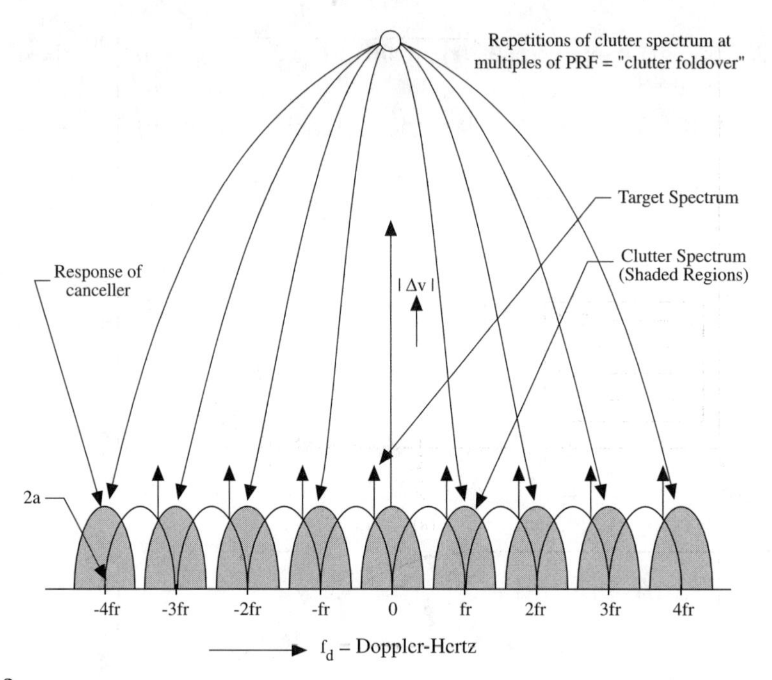

FIGURE 9.2
Output of MTI single delay-line canceller.

$$v_3^{(BPV)}(t) = p[t - (t_0 + 2Tr)] \cos\{(\omega_o + \omega_d)[t - (t_0 + 2Tr)] + \psi\} \tag{9.6a}$$

where $p(t - nTr) = a[u(t - nTr) - u(t - (n + 1)Tr]; n = 1,2.$

From (9.1a,b), (9.5) and (9.6a), where t_o and ψ are both set to zero,

$$\Delta^2 v = v_1^{(BPV)}(t) - 2v_2^{(BPV)}(t) + v_3^{(BPV)}(t) \tag{9.6b}$$

$$= a[\cos(\omega_d t) - 2\cos\{\omega_d(t - Tr)] + \cos[\omega_d(t - 2Tr)]\}$$

$$= a\{\cos(\omega_d t)[1 - 2\cos\omega_d Tr + \cos 2\omega_d Tr]$$

$$+ \sin\omega_d t[-2\sin\omega_d Tr + \sin 2\,\omega_d Tr]\}$$

The amplitude of this "double delay-line canceller" is

$$|\Delta^2 v| = a\sqrt{(1 - 2\cos\omega_d Tr + \cos 2\omega_d Tr)^2 + (-2\sin\omega_d Tr + \sin 2\omega_d Tr)^2}$$

$$= a\sqrt{(1 + 4 + 1 - 4\cos\omega_d Tr + 2\cos 2\omega_d Tr - 4\cos\omega_d Tr\cos 2\omega_d Tr}$$

$$\overline{- 4\sin\omega_d Tr\sin 2\omega_d Tr)} = \sqrt{2}a\sqrt{3 - 4\cos\omega_d Tr + \cos 2\omega_d Tr}$$

$$= 2a\sqrt{1 - 2\cos\omega_d Tr + \cos^2\omega_d Tr} = 2a\left|1 - \cos\left(2\pi\frac{f_d}{fr}\right)\right|$$

$$= 4a\sin^2\left(\frac{\pi f_d}{fr}\right) \tag{9.6c}$$

or

$$|\Delta^2 v| = 4a\sin^2\left(\frac{\pi f_d}{fr}\right) \tag{9.6d}$$

The double delay-line canceller or "three-pulse canceller" has an output with peaks and nulls at exactly the same places as the single delay-line canceller, i.e., nulls at multiples of the PRF and peaks at $(n + 1/2)\, fr$, where $n = 0, \pm 1, \pm 2, \pm 3, \ldots$ But the nulls are broader, and hence provide broader rejection of targets near the blind speeds, i.e., the rejection band around the nulls (blind speeds) increases as the number of pulse differencing operations increases.

One can extend this process of increasing the number of cancellations (or equivalently the number of pulses and delay lines) to shape the MTI response almost as desired. One can also weight the differences in various ways to achieve synthesis of filters with still broader notches or other desired characteristics.[8]

Extension of this process to more differences should be done with complex envelope representations. From (9.2a,b) or (9.6a), we have with $t_o = \psi = 0$

$$\hat{v}_k^{(\mathrm{BPV})}(t) = a e^{j\omega_{\mathrm{d}}(t-kTr)} \tag{9.7a}$$

and hence

$$\begin{aligned} \Delta^1 v_{k,k+1} &= \hat{v}_k^{(\mathrm{BPV})}(t) - \hat{v}_{k+1}^{(\mathrm{BPV})}(t) = a e^{j\omega_{\mathrm{d}}(t-kTr)}(1 - e^{j\omega_{\mathrm{d}}Tr}) \\ &= -2je^{j\omega_{\mathrm{d}}[t-(k-1/2)Tr]}\sin\left(\frac{\pi f_{\mathrm{d}}}{fr}\right) \end{aligned} \tag{9.7b}$$

For the single-pulse canceller, we have an amplitude

$$|\Delta^1 v_{1,2}| = 2a\left|\sin\left(\frac{\pi f_{\mathrm{d}}}{fr}\right)\right| \tag{9.7c}$$

which is consistent with (9.4).

For the double-pulse canceller,

$$\begin{aligned} \Delta^2 v_{1,2,3} &= \Delta^1 v_{1,2} - \Delta^1 v_{2,3} \\ &= -2jae^{j\omega_{\mathrm{d}}[t-(Tr/2)]}\left[\sin\left(\frac{\pi f_{\mathrm{d}}}{fr}\right)e^{-j\omega_{\mathrm{d}}Tr} - \sin\left(\frac{\pi f_{\mathrm{d}}}{fr}\right)e^{-j2\omega_{\mathrm{d}}Tr}\right] \\ &= -2ja\sin\left(\frac{\pi f_{\mathrm{d}}}{fr}\right)e^{j\omega_{\mathrm{d}}(t-(Tr/2))}e^{-j\omega_{\mathrm{d}}(3/2)Tr}\left(e^{j\omega_{\mathrm{d}}Tr/2} - e^{-j\omega_{\mathrm{d}}Tr/2}\right) \\ &= (-2j)(2ja)\sin^2\left(\frac{\pi f_{\mathrm{d}}}{fr}\right)e^{j\omega_{\mathrm{d}}[t-(5Tr/2)]} \end{aligned} \tag{9.7d}$$

whose amplitude is

$$|\Delta^2 v_{1,2,3}| = 4a\sin^2\left(\frac{\pi f_{\mathrm{d}}}{fr}\right) \tag{9.7e}$$

consistent with (9.6c).

For the case of four pulses, it would follow from (9.7b) that

$$\Delta^3 v_{1,2,3,4} = ae^{j\omega_d(t-Tr)}\{[(1 - e^{j\omega_d Tr}) - (e^{j\omega_d Tr} - e^{j\omega_d 2Tr})] - [(e^{j\omega_d Tr} - e^{j\omega_d 2Tr}) - (e^{j\omega_d 2Tr} - e^{j\omega_d 3Tr})]\}, \tag{9.7f}$$

[since $(1 - x)^3 = 1 - 3x + 3x^2 - x^3$]

$$\begin{aligned}\Delta^3 v_{1,2,3,4} &= ae^{j\omega_d(t-Tr)}(1 - 3e^{j\omega_d Tr} + 3e^{j2\omega_d Tr} - e^{j3\omega_d Tr}) \\ &= ae^{j\omega_d(t-Tr)}(1 - e^{j\omega_d Tr})^3\end{aligned} \tag{9.7g}$$

whose amplitude is

$$|\Delta^3_{1,2,3,4}| = 8a\left|\sin^3\left(\frac{\pi f_d}{fr}\right)\right| \tag{9.7h}$$

The five-pulse case yields [since $(1 - x)^4 = 1 - 4x + 6x^2 - 4x^3 + x^4$]

$$\begin{aligned}\Delta^4 v_{1,2,3,4,5} = ae^{j\omega_d(t-Tr)}\{&[[(1 - e^{j\omega_d Tr}) - (e^{j\omega_d Tr} - e^{j\omega_d 2Tr})] \\ &- [(e^{j\omega_d Tr} - e^{j2\omega_d Tr}) - (e^{j2\omega_d Tr} - e^{j3\omega_d Tr})]] \\ &- ([(e^{j\omega_d Tr} - e^{j2\omega_d Tr}) - (e^{j2\omega_d Tr} - e^{j3\omega_d Tr})] \\ &- [(e^{j2\omega_d Tr} - e^{j3\omega_d Tr}) - (e^{j3\omega_d Tr} - e^{j4\omega_d Tr})]]\} \\ &\times ae^{j\omega_d(t-Tr)}(1 - 4e^{j\omega_d Tr} + 6e^{j\omega_d Tr} - 4e^{j3\omega_d Tr} + e^{j4\omega_d Tr}) \\ = ae^{j\omega_d(t-Tr)}&(1 - e^{j\omega_d Tr})^4\end{aligned} \tag{9.7i}$$

whose amplitude is

$$|\Delta^4 v_{1,2,3,4,5}| = 16a\ \sin^4\left(\frac{\pi f_d}{fr}\right) \tag{9.7j}$$

This process can be continued indefinitely, with the general result for the N-pulse case

$$|\Delta^{N-1} v_{1,2,3,4,\ldots,N}| = 2^{N-1}a\left|\sin^{N-1}\left(\frac{\pi f_d}{fr}\right)\right| \tag{9.7k}$$

The effect of the repeated differencing is simply to widen the notches that occur at the integral multiples of the PRF to allow more target energy near the nulls. To show this, we find the value of f_d corresponding to $a - 3$ dB point on the response curve. For the N-pulse case, (9.7k) yields the condition [where this point is denoted by $(f_d)_{-3\text{ dB}}$]

$$\left|\sin^{N-1}\left(\frac{\pi f_d}{fr}\right)\right| = \frac{1}{\sqrt{2}} \tag{9.7l}$$

or equivalently

$$(f_d)_{-3\text{ dB}} = \frac{fr}{\pi}\sin^{-1}\left[\frac{1}{(2)^{1/[2(N-1)]}}\right] \tag{9.7m}$$

The −3 dB bandwidth around a peak (B_N) is given by $2(f_d)_{-3\text{ dB}}$ or

$$B_N = \frac{2fr}{\pi}\sin^{-1}\left(\frac{1}{(2)^{1/[2(N-1)]}}\right) \tag{9.7n}$$

Values of B_N for increasing values of N are given in Table 9.1. As shown in Table 9.1, the bandwidth around a peak decreases with increasing order of pulse cancellation, implying an increase of the width of the notch centered at each null as the order increases. For a large number of pulse cancellations, the width of the notch approaches the PRF, being only about half the PRF for the two-pulse canceller, 67% of the PRF for the three-pulse canceller, and 96% of the PRF for the case $N = 9$. The effect is shown graphically in Figure 9.3.

The ultimate limits on increasing the number of cancellation stages are as follows:

1. If the rejection region around DC is too broad, slow-moving targets that the radar might be trying to detect will also be blocked out along with the undesired clutter.
2. It is pointed out in Skolnick's text[8] that the N-pulse canceller, which has been modeled here as a cascade of $(N-1)$ single-pulse delay line cancellers, is equivalent to a nonrecursive (or FIR) filter of $(N-1)$ stages (shown in Fig. 9.4) in the sense that its response is exactly the same, if the chosen filter weightings are chosen properly.

In the language of digital filtering, a FIR filter has the response characteristic

$$y(n) = \sum_{m=0}^{N-1} b_m x(n - m) \tag{9.8}$$

where $x(k)$ and $y(k)$ are input and output sequences, respectively. The action of the filter whose operation is illustrated in Figure 9.4 is

$$y(t) = \sum_{m=0}^{N-1} b_m x(t - m\,Tr) \tag{9.9}$$

If Tr is the sampling interval and $t = k\,Tr$, i.e., each pulse is viewed as a digital sample and each pulse input or output as a sequence $x(k\,Tr)$ or $y(k\,Tr)$, respectively, then

TABLE 9.1

Notch Width vs. Order of N-pulse Canceller

N	$2^{\{1/[1/2(N-1)]\}}$	$\sin^{-1}\left(\frac{1}{2^{1/2(N-1)}}\right)$	$\sin^{-1}\left(\frac{1}{2^{1/2(N-1)}}\right)$	B_N/f_r	Width of notch
2	$1/\sqrt{2}$	45°	0.785	0.4993	0.5007 fr
3	0.8409	48°	0.8377	0.3333	0.6667 fr
4	0.8909	56.1°	0.9791	0.2296	0.7704 fr
5	0.9170	60.98°	1.064	0.1609	0.8391 fr
6	$1/4\sqrt{2}$	10.18°	0.178	0.1132	0.8868 fr
7	1/8	7.18°	0.125	0.0795	0.9205 fr
8	$1/8;\sqrt{2}$	5.07°	0.088	0.0560	0.9440 fr
9	1/16	3.58°	0.062	0.0394	0.9606 fr

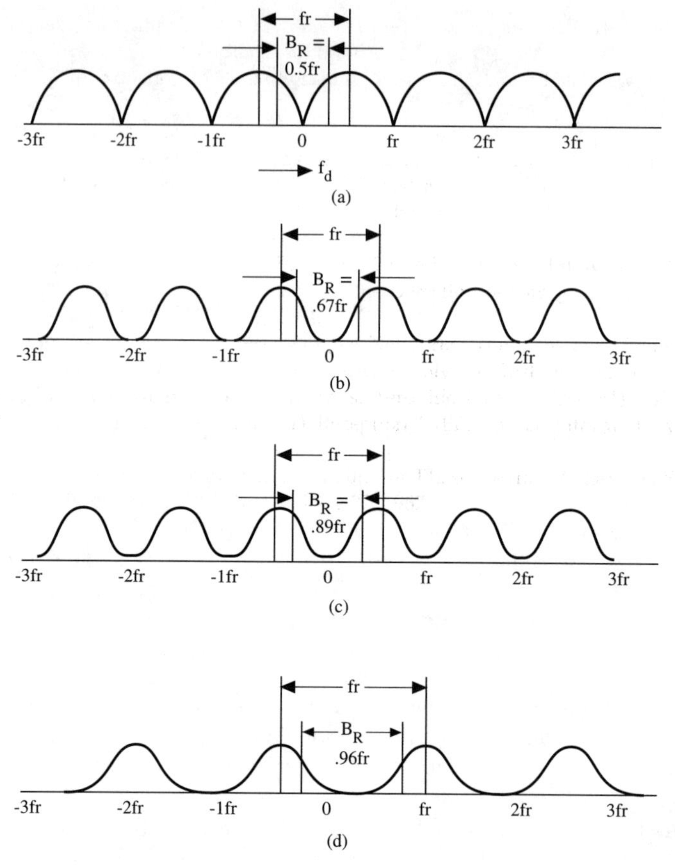

FIGURE 9.3
Blind speed rejection bands for N-pulse cancellers. (a) $N = 2$. (b) $N = 3$. (c) $N = 6$. (d) $N = 9$.

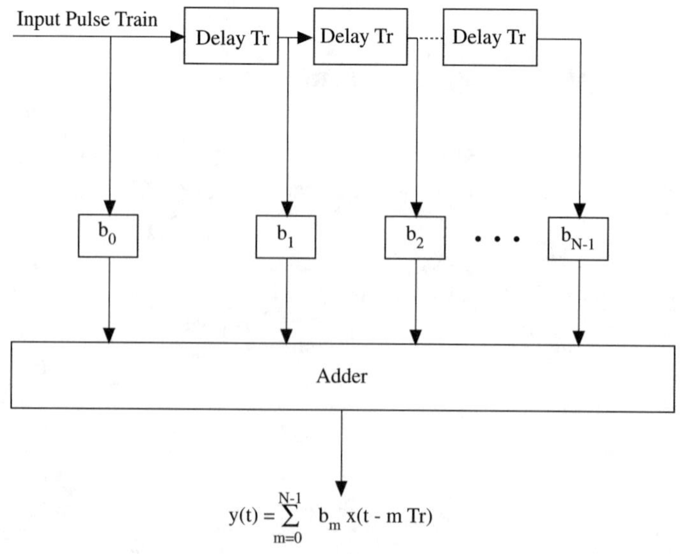

FIGURE 9.4
$(N-1)$-stage FIR filter as MTI device.

$$y(n\,Tr) = \sum_{m=0}^{N-1} b_m x[(n - m)Tr] \tag{9.10}$$

If we adopt time units in which $Tr = 1$, then (9.10) becomes

$$y(n) = \sum_{m=0}^{N-1} b_m x(n - m) \tag{9.11}$$

which is a discrete expression identical to (9.8), the generic input–output characteristic for a FIR filter.[9]

The transfer function of the digital filter (9.11) is

$$H(\Omega) = \frac{Y(\Omega)}{X(\Omega)} = b_0 + b_1 e^{-j\Omega} + b_2 e^{-2j\Omega} + \cdots + b_{N-1} e^{-j(N-1)\Omega} \tag{9.12}$$

where Ω is the discrete frequency.

The two-pulse canceller is implemented by setting $b_o = 1$, $b_1 = -1$, $b_m = 0$ for $m > 1$, in which case (9.11) becomes

$$y(n) = x(n) - x(n - 1) \tag{9.13a}$$

and (9.12) becomes

$$H(\Omega) = 1 - e^{-j\Omega} = 2je^{-j\Omega/2} \sin\left(\frac{\Omega}{2}\right) \tag{9.13b}$$

We now note that the discrete frequency Ω is equivalent to $2\pi f_d\, Tr$ and hence the true frequency response of this filter has amplitude

$$|H(2\pi f_d Tr)| = \left|\sin\left(\frac{\pi f_d}{fr}\right)\right| \tag{9.13c}$$

The three-pulse canceller has weightings $b_o = 1$, $b_1 = -2$, $b_2 = 1$, $b_m = 0$ for $m > 2$. In this case, (from 9.12),

$$H(\Omega) = 1 - 2e^{-j\Omega} + e^{-2j\Omega} = (1 - e^{-j\Omega})^2 \tag{9.14a}$$

and hence

$$|H(2\pi f_d Tr)| = \left|\sin\left(\frac{\pi f_d}{fr}\right)\right|^2 \tag{9.14b}$$

which is consistent with (9.7e) for the three-pulse canceller except for the multiplicative factor 4a.

To implement the equivalent of cascaded two-pulse cancellers discussed above, the selection of weightings for $(N - 1)$ stages should be the binomial coefficients. From (9.12)

$$H(\Omega) = \sum_{m=0}^{N-1} b_m^{(N-1)} e^{-jm\Omega} \tag{9.15a}$$

$$= \sum_{m=0}^{N-1} \frac{(N-1)!(-1)^m e^{-jm\Omega}}{m!(N-1-m)!} = (1 - e^{-j\Omega})^{N-1}$$

where

$$b_m^{(N-1)} = \frac{(-1)^m (N-1)!}{m!(N-1-m)!}$$

The amplitude of the frequency response in this case is

$$H(2\pi f_d Tr) = \left|\sin\left(\frac{\pi f_d}{fr}\right)\right|^{N-1} \tag{9.15b}$$

The cases $N = 2, 3, 4$, and 5, respectively, yield expressions equivalent to (9.4), (9.6c) (or 9.7e), (9.7h), and (9.7j), respectively, and the general result (9.15b) for arbitrarily large N is equivalent to (9.7k), demonstrating that the cascade of two-pulse cancellers is equivalent to the FIR filter with binomial weightings.

Other weightings can be used to shape the FIR filter response to the designer's desires, e.g., to change the Doppler nulls or blind speeds into troughs a specified number of dB down from the peaks. It should be borne in mind that a price must always be paid for these design modifications of the basic MTI concept. The complete null at DC is the desired response of MTI, so any compromise from that may open the door to more unwanted clutter. On the positive side, it may also open the door to a desired target at one of the blind speeds, which could not be seen at all with standard MTI.

Another way to reduce the effect of blind speeds or to remove them completely is to superpose MTI outputs with *different* PRFs.* This will obviously fill in the nulls. The degree to which that can be accomplished depends on how many PRFs one is willing to use. The principle is illustrated in Figure 9.5, which shows the superposition of three two-pulse canceller outputs, each with a different PRF. The sum of the three outputs (amplitude-additive) is

$$|\Delta v| = |\Delta v_1| + |\Delta v_2| + |\Delta v_3| = \left|\sin\left(\frac{\pi f_d}{fr1}\right)\right| + \left|\sin\left(\frac{\pi f_d}{fr2}\right)\right| + \left|\sin\left(\frac{\pi f_d}{fr3}\right)\right| \tag{9.16}$$

The null at $f_d = 0$, which is the *desired* null, is preserved in this output, but the nulls at integral multiples of the PRF are eliminated, *if* the PRFs are not commensurate with one another. Such a scheme does not eliminate the blind speeds entirely and may introduce new ones. For example suppose

$$fr2 = 2fr1, fr3 = \frac{fr1}{2} \tag{9.17a}$$

in which case the output (9.16) is

* This is called staggered PRF-MTI.[10–12]

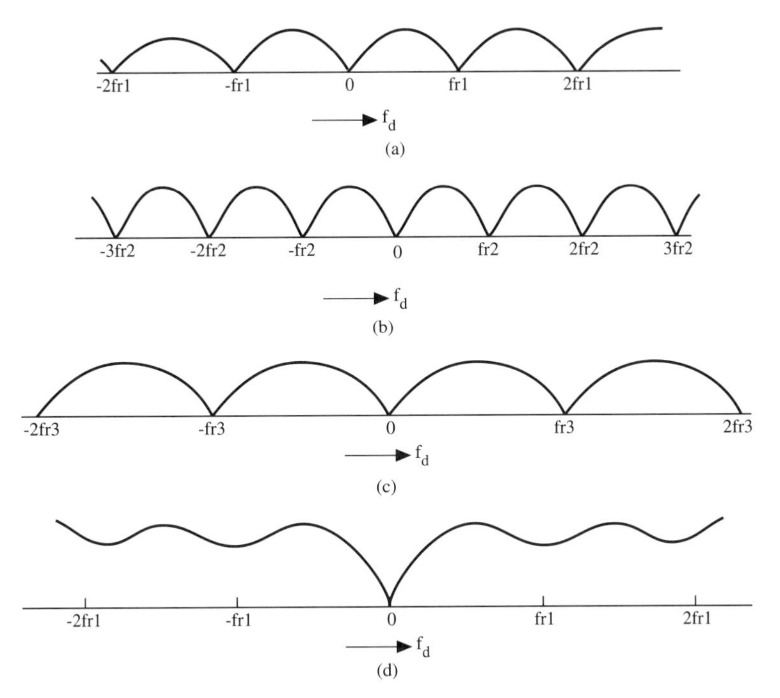

FIGURE 9.5
Two-pulse canceller-type MTI with multiple PRFs. (a) $|\Delta v_1|$. (b) $|\Delta v_2|$. (c) $|\Delta v_3|$. (d) Total: $|\Delta v(f_d)| = |\Delta v_1| + |\Delta v_2| + |\Delta v_3|$.

$$|\Delta v(f_d)| = \left|\sin\left(\frac{\pi f_d}{fr1}\right)\right| + \left|\sin\left(\frac{\pi f_d}{2fr1}\right)\right| + \left|\sin\left(\frac{2\pi f_d}{fr1}\right)\right| \tag{9.17b}$$

At $f_d = nfr1$, the output is

$$|\Delta v(nfr1) = |\sin(n\pi)| + \left|\sin\left(\frac{n\pi}{2}\right)\right| + |\sin(2\pi n)| = \left|\sin\left(\frac{n\pi}{2}\right)\right|$$

$$= \begin{matrix} 1 & \text{if } n \text{ is odd} \\ 0 & \text{if } n \text{ is even} \end{matrix} \tag{9.17c}$$

The outputs at $f_d = nfr2$ and $nfr3$, respectively, are

$$|\Delta v(nfr2)| = |\Delta v(2nfr1)| = |\sin(2\pi n)| + |\sin(n\pi) + |\sin(4\pi n)| = 0 \text{ for all } n \tag{9.17d}$$

and

$$|\Delta v(nfr3)| = \left|\Delta v\left(\frac{nfr1}{2}\right)\right| = \left|\sin\left(\frac{n\pi}{2}\right)\right| + \left|\sin\left(\frac{n\pi}{4}\right)\right| + |\sin n\pi| = \left|\sin\left(\frac{n\pi}{2}\right)\right| + \left|\sin\left(\frac{n\pi}{4}\right)\right| \tag{9.17e}$$

With the relationship between the three PRFs given by (9.17a), nulls (blind speeds) exist at $f_d = nfr2$ for all integral values of n. Nulls also exist at $f_d = nfr1$ for all even integers n and at $f_d = nfr3$ for $n = 4, 8, 12, 16, \ldots$ or in general for $n = 4m$, where m is an arbitrary integer.

Thus by introducing the additional PRFs $fr2$ and $fr3$ we have eliminated blind speeds at *some* but not *all* integral multiples of $fr1$ and have introduced some *new* blind speeds that did not exist before. That is hardly a satisfactory choice for $fr2$ and $fr3$.

The same effect can be attained by variation of the radar *frequency* rather than PRF. The pulse canceller response in terms of Doppler speed v_d is $\sin(\pi 2v_d/\lambda fr)$. For fixed Doppler speed, a variation of the wavelength λ would alter the response in the same way as a change in PRF. A scheme like that described above, involving superposition of three canceller outputs with three different PRFs and fixed wavelength, could also be accomplished with three different wavelengths and fixed PRF.

9.3 PERFORMANCE MEASURES FOR MTI

The usual measure of performance for an MTI system (or any other system attempting to combat the effects of clutter) is the "improvement factor,"which *could* be defined as

$$I_F(v_d) = \left[\frac{(S/C)_o}{(S/C)_i}\right]_{v_d} \tag{9.18a}$$

for a single doppler speed v_d, but is more properly defined as an average over the Doppler spectrum (as in Skolnick,[1], p. 132), i.e.,

$$I_F = \left[\frac{(S/C)_o}{(S/C)_i}\right]_{\text{all } f_d} \tag{9.18b}$$

where "all f_d" denotes averaging over all Doppler frequencies (or equivalently Doppler speeds) of interest, and where $(S/C)_i$ and $(S/C)_o$ denote the signal-to-clutter ratios at the input and output of the MTI circuit, respectively.

To evaluate the improvement factor, we note that the clutter power at the input to the MTI is the integral over the clutter spectrum as given by (8.43a), (8.43b), or some other model for the spectrum, i.e., input clutter is

$$C_i = \int_{-\infty}^{\infty} df\, W(f) \tag{9.19a}$$

The output clutter power integral has its integrand weighted by the frequency response function of the MTI device, whose amplitude-square for an N-pulse canceller is $|2\sin(\pi f/fr)|^{2(N-1)}$. Thus the output clutter power is

$$C_o = \int_{-\infty}^{\infty} df\, W(f) \left|2\sin\left(\frac{\pi f}{fr}\right)\right|^{2(N-1)} \tag{9.19b}$$

The Gaussian form of the doppler spectrum of clutter, as given by (8.43a), is the most convenient form for analysis. Using it to determine C_i and C_o will produce a result that is both easily tractable mathematically and easily interpreted in such a way that the general ideas could be extended to other spectral shapes without much real loss of generality. First, from (8.43a) and (9.19a)*

$$C_i = \int_{-\infty}^{\infty} df\, W(f_o) e^{-5.54((f-f_o)/\Delta f_c)^2} = \frac{W(f_o)\Delta f_c \sqrt{\pi}}{\sqrt{5.54}} \tag{9.20a}$$

and from (8.43a) and (9.19b)

$$C_o = \int_{-\infty}^{\infty} df\, W(f_o) e^{-5.54((f-f_o)/\Delta f_c)^2} \left[\frac{Ae^{j\beta((f-f_o)/\Delta f_c)} - A^* e^{-j\beta((f-f_o)/\Delta f_c)}}{2j}\right]^{2(N-1)} \tag{9.20b}$$

where $A = 2e^{j\pi f_o/fr}$; $\beta = \pi\Delta f_c/fr$.

Using the binomial theorem to expand the bracketed quantity in (9.20b), we obtain

$$C_o = \frac{W(f_o)\Delta f_c 4^{N-1}}{(2j)^{2N-2}} \sum_{k=o}^{2(N-1)} \frac{(2N-2)!(-1)^k}{(2N-2-k)!k!} A^k (A^*)^{2N-2-k} I_k \tag{9.20c}$$

where

$$I_k = \int_{-\infty}^{\infty} dx\, e^{-5.54x^2 - j\beta(2N-2-2k)x}$$

$$= \sqrt{\frac{\pi}{5.54}}\, e^{-(\beta^2/5.54)(N-1-k)^2} \tag{9.20c$'$}$$

Then, from (9.20a, c) and (9.20c)′, the ratio of input to output clutter power for a N-pulse canceller is

$$\left(\frac{C_i}{C_o}\right) = [(-1)^{N-1}] \Big/ \sum_{k=0}^{2(N-1)} \frac{(-1)^k(2N-2)!}{(2N-2-k)!k!} e^{-j(2\pi f_o/fr)(N-1-k)} e^{-(\pi\Delta f_c/fr)^2[(N-1-k)^2/5.54]} \tag{9.21}$$

The argument in the Gaussian functions in our analysis is expressed in terms of Δf_c, the spacing between −3 dB points on the one-way power Doppler spectrum. In some standard texts on radar other measures of spread are used. For example, Skolnick[1] uses the form $\exp(-f^2/2\sigma_c^2)$ where σ_c is the standard deviation of Doppler frequency and obtains results for the improvement factor of the two-pulse canceller [Eq. (4.23) on p. 132 of Skolnick[1]] in terms of σ_c. Nathanson et al.[4] obtain more general results for two- and three-pulse cancellers [Eqs. (9.20) on p. 402 of Nathanson et al.[4]] also in terms of σ_c. Barton[3] [Eqs. (5.3.14) and (5.3.15) on p. 245] obtains results for two- and three-pulse cancellers in terms of σ_v, the rms Doppler velocity spread. To reconcile results obtained independently here with those obtain in those references, we can use the conversion

* Equation (8.43a) applies to two-way voltage or one-way power. To use it here, we double the exponent to make it applicable to two-way power, noting that $W(f_o)$ now means two-way power at $f = f_o$. Δf_c now denotes the separation between 6 dB points on the two-way power Doppler spectrum.

$$\sigma_c = \frac{2}{\lambda}\sigma_v = \frac{\Delta f_c}{\sqrt{11.08}} \tag{9.21a$'$}$$

where σ_c and Δf_c are in Hertz and σ_v is in meters/second assuming that λ is in meters.

For $N = 2$, (9.21) with the aid of (9.21a)′ gives us

$$\left(\frac{C_i}{C_o}\right)_{N=2} = \frac{0.5}{1 - e^{-8\pi^2\sigma_v^2/(\lambda fr)^2}\cos(4\pi v_0/(\lambda fr))} \tag{9.21b$'$}$$

where v_0 is the center of the Doppler spectrum in velocity units (m/s), equal to $(\lambda/2) f_0$.

For $N = 3$

$$\left(\frac{C_i}{C_o}\right)_{N=3} = (1/6)\Big/\left[1 - \frac{4}{3}e^{-8\pi^2\sigma_v^2/(\lambda fr)^2}\cos\left(\frac{4\pi v_0}{(\lambda fr)}\right) + \frac{1}{3}e^{-32\pi^2\sigma_v^2/(\lambda fr)^2}\cos\left(\frac{8\pi v_0}{(\lambda fr)}\right)\right] \tag{9.21c$'$}$$

For $N = 4$

$$\left(\frac{C_i}{C_o}\right)_{N=4} = 0.05\Big/\left[1 - \frac{3}{2}e^{-8\pi^2\sigma_v^2/(\lambda fr)^2}\cos\left(\frac{4\pi v_0}{(\lambda fr)}\right) + \frac{3}{5}\cos\left(\frac{8\pi v_0}{(\lambda fr)}\right)e^{-32\pi^2\sigma_v^2/(\lambda fr)^2} - \frac{1}{10}\cos\left(\frac{12\pi v_0}{\lambda fr}\right)e^{-72\pi^2\sigma_v^2/(\lambda fr)^2}\right] \tag{9.21d$'$}$$

For $N = 5$

$$\left(\frac{C_i}{C_o}\right)_{N=5} = (1/70)\Big/\left[1 - 1.6e^{-8\pi^2\sigma_v^2/(\lambda fr)^2}\cos\left(\frac{4\pi v_0}{\lambda fr}\right) + 0.8e^{-32\pi^2\sigma_v^2/(\lambda fr)^2}\cos\left(\frac{8\pi v_0}{\lambda fr}\right) - 0.22857\,e^{-72\pi^2\sigma_v^2/(\lambda fr)^2}\cos\left(\frac{12\pi v_0}{\lambda fr}\right) + 0.02857e^{-128\pi^2\sigma_v^2/(\lambda fr)^2}\cos\left(\frac{16\pi v_0}{\lambda fr}\right)\right] \tag{9.21e$'$}$$

The next step in evaluating the effectiveness of the N-pulse canceller is to calculate the target signal power at input and output. As an additional feature, let us calculate the input and output noise power values.

The IF signal-plus-noise waveform is

$$\begin{aligned} v_{IF}(t) &= a\cos[(\omega_o + \omega_d)t] + [x_n(t)\cos\omega_o t + y_n(t)\sin\omega_o t] \\ &= \{a + [x_n(t)\cos\omega_d t - y_n(t)\sin\omega_d t]\}\cos[(\omega_o + \omega_d)t] \\ &\quad + \{[x_n(t)\sin\omega_d t + y_n(t)\cos\omega_d t]\}\sin[(\omega_o + \omega_d)t] \end{aligned} \tag{9.22}$$

where a is the input signal amplitude, $x_n(t)$ and $y_n(t)$ the in-phase and quadrature noise components, respectively, and ω_o is the IF. The noise components vary slowly compared to $\cos \omega_o t$ and $\sin \omega_o t$. To simplify the analysis without significant loss of generality, the signal phase is set to zero.

To convert $v_{IF}(t)$ to a bipolar video signal, we multiply by $\cos \omega_o t$ and low-pass filter (or equivalently average over many IF cycles), with the approximate result (justified by the assumption $\omega_d << \omega_o$)

$$
\begin{aligned}
v_i^{(BP)}(t) &= a[\cos(\omega_d t)\,\overline{\cos^2 \omega_o t} - \sin(\omega_d t)\,\overline{\sin \omega_o t \cos \omega_o t}] \\
&\quad + x_n(t)\,\overline{\cos^2 \omega_o t} + y_n(t)\,\overline{\sin \omega_o t \cos \omega_o t} \\
&= \frac{1}{2}[a \cos(\omega_d t) + x_n(t)] = s_i(t) + n_i(t)
\end{aligned}
\tag{9.23}
$$

Confining attention for the moment to the signal only (i.e., set $x_n = 0$), the signal power at the canceller input is

$$
\overline{[s_i(t)]^2} = \frac{1}{4}\frac{a^2}{2}[1 + \overline{\cos 2\omega_d t}] = \frac{1}{4}\frac{a^2}{2}
\tag{9.24a}
$$

Let us model the canceller as a general FIR filter with weighting coefficients $b_0, b_1, b_2, \ldots, b_{N-1}$ [see Eq. (9.11)]. The target signal output of the canceller is

$$
s_o(t) = \frac{a}{2}\sum_{m=0}^{N-1} b_m \cos[\omega_d(t - mTr)]
\tag{9.24b}
$$

where $bm = bm^{(N-1)}$

The time averaged signal power output is

$$
\begin{aligned}
\overline{s_o^2(t)} &= \frac{a^2}{4}\left\{\sum_{m=0}^{N-1}\frac{b_m^2}{2}[1 + \overline{\cos(2\omega_d t - 2\omega_d mTr)}]\right. \\
&\quad + \sum_{\substack{m,n=0 \\ m\neq n}}^{N-1}{}' \frac{b_m b_n}{2}\overline{\{\cos[2\omega_d t - (m+n)\omega_d Tr]} \\
&\quad \left. + \overline{\cos[\omega_d Tr(m - n)]\}}\right\} \\
&= \frac{a^2}{4}\left\{\sum_{m=0}^{N-1}\frac{b_m^2}{2} + \sum_{\substack{m,n=0 \\ m\neq n}}^{N-1}\frac{b_m b_n}{2}\cos[\omega_d Tr(m - n)]\right\}
\end{aligned}
\tag{9.24c}
$$

For the two-pulse canceller, $b_0 = 1$, $b_1 = -1$, $b_m = 0$ for $m > 1$, hence

$$
\overline{s_o^2(t)} = \frac{a^2}{4}\left[\left(\frac{1}{2} + \frac{1}{2}\right) - \left(\frac{1}{2} + \frac{1}{2}\right)\cos(\omega_d Tr)\right]
\tag{9.25a}
$$

The ratio of output to input signal power [from (9.25a) and (9.24a)] is

$$\left(\frac{S_o}{S_i}\right)_{N=2} = \frac{\overline{s_o^2(t)}}{\overline{s_i^2(t)}} = \frac{(a^2/4)}{(a^2/8)}[1 - \cos(\omega_d Tr)] = 2[1 - \cos(\omega_d Tr)] \tag{9.25b}$$

For the three-pulse canceller, $b_0 = 1$, $b_1 = -2$, $b_2 = 1$, $b_m = 0$ for $m > 2$; then from (9.24c) and (9.24a)

$$\left(\frac{S_o}{S_i}\right)_{N=3} = \frac{(a^2/4)}{(a^2/8)}\left[\frac{1}{2}(1 + 4 + 1) + (b_0 b_1 + b_1 b_2)\cos(\omega_d Tr)\right.$$

$$\left. + (b_0 b_2)\cos(2\omega_d Tr)\right] = 2(3 - 4\cos\omega_d Tr + \cos 2\omega_d Tr)$$

$$= 6 - 8\cos\omega_d Tr + 2\cos 2\omega_d Tr \tag{9.25c}$$

For the four-pulse canceller, where $b_0 = 1$, $b_1 = -3$, $b_2 = 3$, $b_3 = -1$,

$$\left(\frac{S_o}{S_i}\right)_{N=4} = 2\left[\frac{1}{2}(1 + 9 + 9 + 1) + (b_0 b_1 + b_1 b_2 + b_2 b_3)\cos(\omega_d Tr)\right.$$

$$\left. + (b_0 b_2 + b_1 b_3)\cos(2\omega_d Tr) + (b_0 b_3)\cos(3\omega_d Tr)\right]$$

$$= 20 - 30\cos(\omega_d Tr) + 12\cos(2\omega_d Tr) - 2\cos(3\omega_d Tr) \tag{9.25d}$$

For $N = 5$, $b_0 = 1$, $b_1 = -4$, $b_2 = 6$, $b_3 = -4$, $b_4 = 1$; then

$$\left(\frac{S_o}{S_i}\right)_{N=5} = 2\left[\frac{1}{2}(1 + 16 + 36 + 16 + 1)\right.$$

$$+ (b_0 b_1 + b_1 b_2 + b_2 b_3 + b_3 b_4)\cos(\omega_d Tr)$$

$$+ (b_0 b_2 + b_1 b_3 + b_2 b_4)\cos(2\omega_d Tr)$$

$$\left. + (b_0 b_3 + b_1 b_4)\cos 3\omega_d Tr + (b_0 b_4)\cos 4\omega_d Tr\right]$$

$$= 70 - 112\cos(\omega_d Tr) + 56\cos(2\omega_d Tr) - 16\cos(3\omega_d Tr)$$

$$+ 2\cos(4\omega_d Tr) \tag{9.25e}$$

Before proceeding, we note that a meaningful answer requires that we average over the probability distribution of target doppler shifts. To simplify that process, we can again use a Gaussian function, this time to model the a priori PDF of target Doppler velocity, i.e.,

$$p(v) = \frac{1}{\sqrt{2\pi\eta_v^2}}e^{-(v-\langle v\rangle)^2/2\eta_v^2} \tag{9.26}$$

where $\eta_v = \langle(v - \langle v\rangle)^2\rangle = \langle v^2\rangle - \langle v\rangle^2$.

Since $\omega_d = 4\pi v/\lambda$, it follows from (9.26) that

$$\langle\cos(n\omega_d Tr)\rangle = \frac{1}{2\sqrt{2\pi\eta_v^2}}\int_{-\infty}^{\infty} dv\, e^{-(v-\langle v\rangle)^2/2\eta_v^2}\left(e^{j4\pi nv/\lambda fr} + e^{-j4\pi nv/\lambda fr}\right)$$

$$= \cos\left(\frac{4\pi n\langle v\rangle}{\lambda fr}\right)e^{-8\pi^2 n^2\eta_v^2/(\lambda fr)^2} \tag{9.27}$$

Based on (9.27) and (9.25b,c,d,e), we have the output-to-input signal power ratios averaged over the distribution of Dopplers for lower-order pulse cancellers, as follows:

$$\left[\left(\frac{S_o}{S_i}\right)_{N=2}\right]_{av} = 2\left[1 - \cos\left(\frac{4\pi\langle v\rangle}{\lambda fr}\right)e^{-8\pi^2\eta_v^2/(\lambda fr)^2}\right] \tag{9.28a}$$

$$\left[\left(\frac{S_o}{S_i}\right)_{N=3}\right]_{av} = 6\left[1 - \frac{4}{3}\cos\left(\frac{4\pi\langle v\rangle}{\lambda fr}\right)e^{-8\pi^2\eta_v^2/(\lambda fr)^2}\right.$$
$$\left. + \frac{1}{3}\cos\left(\frac{8\pi\langle v\rangle}{\lambda fr}\right)e^{-32\pi^2\eta_v^2/(\lambda fr)^2}\right] \tag{9.28b}$$

$$\left[\left(\frac{S_o}{S_i}\right)_{N=4}\right]_{av} = 20\left[1 - \frac{3}{2}\cos\left(\frac{4\pi\langle v\rangle}{\lambda fr}\right)e^{-8\pi^2\eta_v^2/(\lambda fr)^2}\right.$$
$$\left. + \frac{3}{5}\cos\left(\frac{8\pi\langle v\rangle}{\lambda fr}\right)e^{-32\pi^2\eta_v^2/(\lambda fr)^2} - \frac{1}{10}\cos\left(\frac{12\pi\langle v\rangle}{\lambda fr}\right)e^{-72\pi^2\eta_v^2/(\lambda fr)^2}\right] \tag{9.28c}$$

$$\left[\left(\frac{S_o}{S_i}\right)_{N=5}\right]_{av} = 70\left[1 - 1.6\cos\left(\frac{4\pi\langle v\rangle}{\lambda fr}\right)e^{-8\pi^2\eta_v^2/(\lambda fr)^2}\right.$$
$$+ 0.8\cos\left(\frac{8\pi\langle v\rangle}{\lambda fr}\right)e^{-32\pi^2\eta_v^2/(\lambda fr)^2}$$
$$- 0.22857\cos\left(\frac{12\pi\langle v\rangle}{\lambda fr}\right)e^{-72\pi^2\eta_v^2/(\lambda fr)^2}$$
$$\left. + 0.02857\cos\left(\frac{16\pi\langle v\rangle}{\lambda fr}\right)e^{-128\pi^2\eta_v^2/(\lambda fr)^2}\right] \tag{9.28d}$$

For both ground-based and airborne radars and moving targets, the cosine terms in (9.28a–d) are generally nonzero, because there is an average Doppler velocity $\langle v\rangle$ that depends on the average relative speed of target and radar. The parameter $\eta_v = \langle(v - \langle v\rangle)^2\rangle$ is a measure of the maximum deviation from $\langle v\rangle$ that one can expect. The issue of whether or not the terms beyond the first in (9.28a–d) can be neglected, or how many of them can be neglected, depends on the ratio of η_v to (λfr), i.e., if the maximum expected deviation from the average doppler $\langle v\rangle$ is large compared with the PRF, then all of these exponents are negligibly small. If not, they may contribute significantly to a reduction in (S_o/S_i) resulting from inclusion of terms beyond the first.

To calculate the ratio of output to input noise, we return to (9.23) and set $a = 0$, resulting in

$$[v_i^{(BP)}(t)]_{a=0} = \frac{x_n(t)}{2} \tag{9.29a}$$

The counterpart of (9.24a) for noise alone is

$$\overline{\langle[n_i(t)]^2\rangle} = \frac{1}{4}\langle x_n^2(t)\rangle = \frac{1}{4}\langle x_n^2\rangle \tag{9.29b}$$

The counterpart of (9.24b) for this case is

$$n_o(t) = \sum_{m=0}^{N-1} b_m x_n(t - mTr) \tag{9.29c}$$

The mean output noise power is

$$\langle n_o^2(t)\rangle = \frac{1}{4}\left\{\sum_{m=0}^{N-1} b_m^2\langle x_n^2\rangle + \sum_{\substack{l,m=0 \\ l\neq m}}^{N-1}{}' \, b_l b_m \langle x_n(t - lTr)x_n(t - mTr)\rangle\right\} \tag{9.29d}$$

since $\langle x_n^2(t - mTr)\rangle = \langle x_n^2\rangle$ = independent of time.

Since we can always assume that noise correlation time is much lower than the pulse repetition interval, it is always true that

$$\begin{aligned}\langle x_n(t - lTr)x_n(t - mTr)\rangle &= \langle x_n(t)x_n[t - (m - l)Tr]\rangle \\ &= 0 \qquad \text{if } m \neq l\end{aligned} \tag{9.29e}$$

from which it follows that the cross-terms in (9.29d) vanish and

$$\langle n_o^2(t)\rangle = \frac{\langle x_n^2\rangle}{4}\sum_{m=0}^{N-1} b_m^2 \tag{9.29f}$$

The noise-out-to-noise-in ratio for the N-pulse canceller (from 9.29b,f) is

$$\left(\frac{N_o}{N_i}\right)_N = \sum_{m=0}^{N-1} b_m^2 \tag{9.29g}$$

From the same arguments as in (9.25a,b,c,d,e), we obtain

$$\left(\frac{N_o}{N_i}\right)_{N=2} = 1 + 1 = 2 \tag{9.30a}$$

$$\left(\frac{N_o}{N_i}\right)_{N=3} = 1 + 4 + 1 = 6 \tag{9.30b}$$

$$\left(\frac{N_o}{N_i}\right)_{N=4} = 1 + 9 + 9 + 1 = 20 \tag{9.30c}$$

$$\left(\frac{N_o}{N_i}\right)_{N=5} = 1 + 16 + 36 + 16 + 1 = 70 \tag{9.30d}$$

Comparing (9.30a–d) with (9.28a–d), we note that if all terms beyond the first can be neglected in (9.28a–d), then (S_o/S_i) and (N_o/N_i) are identical for all values of N. This means that the N-pulse canceller increases the signal and the noise in equal amounts if the maximum target doppler *relative to its mean value* (i.e., the parameter η_v) can be very large compared to λfr.

The ultimate in performance evaluation of the N-pulse canceller is the determination of signal-to-interference ratio (SIR) as defined by (8.25a, b, or c) at the output and input of the canceller. Using the basic definition as given in (9.25a), the ratio of SIR at the output to that at the input (which we choose to call the "general improvement factor") is

$$\begin{aligned} I_{FG} &= \frac{(\mathrm{SIR})_o}{(\mathrm{SIR})_i} = \frac{[(P_{Rt})_o/(P_{Rt})_i][(P_{Rc})_i + (P_n)_i]}{[(P_{Rc})_o + (P_n)_o]} \\ &= \frac{[(P_{Rt})_o/(P_{Rt})_i]\{1 + [1/[(P_{Rc})_i/(P_n)_i]]\}}{[(P_{Rc})_o/(P_{Rc})_i]\{1 + 1/[(P_{Rc})_o/(P_n)_o]\}} \end{aligned} \tag{9.31}$$

where P_{Rt}, P_{Rc}, and P_n are, respectively, the target signal power, clutter power, and noise power and where subscripts i and o denote input and output, respectively.

Rewriting (9.31) in the notation we have been using in this chapter, where

$$\frac{(P_{Rt})_o}{(P_{Rt})_i} \rightarrow \left(\frac{S_o}{S_i}\right)$$

$$\frac{(P_n)_o}{(P_n)_i} \rightarrow \left(\frac{N_o}{N_i}\right)$$

$$\frac{(P_{Rc})_o}{(P_{Rc})_i} \rightarrow \left(\frac{C_o}{C_i}\right)$$

we have

$$I_{FG} = \frac{(\mathrm{SIR})_o}{(\mathrm{SIR})_i} = \frac{(S_o/S_i)[1 + (N_i/C_i)]}{(C_o/C_i)[1 + (N_o/C_o)]} = I_F \frac{[1 + (N_i/N_o)(N_o/C_o)(C_o/C_i)]}{[1 + (N_i/C_i)(N_o/N_i)(C_i/C_o)]} \tag{9.31'}$$

The values of (C_o/C_i) are obtained from (9.21) for general N and from (9.21b,c,d,e)′ for $N = 2, 3, 4$, and 5. Those of (S_o/S_i) could be obtained for arbitrary N by ensemble averaging $[\cos \omega_d\, Tr(m - n)]$ in (9.24c) over the target Doppler distribution (9.26) For $N = 2, 3, 4$, and 5, (S_o/S_i) can be obtained from (9.28a,b,c,d). The values of (N_o/N_i) are obtained from (9.29g) for arbitrary N and from (9.30a,b,c,d) for $N = 2, 3, 4$, and 5. In using (9.25a–e), (9.28a,b,c,d), (9.21), (9.21a–e)′ or (9.29g) the values of b_m are those of the binomial coefficients with alternating signs [see Eq. (9.15a)], i.e.,

$$b_m = b_m^{(N-1)} = \frac{(-1)^m(N-1)!}{m!(N-1-m)!} \tag{9.31)''$$

Summarizing the results for the N-pulse canceller, we write (9.31)′ in the form

$$I_{FG} = I_F \frac{[1 + (1/\rho_{ci})]}{[1 + (L_N \rho_{cdN}/\rho_{ci})]} \tag{9.32}$$

where

$$L_N = \frac{N_o}{N_i}, \qquad \rho_{ci} = \frac{C_i}{N_i} = \text{input CNR}$$

$$\rho_{cdN} = \frac{C_i}{C_o}$$

and where [from (9.29g) and (9.30a,b,c,d)]

$$L_N = \sum_{m=0}^{N-1} b_m^2 \tag{9.32a)'$$

$$L_2 = 2, \qquad L_3 = 6, \qquad L_4 = 20, \qquad L_5 = 70 \tag{9.32b)'$$

from (9.21) and (9.21c) with the aid of (9.31)″

$$\rho_{cdN} = \frac{(-1)^{N-1}}{\sum_{k=0}^{2(N-1)} b_k^{(2N-2)} e^{-j\xi(N-1-k)-\zeta^2(N-1-k)^2}} \tag{9.32c)'$$

where

$$\xi = \frac{4\pi v_o}{(\lambda fr)}, \qquad \zeta = \frac{2\pi\sqrt{2}\sigma_v}{(\lambda fr)}, \qquad b_k^{(2N-2)} = \frac{(-1)^k(2N-2)!}{(2N-2-k)!k!}$$

$$\rho_{cd2} = \frac{0.5}{1 - \cos\xi e^{-\zeta^2}}$$

$$\rho_{cd3} = \frac{(1/6)}{1 - (4/3)\cos\xi e^{-\zeta^2} + (1/3)\cos 2\xi e^{-4\zeta^2}} \tag{9.32d)'$$

$$\rho_{cd4} = \frac{0.05}{1 - 1.5\cos\xi e^{-\zeta^2} + 0.6\cos 2\xi e^{-4\zeta^2} - 0.1\cos 3\xi e^{-9\zeta^2}}$$

$$\rho_{cd5} = \frac{(1/70)}{1 - 1.6\cos\xi e^{-\zeta^2} + 0.8\cos 2\xi e^{-4\zeta^2} - 0.22857\cos 3\xi e^{-9\zeta^2} + 0.02857\cos 4\xi^{-16\zeta^2}}$$

from (9.20a) and (9.21a)′

$$\rho_{ci} = \frac{W(f_o)}{\langle n_i^2 \rangle} 2\sqrt{2\pi}\,\frac{\sigma_v}{\lambda} \tag{9.32e$'$}$$

and from (9.18b), (9.21), (9.21a–e)$'$, (9.24a,c), (9.27), and (9.28a,b,c,d), with the aid of (9.32c,d)$'$

$$I_F = (I_F)_N = (-1)^{N-1}$$

$$\times \frac{\left[\sum_{m=0}^{N-1} [(b_m^{(N-1)})^2/2] + \sum_{\substack{m,n=0\\ m\neq n}}^{N-1} [b_m^{(N-1)} b_n^{(N-1)}/2] \cos[(m-n)\xi_s] e^{-(m-n)^2\zeta_s^2}\right]}{\sum_{k=0}^{2(N-1)} b_k^{(2N-2)} e^{-j\xi(N-1-k)-\zeta^2(N-1-k)^2}} \tag{9.32f$'$}$$

where

$$\xi_s = \frac{4\pi\langle v\rangle}{(\lambda fr)} \qquad \zeta_s = \frac{2\pi\sqrt{2}\eta_v}{(\lambda fr)}, \qquad b_k^{(N)} = \frac{(-1)^k N!}{(N-k)!k!}$$

and

$$(I_F)_{N=2} = \frac{(1 - \cos\xi_s\, e^{-\zeta_s^2})}{(1 - \cos\xi\, e^{-\zeta^2})}$$

$$(I_F)_{N=3} = \frac{(1 - 1.333\cos\xi_s\, e^{-\zeta_s^2} + 0.333\cos 2\xi_s\, e^{-4\zeta_s^2})}{(1 - 1.333\cos\xi\, e^{-\zeta^2} + 0.333\cos 2\xi\, e^{-4\zeta^2})}$$

$$(I_F)_{N=4} = \frac{(1 - 1.5\cos\xi_s\, e^{-\zeta_s^2} + 0.6\cos 2\xi_s\, e^{-4\zeta_s^2} - 0.1\cos 3\xi_s\, e^{-9\zeta_s^2})}{(1 - 1.5\cos\xi\, e^{-\zeta^2} + 0.6\cos 2\xi\, e^{-4\zeta^2} - 0.1\cos 3\xi\, e^{-9\zeta^2})}$$

$$(I_F)_{N=5} = \frac{\begin{aligned}(1 &- 1.6\cos\xi_s\, e^{-\zeta_s^2} + 0.8\cos 2\xi_s\, e^{-4\zeta^2}\\ &- 0.22857\cos 3\xi_s\, e^{-9\zeta_s^2} + 0.02857\cos 4\xi_s\, e^{-16\zeta_s^2})\end{aligned}}{\begin{aligned}(1 &- 1.6\cos\xi\, e^{-\zeta^2} + 0.8\cos 2\xi\, e^{-4\zeta^2}\\ &- 0.22857\cos 3\xi\, e^{-9\zeta^2} + 0.02857\cos 4\xi\, e^{-16\zeta^2})\end{aligned}} \tag{9.32g$'$}$$

Equations (9.32), (9.32a)$'$, (9.32c)$'$, and (9.32f)$'$, which give the results for an arbitrary value of N, leave the FIR filter coefficients $b_k{}^{(N)}$ unspecified and hence are applicable to an arbitrary FIR filter, not only to the N-pulse canceller. Equations (9.32b)$'$, (9.32d)$'$, and (9.32g)$'$ for $N = 2, 3, 4,$ and 5 apply only to the N-pulse canceller.

Consider the special case where the target Doppler spread is extremely large, i.e.,

$$\zeta_s >> 1 \tag{9.33a}$$

which implies that

$$e^{-\zeta_s^2} << 1 \tag{9.33b}$$

which in turn implies that the terms beyond the first in numerators of all expressions in (9.32g)$'$ are negligible and we can rewrite (9.32g)$'$ in the generic form

$$(I_F)_N = \frac{1}{1 + c_1^{(N)}\cos\xi\, e^{-\zeta^2} + c_2^{(N)}\cos 2\xi\, e^{-4\zeta^2} + \cdots + c_{N-1}^{(N)}\cos[(N-1)\xi]e^{-[(N-1)\zeta]^2}} \tag{9.33c}$$

where $c_1^{(2)} = -1, c_1^{(3)} = -1.333, c_1^{(4)} = -1.5, c_1^{(5)} = -1.6$, and the remaining coefficients $c_k^{(N)}$ for $N \leq 5$ can be inferred from (9.32g)′.

Another way to obtain the result (9.33c) is to evaluate canceller performance by averaging I_F in (9.32g)′ over all possible mean target Doppler speeds $\langle v \rangle$, assuming all speeds between $-V$ and $+V$ are equally probable. Then

$$\langle \cos n\xi_s \rangle = \frac{1}{V}\int_{-V}^{V} d\langle v \rangle \cos n\xi_s = 2\,\mathrm{sinc}\left(\frac{4\pi nV}{\lambda fr}\right) \tag{9.33d}$$

which becomes negligibly small for large argument.

Another possible assumption is that of a Gaussian distribution of Doppler speeds around zero, which is in effect what we have done here. In this case $\langle v \rangle = \xi_s = 0$ in the numerators of Eqs. (9.32g)′. In either case (uniform or Gaussian PDF of target Doppler speeds assuming a large spread), the approximate form (9.33c) for the improvement factor follows.

The form (9.33c) is found for low orders of pulse canceller in a number of standard references (e.g., Nathanson et al.,[4] pp. 402–403). The assumption behind it, as indicated above, is that I_F is being averaged over a range of target Doppler speeds. The general result (9.32g)′ applies to a specific mean target Doppler speed and a specific rms spread around that mean.

Another limiting case is that for which

$$\zeta << 1, \qquad \zeta_s << 1, \qquad \xi = 0 \tag{9.34a}$$

The third condition in (9.34a) is equivalent to setting v_0, the mean Doppler velocity, to zero. However, we are keeping the nonzero value of ξ_s intact, reflecting the case where the target is moving along the LOS while the clutter spectrum is centered at zero Doppler. That case is ideal for examining the effectiveness of MTI, which was conceived for the purpose of separating a moving target from essentially fixed clutter. The first two conditions in (9.34a) are tantamount to assuming that σ_v and η_v, the rms Doppler velocity bandwidths of clutter and target signal, respectively, are small compared with $(\lambda fr)/2\pi\sqrt{2}$.

If the conditions (9.34a) hold, then the exponentials in (9.32c,d,f,g)′ can all be approximated by the first two terms of their power series and the factors $\cos n\xi$ are all equal to unity. Equations (9.32g)′ (excluding the $N = 5$ case) are approximated by

$$(I_F)_{N=2} = \frac{1}{\zeta^2}\left[(1 - \cos\xi_s) + \zeta_s^2(\cos\xi_s)\right] \rightarrow \left(\frac{\eta_v}{\sigma_v}\right)^2 \text{ as } \langle v \rangle \rightarrow 0 \tag{9.34b}$$

$$(I_F)_{N=3} = \frac{1}{2\zeta^4}\left\{\left(1 - \frac{4}{3}\cos\xi_s + \frac{1}{3}\cos 2\xi_s\right) + \zeta_s^2\left[\frac{4}{3}(\cos\xi_s - \cos 2\xi_s)\right]\right.$$
$$\left. + \zeta_s^4\left[\frac{2}{3}(4\cos 2\xi_s - \cos\xi_s)\right]\right\} \rightarrow \left(\frac{\eta_v}{\sigma_v}\right)^4 \text{ as } \langle v \rangle \rightarrow 0 \tag{9.34c}$$

$$(I_F)_{N=4} = \frac{1}{6\zeta^6}\left[\left(1 - \frac{3}{2}\cos\xi_s + \frac{3}{5}\cos 2\xi_s - \frac{1}{10}\cos 3\xi_s\right)\right.$$

$$+ \zeta_s^2\left(\frac{3}{2}\cos\xi_s - \frac{12}{5}\cos 2\xi_s + \frac{9}{10}\cos 3\xi_s\right)$$

$$+ \zeta_s^4\left(-\frac{3}{4}\cos\xi_s + \frac{24}{5}\cos 2\xi_s - \frac{81}{20}\cos 3\xi_s\right)$$

$$+ \zeta_s^6\left(\frac{1}{4}\cos\xi_s - \frac{32}{5}\cos 2\xi_s + \frac{243}{20}\cos 3\xi_s\right)\Bigg] \to \left(\frac{\eta_v}{\sigma_v}\right)^6 \text{ as } \langle v\rangle \to 0 \qquad (9.34\text{d})$$

The results (9.34b–d) in the limiting case of zero target Doppler velocity tell us that the pulse canceller provides no improvement in clutter suppression if the rms spread of the target and clutter Doppler spectra are the same, degrades clutter suppression if the target spread is much lower than the clutter spread, and provides improvement proportional to $2(N-1)$ power of the ratio of target spread to clutter spread if the former exceeds the latter. This is expected, since a large spread of target LOS velocities around zero when the clutter has a very small spread around zero must provide better detection of target motion in a background of stationary clutter.

The families of curves on Figure 9.6a, b, and c show the improvement factor I_F for the N-pulse canceller in dB as given by Eqs. (9.32g)′ for $N = 2,3$, and 4, respectively, under assumptions that (1) the center of the Doppler spectrum of clutter is zero (i.e., $v_0 = 0$), and (2) the rms Doppler spread of the target is small, i.e., $\eta_v << \lambda fr/2\pi\sqrt{2}$. The remaining parameters are σ_v, the rms Doppler spread of clutter, and $\langle v\rangle$, the mean Doppler speed of the target. The curves are presented for a few variations of σ_v and $\langle v\rangle$ to demonstrate the increase of I_F with an increase in mean target Doppler speed and a decrease in clutter Doppler spread. As the Doppler speed of the target increases beyond the limit of the clutter Doppler spectrum, the target signal becomes more distinguishable from the clutter signal. That effect is evident from the curves of Figure 9.6a, b, and c, which show higher values of I_F as $\langle v\rangle$ increases and σ_v decreases.

9.4 MTI, CW-DOPPLER, AND PULSED-DOPPLER RADARS IN GENERAL

The N-pulse canceller, which was discussed at great length in Section 9.3, is only one category of coherent radar processing scheme that attempts to separate target and clutter signals on the basis of Doppler information. In fact, it is not the only possible way to implement an MTI radar, defined in general as a radar that uses Doppler information to separate moving target returns from stationary target returns, the latter assumed to be clutter. There are other filtering schemes that could be used to do this, since the canceller is not necessarily the optimum such processor.[13,14]

A more general concept than the MTI is that of the pulsed Doppler radar (sometimes abbreviated as PD). In general, a PD is a pulsed radar whose receiver processes the incoming signals coherently and extracts Doppler information from them together with the range information inherent in pulsing. A Doppler radar that does not use pulsing but merely extracts the Doppler shift of a continuous-wave (CW) return signal is a "CW-Doppler radar" or "Doppler radar." A CW-doppler radar can acquire range information along a trajectory by integrating the range rate, to which the Doppler shift is proportional, if range is known for at least one instant

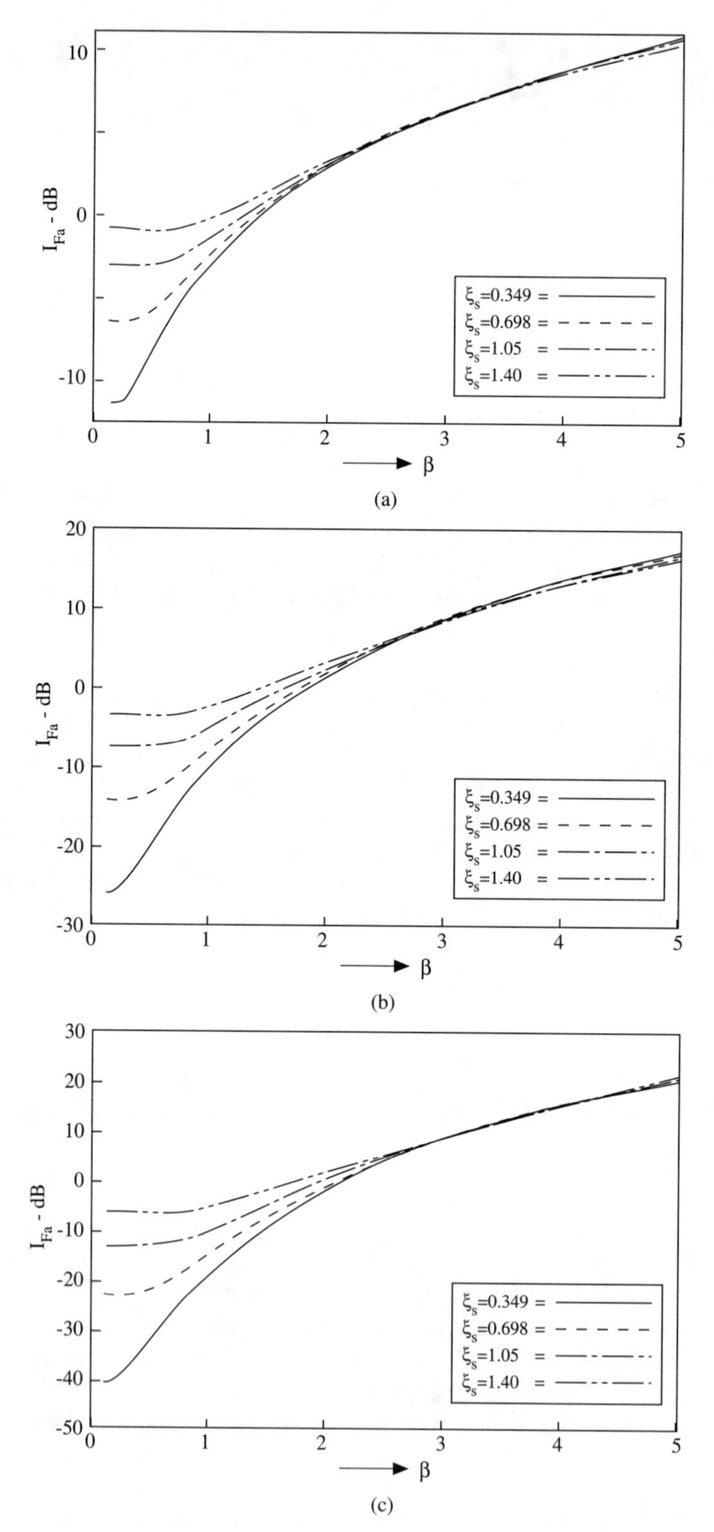

FIGURE 9.6
Improvement factors for N-pulse cancellers. (a) $N = 2$. (b) $N = 3$. (c) $N = 4$. $\beta = \sqrt{2}\langle v \rangle / \sigma_v$; $\xi_s = 4\pi\langle v \rangle / (\lambda f r)$.

of time. It can also use frequency modulation to acquire range information, in which case it is called an FM-CW-doppler radar.

The MTI is really a special class of pulsed-doppler radar, but in radar literature the two designations PD and MTI are sometimes used as if these concepts were distinct. There are different criteria for deciding when to call the radar an MTI and when to call it a PD. One distinction (the most logical one in the author's opinion) can be made by first saying that in general the PD radar separates the returns within a given range gate according to their Doppler frequencies, leaving open the question of how returns in different Doppler regions are classified (e.g., desired target or clutter). A PD becomes an MTI only when the processing of the Doppler information specifically separates moving from stationary targets within a given range gate. In Section 9.3, we tacitly generalized the canceller type of MTI by accounting for a clutter spectrum not necessarily centered at zero Doppler, thereby accounting for a moving platform, i.e., an airborne or other vehicle-borne radar, or clutter with a nonzero mean Doppler velocity v_o, e.g., wind-driven ocean currents or clusters of moving discrete objects such as cars on a busy road. If v_0 is nonzero, then the MTI becomes a processing scheme to distinguish targets moving at a mean Doppler speed $\langle v \rangle$ from clutter moving at a different mean Doppler speed v_0.

9.4.1 Low, Medium, and High PRF Operation

In some sources[15] a pulsed Doppler is called an MTI if it is operated at "low PRF," meaning that it is designed to have unambiguous range and will incur Doppler ambiguities (Chapter 7, Sections 7.4, 7.5, and 7.6) implying the existence of the kind of "blind speeds" discussed in Section 9.2 and illustrated in Figures 9.3 and 9.5.

The radars specifically designated as pulsed Doppler rather than MTI radars are usually designed to minimize blind speeds at a cost of increased range ambiguities, with their attendant clutter foldover. If such a radar is designed to completely eliminate Doppler ambiguities, it is operated at "high PRF" and will incur a high level of range ambiguity problems as the price of the elimination of blind speeds. A pulsed-Doppler radar may also be operated at "medium PRF," which is a compromise between low PRF to eliminate range ambiguities and high PRF to eliminate Doppler ambiguities. Medium PRF operation usually implies the possible existence of both range and Doppler ambiguities.

To put these ideas on a more quantitative basis, we first invoke the minimum PRF that will allow range ambiguities, defined in Eq. (7.49) in Chapter 7 as the maximum PRF for unambiguous range operation. Designating that PRF as $fr1$ in (7.49), we have

$$fr1 = c/2r_{\max} \tag{9.35a}$$

where $r_{\max}$ is the maximum range of targets and clutter to be encountered in the scenario of interest.

The minimum PRF for unambiguous Doppler operation is defined in Eq. (7.56) in Chapter 7. Designating that PRF as $fr2$ in (7.56), we have

$$fr2 = 2|v_{\max}|/\lambda_0 \tag{9.35b}$$

where λ_0 is wavelength corresponding to the transmitted signal frequency and $|v_{\max}|$

is the maximum velocity of targets and clutter to be encountered in the scenario of interest.

Some values of *fr*1 and *fr*2 for a number of radar frequencies and a few realistic values of r_{max} and $|v_{max}|$ are shown in Table 9.2.

Based on (9.35a,b), the categories are

A. If $c/2r_{max} < 2\,|v_{max}|/\lambda_0$
   1. Low PRF: PRF $\leq c/2r_{max}$
   2. Medium PRF: $c/2r_{max} <$ PRF $< 2\,|v_{max}|/\lambda_0$
   3. High PRF: PRF $\geq 2\,|v_{max}|/\lambda_0$

B. If $c/2r_{max} \geq 2\,|v_{max}|/\lambda_0$
   (1) Low PRF: PRF $\leq 2\,|v_{max}|/\lambda_o$
   (2) Medium PRF: $2\,|v_{max}|/\lambda_0 <$ PRF $< c/2r_{max}$
   (3) High PRF: PRF $\geq 2r_{max}$

If multiple PRFs are used, as discussed in Section 9.1, the PRF ranges indicated above apply to the *average* PRF.

9.4.2 DFT Processing of Pulsed-Doppler Signals

In all modern radars, including those designated as MTI or pulsed Doppler, the digital processing techniques that have been improving so rapidly during the past several years are the basis for the treatment of received signals. Even delay line cancellers, which have been used since the early days of radar, are now implemented as FIR digital filters, as pointed out in Section 9.2.

TABLE 9.2

Limiting PRFs for Low, High, and Medium PRF Radars

(a) Upper limit of unambiguous range operation *fr*1 = $c/2r_{max}$ (pps)

r_{max} (km)	*fr*1 (pps)
30	5000
100	1500
300	500
750	200
1000	150
1500	100
2000	75

(b) Lower limit of unambiguous doppler operation *fr*2 = $2|v_{max}|/\lambda_o$

Frequency	$\|v_{max}\|$ = 100 mph = 44.7 m/sec	500 mph = 223.50 m/sec (1 mph = 0.447 m/sec)
300 MHz	89.4	447
1 GHz	298	1490
2 GHz	596	2980
4 GHz	1192	5960
8 GHz	2384	11920
10 GHz	2980	14900
30 GHz	8940	44700
60 GHz	17880	89400

The receiver processing for a pulsed doppler radar is illustrated in the block diagram in Figure 9.7. The incoming signal from a target is a train of uniformly spaced IF pulses, assumed to be of constant amplitude during a dwell. The pulses are coherently integrated and driven through a bank of contiguous range gates covering the range span of interest. The output of each range gate enters a bank of contiguous narrowband Doppler filters covering the span of Doppler speeds from zero to the PRF. The set of outputs of the Doppler filters for each of the range gates enters logic circuitry that sets thresholds and decides which Doppler filter outputs are probably clutter and which outputs are probably from desired targets. We note that an MTI pulse canceller automatically discards signals with Doppler speeds near zero, based on the assumption that they are clutter signals. The processing we are now describing may base its operating logic on that same assumption and discard the outputs of the filters centered near zero Doppler even if they have extremely large amplitudes, retaining only those outputs from filters with large Doppler for further processing on the assumption that they come from moving targets. When implemented this way this processing scheme acts as a moving target indicator, or MTI.

This output can be processed in many other ways as well, depending on the application. For example, it could filter out signals in Doppler bins arising from nonstationary clutter sources moving at LOS velocities different from those of targets to be detected, which might be stationary or moving. An example is detection of a ship on a rough sea surface, where signals from ocean waves could be a major source

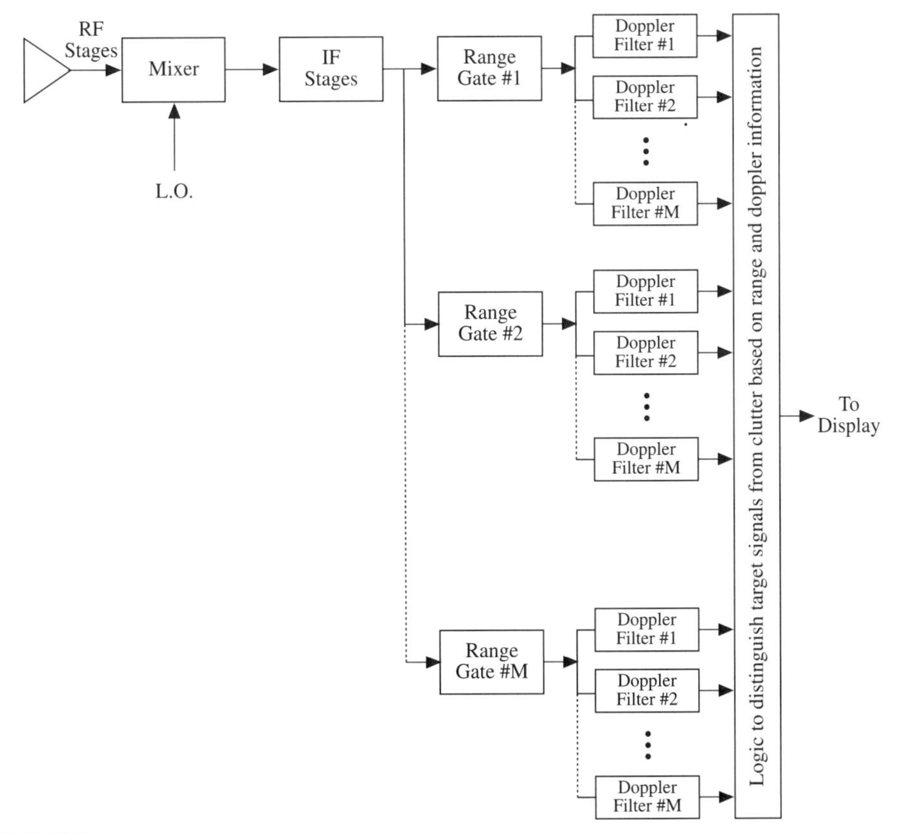

FIGURE 9.7
Receiver for pulsed-doppler radar.

of clutter in Doppler bins far removed from zero and ship speeds might differ enough from those of the waves to allow separation by Doppler. Another example is detection of an aircraft in the vicinity of a moving rainstorm, where the speed of the storm differs significantly from that of the aircraft.

We note that the coherent processing of a train of uniformly spaced IF pulses gives rise to the equivalent of a discrete Fourier transform (DFT). To show this, consider each pulse as occupying a negligibly small time interval compared to the pulse repetition period *Tr*. The pulse train coming from range r_0 is given in the time domain by

$$s(t;\omega) = \sum_{n=0}^{N-1} A_n p\left(t - \frac{2r_0}{c} - nTr\right) e^{j\omega[t-(2r_0/c)-nTr]} \tag{9.36a}$$

where A_n is the amplitude of the nth pulse, $p(t)$ is the normalized waveform (assumed the same except for the delay for all N pulses), and ω is the frequency, which includes the Doppler shift, i.e., $\omega = \omega_o + \omega_d$, where ω_o and ω_d are IFs corresponding to transmitted frequency and Doppler shift, respectively.

To coherently process the pulse train (9.36a), we match it to the reference waveform

$$s_R(t) = \sum_{m=0}^{N-1} p\left(t - \frac{2r_0}{c} - mTr\right) e^{-j\omega_0[t-(2r_0/c)-mTr]} \tag{9.36b}$$

The output of the range gate perfectly matched to the range r_0 but mismatched to the frequency in the amount $\omega_d = \omega - \omega_d$. (See Section 7.3 for a discussion of ambiguity functions for pulse trains. The output of this operation is the frequency ambiguity function of the pulse train (9.36a) with the amplitude A_n stipped off of the nth term in the expression.).

The expression (7.46f,g,h) in Section 7.3 gives us the ambiguity function of a finite pulse train such as (9.36a). Changing notation in (7.46h) from N_p to N and from Ω to ω_d and setting the delay τ to zero, reflecting the fact that the match to the range is assumed perfect, only the first set of terms at the top of (7.46h) (i.e., those corresponding to the *unambiguous* range) will survive. If we reinsert the amplitudes A_n into the output, then write Eq. (7.46h) with the indicated changes, set the summation index m in (7.46h) equal to $n + 1$, and sum on n, we obtain the output of the filter matched to the range r_o as a function of Doppler frequency:

$$s_0(\omega_d) = \hat{\Lambda}_N(0,\omega_d) = \hat{\Lambda}_1(0,\omega_d) \sum_{n=0}^{N-1} A_n e^{jn\omega_d Tr} \tag{9.36c}$$

where $\hat{\Lambda}_1(0,\omega_d)$ is the frequency ambiguity function (AF) of the single-pulse waveform $p(t)$. The single-pulse AF $\hat{\Lambda}_1(0,\omega_d)$ has a spectral spread many times wider than that of the summation in (9.36c) and can be assumed nearly constant over most of the spectral region of the summation. We consider it as a constant scale factor and denote the product $\hat{\Lambda}_1(0,\omega_d)$ by $f(n)$, which we characterize as a discrete sequence with samples at $n = 0,1,\ldots,N-1$. We rename $s_0(\omega_d)$, recognizing it as a function of ω_d and calling it $\hat{F}(\omega_d)$. Then we finally rewrite (9.36a) in the form*

* We have used the complex conjugate of $\hat{F}(\omega_d)$ to follow the convention that the exponents inside the summation on the right-hand side of (9.36e) will be $-j(2\pi nk/N)$ rather than $+j(2\pi nk/N)$.

$$\hat{F}^*(\omega_d) = \sum_{n=0}^{N-1} f(n)e^{jn\omega_d Tr} \tag{9.36d}$$

The next step is to sample $\hat{F}^*(\omega_d)$ at regularly spaced radian frequency intervals $2\pi/NTr$ and rename $\hat{F}^*(\omega_d)$ at $\omega_d = 2\pi k/NTr$, calling the kth sample $F(k)$ where k is one of the integers 0,1,2, . . . $(N - 1)$. The result is

$$F(k) = \sum_{n=0}^{N-1} f(n)e^{-j(2\pi nk/N)}; \qquad k = 0, \ldots, N \tag{9.36e}$$

The expression (9.36e) is the discrete Fourier transform of the incoming pulse train.

The implication of (9.36e) is that the operations shown in Figure 9.7 can be implemented as a computation of the DFT of the sequence of incoming pulse amplitudes at the output of the IF stage of the receiver. The digital output of each range gate is the discrete sequence $F(k)$; $k = 0,1, \ldots (N - 1)$, which contains the distribution of Doppler frequencies in the radar return from targets and clutter at range r_0, from $\omega = 0$ to $\omega = 2\pi fr(1 - 1/N)$. The number of narrowband and digital filters following each range gate is N, the same as the number of incoming pulses. If those filters were perfectly matched to the range gate output, then $F(k)$ as given by (9.36e) could be considered as proportional to the output of the particular member of the filter bank that is tuned to the frequency $f = kfr/N$.

The DFT operation described above required N^2 multiplications, but implementing it as a "fast Fourier transform," or FFT, can reduce the number to somewhere near $N \log N$. In modern radar signal processing, these computations are always executed through FFT algorithms, which are continuously being improved.

If the pulse Doppler radar is to be operated as an MTI, the sample $F(0)$ and possibly $F(1)$ and $F(2)$ can be excluded from the output of the final processing logic stage to filter out returns from stationery clutter. This is tantamount to using a bank of analog Doppler filters covering the span from $f= 0$ to $f = fr$ and excluding the output of the first few filters covering a bandwidth from $\omega_d = 0$ to $\omega_d = W$, where W is well below the Doppler frequency of a desired target. If we are interested in targets with a given narrow range of Doppler speeds and are working against clutter with a larger Doppler spread than the target but that overlaps the target's Doppler region we can set up the logic accordingly, rejecting filter outputs outside the target's Doppler spectrum to maximize the improvement factor I_F realized in the outputs of Doppler filters that contain both clutter and target signals.

We now return to the representation of the incoming pulse train given by (9.36a). If all of the pulses are equally weighted, i.e., if $A_n = 1$ for all n, then the remaining steps in the analysis leading to (9.36e) give us the expression

$$|F(k)| = \left|\sum_{n=0}^{N-1} e^{j(2\pi nk/N)}\right| = \left|\frac{\sin(\pi k)}{\sin(\pi k/N)}\right| \tag{9.36f}$$

the equivalent of which we encountered in Section 7.3 as the Fourier transform of a train of pulses with uniform spacing and amplitude given by Eq. (7.44). To show the equivalence of (9.36f) and (7.44), we note that as indicated below (9.36d), ω becomes $2\pi k/NTr$ when we convert from analog to digital Fourier transforms. Then $\omega Tr/2$ and $N_p\omega Tr/2$ in (7.44) become $\pi k/N$ and πk, respectively, in (9.36a) [where the subscript p on N_p in (7.44) has been removed in (9.36a)]. A pair of rough plots of this FT as a function of analog frequency in Hertz is shown in Figure 7.10, for $N = 5$ and $N = 10$. These plots show the peaks occurring at integral multiples of the PRF, the sidelobes around those peaks, and the periodic nature of this FT, the period being the PRF. The recurring peaks are the Doppler ambiguities as discussed in Section 7.3, and the nulls between the peaks are blind-speeds, discussed for N-pulse cancellers in Section 9.2.

The amplitudes A_n in the incoming pulse train as represented by (9.36a) were allowed to be variable with n. We were accounting for the possibility of time variation of pulse amplitude, e.g., due to the target, where the pulse amplitudes are weighted with antenna pattern values. In the usual treatment of the subject, A_n is assumed to be uniform in the incoming pulse train. If we had used the standard approach, then the A_ns would have represented weightings introduced as processing of the pulse train. Since the processing can be viewed as the action of a FIR digital filter, the A_ns are equivalent to the coefficients of the FIR filter, which we have denoted by b_n in Section 9.2 [see Eqs. (9.8), (9.9), (9.10), and Fig. 9.4].

9.4.3 General Performance Evaluation for MTI and Pulsed-Doppler Radar

Performance evaluation for N-pulse cancellers was discussed in Section 9.3. The ratio (C_i/C_o), called the "clutter attenuation" or CA, the signal output/input ratio or "signal gain" (S_o/S_i), the noise output/input ratio (N_o/N_i), and the improvement factor $I_F = (C_i/C_o)(S_o/S_i)$ for clutter-limited radars were evaluated for the general N-pulse canceller and specifically for the cases $N = 2, 3, 4,$ and 5.

The processing based on the DFT as discussed in Section 9.4 was shown to be equivalent to the use of a FIR digital filter. [See discussions from Eq. (9.8) through (9.13a) and Figure 9.4.] The N-pulse canceller was shown to be equivalent to such a FIR filter with coefficients $b_n^{(N-1)}$ equal to the binomial coefficients. The performance evaluation in Section 9.3 confined itself to that specific choice of the b_ns. Most of the formulation was general for FIR filter processing, but the analysis was quickly specialized to the binomial coefficients applicable to cancellers.

In this section we cover a similar analysis for other choices of the b_ns. The FIR filter with binomial coefficients is not even necessarily the optimum anticlutter filter for the MTI, so it is instructive to consider other possible FIR filter coefficients that might provide some advantages over pulse cancellers for MTI and other pulsed-Doppler radars.

Consider first the clutter attenuation. The expression $H(\Omega)$ in (9.15a) can be expressed in terms of $b_m^{(N-1)}$ without specifying those coefficients. Letting $b_m = b_m^{(N-1)}$ in (9.15a) and with the aid of (9.19a), we can express (9.19b) in the form

$$\left(\frac{C_o}{C_i}\right)_N = \frac{1}{C_1}\int_{-\infty}^{\infty} df\, W(f)\left(\sum_{m=0}^{N-1} b_m^2 + \sum_{\substack{m,n=0\\ m\neq n}}^{N-1}{}' \, b_m b_n e^{-j(m-n)\pi f/fr}\right)$$

$$= \sum_{m=0}^{N-1}\left(b_m^2 + 2\sum_{l=1}^{N-1-m} b_m b_{m+l} \cos l\xi e^{-l^2\zeta^2}\right) \tag{9.37a}$$

We have already used this approach on the output/input signal ratio, beginning with Eq. (9.24c), but then specified the binomial coefficients for the subsequent analysis. If we invoke (9.24a) and (9.24c), then use (9.27) to average over target doppler with the PDF given by (9.26), we obtain

$$\left(\frac{S_o}{S_i}\right)_N = \sum_{m=0}^{N-1}\left(b_m^2 + 2\sum_{l=1}^{N-1-m} b_m b_{m+l} \cos l\xi_s e^{-l\zeta_s^2}\right) \tag{9.37b}$$

If parameter regimes where the system is noise-limited as well as clutter-limited are important, we would also need the noise output/input ratio, which was also calculated in Section 9.2. The result is (9.29f), repeated below.

$$\left(\frac{N_o}{N_i}\right)_N = \sum_{m=0}^{N-1} b_m^2 \tag{9.37c}$$

The improvement factor for a clutter-limited radar, as defined in (9.18b) and obtained from (9.37a,b), is

$$(I_F)_N = \frac{\sum_{m=0}^{N-1}\left(b_m^2 + 2\sum_{l=1}^{N-1-m} b_m b_{m+l} \cos l\xi_s e^{-l^2\zeta_s^2}\right)}{\sum_{m=0}^{N-1}\left(b_m^2 + 2\sum_{l=1}^{N-1-m} b_m b_{m+l} \cos l\xi e^{-l^2\zeta^2}\right)} \tag{9.37d}$$

where

$$\xi = \frac{4\pi v_o}{(\lambda fr)}, \qquad \xi_s = \frac{4\pi\langle v\rangle}{(\lambda fr)}, \qquad \zeta = \frac{2\pi\sqrt{2}\sigma_v}{(\lambda fr)}, \qquad \zeta_s = \frac{2\pi\sqrt{2}\eta_v}{(\lambda fr)}$$

The general improvement factor that includes the possibility of noise-limited operation is obtained from (9.31)′ and is given by

$$(I_{FG})_N = (I_F)_N \frac{[1 + (N_i/C_i)]}{[1 + (N_i/C_i)(N_o/N_i)_N(C_i/C_o)_N]}$$

where $(I_F)_N$ is given by (9.37d), (N_o/N_i) by (9.37c), (C_i/C_o) by the inversion of (9.37a), and (N_i/C_i) is obtained from (9.20a) based on an assumed input noise level N_i.

Let us now consider, as alternatives to the binomial weightings characterizing the N-pulse canceller, the following weightings: (a) unweighted FIR filter

$$b_m^{(N-1)} = 1; \qquad m = 0, 1, \ldots, (N-1) \tag{9.38a}$$

(b) FIR filter having equal amplitude weightings with alternating signs

$$b_m^{(N-1)} = (-1)^m; \qquad m = 0, 1, \ldots, (N-1) \tag{9.38b}$$

It is a matter of simple algebra to determine the coefficients that replace those in the expressions (9.32g)′, which were based on binomial weightings:

The generic expression for the improvement factor, as given by (9.37d) is

$$(I_F)_N = F_N^{(s)}/F_N^{(c)} \tag{9.39}$$

where

$$F_N^{(\nu)} = \sum_{m=0}^{N-1}\left(U_m + 2\sum_{l=1}^{N-1-m} V_{ml}\cos l\xi_\nu e^{-l^2\zeta_\nu^2}\right)$$

and where

$$U_m = b_m^2$$
$$V_{lm} = b_m b_{m+l}$$
$$\nu = \text{s for numerator, c for denominator}$$

It is evident that for these two cases

$$\begin{aligned} U_m &= 1 \text{ for } m = 0, 1, \ldots, (N-1) \\ V_{lm} &= 1 \text{ in Case (a) for } m = 0, 1, \ldots, (N-1), l = 1, 2, \ldots, N-1-m \\ &\quad (-1)^l \text{ in Case (b) for the same range of values of } l \text{ and } m \end{aligned} \tag{9.39$'$}$$

From (9.39) and (9.39)′, the generic form of F_N for these two cases is

$$\begin{aligned} F_2 &= 2(1 \pm \cos\xi\, e^{-\zeta^2}) \\ F_3 &= 3(1 \pm 1.33\cos\xi\, e^{-\zeta^2} + 0.67\cos 2\xi e^{-4\zeta^2}) \\ F_4 &= 4(1 \pm 1.5\cos\xi e^{-\zeta^2} + \cos 2\xi e^{-4\zeta^2} \pm 0.5\cos 3\xi e^{-9\zeta^2}) \\ F_5 &= 5(1 \pm 1.6\cos\xi e^{-\zeta^2} + 1.2\cos 2\xi e^{-4\zeta^2} \pm 0.8\cos 3\xi e^{-9\zeta^2} + 0.4\cos 4\xi e^{-16\zeta^2}) \end{aligned} \tag{9.40}$$

where, in the ± terms, the plus sign applies to Case (a) and the minus sign to Case (b). There are no subscripts on ξ or ζ in (9.40), but it should be understood that ξ and ζ become ξ_s and ζ_s in the numerator of (9.39) and ξ_c and ζ_c in the denominator.

Consider the improvement factor averaged over all possible target Doppler speeds as given by (9.33c), where the coefficients $c_k^{(N)}$ in that particular expression are obtained from the binomial filter coefficients b_n characteristic of the N-pulse canceller. Averaging $(I_F)_N$ over target Doppler speeds for other sets of filter weighting coefficients b_n would also result in the negligibility of all terms beyond the first in $F_N^{(s)}$, the numerator of $(I_F)_N$ in (9.39). The result is (9.33c), applicable to any FIR Doppler filter, including one whose weighting coefficients are those of (9.38a) or (9.38b).

In what follows we will compare the clutter attenuation $(C_i/C_o)_N$, the signal ratio $(S_o/S_i)_N$ and the improvement factor $(I_F)_N$, which is the product of those two ratios, for (1) the N-pulse canceller (i.e., binomial weighting coefficients), (2) the unweighted FIR filter [Eq. (9.38a)], and (3) the filter whose coefficients are given by (9.38b). From (9.37b), where averaging over all target Dopper frequencies results in the negligibility of the $m \neq n$ summation,

$$(S_o/S_i)_N = \sum_{m=0}^{N-1} b_m^2 \tag{9.41a}$$

From (9.37a), we obtain the clutter attenuation (CA) for arbitrary weighting coefficients, i.e.,

$$(\text{CA})_N = (C_i/C_o)_N = \frac{1}{\left(\sum_{m=0}^{N-1} b_m^2\right)\left[1 + \sum_{l=1}^{N-1} c_l^{(N)} \cos l\xi e^{-l^2\zeta^2}\right]} \tag{9.41b}$$

and

$$(I_F)_N = (\text{CA})_N (S_o/S_i)_N \tag{9.41c}$$

where

$$c_l^{(N)} = 2 \sum_{m=0}^{N-1-l} b_m b_{m+l} \Big/ \sum_{m=0}^{N-1} b_m^2$$

Finally, the improvement factor $(I_F)_N$, the product of CA and $(S_o/S_i)_N$, as obtained from (9.41a,b), can be written in the form (9.33c), where the coefficients $c_l^{(N)}$ for the three cases under consideration here and for $N = 2,3,4,5$ are

For Case (a): binomial weightings: N-pulse canceller

$$\begin{aligned}
c_1^{(2)} &= -1 \\
c_1^{(3)} &= -1.333 \\
c_2^{(3)} &= 0.333 \\
c_1^{(4)} &= -1.5 \\
c_2^{(4)} &= 0.6 \\
c_3^{(4)} &= -0.1 \\
c_1^{(5)} &= -1.6 \\
c_2^{(5)} &= 0.8 \\
c_3^{(5)} &= -0.22857 \\
c_4^{(5)} &= 0.02857
\end{aligned} \tag{9.42a}$$

For Case (b) ($b_m = 1$) and Case (c) [$b_m = (-1)^m$]

$$
\begin{aligned}
c_1^{(2)} &= \pm 1 \\
c_1^{(3)} &= \pm 1.33 \\
c_2^{(3)} &= 0.67 \\
c_1^{(4)} &= \pm 1.5 \\
c_2^{(4)} &= 1 \\
c_3^{(4)} &= \pm 0.5 \\
c_1^{(5)} &= \pm 1.6 \\
c_2^{(5)} &= 1.2 \\
c_3^{(5)} &= \pm 0.8 \\
c_4^{(5)} &= 0.4
\end{aligned}
\tag{9.42b}
$$

In the case where the rms Doppler spread of the clutter is sufficiently small to justify neglect of all but the first few terms of the power series for $e^{-l^2\zeta^2}$ in (9.33c), we can invoke an approximation that is particularly convenient for the case of stationary clutter, primarily applicable to a ground-based radar. In this case, for the N-pulse canceller, the terms in the denominator of (9.33c) that do not involve powers of ζ cancel out, as do those involving the lowest powers of ζ^2, leaving only one surviving term always proportional to $1/\zeta^{2N}$. We can see this by examining (9.42a) and noting that

$$
\begin{aligned}
1 + c_1^{(2)} &= 0 \\
1 + c_1^{(3)} + c_2^{(3)} &= 0 \\
1 + c_1^{(4)} + c_2^{(4)} + c_3^{(4)} &= 0 \\
1 + c_1^{(5)} + c_2^{(5)} + c_3^{(5)} + c_4^{(5)} &= 0
\end{aligned}
\tag{9.43a}
$$

It follows through (9.32g)′ and (9.42a) that if $\xi = 0$, $\zeta << 1$, and $\zeta_s >> 1$,

$$
\begin{aligned}
(I_F)_2 &\approx \frac{1}{1 + c_1^{(2)}(1 - \zeta^2)} = \frac{1}{1!\zeta^2} \\
(I_F)_3 &\approx \frac{1}{1 + c_1^{(3)}[1 - \zeta^2 + (\zeta^4/2!)] + c_2^{(3)}[1 - 4\zeta^2 + (16\zeta^4/2!)]} \\
&= \zeta^2(-c_1^{(3)} - 4c_2^{(3)}) + \zeta^4\left(\frac{c_1^{(3)}}{2} + 8c_2^{(3)}\right) = \frac{1}{2!\zeta^4}
\end{aligned}
\tag{9.43b}
$$

$$
\begin{aligned}
(I_F)_4 \approx \frac{1}{1 + c_1^{(4)}[1 - \zeta^2 + (\zeta^4/2!) - (\zeta^6/3!)] + c_2^{(4)}[1 - 4\zeta^2 + (16\zeta^4/2!) - (64\zeta^6/3!)]} \\
+ c_3^{(4)}\left(1 - 9\zeta^2 + \frac{81\zeta^4}{2!} - \frac{729\zeta^6}{3!}\right)
\end{aligned}
$$

$$= \frac{1}{0 + (0)\zeta^2 + (0)\zeta^4 + 3!\zeta^6} = \frac{1}{3!\zeta^6}$$

$$(I_F)_5 \approx \frac{1}{4!\zeta^8}$$

We can infer from (9.43b) that for the N-pulse canceller

$$(I_F)_N \approx \frac{1}{(N-1)!}\left(\frac{\lambda fr}{2\pi\sqrt{2}\sigma_v}\right)^{2(N-1)} \qquad \text{if } v_o = 0 \tag{9.43c}$$

The implication of (9.43c) is that the improvement in performance against stationary clutter due to the N-pulse canceller is inversely proportional to the $2(N-1)$ power of the clutter Doppler spread, i.e., increasing the number of stages of the canceller enhances the clutter suppression significantly. Expressing (9.43c) in dB, we have

$$[(I_F)_2]_{dB} = 20 \log_{10} (\lambda fr/2\pi\sqrt{2}) - 20 \log_{10} \sigma_v \tag{9.43d}$$

and to show the improvement over the two-pulse canceller attained by increasing the number of stages, we have

$$\begin{aligned}[(I_F)_N/(I_F)_2]_{dB} = &-10 \log_{10} [(N-1)!] \\ &+20(N-2) \log_{10} (\lambda fr/2\pi\sqrt{2}) - 20(N-2) \log_{10} \sigma_v\end{aligned} \tag{9.43e}$$

The curves on Figure 9.8a show the improvement factor in dB vs. σ_v for $N = 2, 3, 4$, and 5, demonstrating that the decrease in improvement with N due to the factorial term in (9.43e) is offset by the increase with N due to the remaining terms, resulting in a net enhancement as N increases.

It is instructive to consider the improvement against stationary clutter attainable with the other two FIR filters we have considered here. Analysis analogous to that performed above [i.e., Eqs. (9.43a,b,c,d,e) using the weighting coefficients in (9.42b)] leads to the following results for $N = 2, 3, 4$, and 5.

For uniform weightings

$$\begin{aligned}
(I_F)_2 &\approx \frac{1}{2[1-(\zeta^2/2)]} \approx \frac{1}{2} + \frac{\zeta^2}{4} + O(\zeta^4) \\
(I_F)_3 &\approx \frac{1}{3[1-(4\zeta^2/3)]} \approx \frac{1}{3} + \frac{4\zeta^2}{9} + O(\zeta^4) \\
(I_F)_4 &\approx \frac{1}{[1-2.5\zeta^2]} \approx \frac{1}{4} + \frac{\zeta^2}{1.6} + O(\zeta^4) \\
(I_F)_5 &\approx \frac{1}{5[1-4\zeta^2]} \approx \frac{1}{5} + 0.8\zeta^2 + O(\zeta^4)
\end{aligned} \tag{9.44a}$$

For uniform amplitude-alternating sign weightings: where $\zeta \ll 1$

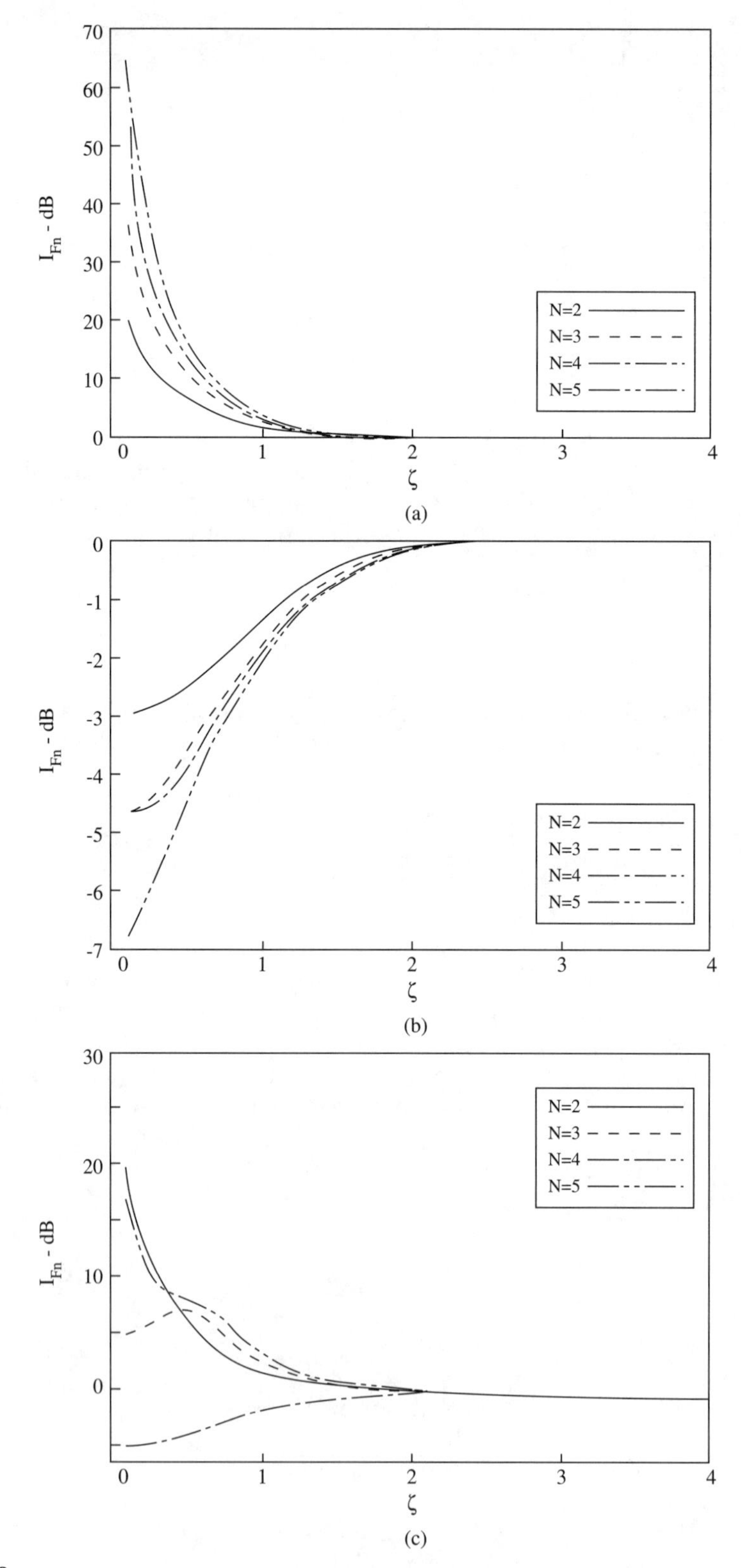

FIGURE 9.8
Improvement factor for moving targets in stationary clutter due to FIR Doppler filters. (a) Binomial weightings (*N*-pulse canceller). (b) Uniform weightings. (c) Uniform amplitude-alternating sign weightings. $\zeta = 2\pi\sqrt{2}\sigma_v/(\lambda fr)$.

$$(I_F)_2 \approx \frac{1}{\zeta^2}$$

$$(I_F)_3 \approx \frac{3}{1 - 4\zeta^2 + \cdots} \approx 3 + 12\zeta^2 + O(\zeta^4) \tag{9.44b}$$

$$(I_F)_4 \approx \frac{1}{2\zeta^2}$$

$$(I_F)_5 \approx \frac{1}{1 - 12\zeta^2} \approx 5 + 60\zeta^2 + O(\zeta^4)$$

From Eqs. (9.44a) and Figure 9.8b, we can immediately see that the FIR filter with uniform weightings offers no improvement in stationary clutter suppression; such a filter actually degrades clutter suppression, and increasing the order of the filter causes further degradation, the opposite of the effect of that measure with binomial weightings.

On the other hand, Eqs. (9.44b) and Figure 9.8c show that the filter with uniform amplitude weightings with alternating signs behaves somewhat more like the N-pulse canceller for *even* values of N. For example, the two provide identical improvement factors for $N = 2$. For $N = 4$, the improvement factor is inversely proportional to the square of ζ, whereas it is proportional to the inverse sixth power of ζ for the four-pulse canceller. For odd values of N, the improvement is nothing like that of the corresponding N-pulse canceller for $\zeta \ll 1$, although for all odd and even values of N, increasing the order does increase the improvement factor.

For an interpretation of these results in terms of the frequency response of the three filters considered, we first note that the binomial weightings give rise to the Doppler frequency response [see Eq. (9.15b)].

$$|F_N(f_d)| = \left|\sin\left(\frac{\pi f_d}{fr}\right)\right|^{(N-1)} \tag{9.45a}$$

The uniform weightings result in [see Eq. (9.36f)] (where $k = Nf_d/fr$)

$$|F_N(f_d)| = \left|\frac{\sin(N\pi f_d/fr)}{\sin(\pi f_d/fr)}\right| \tag{9.45b}$$

For the uniform amplitude weightings with alternating signs

$$|F_N(f_d)| = \left|\sum_{n=0}^{N-1} (-1)^n e^{-jn\omega_d Tr}\right| = \left|\frac{1 - (-1)^N e^{-j\omega NTr}}{1 + e^{-j\omega Tr}}\right|$$

$$= \frac{1}{\left|\cos\left(\frac{\pi f_d}{fr}\right)\right|} \begin{Bmatrix} |\sin(N\pi f_d/fr)| & \text{if } N \text{ is even} \\ |\cos(N\pi f_d/fr)| & \text{if } N \text{ is odd} \end{Bmatrix} \tag{9.45c}$$

From (9.45a) and (9.45c), we note that the responses of the N-pulse canceller for all values of N *and* those of the FIR filter with weightings of equal amplitude and alternating signs for *even* values of N have nulls at $f_d = 0$, implying that returns with zero Doppler are blocked by the filter. The equal amplitude-alternating sign filter for *odd* values of N and the filter with uniform weightings do not have nulls at $f_d = 0$; in fact *peaks* in the response occur at zero Doppler. Such filters cannot exclude stationary clutter and hence cannot perform as MTI filters against clutter with zero mean Doppler.

9.4.4 Optimization of Filters for Clutter Suppression

A conclusion that can be drawn from the discussion in Section 9.4.3 is that the N-pulse canceller type of MTI is more effective than the Nth-order FIR filter with uniform-amplitude and alternating signs with an *even* value of N for detection of a moving target in clutter with mean zero Doppler. The filter with uniform weightings or that with uniform amplitude weightings with alternating signs and an *odd* value of N are counterproductive for that application. It is natural to ask whether there is an "optimum" MTI filter.

The answer is that one could design an FIR filter whose coefficients b_n would maximize the improvement factor, but that the optimum coefficients in general would depend on the form of the Doppler spectrum of the clutter, the target Doppler for which the filter is optimized and the mean Doppler of the clutter. This problem and the practical limitations of the optimization process have been treated by a number of investigators. A good discussion of this topic can be found in Nathanson et al.,[4] pp. 410–412, which cites work by Capon,[16,17] Kretschmer,[18] and Kroszezynski[19] and elaborates in some detail on the work of Ares.[20] We will confine attention here to a few elementary ideas on maximization of the improvement factor attainable with an anticlutter filter.

Consider a second-order FIR filter. The general form of the improvement factor for specific values of $\langle v \rangle$, v_0, η_v, and σ_v (or equivalently the normalized forms of these variables ξ_s, ξ, ζ_s, and ζ) is given by (9.37d) for $N = 2$.

$$(I_F)_2 = \frac{1 + d_1^2 + 2d_1 B}{1 + d_1^2 + 2d_1 A} \tag{9.46a}$$

where $d_1 = b_1/b_0$; $A = \cos \xi \, e^{-\zeta^2}$, $B = \cos \xi_s \, e^{-\zeta_z^2}$.

Differentiating $(I_F)_2$ with respect to d_1 and setting the derivative to zero, we obtain

$$\frac{\partial (I_F)_2}{\partial d_1} = \frac{2(A - B)(d_1^2 - 1)}{(1 + d_1^2 + 2d_1 A)^2} = 0 \tag{9.46b}$$

If $A = B$, the optimum value of d_1 is indeterminate. But assuming that $A \neq B$, then the values of d_1 corresponding to an extremum of $(I_F)_2$ are

$$d_1 = \pm 1 \tag{9.46c}$$

Differentiating again, we have

$$\left[\frac{\partial^2 (I_F)_2}{\partial d_1^2}\right]_{d_1 = \substack{+1 \\ -1}} = \pm 2 \frac{(A - B)}{(1 \pm A)} \tag{9.46d}$$

Specializing to the case where the average over the target Dopplers is taken, such that $B = 0$, it follows from (9.46d) that

$$\left[\frac{\partial^2 (I_F)_2}{\partial d_1^2}\right]_{\substack{d_1=+1\\-1}} = \pm\frac{2A}{1 \pm A} \begin{matrix} > 0 & \text{if } d_1 = 1 \\ < 0 & \text{if } d_1 = -1 \end{matrix} \tag{9.46e}$$

which implies that $(I_F)_2$ has its maximum value when

$$b_1 = -b_0 \tag{9.46f}$$

We note that the two pulse canceller obeys the condition (9.46f) and hence it is the optimum second-order FIR filter. Moreover, this optimization is independent of the spectrum of the clutter except that it holds *exactly* only if $A > 0$, which requires that $\cos \xi > 0$, or equivalently

$$-(4n+1)\left(\frac{\lambda f r}{8}\right) < v_0 < (4n+1)\left(\frac{\lambda f r}{8}\right) \tag{9.46g}$$

where n is an arbitrary positive integer.

For clutter with zero average Doppler, of course, $\cos \xi = 1$ and the optimization holds exactly. Thus it always holds for stationary clutter but not in all cases for airborne MTI (AMTI), for which the average clutter Doppler is not only nonzero, being affected by the radar's velocity relative to the ground, but it also changes during flight for the clutter within a range gate as the line-of-sight component of the relative velocity changes. It also does not always hold for rain clutter, even for a ground-based radar, because of the nonzero velocity of the rainfall, which may have a line-of-sight component with respect to the radar.

Extension of the above analysis to FIR filters of orders exceeding 2 is much more difficult and will not be covered here. A general conclusion is that the optimum coefficients for higher order FIR filters are functions of the factors ($\cos l\xi\ e^{-l^2\zeta^2}$) appearing in the improvement factor.

We will not pursue this topic further here. The reader is referred to some of the sources cited at the end of this chapter for more detail, particularly Chapter 9 of Nathanson et al.,[4] by P. Reilly, and some of the many references cited in that source. Topics covered include the various practical limitations of MTI processors, special problems incurred in airborne MTI, and the limitations of digital MTI, including quantization errors.

REFERENCES

1. Skolnick, M.I., "Introduction to Radar Systems," 2nd ed. McGraw-Hill, New York, 1980, Chapter 4.
2. Eaves, J.L., and Reedy, E.K. (eds.), "Principles of Modern Radar." Van Nostrand Reinhold, New York, 1987, Chapter 14, by C.R. Barrett.
3. Barton, D.K., "Modern Radar System Analysis." Artech, Norwood, MA, 1988, Section 5.3 (on MTI), pp. 232–253; Section 5.4 (on pulsed Doppler), pp. 253–274.
4. Nathanson, F.E., Reilly, J.P., and Cohen, M.N., "Radar Design Principles," 2nd ed. McGraw-Hill, New York, 1991, Chapter 9 (on MTI), by J.P. Reilly; Chapter 11 (on pulse Doppler), by F.E. Nathanson.
5. Barton, D.K., "Modern Radar System Analysis." Artech, Norwood, MA, 1988, Section 5.3.
6. Nathanson, F.E., Reilly, J.P., and Cohen, M.N., "Radar Design Principles," 2nd ed. McGraw-Hill, New York, 1991, Chapter 9.
7. Levanon, N., "Radar Principles." Wiley, New York, 1988; Chapter 11, pp. 222–244.
8. Skolnick, M.I., "Introduction to Radar Systems," 2nd ed. McGraw-Hill, New York, 1980, Sections 4.1 through 4.3, pp. 101–117.

9. Oppenheim, A.V., and Schaefer, R.W., "Discrete-Time Signal Processing." Prentice-Hall, Englewood Cliffs, NJ, 1989, particularly Chapter 7, Section 7.4–7.7, pp. 444–488.
10. Skolnick, M.I., "Introduction to Radar Systems," 2nd ed. McGraw-Hill, New York, 1980, Section 4.3, pp. 114–117.
11. Levanon, N., "Radar Principles." Wiley, New York, 1988, pp. 231–238.
12. Barton, D.K., "Modern Radar System Analysis." Artech, Norwood, MA, 1988, pp. 239–240.
13. Skolnick, M.I., "Introduction to Radar Systems," 2nd ed. McGraw-Hill, New York, 1980, pp. 117–139.
14. Nathanson, F.E., Reilly, J.P., and Cohen, M.N., "Radar Design Principles," 2nd ed. McGraw-Hill, New York, 1991, pp. 388–410 by J.P. Reilly.
15. Skolnick, M.I., "Introduction to Radar Systems," 2nd ed. McGraw-Hill, New York, 1980, p. 107.
16. Capon, J., "Optimum Weighting Functions for the Detection of Sampled Signals in Noise." IEEE Trans. Inform. Thry. IT-10(2), 152–159, 1964.
17. Capon, J., "On the Properties of Time-Variable Networks: The Coherent Memory Filter." Proc. Symp. Active Networks Feedback Systems, 561–581, Polytech. Inst., Brooklyn, April 19–20, 1960.
18. Kretschmer, F.F., "MTI Weightings." IEEE Trans. Aero. Elect. Syst. AES-10(1), 143–156, 1974.
19. Kroszezynski, J., "On the Optimum MTI Reception." IEEE Trans. Inform. Thry. IT-11(3), 451–452, 1965.
20. Ares, M., "Some Anticlutter Waveforms Suitable for Phased Array Radars." G.E. Report, Dept. Heavy Military Electronics," R67EMH20, June 27, 1967.

Chapter 10

Radar Antennas

CONTENTS

The antenna is a very important element in a radar system. Its design controls the angular distribution of radiated energy and, reciprocally, that of received energy. In Chapter 2 and elsewhere in this book, antenna patterns, peak gain of antenna, and reciprocity have been alluded to and used freely without regard to the methods of controlling these antenna properties.

Radar antennas could easily be the topic of an entire course. It is important that radar engineers have *some* knowledge of antennas without necessarily becoming an expert on the subject. Presentation of a brief account of some aspects of antennas that are most important in radar systems is the objective of this chapter. Some references on antennas[1,4–10] and on radar antennas in particular[11–13] are listed at the end of this chapter for readers interested in learning more about the subject.

10.1 THE APERTURE ANTENNA

The preferred way to model microwave antennas is through "aperture theory." Without pursuing the electromagnetic theory required for a thorough understanding of aperture theory, we present below a relationship that follows directly from very general EM-theory laws and that can be used as a basis for calculating the angular distribution of power radiated into the space surrounding the antenna from the field on an aperture at the antenna location. Although this relationship appears in *some* form in many references, it is found in an especially convenient form in "Antennas and Radiowave Propagation," by R.E. Collin[1] on p. 181, Eqs. (4.37a)–(4.39b). Collin derives this relationship through an application of stationary phase, but it can also be derived directly from the very general "Stratton–Chu-Integral" (J.A. Stratton, "Electromagnetic Theory,"[2]), which expresses the fields in an infinite linear homogeneous isotropic medium (free space in this case) in terms of those on a *closed* surface embedded in the medium.[3] The "aperture" is an *open* surface, but the fields on the remaining portion of the surface that closes it can be approximated as zero and the field at any point in space due to radiation from the aperture can be calculated from the fields on the aperture.

The relationship as given by Collin in this most general form is (where definitions relate to Fig. 10.1)

$$E_\theta = \frac{jk_0 e^{-jk_0 r}}{4\pi r}[(f_x \cos\phi + f_y \sin\phi) + Z_0 \cos\theta(g_y \cos\phi - g_x \sin\phi)] \quad (10.1a)$$

$$E_\phi = \frac{jk_0 e^{-jk_0 r}}{4\pi r}[\cos\theta(f_y \cos\phi - f_x \sin\phi) - Z_0(g_x \cos\phi + g_y \sin\phi)] \quad (10.1b)$$

where the observation point is located at the tip of a vector $\mathbf{r}$ originating at the center of the aperture 0. The spherical coordinates of the observation point, designating 0 as the origin of coordinates, are (r, θ, ϕ) and its corresponding rectangular coordinates are (x, y, z). The vector $\mathbf{r}'$ also originates at 0. It is a vector from the center of the aperture to an arbitrary point (x', y', z') on the aperture, defined in a primed rectangular coordinate system (x', y', z'). The aperture is in the (x', y') plane, which is parallel to the $(x–y)$ plane and very close to that plane; E_θ and E_ϕ are the θ and ϕ components of the complex electric fields radiated from the antenna aperture at the observation point (r, θ, ϕ). Z_0 is the wave impedance of free space, equal to 120π or 377 Ω, $k_0 = 2\pi/\lambda_0$ where λ_0 is the free-space wavelength, and $\mathbf{f}$ and $\mathbf{g}$, whose x and y components are (f_x, g_x) and (f_y, g_y), respectively, are the vectors.

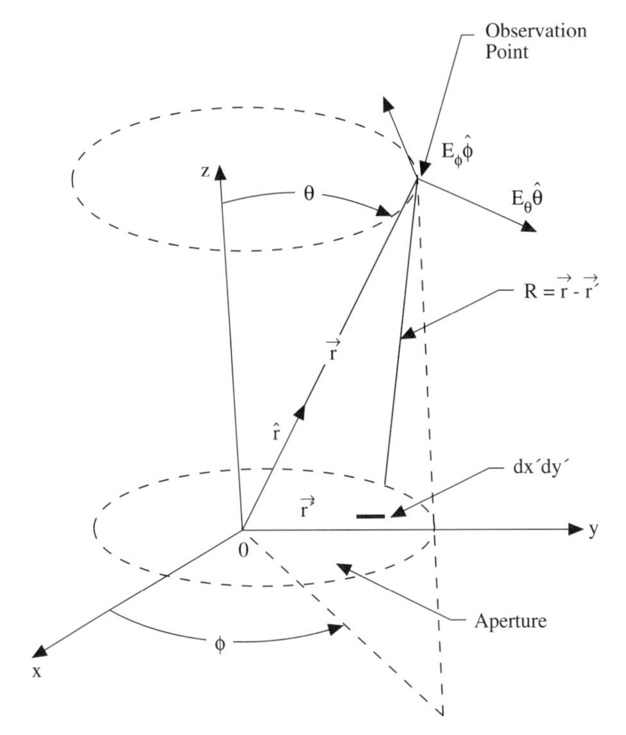

FIGURE 10.1
Aperture antenna.

$$\mathbf{f} = \int_{\Delta A} dx' \int dy' \; e^{jk_0\hat{r}\cdot\mathbf{r}'}\mathbf{E}_a(x', y') \tag{10.2a}$$

$$\mathbf{g} = \int_{\Delta A} dx' \int dy' \; e^{jk_0\hat{r}\cdot\mathbf{r}'}\mathbf{H}_a(x', y') \tag{10.2b}$$

where $\hat{r}$ is the unit vector in the direction of **r**, given by

$$\hat{r} = \hat{x}\sin\theta\cos\phi + \hat{y}\sin\theta\sin\phi + \hat{z}\cos\theta \tag{10.2c}$$

where

$$\mathbf{r}' = \hat{x}x' + \hat{y}y' + \hat{z}z' \tag{10.2d}$$

where $(\hat{x}, \hat{y}, \hat{z})$ are unit base vectors along (z, y, z) directions, where $\mathbf{E}_a(x', y')$ and $\mathbf{H}_a(x', y')$ are, respectively, the electric and magnetic field vectors on the aperture plane, and ΔA is the aperture area. The integrals in (10.2a, b) are spatial Fourier transforms of the aperture fields. The relationships (10.1a, b) apply only when the observation point is in the far zone of the aperture. They must subtend a negligibly small angle at the observation point *and* the distance r must be large compared to wavelength.

The magnetic field on an aperture can generally be expressed in terms of the electric field through the relationships

$$H_{ax} = -Y_w E_{ay} \tag{10.3a}$$

$$H_{ay} = +Y_w E_{ax} \tag{10.3b}$$

where Y_w is the wave admittance for the wave propagating down the transmission line or waveguide feeding power to the antenna. It follows from (10.3a, b) and (10.2a, b) that

$$g_x = -Y_w f_y \tag{10.4a}$$

$$g_y = +Y_w f_x \tag{10.4b}$$

and, in turn, by substitution of (10.4a, b) into (10.1a, b) that

$$E_\theta = \frac{jk_0 e^{-jk_0 r}}{4\pi r}(f_x \cos\phi + f_y \sin\phi)(1 + Z_0 Y_w \cos\theta) \tag{10.5a}$$

$$E_\phi = \frac{jk_0 e^{-jk_0 r}}{4\pi r}(f_y \cos\phi - f_x \sin\phi)(\cos\theta + Z_0 Y_w) \tag{10.5b}$$

The forms (10.5a, b) enable us to determine E_θ and E_ϕ directly from the *electric* fields only on the aperture plane.

We now proceed to specialized aperture distributions that are close to those used in actual radar antennas and provide some insight into the behavior of those fields. First we note that [from (10.2a), (10.2c), and (10.2d)]

$$f_{\substack{x\\y}} = \int_{\Delta A} dx' \int dy' \, e^{jk_0[\sin\theta(x'\cos\phi + y'\sin\phi) + z'\cos\theta]} E_{a\substack{x\\y}}(x', y') \tag{10.6}$$

10.2 THE RECTANGULAR APERTURE WITH UNIFORM ILLUMINATION

Consider an aperture in the (x–y) plane that is rectangular, with its sides along x and y axis (see Fig. 10.2). The aperture extends from $x = -a/2$ to $x = +a/2$ and from $y = -b/2$ to $y = +b/2$. The electric field distribution on the aperture is assumed to be uniform. Calling that distribution the "illumination function," this

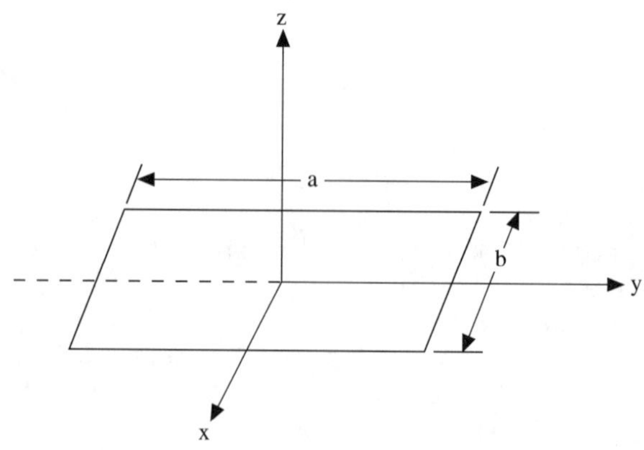

FIGURE 10.2
Rectangular aperture.

case is usually referred to as "uniform illumination." From (10.5a, b) and (10.6), the uniform illumination assumption and the assumption that $Y_w = Y_0 = 1/Z_0$, the wave admittance of free space (valid in this case because uniform illumination would usually be provided by a transmission line in the TEM mode) the functions f_x and f_y are

$$f_{\substack{x\\y}} = E_{a\substack{x\\y}}(1 + \cos\theta) \int_{-a/2}^{a/2} dx'\, e^{jk_0 x' \sin\theta\cos\phi} \ldots$$

$$\int_{-b/2}^{b/2} dy'\, e^{jk_0 y' \sin\theta\sin\phi} \tag{10.7a}$$

$$= abE_{a\substack{x\\y}}(1 + \cos\theta)\operatorname{sinc}\left(\frac{\pi a}{\lambda_0}\sin\theta\cos\phi\right)\operatorname{sinc}\left(\frac{\pi b}{\lambda_0}\sin\theta\sin\phi\right)$$

and the field components E_θ and E_ϕ are

$$E_{\substack{\theta\\\phi}} = \frac{je^{-j2\pi r/\lambda_0}}{2\lambda_0 r}\begin{Bmatrix}(E_{ax}\cos\phi + E_{ay}\sin\phi)\\(E_{ay}\cos\phi - E_{ax}\cos\phi)\end{Bmatrix}\ldots$$

$$(1 + \cos\theta)\operatorname{sinc}\left(\frac{\pi a}{\lambda_0}\sin\theta\cos\phi\right)\operatorname{sinc}\left(\frac{\pi b}{\lambda_0}\sin\theta\sin\phi\right) \tag{10.7b}$$

From (10.7b) we see that aside from the factor $(1 + \cos\theta)$ the field components are products of two functions of θ and ϕ, one due to the phase variation in the x-direction, the other due to that in the y-direction. "Principal planes" are defined as the x–z plane and the y–z plane. Planar cuts in either of these principal planes, neglecting the factor $(1 + \cos\theta)$, both yield the antenna pattern function

$$F(\theta) = \operatorname{sinc}\left(\frac{\pi d}{\lambda_0}\sin\theta\right) \tag{10.8}$$

where $d = a$ for the x–z plane and b for the y–z plane.

The factor $(1 + \cos\theta)$ has a small effect on the pattern, except at large values of θ, where its effect is to reduce the radiated power by a factor near -6 dB. This can be seen by plotting $((1 + \cos\theta)/2)^2$ in dB as a function of θ as in Figure 10.3.

Thus for angles less than about 35°, the effect of this factor is nearly negligible and it is often neglected in standard engineering analysis work.

A rough plot of (10.8) is shown in Figure 10.4.

It is easy to deduce from (10.8) that nulls occur where $\pi d/\lambda_0 \sin\theta = n\pi$, where n is any nonzero integer, that the sidelobe peaks are at $\pi d/\lambda_0 \sin\theta = (|n| + 1/2)\pi$, where $n = \pm1, \pm2, \pm3, \ldots$, and that the amplitude at the nth sidelobe peak, where the amplitude of the sine function is near unity, is approximately $\lambda_0/\pi d \sin\theta(|n| + 1/2)$. The locations and amplitudes (relative to mainlobe peak) in dB of the first few sidelobes are given in Table 10.1. The sidelobe angles are given by $\theta_n = \sin^{-1}((\lambda_0/2d)(2|n| + 1))$. Amplitudes are $\lambda_0/\pi d \sin\theta_n = 1/\pi(|n| + 1/2)$. Thus the amplitude of the nth sidelobe peak, in dB relative to that of the mainlobe peak, is

$$(A_n)_{\mathrm{dB}} = 20\log_{10}\left(\frac{1}{\pi(|n| + 1/2)}\right); \qquad |n| = 1, 2, 3, 4, \cdots \tag{10.9a}$$

and the angles at which these peaks occur are approximately

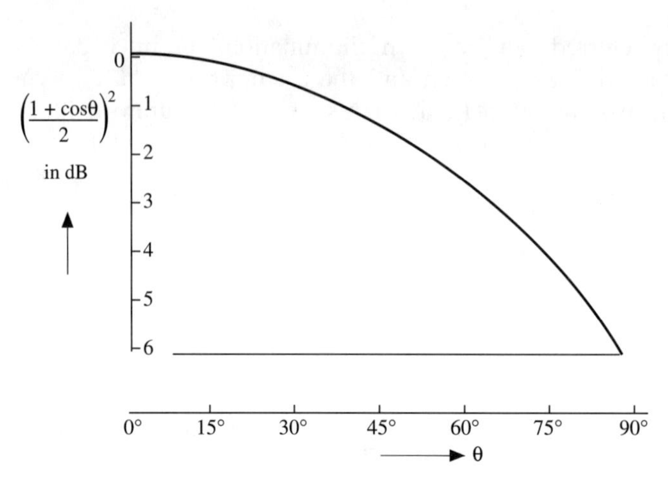

FIGURE 10.3
Effect of factor (1 + cos θ) on radiated power.

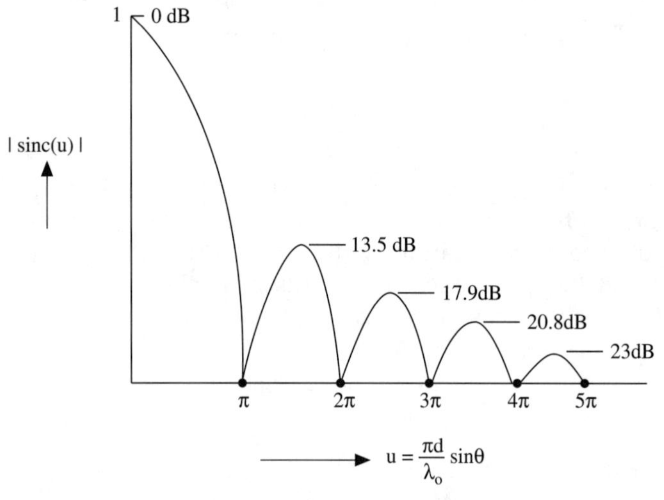

FIGURE 10.4
Principal plane pattern. Rectangular aperture-dimension d-uniform illumination.

TABLE 10.1

Sidelobe Levels and Locations for Uniformly Illuminated Rectangular Aperture

			θ_n degrees			
$\|n\|$	$\frac{1}{\pi}\frac{1}{(\|n\| + 1/2)}$	$(A_n)_{dB}$	$\frac{\lambda_0}{2d} = 0.001$	0.01	0.1	0.2
1	$2/3\pi = 0.2122$	−13.47	0.17°	1.72°	17.5°	36.9°
2	$2/5\pi = 0.1273$	−17.90	0.286°	2.87°	30°	90°
3	$2/7\pi = 0.0910$	−20.82	0.401°	4.01°	44.4°	Outside visible space
4	$2/9\pi = 0.0707$	−23.01	0.516°	5.16°	64.2°	Outside visible space
5	$2/11\pi = 0.0578$	−24.75	0.688°	6.89°	Outside visible space	Outside visible space

$$\theta_n = \sin^{-1}\left(\frac{\lambda_0}{2d}(2|n| + 1)\right); \qquad |n| = 1, 2, 3, 4, \cdots \tag{10.9b}$$

Table 10.1 gives values of $(A_n)_{dB}$ and θ_n for the first few values of $|n|$.

Table 10.1 shows the first few sidelobes at extremely small angles off boresight (less than 1°) if the aperture dimension d is a thousand times a half-wavelength. If d is only 100 times a half-wavelength; the first few sidelobes are still only a few degrees from boresight. If d is only 10 times a half-wavelength, then sidelobes 1, 2, 3, and 4 occur at about 18°, 30°, 45°, and 64° off boresight and sidelobe 5 *does not occur at all*. That means no sidelobe exists in "visible space," i.e., the sine of the angle θ_n in the argument of the pattern function exceeds unity, which means the sidelobe cannot exist. If d is only five times $\lambda_0/2$, then only the first two sidelobes exist and they are at 37° and 90°.

Consider next the *beamwidth* of this antenna, which we define as the spacing between its half-power (−3 dB) points. The condition is (where θ_B is the beamwidth)

$$\left|\text{sinc}\left[\frac{\pi d}{\lambda_0}\sin\left(\frac{\theta_B}{2}\right)\right]\right| = \frac{1}{\sqrt{2}} \tag{10.10a}$$

For small values of the argument of the sinc function, denoted by u, (10.10a) gives us

$$1 - \frac{u^2}{3!} + \cdots = \frac{1}{\sqrt{2}} \tag{10.10b}$$

from which it follows that

$$u = \frac{\pi d}{\lambda_0}\sin\left(\frac{\theta_B}{2}\right) = \sqrt{6\left(1 - \frac{1}{\sqrt{2}}\right)} = 1.326 \tag{10.10c}$$

or equivalently

$$\theta_B = 2\sin^{-1}\left(0.422\frac{\lambda_0}{d}\right) = 0.844\frac{\lambda_0}{d} \tag{10.10d}$$

assuming that $\lambda_0/d << 1$, which is a condition required for the validity of the analysis on which this discussion is based.

If we were to define the beamwidth as the spacing between the first nulls on either side of boresight, then since the first null occurs where $(\pi d/\lambda_0)\sin(\theta_B/2) = \pi$, then the beamwidth so-defined would be $2\sin^{-1}(\lambda_0/d)$ or about twice the value obtained through the half-power point definition. Since the latter is the more conventional definition, and the spacing between nulls is easier to obtain exactly, we can use as a general rule the idea that the beamwidth is roughly half the spacing between nulls for the uniform illumination case, and we can then approximate the beamwidth as the ratio (λ_0/d).

The peak gain of this antenna can be obtained from the basic definition (2.2)

$$G_0 = \frac{4\pi r^2 (dP/dA)_{\text{peak}}}{P_T} \tag{10.11a}$$

where $(dP/dA)_{\text{peak}}$ is the peak value of power density of the radiation and P_T is the total radiated power. For an aperture antenna, there is a convenient way to approximate P_T *without* integrating the pattern function over the surrounding sphere. The total radiated power is the power entering the aperture from the transmission line that feeds it. The aperture fields $\mathbf{E}_a(x, y)$ and $\mathbf{H}_a(x, y)$ [see Eqs. (10.2a, b)] are uniform, as they must be in this case. In general, the power coming through the transmission line is given by

$$P_a = \iint_{\Delta A} dx\, dy \frac{Y_w}{2} |\mathbf{E}_a(x, y)|^2 \tag{10.11b}$$

(where Y_w is the wave admittance of the mode propagating in the line, free space in this case; hence $Y_w = Y_0 = 1/Z_0 = 1/377\ \Omega^{-1}$).

Since the field $\mathbf{E}_a(x, y)$ is uniform on the aperture, then (10.11b) yields

$$P_a = \frac{Y_0}{2} |\mathbf{E}_a|^2 (ab) \tag{10.11c}$$

From (10.7a) and the fact that the peak radiation occurs at $\theta = 0°$, we obtain

$$f_{\substack{x\\y}} = E_{a\substack{x\\y}}(ab) \tag{10.11d}$$

and from (10.11b), equating P_a with the total radiated power P_T

$$P_a = P_T = \frac{Y_w}{2} \Delta A = \frac{Y_0}{2} |\mathbf{E}_a|^2 (ab) \tag{10.11e}$$

$$= \frac{Y_0}{2} (ab)(|E_{ax}|^2 + |E_{ay}|^2)$$

The power density dP/dA at $\theta = 0°$ for the field components E_θ and E_ϕ is obtained from (10.5a, b). The result is

$$\left[\left(\frac{dP}{dA}\right)_\theta\right]_{\text{peak}} = \frac{Y_0}{2} |E_\theta|^2_{\theta=0°} = (2)^2 \left(\frac{k_0}{4\pi r}\right)^2 \frac{Y_0}{2} [|f_x|^2 \cos^2\phi + |f_y|^2 \sin^2\phi$$

$$+ 2\, Re(f_x f_y^*) \cos\phi \sin\phi] \tag{10.11f}$$

$$\left[\left(\frac{dP}{dA}\right)_\phi\right]_{\text{peak}} = \frac{Y_0}{2} |E_\phi|^2_{\theta=0°} = (2)^2 \left(\frac{k_0}{4\pi r}\right)^2 \frac{Y_0}{2} [|f_x|^2\, sin\, \phi^2 + |f_y|^2 \cos^2\phi$$

$$- 2\, Re(f_x f_y^*) \cos\phi \sin\phi] \tag{10.11g}$$

The total radiated power density is the sum of power densities for θ and ϕ polarized fields, thus from (10.11f, g)

$$\left(\frac{dP}{dA}\right)_{\text{peak}} = \frac{Y_0}{2(\lambda_0 r)^2}(|f_x|^2 + |f_y|^2) = \frac{1}{\lambda_0^2 r^2}(ab)^2 \frac{Y_0}{2}(|E_{ax}|^2 + |E_{ay}|^2) \tag{10.11h}$$

where we have invoked the specialization (10.11d) and noted that $k_0 = 2\pi/\lambda_0$.

From (10.11a), (10.11e), and (10.11h)

$$G_0 = 4\pi r^2 \frac{Y_0}{2} \frac{(ab)^2}{\lambda_0^2 r^2} \frac{[|E_{ax}|^2 + E_{ay}|^2]}{\frac{Y_0}{2}(ab)(|E_{ax}|^2 + |E_{ay}|^2)} = \frac{4\pi}{\lambda_0^2}(ab) \tag{10.11i}$$

If we repeat Eq. (2.11) here, i.e.,

$$G_0 = \frac{4\pi}{\lambda_0^2} A_{e0} \tag{10.12}$$

we realize that the peak *effective* aperture area in the uniform illumination case is exactly equal to the *actual* aperture area. An analysis equivalent to the one above for the uniformly illuminated circular aperture would yield the analog of (10.11d)

$$f_{\substack{x\\y}} = E_{\substack{ax\\y}}(\pi\rho)^2 \tag{10.13}$$

where ρ = radius of aperture. Invoking (10.12), the form of (10.11e) applicable to the uniform circular aperture (where $\Delta A = \pi\rho^2$) and (10.11f, g) for this case, we would obtain the analog of (10.11i) and finally the result

$$A_{e0} = \Delta A \text{ for uniformly illuminated aperture of any shape,} \tag{10.14a}$$

where ΔA is the actual physical aperture area,

and consequently

$$A_{e0} = ab \text{ for uniform rectangular aperture of dimensions } a, b \tag{10.14b}$$

$$= \pi\rho^2 \text{ for uniform circular aperture of radius } \rho \tag{10.14c}$$

The uniform aperture is a very convenient reference case for evaluating the effect of tapering the illumination function on the peak gain of an aperture antenna. For radar antennas, this is particularly important because one of the major considerations in radar antenna design is the attempt to design antennas with very low sidelobes to avoid sidelobe clutter. That is accomplished through some form of taper, and a price must be paid for that measure in reduced gain and increased mainlobe beamwidth. It is important to be able to evaluate that price quantatively.

Before closing this section, we should mention that there is usually a simple way to estimate the gain of an antenna. An expression for the gain, given by (2.29), is

$$G_0 = \frac{4\pi}{\Omega_b} \tag{10.15}$$

where Ω_b is the solid angle of coverage. An approximation of the gain is achieved by estimating the solid angle of coverage as the product of -3 dB beamwidths in the two principal planes. For the rectangular aperture, these are the $(x\text{–}z)$ and $(y\text{–}z)$

planes. It was determined from (10.10d) that these beamwidths would be $[0.844(\lambda_0/a)]$ and $[0.844(\lambda_0/b)]$. Thus

$$\Omega_b = (0.844)^2 \frac{\lambda_0^2}{(ab)} \tag{10.16a}$$

and from (10.15)

$$G_0 = \frac{4\pi(ab)}{\lambda_0^2(0.844)^2} = \frac{4\pi}{\lambda_0^2}(A_{e0})_{est} \tag{10.16b}$$

where $(A_{e0})_{est}$ = estimated value of $A_{e0} = 1.4(ab)$, which is about 1.5 dB above the correct value.

10.3 THE CIRCULAR APERTURE WITH UNIFORM ILLUMINATION

Consider an aperture of circular shape, as shown in Figure 10.5. Its radius is ρ_0 and it is in the x–y plane.

The vector $\mathbf{r}'$ to the source-point in this geometry should be expressed in cylindrical coordinates (Fig. 10.6), as should the unit vector $\hat{r}$ toward the observation point. We note that (where r', ϕ', 0) are the source-point coordinates)

$$\mathbf{r}' = \hat{x}\rho' \cos\phi' + \hat{y}\rho' \sin\phi' \tag{10.17a}$$

and

$$\hat{r} = \hat{x}\sin\theta\cos\phi + \hat{y}\sin\theta\sin\phi + \hat{z}\cos\theta \tag{10.17b}$$

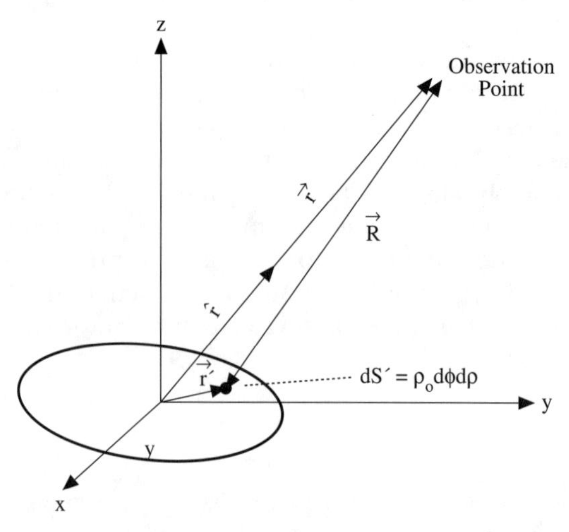

FIGURE 10.5
Circular aperture antenna.

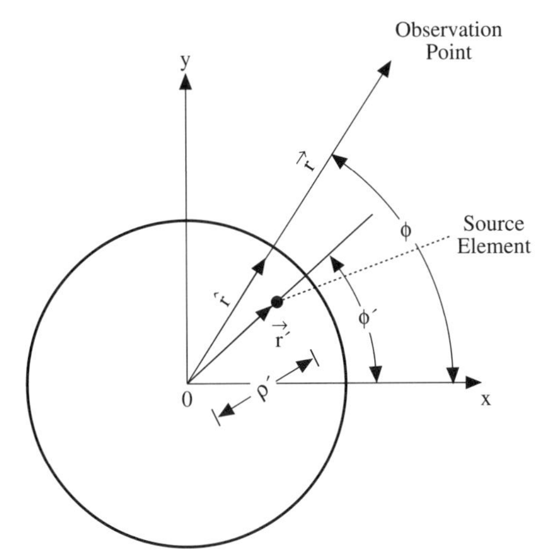

FIGURE 10.6
Coordinates of source and observation point.

The phase factor in the integrals is

$$e^{jk_0\hat{r}\cdot\mathbf{r}'}e^{jk_0\rho' \sin\theta(\cos\phi\cos\phi' + \sin\phi\sin\phi')} = e^{jk_0\rho' \sin\theta\cos(\phi'-\phi)} \tag{10.17c}$$

From (10.2a) in this case

$$\begin{aligned}\mathbf{f} &= \int_0^{2\pi} d\phi' \int_0^{\rho_0} d\rho'\ \rho' e^{jk_0\rho' \sin\theta\cos(\phi'-\phi)}\ \mathbf{E}_a(\rho', \phi') \\ &= 2\pi\mathbf{E}_a \int_0^{\rho_0} d\rho'\ \rho' \left(\frac{1}{2\pi}\int_0^{2\pi} d\phi''\ e^{jk_0\rho' \sin\theta\cos\phi''}\right)\end{aligned} \tag{10.17d}$$

where $\phi'' = \phi' - \phi$ and it is recognized that the limits on the ϕ' integral, since it extends over 2π, can be set arbitrarily.

From a well-known integral formula (follows from Spiegel,[14] #24.98, p. 143)

$$J_0(A) = \frac{1}{2\pi}\int_0^{2\pi} d\phi\ e^{jA\cos\phi} \tag{10.17e}$$

where $J_0(A)$ is the zero-order Bessel function of the first kind with argument A, we obtain

$$\begin{aligned}\mathbf{f} &= 2\pi\mathbf{E}_a \int_0^{\rho_0} d\rho'\ \rho' J_0(k_0\rho' \sin\theta) \\ &= \frac{2\pi\mathbf{E}_a}{(k_0 \sin\theta)^2}\int_0^{k\rho_0 \sin\theta} dx\ xJ_0(x)\end{aligned} \tag{10.17f}$$

(from Spiegel,[14] #2474, p. 142)

$$= \frac{2\pi \mathbf{E}_a}{(k_0 \sin \theta)^2} (k_0 \rho_0 \sin \theta) J_1(k_0 \rho_0 \sin \theta)$$

$$= 2k_0(\pi \rho_0^2)\mathbf{E}_a \frac{J_1(k_0 \rho_0 \sin \theta)}{k_0 \rho_0 \sin \theta}$$

The radiation pattern given by (10.17f) looks very much like that of the rectangular aperture with uniform illumination when plotted against $k\rho_0 \sin \theta$. A rough sketch showing sidelobe locations is shown in Figure 10.7. The nulls are, of course, at the zeros of the Bessel function $J_1(u)$ and the sidelobes about midway between them. The sidelobe amplitudes are determined from Bessel function tables. An approximation can be made to estimate the beamwidth if it is sufficiently small. For small argument x, the power series approximation for the Bessel function $J_1(x)$ is

$$J_1(x) = \frac{x}{2} - \frac{x^3}{2^2 4} + \cdots \tag{10.18a}$$

(Spiegel,[14] Eq. (24.6), p. 136).

From (10.18a) if $k_0 \rho_0 \sin \theta = 2\pi \rho_0 / \lambda_0 \sin \theta << 1$,

$$\frac{J_1(k_0 \rho_0 \sin \theta)}{k_0 \rho_0 \sin \theta} = 1 - \frac{(k_0 \rho_0 \sin \theta)^2}{16} + \cdots \tag{10.18b}$$

The beamwidth θ_B, defined as the spacing between -3 dB points, can be determined approximately from (10.18b) through the condition

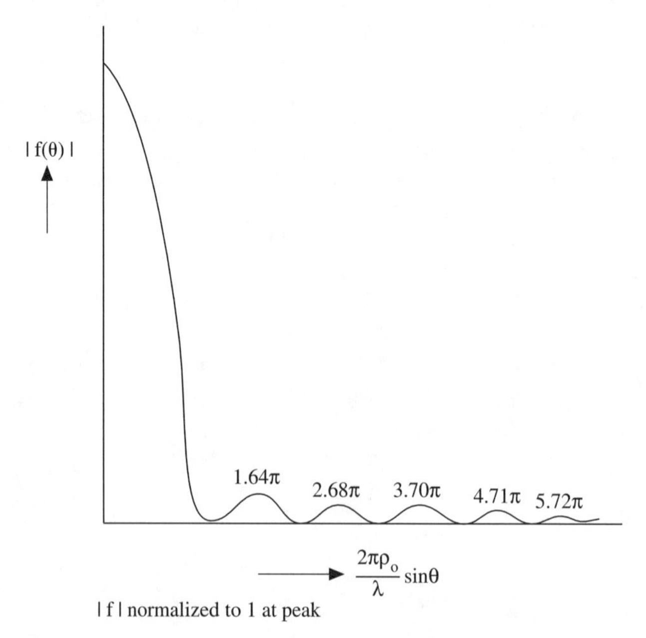

FIGURE 10.7
Radiation pattern of circular aperture:uniform illumination.

$$1 - \frac{\left(k_0\rho_0 \sin(\theta_B/2)\right)^2}{16} = \frac{1}{\sqrt{2}} \tag{10.18c}$$

or equivalently

$$\begin{aligned}\theta_B &= 2\sin^{-1}\left(\frac{2\lambda_0}{\pi\rho_0}\sqrt{1-\frac{1}{\sqrt{2}}}\right) = 2\sin^{-1}\left(1.08\,\frac{\lambda_0}{\pi\rho_0}\right)\\ &= 2\sin^{-1}\left(0.689\,\frac{\lambda_0}{D}\right) = 1.378\,\frac{\lambda_0}{D} \qquad \text{if } \frac{\lambda_0}{D} << 1\end{aligned} \tag{10.18d}$$

where D is the aperture diameter.

From the argument above that showed the effective aperture area of an antenna with a uniformly illuminated aperture to be the same as the true physical aperture area, we know that

$$A_{e0} = \pi\rho_0^2 \tag{10.19a}$$

from which it follows, since $G_0 = 4\pi A_{e0}/\lambda_0^2$, that the exact peak gain of this antenna is $(4\pi/\lambda_0^2)(\pi\rho_0^2)$ or

$$G_0 = \left(\frac{2\pi\rho_0}{\lambda_0}\right)^2 = \left(\frac{\pi D}{\lambda_0}\right)^2 \tag{10.19b}$$

An estimate of the gain for a circularly symmetric antenna through the definition of peak gain given by (2.29) is

$$G_0 = \frac{4\pi}{\int_0^{2\pi} d\phi \int_0^{\pi} d\theta \sin\theta g(\theta)} = \frac{2}{\int_0^{\pi} d\theta \sin\theta g(\theta)} \tag{10.19c}$$

where $g(\theta)$ is the normalized one-way power pattern function.

If the pattern were approximated as unity between $\theta = 0$ and $\theta = \theta_B/2$, and as zero for $\theta > \theta_B/2$, then (10.19c) would imply [with the aid of (10.18d) and the assumption that $\theta_B << \pi/2$] that the estimated gain is

$$\begin{aligned}(G_0)_{est} &= \frac{2}{-\int_{\cos(0)}^{\cos(\theta_B/2)} d(\cos\theta)} = \frac{2}{1-\cos(\theta_B/2)}\\ &= \frac{2}{1-\left\{1-\frac{((\theta_B/2))^2}{2}\right\}} = \frac{4(2)^2}{\theta_B^2} = \frac{16\pi^2\rho_0^2}{(2.16)^2\lambda_0^2}\\ &= \left(\frac{1.85\pi\rho_0}{\lambda_0}\right)^2 = 0.8556\left(\frac{\pi D}{\lambda_0}\right)^2\end{aligned} \tag{10.19d}$$

which is approximately 0.677 dB below the true gain as given in (10.19b).

TABLE 10.2

Sidelobe Levels and Locations for Uniformly Illuminated Circular Aperture

			$\lambda_0/2D$			
$\lvert n\rvert$	x	$(A_n)_{dB}$	**0.001**	**0.01**	**0.1**	**0.2**
1	5.15	−11.8	0.188°	1.879°	19.14°	41.2°
2	8.42	−14.92	0.307°	3.0725°	32.41°	Ouside visible space
3	11.6	−16.99	0.424°	4.2357°	47.60°	Outside visible space
4	14.8	−18.54	0.5399°	5.407°	70.44°	Outside visible space

The locations and amplitudes of the first few sidelobes, as obtained from Bessel function tables [e.g., Abramowitz and Stegun (A&S),[15] Section 9] are shown in Table 10.2 for values of $\lambda_0/2D$ comparable to those of $\lambda_0/2d$ used in the similar tabulation for the rectangular aperture with uniform illumination (Table 10.1). From A&S, Section 9, p. 409, the first five zeros of $J_1(x)$ are shown to be at $x = 3.83171$, 7.10559, 10.17347, 13.32369, and 16.47063. The first four sidelobe peaks occur near the values where $J_1'(x) = 0$. From the tables on pp. 390–396 of A&S and inspection of Figure 10.7, locations of these peaks are roughly at $x = 5.15$, where $|J_1(5.15)| = 0.340$, $x = 8.42$, where $|J_1(8.42)| = 0.271$, $x = 11.6$, where $|J_1(11.6)| = 0.232$, and $x = 14.8$, where $|J_1(14.8)| = 0.207$.

The argument x is $(2\pi\rho_0/\lambda_0)\sin\theta$.

10.4 LINEAR ANTENNA ARRAY: UNIFORM EXCITATION

A linear antenna array consists of an aggregate of individual antennas arranged in a line, usually (but not necessarily with equal spacing between antennas), as shown in Figure 10.8. The individual antennas, called "elements" when they are

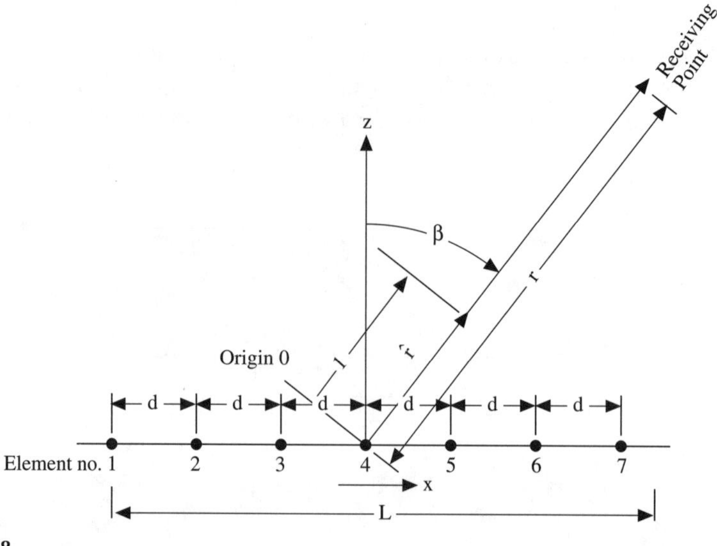

FIGURE 10.8
Linear antenna array—seven elements.

in an array, are assumed to have very broad angular coverage, so that the radiation pattern is determined primarily by the arrangement, excitation and spacing of the elements and is not much affected by the pattern of the individual antennas.[16–21]

For an arbitrarily arranged aggregate of elements, the complex amplitude of the superposition of contributions from these elements is determined with the aid of the assumption that the observation point is far enough away from the array to allow the total field to be regarded as that of a sum of plane wave fields. Two assumptions are required to justify the analysis, as follows:

$$r >> L \tag{10.20a}$$

$$r >> \lambda_0 \tag{10.20b}$$

where r is the distance from the array center (designated as origin), L is the array length, and λ_0 is wavelength. Assumption (10.20a) assures us that the array will subtend a very small angle at the observation point and hence we see all the radiation as emanating from the same point. That assumption was already made for single antennas, and it is merely being extended to the entire array. Assumption (10.20b) merely assures us that we are in the electromagnetic "far zone" of the array. With these assumptions, we can express the total complex field (neglecting polarization) as

$$f(r, \theta, \phi) = \sum_{n=1}^{N} C_n e^{j\psi_n} \frac{e^{-jk_0 r}}{r} e^{jk_0 \hat{r}\cdot\mathbf{r}_n'} f_n(\theta, \phi) \tag{10.21a}$$

where $k_0 = 2\pi/\lambda_0$, where $f_n(\theta, \phi)$ is the complex field radiation pattern of the nth element, $\mathbf{r}_n'$ the position of the nth element relative to the array center, $\hat{r}$ the unit vector in the direction of the observation point and emanating at the origin 0, and C_n and ψ_n the amplitude and phase, respectively, of the excitation on the nth element.

To render array theory useful for ordinary engineering system analysis, we assume that all elements of the array have the same field radiation pattern, which simplifies (10.21a) to the following form

$$f(r, \theta, \phi) = \frac{e^{-jk_0 r}}{r} f_0(\theta, \phi) F_N(\theta, \phi) \tag{10.21b}$$

where $f_0(\theta, \phi)$ is the complex field radiation pattern of a single element of the array and $F_N(\theta, \phi)$ is the "array factor," where the subscript N indicates the number of elements and

$$F_N(\theta, \phi) = \sum_{n=1}^{N} C_n e^{j\psi_n + jk_0 \hat{r}\cdot\mathbf{r}_n'} \tag{10.21c}$$

If all of the array spacings are the same and denoted by d, as shown in Figure 10.8 for the seven-element array portrayed in that figure, and the elements are along the x-axis, then

$$\mathbf{r}_n' = \hat{x}[x_0 + (n - 1)d] \tag{10.21d}$$

where x_0 is the designated reference point for the element positions. In the case shown in Figure 10.8, x_0 is at $-L/2$, the position of the first element. Since x_0 is not important in the end result, it can be chosen wherever it is convenient.

If β is designated as the angle between the array axis and the vector $\hat{r}$, then

$$\hat{r} \cdot \mathbf{r}'_n = [x_0 + (n - 1)d] \cos \beta \tag{10.21e}$$

which, when substituted into (10.21b, c), gives us the result

$$f(r, \theta, \phi) = \frac{e}{r}^{-jk_0 r} f_0(\theta, \phi)\, F_N(\theta, \phi) \tag{10.21f}$$

where

$$F_N(\theta, \phi) = \sum_{n=1}^{N} C_n e^{j[\psi_n + [k_0 x_0 + k_0(n-1)d]\cos \beta]} \tag{10.21g}$$

We now assume that the phase variation along the array is linear, i.e.,

$$\psi(x) = \psi_0 + \alpha x \tag{10.21h}$$

where ψ_0 is an arbitrary reference phase and α is the rate of variation of phase along the array axis in radians/meter.

It follows from (10.21h) that

$$\psi_n = \psi_0 + \alpha(n - 1)d \tag{10.21i}$$

where ψ_0 is the phase at $x = x_0$.

From (10.21h, i), the array factor of a linear array with uniform spacing d and linear phase variation α is given by

$$F_N(\theta, \phi) = \sum_{n=1}^{N} C_n e^{j(kx_0 \cos \beta + \psi_0)} e^{j(n-1)(k_0 \cos \beta + \alpha)d} \tag{10.22}$$

We now assume uniform amplitude excitation on all the array elements, i.e.,

$$C_n = C_0; \qquad n = 1, 2, \ldots, N \tag{10.23a}$$

Substituting (10.23a) into (10.22) and invoking the truncated geometric series $[\Sigma_{k=0}^{N-1} x^k = (1 - x^N)/(1 - x)]$, we obtain the array factor of a linear array with uniform spacing d, linear phase variation α, and uniform amplitude excitation $C_n = C_0$; $n = 1, 2, \ldots, N$, given by

$$F_N(\theta, \phi) = C_0 e^{j(k_0 x_0 \cos \beta + \psi_0)} \ldots \sum_{m=0}^{N-1} (e^{j(k_0 \cos \beta + \alpha)d})^m = C_0 e^{(k_0 x_0 \cos \beta + \psi_0)} \ldots \left[\frac{1 - e^{j(k_0 \cos \beta + \alpha)dN}}{1 - e^{j(k_0 \cos \beta + \alpha)d}}\right] \tag{10.23b}$$

and whose amplitude is

$$|F_N(\theta, \phi)| = |C_0| \left| \frac{\sin\left(\frac{Nd(k_0 \cos \beta + \alpha)}{2}\right)}{\sin\left(\frac{d(k_0 \cos \beta + \alpha)}{2}\right)} \right| \tag{10.23c}$$

Assuming that the pattern of the individual array element is isotropic (this simplifying assumption enables us to isolate the pattern specifically due to the array factor from any effects due to the individual elements), we can determine the beamwidth and sidelobe levels of the pattern given by (10.23c). This will be shown in Section 10.6. If we set α to zero in (10.23c), we might recognize similarities between (10.23c) and (10.8). These similarities will be elaborated upon in Section 10.6. Also in that section we will further discuss the uniformly excited linear array.

10.5 NONUNIFORM APERTURE DISTRIBUTIONS

A very important goal in design of radar antennas to operate in clutter environments is sidelobe reduction. A significant amount of ground clutter power can appear in the antenna sidelobes (see Section 8.9). In particular, at low grazing angles, the main elevation beam will intercept ground clutter only at long ranges, but high elevation beam sidelobes can cause ground clutter contributions at short ranges. The higher intensity of the short-range sidelobe contributions can sometimes compete with the longer range (and therefore lower intensity) mainbeam contributions. Consequently, it is considered a worthwhile antenna design goal in clutter-limited radar systems to strive for sidelobe reduction.

Tapering the aperture illumination can bring about sidelobe reduction at the price of a wider mainbeam. To illustrate the point, consider a rectangular aperture whose radiated field components are given by (10.5a,b) where $E_{\substack{ax \\ y}}(x', y')$ in (10.6) is given by

$$\begin{aligned} E_{\substack{ax \\ y}}(x', y') &= E_{\substack{ax0 \\ y}}\left(1 - \frac{2|x'|}{a}\right); \qquad |x'| \le a/2, \qquad |y'| \le b/2 \\ &= 0; \qquad |x'| > a/2 \end{aligned} \tag{10.24}$$

Evaluation of the integral in (10.6) results in

$$f_{\substack{x \\ y}} = \frac{ab}{2} \operatorname{sinc}\left(\frac{k_0 b}{2} \sin \theta \sin \phi\right)\left[\operatorname{sinc}\left(\frac{k_0 a}{4}\right) \sin \theta \cos \phi\right]^2 \tag{10.25}$$

The sidelobe levels for the uniformly illuminated aperture are given in Table 10.1. These levels $(A_n)_{\text{dB}}$ in the third column of the table are based on principal plane cuts of the pattern as given by (10.8). A y–z plane cut based on (10.25) will yield the same results as in the table, since the y-directed aperture illumination in our example is uniform. But an x–z plane cut based on (10.25) results in a version of (10.8) given by

$$F(\theta) = \left[\operatorname{sinc}\left(\frac{\pi a}{2\lambda_0} \sin \theta\right)\right]^2 \tag{10.26}$$

The sidelobe levels are approximately the values of $F(\theta)$ at $(\pi a/2\lambda_0) \sin \theta_n = \pm\pi(n + \frac{1}{2})$ for $n = 1, 2, 3, \ldots$.

Extending (10.9a) to this case, we have

$$(A_n)_{\mathrm{dB}} = 40 \log_{10}\left[\frac{1}{\pi(|n| + \frac{1}{2})}\right]; \qquad |n| = 1, 2, 3, 4, \ldots \tag{10.27}$$

This results in a doubling of the dB values of the sidelobe amplitudes, i.e., the first sidelobe is 26.94 dB below the mainlobe peak, an increase of 13.47 dB relative to the uniform illumination case. Values for the next few sidelobes are 35.8, 41.64, 46.02, and 49.5 dB for sidelobes 2, 3, 4, and 5, respectively. Referring back to Table 10.1, this indicates sidelobe level decreases of 13.47, 17.90, 20.82, 23.01, and 24.75 dB below the uniform illumination case for sidelobes 1, 2, 3, 4, and 5, respectively.

The price to be paid for these reductions is a widening of the mainbeam and a reduction in gain. To examine the first of these modifications quantitatively, we invoke the counterpart of (10.10a), which is

$$\left|\mathrm{sinc}\left[\frac{\pi a}{2\lambda_0}\sin\left(\frac{\theta_{\mathrm{B}}}{2}\right)\right]\right| = \left[\frac{1}{\sqrt{2}}\right]^{1/2} \tag{10.28a}$$

The counterparts of (10.10b, c, d) for this case are (where $d = a$ and u is the argument of the sinc function)

$$1 - \frac{u^2}{3!} + \cdots = \frac{1}{2^{1/4}} = 0.8408 \tag{10.28b}$$

$$u = \frac{\pi a}{2\lambda_0}\sin\left(\frac{\theta_{\mathrm{B}}}{2}\right) = \sqrt{6(1 - (1/2^{1/4}))} = 0.9772 \tag{10.28c}$$

$$\theta_{\mathrm{B}} = 2\sin^{-1}\left(0.622\,\frac{\lambda_0}{a}\right) = 1.244\,\frac{\lambda_0}{a} \tag{10.28d}$$

From (10.28d) we conclude that the half-power beamwidth has been increased by a factor $1.244/0.844 = 1.47$ as a result of the linear taper, where the aperture dimension in wavelengths is the same in both cases.

Let us now examine the effect of the linear taper on the antenna gain and effective aperture area. Following the analysis from Eqs. (10.11a) through (10.11i), we note a deviation at (10.11b) due to the aperture illumination assignment (10.24). The integral of (10.11b) in this case is

$$\begin{aligned} P_{\mathrm{a}} &= \int_{-b/2}^{b/a}\int_{-a/2}^{a/2} dx\,dy\,\frac{Y_{\mathrm{w}}}{2}\,|\mathbf{E}_{\mathrm{a}0}|^2\left(1 - \frac{2|x|}{a}\right)^2 \\ &= \frac{Y_{\mathrm{w}}}{2}\,|\mathbf{E}_{\mathrm{a}0}|^2 b^2\left(\frac{a}{2}\right)\int_0^1 du(1-u)^2 = \frac{Y_{\mathrm{w}}}{2}\,|\mathbf{E}_{\mathrm{a}0}|^2\,\frac{ab}{3} \end{aligned} \tag{10.29a}$$

The counterpart of (10.11c) is

$$P_{\mathrm{a}} = \frac{Y_0}{2}\,|\mathbf{E}_{\mathrm{a}0}|^2\,\frac{(ab)}{3} \tag{10.29b}$$

From (10.25), the counterpart of (10.11d) is

$$f_{\substack{x\\y}} = E_{\substack{ax0\\y}} \frac{(ab)}{2} \tag{10.29c}$$

Executing the steps from (10.11d) through (10.11i) for the linear taper case would change the result by a factor $[(ab/2)^2/(ab)^2][(ab)/(ab/3)] = 3/4$.

Thus from the counterparts of (10.11i) and (10.12) for this case, we obtain the result

$$\frac{(G_0)_{\text{L.T.}}}{(G_0)_{\text{U.I.}}} = \frac{(A_{eo})_{\text{L.T.}}}{(A_{eo})_{\text{U.I.}}} = 0.75 \text{ or } -1.25 \text{ dB} \tag{10.29d}$$

where L.T. and U.I. denote "linear taper" and "uniform illumination," respectively.

The 1.25 dB decrease in gain indicated by (10.29d) is a small price to pay for the 13.47, 17.90, and 20.82 dB decreases in levels of the first three sidelobes. This gain reduction is due to the fact that the length of the region (in the x-direction) being illuminated is effectively reduced relative to the case of uniform illumination. That effective illumination dimension is not actually a, the true aperture length in the x-direction, but a number less than a. The effective and true aperture lengths in the y-direction are the same because the taper is entirely x-directed. If we think of the "effective aperture area" A_{e0} as the product of the effective aperture lengths in x and y directions, then the 1.25 dB reduction in A_{e0} indicated by (10.29d) is largely due to the fact that only a fraction of the actual x-aperture length is being illuminated.

In some radar applications, the sidelobe reduction attainable by linear tapering is inadequate. In severe ground clutter conditions, first sidelobe levels as far down as 50 or 60 dB might be a very important design goal. Let us examine a more extreme taper to see what is required to produce that level of reduction and what price must be paid in reduced gain.

Consider as an illustrative example an aperture distribution that produces an x–z plane pattern of the form

$$F(\theta) = \left[\operatorname{sinc}\left(\frac{\pi a}{N\lambda_0}\sin\theta\right)\right]^N \tag{10.30}$$

where N is a positive integer exceeding 2. Note that the aperture distribution required to obtain this pattern is not necessarily realistic. Note also that $N = 1$ gives us the uniform illumination case and $N = 2$ the linearly tapered aperture case.

The nulls of (10.30) occur where

$$\frac{\pi a}{N\lambda_0}\sin\theta = \pm n\pi \tag{10.31a}$$

where n is any positive integer. The sidelobe peaks, then, are roughly midway between the nulls, i.e., where

$$\frac{\pi a}{N\lambda_0}\sin\theta_n = \pm\left(n + \frac{1}{2}\right)\pi \tag{10.31b}$$

The analog of (10.27) is

$$(A_n)_{\text{dB}} = 20\,N \log_{10}\left[\frac{1}{\pi(|n| + 1/2)}\right]; \qquad |n| = 1, 2, 3, 4, \ldots \tag{10.31c}$$

If $N = 3$, for example, the first three sidelobe levels are -40.4, -53.7, and -62.5 dB. For $N = 4$, the levels are -53.9, -71.6, and -83.3 dB, and for $N = 5$, the levels are -67.4, -90, and -104 dB. If the specification were for sidelobe levels of -60 dB or better, we would require that N be 5 or greater.

To examine the -3 dB beamwidth as a function of N, we can generalize the approximation leading to Eqs. (10.28a–d) for this case, leading to

$$\left|\text{sinc}\left[\frac{\pi a}{N\lambda_0}\sin\left(\frac{\theta_B}{2}\right)\right]\right| = \left(\frac{1}{\sqrt{2}}\right)^{1/N} \tag{10.32a}$$

from which it follows for small argument that

$$1 - \frac{u^2}{3!} + \cdots = \frac{1}{2^{1/2N}} \tag{10.32b}$$

and hence that

$$u = \frac{\pi a}{N\lambda_0}\sin\left(\frac{\theta_B}{2}\right) = \sqrt{6\left(1 - \frac{1}{2^{1/2N}}\right)} \tag{10.32c}$$

and

$$\theta_B = 2\sin^{-1}\left(K_N\frac{\lambda_0}{a}\right) = 2K_N\frac{\lambda_0}{a} \tag{10.32d}$$

where

$$K_N = \frac{N}{2}\sqrt{6(1 - 2^{-1/2N})}$$

The peak gain can be calculated from (2.29), where the solid angle of beam coverage Ω_b in the denominator of (2.29) is approximated by

$$\Omega_b = (\theta_B)_E(\theta_B)_H \tag{10.33}$$

where θ_B is the -3 dB beamwidth in radians and the subscripts E and H refer to the two principal planes designated as "E-plane" and "H-plane," respectively. Without elaborating on those definitions, we can assume that the aperture distribution generating the field in one of the principal planes, say the H-plane, is independent of N (e.g., uniform illumination) and that the pattern in the other principal plane, now designated as the E-plane, is that given by (10.30). Then the peak gain for a given value of N relative to that for $N = 1$ (uniform illumination) is

$$\frac{(G_0)_N}{(G_0)_1} = \frac{\sin^{-1}(K_1)}{\sin^{-1}(K_N)} = \frac{1}{N}\sqrt{\frac{1-2^{-1/2}}{1-2^{-1/2N}}} \tag{10.34a}$$

while the ratio of half-power beamwidths in the relevant principal planes is the reciprocal of (10.34a), i.e.,

$$\frac{(\theta_B)_N}{(\theta_B)_1} = \frac{\sin^{-1}(K_N)}{\sin^{-1}(K_1)} = N\sqrt{\frac{1-2^{-1/2N}}{1-2^{-1/2}}} \tag{10.34b}$$

The two ratios in (10.34a, b) are shown plotted vs. N in Figure 10.9. The ratio of gains in (10.34a) is given in dB. The level of the first and highest sidelobe obtained from (10.31) and denoted by SLL is also given in dB relative to the uniform illumination case.

More common tapered aperture illumination functions that could be used to further illustrate this point are the cosine function (the field along one axis of the cross section of a rectangular waveguide propagating the TE_{10} mode) and the cosine-squared function. These are

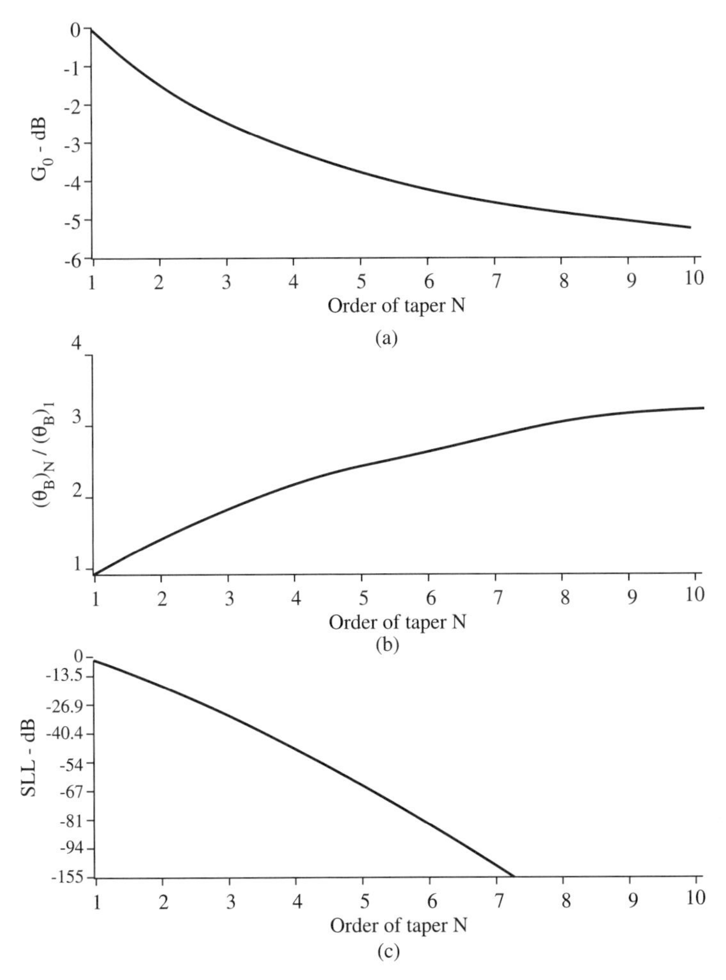

FIGURE 10.9
Variation of peak gain, beamwidth, and SLL with order of taper. (a) Peak gain ratio. (b) Half-power beamwidth ratio. (c) First sidelobe level below $N = 1$ case.

$$f(x) = \cos\left(\frac{\pi x}{a}\right); \qquad |x| \le a/2$$
$$= 0; \qquad |x| > a/2 \tag{10.35a}$$

and

$$f(x) = \cos^2\left(\frac{\pi x}{a}\right); \qquad |x| \le a/2$$
$$= 0; \qquad |x| > a/2 \tag{10.35b}$$

The pattern for cosine illumination, normalized to unity at its peak, which occurs at $u = 0$, where $u = k_x a$, is given by (where $\pi/2$ is the normalizing factor).

$$\frac{2}{\pi} F(u) = \frac{1}{a}\int_{-a/2}^{a/2} dx \cos\left(\frac{\pi x}{a}\right) e^{jk_x x}$$
$$= \frac{2}{a}\int_{0}^{a/2} dx \cos\left(\frac{\pi x}{a}\right)\cos(k_x x)$$
$$= \frac{1}{a}\left\{\int_{0}^{a/2} dx \cos\left[\left(k_x + \frac{\pi}{a}\right)x\right]\right.$$
$$\left. + \int_{0}^{a/2} dx \cos\left[\left(k_x - \frac{\pi}{a}\right)x\right]\right\}$$
$$= \frac{1}{a}\left\{\left[\sin\left(\left(k_x + \frac{\pi}{a}\right)\frac{a}{2}\right)\Big/\left(k_x + \frac{\pi}{a}\right)\right] + \left[\sin\left(\left(k_x - \frac{\pi}{a}\right)\frac{a}{2}\right)\Big/\left(k_x - \frac{\pi}{a}\right)\right]\right\}$$
$$= \frac{2\pi \cos(u/2)}{(\pi^2 - u^2)} \tag{10.36a}$$

and that for cosine-squared illumination is (where $1/2$ is the normalizing factor)

$$\frac{1}{2} F(u) = \frac{1}{a}\int_{-a/2}^{a/2} dx \cos^2\left(\frac{\pi x}{a}\right) e^{jk_x x} = \frac{1}{2a}\int_{-a/2}^{a/2} dx\left[1 + \cos\left(\frac{2\pi x}{a}\right)\right] e^{jk_x x}$$
$$= \frac{1}{a}\left\{\int_{0}^{a/2} dx \cos(k_x x) + \int_{0}^{a/2} dx \cos\left(\frac{2\pi x}{a}\right)\cos(k_x x)\right\}$$
$$= \frac{1}{k_x a}\sin\left(\frac{k_x a}{2}\right) + \frac{1}{2a}\left\{\frac{\sin\left(\left(k_x + \frac{2\pi}{a}\right)\frac{a}{2}\right)}{\left(k_x + \frac{2\pi}{a}\right)} + \frac{\sin\left(\left(k_x - \frac{2\pi}{a}\right)\frac{a}{2}\right)}{\left(k_x - \frac{2\pi}{a}\right)}\right\}$$
$$= \frac{1}{2}\left\{\operatorname{sinc}\left(\frac{u}{2}\right) + \left(\frac{u}{2}\right)\frac{\sin(u/2)}{\pi^2 - (u/2)^2}\right\} = \frac{1}{2}\frac{\operatorname{sinc}(u/2)}{1 - (u/2\pi)^2} \tag{10.36b}$$

Another simple tapered aperture function is the "cosine on a pedestal," where

$$f(x) = e + (1 - e)\cos\left(\frac{\pi x}{a}\right); \qquad |x| \le a/2 \tag{10.37a}$$

where e is the "edge illumination," a positive number between 0 and 1. The closer the edge illumination to unity the closer the pattern to that for a uniformly illuminated aperture. It is evident that if there is no edge illumination, we have the cosine pattern as given by (10.36a). Edge illumination acts as an additional control on the behavior of the pattern through its effect on the size of the "pedestal."* The normalized pattern from (10.37a) is given by

$$F(u) = K_n\left[e\,\mathrm{sinc}(u) + \frac{(1-e)2\pi}{\pi^2 - u^2}\cos\left(\frac{u}{2}\right)\right]$$

$$\text{(where } K_n \text{ is the normalizing factor)} \tag{10.37b}$$

No further discussion will be presented here on these forms of tapered illumination. The reader is referred to antenna references listed at the end of Chapter 10 for elaboration of these ideas[22–24]. In Figure 10.10, we show sketches of illumination

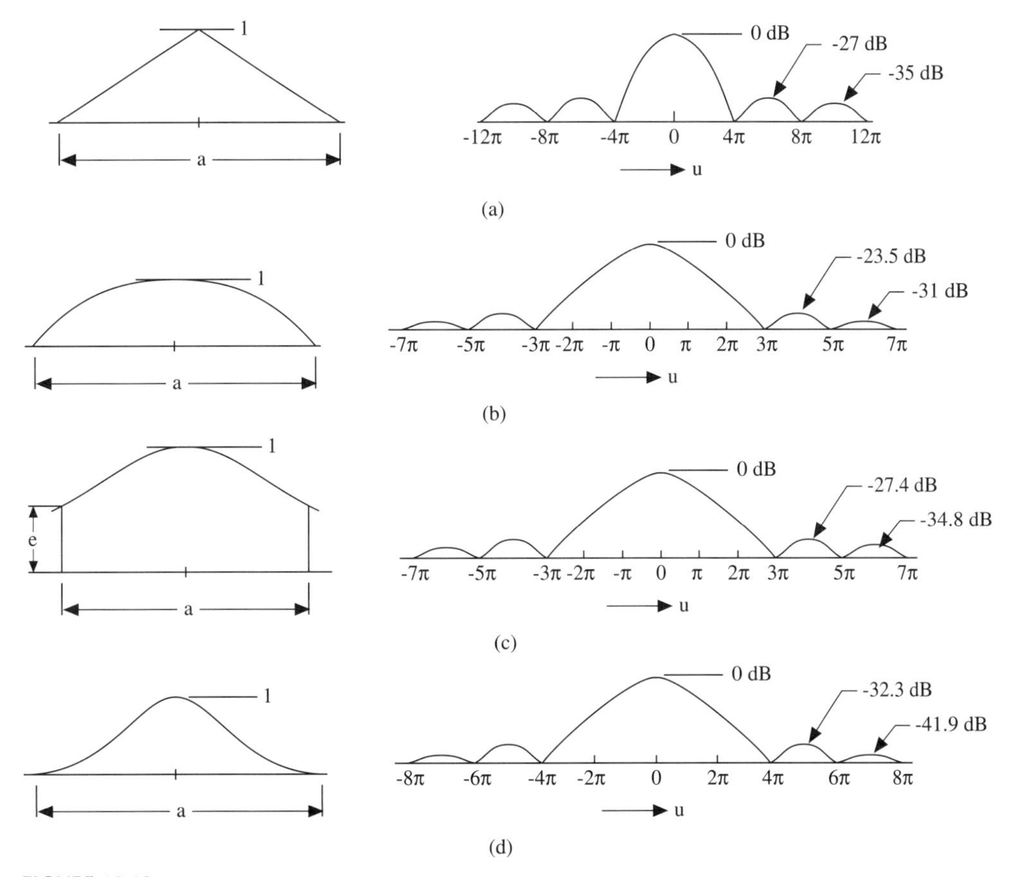

FIGURE 10.10
A few examples of tapered aperture antennas. (a) Linear taper. (b) Cosine taper. (c) Cosine-on-a-pedestal taper; $e = 0.5$. (d) Cosine-squared taper; $u = (2\pi a \sin\theta)/\lambda$.

* The cosine squared antenna is a cosine on a pedestal where $e = 0.5$ and $a \rightarrow a/2$.

function and patterns, for a few of the simple types of tapered apertures we have discussed here.

10.6 UNIFORM AND NONUNIFORM ARRAY EXCITATION[16–21]

The analogy between uniform linear aperture illumination and uniform linear array excitation was alluded to in Section 10.4. Before we launch a discussion of nonuniform array excitation, we should recognize that the pattern functions for uniform linear aperture illumination and uniform linear array excitation are very similar and exhibit the same general features. From (10.7b) and (10.23c), we recognize that the pattern functions, normalized to unity at the peak, for these two cases in the x–z principal plane can be written in the following forms:

$$F_A(u) = \text{sinc}(u)$$

$$F_B(u) = \frac{1}{N}\frac{\sin(Nu)}{\sin u} \tag{10.38}$$

where A and B denote "linear aperture" and "linear array," respectively, the linear phase variation in the array case has been set to zero, where $u = (\pi d/\lambda_0)\sin\theta$, d being the interelement spacing in the array case and the aperture dimension in the aperture case. Rough-sketch plots of $F_A(u)$ and $F_B(u)$ vs. u are shown in Figure 10.11.

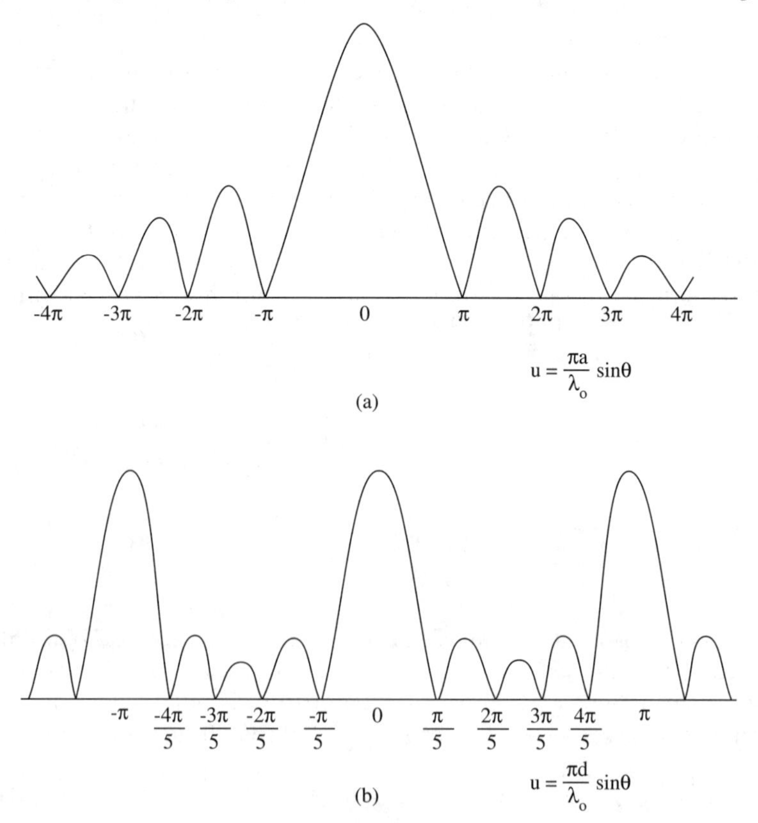

FIGURE 10.11
Comparison between uniform linear aperture illumination and uniform linear array excitation. (a) Pattern for uniformly illuminated linear aperture. (b) Pattern for linear array with uniform excitation, 5 elements.

To construct these plots, we note that the array pattern repeats itself at intervals of $u = \pi$, whereas the aperture pattern decays to negligibly small values as $|u|$ increases. Within a repetition period of the array pattern, it behaves very much like the aperture pattern.

Referring to (10.21g), let us assume an odd number of array elements $N = 2M + 1$ with symmetrical amplitude excitation about the central element. The tapering assumption is

$$C_m = C_{-m} = C_0\gamma^m; \qquad \text{where } 0 < \gamma < 1 \tag{10.39a}$$

Then (10.22) can be rewritten in the form

$$F_N(w) = [C_0 e^{j(k_0 x_0 \cos\beta + \psi_0)}][1 + 2\sum_{m=1}^{M} \gamma^m e^{jmw}] \tag{10.39b}$$

where $w = (k_0 \cos\beta + \alpha)d$.

Applying the truncated geometric series to the summation in (10.39b) and* setting C_0 so that $|F_N(0)| = 1$, we obtain

$$|F_N(w)| = |C_0|\left|1 + 2\gamma e^{jw}\left(\frac{1 - \gamma^M e^{jwM}}{1 - \gamma e^{jw}}\right)\right| = \tag{10.39c}$$

$$|C_0|\sqrt{\frac{1 + 2\gamma\cos w + \gamma^2\{1 - 4\gamma^{M-1}\cos[w(M+1)]\} - 4\gamma^{M+2}\cos(Mw) + 4\gamma^{2(M+1)}}{1 + \gamma^2 - 2\gamma\cos w}}$$

Figure 10.12 shows a family of plots of $|F_N(w)|$ vs. w based on (10.39c) for a nine-element array with various values of γ from 0 to 1. It is evident that the result (10.39c) degenerates to (10.23c) (the uniform excitation case) if $\gamma = 1$ and to the case

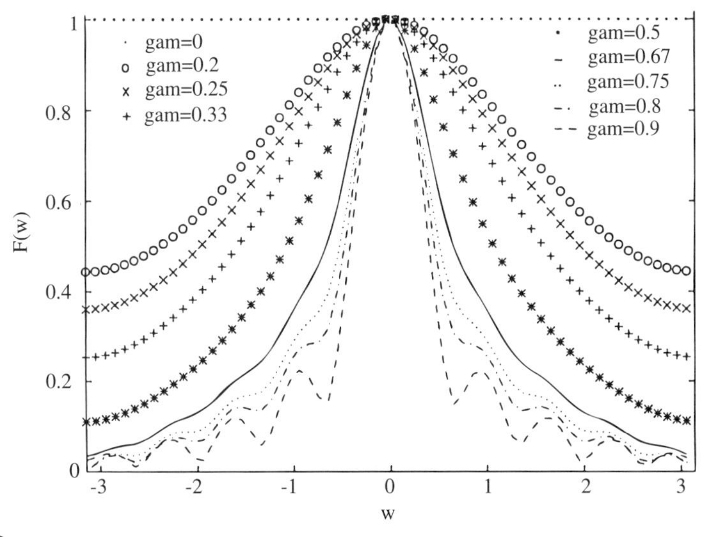

FIGURE 10.12
Effect of tapering on pattern of linear array.

* $|C_0| = \left|\dfrac{1 - \gamma}{1 + \gamma - 2\gamma^{M+1}}\right|$.

of a single isotropic radiating element if $\gamma = 0$. As γ increases from 0 to 1, the sidelobe levels increase toward those of the uniformly excited array and the mainlobe beamwidth measured in w-space tends to decrease. These are the expected effects, and serve to illustrate that tapering array element excitation has the same effect as tapering a continuous aperture distribution.

The effect of variation of γ on peak gain, beamwidth, and "effective array length relative to true array length" is shown in Table 10.3 for a nine-element array.*

The "effective array length" L_{eff} is defined in terms of the "true array length" L_{true} by

$$L_{\text{eff}} = L_{\text{true}}\left(\frac{1}{N}\sum_{n=1}^{N} |C_n|^2\right) \tag{10.40}$$

where the largest $|C_n|$ has the value of unity.

This quantity is analogous to the "effective aperture length" discussed in Section 10.5 and has the same meaning.

There is a great deal of latitude in excitation of array elements to realize design goals. However, the concept of "visible space" is an important constraint in array design. Consider the argument of the array factor $w = d(k_0 \cos\beta + \alpha)$. We can obtain insights into array behavior by plotting $|F_N(w)|$ vs. w as in Figure 10.11. But there is a region in w-space that corresponds to reality and another region that cannot exist in the real world. The former region, called "visible space," is that for which

$$-1 \leq \cos\beta \leq 1 \tag{10.41a}$$

reflecting the fact that the cosine (or sine) of a real angle cannot be outside the range −1 to 1. This condition translates into

TABLE 10.3

Effect of Excitation Tapering on Gain and Beamwidth of Array

Value of γ	Effective array length relative to true array length	−3 dB beamwidth relative to uniform excitation case	Peak gain dB relative to uniform excitation case
0	0.1111	9.000	−9.542
1/5	0.1204	8.306	−9.194
1/4	0.1259	7.943	−8.999
1/3	0.1389	7.199	−8.573
1/2	0.1480	6.757	−8.298
2/3	0.1965	5.089	−7.066
3/4	0.2397	4.172	−6.203
4/5	0.2755	3.629	−5.598
0.9	0.3808	2.626	−4.193
1	1	1	0

* This is only an illustrative example and this taper is not realistic. It has an exaggerated effect on gain and beamwidth, reducing the former by more than 8 dB and increasing the latter by a factor exceeding 6 when $\gamma = 0.5$, which is implied by the fact that only 15% of the aperture length is used in this case.

$$\alpha d - 2\pi\left(\frac{d}{\lambda}\right) \leq w \leq \alpha d + 2\pi\left(\frac{d}{\lambda}\right) \tag{10.41b}$$

If the interelement spacing d is set at $\lambda_0/2$ and $\alpha = 0$, for example

$$-\pi \leq w \leq \pi \tag{10.41c}$$

If d is set at λ_0 and α at π/λ_0, then

$$-\pi \leq w \leq 3\pi \tag{10.41d}$$

and if d is $3\lambda_0/4$ and $\alpha = \pi/2\lambda_0$

$$\frac{-9\pi}{8} \leq w \leq \frac{15\pi}{8} \tag{10.41e}$$

This implies that when we examine a curve like one of those in Figure 10.11, we must determine which part of the curve lies in visible space, because that is the only part that is meaningful. As an example, a five-element array with uniform excitation is shown in Figure 10.13. Visible space is indicated for the three cases (10.41c, d, e). It is seen that the only sidelobes that are real are those in visible space; hence design for low sidelobes must take that into account to be meaningful.

10.7 COSECANT SQUARE ANTENNA

The cosecant square antenna, sometimes used in ground-based or ship-borne search and surveillance radars looking for flying targets,[25–28] has an elevation pattern (one-way power) of the form

$$|f(\alpha)|^2 = |f(\alpha_1)|^2 \frac{\csc^2\alpha}{\csc^2\alpha_1}; \qquad \alpha_1 \leq \alpha_2 \tag{10.42}$$

where α is the elevation angle, measured from the horizon, as shown in Figure 10.14 (which is similar to Rohan's[28] illustration, Fig. 10.6, p. 230, or Barton's [12] illustration, Fig. 4.2.9, p. 163).

The reason for choice of the cosecant-square antenna is to provide a nearly constant return signal power for an approaching target at fixed altitude over a wide but limited range of typical aircraft altitudes.

An antenna with this pattern points its beam at an angle somewhat above the horizon. In the region $0 \leq \alpha \leq \alpha_1$, there is a fan beam with a null at the horizon, which helps eliminate ground clutter, and whose peak is at an angle α_m below α_1 (α_m is close to half of α_1) and with a null near $\alpha = 0$, thus eliminating much of the ground clutter that can be an important undesirable feature of ground based search radar antennas. For $\alpha > \alpha_2$, the coverage decays rapidly with increasing altitude, which limits its response to high-altitude targets. It is designed to look out over a limited range of elevation angles, although there is some flexibility in its elevation beamwidth through specification of values of α_1 and α_2.

Blake[29] gives the values of $f(\alpha_1)$ as equivalent to

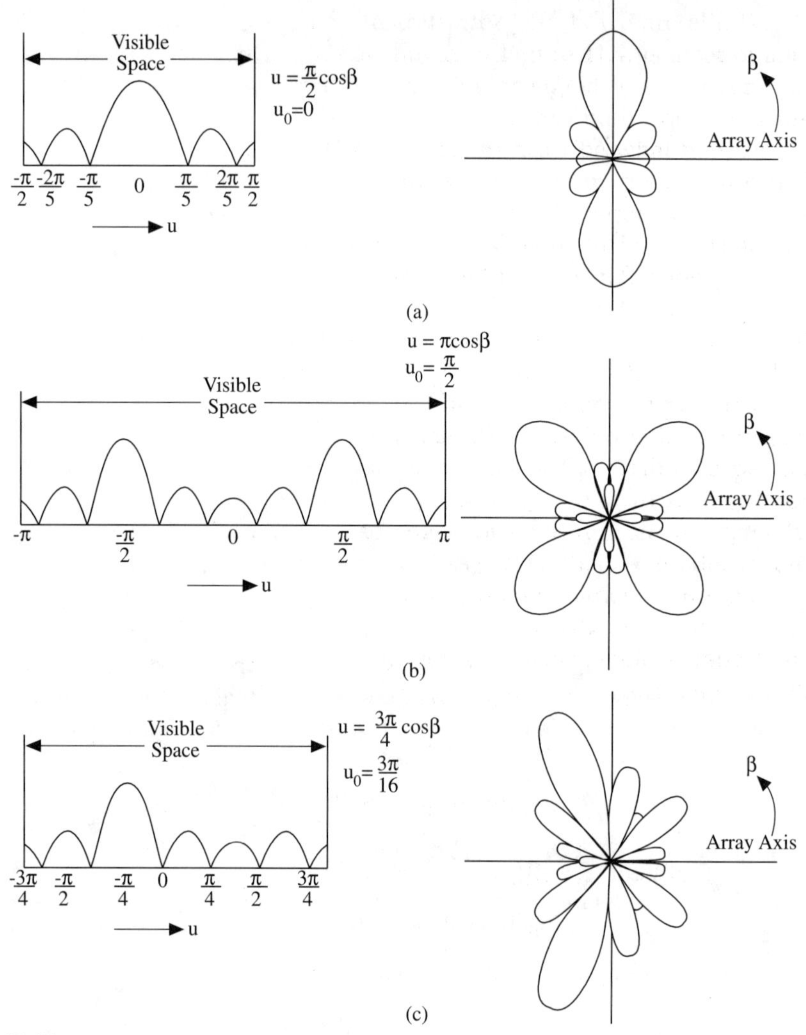

FIGURE 10.13
Meaning of visible space for arrays: 5 element array pattern. (a) Case 1: $d = \lambda_0/2$, $\alpha = 0$. (b) Case 2: $d = \lambda_0$, $\alpha = \pi/\lambda_0$. (c) Case 3: $d = 3\lambda_0/4$, $\alpha = \pi/2\lambda_0$.

$$|f(\alpha_1)| = \frac{|\sin(\alpha_{b/2} + \alpha_m)|}{\sqrt{2}} \tag{10.43}$$

where α_b is the -3 dB beamwidth of the pattern for $\alpha < \alpha_1$ and α_m is the upward tilt angle of the beam, relative to horizontal, typically about half a beamwidth, such that the lower -3 dB point is roughly in the horizontal direction and α_1 is the upper -3 dB point, as shown on Figure 10.14.

The return from a target at altitude h and slant range r will be weighted by $|F(\theta, \phi)|^4/r^4$, where $|F(\theta, \phi)|$ is the one-way field strength antenna pattern. Given omnidirectional azimuthal coverage, $F(\theta, \phi)$ is independent of ϕ and hence with the flat earth approximation and the assumption of constant RCS, the cosecant-squared pattern results in a power weighting as a function of r and α (where $\alpha = \pi/2 - \theta$)

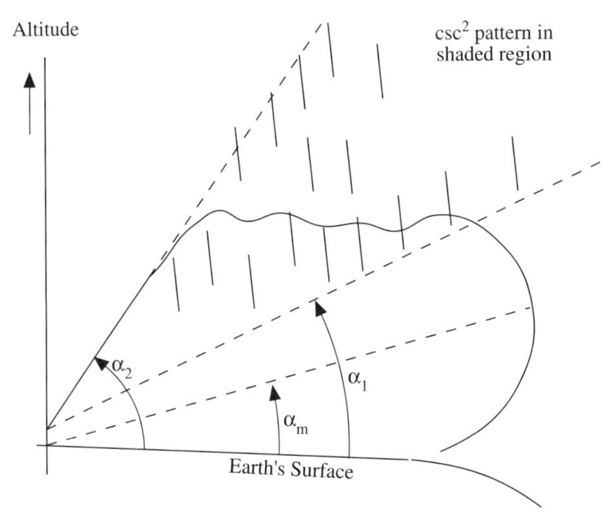

FIGURE 10.14
Cosecant-squared elevation pattern. From D. Barton,[12] "Modern Radar Systems Analysis," by permission of Artech House.

$$P(\alpha, r) = K\frac{(\csc\alpha)^4}{r^4} = K\left(\frac{r}{h}\right)^4\frac{1}{r^4} = \frac{K}{h^4} \tag{10.44}$$

which implies constant received power from an approaching target flying at constant altitude. This is somewhat in error for ranges sufficiently long to require accounting for earth curvature, but the approximate result (10.44) is the rationale for the cosecant-squared antenna.

The peak gain of the cosecant-square antenna can be calculated from (2.29). The quantity Ω_b in (2.29), the solid angle of coverage, is

$$\Omega_b = \int_0^{2\pi} d\phi\left\{\int_{\pi/2}^{(\pi/2)-\alpha_1} d\theta \sin\theta|f_1(\alpha)|^2 + \int_{(\pi/2)-\alpha_1}^{(\pi/2)-\alpha_2} d\theta\frac{\sin\theta}{\sin^2[(\pi/2)-\theta]}\right\} \tag{10.45a}$$

Because of the omnidirectional azimuthal coverage

$$\begin{aligned}\frac{\Omega_b}{2\pi} &= \int_0^{\alpha_1} d\alpha \cos\alpha|f_1(\alpha)|^2 + \int_{\alpha_2}^{\alpha_1} d\alpha\frac{\cos\alpha}{\sin^2\alpha} \\ &= \int_0^{\alpha_1} d\alpha\cos\alpha|f_1(\alpha)|^2 + \left(\frac{1}{\sin\alpha_1} - \frac{1}{\sin\alpha_2}\right)\end{aligned} \tag{10.45b}$$

where $|f_1(\alpha)|^2$ is the pattern approximation used in the region $0 < \alpha < \alpha_1$, which is variable for different antennas but may be approximated by a Gaussian function or a sinc function pattern.

The important point is that the major contribution to the solid angle of coverage should come from the region between α_1 and α_2, which means that a rough estimate of the largest possible value of the peak gain is

$$G_0 = \frac{2 \sin \alpha_1 \sin \alpha_2}{\sin \alpha_2 - \sin \alpha_1} \tag{10.45c}$$

For $\alpha_1 = 30°$ and $\alpha_2 = 50°$, $G_0 = 2.88$ or about 4.6 dB. That is low, as expected, because the cosecant-square antenna is not intended to be highly directive.

10.8 HORNS, LENS, REFLECTOR, AND DIPOLE AND SLOT ANTENNAS

Up to this point, we have said little of the actual specific antennas used in radars, except to discuss their beam patterns, directivities or gains, and beamwidths in the abstract. We cannot cover much detail on the specifics of radar antennas because it is an enormously broad subject. References at the end of this chapter contain material of interest to readers who want to pursue the subject further.

In this section, there are brief discussions of a few general ideas about four general classes of antennas used in radar, the microwave horn,[30–32] the microwave lens,[33–35] reflector antennas,[36–40] and dipole and slot antennas,[41,42] to be covered in Subsections 10.8.1, 10.8.2, 10.8.3, and 10.8.4, respectively.

10.8.1 The Microwave Horn

The horn antenna,[30–32] which can have a rectangular or circular cross section, is illustrated in four forms in Figure 10.15a–d. The first, in (a), is the "H-plane horn"; (b) shows the "E-plane horn," (c) the "pyramidal horn," and (d) the "conical horn." (d) has a circular cross section, while (a), (b), and (c) have rectangular cross sections, based on the fact that the horn is fed by a rectangular waveguide propagating the dominant TE_{10} mode, with a flare in the H-plane (a), the E-plane (b), or both [(c), the pyramidal horn]. Obviously the horn shown in (d) is a flared open-ended circular waveguide. The use of horns is based on the fact that an open-ended waveguide whose end acts as the aperture does not generally have a very high gain. That is because the dimensions of the guide cannot be too large or higher order modes will propagate. If we want single dominant mode propagation only in a waveguide the x-dimension a must be $<\lambda$. The purpose of the flare is to allow a very gradual taper in the guide such that there is a minimal higher order mode generation and yet the aperture dimension is significantly larger than that of the untapered waveguide mouth. This requires that L, the length of the flared section, is large compared with a', i.e., the flare angle cannot exceed a given value before the cylindrical phase front begins to differ significantly from the planar phase front that would exist on the aperture without the flare.

The criterion for a satisfactory horn is that the phase difference between the cylindrical and plane wave fronts on the aperture differ by less than 45°, i.e.,

$$\frac{2\pi \Delta d}{\lambda} \leq \frac{\pi}{4}, \qquad \text{or equivalently}$$

$$M - L = \Delta d \leq \frac{\lambda}{8} \tag{10.46}$$

If that criterion is met, then the ratio of flared to unflared aperture dimension (from Fig. 10.16) is obtained from the geometric relationships

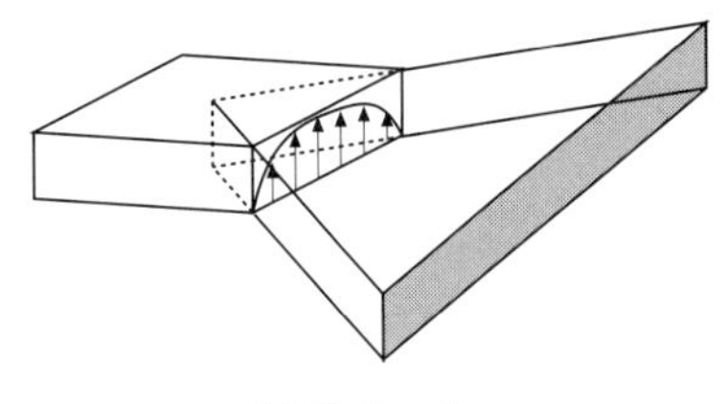

(a) H-plane horn

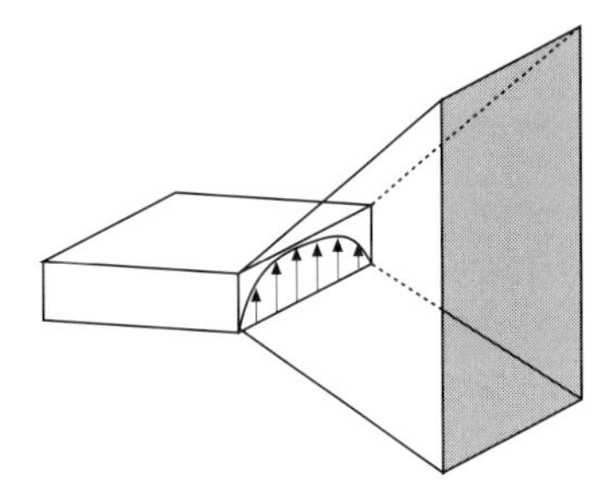

(b) E-plane horn

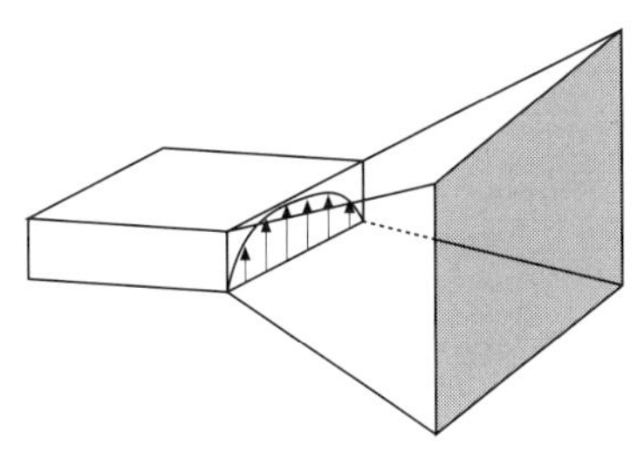

(c) Pyramidal horn

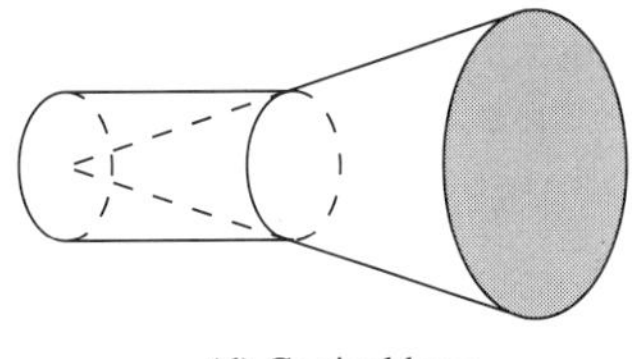

(d) Conical horn

FIGURE 10.15
The microwave horn. From Balanis,[4] "Antenna Theory," (Fig. 12.1, p. 533), Copyright © John Wiley and Sons, Inc. 1982, by permission of John Wiley and Sons, Inc.

$$\left(\frac{a'}{2}\right)\Big/L = \left(\frac{a}{2}\right)\Big/b = \tan\frac{\psi}{2} \tag{10.47a}$$

or

$$\left(\frac{a'}{2}\right)\Big/M = \left(\frac{a}{2}\right)\Big/\sqrt{b^2 + \left(\frac{a}{2}\right)^2} = \sin\frac{\psi}{2} \tag{10.47b}$$

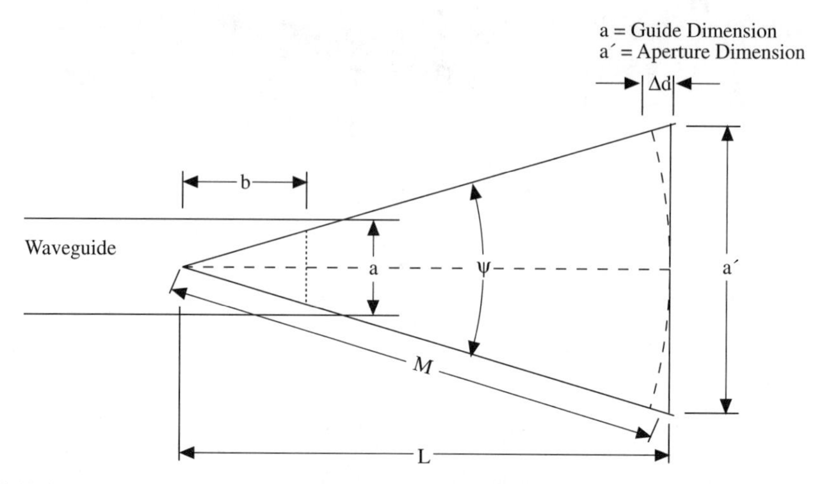

FIGURE 10.16
Principle of microwave horn. Similar to Figure 4.8 on p. 187 of Collin,[1] "Antennas and Radiowave Propagation," by permission of McGraw-Hill.

From (10.46) and (10.47a) or (10.47b) it is easily shown that the required condition is*

$$\psi \leq 4 \tan^{-1}\left(\frac{\lambda}{4a'}\right) \tag{10.48a}$$

and hence that

$$L \geq a'/4 \tan^{-1}\left(\frac{\lambda}{4a'}\right) \approx (a')^2/\lambda \tag{10.48b}$$

which implies that L must exceed a value that might be prohibitively high for practical purposes in some cases.

These conditions (10.48a, b) are required for a minimally distorted aperture distribution and hence a beam of nearly the same shape as that attainable with an open waveguide aperture of dimension a but with a higher directivity, roughly in the amount (a'/a).

For the H-plane horn, the flare is in the H-plane, whose aperture distribution is given by

$$\begin{aligned} f(x) &= \cos\left(\frac{\pi x}{a}\right); \qquad |x| \leq \frac{a}{2} \\ &= 0; \qquad \text{otherwise} \end{aligned} \tag{10.49}$$

(This is the x-illumination of a rectangular waveguide propagating the TE_{10} mode, where the E-vector is in the y-direction.)

* From (10.47a, b), $L/M = \cos(\psi/2)$ and from (10.47b), $M = a'/(2\sin(\psi/2))$; hence from (10.46), $[a'(1-\cos(\psi/2))]/(2\sin(\psi/2)) = (a'/2)\tan(\psi/4) \leq \lambda/8$, which implies (10.48a), since $\psi/4$ is in the first quadrant. Because $\psi << \pi$, $\tan(\psi/2) = \psi/2$ and (10.47a) implies that $L \approx a'/\psi$, which in turn implies (10.48b) with the aid of (10.48a).

In the E-plane the aperture distribution is uniform. In the H-plane, it is that given in (10.49). The pattern resulting from these distributions applied to a component of (10.5a) or (10.5b) with the aid of (10.6) is*

$$F(\theta, \phi) = \text{sinc}\left(\frac{\pi b}{\lambda} \sin\theta \sin\phi\right) \frac{\cos[(\pi a/\lambda)\sin\theta\cos\phi]}{\{1 - [(2a/\lambda)\sin\theta\cos\phi]^2\}} \tag{10.50}$$

The directivity or peak gain for a TE_{10} mode rectangular waveguide aperture antenna can be calculated from (10.11a), as follows:

$$G_0 = \frac{4\pi r^2 (dP/dA)_{\text{peak}}}{P_a} \tag{10.51a}$$

where P_a, as given by (10.11b) with the aid of (10.49), is

$$P_a = \frac{Y_0}{2} |\mathbf{E}_{a0}|^2 \int_{-b/2}^{b/2} dy \int_{-a/2}^{a/2} dx \cos^2\left(\frac{\pi x}{a}\right) = \frac{Y_0}{2} |E_{a0y}|^2 \frac{ab}{2} \tag{10.51b}$$

where $|E_{a0y}|^2$ is the peak value of the aperture field and it is assumed that $Y_w = Y_0$, i.e., that the waveguide medium is free space. The value of $(dP/dA)_{\text{peak}}$ is obtained from the arguments in Section (10.2) from (10.11f) through (10.11h) where the field is entirely y-directed in this case; hence from the generalization of (10.11h) to allow for a nonuniform aperture field

$$\left(\frac{dP}{dA}\right)_{\text{peak}} = \frac{1}{\lambda^2 r^2} \frac{Y_0}{2} |E_{a0y}|^2 \left| \int_{-b/2}^{b/2} dy \int_{-a/2}^{a/2} dx \cos\left(\frac{\pi x}{a}\right) \right|^2 \tag{10.51c}$$

$$= \frac{Y_0}{2\lambda^2 r^2} \left(\frac{2}{\pi}\right)^2 (ab)^2 |E_{a0y}|^2$$

Substituting (10.51b, c) into (10.51a), we obtain

$$G_0 = \left[\frac{4\pi}{\lambda^2}(ab)\right]\left[\frac{8}{\pi^2}\right] \tag{10.51d}$$

The square-bracketed quantity in (10.51d) is the value of G_0 for a uniformly illuminated aperture. The factor on that quantity demonstrates the reduction in peak gain relative to the uniform aperture case due to the cosine taper. For TE_{10} mode propagation a is about $\lambda/2$ and b is about half of a; thus the gain for this waveguide aperture antenna is about 1.27 or 1.04 dB, which renders this kind of antenna not sufficiently directive. The flare due to the horn increases the gain by a factor (a'/a) for the H-plane horn, (b'/b) for the E-plane horn, and (a'/a) (b'/b) for the pyramidal horn; hence the larger the flare the better the gain improvement. However, that is limited by the requirement for minimally distorted phase front as discussed earlier.

A more rigorous treatment of horns is given by Balanis[4] (Chapter 12, pp. 532–590) in which the effect of the phase front on the pattern is accounted for

* Note that (10.49) is equivalent to (10.35a). The calculation leading from (10.35a) to (10.36a) results in (10.50).

accurately, giving rise to Fresnel integrals and resulting in more accurate patterns and peak gain values for E-plane horns [see Balanis,[4] Eqs. (12.11b, c) on pp. 538–539 and (12.19) on p. 546], H-plane horns [see Balanis,[4] Eqs. (12.30b, c) on p. 554 and (12.42c) on p. 562], and pyramidal horns [see Balanis,[4] Eqs. (12.48a, b, c) on p. 567 and (12.54e) on p. 572]. Horns are used in some radars as primary radiators. They can also be used as elements of arrays or as feeding elements for reflector antennas.

10.8.2 Lens Antennas

The elementary principle of operation of a microwave lens antenna (fundamentally the same as an optical lens) is best illustrated as in Figure 10.17 by a planar cut of a lens with a curved rear surface and a flat front surface. In the case shown in (c), the field pattern and the gain are obtained from the theory of the rectangular aperture antenna as described in Section 10.2. In the case shown in (d), the applicable theory is that of the circular aperture as discussed in Section 10.3. In both cases the "aperture" is the front of the lens and the field distribution on the aperture is uniform if the lens behaves according to the idealized theory, but in practice would differ from uniform depending on properties of the feed and the lens structure.

Following a treatment by Balanis[4] (Section 13.6.1, pp. 648–654), and referring to Figure 10.17a, we assume the lens material to be a dielectric with relative permittivity ϵ_R. The refractive index is $\nu = \sqrt{\epsilon_R}$, assumed to be greater than unity. That implies that entering rays bend inward, implying a convex surface. The distance *sa* is the focal length *f*. The source of the radiation is at point *s*. The horizontal ray from *s* to *d* strikes the lens at the point *a* and since it is normal to the lens it continues to travel horizontally inside the lens. The *electrical* path length (EPL) is the product of the *actual* path length and the refractive index; hence the EPL from *s* to *d* is

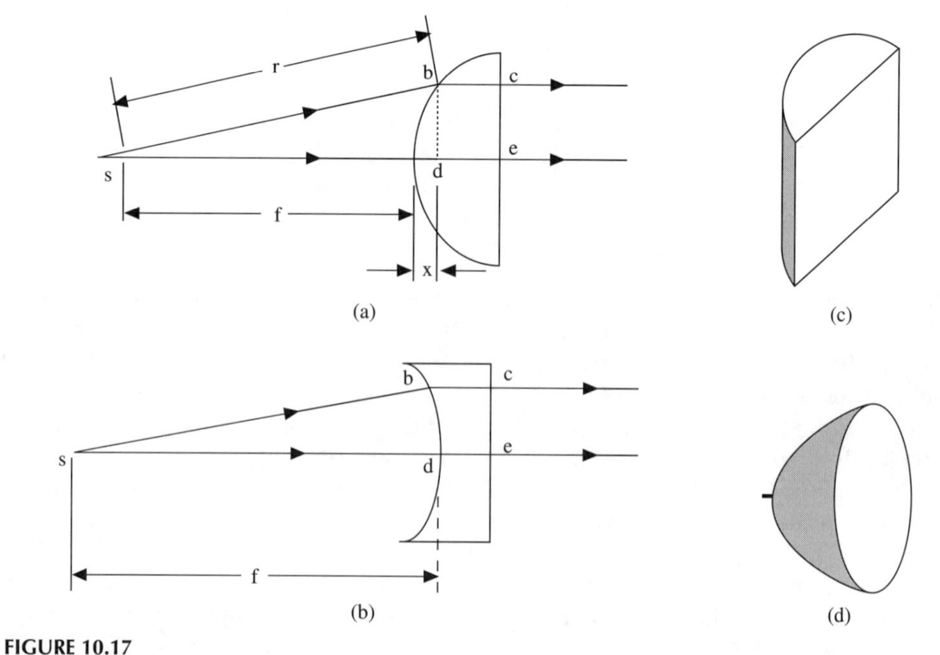

FIGURE 10.17
Lens antennas. (a) $\nu > 1$; convex planar. (b) $\nu < 1$; concave planar. (c) Cylindrical lens-convex and planar surfaces. (d) Circular lens-convex and planar surfaces.

$$D_1 = k_0(f + \nu x) \tag{10.52a}$$

while the EPL of the oblique ray, which travels entirely in free space from s to b, is

$$D_2 = k_0 r = k_0\sqrt{(f + x)^2 + y^2} \tag{10.52b}$$

To collimate the beam when it emerges from the lens, we want all rays to be in phase and propagating horizontally when they emerge from the right-hand side of the lens. Since the EPLs of the two paths de and bc are the same, the condition for collimation is that the two ray segments sad and sb be in-phase at points d and b, respectively. The required condition, then, is $D_1 = D_2$, or equivalently, from (10.52a, b)

$$f + \nu x = \sqrt{(f + x)^2 + y^2} \tag{10.52c}$$

By squaring both sides of (10.52c) and rearranging terms, we obtain the equation of the lens surface on the left, which turns out to be that of a hyperbola centered at $x = f/(\nu + 1)$, $y = 0$, with x and y dimensions $f/\sqrt{\nu + 1}$ and $f\sqrt{(\nu - 1)/(\nu + 1)}$, respectively.

$$\frac{\{x + [f/(\nu + 1)]\}^2}{(f/\sqrt{\nu + 1})^2} - \frac{y^2}{(f\sqrt{(\nu - 1)/(\nu + 1)})^2} = 1; \qquad \nu > 1 \tag{10.52d}$$

If $\nu < 1$, which will result in a concave surface as in Figure 10.17b, the collimation condition analogous to (10.52c) is [see Balanis,[4] Eq. (13.76) on p. 651]

$$f - \nu x = \sqrt{(f - x)^2 + y^2} \tag{10.52e}$$

from which the equation analogous to (10.52d) can be obtained by squaring both sides and rearranging terms, i.e.,

$$\frac{\{x - [f/(\nu + 1)]\}^2}{(f/\sqrt{\nu + 1})^2} + \frac{y^2}{(f\sqrt{(1 - \nu)/(1 + \nu)})^2} = 1; \qquad \nu < 1 \tag{10.52f}$$

which is the equation of an ellipse centered at $x = f/(\nu + 1)$, $y = 0$ with x and y dimensions $f/\sqrt{\nu + 1}$ and $f\sqrt{(1 - \nu)/(1 + \nu)}$, respectively.

If the lens is made of a material with index of refraction $\nu > 1$ and (10.52d) is the basis of the design of the convex curved left-hand surface of the lens and the right-hand surface is perfectly flat and vertical, a source at the focal point s will produce a perfectly collimated beam propagating horizontally to the right as shown in Figure 10.17a. The same result can be obtained with a lens whose left-hand concave surface is based on (10.52f) and with an index of refraction $\nu < 1$. A natural dielectric material has a refractive index exceeding unity, hence the lens with convex surface based on (10.52d) is easier to construct and more commonly used. The lens with a concave surface based on (10.52f) requires an artificial dielectric in order to synthesize a material with $\nu < 1$ (see Balanis,[4] pp. 651–652, Kraus,[6] Section 14.2, pp. 663–684, or Skolnick,[11] pp. 249–252).

Figure 10.17 shows planar cuts of this type of lens antenna operating in its transmitting mode, where a diverging cylindrical or spherical wave beam is launched at the focal point on the left and converted by the lens into a collimated beam propagating horizontally to the right. In the receiving mode, a plane wave (collimated) beam propagating horizontally to the left will be converted by the lens into a beam that converges onto the focal point s, where a transducer that delivers the signal to the receiver is placed.

The planar cuts shown can either be cross sections of a segment of a cylinder or of a sphere. If they are the latter, then all planar cuts are the same. The views perpendicular to those of (a) and (b) on Figure 10.17 are shown as either rectangular [in (c)] or circular [in (d)].

There are microwave lens designs other than the simple ones illustrated in Figure 10.17, e.g., lenses with both surfaces curved (see Balanis,[4] Fig. 13.32, p. 646), lenses with variable refractive index such as the Luneberg lens (see Balanis,[4] Section 13.6.3, pp. 652–654, or Skolnick,[11] pp. 252–253), and designs of the types illustrated in Figure 10.17 with both cylindrical and spherical geometries (see Balanis,[4] Fig. 13.33, p. 649). In general, lenses can be useful in radar as a means of generating very narrow beams and in beam-steering applications. However, lens antennas are less commonly used in radar than reflector antennas, which have some significant advantages and which will be briefly discussed in the next subsection.

10.8.3 Reflector Antennas

The operating principle of a reflector antenna,[36–40] based on geometric optics (G.O.), is illustrated in Figure 10.18.

The idealized G.O. theory predicts that (referring to Fig. 10.18a), a cylindrical or spherical wave launched at the focal point s and propagating in the region to the left of s will impinge on the reflector and be reflected as a collimated beam propagating horizontally to the right if the reflector surface is parabolic. The cylindrical wave case is that illustrated in (c) of the figure and the spherical wave case is illustrated in (d). The planar cuts in (a) or (b) can apply to either case (c) or (d). In case (c), the reflector surface ideally follows the equation of a parabola in the plane of the paper. In case (d) it follows the equation of a paraboloid.

In the receiving mode, by reciprocity, a collimated beam propagating horizontally to the left will be reflected back toward the focal point as a converging beam. All the rays striking the reflector will be in-phase at the focal point s. The feeding element at s, which may be, for example, a small horn, as shown in the diagram, sends radiation to the reflector in the transmit mode and gathers radiation from the reflector in the receive mode.

To show that the parabolic shape ensures that the reflected wave in the transmit mode is a collimated beam, consider the two rays sab and scs. Both of these rays must have the same electrical path length for the reflected beam to be collimated.

The EPL of the ray scs is

$$D_1 = 2f \tag{10.53a}$$

That of the ray sab is

$$D_2 = \sqrt{(f - x)^2 + y^2} + (f - x) \tag{10.53b}$$

We express the condition $D_1 = D_2$ in the form

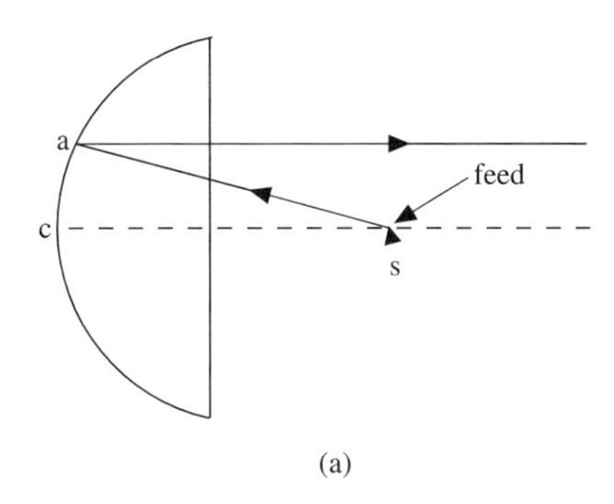

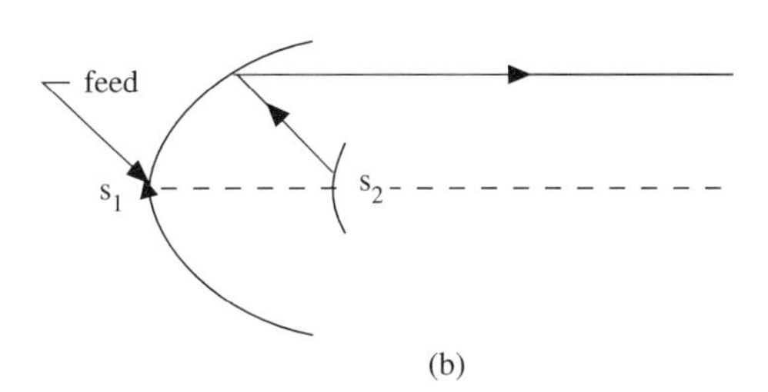

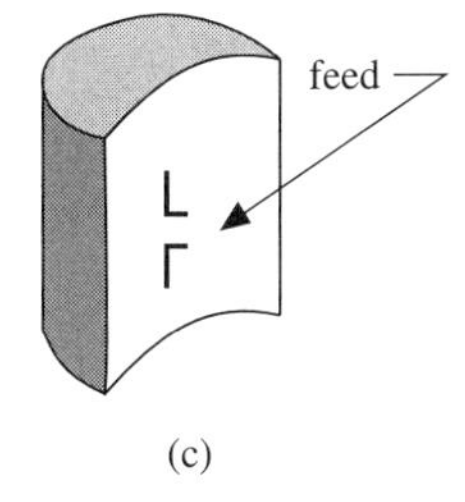

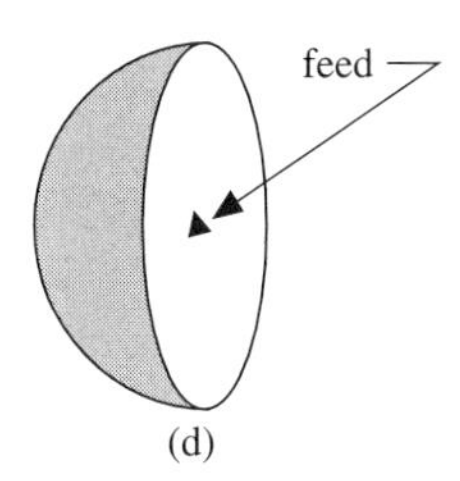

FIGURE 10.18
Reflector antennas. (a) Paraboloidal or spherical reflector-feed at focal point. (b) Paraboloidal or spherical reflector-cassegrain feed. (c) Cylindrical wave case. (d) Spherical wave case.

$$f + x = \sqrt{(f - x)^2 + y^2} \tag{10.53c}$$

Squaring both sides of (10.53c) and rearranging terms, we obtain

$$x = \frac{y^2}{2f} \tag{10.53d}$$

which is the equation of a parabola or that of a paraboloid if symmetry about the x-axis holds, i.e., case (d) in Figure 10.18.

The cassegrain type of reflector antenna is shown in Figure 10.19b. In this configuration, in the transmit mode the primary feed located at c sends a diverging beam toward the secondary reflector, hyperbolic in shape and centered at s. A reflected ray from s is then rereflected at a, resulting in a collimated beam propagating horizontally to the right.

In the cassegrain configuration, the condition analogous to (10.53c) is

$$3(f - f_2) = \sqrt{(f - x')^2 + (y')^2} + \sqrt{(x - x')^2 + (y - y')^2} + x \tag{10.54}$$

The condition (10.54) will lead to a hyperbolic equation for the secondary reflector.

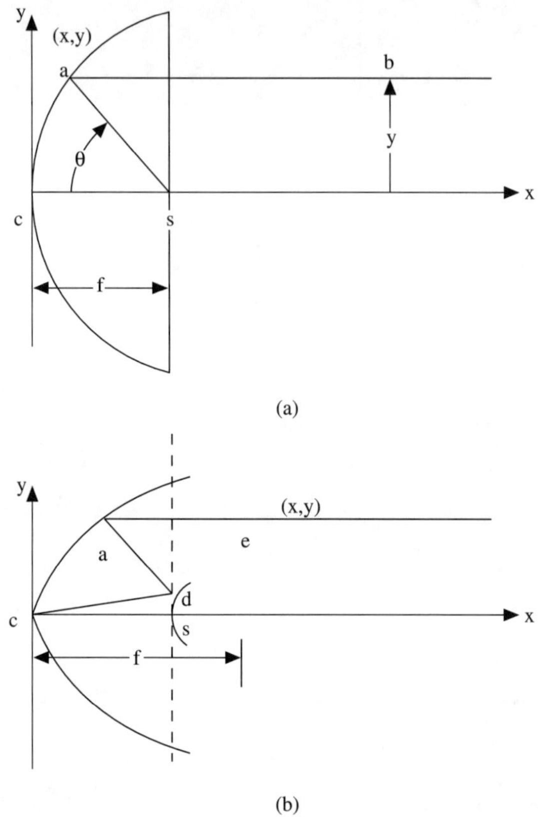

FIGURE 10.19
Principle of the parabolic reflector. (a) Feed at focal point. (b) Cassegrain feed.

Spherical rather than paraboloidal "dish" reflectors are sometimes used in reflector antennas. In spite of the fact that the beam is not perfectly collimated, angular scanning is easier to accomplish with a spherical reflector than with a paraboloidal reflector (Balanis,[4] p. 642), because of its inherent symmetry. Coverage on the spherical reflector antenna can be found in Balanis[4] (pp. 642–646).

In discussing aperture antennas in Sections 10.1, 10.2, and 10.3 we did not refer to the specifics of the physical antennas that lead to that type of theoretical modelling. An "aperture antenna" is very often either a parabolic reflector as pictured in Figure 10.18c, whose face is a rectangular aperture (and with a perfectly collimated beam and no phase distortion it is a uniformly illuminated rectangular aperture), or a paraboloidal ("dish") reflector, as in Figure 10.18d, whose face is a circular aperture. An E-plane, H-plane, or pyramidal horn is also modelled as a rectangular aperture antenna and a conical horn as a circular aperture antenna, as evidenced by our discussions in Section 10.8.1.

Obviously, lens antennas (Section 10.3.2) are also modeled as aperture antennas, rectangular or circular, depending on the configuration. The cosecant square antenna discussed in Section 10.7 can be constructed using a cylindrical reflector antenna of the type shown in Figure 10.18c, where the parabolic surface is in the elevation plane. By using multiple feeds or a reshaping of the reflector surface the beam can be modified to provide the desired coverage at the large elevation angles, which would not be covered with the standard single-feed parabolic reflector.

A major problem with reflector antennas is aperture blocking by the feed. In a cassegrain reflector antenna, that includes aperture blocking by the secondary reflector. Such blocking results in distortion of the amplitude and phase distribution of the fields on the aperture, leading in turn to departures of the far-zone patterns from their theoretical ideals. Accounting for errors due to blocking constitutes an important problem area in the design of reflector antennas.

10.8.4 Dipole and Slot Antennas

Dipole and slot antennas[41,42] are members of the general category of "linear antennas," those driven by a linear distribution of current (for a dipole, we mean electric current; for a slot, "magnetic current," a useful fictional abstraction in EM theory[43]). At microwave frequencies a dipole is most likely to be an element of an array, rather than the sole antenna for a radar transceiver. Two types of dipole antenna, the "short dipole" and the "half-wave dipole," are illustrated in Figure 10.20. A short dipole, as shown in Figure 10.20a, is one whose length is very much smaller than a wavelength and the half-wave dipole shown in (b) is fed at its center and has two current carrying wires a quarter-wave long and with current maximized at the feed gap and zero at the ends.

Consider first the short dipole whose current is in the z-direction. The electric field pattern, shown as part of Figure 10.20a, can be shown to be entirely in the θ-direction and the magnetic field in the ϕ-direction at any far-zone point around the

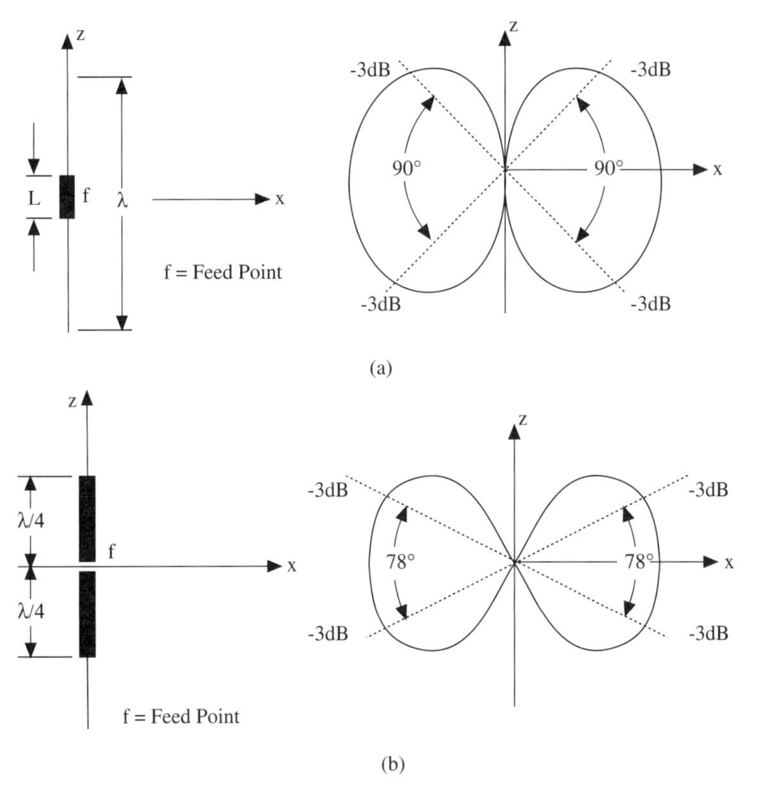

FIGURE 10.20
Dipole antennas. (a) Short dipole; $L << \lambda$. (b) Half-wave dipole.

antenna. The standard electromagnetic analysis by which we can obtain the results[44–46] begins with the magnetic vector potential

$$\mathbf{A} = \hat{z}A = \hat{z}\,\frac{\mu_0 e^{-jkr}}{4\pi r}\,I\Delta z \tag{10.55a}$$

where Δz is the dimension of the dipole current element and I the dipole current.

By using the far-zone approximation $\nabla = jk\hat{r}$ for the del-operator and the expression for the magnetic field as obtained from the magnetic vector potential, we obtain

$$\mathbf{H} = \frac{1}{\mu_0}\nabla \times \mathbf{A} = \frac{jk}{\mu_0}(\hat{r} \times \mathbf{A}) = \frac{jkA}{\mu_0}\sin\theta\,\hat{\phi} \tag{10.55b}$$

Then from one of the Maxwell equations, again invoking the far zone approximation

$$\mathbf{E} = \frac{1}{j\omega\epsilon_0}\nabla \times \mathbf{H} = -Z_w(\hat{r} \times \mathbf{H}) = \frac{jkA}{\mu_0}\sqrt{\frac{\mu_0}{\epsilon_0}}\sin\theta\hat{\theta} \tag{10.55c}$$

From (10.55a, b, c), the electric and magnetic field patterns both have the sin θ angular dependence, being symmetrical around the z-axis and therefore independent of ϕ. The field pattern is doughnut-shaped as shown in Figure 10.20a. The radiated power density dP/dA is

$$\frac{dP}{dA} = \frac{1}{2}Re[(\mathbf{E} \times \mathbf{H}^*) \cdot \hat{r}] = (\hat{\theta} \times \hat{\phi}) \cdot \hat{r}\,\frac{1}{2}\frac{k^2}{\mu_0^2}\sqrt{\frac{\mu_0}{\epsilon_0}}\,|A|^2\sin^2\theta \tag{10.55d}$$

$$= \frac{|I|^2}{8r^2}\left(\frac{\Delta z}{\lambda}\right)^2 \sin^2\theta$$

which shows the field to have its peak value at $\theta = 90°$ and nulls at $\theta = 0$.

Integration of dP/dA over the entire solid angle surrounding the dipole results in

$$P_{tot} = \frac{|I|^2}{8}\left(\frac{\Delta z}{\lambda}\right)^2 \int_0^{2\pi} d\phi \int_0^{\pi} d\theta \sin\theta \sin^2\theta = \frac{|I|^2}{6r^2}\left(\frac{\Delta z}{\lambda}\right)^2 \pi \tag{10.55e}$$

where P_{tot} is the total power radiated into space by the short dipole antenna at the radial distance r. It is evident that the total radiated power is proportional to the square of the dipole length relative to wavelength, which has been assumed small; hence not much power is radiated by a short dipole.

Equations (10.55d, e) can be used to calculate the peak gain through (2.27), as follows:

$$G_0 = 4\pi r^2\,\frac{(dP/dA)_{\theta=(\pi/2)}}{P_{tot}} = \frac{3}{2} \to 1.76\text{ dB} \tag{10.55f}$$

This is a very small gain and might help to explain why a short dipole is virtually never used as the sole antenna on a radar. There would be a reason to use it if very broad angular coverage were desired. The half-power beamwidth is 90° as shown in Figure 10.20a ($|\sin\theta| = 1/\sqrt{2}$ when $\theta = 45°$ or 135°, hence the separation between these angles around the 90° peak broadside to the dipole is 135° − 45° = 90°). Another reason it could not be used except at the extremely low end of the radar spectrum is that it would have to be impractically small to qualify as a "short" dipole, e.g., no longer than about 3 mm at X-band and no longer than about 3 cm at L-band.

However, the "short" dipole is a useful abstraction as a starting point for discussion of longer dipole antennas which overcome some of these shortcomings. The analysis is extended by replacing the constant current I in (10.55a) by an integral that accounts for the z-variation of the current along the dipole and the phase factor $e^{jk\mathbf{r}'\cdot\hat{r}}$, where $\mathbf{r}' = \hat{z}z'$, z' being the distance of an incremental element dz' from the origin. Since the z-component of $\hat{r}$ is $\cos\theta$, the integral takes the form

$$\hat{F}(\theta) = \frac{1}{L}\int_{-L/2}^{L/2} dz' I_0 i(z') e^{jkz'\cos\theta} \tag{10.56a}$$

where L is the dipole length, I_0 is the current at the origin, and $i(z')$ is the normalized current as a function of z' along the dipole. The only other angular dependence is through the factor $\sin\theta$; therefore the radiation pattern is

$$F(\theta) = \sin\theta \hat{F}(\theta)/\hat{F}(\pi/2) \tag{10.56b}$$

The most widely used antenna of this type is the half-wave dipole shown in Figure 10.20b, where $L = \lambda/2$. At X-band the required length is still impractically small (about 1.5 cm), but at the low end of the radar spectrum, e.g., L-band, it is about 15 cm and at UHF, as large as 50 cm. The principal use of a dipole in radar is as an element of an array.

The normalized current distribution for the half-wave dipole (see Fig. 10.20b) required to be symmetrical about the origin, zero at the end-points and maximized at the origin, is

$$i(z) = \sin\left(\frac{\pi}{2} - k|z|\right), \qquad |z| \leq \lambda/4 \tag{10.56c}$$

It can be shown* that the distribution (10.56c) applied to (10.56b) results in

$$F(\theta) = \frac{\cos[(\pi/2)\cos\theta]}{\sin\theta} \tag{10.56d}$$

and that the field polarization directions, the location of the peak at $\theta = 90°$ and the null at $\theta = 0$ and 180°, are the same as those of the short dipole. Calculations analogous to those in Eqs. (10.55d, e, f) lead to a peak gain of 1.64 or 2.15 dB and a half-power beamwidth in the E-plane of about 78°, which provides only slightly less angular coverage in the E-plane than the short dipole and also has omnidirectional coverage in the H-plane. The pattern is doughnut-shaped like that of the short dipole, but the "doughnut" is thinner in the half-wave dipole case.

* From (10.56a) with $L = \lambda/2$, $\hat{F}(\theta) = (I_0/\pi)\int_{-\pi/2}^{\pi/2} dy \cos y e^{jy\cos\theta} = (2I_0/\pi)\{\cos[(\pi/2)\cos\theta]/\sin^2\theta\}$. It follows that $\hat{F}(\pi/2) = 2I_0/\pi$ and hence (10.56b) implies (10.56d).

In Figure 10.21b, the use of half-wave dipoles as elements in a linear antenna array is illustrated. If this is to be a broadside array, the important feature of the dipoles is their orientation relative to the array axis. They are normal to the array axis, which ensures that the overall array pattern as given by (10.21b) and the peak gain will be dominated by the array factor $F_N(\theta, \phi)$ and will be only slightly influenced by the pattern of the dipole. The pattern cut in the x–y plane, as illustrated in Figure 10.21c, is that of the array factor, as if the elements were isotropic radiators. This is valid because the view is along the dipole axis and the element patterns are omnidirectional in the x–y plane, hence the overall pattern cut in that plane is exactly that of the array factor.

In the x–z plane (as illustrated in Fig. 10.21b) where the view is broadside to the dipoles, the pattern is dominated by that of the dipoles for smaller values of θ, particularly when θ is near 0° or 180°, since the dipole pattern has nulls at $\theta = 0°$ and 180°. When the array is operated as a broadside array, and targets of interest are not far off the H-plane of the dipoles (the x–y plane on Figure 10.21a and c) then the beamwidth is primarily determined by the array factor and is inversely proportional to the ratio (L/λ), the array length in wavelengths. If it is operated as an "endfire" array, as shown in Figure 10.21d, where the E-plane pattern peak is at $\theta = 90°$, then the dipoles play a major role in determination of the beamwidth, which is in that case inversely proportional to the square root of (L/λ). The directivity or peak gain is about 3.6 dB higher for the broadside dipole array than for a similarly configured broadside array of the same number of isotropic elements, and about 3 dB higher than that corresponding to an endfire array of isotropic elements.[47]

A "slot radiator" is the "dual" of a dipole radiator.[48] To construct such a radiator, a thin slot is cut in a perfectly conducting material, as shown in Figure 10.22a. An electric field is set up normal to the slot direction (the x-direction in the diagram) by applying a voltage in that direction. This gives rise to a "magnetic current" along the slot axis (the z-axis in the diagram), given by $(\hat{n} \times \mathbf{E})$, where $\hat{n}$ is the outward normal to the slot (in the direction normal to the paper in the diagram). The far zone pattern of the slot radiator is obtained in the same way as that for a dipole by replacing the *magnetic* vector potential $\mathbf{A}$ in (10.55a) by the *electric* vector potential $\mathbf{A}_m$, μ_0 by ϵ_0, the *electric* current I by a "magnetic current" I_m, the magnetic field $\mathbf{H}$ in (10.55b) by the electric field $\mathbf{E}$, which is now in the ϕ-direction, and the electric field $\mathbf{E}$ in (10.55c) by $\mathbf{H}$, which is now in the θ direction. This procedure is an example of the "duality principle," by which many EM problems can be attacked.[49] Equations (10.55d, e, f) and the remaining equations (10.56a, b, c, d) are unchanged except that I_0 and $i(z)$ now refer to magnetic rather than electric current. A short slot or half-wave slot show the exact same scalar behavior as a short dipole or half wave dipole, respectively, the only difference being that the E and H field polarization directions are interchanged.

Slot antennas are used in radar primarily as elements of arrays. In Figure 10.22b a waveguide slot array is shown.[50] In this type of antenna array, used at the higher microwave frequencies, slots are cut on the top surface of the waveguide on alternate sides of the center line and spaced such that their relative phase differences can be controlled to cause the beam to point in a desired direction. Such an array used at X-band or above can provide a radar system with a small antenna that can scan electronically by time varying the phase on the elements.

10.9 SCANNING OF RADAR ANTENNAS

It is customary in radar practice to scan the antenna over an angular region to locate and detect targets. For an antenna with a narrow beam, it is obvious that

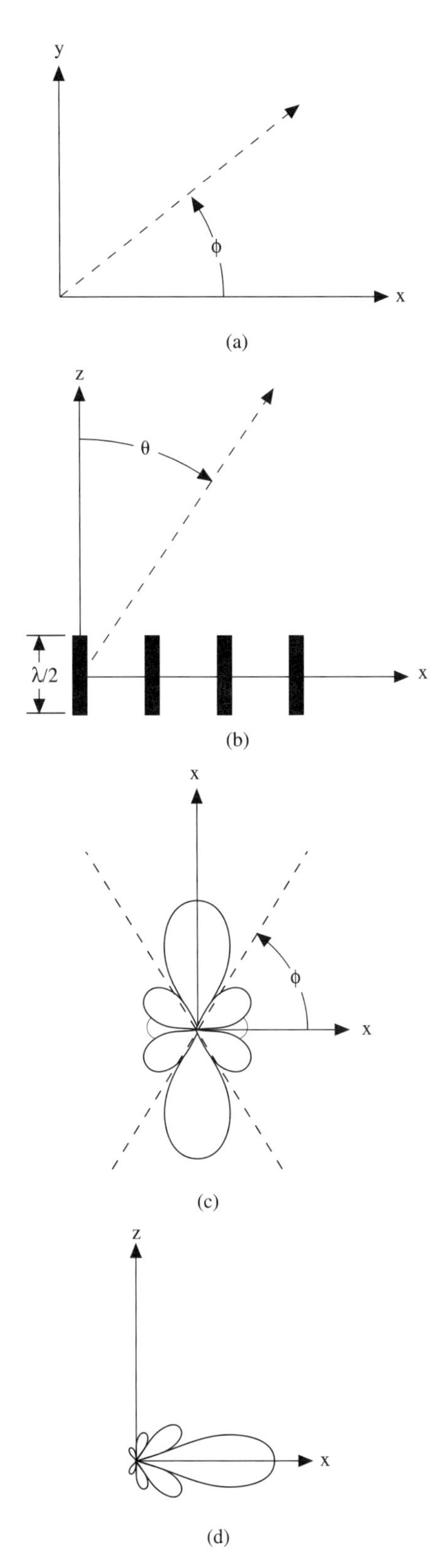

FIGURE 10.21
Dipoles as elements of an array. (a) x-y plane view (H-plane). (b) x-z plane view (E-plane). (c) H-plane pattern cut-broadside array. (d) E-plane pattern cut-endfire array.

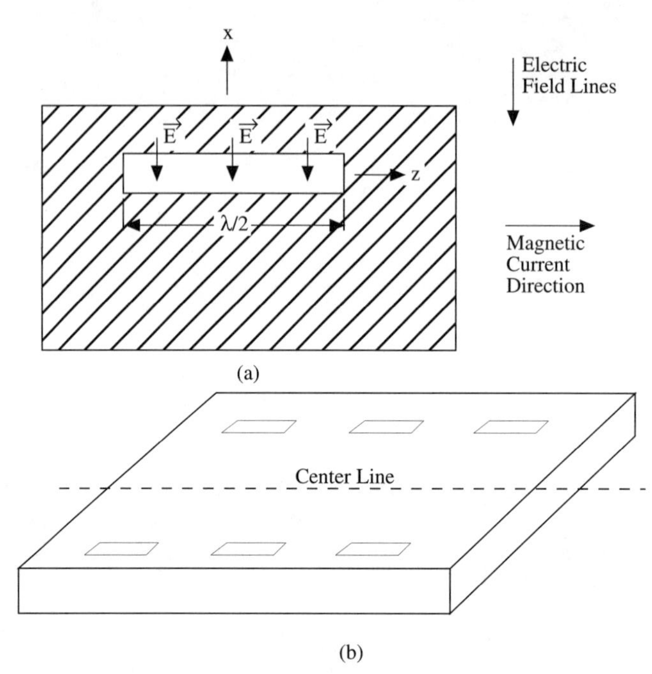

FIGURE 10.22
Slot antennas. (a) A Half-wave slot (shaded region is conducting sheet). (b) Waveguide slot array.

the time interval during which the beam is "on target" provides an indirect measure of the target's angular location. This has been alluded to in various ways throughout this book, e.g., in Section 5.8, particularly the discussion of "Effects of antenna scanning" in detection of targets; Eqs. (5.95) through (5.109) and Figure 5.27. The concept of the Swerling models discussed in Section 5.8 is based on the relationship between the antenna scanning rate and the PRF.

In Subsections 10.9.1 and 10.9.2, brief discussions of mechanical and electronic scanning of radar antennas will be presented.

10.9.1 Mechanical Scanning

Mechanical scanning is accomplished by physically rotating the aperture plane* in such a way that the beam changes its pointing direction as a function of time. When scanning in azimuth, if there is complete uncertainty about the azimuthal angle of a target or targets, the procedure is to scan continuously through 360°. If there is some prior knowledge about the target's probable range of azimuthal angles, then back-and-forth scanning over the range of angles is advisable. The strategy should be to waste as little transmitted power as possible by scanning only over those regions in which the presence of targets of interest is highly probable.

The scanning over elevation angles is nearly always back and forth, because the elevation angle can only vary from $-90°$ to $+90°$.

Figure 10.23a illustrates continuous linear scanning in azimuth. Figure 10.23b and c illustrates back-and-forth scanning in azimuth and elevation, respectively.

An alternative to mechanical scanning in elevation is the use of multiple beams of nearly equal amplitude whose aggregate covers a large region of elevation

* This can be accomplished by rotation of the complete antenna assembly or by motion of the feed in a reflector antenna.

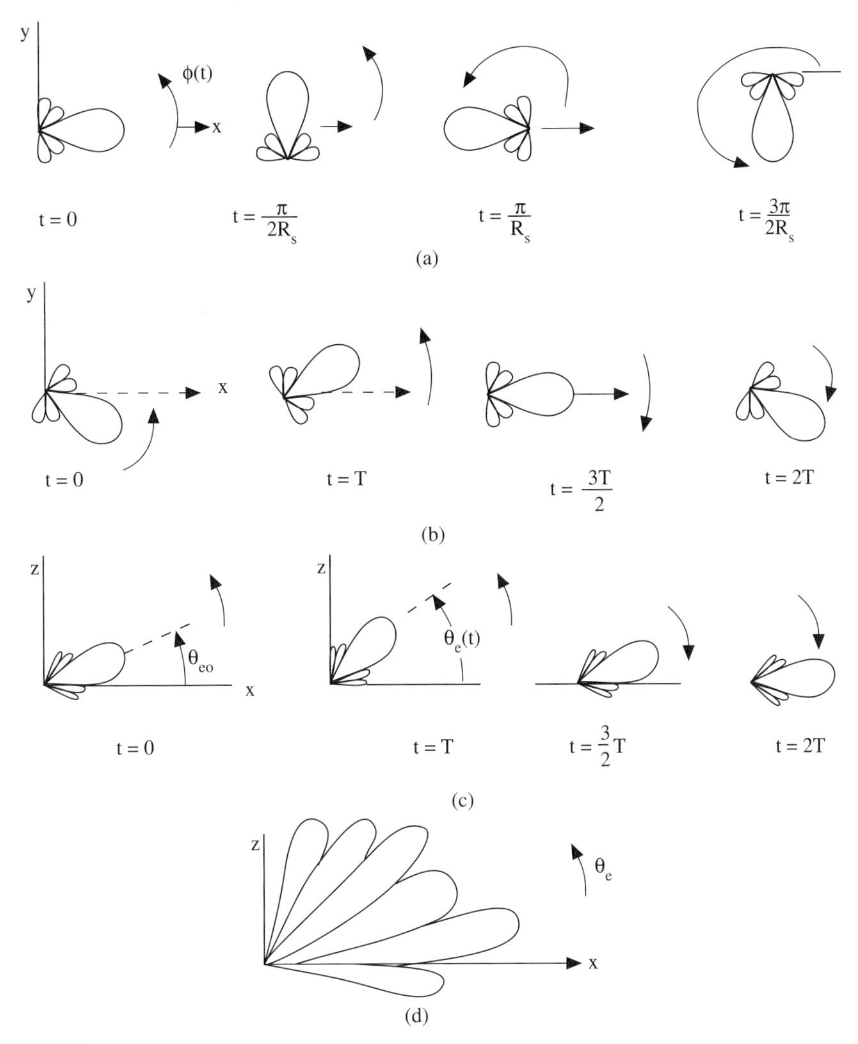

FIGURE 10.23
Mechanical scanning. (a) Continuous linear scanning in azimuth. (b) Back-and-forth scanning in azimuth. (c) Back-and-forth scanning in elevation. (d) Multiple beams covering 90 degrees of elevation angle.

angle and the separate returns to which can be resolved. In principle the same scheme could also be used for azimuth angle. This idea is illustrated in Figure 10.23d.

Consider an antenna on a search radar with a one-way power pattern $F(\theta, \phi)$(same for transmission and reception) scanning continuously and linearly past a target in the azimuth angle ϕ where the peak of the beam has a constant elevation angle θ_{B0}, as shown in Figure 10.23a. The scanning rate is R_s degrees per second. The target is assumed roughly stationary over several scans and has azimuth angle ϕ_T. The target return signal voltage envelope traced out as the peak of the beam scans past the target is

$$v_R(t) = K_0 F(\theta - \theta_{B0}, \phi - \phi_{B0}) \tag{10.57a}$$

where

$$K_0 = \sqrt{\frac{P_T G_0^2 \lambda^2 \sigma_0}{(4\pi)^3 r^4}}$$

and ϕ_B = azimuth angle of the peak of the beam, equal to $\phi_{B0} + R_s t$.

Expanding $F(\theta, \phi)$ in a Taylor series in ϕ about the target's azimuth angle ϕ_T, we obtain from (10.57a)

$$F(\theta, \phi) = F(\theta, \phi_T) + \left(\frac{\partial F(\theta, \phi)}{\partial \phi}\right)_T \Delta\phi + \frac{1}{2}\left(\frac{\partial^2 F(\theta, \phi)}{\partial \phi^2}\right)_T (\Delta\phi)^2 + \cdots \tag{10.57b}$$

where $(\partial^n F(\theta, \phi)/\partial\phi^n)_T$ is the nth derivative of $F(\theta, \phi)$ at $\phi = \phi_T$ and $\Delta\phi = \phi = \phi_T$.

To be specific about the pattern function, consider a rectangular aperture antenna with uniform illumination in the horizontal direction. Assuming a horizontal aperture dimension a, the horizontal pattern cut is

$$F(\theta, \phi) = \left\{\text{sinc}\left[\frac{\pi a}{\lambda}\cos(\theta - \theta_{B0})\sin(\phi - \phi_B)\right]\right\}^2 \tag{10.57c}$$

which must reach its maximum when $\phi_B = \phi_T$. It is evident from the form of (10.57c) that the peak of the pattern is reached when the argument of the sinc function is zero, which occurs when $\phi = \phi_B = \phi_{B0} + R_s t_T$, where t_T must coincide with the time at which the azimuth angle of the peak of the beam equals that of the target, or equivalently

$$t_T = \frac{(\phi_T - \phi_{B0})}{R_s} \tag{10.57d}$$

We note from (10.57b) that the expansion should be about ϕ_B, since $\phi_B = \phi_T$. For the specific pattern in (10.57c), under the assumption that the beam is very narrow, we can approximate the sinc function of x by $1 - (x^2/6)$ and hence, if we assume $\theta = \theta_{B0}$ (i.e., scanning in the principal elevation plane)

$$F(\theta, \phi) = 1 - \frac{1}{3}\left[\frac{\pi a}{\lambda}\sin(\phi - \phi_B)\right]^2 \tag{10.57e}$$

The terms of (10.57b) take the form

$$\begin{aligned} F(\theta, \phi_T) &= 1 \\ \left(\frac{\partial F}{\partial \phi}(\theta, \phi)\right)_T &= -\frac{2}{3}\left[\frac{\pi a}{\lambda}\sin(\phi - \phi_B)\right]\cos(\phi - \phi_B) = 0 \\ \left(\frac{\partial^2 F(\theta, \phi)}{\partial \phi^2}\right)_T &= -\frac{2}{3}\frac{\pi a}{\lambda} \end{aligned} \tag{10.57f}$$

From (10.57a, e, f) where $\phi = \phi_B + R_s(t - t_T)$, the voltage envelope traced out by the beam as it scans past the target is

$$v_R(t) = K_0\left[1 - \frac{\pi^2}{3}\left(\frac{a}{\lambda}\right)^2 R_s^2(t - t_T)^2\right] \quad (10.57g)$$

where t_T is given by (10.57d) and $\sin(\phi - \phi_B)$ is approximated by $(\phi - \phi_B)$ since $(\phi - \phi_B)$ is a small angle in the region of interest.

In Figure 10.24a, rough plots of $v_R(t)$ vs. $(t - t_T)$ are shown for a few values of (a/λ) for fixed R_s and in Figure 10.24b, sketches are shown for a few values of R_s for fixed a/λ. A constraint is that (a/λ) must be significantly greater than unity to uphold the assumption of a narrow beam.

The important parameter with a scanning antenna is the "dwell time" roughly defined in Chapter 2, Section 2.11, as the time interval during which the beam is "on-target." There are various ways in which we could define the dwell time, but a way to view its meaning is to conceive a perfectly flat beam of beamwidth B (angle in degrees between -3 dB points) scanning past the target. For that case, there is a nonzero signal return between $t = t_T - B/2R_s$ and $t = t_T + B/2R_s$ and a zero return if $|t - t_T| > B/2R_s$. Then the beam is "on target" for a time interval B/R_s and returns a zero signal outside that interval. The dwell time in this case is unequivocally defined as

$$T_d = \frac{B}{R_s} \quad (10.57h)$$

For a realistic antenna pattern, the "on target" state can be defined, for example, as that fulfilling the condition that the signal power is greater than or equal

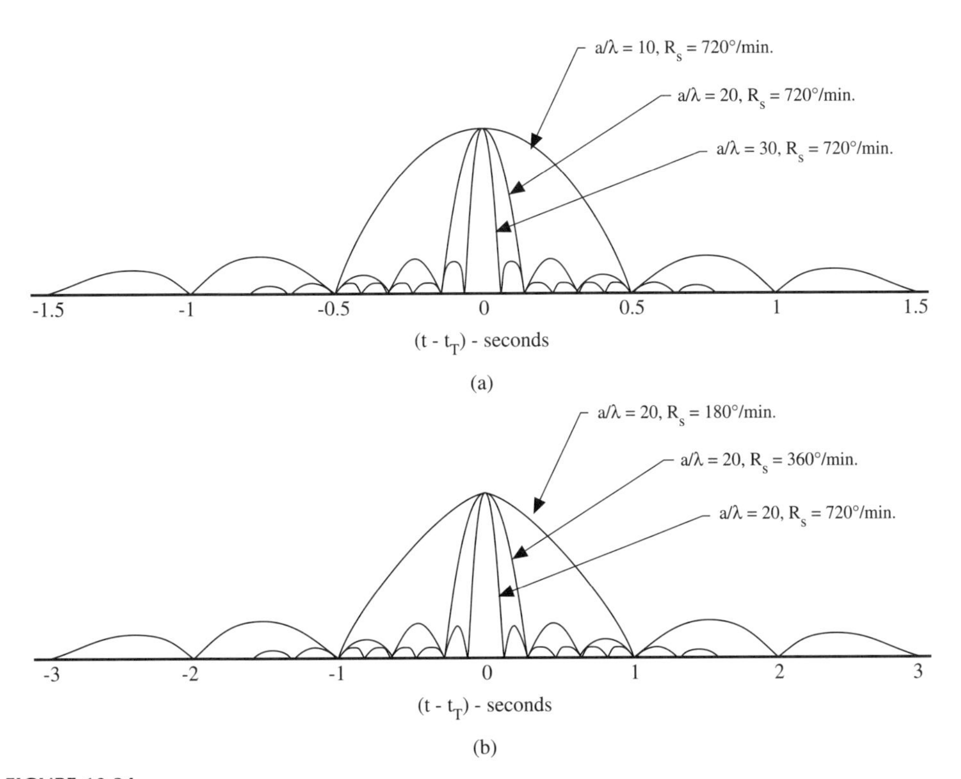

FIGURE 10.24
Time variation of signal voltage due to scanning.

to one-half of its value at the instant when the beam traverses the target. If the dwell time is defined this way, a 3-dB "scanning loss" will be incurred in the received power.

The basis of the scanning loss is the fact that during the dwell, the target signal level is between that at the peak of the beam and that at 3 dB below the peak, i.e., it is always above the latter, but not as large as the former except at the instant $t = t_T$. Therefore a 3-dB "safety factor" should be included when calculating the SNR for purposes of evaluating detection probability or parameter measurement accuracy.

We could also define the dwell time differently, e.g., by defining the beamwidth as spacing between 6 dB points, in which case T_d will increase and the scanning loss must be redefined as 6 dB.

If we define B as the -3 dB beamwidth, then the dwell time is obtained through (10.57g) from the condition

$$\frac{v_R(T_d/2)}{K_0} = 1 - \frac{\pi^2}{3}\left(\frac{a}{\lambda}\right)^2 R_s^2 \left(\frac{T_d}{2}\right)^2 = \frac{1}{\sqrt{2}} \tag{10.57i}$$

from which it follows that

$$T_d = \frac{2.205}{R_s}\left(\frac{\lambda}{a}\right) \tag{10.57j}$$

with a scanning loss of 3 dB.

If we are willing to accept a 6 dB scanning loss, then $1/\sqrt{2}$ is replaced by $1/2$ on the right-hand side of (10.57i) and the recalculated value of T_d exceeds that in (10.57j) by a factor 1.306, i.e., a 31% increase in dwell time.

If we were to invoke (10.10d) in (10.57j) we could express T_d in terms of the half-power beamwidth $B_{1/2}$, which is about $0.844\,\lambda/a$ according to (10.10d). Rewriting (10.57j) in terms of half-power beamwidth results in

$$T_d = 2.5\left(\frac{B_{1/2}}{R_s}\right) \tag{10.57k}$$

For a different beamshape, as long as the beam is very narrow, the dwell time should be quite close to that given in (10.57k), although changes in shape would result in some changes in the coefficient.

10.9.2 Electronic Scanning

Electronic scanning can be accomplished by time varying the spatial phase or amplitude distribution over an aperture or on the elements of an array or by varying the frequency. It does not require physical motion of the antenna and hence offers enormous practical advantages over mechanical scanning. This is particularly advantageous in airborne or satellite-borne radars, where size and weight are highly significant.

It also has a few *dis*advantages, for example, it changes the shape of the antenna pattern and usually the gain, beamwidth, and sidelobe levels as it points the peak of the beam in different directions.

Let us concentrate on electronic scanning through variation of the phase distribution along a linear array axis (a "phased array"). From (10.22), the argument of the array factor is

$$u = \frac{2\pi d}{\lambda} \cos \beta + \alpha d \tag{10.58a}$$

The peak of the beam occurs when $u = 0$. This applies to both uniform and tapered array excitation. The condition $u = 0$ occurs when

$$\beta = \beta_{peak} = \cos^{-1}\left(-\frac{\alpha\lambda}{2\pi}\right) \tag{10.58b}$$

To be specific about the array axis direction, we place it in the x-direction, as shown in Figure 10.25. This implies that

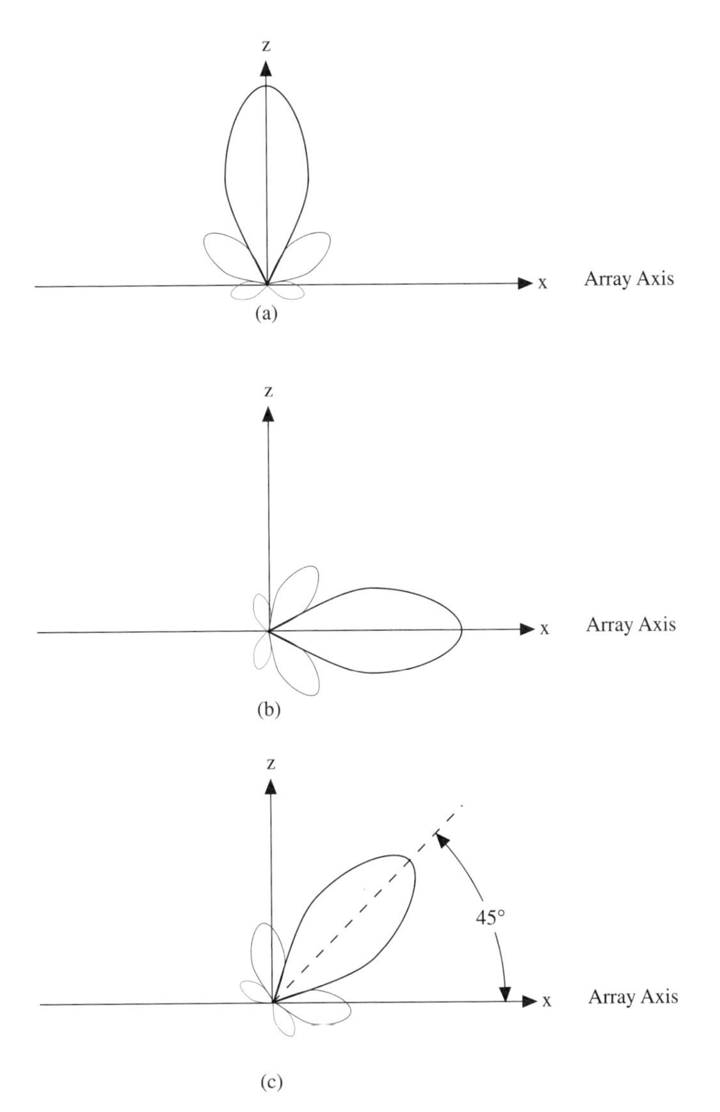

FIGURE 10.25
Effect of phase rate on beam direction. (a) Broadside array; $\alpha = 0$. (b) Endfire array; $\alpha = -2\pi/\lambda$. (c) $\alpha = -\sqrt{2}\pi/\lambda$; beam at 45°.

$$\cos \beta_m = \sin \theta \cos \phi = -\frac{\alpha\lambda}{2\pi} \tag{10.58c}$$

where β_m is the value of β at the peak of the beam.

Consider a cut in the x–z principal plane, as shown in Figure 10.25a, such that $\phi = 0$. From (10.58b, c)

$$\theta_m = \sin^{-1}\left(-\frac{\alpha\lambda}{2\pi}\right) \tag{10.58d}$$

If $\alpha = 0$, then $\theta_{peak} = 0$ (Fig. 10.25a), implying a broadside array. If $\alpha = -2\pi/\lambda$, then from (10.58c), $\theta_{peak} = 90°$, indicating an endfire array (Fig. 10.25b). If $\alpha = -\sqrt{2}\pi/\lambda$, the beam points at 45° relative to the array axis (Fig. 10.25c). It is clear from this simple example that the beam can be pointed in an arbitrary direction between the two extremes of broadside and endfire by varying the rate of linear phase variation along the array axis. We can also apply this concept to a planar array, with two mutually orthogonal array axes, say x- and y-axes, and a two-dimensional grid of elements distributed at intervals d_x and d_y in x- and y-directions, respectively, as illustrated in Figure 10.26. The array factor can be generalized as the product of x and y array factors, i.e.,

$$F_{MN}(u, v) = F_M(u)F_N(v) \tag{10.59a}$$

where

$$u = \frac{2\pi d_x}{\lambda} \sin \theta \cos \phi + \alpha_x d_x$$

$$v = \frac{2\pi d_y}{\lambda} \sin \theta \sin \phi + \alpha_y d_y$$

where α_x and α_y are the linear phase variation rates on the x- and y-axes, respectively. The peak of the beam can be defined as occurring at $u = v = 0$, hence the generalization of (10.58c) for this case is

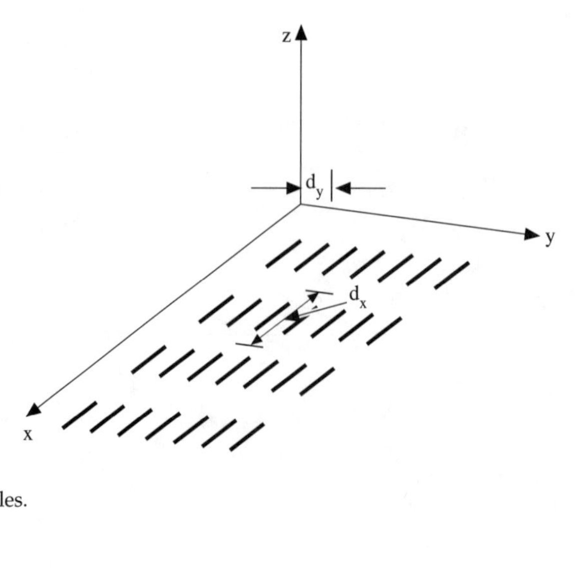

FIGURE 10.26
Planar array of dipoles.

$$\sin\theta_{\mathrm{m}}\cos\phi_{\mathrm{m}} = -\frac{\alpha_x\lambda}{2\pi}$$

$$\sin\theta_{\mathrm{m}}\sin\phi_{\mathrm{m}} = -\frac{\alpha_y\lambda}{2\pi} \tag{10.59b}$$

From Eqs. (10.59b), we can infer that the peak of the beam in elevation and azimuth directions occurs when

$$\theta_{\mathrm{m}} = \sin^{-1}\left(\frac{\lambda}{2\pi}\sqrt{\alpha_x^2+\alpha_y^2}\right) \qquad \text{(elevation peak)}$$

$$\phi_{\mathrm{m}} = \tan^{-1}\left(\frac{-\alpha_y}{-\alpha_x}\right) = \sin^{-1}\left(\frac{-\alpha_y}{\sqrt{\alpha_x^2+\alpha_y^2}}\right) = \cos^{-1}\left(\frac{-\alpha_x}{\sqrt{\alpha_x^2+\alpha_y^2}}\right)$$

$$\text{(azimuth peak)} \tag{10.59c}$$

The planar array behaves very much like a rectangular aperture; hence we could apply these same ideas to a rectangular aperture in the x–y plane with linear phase variations α_x and α_y radians/meter in x- and y-directions, respectively. The argument parameters u, v for that case are the same as those of (10.59a) except that d_x and d_y now represent the x and y aperture dimensions rather than the interelement spacings in these two directions as in an array. Hence, since d_x and d_y are not required in (10.59c), it follows that it is applicable to either a planar array or a rectangular aperture. If we are interested in the variation of the boresight direction in only one principal plane, then the one-dimensional form (10.58d) is equally applicable to a linear array or a linear illumination function in one direction of an aperture.

To accomplish electronic scanning through phase variation with either an aperture antenna or an array, it is necessary to establish a time variation of α_x and/or α_y. Using (10.59c) and letting α_x and α_y be linear functions of time of the form

$$\alpha_{\substack{x\\y}}(t) = \alpha_{\substack{0x\\y}} + \dot{\alpha}_{\substack{x\\y}}t \tag{10.60a}$$

we vary elevation and azimuth angles as *nonlinear* functions of time. In radar antenna scanning, the time variation of θ_{peak} or ϕ_{peak} should be linear. It is evident that linear variation of these angles cannot necessarily be achieved by linear variation of α_x or α_y.

To examine the rates of change of θ_{m} and ϕ_{m} as functions of α_x, α_y, and their respective rates of change $\dot{\alpha}_x$, $\dot{\alpha}_y$, we differentiate Eqs. (10.59c) with respect to time, obtaining the results

$$\dot{\theta}_{\mathrm{m}} = \pm\left(\frac{\lambda}{2\pi}\right)\frac{(\alpha_x\dot{\alpha}_x+\alpha_y\dot{\alpha}_y)}{\sqrt{\alpha_x^2+\alpha_y^2}\sqrt{1-(\lambda/2\pi)^2(\alpha_x^2+\alpha_y^2)}}, \qquad \begin{array}{l} +\text{ if } 0\le\theta_{\mathrm{m}}\le\dfrac{\pi}{2} \\ -\text{ if } \dfrac{\pi}{2}\le\theta_{\mathrm{m}}\le\pi \end{array} \tag{10.60b}$$

where $(\alpha_x^2+\alpha_y^2) < (2\pi/\lambda)^2$

$$\dot{\phi}_{\mathrm{m}} = \frac{\alpha_x\dot{\alpha}_y-\alpha_y\dot{\alpha}_x}{(\alpha_x^2+\alpha_y^2)} \tag{10.60c}$$

From (10.60b, c), we see that if $\dot{\alpha}_x$ and $\dot{\alpha}_y$ are constant, then $\dot{\theta}_{\mathrm{m}}$ and $\dot{\phi}_{\mathrm{m}}$ are still not constant, because α_x and α_y are functions of time. This implies that linear scanning in θ_m and ϕ_m are not attainable through linear variations of α_x and α_y.

Further coverage of electronic scanning can be found in some standard references on radar systems.[51–53]

REFERENCES

1. Collin, R.E., "Antennas and Radiowave Propagation." McGraw-Hill, New York, 1985.
2. Stratton, J.A., "Electromagnetic Theory." McGraw-Hill, New York, 1941.
3. Stratton, J.A., "Electromagnetic Theory." McGraw-Hill, New York, 1941, p. 466.
4. Balanis, C.A., "Antenna Theory—Analysis and Design." Harper & Row, New York, 1982.
5. Silver, S. (ed.), "Microwave Antenna Theory and Design." McGraw-Hill, New York, 1949.
6. Kraus, J.D., "Antennas," 2nd ed. McGraw-Hill, New York, 1988.
7. Griffiths, J., "Radiowave Propagation and Antennas." Prentice-Hall, Englewood Cliffs, NJ, 1987.
8. Elliott, R.S., "Antenna Theory and Design." Prentice-Hall, Englewood Cliffs, NJ, 1981.
9. Stutzman, W.L., and Thiele, G.A., "Antenna Theory and Design." Wiley, New York, 1987.
10. Wolff, E.A., "Antenna Analysis." Artech, Norwood, MA, 1988.
11. Skolnick, M.I., "Introduction to Radar Systems," 2nd ed. McGraw-Hill, New York, 1980, Chapters 7 and 8.
12. Barton, D.K., "Modern Radar System Analysis." Artech, Norwood, MA, 1988, Chapter 4.
13. Eaves, J.L., and Reedy, E.K. (eds.), "Principles of Modern Radar." Van Nostrand Reinhold, New York, 1987; Chapter 6, by D. Bodnar.
14. Spiegel, M., "Mathematical Handbook." McGraw-Hill, New York, 1968.
15. Abramowitz, M., and Stegun, I.A., "Handbook of Mathematical Functions." Dover, New York, 1965.
16. Collin, R.E., "Antennas and Radiowave Propagation." McGraw-Hill, New York 1985, Chapter 3, Sections 3.6, 3.7.
17. Balanis, C.A., "Antenna Theory—Analysis and Design." Harper & Row, New York, 1982, Chapter 6.
18. Kraus, J.D., "Antennas," 2nd ed. McGraw-Hill, New York, 1988, Chapter 4.
19. Griffiths, J., "Radiowave Propagation and Antennas." Prentice-Hall, Englewood Cliffs, NJ, 1987, Chapter 5.
20. Skolnick, M.I, "Introduction to Radar Systems," 2nd ed. McGraw-Hill, New York, 1980, Chapter 8.
21. Barton, D.K., "Modern Radar System Analysis." Artech, Norwood, MA, 1988, Section 4.3, pp. 164–187.
22. Collin, R.E., "Antennas and Radiowave Propagation." McGraw-Hill, New York, 1985, Chapter 4; especially pp. 174, 175, 206–215.
23. Skolnick, M.I., "Introduction to Radar Systems," 2nd ed. McGraw-Hill, New York, 1980, Section 7.2, pp. 228–235, especially Table 7.1, p. 232.
24. Barton, D.K., "Modern Radar System Analysis." Artech, Norwood, MA, 1988, pp. 151–157, especially Table 4.1, p. 154; also Section 4.4, pp. 188–196.
25. Barton, D.K., "Modern Radar System Analysis." Artech, Norwood, MA, 1988, pp. 162–164.
26. Skolnick, M.I., "Introduction to Radar Systems," 2nd ed. McGraw-Hill, New York, 1980, Section 7.7, pp. 258–261.
27. Blake, L.V., "Radar Range Performance Analysis." Artech, Norwood, MA, 1986, pp. 375–376.
28. Rohan, P., "Surveillance Radar Performance Prediction." Peter Peregrinus, 1983, pp. 229–238, especially Figure 10.6, p. 230 and Figure 10.7, p. 232.
29. Blake, L.V., "Radar Range Performance Analysis." Artech, Norwood, MA, 1986, Eq. (8.23), p. 376.
30. Collin, R.E., "Antennas and Radiowave Propagation." McGraw-Hill, New York, 1985, Section 4.4, pp. 179–189.
31. Balanis, C.A., "Antenna Theory—Analysis and Design." Harper & Row, New York, 1982, Chapter 12.
32. Kraus, J.D., "Antennas," 2nd ed. McGraw-Hill, New York, 1988, Sections 13-8 to 13-15, pp. 644–659.
33. Collin, R.E., "Antennas and Radiowave Propagation." McGraw-Hill, New York, 1985, Section 4.5, pp. 194–199.
34. Balanis, C.A., "Antenna Theory—Analysis and Design." Harper & Row, New York, 1982, Section 13.6, pp. 646–654.
35. Kraus, J.D., "Antennas," 2nd ed. McGraw-Hill, New York, 1988, Chapter 14.
36. Collin, R.E., "Antennas and Radiowave Propagation." McGraw-Hill, New York, 1985, Sections 4.6–4.11, pp. 199–260.
37. Balanis, C.A., "Antenna Theory—Analysis and Design." Harper & Row, New York, 1982, Sections 13.4, 13.5, pp. 604–645.

38. Kraus, J.D., "Antennas," 2nd ed. McGraw-Hill, New York, 1988, Chapter 12.
39. Skolnick, M.I., "Introduction to Radar Systems," 2nd ed. McGraw-Hill, New York, 1980, Sections 7.3, 7.4, pp. 235–248.
40. Barton, D.K., "Modern Radar System Analysis." Artech, Norwood, MA, 1988, pp. 158–161.
41. Collin, R.E., "Antennas and Radiowave Propagation." McGraw-Hill, New York, 1985, Sections 4.12, 4.13, pp. 261–273.
42. Kraus, J.D., "Antennas," 2nd ed. McGraw-Hill, New York, 1988, Chapter 13, Sections 13-2 to 13-7, pp. 624–644.
43. Collin, R.E., "Antennas and Radiowave Propagation." McGraw-Hill, New York, 1985, pp. 175–182.
44. Collin, R.E., "Antennas and Radiowave Propagation." McGraw-Hill, New York, 1985, pp. 19–25.
45. Shen, C.S., and Kong, J.A., "Applied Electromagnetism," 2nd ed. PWS Publishers, Boston, 1987, Sections 7.1–7.2, pp. 183–193.
46. Balanis, C.A., "Antenna Theory—Analysis and Design." Harper & Row, New York, 1982, Chapter 4, especially pp. 100–103.
47. Collin, R.E., "Antennas and Radiowave Propagation." McGraw-Hill, New York, 1985, pp. 111–118.
48. Collin, R.E., "Antennas and Radiowave Propagation." McGraw-Hill, New York, 1985, Section 4.12, pp. 261–265.
49. Balanis, C.A., "Antenna Theory—Analysis and Design." Harper & Row, New York, 1982, Section 3.7, pp. 93–94.
50. Collin, R.E., "Antennas and Radiowave Propagation." McGraw-Hill, New York, 1985, Section 4.13, pp. 265–271.
51. Skolnick, M.I., "Introduction to Radar Systems," 2nd ed. McGraw-Hill, New York, 1980, Section 7.4, pp. 244–248 and Chapter 8, pp. 278–342.
52. Barton, D.K., "Modern Radar System Analysis." Artech, Norwood, MA, 1988, pp. 165–187.
53. Carpentier, M., "Principles of Modern Radar Systems." Artech, Norwood, MA, 1988, Chapter 8.

Chapter 11

Topics in Radar Propagation

CONTENTS

One should not forget that radar involves the propagation of radio waves at microwave frequencies in the vicinity of the earth. The science of terrestrial radio propagation has a great many important ideas to contribute to the art of radar engineering, and the effects of certain propagation phenomena on radar performance can be extremely important.

Propagation phenomena of importance in radar are, among others,

1. atmospheric refraction (ray-bending),
2. atmospheric- and precipitation-induced attenuation,
3. interference due to earth surface reflections (source of multipath),

4. the radio horizon,
5. diffraction,
6. effects of rough surface terrain, and
7. effect of polarization.

We will consider briefly each of these phenomena and some of their implications for radar system analysis. A classical reference on terrestrial microwave radio propagation is "Propagation of Short Radio Waves," edited by D.E. Kerr, which is Volume 13 of the MIT Radiation Laboratory Series.[1] Although published in 1951, it is still used as a basic reference on the subject. Other much more recent references were consulted by the author in developing this material, a few of them written largely in the context of microwave communications technology.[2–6] Also, some of the standard texts on radar contain chapters or sections on propagation concerns in radar systems,[7–11] also consulted in development of this chapter.

11.1 REFRACTION-RAY BENDING

In all of our thinking about the propagation of a radar wave in this book, we have conceived of the waves as traveling in straight lines ("line-of-sight" propagation). In reality, radio waves traveling over the earth's surface are propagating in a stratified medium (air), which is decreasing in density with altitude. If a wave propagates over a long distance, it is continuously refracted. Snell's law of refraction predicts that a wave propagating from a dense to a less dense medium will be bent away from the normal to the interface between the media and a wave propagating into a denser medium will be bent toward the normal. Figure 11.1 illustrates this for a layered medium model consisting of infinite slabs of material gradually changing their refractive index.

A wave propagating in a medium numbered 1 with wave number k_1 incident on a medium numbered 2 with wave number k_2 undergoes a refraction or bending process. Assuming that the angle of incidence [i.e., angle relative to normal in medium (1)] is θ_1 and the angle of refraction [the angle, relative to the normal, of the wave propagating in medium (2)] is θ_2, Snell's law tells us that

$$k_1 \sin \theta_1 = k_2 \sin \theta_2 \tag{11.1}$$

Equation (11.1) written in the form

$$\sin \theta_2 = \left(\frac{k_1}{k_2}\right) \sin \theta_1 \tag{11.2}$$

implies that if k_1 exceeds k_2, the angle θ_2 exceeds θ_1, i.e., the wave is bent *away from* the normal, and if k_2 exceeds k_1, then θ_1 exceeds θ_2, i.e., the wave is bent *toward* the normal. The former situation is illustrated in Figure 11.1a where the wave propagates *upward* and the latter is illustrated in Figure 11.1b where the wave propagates *downward*.

The net effect of the above is that, even over a short range where the earth could be considered flat, a wave from the radar to a target at a higher altitude travels in a curved path that is concave upward as shown in Figure 11.2, and the return wave from the target also travels that same path in the reverse direction. This is due

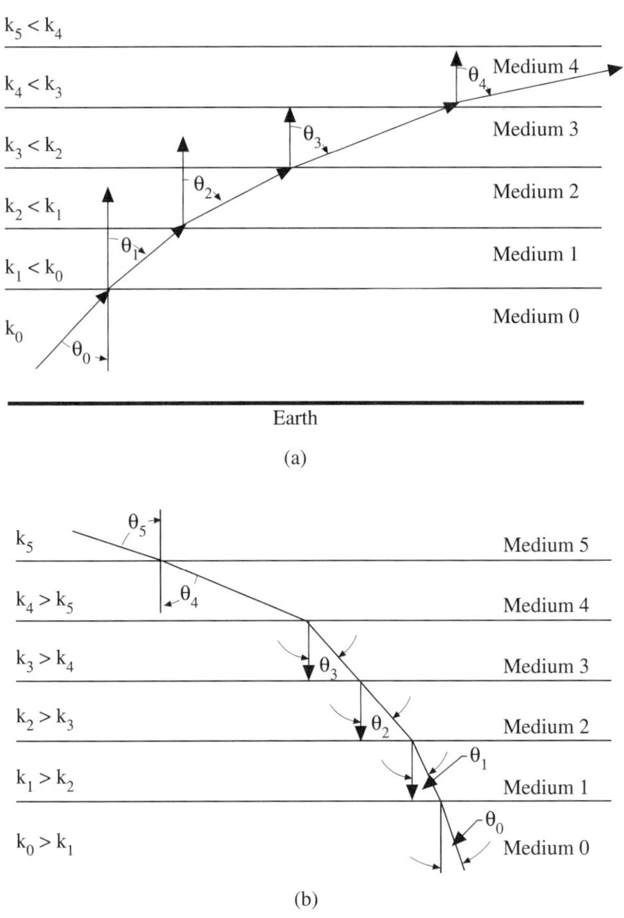

FIGURE 11.1
Simple ideas of refraction. (a) Upward propagation. (b) Downward propagation.

to the continuous *decrease* of refractive index [the ratio k_2/k_1] as the altitude decreases. The effect can be modelled with a layered medium as in Figure 11.1, where the layers are allowed to become progressively thinner until one approaches a continuous variation of refractive index with altitude.

The curvature (reciprocal radius of curvature) of a radio wave propagating at an oblique angle relative to the horizontal can be shown to be given by[12]

$$\rho = -\frac{d\nu(z)}{dz} \tag{11.3}$$

where $\nu(z)$ is the atmospheric refractive index, which generally decreases with increasing altitude z. The experimentally derived functional dependence of $\nu(z)$ on z can be obtained from an exponential model for the "refractivity" of the troposphere, the part of the atmosphere where most common radar propagation takes place. The refractivity N and refractive index ν are related by the expression

$$N = (\nu - 1) \times 10^6 \tag{11.4}$$

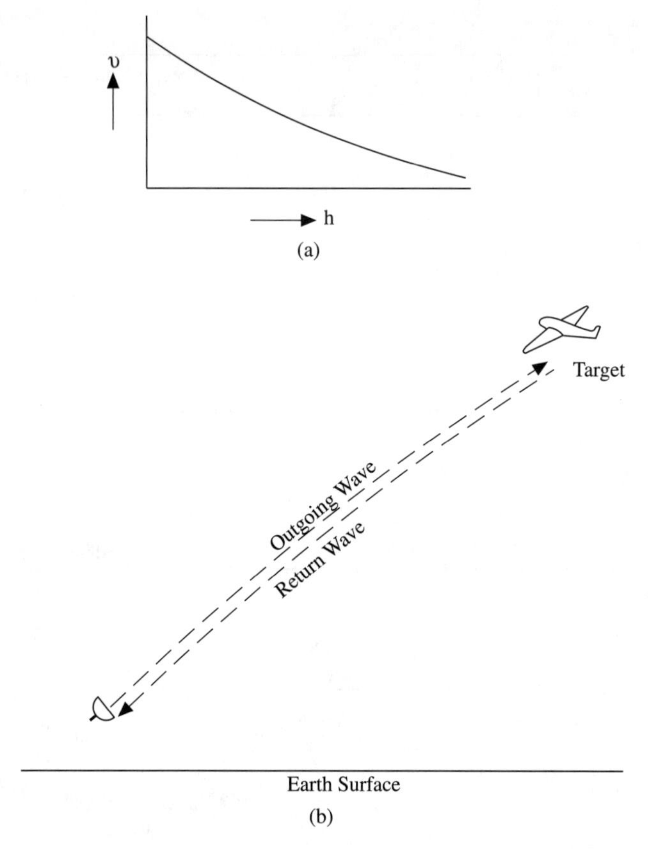

FIGURE 11.2
The curved path of a radar wave. (a) Refractive index ν vs. altitude h. (b) Ray paths on transmission and return.

and the exponential law for $N(z)$ is

$$N(z) = N(z_0)e^{-\alpha(z-z_0)} \tag{11.5}$$

where Barton[13] gives $N(z_0)$ as 313 and α as 0.0001439 m^{-1} and z_0 is sea level.
From (11.4) and (11.5)

$$-\frac{d\nu(z)}{dz} = -10^{-6} \times \frac{dN(z)}{dz} = \alpha N(z) \times 10^{-6} = 4.504(10^{-8})\ \text{m}^1 \text{ at } z = z_0 \tag{11.6}$$

using Barton's numbers.

Looking at Figure 11.3 for the spherical earth, we note that the curved ray-path delivers power to points below the horizon that would be invisible to the radar in the absence of refraction.

The actual radius of the earth is about 3440 nmi or equivalently about 6340 km. The "earth curvature," the reciprocal of the radius of curvature, is about 15.8 $\times\ 10^{-8}\ \text{m}^{-1}$. The difference between the curvature of the ray-path and that of the earth, i.e., the curvature of the ray *relative* to that of the earth is $(-d\nu/dz) - (1/r_a)$, where r_a is the actual earth radius. That difference is $(4.78 - 15.8)(10^{-8}) = -11.02(10^{-8})$. Suppose a hypothetical earth with a smaller curvature $1/r_e$ were used to model radar propagation. If that curvature were equal to $-[(-d\nu/dz) - (1/r_a)]$

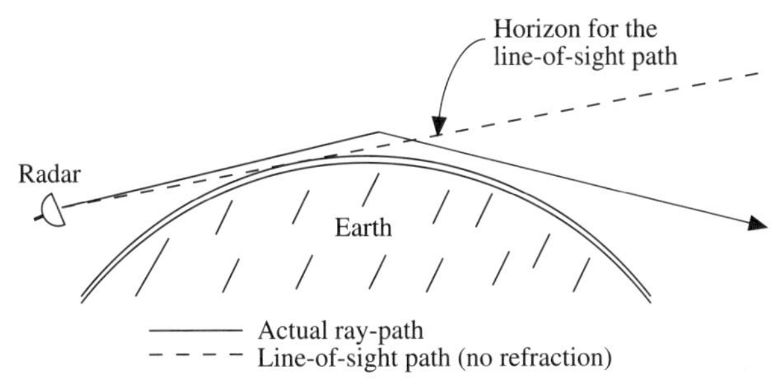

FIGURE 11.3
Radar wave propagating over spherical earth.

$= (dv/dz) + (1/r_a) = 11.02(10^{-8})$, then the ratio of the *actual* earth curvature to this new hypothetical earth's curvature, or equivalently the ratio of the radius of the hypothetical earth to that of the actual earth, is

$$\frac{r_e}{r_a} = \frac{15.8(10^{-8})}{11.3(10^{-8})} = 1.398 \tag{11.7}$$

The usual number used to allow consideration of the radar wave as propagating in a straight line is 4/3, which is about 1.333. The number arrived at in (11.7) is about 4.9% higher than 4/3. However, since the number is dependent on the parameter dv/dz, which is in turn dependent on location and on various geophysical parameters, the use of the "four-thirds earth radius" is usually satisfactory in radar analysis unless extreme accuracy is required. That might be the case, for example, if we were analyzing the effect of ray-bending on target angle measurement and required very high precision.

11.2 INTERFERENCE DUE TO EARTH'S SURFACE: MULTIPATH AND THE PROPAGATION FACTOR

A propagation effect that is sometimes extremely important in radar is that of interference due to the presence of the earth. That is the phenomenon that gives rise to the most common form of "multipath," the existence of indirect paths between radar and target due to earth reflections.[14–19]

11.2.1 Flat Smooth Earth

In standard radar system modeling we view the radar and the target as if they were both immersed in infinite free space. But in virtually all terrestrial radar environments, there are earth-based effects, i.e., reflection of the transmitted radar wave from the earth's surface in the direction of the target and reflection of the target-scattered wave from the earth's surface in the direction of the radar. If the flat earth approximation is used and the earth can be modeled as a perfectly smooth flat plane reflector, then Figure 11.4 illustrates the effects we are alluding to here. The result of these earth reflections, as shown in Figure 11.4, is a set of additional indirect paths undergone by the transmitted radar signal before it returns to the receiver. The fact that these are superposed on the direct path from radar-to-target-to-radar leads to the nomenclature "multipath" to describe what happens.

In Figure 11.4a, the direct (radar-to-target-to-radar) path is shown. That is the only path we have been dealing with up to now. In Figure 11.4b the radar-to-earth reflection point-to-target-to-radar path is shown. In Figure 11.4c we show the reversed traversal of path (b), radar-to-target-to-earth reflection point-to-radar. Finally in Figure 11.4d we see the radar-to-earth-to-target-to-earth-to-radar path. Paths (b) and (c) are "two-bounce" paths and path (d) is a "three-bounce" path, i.e., if in each case we assume a *single* reflection point on the earth's surface, then there are two reflections on paths (b) and (c) and three reflections on path (d).

The assumption that there is only a single reflection point on the earth for each indirect path follows from the approximation of the earth as a perfectly flat infinite plane that is electrically homogeneous throughout. In reality, due to the earth's roughness and electrically nonhomogeneous nature, there are actually many earth reflection points. However, experience has shown that the model we are using gives very satisfactory results in many practical radar situations, since the dominant reflection comes from the "specular point" on the earth's surface. That point is defined as the one where the law of reflection (incidence angle = reflection angle) holds. With that simplification, we can express the total return signal voltage as follows (including all four paths):

$$V_r = \frac{C_0}{r_{RT}^2}\sqrt{\frac{P_T G_0^2 \lambda^2}{(4\pi)^3}}\,\{e^{-j2kr_{RT}} f^2(\theta_{RT}) R_T(\theta_{TR}/\theta_{TR})$$
$$+ e^{-jk(r_{RP}+r_{PT}+r_{TR})}[f(\theta_{RP}) R_P(\theta_{PT}/\theta_{PR}) R_T(\theta_{TR}/\theta_{TP}) f(\theta_{RT})$$
$$+ f(\theta_{RT}) R_T(\theta_{TP}/\theta_{TR}) R_P(\theta_{PR}/\theta_{PT})\, f(\theta_{RP})]$$
$$+ e^{-2jk(r_{RP}+r_{PT})}[f(\theta_{RP}) R_P(\theta_{PT}/\theta_{PR}) R_T(\theta_{TP}/\theta_{TP}) \cdots$$
$$R_P(\theta_{PR}/\theta_{PT})\, f(\theta_{RP})]\} \tag{11.8}$$

where

$r_{AB} = r_{BA}$ = distance between points A and B
θ_{AB} = elevation angle of B with respect to A
θ_{BA} = elevation angle of A with respect to B

(*Note*: θ_{AB} and θ_{BA} are not necessarily equal, but in a flat earth geometry $\theta_{BA} = \pi - \theta_{AB}$ if θ denotes the spherical polar angle, defined relative to the vertical axis, and $\theta_{BA} = -\theta_{AB}$ if θ denotes the elevation angle, defined relative to the horizontal axis.)

$f(\theta_{RL})$ = one-way complex field antenna radiation pattern in the direction of point L, normalized to unity at its peak (also the receptivity pattern assuming reciprocity)*

Before proceeding, we note that the earth-*reflection* point *P* (ERP) depicted

* This is the elevation plane pattern. The azimuth angle is fixed for all ray-paths considered in this process and hence is not indicated on the pattern function.

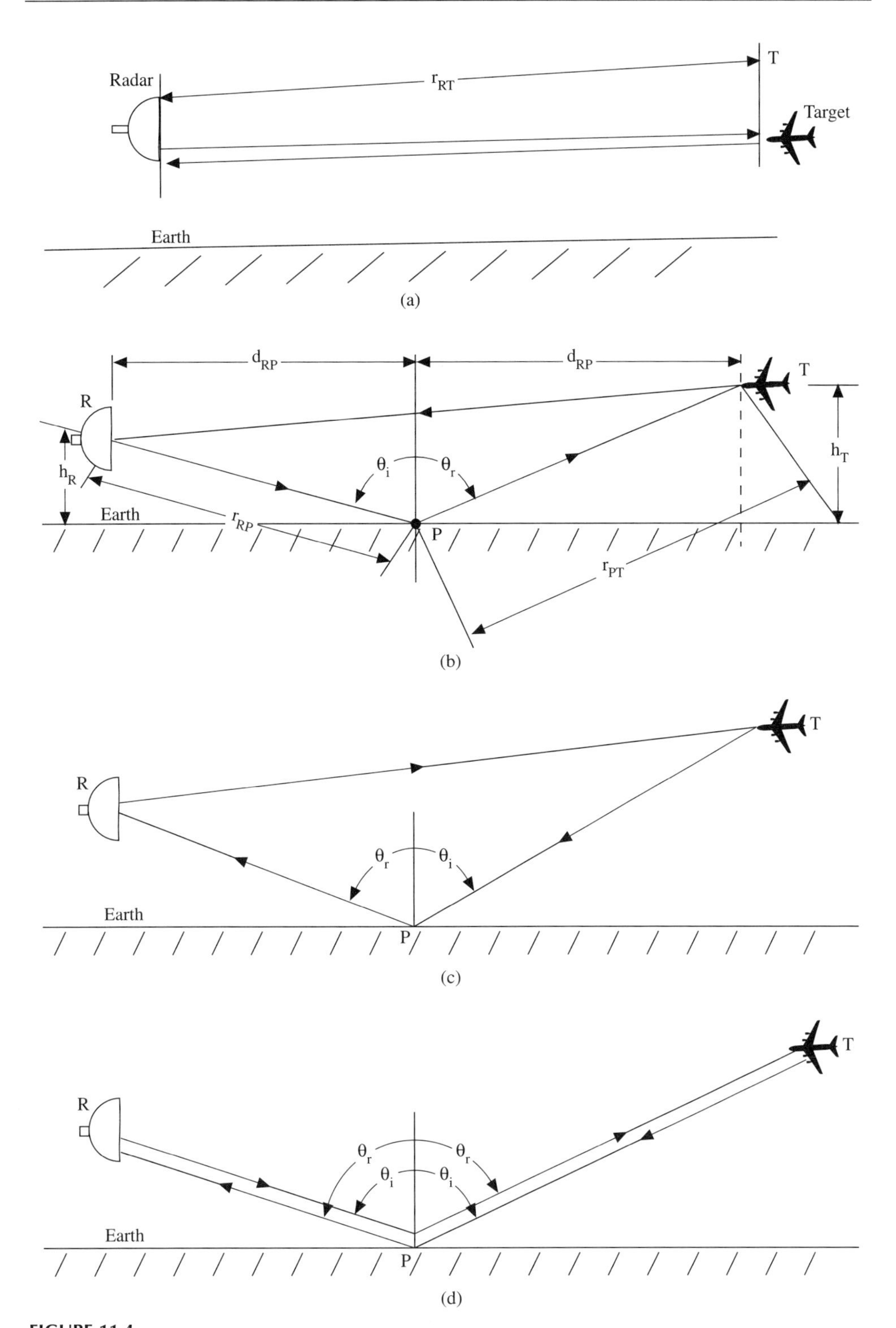

FIGURE 11.4
Simplest form of multipath. (a) Direct path. (b) Indirect-direct path. (c) Direct-indirect path. (d) Indirect-indirect path.

in Figure 11.4 is the point that obeys the law of reflection and hence can be found by the relationships

$$\operatorname{ctn} \theta_i = \frac{h_R}{d_{RP}} = \frac{h_T}{d_{PT}} = \operatorname{ctn} \theta_r \tag{11.9a}$$

$$d_{RT} = d_{RP} + d_{PT} \tag{11.9b}$$

$$r_{\substack{RP\\PT}} = \sqrt{d^2_{\substack{RP\\PT}} + h^2_{\substack{R\\T}}} \tag{11.9c}$$

where d_{RP} and d_{PT} are distances between the projection of the radar on the ground and the reflection point and the projection of the target on the ground and the reflection point, respectively, (i.e. d_{RP} and d_{PT} are ground ranges of ERP with respect to radar and target, respectively), and h_R and h_T are altitudes of radar and target, respectively.

The location of the reflection point relative to the radar, as obtained from (11.9a, b), is

$$d_{RP} = \frac{d_{RT}}{1 + (h_T/h_R)} \tag{11.10a}$$

or the distance relative to the target is

$$d_{PT} = \frac{d_{RT}}{1 + (h_R/h_T)} \tag{11.10b}$$

11.2.2 Effect of Earth Curvature

For situations in which the range-to-target is more than a few kilometers, the error in the propagation factor as calculated with the flat-earth approximation is unacceptable. This is particularly true for frequencies sufficiently high to render the phase differences between direct and reflected waves change appreciably due to earth curvature. To treat the effect correctly if we want to model it with straight rays, we should account for refraction and invoke the equivalent earth radius concept discussed in Section 11.1, culminating in 11.7. Having done that, locating the earth reflection point and determining the earth-curvature correction to the propagation factor ("divergence factor") cannot be accomplished through the simple formulas (11.9a, b) and (11.10a, b). A much more involved analysis is necessary.

Before discussing those issues we will define the range to the "radio horizon" as that range beyond which a point that should be illuminated by the radar is obscured from its view by the earth. The radio horizon, illustrated in Figure 11.5, is defined in terms of the equivalent earth radius r_e, i.e., refraction is thereby accounted for and straight-line paths for LOS transmission are assumed.

The "regions" usually defined in discussions of the propagation factor are illustrated in Figure 11.5. The figure shows the three points P_1, P_2, and P_3, where P_1 is well above the radio horizon and therefore in the "interference region," which is the *only* region if the flat earth approximation is used. Point P_2 is well below the radio horizon and defined as being in the "diffraction region," implying that it can be seen by the radar only through the phenomenon of diffraction around the earth. Finally, P_3 is in the "intermediate region," defined as a region just short of the radio horizon but very close to it.

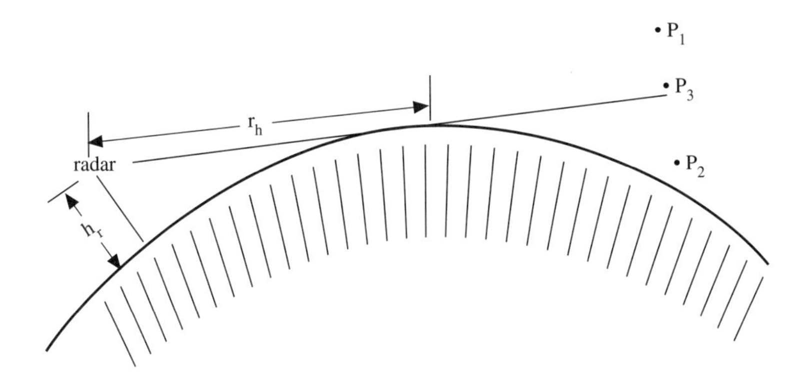

FIGURE 11.5
The radio horizon and propagation regions.

The range to the radio horizon (slant range, not ground range) is easily seen from Figure 11.5 to be given by

$$r_h = \sqrt{(r_e + h_r)^2 - r_e^2} \tag{11.11}$$

Since $h_r << r_e$ in all cases of interest, we can approximate r_h as

$$\begin{aligned} r_h &= \sqrt{r_e^2 + 2r_e h_r + h_r^2 - r_e^2} = \sqrt{2r_e h_r}\sqrt{1 + \frac{h_r}{2r_e}} \\ &\simeq \sqrt{2r_e h_r}\left(1 + \frac{h_r}{4r_e}\right) \simeq \sqrt{2r_e h_r} \end{aligned} \tag{11.12}$$

Figure 11.6 indicates the range to the radio horizon as a function of radar altitude h_R for a "four-thirds earth" [equivalent earth radius $= 8.453(10^3)$ km]. We see from the curve that a ground-based radar at 10 m altitude has a radio horizon range of about 13 km while a high-flying airborne radar at a 10.6 km (about 35,000 feet) altitude has a radio horizon range of about 423 km, or 264 miles. A satellite-borne radar at 200 miles altitude has a radio horizon range of about 1453 miles.

We now turn our attention to determination of the "divergence factor,"[20–23] which models the effect of earth curvature on the field strength of the radar signal received at a point residing in the interference region, i.e., a point P well above the radio horizon.

The classical treatment of this problem is in Kerr,[1] Section 5.2, pp. 404–406. The result of that analysis is the "divergence factor," which is a factor used as a weighting on the indirect wave's reflection coefficient. This is to account for the spreading of the reflected wave in a wide cone due to the presence of more than a single specular reflection point as in the flat earth case. The result given in Kerr,[1] [Eq. (16) on p. 406], attributed to Van der Pol and Bremmer,[24] is

$$D = \frac{a(R_1 + R_2)\sqrt{\sin\tau_2 \cos\tau_2}}{\sqrt{[(a + z_2)R_1 \cos\tau_3 + (a + z_1)R_1 \cos\tau_1](a + z_1)(a + z_2)\sin\theta}} \tag{11.13a}$$

where, in the notation used in Kerr,[1] a = earth's radius, τ_2 is the reflection angle

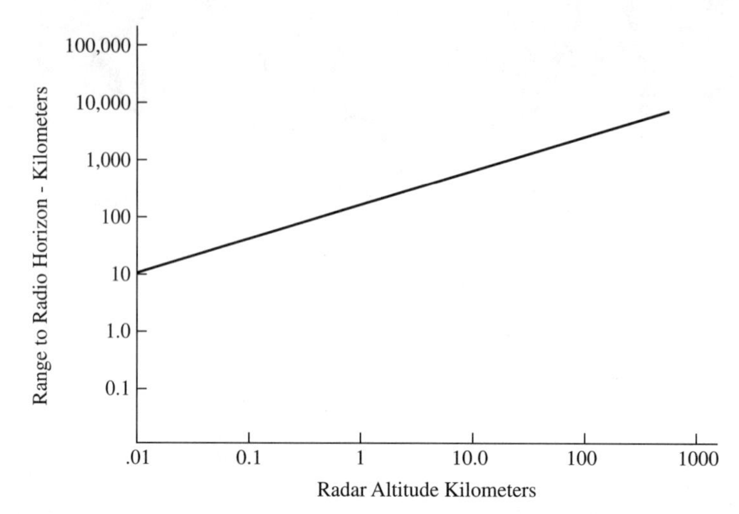

FIGURE 11.6
Range to radio horizon vs. radar altitude.

(which is equal to the incidence angle or equivalently the complement of the grazing angle at the reflection point), R_1 and R_2 are slant ranges between radar and ERP and between terminal point and ERP, respectively, τ_1 is angle between the ray from the transmit point to the ERP and the ray from the transmit point to the earth center, τ_3 is the counterpart of τ_1 for the terminal point, z_1 and z_2 are altitudes of transmit and terminal points, respectively, and θ is the angle between transmitting and terminal points with respect to the earth center. The geometry is shown in Figure 5.7 on p. 404 of Kerr[1] and the analysis to reach (11.13a) is in Eqs. (8) through (15) on pp. 404 and 405 of Kerr.[1]

A simplified approximation to (11.13a) is given in Eq. (17) on p. 406 of Kerr,[1] which is adequate for virtually any radar situation we might encounter (i.e., heights of transmitting and terminal points are small compared with earth radius, true in all terrestrial radars and even satellite-based radars. Above all, that condition is met in nearly all situations where the propagation factor is an important effect). Changing notation from that used in Kerr,[1] the equivalent of that approximation is

$$D \cong \sqrt{\frac{(r_{13} + r_{23})}{\left\{\left[1 + \frac{(z_1)}{r_e}\right]\left[1 + \frac{(z_2)}{r_e}\right] r_e \sin\theta_{012}\left[1 + \frac{2r_{13}r_{23}}{r_e(r_{13} + r_{23})\cos\theta_i}\right]\right\}}} \tag{11.13b}$$

where point 1 = radar, point 2 = terminal point, point 3 = ERP

r_{13} = slant range between radar and ERP
r_{23} = slant range between terminal point and ERP
z_1 = altitude of radar
z_2 = altitude of terminal point
r_e = equivalent earth radius (e.g., 4/3 times true earth radius)
θ_i = local angle of incidence at the ERP; the angle θ_{012} is the angle between the radar and the terminal point relative to the center of the earth.

From Figure 11.5, it is easily seen that if $z_1 << r_e$, $z_2 << r_e$, and the path length between points 1 and 2 is very small compared to earth radius, then it is permissible, with negligible error, to invoke the approximations

$$1 + \frac{z_k}{r_e} \approx 1 \qquad \text{for } k = 1, 2 \tag{11.13c}$$

$$r_e \sin \theta_{012} \approx d_{13} + d_{23} \tag{11.13d}$$

where d_{13} and d_{23} are ground ranges between radar and ERP and ERP and terminal point, respectively. With the aid of (11.13c, d), Eq. (11.13b) is further simplified to the form [Eq. (18) on p. 406 of Kerr[1]]

$$D \approx 1 \Big/ \sqrt{1 + \frac{2d_{13}d_{23}}{r_e(d_{13} + d_{23}) \sin \alpha_G}} \tag{11.14}$$

where α_G, the grazing angle of the wave impinging on the ERP from the radar, is used in lieu of its complement θ_i, the angle of incidence.

The form (11.14) is that which is generally used for the divergence factor in radar propagation and appears in many standard treatises on radar and microwave propagation.[20–23]

It remains to determine the position of the ERP relative to radar and terminal points, i.e., to obtain expressions that are curved earth generalizations of (11.10a, b) for the flat earth case. Referring to Figure 11.7, we can write laws of cosines for the triangles 130, 230, and 120 and the relationships between the various angles in the diagram, as follows:

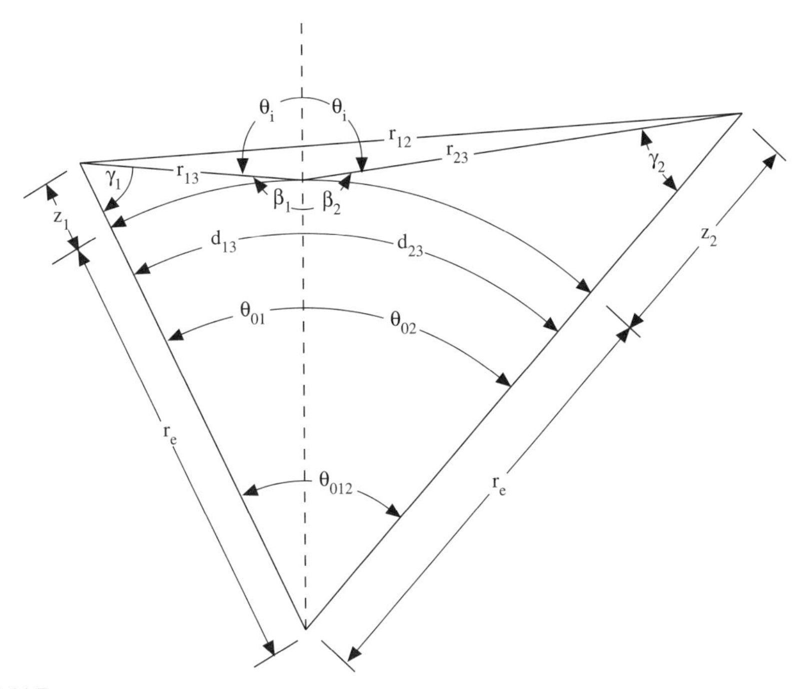

FIGURE 11.7
Location of ERP for curved earth case.

Triangle 130:

$$(r_e + z_1)^2 = r_e^2 + r_{13}^2 + 2r_e r_{13} \cos \theta_i \tag{11.15a}$$

Triangle 230 (note $\theta_r = \theta_i$)

$$(r_e + z_2)^2 = r_e^2 + r_{23}^2 + 2r_e r_{23} \cos \theta_i \tag{11.15b}$$

Triangle 120:

$$r_{12}^2 = (r_e + z_1)^2 + (r_e + z_2)^2 - 2(r_e + z_1)(r_e + z_2)\cos \theta_{012} \tag{11.15c}$$

Triangle 123:

$$r_{12}^2 = r_{13}^2 + r_{23}^2 - 2r_{13}r_{23} \cos(2\theta_i) \tag{11.15.d}$$

Angle relationships:

$$\theta_{01} + \gamma_1 + \beta_1 = \theta_{01} + \gamma_1 + \pi - \theta_i = \pi; \qquad \theta_{01} + \gamma_1 = \theta_i$$
$$\theta_{02} + \gamma_2 + \beta_2 = \theta_{02} + \gamma_2 + \pi - \theta_i = \pi; \qquad \theta_{02} + \gamma_2 = \theta_i \tag{11.15e}$$

$$r_e\theta_{01} = d_{13} = \text{ground range—radar to ERP}$$
$$r_e\theta_{02} = d_{23} = \text{ground range—terminal point to ERP}$$
$$r_e\theta_{012} = d_{13} + d_{23} = d_{12} = \text{ground range—radar to target}$$

From (11.15a, b), with the aid of (11.13c)

$$\cos \theta_i \approx \frac{2r_e z_1 - r_{13}^2}{2r_e r_{13}} \approx \frac{2r_e z_2 - r_{23}^2}{2r_e r_{23}} \tag{11.16a}$$

From (11.15c, d, e), again with the aid of (11.13c)

$$r_{12}^2 \approx 2r_e^2\left[1 + \left(\frac{z_1 + z_2}{r_e}\right)\right](1 - \cos \theta_{012}) = 4r_e^2 \sin^2\left(\frac{d_{12}}{r_e}\right)$$
$$\approx r_{13}^2 + r_{23}^2 - 2r_{13}r_{23} \cos(2\theta_i) \tag{11.16b}$$

Equating both expressions for $\cos \theta_i$ in (11.16a), invoking the last expression in (11.15e) (i.e., $d_{23} = d_{12} - d_{13}$) and noting, with the aid of (11.13c), that

$$r_{\substack{13\\23}} \approx d_{\substack{13\\23}} (1 + z_{\substack{1\\2}}/2r_e) \approx \begin{matrix} d_{13} \\ (d_{12} - d_{13}) \end{matrix} \tag{11.17}$$

we obtain the following cubic equation in d_{13},

$$2d_{13}^3 - 3d_{13}^2 d_{12} + d_{13}[d_{12}^2 - 2r_e(z_1 + z_2)] + 2r_e z_1 d_{12} = 0 \tag{11.18}$$

This result is identical with the first equation (not numbered) on p. 113 of Kerr[1] where $r_1 \rightarrow d_{13}$, $r \rightarrow d_{12}$, $a_e \rightarrow r_e$. The formal solution of that equation, also given on p. 113 of Kerr[1] with these same notational changes, is

$$d_{13} = \frac{d_{12}}{2} + p\cos\left(\frac{\Phi + \pi}{3}\right) \tag{11.19}$$

where

$$p = \frac{2}{\sqrt{3}}\sqrt{r_e(z_1 + z_2) + \left(\frac{r_{12}}{2}\right)^2}$$

$$\Phi = \cos^{-1}\left[\frac{2r_e(z_2 - z_1)r_{12}}{p^3}\right]$$

Using d_{13}, given by (11.19) as a function of the known altitudes z_1 and z_2, the known slant range r_{12}, and the effective earth radius r_e, we can invoke (11.17) to solve for r_{13} and r_{23} in terms of d_{13}, d_{23}, z_1, z_2, and r_e and the first expression in (11.16a) to solve for the angle of incidence θ_i in terms of z_1, r_e, and r_{13}.

11.2.3 Effect of Surface Roughness

When the wave from the transmitter impinges on the earth's surface near the ERP, the idealized reflection from a smooth locally planar surface, which is depicted in the discussion above, is modified by height fluctuations in that region. It is easy to see from Figure 11.8 that the point designated as the ERP in the previous discussions is based on the assumption of a perfectly smooth locally horizontal earth, i.e., the earth's curvature is sufficiently large compared to the illuminated region to assume that the reflection process is equivalent to that occurring when a plane wave in air impinges on an infinite flat horizontal plane interface between earth and air. But if there are appreciable surface height variations in the illuminated region, there are hardly any points within that region where the law of reflection is valid; hence the ERP we have defined earlier is really based on an *average* locally horizontal flat plane with random height variations around the average height of the plane. If the height variations are

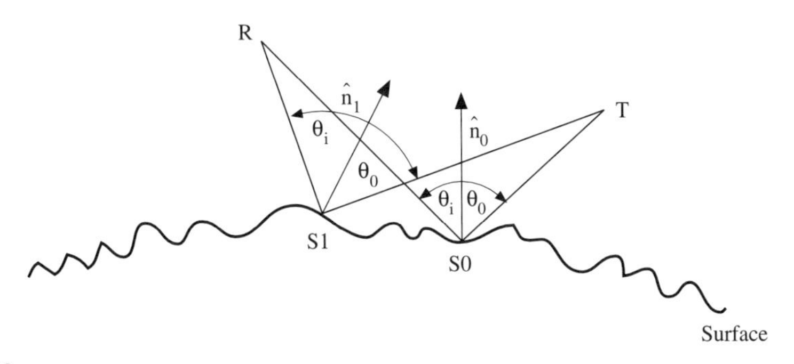

FIGURE 11.8
Effect of surface roughness on propagation factor. R = radar, T = target. SO = ERP for horizontal surface; $\hat{n}_o$ = normal at S0; θ_o = angle of incidence at S0. S1 = ERP for sloping surface; $\hat{n}_1$ = normal at S1; θ_i = angle of incidence at S1.

sufficiently large, the slopes of local points will differ so much from zero that the law of reflection would apply to receiving points far removed from the designated terminal point, but almost never to that point. In fact, given a designated terminal point, there would be many ERPs for that point all over the earth's surface due to the roughness that we know exists everywhere on earth.

Since the accurate modeling of the true situation for appreciable roughness is virtually impossible, we must confine ourselves to a small local region near the ERP whose location we have inferred from the assumption of a locally horizontal earth's surface.* Then we must assume that the large-scale roughness in that region is sufficiently small to justify the assumption that the local slopes are well below 45°. Since there is a continuum of scales of roughness, it is evident that extremely small scale roughness may have very large slopes but they are randomly distributed and will average out to zero over the illuminated region. This allows construction of a multiplicative factor modifying the propagation factor obtained with the locally smooth earth assumption.

A simple theory neglecting polarization that gives us this correction factor has been expounded in some classical references, particularly the well-known treatise on rough surface scattering by Beckmann and Spizzichino,[25] and the same result is obtained in other more recent references using theory that includes polarization.[26–28]

A field component of a wave scattered from a small patch of area $\Delta S = \Delta x \Delta y$ on a rough horizontal surface is given by

$$U(\mathbf{r}_s) = \iint_{\Delta S} dx'\, dy'\, f(x', y', z') e^{jk\mathbf{r}' \cdot (\hat{r}_i + \hat{r}_s)} \tag{11.20}$$

where $\mathbf{r}' = \hat{x}x' + \hat{y}y' + \hat{z}z'$, $\hat{r}_i = \sin\theta_i(\hat{x}\cos\phi_i + \hat{y}\sin\phi_i) + \hat{z}\cos\theta_i$, $\hat{r}_s = \sin\theta_s(\hat{x}\cos\phi_s + \hat{y}\sin\theta_s) + \hat{z}\cos\theta_s$ where $\hat{r}_i$ and $\hat{r}_s$ are the unit vectors directed from the scattering point toward the source of the incident wave and the scattered wave receiving point, respectively, (θ_i, ϕ_i) and (θ_s, ϕ_s) are their spherical coordinate angles, and $\mathbf{r}'$ is a vector from the center of the patch to an arbitrary point on the patch. The vector $\mathbf{r}_s$ is $\hat{r}_s r_s$, where r_s is distance from patch center to observation point. The function $f(x', y', z')$ is a function on the patch whose details are not important for this argument but would be obtained from the Stratton–Chu integral[29] and involve the reflection coefficients of the surface.

We will express the exponential in (11.20) in the form

$$e^{jk(\hat{r}_i + \hat{r}_s)\cdot\mathbf{r}'} = e^{jk[\alpha_1 x' + \alpha_2 y' + \alpha_3 z'(x', y')]} \tag{11.21}$$

where $\alpha_1 = \sin\theta_i \cos\theta_i + \sin\theta_s \cos\phi_s$, $\alpha_2 = \sin\theta_i \sin\phi_s + \sin\theta_s \sin\phi_s$, $\alpha_3 = \cos\theta_i + \cos\theta_s$ and where $z'(x', y')$ is the height as a function of the horizontal coordinates of a point on the patch.

It is evident from (11.21) that the surface roughness in the patch is contained in the function $z'(x', y')$, present in the exponent given in (11.21) and also in the function $f[x', y', z'(x', y')]$.

If we assume $z'(x', y')$ to be a sample function of a zero-mean Gaussian random process then from (11.20) and (11.21), the average of $U(\mathbf{r}_s)$ is

* The usual assumption is that the local region alluded to is that within the first Fresnel zone around the ERP.

$$\langle U(\mathbf{r}_s)\rangle = \iint_{\Delta S} dx'\, dy'\, e^{jk(\alpha_1 x' + \alpha_2 y')} \langle f[x', y', z'(x', y')] e^{jk\alpha_3 z'(x', y')}\rangle \tag{11.22}$$

If we now further assume that the variation of $z'(x', y')$ in $f(x', y', z')$ has a negligible effect on the averaging process compared to that of $e^{jk\alpha_3 z'(x', y')}$, we can approximate $f[x', y', z'(x', y')]$ by its mean value $f(x', y', 0)$ and remove it from the averaging brackets. We then assume that the statistics of $z'(x', y')$ are spatially uniform throughout the patch. The result of these assumptions is that (11.22) takes the form

$$\langle U(\mathbf{r}_s)\rangle = I_0 \langle e^{jk\alpha_3 z'}\rangle \tag{11.23}$$

where it is indicated notationally in (11.23) that the statistics of z' are independent of x' and y' and where

$$I_0 = \iint_{\Delta S} dx' dy'\, e^{jk(\alpha_1 x' + \alpha_2)} f(x', y', 0) \tag{11.24}$$

The average in (11.23) is

$$\begin{aligned} \langle e^{jk\alpha_3 z'}\rangle &= \frac{1}{\sqrt{2\pi\langle z^2\rangle}} \int_{-\infty}^{\infty} dz'\, e^{-[(z')^2/2\langle z^2\rangle] + jk\alpha_3 z'} \\ &= \frac{1}{\sqrt{\pi}} \int_{-\infty}^{\infty} du\, e^{-[u^2 - 2(jk/2)\sqrt{2\langle z^2\rangle}\alpha_3)u]} \end{aligned} \tag{11.25}$$

Completion of the square in the exponential and integration results in

$$\langle e^{jk\alpha_3 z'}\rangle = e^{-(8\pi^2/\lambda^2)\langle z^2\rangle \cos^2\theta_i} \tag{11.26}$$

(noting that, the reflection being specular, $\theta_s = \theta_i$ and hence $\alpha_3 = 2\cos\theta_i$).

The integral in (11.24) is that from which the reflection from the smooth, locally flat earth's surface is obtained. The average in (11.26) is the factor that accounts in an approximate way for the effect of surface roughness. A more sophisticated way to obtain the roughness factor circumvents the assumption of negligibility of the factor $f(x', y', z')$ in the averaging process. It is based on integration by parts and on the fact that $f(x', y', z')$ is linearly dependent on the surface normal unit vector. This derivation is given in Beckman and Spizzichino,[25] Chapter 5, Section 5.3, and in Tsang et al.,[27] and will not be presented here.

Since the reflection from the smooth surface is assumed specular and the reflecting plane is approximated as infinite, the factor $f(x', y', 0)$ includes the Fresnel reflection coefficient.

For vertical polarization, where the electric field vector of the incoming wave is in the vertical plane, the Fresnel reflection coefficient is given by

$$R_v = \frac{\epsilon_{cR}\cos\theta_i - \sqrt{\epsilon_{cR} - \sin^2\theta_i}}{\epsilon_{cR}\cos\theta_i + \sqrt{\epsilon_{cR} - \sin^2\theta_i}} \tag{11.27a}$$

For horizontal polarization, where the electric field vector of the incoming wave is in the horizontal plane, the coefficient is given by

$$R_h = \frac{\cos\theta_i - \sqrt{\epsilon_{cR} - \sin^2\theta_i}}{\cos\theta_i + \sqrt{\epsilon_{cR} - \sin^2\theta_i}} \tag{11.27b}$$

where ϵ_{cR} is the complex relative permittivity of the earth $= (\epsilon/\epsilon_0) - (j\sigma/\omega\epsilon_0)$, where ϵ is permittivity, σ the conductivity, and ω the radian frequency.

The derivation of (11.27a, b) can be found in any standard text on electromagnetic theory.[30–32,34] It is based on the assumption that the interface between free space and the medium (the earth in this case) is a flat plane of infinite extent, that the incoming wave is a plane wave and the medium is linear, homogeneous, and isotropic (LHI) but may be partially conducting, as evidenced by a nonzero conductivity σ and hence a complex relative permittivity ϵ_{cR}. In the present context, the Fresnel reflection coefficient is only a rough approximation to reality. The assumption of a plane incident wave can be considered as a reasonable approximation within a small local region around the ERP, although the *actual* incident wave is spherical and weighted by the antenna pattern. The notion of a surface that is smooth and a medium that is uniform *on the average* is one whose validity depends on the size of the region around the ERP that is the major contributor to the ground reflections. The reflection coefficient could be generalized to account for at least *vertical* nonuniformity of the earth in the vicinity of the ERP as a medium composed of infinite horizontal layers, e.g., a top layer of vegetation or snow and a second layer of soil for certain types of terrain.[33] This would be important only if the wave penetrates the top layer and enters the second layer, a phenomenon dependent on the "skin depth" of the medium of which the top layer is composed. Skin depth is defined in a different context in Section 11.4. The skin depth is defined as the reciprocal of the attenuation in nepers/meter, i.e., it is the depth of penetration of the wave into the medium required for the wave amplitude to decay to $1/e$ of its value when it enters the medium at normal incidence. If the thickness of the top layer is significantly exceeded by the skin depth, then reflection coefficients from a layered medium model might provide better accuracy than (11.27a, b).

Table 11.1 shows some approximate values of $\epsilon_r = \epsilon/\epsilon_0$ and the "loss-tangent" $\sigma/\omega\epsilon$ for some common terrestrial media in radar bands from 1 GHz to the millimeter wave bands, obtained from a number of well-known sources.[35–38,*]

The integral I_0 in (11.24) is that over a patch, which has dimensions in both directions that span many wavelengths. If we then assume $f(x', y', 0)$ to be constant over the patch and note that, for specular reflection, $\theta_s = \theta_i$, $\phi_s = \phi_i + \pi$, $\cos\phi_s = -\cos\phi_i$, $\sin\phi_s = -\sin\phi_i$, then $\alpha_1 = \alpha_2 = 0$ and hence the phase factor in the integrand is unity and

$$I_0 = f(x_0, y_0, 0)\Delta S \tag{11.28}$$

where (x_0, y_0) is the ERP and ΔS is the patch area.

11.3 DIFFRACTION EFFECTS

There are two classes of diffraction effects that are particularly important in radar wave propagation in a terrestrial environment. The first is the propagation of

* The effects of parameters such as temperature, moisture content, packing density, etc., which can change constitutive parameters of some of these media significantly, are not necessarily accounted for in values reported in the cited references. These are merely typical rough values to be used in radar system studies and any one of them could change by factors of 2, 3, or even more due to varying conditions.

TABLE 11.1

Values of Permittivity and Conductivity for Earth Media

Material	ϵ/ϵ_0	$\sigma/\omega\epsilon$
Sea water	L and S bands ≈ 70 C, X, Ku, Ka bands ≈ 52	L band ≈ 2.4 S band ≈ 0.435 Other bands ≈ 0.605
Fresh water	L and S bands ≈ 78 C and X bands ≈ 55 Ku and Ka ≈ 34	L band ≈ 0.016 S band ≈ 0.157; C and X bands ≈ 0.54 Ku and Ka bands ≈ 0.265
Snow, freshly packed	1.2–1.26, all bands	L band ≈ 0.0012 S band ≈ 0.0029 Other bands ≈ 0.00042
Snow tightly packed	≈ 1.5, all bands	≈ 0.0009, all bands
Sea ice	L and S bands ≈ 3.2 Other bands ≈ 3.17	L and S bands ≈ 0.0009 Other bands ≈ 0.0007
Sandy soil	L and S bands ≈ 2.55 Other bands ≈ 2.53	L band ≈ 0.01 S band ≈ 0.062 Other bands ≈ 0.0036
Loamy soil	L band ≈ 2.47 Other bands ≈ 2.44	L band ≈ 0.0065; S band ≈ 0.0011 Other bands ≈ 0.0014
Clay soil	L band ≈ 2.38; S band ≈ 2.27; Other bands ≈ 2.16	L band ≈ 0.02 S band ≈ 0.015 Other bands ≈ 0.013
Cement and asphalt	L and S bands ≈ 2.4 Other bands ≈ 2.35	L and S bands ≈ 0.00078 Other bands ≈ 0.00068

the wave beyond the radio horizon. The second is the type of diffraction encountered when the wave propagates at low grazing angles over very rough terrain (e.g., rough sea surface or mountains), e.g., where diffraction effects allow some radio wave energy to appear behind protrusions, such as hills or ocean waves, where LOS transmission is blocked.

First consider diffraction due to an obstacle in the ray path between two points. This effect can be modeled by knifedge diffraction theory.[39–45] Figure 11.9 shows a radio wave traveling between two points A and B with a wedge obstructing the line-of-sight between A and B. In Figure 11.9a the LOS between the two points is shown to be far above the obstruction. In Figure 11.9b it just barely clears the obstruction. In Figure 11.9c it is directly blocked by the obstruction.

From Huyghen's principle, the field at B is the superposition of contributions from points on a closed surface around B. If we construct an infinite vertical plane normal to the LOS at the wedge location, that plane can be considered as a surface that closes around B at infinity. The wedge and the earth beneath it are assumed opaque to the radio wave propagating from A. The plane in free-space above the top of the wedge contributes to the received wave at B. To simplify the analysis, the wedge is considered as extending to infinity in the direction normal to the plane of the paper.

The superposition of fields on the plane that contribute to the field at B, ignoring polarization, is given by

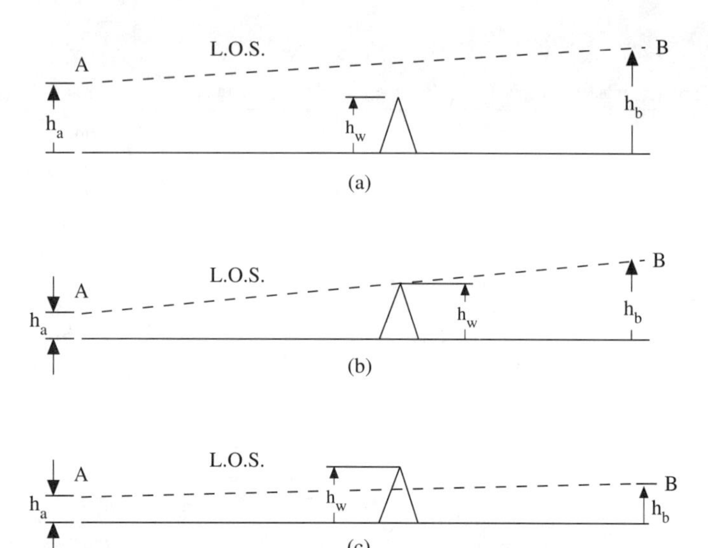

FIGURE 11.9
Diffraction due to an obstacle.

$$E_B = \int_{-\infty}^{\infty} dy' \int_{h_W}^{\infty} dz' \, A(y', z') e^{j\phi_{APB}(y', z')} \tag{11.29}$$

where h_w is the height of the wedge, and $A(y, z)$ and $\phi_{APB}(y, z)$ are the amplitude and phase of the field on the plane as a function of height.

Let us focus for the moment on the phase $\phi_{APB}(y, z)$, which is related to the path length between A and B by the relationship

$$\phi_{APB}(y, z) = \frac{2\pi}{\lambda}(AP + PB) \tag{11.30}$$

where P is a point on the vertical plane. We use rectangular coordinates and subscripts a, b corresponding to points A, B. We take the origin at $x = x_a$, $y = 0$, $z = 0$, an arbitrary point on the plane as x, y, z, and let both A and B lie on the y–z plane. Then

$$\phi_{APB}(y, z) = \frac{2\pi}{\lambda}[\sqrt{x^2 + y^2 + (z - z_a)^2} + \sqrt{(x - x_b)^2 + y^2 + (z - z_b)^2} \tag{11.31}$$

The direct electrical path length between A and B is

$$\phi_{AB} = \frac{2\pi}{\lambda}\sqrt{(x_b - x_a)^2 + (z_b - z_a)^2} \tag{11.31a)'}$$

From (11.31) and (11.31a)′, the difference in electrical path lengths is

$$\Delta\phi(y, z) = \phi_{APB}(y, z) - \phi_{AB}$$
$$= \frac{2\pi}{\lambda}\left[\sqrt{d_1^2 + y^2 + (z - h_a)^2} + \sqrt{d_2^2 + y^2 + (z - h_b)^2} - \sqrt{d^2 + (\Delta h)^2}\right] \qquad (11.31b)'$$

where $x = d_1$, $x_b - x = d_2$, $z_a = h_a$, $z_b = h_b$, $d = d_1 + d_2$, $\Delta h = h_b - h_a$.

Assuming that $(d_1^2 + y^2) >> (z - h_a)^2$, $(d_2^2 + y^2) >> (z - h_a - \Delta h)^2$, and $d >> |\Delta h|$, we approximate (11.31b)′ as

$$\Delta\phi(y, z) \approx \frac{2\pi}{\lambda}\left\{\left[\sqrt{d_1^2 + y^2}\left(1 + \frac{1}{2}\frac{(z - h_a)^2}{(d_1^2 + y^2)}\right)\right] + \left[\sqrt{d_2^2 + y^2}\left(1 + \frac{1}{2}\frac{(z - h_b)^2}{(d_2^2 + y^2)}\right)\right] - d\left(1 + \frac{1}{2}\left(\frac{\Delta h}{d}\right)^2\right)\right\} \qquad (11.31c)'$$

At this point we can simplify matters by modeling this as a two-dimensional problem, in effect assuming that the transmitting point A, the wedge and the terminal point B are uniform in the y-direction, which implies that there is no y-dependence in the path length difference, allowing us to set $y = 0$ in (11.31c)′ and eliminate the y-integration in (11.29). Then we obtain the result

$$\Delta\phi(z) = \Delta\phi(0, z) \simeq \frac{2\pi}{\lambda}\left\{(d_1 + d_2 - d) + \frac{1}{2}\left[\frac{(z - h_a)^2}{d_1} + \frac{(z - h_b)^2}{d_2} - \frac{(\Delta h)^2}{d}\right]\right\}$$
$$= \frac{\pi}{\lambda}\left\{\left(\frac{d_1 + d_2}{d_1 d_2}\right)(z - h_{av})^2 + \left(\frac{d_2 - d_1}{d_1 d_2}\right)(z - h_{av})\Delta h + \left[\frac{(d_2 - d_1)^2}{d_1 d_2 (d_1 + d_2)}\right]\left(\frac{\Delta h}{2}\right)^2\right\} \qquad (11.31d)'$$

where $h_{av} = (h_a + h_b)/2$ = average height of transmitting and receiving points.

In the usual form of (11.31d)′, found in a number of standard treatises, points A and B are assumed to be at the same height and hence $\Delta h = 0$, and $h_{av} = h_a = h_b$, which eliminates the last two terms of (11.31d)′. We note that these terms would also be eliminated if $d_1 = d_2$, even if $\Delta h \neq 0$.

To determine E_B, the further assumption must be made that the variation of $A(y, z)$ with z is very slow compared to that of the exponential factor and can be considered constant in the integrand. By recognizing that the quantity in (11.31d)′ is

$$\frac{\pi}{\lambda}\left(\frac{d_1 + d_2}{d_1 d_2}\right)\left[(z - h_{av}) + \frac{\Delta h}{2}\left(\frac{d_2 - d_1}{d_2 + d_1}\right)\right]^2$$

and setting $A(y, z)$ to $A_0\delta(y)$, we can express (11.29) in the form

$$E_B \cong \frac{A_0}{\eta}\int_X^{\infty} du\, e^{j(\pi/2)u^2} \tag{11.32}$$

where $X = \eta[(h_w - h_{av}) + (\Delta h/2)\xi]$,

$$\eta = \sqrt{\frac{2}{\lambda}\left(\frac{d_1 + d_2}{d_1 d_2}\right)}, \qquad \xi = \left(\frac{d_2 - d_1}{d_2 + d_1}\right)$$

The integral in (11.32) is expressed in terms of Fresnel cosine and sine integrals $C(x)$ and $S(x)$, respectively, where

$$C(x) = \int_0^x du \cos\left(\frac{\pi}{2}u^2\right) \tag{11.32a)'}$$

$$S(x) = \int_0^x du \sin\left(\frac{\pi}{2}u^2\right)$$

Thus from (11.32) and (11.32a)′

$$E_B = \frac{A_0}{\eta}\left\{\left(\frac{1}{2} - C(X)\right) - j\left(\frac{1}{2} - S(X)\right)\right\} \tag{11.32b)'}$$

where $X = \eta[(h_w - h_{av}) + (\Delta h/2)\xi]$.

The curve in Figure 11.10 shows the loss relative to free space, given by the quantity $(E_B\eta/A_0)$ (in dB) in (11.32b)′, vs. the parameter X, to show that the loss is initially negligibly small, oscillating about 0 dB for $X \leq -1$ and then increasing rapidly with X beyond $X = -1$, approaching 20 dB when X reaches 2. This behavior, to give us insight in the radar context, must be related to the parameter $(h_{av} - h_w)$, the average height of transmit and receive points (e.g., radar and target) relative to the height of the wedge. Intuitively, we expect line-of-sight transmission with no

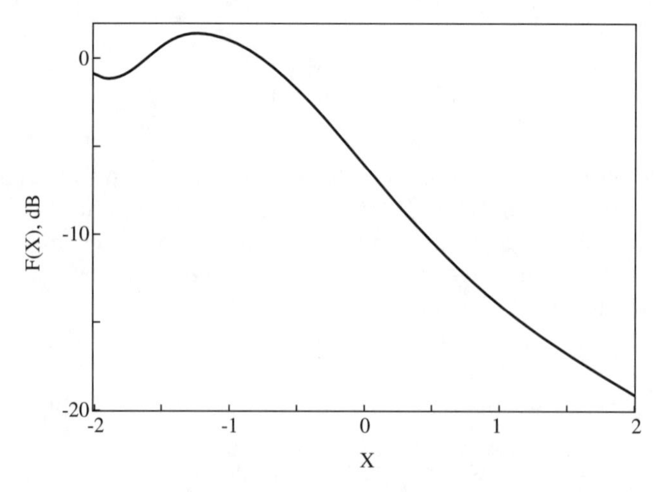

FIGURE 11.10
Diffraction loss vs. parameter X.

significant loss when the average height of the two terminal points far exceeds the height of the wedge. When that average is below the wedge height, we would expect the loss to increase rapidly as it descends further because one or both of the terminal points is in the "shadow zone" of the wedge, where the LOS path for either point is obscured (or blocked in the extreme case) from the other point by the wedge.

In the radar context, we must multiply the number of dB by 2 because both the illumination of the target by the radar transmitter and the return from the target to the receiver are obscured by the wedge.

Figure 11.11a, b, and c shows the two-way diffraction loss in dB, relative to that of the free-space path vs. $(h_{av} - h_w)$, the average terminal height above the wedge height for $d_1 = d_2$, $d_1 = 5d_2$, and $d_2 = 5d_1$, respectively, all for $\Delta h = 0$, i.e., where both terminal points are at the same height. Figure 11.12a, b, and c shows the loss vs. $(h_{av} - h_w)$ for $\Delta h/h_{av} = -0.5$, 0, and $+0.5$, respectively.

The trends shown on those curves are as expected and show how the losses become prohibitively high as either terminal point, radar, or target descends into the shadow zone of the other.

We now consider the other form of diffraction mentioned earlier, that due to the earth when either terminal point (radar or target) is beyond the radio horizon with respect to the other, or equivalently, within the other's shadow zone. The analysis of this effect, appearing in classical references (e.g., Kerr,[1] Section 2.12, pp. 109–112) is beyond the scope of this book. The important result for our purposes, based on the theory of wave propagation in a horizontally stratified atmosphere over a spherical earth as expounded in Kerr[1] [Chapter 2, pp. 27–180; given by Eqs. (4.27), (4.27a), and (4.28) on p. 109], wherein the propagation factor (or "height-gain function") in the diffraction zone is approximated by

$$F \approx 2\sqrt{\pi X}\left|\sum_{m=1}^{\infty} e^{jA_m X} U_m(Z_1) U_m(Z_2)\right| \tag{11.33}$$

where the functions U_m for the mth mode of the expansion are functions of Z_1 and Z_2, the normalized altitudes of the two terminal points, e.g., radar and target, and X is the normalized distance between the two points along the earth's surface. A good approximation to (11.33) is to use only the first mode, which results in a standard expression of the form[46]

$$F \approx V_1(X) U_1(Z_1) U_2(Z_2) \tag{11.34}$$

where

$$V_1(X) = 2\sqrt{\pi X} e^{-2.02X}$$

and where

$$Z_{\frac{1}{2}} = \frac{z_{\frac{1}{2}}}{H}$$

$$X = \frac{d}{L}$$

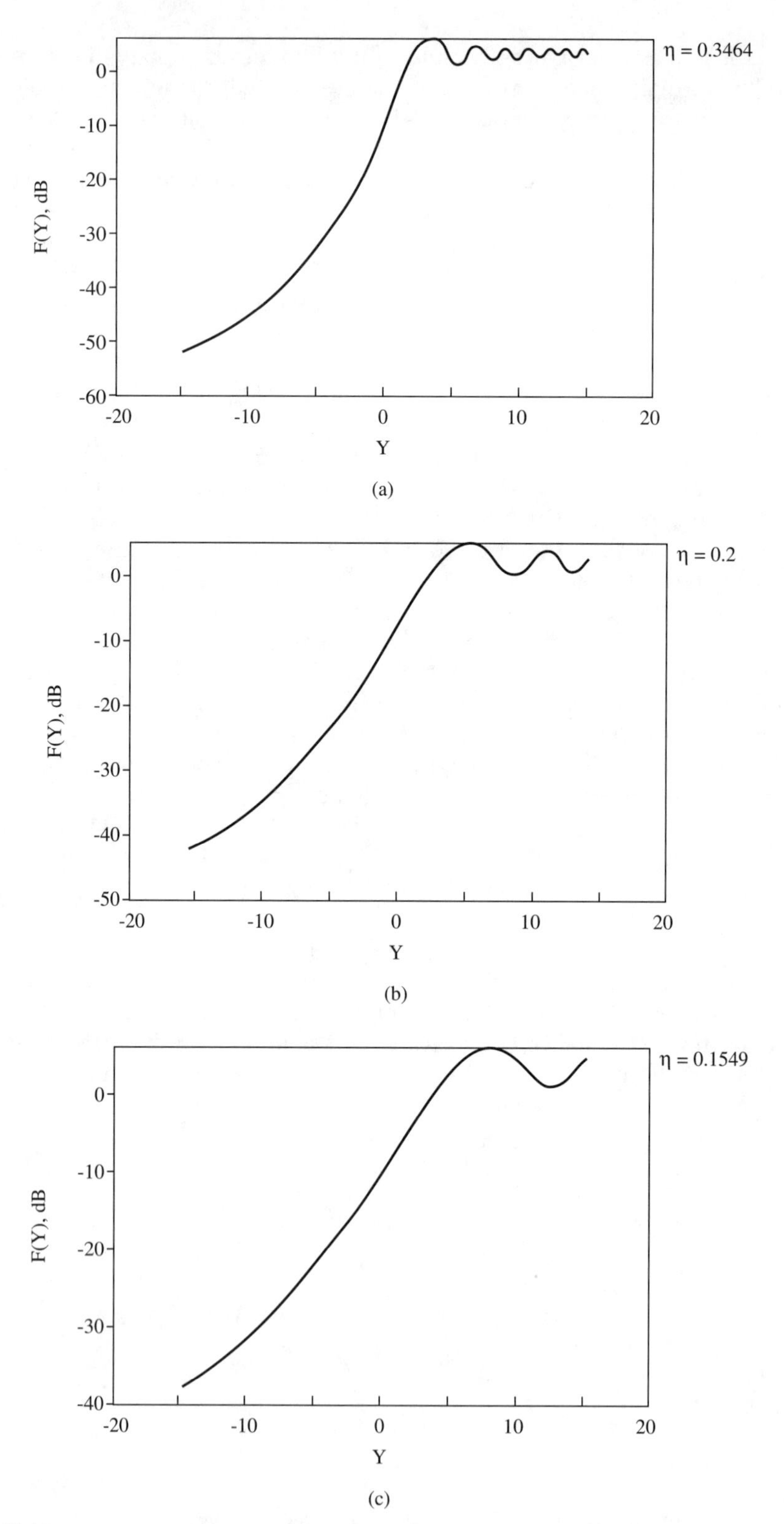

FIGURE 11.11
Two-way diffraction loss for three different wedge positions. $Y = \eta(h_{av} - h_w)$.

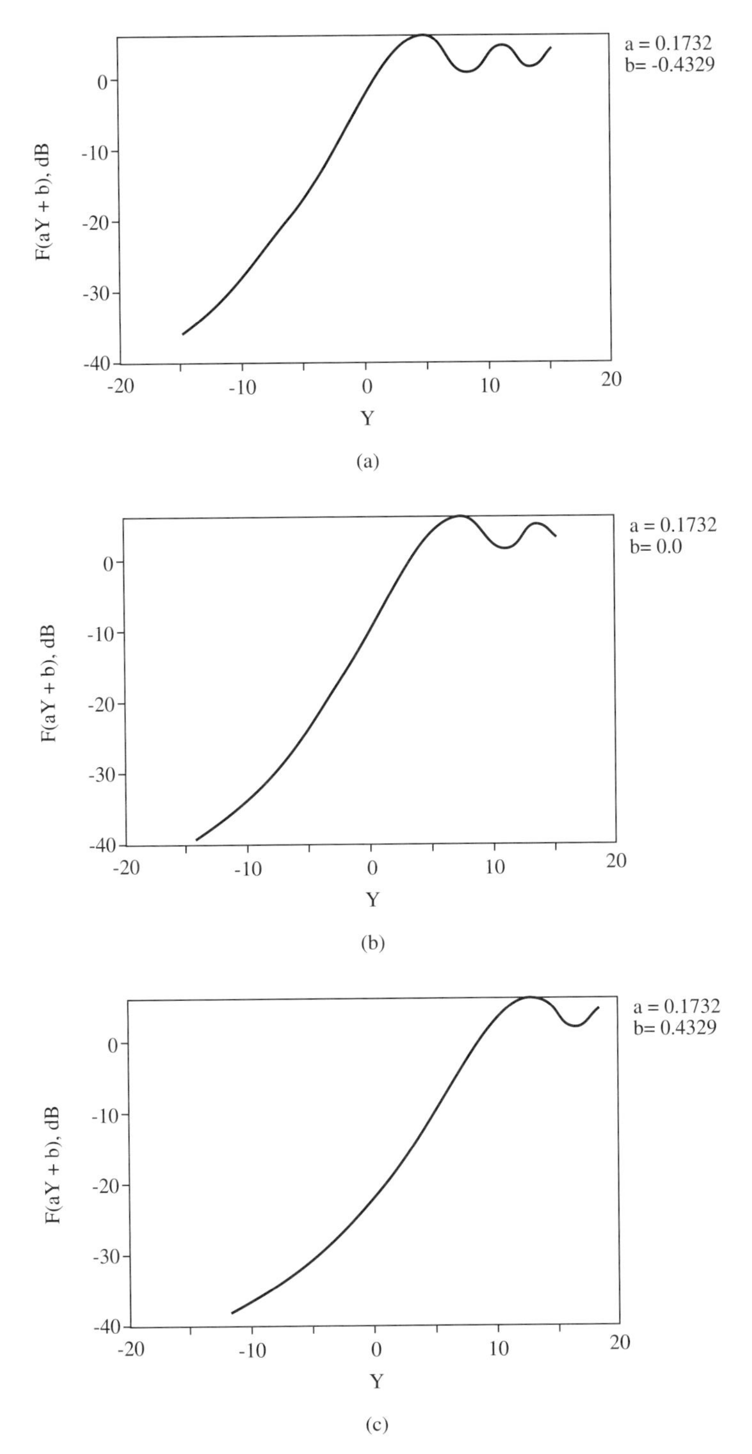

FIGURE 11.12
Two-way diffraction loss for three transmitter–receiver height differences. $Y = h_{av} - h_w$. a = η. b = $-\eta\xi(\Delta h/2)$.

$z_{1,2}$ being terminal point altitudes in meters and the normalization factors being

$$L = 2\left(\frac{r_e^2}{4k}\right)^{1/3} = 28.41\lambda^{1/3}(10^3)\text{ m}$$

$$H = \left(\frac{r_e}{2k^2}\right)^{1/3} = 47.55\lambda^{2/3}\text{ m}$$

The expression (11.34) can be used to roughly estimate the losses due to diffraction around the earth for radar systems studies in which this phenomenon is sufficiently important to warrant its consideration.

11.4 ATMOSPHERIC ATTENUATION

There is a natural attenuation[47–51] of the field strength of an electromagnetic wave propagating through any medium that has nonzero conductivity. Consider a plane wave propagating in the z-direction in a linear, homogeneous, isotropic medium with wave vector **k**. If the medium has permeability μ_0, permittivity $\epsilon_0\epsilon_R$ and conductivity σ, then the complex amplitude of the electric field vector has the form

$$\mathbf{E} = \mathbf{E}_0 e^{-jk_0\nu_c z} \tag{11.35}$$

where $k_0 = \omega\sqrt{\mu_0\epsilon_0}$ and ν_c is the complex refractive index, given by

$$\nu_c = \sqrt{\epsilon_R - j\frac{\sigma}{\omega\epsilon_0}} \qquad (11.35a)'$$

If the medium is "slightly lossy" defined by the condition

$$\frac{\sigma}{\omega\epsilon_0\epsilon_R} << 1 \qquad (11.35b)'$$

then

$$\sqrt{1 - \frac{j\sigma}{\omega\epsilon_0\epsilon_R}} \approx 1 - j\frac{\sigma}{2\omega\epsilon_0\epsilon_R} \qquad (11.35c)'$$

from which it follows that we can express (11.35) in the form

$$\mathbf{E} = \mathbf{E}_0 e^{-jkz} e^{-\alpha z} \qquad (11.35d)'$$

where $k = k_0\sqrt{\epsilon_R}$ and the "attenuation coefficient" α is given by

$$\alpha = \frac{Z_w\sigma}{2} \qquad (11.35e)'$$

where $Z_w = \sqrt{\mu_0/\epsilon_0\epsilon_R}$ is the wave impedance of the medium in ohms and σ the conductivity in Siemens/meter, or equivalently reciprocal ohms/meter. The unit of α is "nepers/meter" when it is used in field strength rather than power. It is usually expressed in dB/meter, in which case the value is the same whether we are dealing with field strength or power. The power density (power per unit cross-sectional area) of the wave is given by

$$\frac{dP}{dA} = \frac{|\mathbf{E}|^2}{2Z_w} = \frac{|\mathbf{E}_0|^2}{2Z_w} e^{-2\alpha z} \tag{11.36}$$

The power density in dB is

$$\left(\frac{dP}{dA}\right)_{dB} = 10 \log_{10}\left(\frac{|\mathbf{E}_0|^2}{2Z_w}\right) - (20 \log_{10} e)\alpha z \qquad (11.36a)'$$

The conversion between nepers/meter and dB meter, by virtue of (11.36a)′ is

$$\begin{aligned} \gamma &= \text{attenuation in dB/meter} \\ &= (20 \log_{10} e)\alpha = 8.68\alpha \end{aligned} \qquad (11.36b)'$$

where α is attenuation in nepers/meter.

The real part of the refractive index in air, as noted in Section 11.1, is very close to unity. The conductivity that would give rise to an imaginary part of the refractive index and thereby cause attenuation is due to moisture content. Perfectly dry air has no conductivity and would not attenuate a propagating wave through the mechanism discussed above.

Fog and clouds consist of tiny water droplets. As an electromagnetic wave propagates through a region containing these particles, it scatters in all directions from each particle, thus losing energy continuously. Also, each particle, having a finite conductivity characteristic of water, absorbs energy. From these two mechanisms of absorption and scattering, there is an attenuation which can be predicted quantitatively using very simple and well-known physics principles.

The water droplets are depicted as "Rayleigh spheres," meaning that their diameters are extremely small relative to wavelength. At frequencies below about 60 GHz that condition is always fulfilled so the theory of scattering and absorption from small spheres can always be used to predict these effects for a single sphere. Since the spheres are sparsely distributed throughout a volume, multiple scattering between them is a very small effect and can usually be safely neglected.

The theory of scattering and absorption by spheres based on spherical vector wave functions has been thoroughly investigated for decades and appears in many classical references (e.g., Stratton,[29] pp. 563–573, Kerr,[1] pp. 445–454, and others). The key results for purposes of our discussion here are the expressions for the total scattering and absorption cross-sections for a Rayleigh sphere given in Kerr,[1] pp. 672–673:

$$\sigma_s = \frac{K(2\pi a)^6}{\lambda^4} \tag{11.37a}$$

$$\sigma_a = \frac{K(2\pi a)^3}{\lambda} \tag{11.37b}$$

where a is the radius of the spheres, λ the wavelength, and K a constant. To obtain (11.37a) we calculate $\mathbf{E}_s(r, \theta, \phi)$, the electric field of a wave scattered by the sphere in a given direction (θ, ϕ) and at a given distance r from the sphere. We then calculate the Poynting vector $\mathbf{S}(r, \theta, \phi)$, which points in the radial direction and whose magnitude is the power density in the wave scattered in the direction (θ, ϕ) at a distance r from the sphere. That result is

$$\hat{r} \cdot \mathbf{S}(r, \theta, \phi) = \frac{dP}{dA}(r, \theta, \phi) \tag{11.38}$$

where

$$\begin{aligned} \frac{dP}{dA}(r, \theta, \phi) &= \frac{1}{2}\sqrt{\frac{\epsilon_0}{\mu_0}}\,|\mathbf{E}_s(r, \theta, \phi)|^2 \\ &= \frac{1}{2}\sqrt{\frac{\epsilon_0}{\mu_0}}\frac{1}{r^2}\,|\mathbf{F}_s(\theta, \phi|^2 \end{aligned} \tag{11.38a)'$$

where, for a scattered wave in the far zone, $\mathbf{E}_s$ has the form $\mathbf{E}_s(r, \theta, \phi) = (e^{-jkr}/r)\mathbf{F}_s(\theta, \phi)$, and where $\hat{r}$ is the unit vector in the radial direction.

Integration of (11.38a)′ over all directions around the scatterer gives us the total power scattered out from the sphere, i.e.,

$$\begin{aligned} P_s &= r^2 \int_0^{\pi} d\theta \sin\theta \int_0^{2\pi} d\phi \frac{dP}{dA}(r, \theta, \phi) \\ &= \frac{1}{240\pi}\int_0^{\pi} d\theta \sin\theta \int_0^{2\pi} d\phi\, |\mathbf{F}_s(\theta, \phi)|^2 \end{aligned} \tag{11.38b)'$$

It can be shown (Kerr,[1] pp. 402–406) that a Rayleigh water sphere scatters like an electric dipole. The dipole polarization per unit volume is given by

$$\frac{d\mathbf{D}}{d\mathbf{V}} = 3\left(\frac{\nu - 1}{\nu + 2}\right)\epsilon_0 E_0 \hat{z} = \frac{\mathbf{J}}{j\omega} \tag{11.38c)'$$

where $\mathbf{J}$ is the dipole current density and where the dipole is assumed oriented in the z-direction, ν is the refractive index of the sphere, and E_0 is the exciting field. Multiplying this by the volume of the sphere, we obtain the dipole moment

$$\mathbf{D} = \left(\frac{4\pi a^3}{3}\right)\left(\frac{d\mathbf{D}}{dV}\right) = \hat{z}\epsilon_0 E_0 4\left(\frac{\nu - 1}{\nu + 2}\right)\pi a^3 \tag{11.38d)'$$

The magnetic field in the far-zone of a z-directed electric dipole is

$$\mathbf{H}_s = \frac{jke^{-jkr}}{4\pi r} j\omega|\mathbf{D}|\,(\hat{r} \times \hat{z}) = \frac{-k\omega e^{-jkr}}{4\pi r}|\mathbf{D}|\hat{\phi} \sin\theta \tag{11.38e)'$$

and the resulting electric field is

$$\mathbf{E}_s = -Z_0(\hat{r} \times \mathbf{H}_s) = -\left(cZ_0 \frac{k^2|\mathbf{D}|e^{-jkr}}{4\pi r} \sin\theta\right)\hat{\theta} \tag{11.38f)'}$$

Taking the absolute square of $\mathbf{E}_s$ in (11.38f)', multiplying by $\frac{1}{2}\sqrt{\epsilon_0/\mu_0}$, integrating over the sphere of radius r surrounding the dipole, and invoking (11.38d)', we obtain P_s as given by (11.38b)'. The result is

$$P_s = \frac{4\pi}{3} Y_0|E_0|^2k^4a^6 \left|\frac{\nu - 1}{\nu + 1}\right|^2 \tag{11.38g)'}$$

where Y_0 is the wave admittance of free space.

The scattering cross section can be defined as the ratio of total scattered power to the incident wave power density, the latter being given by $(Y_0/2)|E_0|^2$. Thus from (11.38g)'

$$\sigma_s = \frac{P_s}{(Y_0/2)\,|E_0|^2} = \frac{8}{3}\pi k^4a^6\left|\frac{\nu - 1}{\nu + 1}\right|^2 \tag{11.38h)'}$$

The power lost by absorption inside the water sphere is derived from the ohmic loss per unit volume, given by

$$\frac{dP_L}{dV} = \frac{1}{2} Re(\mathbf{E}_0 \cdot \mathbf{J}^*) \tag{11.39}$$

where $\mathbf{E}$ is the electric field of the propagating wave and $\mathbf{J}$ the current density induced within the sphere by that field. From (11.38c)', the current density is related to the dipole moment per unit volume by the expression

$$\mathbf{J} = j\omega\left(\frac{d\mathbf{D}}{dV}\right) = 3j\omega\epsilon_0E_0\left(\frac{\nu - 1}{\nu + 2}\right) \tag{11.39a)'}$$

and hence (11.39) becomes

$$\frac{dP_L}{dV} = \frac{1}{2} Re\left[3j\omega\epsilon_0|E_0|^2\left(\frac{\nu - 1}{\nu + 2}\right)^*\right] \tag{11.39b)'}$$

From (11.39b)' it follows that P_L will be zero unless $[(\nu - 1)/(\nu + 2)]$ has an imaginary part, e.g., unless it is lossy.

$$\text{Im}\left(\frac{\nu - 1}{\nu + 2}\right) = \frac{3\nu_I}{(\nu_R + 2)^2 + \nu_I^2} \tag{11.39c)'}$$

(where ν_R and ν_I are the real and imaginary parts of the refractive index) which, when substituted into (11.39b)', yields

$$\frac{dP_L}{dV} = \frac{9\omega\epsilon_0}{2} \frac{|E_0|^2 \nu_I}{[(\nu_R + 2)^2 + \nu_I^2]} \tag{11.39d)'$$

Assuming σ and ϵ_R to be uniform within the sphere, we integrate dP_L/dV over the sphere to obtain the total power loss P_L, with the result

$$P_L = \frac{6\pi a^3 \omega\epsilon_0 |E_0|^2 \nu_I}{[(\nu_R + 2)^2 + \nu_I^2]} \tag{11.39e)'$$

The absorption cross section σ_a is given by

$$\sigma_a = \frac{P_L}{(Y_0/2)|E_0|^2} = \frac{24\pi^2 a^3 \nu_I}{\lambda[(\nu_R + 2)^2 + \nu_I^2]} \tag{11.39f)'$$

The "extinction cross section" is the sum of scattering and absorption cross sections and is given by

$$\sigma_e = \sigma_s + \sigma_a \tag{11.40}$$

where σ_s and σ_a are given by (11.38h)′ and (11.39f)′, respectively.

To find the average extinction cross section we must integrate over the distribution of sphere radii a, given by $n(a)$, the number of spheres per unit volume with radii between a and $(a + da)$. That average is

$$\sigma_{eav} = \int_0^\infty da\, \sigma_e(a) n(a) \tag{11.41}$$

A wave with electric field amplitude E_0 propagating in the x-direction in a lossless medium has power density

$$P_d = \frac{Y_0}{2} |E_0|^2 \tag{11.42}$$

The power lost as this wave propagates a distance dx through a lossy medium, where the losses are due to scattering and absorption by spherical particles, is

$$\frac{dP_d}{P_d} = -\alpha\, dx \tag{11.43}$$

where α is the one-way attenuation loss in nepers/meter and is given by $2\sigma_{eav}$. Curves showing variation of attenuation by fog and clouds (which is due to scattering and absorption by small spheres) with frequency are presented in Figure 11.13a.

An important source of attenuation at frequencies above 1 GHz is that due to atmospheric gases. The worst offenders are water vapor, which has resonance peaks at wavelengths of 1.35 cm (22.24 GHz) and 1.63 mm (184 GHz) and oxygen, which has resonances at wavelengths of 0.5 cm (60 GHz) and 2.54 mm (118 GHz). The curves shown in Figure 11.13b, the equivalent of which appears in many standard references,[52–54] show the resonances and the general increases in attenuation with increasing frequency. The values are at sea level, at a temperature of 20°C and at pressure of 1 atmosphere (760 mm Hg or 1013 mbar).

The mechanism for attenuation by atmospheric gases is not necessarily the same as that for fog and clouds. Some of the electromagnetic energy in the propagating wave is absorbed through conversion to thermal energy due to collisions between the molecules that are induced by the wave's electric fields. This collision-induced absorption of wave energy gives rise to quantum mechanical resonances at certain

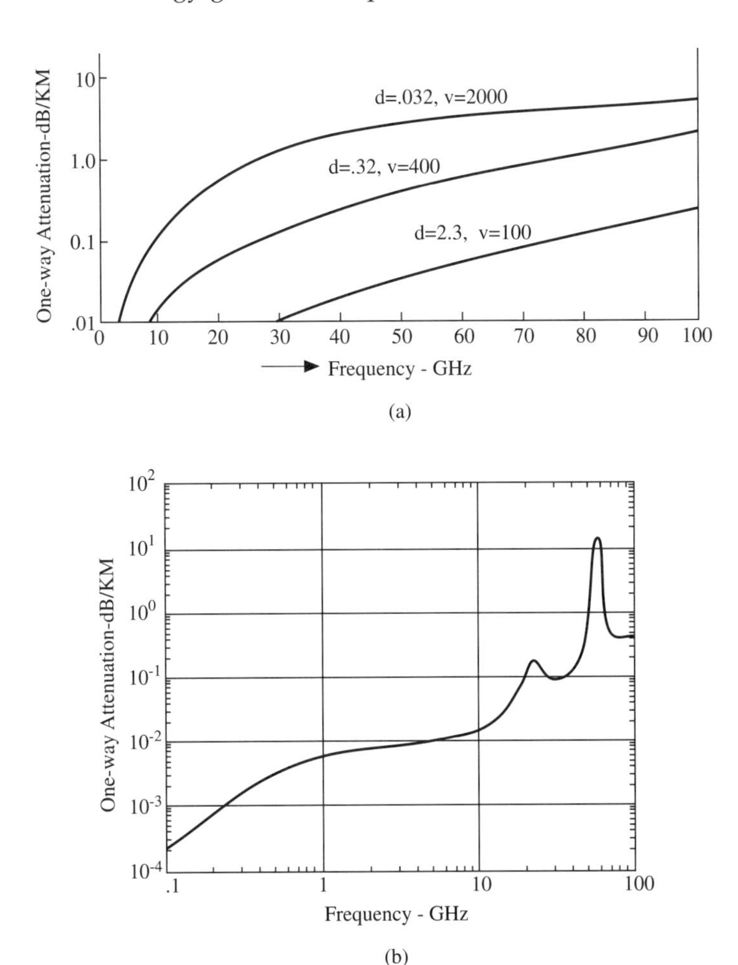

FIGURE 11.13
Attenuation by fog, clouds, and atmospheric gases. (a) Attenuation by fog and clouds; d = density of water droplets-g/m^3, v = visibility, feet. (b) Attenuation by atmospheric gases; Patterned after Figure 5.10 on page 207 of Blake,[9] based on NRL Report 7461, by L. V. Blake, October, 1967, by permission of Artech House.

frequencies. These resonances are not sharply peaked, but rather undergo collision broadening. The theory that predicts the broad resonances that appear on the curves was expounded by Van Vleck in 1947.[55]

It is evident from Figure 11.13 that attenuation due to oxygen absorption is negligible (less than 0.01 dB/km) at frequencies below about 20 GHz (i.e., below K-band) and becomes troublesome (exceeds 1 dB/km) between about 50 and 70 GHz, and extremely troublesome near 60 GHz, and unimportant again until the next resonance at 100 GHz. Water vapor attenuation, while it stays below about 0.2 dB/km until the frequency exceeds 100 GHz, can be important for long range propagation paths between about 20 and 80 GHz. For example, for a 200 km propagation path, water vapor attenuation can cause 20 dB of loss at 20 GHz.

11.5 ATTENUATION DUE TO PRECIPITATION

All forms of precipitation, e.g., rain, snow, and hail, have the potential for corrupting radar signals. This occurs through two mechanisms: (1) attenuation as the wave propagates through the precipitation region and (2) scattering from the precipitation region back to the radar, which is a form of clutter and will not be discussed here.

In the discussion of attenuation by fog and clouds in Section 11.4, we invoked the theory of scattering and absorption by a Rayleigh sphere, which is used as the model for a water droplet. The difference between a water droplet and a raindrop is primarily that of size. While we can safely assume that a water droplet is small compared to wavelength at all radar frequencies up through the millimeter bands, raindrop diameters become large enough to be considered as comparable to a wavelength at the high end of the radar spectrum.

The important dimension parameter is $(2\pi a/\lambda)$, where a is the radius. Throughout the X-band, free-space wavelengths are between 2.5 and 3.75 cm. In Ku-band, they range from 1.67 to 2.5 cm, in K-band from 1.11 to 1.67 cm, and in Ka-band, from 7.5 mm to 1.11 cm. Raindrop radii tend to be distributed between about 1 and 4 mm and hence their orders of magnitude become comparable to wavelength in the millimeter bands. This poses a serious problem in radars required to operate in all kinds of weather.

Attenuation and scattering by rain and snow have been investigated extensively for many decades and there is a spate of literature on the subject.[56–63] The theory used to calculate attenuation due to rain is similar to the development in Section 11.4, culminating in the expression (11.41) for the mean extinction cross section from which attenuation in dB/km follows. The integrand in (11.41) is the product of $\sigma_e(a)$, the cross section of a single raindrop of radius a, and $n(a)$, the number of spheres per unit volume with radii between a and $(a + da)$. The factor $\sigma_e(a)$ is obtained from Rayleigh theory if the drop is small compared to wavelength and from the more extensive Mie theory[64–66] of absorption and scattering by spheres of arbitrary radii if the drop size is comparable to wavelength, i.e., at the high frequencies.

The drop size distribution $n(a)$ is obtained empirically from measurements of rainfall rate R, traditionally measured in mm/hr. Compilations of rain attenuation results were published many years ago by Ryde and Ryde,[67] using drop-size distributions obtained by Laws and Parsons.[68] These results are discussed extensively in Kerr[1] and are summarized for various frequencies and rainfall rates in Figure 11.14, based on Table 8.15, Figure 8.12, and Figure 8.13 on pp. 682, and 683, respectively. On p. 406 of Collin,[3] a simple formula for attenuation as a function of rainfall rate and frequency is presented. Collin cites an IEEE paper by Olsen et al.,[69] in which

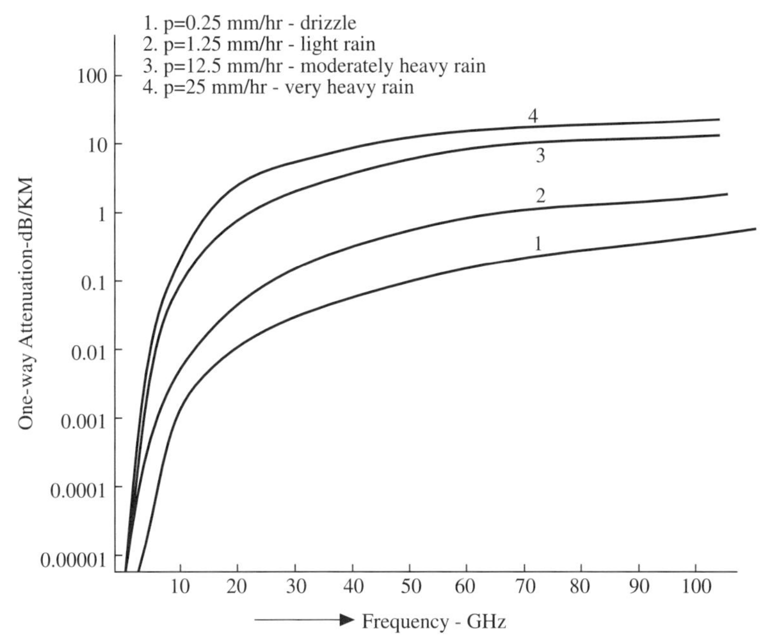

FIGURE 11.14
Attenuation due to rainfall.

empirical formulas for constants arising in this formula are reviewed. The formula, which could be used by a radar systems engineer to roughly estimate the amount of rain attenuation to expect in a given scenario, is

$$2\alpha = aR^b \text{ dB/km} \tag{11.44}$$

where α is the one-way attenuation, R is rainfall rate in mm/hr, and a and b are as follows:

$$a = G_a f^{E_a}, \qquad b = G_b f^{E_b}$$

where f = frequency in Gigahertz and where G_a, E_a, G_b, E_b for various frequency regions are given in Table 11.2.

In addition to rain, attenuation is also caused by other forms of precipitation, i.e., snow and hail, but these topics will not be discussed here. A few literature sources including these topics appear in the reference list at the end of this chapter.[70,71]

TABLE 11.2

Frequency Constants for Empirical Rain Attenuation Formula

Region (GHz)	G_a	E_a	G_b	E_b
$f < 2.9$	$6.39(10^{-5})$	2.03	0.851	0.158
$2.9 \le f < 8.5$	$4.21(10^{-5})$	2.42	0.851	0.158
$8.5 \le f < 25$	$4.21(10^{-5})$	2.42	1.41	−0.0779
$25 \le f < 54$	$4.21(10^{-5})$	2.42	2.63	−0.272
$54 \le f < 164$	$4.09(10^{-2})$	0.699	2.63	−0.272
$164 \le f \le 180$	$4.09(10^{-2})$	0.699	0.616	0.0126
$f > 180$	3.38	−0.151	0.616	0.0126

From Eqs. (6.108 a,b) on page 406 of Collin,[1] "Antennas and Radiowave Propagation," by permission of McGraw-Hill.

11.6 SHADOWING AT LOW GRAZING ANGLES

A radar whose antenna beam points in a nearly horizontal direction over hilly land terrain or a rough sea surface illuminates only a portion of the terrain. This is illustrated by Figure 11.15, which shows how ray paths are obscured or possibly completely blocked by height variations on the surface. Each of the small patches P_1, P_2, P_3, P_4, and the target T are at the peak of Radar 1's antenna beam, but only one of the patches, P_1, is both on the side of the nearest hill facing that radar and not blocked by that hill or the one behind it. Patches P_2, P_3, and P_4 and the target T are not visible to Radar 1 and hence if we rely entirely on line-of-sight (LOS) propagation, we get no return from those patches or from the target. The only contributions would be from diffraction around the hills, which was discussed in Section 11.3. This adversely affects the probability of target detection (and also may decrease the clutter return from the terrain surface).

Radar 2 illuminates nearly all four of the patches and target, although it sees some at very low *local* grazing angles and hence gets little return from them.

The obvious difference between the two radars is in their altitudes and their grazing angles as defined with respect to a horizontal earth's surface. For a flat earth approximation, that is the same as the beam depression angle which is about 10° for Radar 1 and about 60° for Radar 2. For small grazing angles, characteristic of ground based or low-flying airborne radars, the shadowing phenomenon (which is sometimes called "terrain masking") assumes much greater importance than for high-flying airborne radars.

Unless there is a terrain-specific database, in which case it would be possible to calculate the shadowing deterministically, in most radar analysis it is necessary to assume a statistical height variation on the terrain and obtain a shadowing function based on the mean and variance of height fluctuations and the radar's depression angle. Some well-known analysis based on these ideas is discussed in what follows.

The shadowing function to be discussed is that due to Smith[72] and Sancer,[73] referenced in Tsang et al.[27] pp. 93–96, Eqs. (113)–(127). Another reference on this topic is Ulaby et al.,[28] Appendix 12k, p. 1024, Eqs. (12.k.1)–(12.k.4), in which the

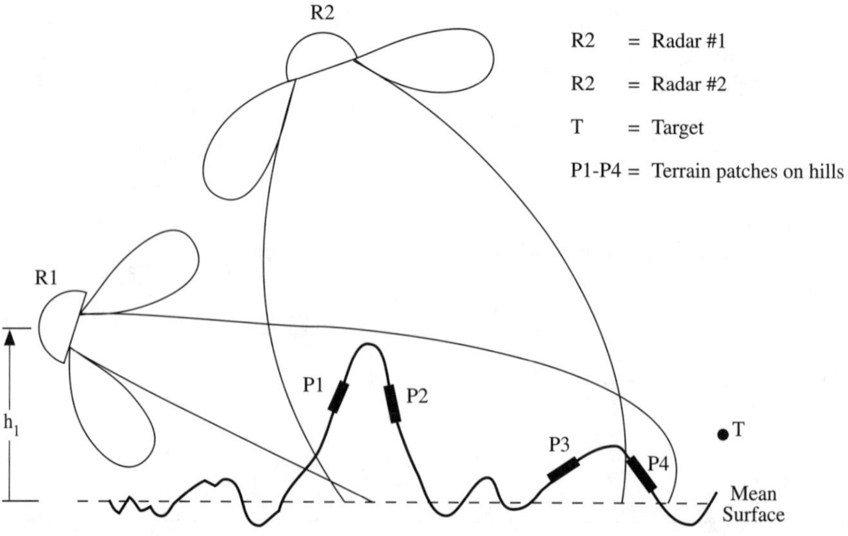

FIGURE 11.15
Illustration of shadowing by terrain.

shadowing function credited to Smith[72] is cited, with contributions by Wagner[74] applicable to the Kirchoff approximation with stationary phase.

The Sancer–Smith shadowing function given in Tsang et al.,[27] which is for the general (including bistatic) case, is based on the insertion of a factor $L(\hat{k}_i, \hat{k}_s, \mathbf{r}')$ into the integrand of the diffraction integral from which the fields are evaluated (which is equivalent to the Stratton–Chu integral). The vector $\mathbf{r}'$ denotes the surface coordinates (x', y'). The unit vectors $\hat{k}_i$ and $\hat{k}_s$ are the unit wave vectors for the incident and scattered waves, respectively. The factor $L(\hat{k}_i, \hat{k}_s, \mathbf{r}') = 1$ if a ray in the direction $\hat{k}_i$ illuminates the point $\mathbf{r}'$ without intersecting the surface at some point earlier on its path, *and* if the ray from $\mathbf{r}'$ along $\hat{k}_s$ does not strike the surface elsewhere. This means that if these conditions are met, the incident ray fully illuminates the point and the scattered wave from the point is not blocked by another part of the surface on its way to the receiving point, and hence the point fully contributes to the integral. If those conditions are *not* met at that point, then $L(\hat{k}_i, \hat{k}_s, \mathbf{r}') = 0$, implying no contribution to the integral from the surface point $\mathbf{r}'$. The surface roughness height function is assumed to be a sample function of a Gaussian random process $z(x, y)$. Shadowing must be formulated in terms of the bivariate Gaussian PDF of the surface slopes (z_x, z_y) given by

$$P(z_x, z_y) = \frac{1}{2\pi\sigma^2 |C''(0)|} e^{[(z_x^2 + z_y^2)/(2\sigma^2 |C''(0)|)]} \tag{11.45}$$

where $C''(0)$ is the second derivative of the normalized correlation function at $|\mathbf{r} - \mathbf{r}'| = 0$ and $\sigma^2 = \langle z^2 \rangle$. Using the stationary phase approximation, implying that $z_x = -k_{dx}/k_{dz}$, $z_y = -k_{dy}/k_{dz}$, where $\mathbf{k}_d = \hat{k}_s - \hat{k}_i$, we obtain from (11.45).

$$P\left(\frac{-k_{dx}}{k_{dz}}, \frac{-k_{dy}}{k_{dz}}\right) = \frac{1}{2\pi\sigma^2 |C''(0)|} e^{-(k_{dx}^2 + k_{dy}^2)/(2\sigma^2 k_{dz}^2 |C''(0)|)} \tag{11.46}$$

The incoherently scattered intensity, as given by Eq. (114) on p. 93 of Tsang et al.,[27] contains a factor $\langle II^* \rangle$, which when modified to account for shadowing contains the products $L(\hat{k}_i, \hat{k}_s, \mathbf{r}) L(\hat{k}_i, \hat{k}_s, \mathbf{r}')$ within the integrand of its double surface integral, where $\mathbf{r}$ and $\mathbf{r}'$ are points on the surface. Using the method of asymptotics to evaluate this double integral, in the limit as $k \to \infty$, the contributions to the integral come from regions where (x, y) is close to (x', y'). The end result of this process (see pp. 85 and 86 of Tsang et al.,[27] where the process is carried out without the shadowing functions, and carry out the same steps with the inclusion of the shadowing functions in the integrands) is (see Eq. (117) on p. 94 of Tsang et al.[27])

$$\left\langle \lim_{k \to \infty} II^* \right\rangle = \frac{4\pi^2 A_0}{k_{dz}^2} \int dL \, p(z_x, z_y, L) \Big|_{z_x = -k_{dx}/k_{dz}, z_y = -k_{dy}/k_{dz}} L^2(\hat{k}_i, \hat{k}_s, \mathbf{r}_-) \tag{11.47}$$

where $\mathbf{r}_- = (x, y)$ and where $P(z_x, z_y, L)$ is the joint PDF of z_x, z_y, and L.

From Bayes' theorem

$$p(z_x, z_y, L) = p(z_x, z_y) p(L/z_x, z_y) \tag{11.48}$$

where $p(L/z_x, z_y)$ is the conditional PDF of L with given values of z_x and z_y.

From the definition of L as a function of $\hat{k}_i$, $\hat{k}_s$, and $\mathbf{r}$ that has the value 1 if $\mathbf{r}$ is illuminated by rays in directions $\hat{k}_i$ and $-\hat{k}_s$ and zero otherwise, we have (see Eq. (119) on p. 94 of Tsang et al.[27]).

$$P(L/z_x, z_y) = P_L(\hat{k}_i, \hat{k}_s/z_x, z_y)\delta(L-1) + [1 - P_L(\hat{k}_i, \hat{k}_s/z_x, z_y)]\delta(L) \tag{11.49}$$

where $P_L(\hat{k}_i, \hat{k}_s/z_x, z_y)$ is the conditional probability that a point is illuminated by rays in the directions $\hat{k}_i$ and $-\hat{k}_s$ given the values of z_x and z_y at that point. Equation (11.49) may be expressed in words as follows: The conditional probability that $L = 1$ at a point given slopes (z_x, z_y) at that point is the probability that rays in directions $\hat{k}_i$ and $-\hat{k}_s$ illuminate the point. The conditional probability that $L = 0$ at a point given slopes (z_x, z_y) at that point is the probability that rays in directions $\hat{k}_i$ and $-\hat{k}_s$ *do not* illuminate the point [equal to $1 - P_L(\hat{k}_i, \hat{k}_s/z_x, z_y)$]. The conditional probability of L given slopes (z_x, z_y) is the sum of these two conditional probabilities.

When (11.49) is substituted into (11.47) with the aid of (11.48), we obtain in the limit as $k \to \infty$

$$\langle \text{II}^* \rangle = \frac{4\pi^2 A_0}{k_{dz}^2} P\left(\frac{-k_{dx}}{k_{dz}}, \frac{-k_{dy}}{k_{dz}}\right) P\left(1 \Big/ \frac{-k_{dx}}{k_{dz}}, \frac{-k_{dy}}{k_{dz}}\right)$$

$$= \frac{4\pi^2 A_0}{k_{dz}^2} P\left(\frac{-k_{dx}}{k_{dz}}, \frac{-k_{dy}}{k_{dz}}\right) P_L\left(\hat{k}_i, \hat{k}_s \Big/ \frac{-k_{dx}}{k_{dz}}, \frac{-k_{dy}}{k_{dz}}\right) \tag{11.50}$$

Without the shadowing function only the factor P_L on the right is missing [Eq. (81) on p. 86 of Tsang et al.[27]]. Hence the shadowing function by which the scattering coefficient (or equivalently the NRCS) is multipled *is* the P_L function in (11.50). Equation (122) on p. 95 of Tsang et al.[27] is the final result for the shadowing function $S(\theta_s, \theta_i)$, given by

$$S(\theta_s, \theta_i) = P_L\left(\hat{k}_i, \hat{k}_s \Big/ \frac{-k_{dx}}{k_{dz}}, \frac{-k_{dy}}{k_{dz}}\right) \tag{11.51}$$

where

$$S(\theta_s, \theta_i) = \begin{cases} \dfrac{1}{1+\Lambda(\mu_s)}, & \phi_s = \phi_i + \pi,\ \theta_s \geq \theta_i \\[2ex] \dfrac{1}{1+\Lambda(\mu_i)}, & \phi_s = \phi_i + \pi,\ \theta_s \leq \theta_i \\[2ex] \dfrac{1}{1+\Lambda(\mu_s)+\Lambda(\mu_i)}; & \text{otherwise} \end{cases} \tag{11.52}$$

where $\mu = \cot\theta$

$$\Lambda(\mu) = \frac{1}{2}\left[\sqrt{\frac{2}{\pi}}\frac{s}{\pi} e^{-\mu^2/2s^2} - \operatorname{erfc}\left(\frac{\mu}{\sqrt{2}s}\right)\right] \tag{11.52$'$}$$

where s = rms surface slope = $\sigma^2|C''(0)|$.

The final form of P_L as given by (11.52) is based on the assumption of Gaussian slope statistics [Eq. (11.45)], which, in turn, is based on that of Gaussian height statistics.

Note that, if a Gaussian PDF of the form $(1/\sqrt{2\pi s^2})e^{-(x-\mu)^2/2\sigma^2}$ were integrated from 0 to ϵ, where $\epsilon << \mu$, the result would be

$$I = \int_0^{\epsilon} \frac{dx}{\sqrt{2\pi\sigma^2}} e^{-(x-\mu)^2/2\sigma^2} = \frac{1}{2}\left(\frac{2}{\sqrt{\pi}}\int_{-\mu/\sqrt{2s}}^{0} dy\, e^{-y^2} + \frac{2}{\sqrt{\pi}}\int_0^{(\epsilon-\mu)/\sqrt{2s}} dy\, e^{-y^2}\right)$$

$$= \frac{1}{2}\left[\operatorname{erfc}\left(\frac{(\mu-\epsilon)}{\sqrt{2s}}\right) - \operatorname{erfc}\left(\frac{\mu}{\sqrt{2s}}\right)\right] \tag{11.53}$$

An asymptotic expansion of erfc(x) for $|x| >> 1$ [ ] is

$$\operatorname{erfc}(x) \simeq \frac{e^{-x^2}}{\sqrt{\pi}x}\left[1 - \frac{1}{2x^2} + 0\left(\frac{1}{x^4}\right)\right] \tag{11.54}$$

Applying this expansion to (11.53) and truncating it after the first term [since $|(\mu-\epsilon)/\sqrt{2s}| >> 1$] yields the result

$$I = \frac{1}{2}\left[\sqrt{\frac{2}{\pi}}\frac{s}{(\mu-\epsilon)} e^{-(\mu-\epsilon)^2/2s^2} - \operatorname{erfc}\left(\frac{\mu}{\sqrt{2s}}\right)\right] \tag{11.55}$$

where for $|\mu| >> \epsilon$, I is identical to Λ as given by (11.52).

For monostatic radar, where $\hat{k}_s = -\hat{k}_i$, the shadowing function of (11.51), which is a factor on the radar cross section, is given by

$$S(\theta_i) = S(\theta_i, \theta_i) = \frac{1}{1 + 2\Lambda(\mu_i)} \tag{11.56}$$

where

$$\Lambda(\mu_i) = \frac{1}{2}\left[\sqrt{\frac{2}{\pi}}\frac{s}{\pi} e^{-\mu^2/2s^2} - \operatorname{erfc}\left(\frac{\mu}{\sqrt{2s}}\right)\right]$$

where $\mu_i = \cot\theta_i$, s = rms surface slope = $\langle z^2\rangle\,|C''(0)|$, $C''(\rho)$ = the second derivative of the correlation function of the heights z_1 and z_2, and $\rho = |z_1 - z_2|$, $\sqrt{\langle z^2\rangle}$ = rms surface height.

REFERENCES

1. Kerr, D.E. (ed.), "Propagation of Short Radio Waves." McGraw-Hill, New York, 1951.
2. Meeks, M.L., "Radar Propagation at Low Altitudes." Artech, Norwood, MA, 1982.
3. Collin, R.E., "Antennas and Radiowave Propagation." McGraw-Hill, New York, 1985, Chapter 6.
4. Griffiths, J., "Radiowave Propagation and Antennas." Prentice-Hall, Englewood Cliffs, NJ, 1987.
5. Freeman, R.L., "Radio System Design for Telecommunications (1 –100 GHz)." Wiley, New York, 1987, Chapter 1; also Section 3.3.
6. Parsons, D., "The Mobile Radio Propagation Channel." Halsted Press, New York (Division of J. Wiley & Sons), 1992.
7. Skolnick, M.I., "Introduction to Radar Systems," 2nd ed. McGraw-Hill, New York, 1980, Chapter 12.
8. Rohan, P., "Surveillance Radar Performance Prediction." Peter Peregrinus, London, UK, 1983, Chapters 5, 6.

9. Blake, L.V., "Radar Range Performance Analysis." Artech, Norwood, MA, 1986, Chapters 5 and 6.
10. Eaves, J.L., and Reedy, E.K. (eds.), "Principles of Modern Radar." Van Nostrand Reinhold, New York, 1987, Chapter 3, by D.G. Bodnar.
11. Barton, D.K., "Modern Radar System Analysis." Artech, Norwood, MA, 1988, Chapter 6.
12. Parsons, D., "The Mobile Radio Propagation Channel." Halsted Press, New York (Division of J. Wiley & Sons), 1992, pp. 28–31.
13. Barton, D.K., "Modern Radar System Analysis." Artech, Norwood, MA, 1988, p. 304.
14. Kerr, D.E. (ed.), "Propagation of Short Radio Waves." McGraw-Hill, New York, 1951, pp. 34–41, also Chapter 5, by D.E. Kerr, W.T. Fishback, and H. Goldstein.
15. Blake, L.V., "Radar Range Performance Analysis." Artech, Norwood, MA, 1986, Chapter 6.
16. Skolnick, M.I., "Introduction to Radar Systems," 2nd ed. McGraw-Hill, New York, 1980, Chapter 12, pp. 442–447.
17. Barton, D.K., "Modern Radar System Analysis." Artech, Norwood, MA, 1988, Section 6.2, pp. 288–296.
18. Eaves, J.L., and Reedy, E.K. (eds.), "Principles of Modern Radar." Van Nostrand Reinhold, New York, 1987, Chapter 4, by H.A. Corriher, Jr.
19. Griffiths, J., "Radiowave Propagation and Antennas." Prentice-Hall, Englewood Cliffs, NJ, 1987, pp. 100–112.
20. Kerr, D.E. (ed.), "Propagation of Short Radio Waves," McGraw-Hill, New York, 1951, pp. 112–122 (by W.T. Fishback); also pp. 404–406 in Chapter 5.
21. Griffiths, J., "Radiowave Propagation and Antennas." Prentice-Hall, Englewood Cliffs, NJ, 1987, Sections 4.2.6, 4.2.7, pp. 108–112.
22. Rohan, P., "Surveillance Radar Performance Prediction." Peter Peregrinus, London, UK, 1983, pp. 152–153.
23. Blake, L.V., "Radar Range Performance Analysis." Artech, Norwood, MA, 1986, pp. 269–271.
24. Van der Pol and Bremmer, "Further Notes on the Propagation of Radio Waves over a Finitely Conducting Spherical Earth." Phil. Mag. 27(Ser. 7, No. 182), 1939.
25. Beckmann, P., and Spizzichino, A., "The Scattering of Electromagnetic Waves from Rough Surfaces." Artech, Norwood, MA, 1987.
26. Ruck, G.T., Barrick, D.E., Stuart, W.D., and Krichbaum, C.K., "Radar Cross-Section Handbook" (2 volumes). Plenum Press, New York, 1970.
27. Tsang, L., Kong, J.A., and Shin, R.T., "Theory of Microwave Remote Sensing." Wiley, New York, 1985.
28. Ulaby, F.T., Moore, R.K., and Fung, A.K., "Microwave Remote Sensing." Volume I, Addison-Wesley, Reading, MA, 1981; Volume II, Addison-Wesley, Reading, MA, 1982; Volume III, Artech, Norwood, MA, 1986.
29. Stratton, J.A., "Electromagnetic Theory." McGraw-Hill, New York, 1941, p. 466.
30. Shen, L.C., and Kong, J.A., "Applied Electromagnetism," 2nd ed. P.W.S. Engineering, Boston, 1987, Chapter 4.
31. Rao, N.N., "Elements of Engineering Electromagnetics," 3rd ed. Prentice-Hall, Englewood Cliffs, NJ, 1991, Section 9.5, pp. 538–547.
32. Jordan, E.C., and Balmain, K.G., "Electromagnetic Waves and Radiating Systems," 2nd ed. Prentice-Hall, Englewood Cliffs, NJ, 1968, Sections 5.09–5.13, pp. 136–152.
33. Tsang, L., Kong, J.A., and Shin, R.T., "Theory of Microwave Remote Sensing." Wiley, New York, 1985, Chapters 2 and 5.
34. Ishimaru, A., "Electromagnetic Wave Propagation, Radiation, and Scattering." Prentice-Hall, Englewood Cliffs, NJ, 1991, pp. 36–40.
35. Von Hippel, A.R., "Microwave Materials and Applications." MIT Press, Cambridge, MA, 1954.
36. Kerr, D.E. (ed.), "Propagation of Short Radio Waves." McGraw-Hill, New York, 1951.
37. Hipp, J.A., "Soil Electromagnetic Parameters as Functions of Frequency, Soil Density and Soil Moisture." Proc. IEEE 62(1), 98–108, 1974.
38. Ulaby, F.T., Moore, R.K., and Fung, A.K., "Microwave Remote Sensing," Vol. III, Artech, Norwood, MA, 1986, Appendix E, pp. 2017–2119.
39. Ishimaru, A., "Electromagnetic Wave Propagation, Radiation, and Scattering." Prentice-Hall, Englewood Cliffs, NJ, 1991, pp. 388–389.
40. Barton, D.K., "Modern Radar System Analysis." Artech, Norwood, MA, 1988, pp. 299–302.
41. Ulaby, F.T., Moore, R.K., and Fung, A.K., "Microwave Remote Sensing," Vol. I, Addison-Wesley, Reading, MA, 1981, pp. 195–121.
42. Meeks, M.L., "Radar Propagation at Low Altitudes." Artech, Norwood, MA, 1982, Section 5, pp. 26–35.

43. Griffiths, J., "Radiowave Propagation and Antennas." Prentice-Hall, Englewood Cliffs, NJ., 1987, pp. 127–130.
44. Collin, R.E., "Antennas and Radiowave Propagation," McGraw-Hill, New York, 1985, Section 6.4, pp. 372–377.
45. Parsons, D., "The Mobile Radio Propagation Channel." Halsted Press, New York (Division of J. Wiley & Sons), 1992, Sections 3.3, 3.4, pp. 37–51.
46. Collin, R.E., "Antennas and Radiowave Propagation." McGraw-Hill, New York, 1985, Section 6.3, pp. 369–372.
47. Kerr, D.E. (ed.), "Propagation of Short Radio Waves." McGraw-Hill, New York, 1951, Chapter 8, by J.H. Van Vleck, E.M. Purcell, and H. Goldstein.
48. Skolnick, M.I., "Introduction to Radar Systems," 2nd ed. McGraw-Hill, New York, 1980, Section 12.7, pp. 459–461.
49. Eaves, J.L., and Reedy, E.K. (eds.), "Principles of Modern Radar." Van Nostrand Reinhold, New York, 1987, Section 3.4.1, pp. 66–70 (by D. Bodnar).
50. Barton, D.K., "Modern Radar System Analysis." Artech, Norwood, MA, 1988, pp. 278–282.
51. Blake, L.V., "Radar Range Performance Analysis." Artech, Norwood, MA, 1986, pp. 197–214.
52. Kerr, D.E. (ed.), "Propagation of Short Radio Waves." McGraw-Hill, New York, 1951, p. 663.
53. Barton, D.K., "Modern Radar System Analysis." Artech, Norwood, MA, 1988, p. 279.
54. Collin, R.E., "Antennas and Radiowave Propagation." McGraw-Hill, New York, 1985, p. 409.
55. Van Vleck, J.H., "The Absorption of Microwaves by Oxygen," and "The Absorption of Microwaves by Uncondensed Water Vapor." Phys. Rev. 71(7), 413–433, 1947.
56. Kerr, D.E. (ed.), "Propagation of Short Radio Waves." McGraw-Hill, New York, 1951, Chapter 7, by H. Goldstein, D.E. Kerr, and A.E. Bent.
57. Ulaby, F.T., Moore, R.K., and Fung, A.K., "Microwave Remote Sensing," Volume I, Addison-Wesley, Reading, MA, 1981, Sections 512–514, pp. 316–330.
58. Skolnick, M.I., "Introduction to Radar Systems," 2nd ed. McGraw-Hill, New York, 1980, Section 13.7, pp. 409–503 and Section 13.8, pp. 504–506.
59. Blake, L.V., "Radar Range Performance Analysis." Artech, Norwood, MA, 1986, pp. 214–221.
60. Collin, R.E., "Antennas and Radiowave Propagation," McGraw-Hill, New York, 1985, pp. 401–407.
61. Nathanson, F.E., Reilly, J.P., and Cohen, M.N., "Radar Design Principles," 2nd ed. McGraw-Hill, New York, 1991, Chapter 6, by F.E. Nathanson and J.P. Reilly.
62. Barton, D.K., "Modern Radar System Analysis." Artech, Norwood, MA, 1988, pp. 38–41 and 279–284.
63. Skolnick, M.I. (ed.), "Radar Handbook," 1st ed. McGraw-Hill, New York, 1970, Section 24.14, pp. 24-23 to 24-27.
64. Stratton, J.A., "Electromagnetic Theory." McGraw-Hill, New York, 1941, pp. 414–420 and 563–573.
65. Kerr, D.E. (ed.), "Propagation of Short Radio Waves." McGraw-Hill, New York, 1951, pp. 445–453 and 608–615.
66. Ulaby, F.T., Moore, R.K., and Fung, A.K., "Microwave Remote Sensing," Vol. I, Addison-Wesley, Reading, MA, 1981, pp. 290–298.
67. Ryde, J.W., and Ryde, D., "Attenuation of Centimeter Waves by Rain, Hail, Fog and Clouds." Report, General Electric Company, Wembly, England, 1945.
68. Laws, J.O., and Parsons, D.A., "The Relationship of Raindrop Size to Intensity." Transact. Am. Geophys. Union, 24th Annu. Meeting, 452–460, 1943.
69. Olsen, R.L., Rodgers, D.V., and Hodge, D.B., "The aR^b Relation in the Calculation of Rain Attenuation." IEEE Trans. AP-26, 318–329, 1978.
70. Kerr, D.E. (ed.), "Propagation of Short Radio Waves." McGraw-Hill, New York, 1951, pp. 685–688.
71. Ulaby, F.T., Moore, R.K., and Fung, A.K., "Microwave Remote Sensing," Vol. I, Addison-Wesley, Reading, MA, 1981, pp. 326–329.
72. Smith, B.G., "Geometrical Shadowing of a Random Rough Surface." IEEE Trans. Ant. Prop. AP-15(5), 668–671, 1967.
73. Sancer, M.I., "Shadow-Corrected Electromagnetic Scattering from a Randomly Rough Surface." IEEE Trans. Ant. Prop. AP-17(5), 577–585, 1969.
74. Wagner, R.J., "Shadowing of Randomly Rough Surfaces." J. Acoust. Soc. Am. 41, 138–147, 1967.

Chapter 12

Radar Parameter Measurement Theory

To attack the problem of optimal measurement of radar parameters (as opposed to mere detection of targets), we can use maximum likelihood estimator (MLE) theory as a theoretical base.* Then we can apply MLE theory to the measure-

* Most of the theory in this chapter is derived here from first principles, the approach being close to that followed by the author in a previous book,[1] but a few general sources of MLE theory are provided in the reference list,[2–6], as well as locations of the sections on radar parameter measurement and tracking in a few well-known radar texts.[7–10]

ment of amplitude, frequency, phase, pulse position, or (in principle) any other radar signal parameter we would like to measure.

12.1 MAXIMUM LIKELIHOOD ESTIMATION OF RADAR SIGNAL PARAMETERS

To find the optimal measurement of a parameter β contained in a signal waveform $s(t;\, \beta)$ immersed in additive noise $n(t)$, we formulate the probability that the total voltage $v(t)$, the superposition of signal and noise, has a value between $v(t)$ and $v(t) + dv$, i.e., the PDF of $v(t)$ at time t. Since the parameter β is to be measured, that implies that it is unknown prior to the measurement. However, from the viewpoint of the party doing the measuring, it has an a-priori PDF of the form $p(\beta)$ (a priori in this context means "before the measurement is made"). The joint PDF of β and v (where we have omitted the argument t on $v(t)$ and just consider it as a random variable) before the measurement is made ("a priori PDF") is

$$p(v, \beta) = p(\beta)p(v/\beta) \tag{12.1a}$$

where

$p(\beta)$ = PDF of unknown parameter β
$p(v/\beta)$ = conditional PDF of v given a fixed value of β

According to Bayes' theorem (or our intuition about probability), the joint PDF of the two variables β and v can be expressed as in (12.1a) or alternatively as

$$p(v, \beta) = p(v)p(\beta/v) \tag{12.1b}$$

where

$p(v)$ = PDF of v (averaged over all possible values of β)
$p(\beta/v)$ = conditional PDF of β given a fixed value of v

From (12.1a, b) we can write

$$p(\beta/v) = \frac{p(\beta)p(v/\beta)}{p(v)} \tag{12.1c}$$

Before we have observed the voltage waveform $v(t)$ at time t, $p(v/\beta)$, $p(v)$ and $p(\beta)$ must be specified theoretically, e.g., if the signal portion of $v(t)$ contains the parameter β, we can write

$$v(t;\, \beta) = s(t;\, \beta) + n(t) \tag{12.2}$$

Fixing time and removing it in (12.2) we can think of $v(t;\, \beta)$, $s(t;\, \beta)$, and $n(t)$ as random variables $v(\beta)$, $s(\beta)$, and n respectively, and rewrite (12.2) as

$$v(\beta) = s(\beta) + n \tag{12.2}'$$

The information about the parameter β, of course, is contained in the signal waveform at time t, designated as $s(\beta)$ in (12.2)′.

We note that the LHS of (12.1c) multiplied by an increment $d\beta$ is the probability that the unknown parameter β falls within the range $\beta - (\beta + d\beta)$. If we were to find the maximum value of $p(\beta/v)$, then the value of β at that maximum value of $p(\beta/v)$ would be the *most probable* value of β, or equivalently the *most likely* value of β. The estimation of the value of β through this theory is aptly named "maximum likelihood estimation."

The maximum value of $p(\beta/v)$ occurs at the value of $\beta = \beta_{est}$, where

$$\left[\frac{\partial p}{\partial \beta}(\beta/v)\right]_{\beta=\beta_{est}} = 0 \tag{12.3}$$

From (12.3) and (12.1c),

$$\left[\frac{\partial p}{\partial \beta}(\beta/v)\right]_{\beta=\beta_{est}} = \left[\frac{\partial p(\beta)}{\partial \beta}\,p(v/\beta) + p(\beta)\,\frac{\partial p}{\partial \beta}(v/\beta)\right]_{\beta=\beta_{est}}$$
$$= 0 \tag{12.3}'$$

or equivalently

$$\left[p(\beta)\,\frac{\partial p}{\partial \beta}(v/\beta)\right]_{\beta=\beta_{est}} = -\left[p(v/\beta)\,\frac{\partial p(\beta)}{\partial \beta}\right]_{\beta=\beta_{est}} \tag{12.3}''$$

Before interpreting (12.3)″, we will generalize the above theory by including another set of parameters $\beta_2, \beta_3, \ldots, \beta_M$ contained in the waveform $s(t)$. Renaming the variable β, i.e., calling it β_1, we now have a vector of unknown parameters

$$\boldsymbol{\beta} = \begin{bmatrix} \beta_1 \\ \beta_2 \\ \vdots \\ \beta_M \end{bmatrix} \tag{12.4a}$$

We also consider $v(t)$, $s(t)$, and $n(t)$ as vectors of time samples of these respective waveforms, at time instants $t_1, t_2, \ldots, t_N$. The vector representations of these sets of samples are

$$\mathbf{v} = \begin{bmatrix} v_1 \\ v_2 \\ \vdots \\ v_N \end{bmatrix}, \quad \mathbf{s} = \begin{bmatrix} s_1 \\ s_2 \\ \vdots \\ s_N \end{bmatrix}, \quad \mathbf{n} = \begin{bmatrix} n_1 \\ n_2 \\ \vdots \\ n_N \end{bmatrix} \tag{12.4b}$$

where $v_k = v(t_k)$, $s_k = s(t_k)$, $n_k = n(t_k)$.

The generalization of (12.1c) for this multidimensional case is

$$p(\boldsymbol{\beta}/\mathbf{v}) = \frac{p(\boldsymbol{\beta})p(\mathbf{v}/\boldsymbol{\beta})}{p(\mathbf{v})} \tag{12.5a}$$

where all the PDFs are multivariate PDFs, i.e.,

$p(\boldsymbol{\beta}/\mathbf{v})$ = PDF for the vector $\boldsymbol{\beta}$ given the vector $\mathbf{v}$ of received voltage samples
$p(\boldsymbol{\beta})$ = a priori PDF of the vector $\boldsymbol{\beta}$
$p(\mathbf{v}/\boldsymbol{\beta})$ = PDF of $\mathbf{v}$ given $\boldsymbol{\beta}$
$p(\mathbf{v})$ = PDF of $\mathbf{v}$ averaged over all possible values of $\boldsymbol{\beta}$

The generalization of (12.3)″ for this case is

$$\left[p(\boldsymbol{\beta})\frac{\partial p}{\partial \beta_k}(\mathbf{v}/\boldsymbol{\beta})\right]_{\beta_k=\beta_{k\text{est}}} = -\left[p(\mathbf{v}/\boldsymbol{\beta})\frac{\partial p(\boldsymbol{\beta})}{\partial \beta_k}\right]_{\beta_k=\beta_{k\text{est}}}; \quad k = 1, 2, \ldots M \tag{12.5b}$$

The usual situation in radar is that we would like to estimate one of the parameters and average out all of the others. In the form (12.5b), the other parameters $\beta_j \ldots, j = 1, \ldots, M, j \neq k$ are assumed to be precisely known a priori, i.e., the maximum likelihood estimator is being found for β_k given known values of all the other β_js.

We now assume that all the unknown variables β_j have a priori PDFs that are nearly constant over a significant region and zero outside that region, as shown in Figure 12.1

The implication of the figure is that, prior to the measurement of β_k, we know that β_k lies somewhere between $\beta_{k0} - \Delta\beta_k/2$ and $\beta_{k0} + \Delta\beta_k/2$ and is equally likely to be anywhere in that region. The trail-off of the curve implies that, at some value of $|\beta_k - \beta_{k0}|$, the likelihood of β_k being near that value becomes smaller as $|\beta_k - \beta_{k0}|$ increases. Beyond a still greater value of $|\beta_k - \beta_{k0}|$, it is considered next to impossible that $|\beta_k - \beta_{k0}|$ could be that large. An obvious example is Doppler velocity of a target V_d, which we usually have to assume may be anywhere between $-V_{max}$ and $+V_{max}$, where V_{max} is the largest possible speed of a target and, since we do not know the target's direction relative to the radar, the central value of V_d must be assumed to be zero (or the velocity of the radar in the case of airborne radars). In that case, we might assume that extremely high speeds (e.g., above Mach 1) become less likely as the speed increases, while speeds below Mach 1 are about equally likely, thus producing the kind of a priori PDF shown in Figure 12.1.

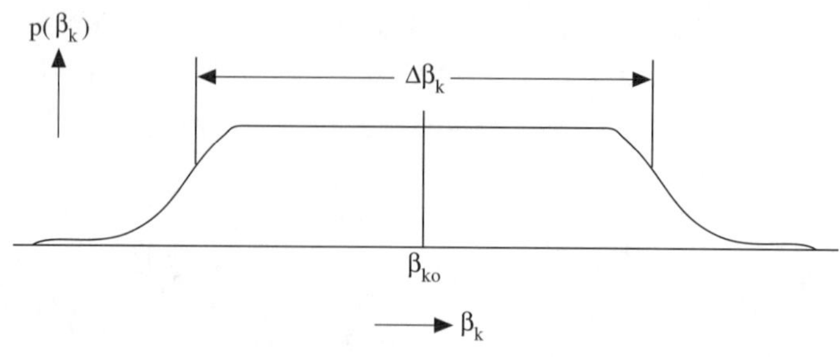

FIGURE 12.1
A priori PDF of a variable β_k.

Given a priori PDFs of all the parameters β, like that depicted in Figure 12.1, it follows that $\partial p(\boldsymbol{\beta})/\partial\beta_k$ on the right-hand side of (12.5b) vanishes over nearly all of the region in which β_k might lie prior to its measurement, and hence (12.5b) becomes [since $p(\boldsymbol{\beta}) \neq 0$]

$$\left[\frac{\partial p}{\partial \beta_k}(\mathbf{v}/\boldsymbol{\beta})\right]_{\beta_k=\beta_{kest}} = 0 \tag{12.6}$$

The next step in the analysis is to assume $p(\mathbf{v}/\boldsymbol{\beta})$ to be a multivariate Gaussian PDF (the situation for a radar signal in additive Gaussian noise at the IF stage of the receiver) and to further assume that the interval between time samples is large compared with the correlation time of the noise so that

$$p(\mathbf{v}/\boldsymbol{\beta}) = \prod_{l=1}^{N} p(v_l/\boldsymbol{\beta}) \tag{12.7a}$$

where $p(v_l/\boldsymbol{\beta})$ is the PDF of the lth sample.

Using (12.7a) in (12.6), we obtain

$$\left[\sum_{l=1}^{N} \frac{\partial p}{\partial \beta_k}(\ln p(v_l/\boldsymbol{\beta})\right]_{\beta_k=\beta_{kest}} = 0 \tag{12.7b}$$

$$\sum_{l=1}^{N}\left\{\frac{\partial}{\partial \beta_k}[v_l^2 - 2v_l s_l(\boldsymbol{\beta}) + s_l^2(\boldsymbol{\beta})]\right\}_{\beta_k=\beta_{kest}} = 0 \tag{12.7c}$$

or equivalently

$$\left[\sum_{l=1}^{N} v_l \frac{\partial s_l(\boldsymbol{\beta})}{\partial \beta_k}\right]_{\beta_k=\beta_{kest}} = \left[\sum_{l=1}^{N} s_l(\boldsymbol{\beta}) \frac{\partial s_l(\boldsymbol{\beta})}{\partial \beta_k}\right]_{\beta_k=\beta_{kest}} \tag{12.7d}$$

Taking the differentiation with respect to β_k outside the summation signs in (12.7d), we obtain

$$\left[\frac{\partial}{\partial \beta_k}\left(\sum_{l=1}^{N} v_l s_l(\boldsymbol{\beta})\right) = \frac{\partial}{\partial \beta_k}\left(\frac{1}{2}\sum_{l=1}^{N} s_l^2(\boldsymbol{\beta})\right)\right]_{\beta_k=\beta_{kest}} \tag{12.7e}$$

The left-hand side of (12.7e) contains within the parenthesis the output of a discrete matched filter which was shown in Chapter 5 to be equivalent to an analog matched filter if the noise is bandlimited white. The same considerations apply on the RHS of (12.7e); thus we can rewrite it in the form

$$\left[\frac{\partial}{\partial \beta_k}\left(\int_{-\infty}^{\infty} dt\, v(t)s(t;\boldsymbol{\beta})\right) = \frac{\partial}{\partial \beta_k}\left(\frac{1}{2}\int_{-\infty}^{\infty} dt\, s^2(t;\boldsymbol{\beta})\right)\right]_{\beta_k=\beta_{kest}} \tag{12.7f}$$

For an expected IF signal pulse of the form

$$s_{\text{exp}}(t) = a\cos(\omega t + \psi); \qquad t_0 - \frac{\tau_p}{2} \le t \le t_0 + \frac{\tau_p}{2} \qquad (12.8)$$

$$0; \qquad |t - t_0| > \frac{\tau_p}{2}$$

12.1.1 Estimation of Signal Amplitude

Suppose β_1 is the amplitude a and the other β-parameters are $\beta_2 = \omega$, $\beta_3 = \psi$, $\beta_4 = t_0$, $\beta_5 = \tau_p$. We assume that β_2, β_3, β_4, and β_5 are a priori known and β_1, the amplitude a, is to be estimated. Then (12.7f) and (12.8) yield the following, for $a = a_{\text{est}}$

$$\frac{\partial}{\partial a}\left(a\int_{t_0-\tau_p/2}^{t_0+\tau_p/2} dt\, v(t)\cos(\omega t + \psi)\right.$$

$$= \frac{\partial}{\partial a}\left(\frac{a^2}{4}\int_{t_0-\tau_p/2}^{t_0+\tau_p/2} dt + \frac{1}{4}\int_{t_0-\tau_p/2}^{t_0+\tau_p/2} dt\cos(2\omega t + 2\psi)\right) \qquad (12.9a)$$

The second integral on the right-hand side vanishes for the usual reasons and the first has the value τ_p, leading to the result

$$a_{\text{est}} = \frac{2}{\tau_p}\int_{t_0-\tau_p/2}^{t_0+\tau_p/2} dt\, v(t)\cos(\omega t + \psi) \qquad (12.9b)$$

The optimally estimated signal amplitude, according to (12.9b), is the output of the coherent matched filter matched exactly to the expected signal waveform.

To show the reasonableness of the result (12.9b), let us suppose that the actual signal-plus-noise waveform consists of the expected signal as given by (12.8) plus the noise $n(t)$. Then from (12.9b), the output of the amplitude estimation process is

$$a_{\text{est}} = \frac{2a}{\tau_p}\int_{t_0-\tau_p/2}^{t_0+\tau_p/2} dt\cos^2(\omega t + \psi) \qquad (12.9c)$$

$$+ \frac{2}{\tau_p}\int_{t_0-\tau_p/2}^{t_0+\tau_p/2} dt\, n(t)\cos(\omega t + \psi)$$

Since the value of the first of the integrals in (12.9c) is $\tau_p/2$, the difference between the estimated and true values of the signal amplitude is

$$a_{\text{est}} - a = \frac{2}{\tau_p}\int_{t_0-\tau_p/2}^{t_0+\tau_p/2} dt\, n(t)\cos(\omega t + \psi) \qquad (12.9d)$$

The rms value of this difference, called the "rms error" in amplitude measurement, is

$$\epsilon_{\text{rms}} = \sqrt{\langle (a_{\text{est}} - a)^2 \rangle}$$
$$= \sqrt{\frac{4}{\tau_p^2} \int_{t_0-\tau_p/2}^{t_0+\tau_p/2} dt_1 \int_{t_0-\tau_p/2}^{t_0+\tau_p/2} dt_2 \, \langle n(t_1)n(t_2) \rangle \cos(\omega t_1 + \psi)\cos(\omega t_2 + \psi)} \quad (12.9e)$$

It was already given in Eq. (4.67) that the ACF of bandlimited white noise can be approximated by

$$\langle n(t_1)n(t_2) \rangle \simeq \frac{\sigma_n^2}{2B_n} \delta(t_1 - t_2) \quad (12.9f)$$

where B_n is the "lowpass" noise bandwidth in Hertz and σ_n^2 the mean noise power.
Substituting (12.9f) into (12.9e) produces the result

$$\epsilon_{\text{rms}} = \frac{2}{\tau_p} \sqrt{\left(\frac{\sigma_n^2}{2B_n}\right) \int_{t_0-\tau_p/2}^{t_0+\tau_p/2} dt \, \frac{1}{2} [1 + \cos(2\omega t + 2\psi)]}$$
$$\simeq \frac{\sigma_n}{\sqrt{B_n \tau_p}} \quad (12.9g)$$

The *fractional* rms error in optimal amplitude estimation is

$$(\epsilon_{\text{rms}})_{\text{frac}} = \frac{\epsilon_{\text{rms}}}{a} = \frac{1}{\sqrt{\rho_0 B_n \tau_p}} \quad (12.9h)$$

where

$\rho_0 = a^2/\sigma_n^2 =$ *peak* SNR at input to the estimation process (as differentiated from $a^2/2\sigma_n^2$, which is the value of SNR referred to the rms amplitude $a/\sqrt{2}$)

The result (12.9h) shows a fractional rms error that is inversely proportional to the voltage SNR and the square root of the product $(B_n\tau_p)$. These parameters will be prominent in estimation of errors in all radar parameters of interest.

12.1.2 Estimation of Signal Phase

Let β_1 be the phase ψ; the other parameters are $\beta_2 = \omega$, $\beta_3 = a$, $\beta_4 = t_0$, $\beta_5 = \tau_p$. Again we assume that β_2, β_3, β_4, and β_5 are a priori known and ψ is to be estimated. Equations (12.7f) and (12.8) give us

$$\frac{\partial}{\partial \psi} \left(a \int_{t_0-\tau_p/2}^{t_0+\tau_p/2} dt \, v(t)\cos(\omega t + \psi) \right) = 0 \quad (12.10a)$$

since the ψ-dependent part of the quantity on the right-hand side of (12.7f) is proportional to $\partial/\partial\psi(\int_{t_0-\tau_p/2}^{t_0+\tau_p/2} dt \cos(2\omega t + 2\psi))$, which is negligibly small for all practical purposes. Equation (12.10a) leads to

$$V_c \frac{\partial}{\partial \psi}(\cos \psi) - V_s \frac{\partial}{\partial \psi}(\sin \psi) = 0 \tag{12.10b}$$

where

$$V_c = \int_{t_0-\tau_p/2}^{t_0+\tau_p/2} dt\, v(t)\cos \omega t, \qquad V_s = \int_{t_0-\tau_p/2}^{t_0+\tau_p/2} dt\, v(t)\sin \omega t$$

Finally, (12.10b) implies that the optimal estimator of phase is

$$\psi_{est} = -\tan^{-1}\left(\frac{V_s}{V_c}\right) \tag{12.10c}$$

Equation (12.10c) yields the following prescription for estimating phase:

1. Perform quadrature matched filtering on the signal-plus-noise waveform.
2. Take the negative ratio of sine-filter output to cosine-filter output.
3. Take the arctangent of that ratio.

The outputs of cosine and sine filters are

$$V_{\substack{s\\c}} = a \int_{t_0-\tau_p/2}^{t_0+\tau_p/2} dt \cos(\omega t + \psi) \begin{matrix} \cos \omega t \\ \sin \omega t \end{matrix} + \int_{t_0-\tau_p/2}^{t_0+\tau_p/2} dt\, n(t) \begin{matrix} \cos \omega t \\ \sin \omega t \end{matrix} \tag{12.10d}$$

The signal portions of these outputs are

$$S_{\substack{s\\c}} = \frac{a}{2}\left\{ \begin{matrix} \cos \psi \\ -\sin \psi \end{matrix} \int_{t_0-\tau_p/2}^{t_0+\tau_p/2} dt + \int_{t_0-\tau_p/2}^{t_0+\tau_p/2} dt \begin{matrix} \cos \\ \sin \end{matrix} (2\omega t + 2\psi) \right\}$$

$$= \frac{a\tau_p}{2}\left\{ \begin{matrix} \cos \psi \\ -\sin \psi \end{matrix} \right\} \tag{12.10e}$$

since the second term is negligibly small in both expressions.

The negative ratio of the signal portions of these outputs is

$$-\frac{S_s}{S_c} = -\frac{(-\sin \psi)}{\cos \psi} = \tan \psi \tag{12.10f}$$

which confirms that the operation dictated by (12.10c) is a direct computation of phase in the absence of noise.

The MLE estimation operation output, then, is

$$\psi_{est} = -\tan^{-1}\left(\frac{\sin \psi_{true} + \dfrac{2}{a\tau_p} n_{os}}{\cos \psi_{true} + \dfrac{2}{a\tau_p} n_{oc}} \right) \tag{12.10g}$$

where n_{os} and n_{oc} are noise outputs of sine and cosine filters, respectively.

In principle, this estimate could be made at an arbitrary value of ψ; hence let us specialize to a value $\psi_{\text{true}} = 0$ and hence $\cos\psi_{\text{true}} = 1$, $\sin\psi_{\text{true}} = 0$.

For the case $\psi_{\text{true}} = 0$, (12.10g) gives us

$$\Delta\psi = \psi_{\text{est}} - \psi_{\text{true}} = -\tan^{-1}\left(\frac{2n_{\text{os}}}{a\tau_p}\bigg/\left(1 + \frac{2n_{\text{oc}}}{a\tau_p}\right)\right) \tag{12.10h}$$

For large SNR, we assume

$$\left|\frac{2n_{\substack{\text{oc}\\\text{s}}}}{a\tau_p}\right| << 1 \tag{12.10i}$$

in which case we can expand the arctangent in a Taylor series with the result (Spiegel,[11] #20.29, p. 111)

$$\Delta\psi = -x + \frac{x^3}{3} - \frac{x^5}{5} + \frac{x^7}{7} - \cdots \tag{12.10j}$$

where

$$x = \frac{2n_{\text{os}}}{a\tau_p}\bigg/\left(1 + \frac{2n_{\text{oc}}}{a\tau_p}\right)$$

For sufficiently small values of $2n_{\text{oc}}/a\tau_p$ and $2n_{\text{os}}/a\tau_p$, we can approximate x as $2n_{\text{os}}/a\tau_p$ and truncate the series (12.10j) after the first term, with the result

$$\Delta\psi \simeq \frac{2n_{\text{os}}}{a\tau_p} \tag{12.10k}$$

and hence the rms error is

$$\begin{aligned} \epsilon_\psi &= \sqrt{\langle(\Delta\psi)^2\rangle} \simeq \frac{2}{a\tau_p}\sqrt{\langle n_{\text{os}}^2\rangle} \\ &= \sqrt{\frac{2\langle n_{\text{os}}^2\rangle}{(a^2\tau_p^2/2)}} = \frac{1}{\sqrt{\rho_0 B_{\text{n}}\tau_p}} \end{aligned} \tag{12.10l}$$

which, according to (12.9h), is the same as the fractional error in optimal measurement of amplitude.

12.1.3 Estimation of Pulse Arrival Time

The estimation of range in a monostatic radar is usually that of pulse arrival time.* To treat the estimation of that parameter from (12.7f), we note that the parameter, which we will call τ, is contained in the signal waveform in the form

* Except for the case when frequency modulation of a CW signal allows estimation of range through that of instantaneous frequency ("FM-CW radar").

$$s(t;\tau) = s(t-\tau)$$

Equation (12.7f), with the aid of (12.11a), becomes

$$\frac{\partial}{\partial\tau}\left(\int_{-\infty}^{\infty} dt\ v(t)s(t-\tau)\right)_{\tau=\tau_{\text{est}}} = \frac{\partial}{\partial\tau}\left(\frac{1}{2}\int_{-\infty}^{\infty} dt\ s^2(t-\tau)\right)_{\tau=\tau_{\text{est}}} \tag{12.11b}$$

$$= \frac{1}{2}\frac{\partial}{\partial\tau}\left(\int_{-\infty}^{\infty} dt'\ s^2(t')\right) = 0$$

reflecting the fact that the translation of the integrand on the right-hand side does not render the integral a function of τ and hence the right-hand side vanishes.

The best way to treat (12.11b) is through the Fourier transforms (FT) of $v(t)$ and $s(t-\tau)$, i.e.,

$$\frac{\partial}{\partial\tau}\left(\int_{-\infty}^{\infty} dt\ v(t)s(t-\tau)\right)$$

$$= \frac{\partial}{\partial\tau}\left(\int_{-\infty}^{\infty} d\omega_1\ V(\omega_1)\int_{-\infty}^{\infty} d\omega_2\ S(\omega_2)\left[\frac{1}{2\pi}\int_{-\infty}^{\infty} dt\ e^{j(\omega_1+\omega_2)t}\right]e^{-j\omega_2\tau}\right)$$

$$= \frac{\partial}{\partial\tau}\left[\frac{1}{2\pi}\int_{-\infty}^{\infty} d\omega\ V(\omega)S^*(\omega)e^{j\omega\tau}\right]_{\tau=\tau_{\text{est}}}$$

$$= \frac{j}{2\pi}\int_{-\infty}^{\infty} d\omega\ \omega V(\omega)S^*(\omega)e^{j\omega\tau_{\text{est}}} = 0 \tag{12.11c}$$

where $S(\omega)$ is the FT of the *expected* signal and $V(\omega)$ is the FT of the *actual* incoming signal plus noise, i.e.,

$$V(\omega) = S(\omega)e^{-j\omega\tau_{\text{act}}} + N(\omega) \tag{12.11d}$$

where τ_{act} is the *actual* time of arrival, where $S(\omega)$, the FT of the *actual* signal, is assumed to be that of the *expected* signal (i.e., the filter is assumed perfectly matched to the true waveshape, consistent with the optimality of the estimation process) and where $N(\omega)$ is the FT of the noise.

Substitution of (12.11d) into (12.11c) yields

$$\int_{-\infty}^{\infty} d\omega\ \omega|S(\omega)|^2 e^{j\omega\Delta\tau} = -\int_{-\infty}^{\infty} d\omega\ \omega N(\omega)S^*(\omega)e^{j\omega\tau_{\text{est}}} \tag{12.11d$'$}$$

where $\Delta\tau = \tau_{\text{est}} - \tau_{\text{act}}$. We now take the ensemble average of the absolute square of both sides of (12.11d). We call the left-hand side of the resulting expression the signal power, and the right-hand side the mean noise power.

The mean power in the noise portion of (12.11d)′ is (where T is a time duration that must approach infinity to allow definition of the noise spectrum)

$$\langle P_N \rangle = T \int_{-\infty}^{\infty} d\omega_1 \int_{-\infty}^{\infty} d\omega_2 \; \omega_1 \omega_2 S^*(\omega_1) S(\omega_2) \cdots \tag{12.11e}$$
$$e^{j(\omega_1 - \omega_2)\tau_{\text{est}}} \frac{\langle N(\omega_1) N^*(\omega_2) \rangle}{T}$$

For noise which has an approximately flat spectrum over the signal band

$$T \lim_{T \to \infty} \frac{\langle N(\omega_1) N^*(\omega_2) \rangle}{T} = N_0 \delta(\omega_1 - \omega_2) \tag{12.11f}$$

where N_0 is the noise power density (power per Hertz).

With the aid of (12.11f), Eq. (12.11e) becomes

$$\langle P_N \rangle = N_0 \int_{-\infty}^{\infty} d\omega \; \omega^2 |S(\omega)|^2 \tag{12.11g}$$

The signal power (from (12.11d)′) is

$$P_s = \left| \int_{-\infty}^{\infty} d\omega \; \omega |S(\omega)|^2 \sin(\omega \Delta\tau) \right|^2 \tag{12.11h}$$

(because the real part of the signal integral, equal to $\int_{-\infty}^{\infty} d\omega \; \omega |S(\omega)|^2 \cos(\omega\Delta\tau)$, vanishes since its integrand is a product of an even function and an odd function of ω).

If the estimation error $\Delta\tau = (\tau_{\text{est}} - \tau_{\text{act}})$ is small enough so that $\sin(\omega\Delta\tau)$ can be approximated by the first term of its power series within the spectral range of $|S(\omega)|^2$, then from Eqs. (12.11d)′ and (12.11g, h),

$$\langle P_s \rangle = \langle (\Delta\tau)^2 \rangle \left| \int_{-\infty}^{\infty} d\omega \; \omega^2 |S(\omega)|^2 \right|^2 = 2\pi N_0 \int_{-\infty}^{\infty} d\omega \; \omega^2 |S(\omega)|^2 = 2\pi \langle P_N \rangle \tag{12.11i}$$

from which it follows that the rms error in estimation of τ due to receiver noise is

$$\epsilon_\tau = \sqrt{\langle (\Delta\tau)^2 \rangle} = \sqrt{\frac{2\pi N_0}{\int_{-\infty}^{\infty} d\omega \; \omega^2 |S(\omega)|^2}} = \frac{1}{\beta_\tau \sqrt{\rho_0}} \tag{12.11j}$$

where

$$\beta_\tau = \sqrt{\frac{\int_{-\infty}^{\infty} d\omega \; \omega^2 |S(\omega)|^2}{\int_{-\infty}^{\infty} d\omega |S(\omega)|^2}} = \text{(a definition of) signal bandwidth in radians/sec}$$

$$E_s = \frac{1}{2\pi} \int_{-\infty}^{\infty} d\omega \; |S(\omega)|^2 = \text{signal energy}$$

$$\rho_0 = \frac{E_s}{N_0} = \text{SNR output of matched filter}$$

12.1.4 Estimation of Signal Frequency

A measurement of range rate in a monostatic radar is equivalent to a measurement of doppler frequency shift. The two are related by the expression [equivalent to Eq. (1.3)]

$$f_d = -\frac{2f_0\dot{r}}{c} = -\frac{2\dot{r}}{\lambda_0} \tag{12.12}$$

where fd is Doppler frequency shift, f_0 the unshifted frequency, and λ_0 the corresponding wavelength, and $\dot{r}$ is range-rate or equivalently negative line-of-sight velocity. The minus sign indicates that an approaching target has a positive Doppler shift and a receding target has a negative Doppler shift.

Another application of frequency estimation is in a FM-CW radar in which instantaneous frequency is used to measure range.

The frequency ω is contained in the IF signal in the form

$$s(t, \omega) = p(t)\cos(\omega t + \psi) \tag{12.13a}$$

where the duration of $p(t)$ is T and where $T >> 2\pi/\omega$.

Equation (12.7f), with the aid of (12.13a), is

$$\frac{\partial}{\partial\omega}\left[\int_{-\infty}^{\infty} dt\ v(t)p(t)\cos(\omega t + \psi)\right]_{\omega=\omega_{est}}$$
$$= \frac{\partial}{\partial\omega}\left[\frac{1}{4}\int_{-\infty}^{\infty} dt\ p^2(t)(1 + \cos(2\omega t + 2\psi)\right]_{\omega=\omega_{est}} \tag{12.13b}$$

The second term on the right-hand side (12.13b) can be approximated as zero for the usual reason that $T >> 2\pi/\omega$; hence, since the first term is independent of frequency, the entire right-hand side of (12.13b) is zero.

Then (12.13b) takes the approximate form

$$\int_{-\infty}^{\infty} dt\ v(t)tp(t)\sin(\omega_{est}t + \psi) = 0 \tag{12.13c}$$

or equivalently

$$U_s(\omega_{est})\cos\psi + U_c(\omega_{est})\sin\psi = 0 \tag{12.13d}$$

where

$$U_{\substack{c\\s}}(\omega) = \int_{-\infty}^{\infty} dt\ tv(t) \begin{matrix}\cos\\ \sin\end{matrix}(\omega t)$$

Given an exact knowledge of the phase ψ, the method of optimal estimation of the frequency ω, based on (12.13d), is to find the value of ω for which

$$\left[\frac{U_s(\omega)}{U_c(\omega)}\right]_{\omega=\omega_{est}} = -\tan\psi \tag{12.13e}$$

If we use (12.13a) to represent the signal waveform where ω is the *actual* signal frequency, denoted by ω_{act}, we can add the noise to the signal waveform and write

$$v(t) = p(t)\cos(\omega_{act}t + \psi) + n(t) \tag{12.14a}$$

and then substitute (12.14a) into (12.13c) with the result

$$\frac{1}{2}\int_{-\infty}^{\infty} dt\; tp^2(t)[\sin(\Delta\omega t) + \sin((\omega_{est} + \omega_{act})t + 2\psi)] = -\int_{-\infty}^{\infty} dt\; tp(t)n(t)\sin(\omega_{est}t + \psi) \tag{12.14b}$$

where $\Delta\omega = \omega_{est} - \omega_{act}$.

The second integral on the left-hand side (12.14b) is negligibly small because of the assumption $T >> 2\pi/\omega$, where $\omega = \omega_{est}$ or ω_{act}. When that term is neglected, and the absolute squares of both sides of the resulting expression are ensemble-averaged, we obtain

$$\frac{1}{4}\left\langle\left|\int_{-\infty}^{\infty} dt\; tp^2(t)\sin(\Delta\omega t)\right|^2\right\rangle = 2\int_{-\infty}^{\infty} dt_1 \int_{-\infty}^{\infty} dt_2\; t_1t_2p(t_1)p(t_2)\langle n(t_1)n(t_2)\rangle \cdots \sin(\omega_{est}t_1 + \psi)\sin(\omega_{est}t_2 + \psi) \tag{12.14c}$$

Invoking (4.67) to evaluate the noise term inside the integral on the right-hand side of (12.14c) and assuming that the estimation error $\Delta\omega$ is sufficiently small to allow $\sin(\Delta\omega t)$ to be represented by the first term of its power series, we obtain

$$\langle(\Delta\omega)^2\rangle\left|\int_{-\infty}^{\infty} dt\; t^2p^2(t)\right|^2 = 2N_0\int_{-\infty}^{\infty} dt\; t^2p^2(t) \tag{12.14d}$$

where N_0 is the noise power density.

Regrouping terms in (12.14d) and taking the square root of both sides, we obtain

$$\epsilon_\omega = \sqrt{\langle(\Delta\omega)^2\rangle} = \frac{N_0}{\left(\frac{1}{2}\int_{-\infty}^{\infty} dt\; t^2p^2(t)\right)} = \frac{1}{\beta_\omega\sqrt{\rho_0}} \tag{12.14e}$$

where

$$\beta_\omega = \sqrt{\frac{\int_{-\infty}^{\infty} dt\; t^2p^2(t)}{\int_{-\infty}^{\infty} dt\; p^2(t)}}$$

$$E_s = \frac{1}{2}\int_{-\infty}^{\infty} dt\; p^2(t) = \text{signal energy}$$

$$\rho_0 = \frac{E_s}{N_0} = \text{matched filter output SNR}$$

This expression is the frequency estimation analog of (12.11j), which applies to time estimation. The quantity β_ω in (12.13j) is analogous to β_τ in (12.11j).

To derive a result equivalent to (12.14e) for the case where the phase is a priori unknown, a much more realistic situation, we return to the step before (12.14b) and assume that the actual signal phase is ψ and the phase assumed in the reference signal is $\psi + \Delta\psi$, where $\Delta\psi$ is a uniformly distributed random variable. Under that assumption, (12.14b) changes to

$$\frac{1}{2}\int_{-\infty}^{\infty} dt\, tp^2(t)[\sin(\Delta\omega t + \Delta\psi) + \sin((\omega_{est} + \omega_{act})t + 2\psi + \Delta\psi)]$$

$$= -\int_{-\infty}^{\infty} dt\, tp(t)n(t)\sin(\omega_{est}t + \psi + \Delta\psi) \quad (12.15a)$$

The second term on the left-hand side is still negligible. The operation of absolute-squaring, ensemble averaging both sides and invoking (4.67) leads to the generalization of (12.14d) for nonzero $\Delta\psi$, as follows:

$$\left\langle \left| \int_{-\infty}^{\infty} dt\, tp^2(t)\sin(\Delta\omega t + \Delta\psi) \right|^2 \right\rangle = 2N_0 \int_{-\infty}^{\infty} dt\, t^2p^2(t) \quad (12.15b)$$

where the averaging process now applies to both random variables $\Delta\omega$ and $\Delta\psi$.

The left-hand side of (12.14c) can be expressed in the form

$$\left| (\cos\Delta\psi) \int_{-\infty}^{\infty} dt\, tp^2(t)\sin(\Delta\omega t) + \sin\Delta\psi \int_{-\infty}^{\infty} dt\, tp^2(t)\cos(\Delta\omega t) \right|^2$$

The second term inside the brackets vanishes based on the assumption that $p^2(t)$ is an even function of t. If we assume that $\Delta\omega$ and $\Delta\psi$ are mutually statistically independent and again assume $\Delta\omega$ to be sufficiently small to use the approximation $\sin(\Delta\omega t) \simeq \Delta\omega t$, it follows from (12.15b) that

$$\langle(\Delta\omega)^2\rangle\langle\cos^2\Delta\psi\rangle \left| \int_{-\infty}^{\infty} dt\, t^2p^2(t) \right|^2 = 2N_0 \int_{-\infty}^{\infty} dt\, t^2p^2(t) \quad (12.15c)$$

which, since $\langle\cos^2\Delta\psi\rangle = 1/2$, differs from (12.14d) by a factor 2 and leads to the result

$$\epsilon_\omega = \sqrt{\langle(\Delta\omega)^2\rangle} = \frac{2N_0}{\left(\frac{1}{2}\int_{-\infty}^{\infty} dt\, t^2p^2(t)\right)} = \frac{2}{\beta_\omega\sqrt{\rho_0}} \quad (12.15d)$$

implying an rms estimation error twice as high as that given by (12.14e).

The discussion above implies that in the absence of a priori knowledge of phase, the optimal frequency estimation must not be phase dependent. That means that we must average over the possible values of phase from 0 to 360° in implementing the estimation process. That is accomplished through quadrature filtering, which produces the amplitude of the matched filter output. The cost of the absence of a priori phase information is a twofold increase in the minimum attainable rms error.

In summary, the optimal estimation of frequency gives us the following results; based on (12.14e) and (12.15d).

$$\epsilon_\omega = \sqrt{\langle(\Delta\omega)^2\rangle} = \frac{1}{\beta_\omega\sqrt{\rho_0}} \quad \text{if phase is known} \tag{12.16}$$

$$= \frac{2}{\beta_\omega\sqrt{\rho_0}} \quad \text{if phase is unknown}$$

where

$$\rho_0 = \text{matched filter output SNR}$$

$$\beta_\omega = \sqrt{\frac{\int_{-\infty}^{\infty} dt\ t^2 p^2(t)}{\int_{-\infty}^{\infty} dt\ p^2(t)}}$$

$p(t)$ being the signal pulse envelope.

12.1.5 Estimation of Angle of Arrival

The estimation of angle of arrival of a received signal can be accomplished by exploiting the angle information contained in the complex antenna pattern (see Chapter 10, Section 10.9 on antenna scanning for background on the discussion to follow).

To investigate this, we note from Eq. (10.6) that the far-zone radiation pattern for a transmitting antenna is related to the field distribution on a planar aperture through a two-dimensional Fourier transformation.

The aperture surface is assumed to be on the $(x - y)$ plane of a cartesian coordinate system. (There is no loss of generality in this assumption, because we can construct the antenna coordinate system to realize the assumption and transform back to another coordinate system to treat the entire radar system problem). The radiation pattern functions can be obtained from (10.6). Whatever the directions of the fields on the aperture plane, the scalar radiation pattern corresponding to an aperture field component $E_s(x', y')$ is (from (10.6) with small modifications)

$$F(u_x, u_y) = \int_{-\infty}^{\infty} dx' \int_{-\infty}^{\infty} dy'\ e^{j(u_x x' + u_y y')} E_s(x', y') \tag{12.17a}$$

where the finite aperture is defined as the region wherein $E_s(x', y')$ is nonzero and where

$$u_x = k_0 \sin\theta \cos\phi$$

$$u_y = k_0 \sin\theta \sin\phi$$

Because of reciprocity the receptivity pattern when the antenna is used in the receiving mode is identical to the radiation pattern when it is used in the transmitting mode.

In the latter case, on which we want to focus, the aperture distribution $E_s(x', y')$ is the distribution of the field of the incoming wave along the aperture plane.

Let us concentrate on the pattern in one of the principal planes, i.e., the $(x - z)$ plane or the $(y - z)$ plane. Choosing the $(x - z)$ plane, the azimuthal angle ϕ in the plane is zero and hence $\sin \phi = 0$, $\cos \phi = 1$. Then $u_x = \sin \theta$, $u_y = 0$ and we can now treat (12.17a) as a one-dimensional Fourier transform relationship, i.e., with a small change in notation, we have

$$F(u) = \int_{-\infty}^{\infty} dx' \; e^{jk_0 x' \sin \theta} f(x') \tag{12.17b}$$

where $f(x') = E_s(x', 0)$.

Superposed on the deterministic distribution $f(x)$ is a random noise along the aperture, which can be denoted by $n(x)$ and gives rise to a random part of the pattern $N(u)$.

The incoming signal voltage may be modeled as a function of u in the form

$$V(u) = F(u) + N(u) \tag{12.17c}$$

The ideal reference signal as dictated by MLE theory is

$$S_R(u + \Delta u) = F^*(u + \Delta u) \tag{12.17d}$$

where Δu is the difference in u between true signal and reference signal. The optimal processing algorithm is based on

$$\frac{\partial}{\partial(\Delta u)} = \int_{-\infty}^{\infty} du \; F^*(u + \Delta u)[F(u) + N(u)] = 0 \tag{12.17e}$$

Combining (12.17e) with (12.17b) and obtaining the noise portion from (12.17c) and changing order of integration, we arrive at the expression

$$\frac{\partial}{\partial(\Delta u)} \int_{-\infty}^{\infty} du \int_{-\infty}^{\infty} dx' \int_{-\infty}^{\infty} dx'' \; e^{j[ux' - (u+\Delta u)x'']} [f(x') + n(x')] f^*(x'')$$

$$= -j2\pi \int_{-\infty}^{\infty} dx' \int_{-\infty}^{\infty} dx'' \; x'' \left[\frac{1}{2\pi} \int_{-\infty}^{\infty} du \; e^{ju(x' - x'')} \right] e^{-j\Delta u x''} \ldots$$

$$[f(x') f^*(x'') + n(x') f^*(x'')] = 0 \tag{12.17f}$$

The square-bracketed expression in the integrand is well known to be the unit impulse $\delta(x' - x'')$; hence (12.17f) becomes

$$\int_{-\infty}^{\infty} dx' \; x' |f(x')|^2 e^{-j\Delta u x'} - \int_{-\infty}^{\infty} dx' \; x' n(x') f^*(x') e^{-j\Delta u x'} \tag{12.17g}$$

Taking the absolute square of both sides of (12.17g), assuming that $|f(x')|^2$ is an even function of x', and ensemble-averaging the noise term, we obtain

$$\left|\int_{-\infty}^{\infty} dx'\ x'|f(x')|^2\sin(\Delta u x')\right|^2$$

$$= \int_{-\infty}^{\infty} dx'\ x'f^*(x') \int_{-\infty}^{\infty} dx''\ x''f(x'')e^{-j\Delta u(x'-x'')}\langle n(x')n^*(x'')\rangle \quad (12.17h)$$

Assuming bandlimited white noise as in previous demonstrations, we obtain

$$\langle n(x')n^*(x'')\rangle = N_0\delta(x' - x'') \quad (12.17i)$$

where N_0 = noise power density.

Equation (12.17h) becomes

$$\left|\int_{-\infty}^{\infty} dx'\ x'|f(x')|^2\langle\sin(\Delta u x')\rangle\right|^2 = N_0 \int_{-\infty}^{\infty} dx'(x')^2|f(x')|^2 \quad (12.17j)$$

Assuming that $(\Delta u x')$ is sufficiently small to allow the approximation $\sin(\Delta u x) \simeq (\Delta u x)$ throughout the nonzero region of $|f(x)|^2$ we obtain

$$\langle(\Delta u)^2\rangle = \frac{N_0}{\int dx\ x^2|f(x)|^2} \quad (12.17k)$$

which can be expressed in the form

$$\epsilon_u = \sqrt{\langle(\Delta u)^2\rangle} = \frac{\sqrt{N_0}}{\sqrt{\int_{-\infty}^{\infty} dx\ x^2|f(x)|^2}} = \frac{1}{\beta_u\sqrt{\rho_0}} \quad (12.17l)$$

where

$$\rho_0 = \frac{\int_{-\infty}^{\infty} dx|f(x)|^2}{N_0} = \text{output SNR of matched filter}$$

$$\beta_u = \sqrt{\frac{\int_{-\infty}^{\infty} dx\ x^2|f(x)|^2}{\int_{-\infty}^{\infty} dx|f(x)|^2}}$$

This result is analogous to (12.11j) for time delay and (12.16) for frequency and was derived in essentially the same way.

12.2 MEASUREMENT ACCURACY

In measuring a parameter by radar using optimal estimation techniques (e.g., maximum likelihood estimation), we arrive at a generic formula for the rms error in measurement of that parameter.[12] In general, the formula is

$$(\epsilon_{\text{rms}})_x = \frac{1}{\beta_x\sqrt{\rho_0}} = \sqrt{\langle(x_{\text{meas}} - x_{\text{true}})^2\rangle} = \sqrt{\frac{N_0}{\int_{-\infty}^{\infty} du\ u^2|S(u)|^2}} \tag{12.18}$$

where N_0 is noise power density, ρ_0 = SNR of matched filter output, x_{meas} = measured value of x, x_{true} = value of x, β_x = rms width of the Fourier transform of a function used to measure the parameter x. Specializations of (12.18) were derived in Section 12.1.3 for pulse arrival time [Eq. (12.11j)], in Section 12.1.4 for signal frequency [Eq. (12.16)], and in Section 12.1.5 for angle of arrival [Eq. (12.17l)].

The parameter β_x can be characterized mathematically as

$$\beta_x = \sqrt{\frac{\int_{-\infty}^{\infty} du\ |S(u)|^2 u^2}{\int_{-\infty}^{\infty} du\ |S(u)|^2}} \tag{12.19}$$

where $s(x)$ is the function of x on which the measurement of x is based, and $S(u)$ is its Fourier transform, i.e.

$$S(u) = \int_{-\infty}^{\infty} dx\ s(x)e^{-jux} \tag{12.19a)'}$$

and

$$s(x) = \frac{1}{2\pi}\int_{-\infty}^{\infty} du\ S(u)e^{jux} \tag{12.19b)'}$$

By recognizing that

$$\frac{ds(x)}{dx} = \frac{1}{2\pi}\int_{-\infty}^{\infty} du(ju)S(u)e^{jux} \tag{12.20a}$$

we can write [with a change in order of integration, and with the aid of (12.20a)]

$$\int_{-\infty}^{\infty} dx\left|\frac{ds(x)}{dx}\right|^2 = \frac{1}{2\pi}\int_{-\infty}^{\infty} du \int_{-\infty}^{\infty} du'\ S(u)S(u')(ju)(-ju') \cdots$$

$$\left(\frac{1}{2\pi}\int_{-\infty}^{\infty} dx\ e^{-j(u-u')x}\right) = \frac{1}{2\pi}\int_{-\infty}^{\infty} du\ u^2|S(u)|^2 \tag{12.20b}$$

From (12.19) and (12.20b) and Parseval's theorem, it follows that

$$\beta_x = \sqrt{\frac{\int_{-\infty}^{\infty} dx|ds(x)/dx|^2}{\int_{-\infty}^{\infty} dx|s(x)|^2}} \tag{12.20c}$$

which provides us with an alternative means of defining β_x.

If we were measuring u rather than x, the rms error would be

$$(\epsilon_{\text{rms}}) = \frac{1}{\beta_u\sqrt{\rho_0}} = \sqrt{\langle(u_{\text{meas}} - u_{\text{true}})^2\rangle} \tag{12.21}$$

where the counterpart of (12.19) is

$$\beta_u = \sqrt{\int_{-\infty}^{\infty} dx\, x^2|s(x)|^2 \Big/ \int_{-\infty}^{\infty} dx|s(x)|^2} \tag{12.21$'$}$$

where, from (12.19a, b)′, we note that the counterpart of (12.20a) is

$$\frac{dS(u)}{du} = -\int_{-\infty}^{\infty} dx\, s(x)(jx)e^{-jux} \tag{12.22a}$$

and that of (12.20b) is

$$\frac{1}{2\pi}\int_{-\infty}^{\infty} du\left|\frac{dS(u)}{du}\right|^2 = \int_{-\infty}^{\infty} dx\, x^2|s(x)|^2 \tag{12.22b}$$

From (12.21a)′, (12.22b), and Parseval's theorem we obtain the counterpart of (12.20c) where x and u are interchanged, i.e.,

$$\beta_u = \sqrt{\frac{\int_{-\infty}^{\infty} du|dS(u)/du|^2}{\int_{-\infty}^{\infty} du|S(u)|^2}} \tag{12.22c}$$

The expressions (12.18) and (12.19) or (12.20c) or their alternate forms (12.21) and (12.21a)′ or (12.22c) provide us with a means of evaluating the accuracy in measurement of the parameters that can be measured by radar and whose Fourier transforms can be evaluated. In the next three sections we will discuss some of the results of application of these expressions to measurement of delay, frequency and angle for specific radar signal pulse shapes and antenna pattern shapes.

12.2.1 Accuracy of Pulse Arrival Time Measurements

The application of MLE theory to this problem gave us the results presented in Section 12.1.3 and, in particular, Eqs. (12.11.j), from which we can determine the rms error due to additive noise in the measurement of pulse time of arrival.

The measurement can be made coherently or noncoherently, the former being more accurate than the latter, but the latter usually being more practical because of the lack of phase information. The block diagrams in Figure 12.2a and b show the measurement procedures dictated by MLE theory for both coherent and noncoherent measurements. Similar diagrams can be found in Barton.[13]

Describing the process illustrated in Figure 12.2, we begin with the incoming signal waveform

$$s(t, \tau_0) = p(t - \tau_0)\cos[\omega_0(t - \tau_0) + \psi] \tag{12.23a}$$

and the reference signal waveform

$$s_R(t, \tau) = p(t - \tau)\cos[\omega_0(t - \tau) + \theta_R] \tag{12.23b}$$

where ψ_R is the phase of the reference signal. Matched filter theory (Chapter 4, Sections 4.3 and 4.4) gives us the filter output signal as

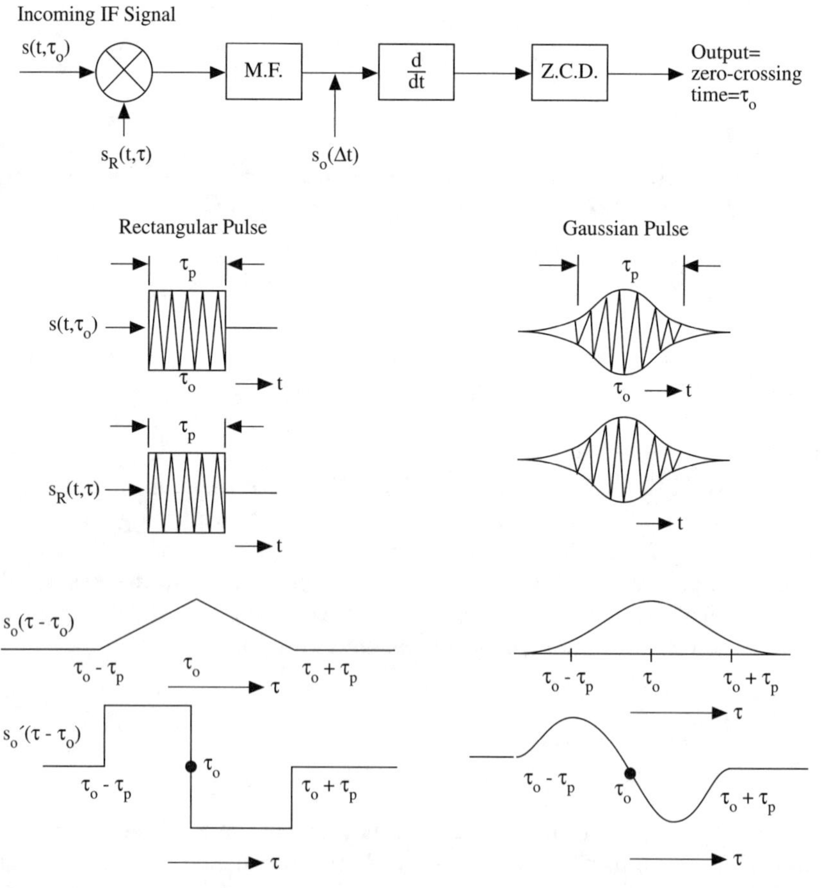

FIGURE 12.2
Optimal pulse time-of-arrival measurement. M.F.: matched filter; Z.C.D.: zero-crossing detector.

$$s_0(\Delta\tau, \Delta\psi) = \left[\frac{1}{2}\int_{-\infty}^{\infty} dt\, p(t)\, p(t - \Delta\tau)\right]\left[\cos(\omega_0\Delta\tau + \Delta\psi)\right.$$

$$+ \cos(\omega_0\Delta\tau + \Delta\psi)\cos(2\omega_0 t + \psi)$$

$$\left. + \sin(\omega_0\Delta\tau + \Delta\psi)\sin(2\omega_0 t + \psi)\right] \tag{12.23c}$$

where $\Delta\tau = \tau - \tau_0$, $\Delta\psi = \psi_R - \psi$.

The last two terms are negligible for reasons cited in earlier sections; hence the result is

$$s_0(\Delta\tau, \Delta\psi) = \frac{1}{2}\Lambda(\Delta\tau)\cos(\omega_0\Delta\tau + \Delta\psi) \tag{12.23d}$$

where

$$\Lambda(\Delta\tau) = \Lambda_p(\Delta\tau)\int_{-\infty}^{\infty} dt\, p^2(t)$$

$$\Lambda_p(\Delta\tau) = \frac{\int_{-\infty}^{\infty} dt\, p(t)p(t - \Delta\tau)}{\int_{-\infty}^{\infty} dt\, p^2(t)}$$

For the coherent measurement process, $\Delta\psi = 0$ because reference phase is assumed matched to the incoming signal phase. For the "noncoherent" or quadrature filtering technique, the phase is eliminated by the process.

An envelope detector applied to the matched filter output signal in (12.23d) results in

$$|s_0(\Delta\tau, \Delta\psi)| = \frac{1}{2}|\Lambda(\Delta\tau)| \tag{12.23e}$$

The correct value of τ for an optimal measurement, according to MLE theory, is that for which $d|\Lambda(\tau)|/d\tau = 0$, i.e., the value of τ for which the time differentiation process as illustrated in Figure 12.2 results in a zero output.

Now let us consider a rectangular pulse of duration τ_p centered at $t = t_0$, i.e.,

$$p(t) = u\left[t - \left(t_0 - \frac{\tau_p}{2}\right)\right] - u\left[t - \left(t_0 + \frac{\tau_p}{2}\right)\right] \tag{12.24a}$$

The FT of this pulse is

$$S(\omega) = \int_{t_0-\tau_p/2}^{t_0+\tau_p/2} dt\, e^{-j\omega t} = \tau_p e^{-j\omega t_0}\mathrm{sinc}\left(\frac{\omega\tau_p}{2}\right) \tag{12.24b}$$

The "effective bandwidth" β_τ, as defined in (12.11j), is

$$\beta_\tau = \sqrt{\frac{\int_{-\infty}^{\infty} d\omega\ \omega^2[\mathrm{sinc}(\omega\tau_p/2)]^2}{\int_{-\infty}^{\infty} d\omega[\mathrm{sinc}(\omega\tau_p/2)]^2}} = \sqrt{\frac{1}{2\pi}\int_{-\infty}^{\infty} d\omega\ \sin^2\!\left(\frac{\omega\tau_p}{2}\right)} = \infty \tag{12.24c}$$

This troubling result can be confirmed through the alternative definition of β_τ as given by (12.20c) where $x \to t$. From that expression and the differentiation of (12.24a), we have

$$\beta_\tau = \sqrt{\frac{1}{\tau_p}\int_{-\infty}^{\infty} dt\left\{\delta\left[t - \left(t_0 - \frac{\tau_p}{2}\right)\right]\right\}^2 - \frac{1}{\tau_p}\int_{-\infty}^{\infty} dt\left\{\delta\left[t - \left(t_0 + \frac{\tau_p}{2}\right)\right]\right\}^2} \tag{12.24d}$$

While the integral of $\delta(t)$ over infinite limits is finite, that of $[\delta(t)]^2$ is infinite, so again we are confronted with an infinite effective bandwidth, which implies through (12.11j) that the rms error is zero regardless of the noise level.

Since this result seems physically impossible, a little thought should convince us that although mathematically correct, it is due to an idealization that is not realistic. The idealization is that of a pulse with infinitely sloping edges, which can usually be used with impunity in analysis. If such a pulse could exist, it would seem that one could locate either leading or trailing edge precisely and thereby determine the delay precisely for any noise level.

Examining the concept of "effective bandwidth" β_τ we note that the *effective* bandwidth of a perfectly rectangular pulse *is* infinite, but the bandwidth as defined in other ways is finite. The usual practical definition of bandwidth is spacing between -3 dB points on the spectrum, which we can determine from (12.24b) to be approximately

$$B_{-3\mathrm{dB}} = \frac{0.844}{\tau_p}\ \text{Hertz} \tag{12.24e}$$

Another definition of bandwidth in the case where the peak value occurs at $\omega = \omega_0$ is (in Hertz)

$$B_A = \frac{\frac{1}{2\pi}\int_{-\infty}^{\infty} d\omega\ |S(\omega)|^2}{|S(\omega_0)|^2} \tag{12.24f}$$

which would obviously be the bandwidth for a perfectly rectangular spectrum.

For the spectrum of (12.242b), (12.24f) yields

$$B_A = \frac{\tau_p}{(\tau_p)^2} = \frac{1}{\tau_p}\ \text{Hertz} \tag{12.24g}$$

If we were to use $B_{-3\mathrm{dB}}$ or B_A to replace the "effective bandwidth" β_τ in (12.11j), we would obtain

$$\begin{aligned} \epsilon_\tau &= \frac{1.185\tau_\mathrm{p}}{\sqrt{\rho_0}} \qquad \text{if } \beta_\tau \to B_{-3\mathrm{dB}} \\ &= \frac{\tau_\mathrm{p}}{\sqrt{\rho_0}} \qquad \text{if } \beta_\tau \to B_\mathrm{A} \end{aligned} \tag{12.24h}$$

To obtain a realistic rms error with a transmitted rectangular pulse, we note that the received pulse at the point where optimal processing occurs has been passed through RF and IF filtering stages and has reduced bandwidth. The baseband equivalent of this situation, with the assumption of a flat filter with Hertz bandwidth B, yields

$$\beta_\tau^2 = \frac{\int_{-\pi B}^{\pi B} d\omega\ \omega^2 |S(\omega)|^2}{\int_{-\pi B}^{\pi B} d\omega |S(\omega)|^2} = \frac{4\int_{-\pi B}^{\pi B} d\omega\ \sin^2(\omega\tau_\mathrm{p}/2)}{\tau_\mathrm{p}^2 \int_{-\pi B}^{\pi B} d\omega \left[\operatorname{sinc}(\omega\tau_\mathrm{p}/2)\right]^2} \tag{12.24i}$$

Let us now consider a pulse shape with the greatest imaginable departure from steeply sloping edges, the Gaussian-shaped pulse with the same total energy as the rectangular pulse of duration τ_p (Chapter 7, Eq. (7.25a)′, and Fig. 7.5)

$$p(t) = e^{-(\pi/2)(t/\tau_\mathrm{p})^2} \tag{12.25a}$$

From the baseband version of (7.25b)

$$S(\omega) = e^{-(\omega\tau_\mathrm{p}/\sqrt{2\pi})^2} \tag{12.25b}$$

It follows from (12.19) with $u \to \omega$ and (12.25b) that

$$\beta_\tau^2 = \frac{\int_{-\infty}^{\infty} d\omega\ \omega^2 e^{-(\omega\tau_p)^2/\pi}}{\int_{-\infty}^{\infty} d\omega\ e^{-(\omega\tau_p)^2/\pi}} = \frac{\pi}{2\tau_\mathrm{p}^2} \tag{12.25c}$$

and hence through (12.11j) that

$$\epsilon_\tau = \sqrt{\frac{2}{\pi}}\frac{\tau_\mathrm{p}}{\sqrt{\rho_0}} = 0.7979\frac{\tau_\mathrm{p}}{\sqrt{\rho_0}} \tag{12.25d}$$

The rms error, as given by (12.25d), is proportional to the pulse duration and inversely proportional to the *voltage* SNR of the matched filter. The proportionality constant in this case is about 0.8.

To focus on the issue of effect of slope of the pulse edges on accuracy, we will compare the Gaussian-shaped pulse with another simple pulse shape, the *nearly* rectangular pulse with linearly sloping edges as shown in Figure 12.3. The pulse waveform and its derivative are

$$
\begin{aligned}
&p(t) = 1 \\
&\frac{dp(t)}{dt} = 0 && \text{if } -\left(\frac{\tau_P - \delta}{2}\right) \le t \le \left(\frac{\tau_P - \delta}{2}\right) \\
&p(t) = \frac{1}{2} + \frac{1}{\delta}\left(t + \frac{\tau_P}{2}\right) \\
&\frac{dp(t)}{dt} = \frac{1}{\delta} && \text{if } -\left(\frac{\tau_P + \delta}{2}\right) < t < -\left(\frac{\tau_P - \delta}{2}\right) \\
&p(t) = \frac{1}{2} - \frac{1}{\delta}\left(t - \frac{\tau_P}{2}\right) \\
&\frac{dp(t)}{dt} = -\frac{1}{\delta} && \text{if } \left(\frac{\tau_P - \delta}{2}\right) < t < \left(\frac{\tau_P + \delta}{2}\right) \\
&p(t) = \frac{dp(t)}{dt} = 0 && \text{if } t < -\left(\frac{\delta_P + \delta}{2}\right) \text{ or } t > \left(\frac{\tau_P + \delta}{2}\right)
\end{aligned}
\tag{12.26a}
$$

Using (12.26a) and (12.20c) with $x \to t$, $s(x) \to p(t)$,

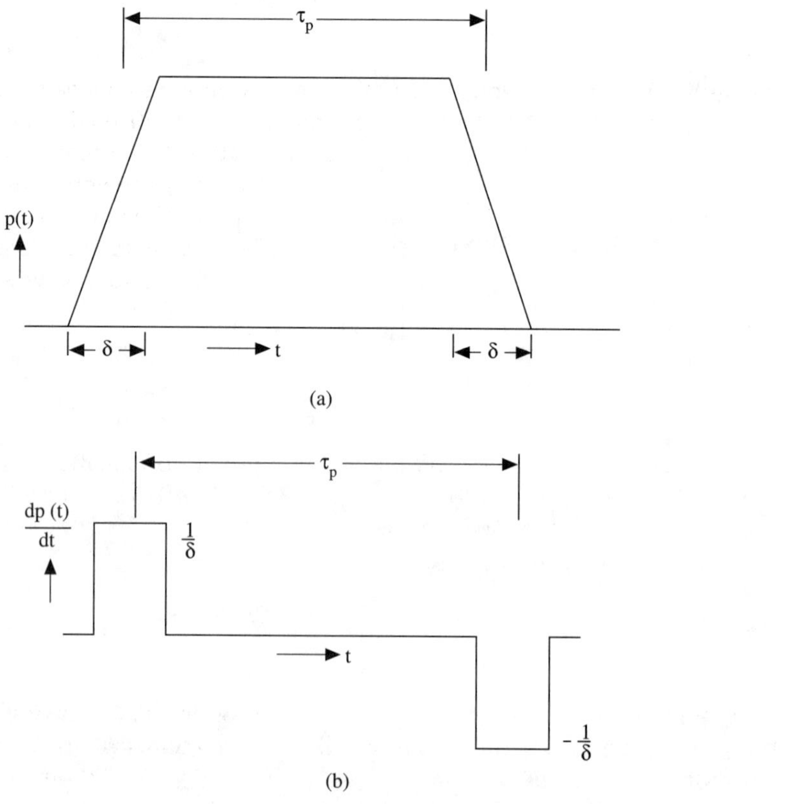

FIGURE 12.3
Pulse with linearly sloping edges. (a) Pulse waveform. (b) Derivative of pulse waveform.

$$\beta_\tau = \sqrt{\frac{2}{\delta\tau_p[1 - (\delta/3\tau_p)]}} \tag{12.26b}$$

Substitution of (12.26b) into (12.11j) yields the rms error

$$\epsilon_\tau = \frac{\tau_p}{\sqrt{\rho_0}} \sqrt{\xi\left(1 - \frac{\xi}{3}\right)} \tag{12.26c}$$

where $\xi = \delta/\tau_p = 1/s\tau_p$ and where $s = 1/\delta$ = slope of edges.

The rms error is again proportional to $\tau_p/\sqrt{\rho_0}$ as for the Gaussian pulse. As the slope of the edges approaches infinity (i.e., the rectangular pulse is approached), the rms error approaches zero, as would be expected from (12.24c). As the slope decreases to the value $2/\tau_p$ (the pulse becomes the triangular function), the proportionality factor is $\sqrt{1/6} = 0.4082$, indicating an rms error about 49% below that with the Gaussian pulse.

It is instructive to see how the rms error relates to bandwidth, which should be the dominant parameter influencing the error. The FT of $p(t)$, through integration by parts, can be written in the form

$$P(\omega) = \frac{1}{j\omega} \int_{-\infty}^{\infty} dt \, \frac{dp(t)}{dt} \, e^{-j\omega t} = \tau_p \, \mathrm{sinc}\left(\frac{\omega\tau_p}{2}\right) \mathrm{sinc}\left(\frac{\omega\delta}{2}\right) \tag{12.27a}$$

The -3 dB bandwidth in Hertz, denoted by B in this discussion, is determined by setting $|P(\omega)|/\tau_p$ in (12.27a) to $1/\sqrt{2}$, i.e.,

$$\left|\mathrm{sinc}\left(\frac{\pi B\tau_p}{2}\right)\mathrm{sinc}\left(\frac{\pi B\delta}{2}\right)\right| = \frac{1}{\sqrt{2}} \tag{12.27b}$$

The first null of $|\mathrm{sinc}(\pi B\tau_p/2)|$ occurs at $B = 2/\tau_p$. This can be used as an approximation for the bandwidth of a perfectly rectangular pulse and implies that the argument of the sinc function is sufficiently small close to the point where $f = B/2$ to justify the approximation

$$\mathrm{sinc}\left(\frac{\pi B\tau_p}{2}\right)\mathrm{sinc}\left(\frac{\pi B\delta}{2}\right) \simeq \left[1 - \frac{1}{6}\left(\frac{\pi B\tau_p}{2}\right)^2\right]\left[1 - \frac{1}{6}\left(\frac{\pi B\delta}{2}\right)^2\right] \tag{12.27c}$$

(since δ is always smaller than τ_p).

Applying (12.27c) in (12.27b) and neglecting the term proportional to $\tau_p^2\delta^2$, we obtain the approximate bandwidth

$$B \simeq \frac{0.844}{\sqrt{\tau_p^2 + \delta^2}} \tag{12.27d}$$

which is the same as (12.24e) where τ_p is replaced by $\sqrt{\tau_p^2 + \delta^2}$. Comparing the bandwidth of the pulse with linearly sloping edges as given by (12.27d) and denoted by B_1 and that of the rectangular pulse as given by (12.24e), and denoted by B_0, we have

$$R = \frac{B_1}{B_0} = \frac{1}{\sqrt{1 + \xi^2}} = \frac{s\tau_p}{\sqrt{(s\tau_p)^2 + 1}} \tag{12.28a}$$

where $\xi = \delta/\tau_p$ and $s = 1/\delta$, the magnitude of the slope of the leading and trailing edges of the pulse. The form (12.28a) points up the role of the slope of the edges in reducing the bandwidth from that of a rectangular pulse. If the slope is infinite then R is unity and if it has its smallest possible value, $2/\tau_p$, then $R = 2/\sqrt{5} = 0.8944$.

Expressing the rms error in (12.26e) in terms of R, we have

$$\epsilon_\tau = \frac{\tau_p}{\sqrt{\rho_0}}\left(\frac{1 - R^2}{R^2}\right)^{1/4}\sqrt{1 - \frac{\sqrt{1 - R^2}}{3R}} \tag{12.28b}$$

For very small departures from a rectangular pulse, where the bandwidth ratio R is very close to 1, i.e., $R = 1 - \Delta R$, (12.28b) takes the approximate form

$$\epsilon_\tau = \frac{1.189\tau_p(\Delta R)^{1/4}}{\sqrt{\rho_o}}, \qquad \text{if } |\Delta R| << 1 \tag{12.28c}$$

for fixed pulse width and SNR, i.e., the rms error varies roughly as the fourth root of $|\Delta R|$ for very small values of $|\Delta R|$.

If the slope s of the edges is the parameter of interest, then the result equivalent to (12.28c) for very steep slopes, from (12.26c) is

$$\epsilon_\tau = \sqrt{\frac{\tau_p}{\rho_0 s}}, \qquad \text{if } s >> \frac{1}{3\tau_p} \tag{12.28d}$$

indicating that the rms error is roughly proportional to the inverse square root of the slope for a sufficiently steep slope.

12.2.2 Accuracy of Frequency Measurements

The optimal frequency measurement is illustrated in Figure 12.4 where a shows the completely coherent measurement procedures and b shows the procedures when due to unknown phase, quadrature filtering must be used.

Consider the coherent process (a), on which the idealized MLE theory is based. The incoming signal is modelled as usual as an RF or IF pulse of the form

$$s(t, \omega_0) = p(t)\cos(\omega_0 t + \psi) \tag{12.29a}$$

The reference signal, assuming phase is a-priori known perfectly, is

$$s_R(t, \omega) = p(t)\cos(\omega t + \psi) \tag{12.29b}$$

The matched filter output, assuming perfect matching in pulse time of arrival, is

$$s_0(\Delta\omega) = \frac{1}{2}\int_{-\infty}^{\infty} dt\, p^2(t)[\cos(\Delta\omega t) + \cos(2\omega_0 t + 2\psi)] \tag{12.29c}$$

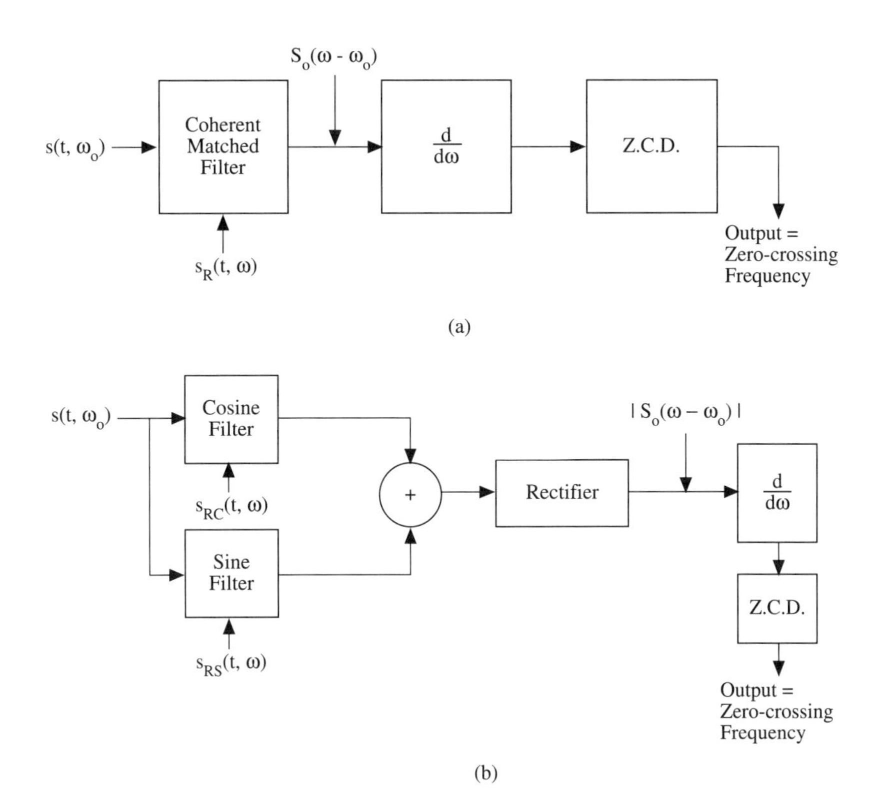

FIGURE 12.4
Optimal frequency measurement process. (a) Coherent measurement; $s(t, \omega_0) = a\cos(\omega_0 t + \psi)$; $s_R(t, \omega)$ = reference signal = $\cos(\omega t + \psi)$; Z.C.D. = zero crossing detector. (b) Quadrature filtering due to unknown phase.

where $\Delta\omega = \omega - \omega_0$, whose second term effectively integrates to a negligibly small value, as has been pointed out many times throughout this book, leaving the form

$$s_0(\Delta\omega) = \frac{1}{2}\int_{-\infty}^{\infty} dt\, p^2(t)\cos(\Delta\omega t) \tag{12.29d}$$

The error in measurement of frequency, as obtained from (12.21) where $x = t$, $u = \omega$, is

$$\epsilon_\omega = \sqrt{\langle(\Delta\omega^2\rangle} = \frac{1}{\sqrt{\rho_0}\beta_\omega} \tag{12.30a}$$

where

$$\rho_0 = \int_{-\infty}^{\infty} dt\, s^2(t)\Big/N_0 = \frac{1}{2\pi}\int_{-\infty}^{\infty} d\omega |S(\omega)|^2\Big/N_0$$

where N_0 is noise-power density and

$$\beta_\omega = \sqrt{\frac{\int_{-\infty}^{\infty} dt\, t^2 s^2(t)}{\int_{-\infty}^{\infty} dt\, s^2(t)}} = \sqrt{\frac{\int_{-\infty}^{\infty} dt\, t^2 p^2(t)}{\int_{-\infty}^{\infty} dt\, p^2(t)}} \tag{12.30b}$$

or equivalently in terms of the signal spectrum (from (12.22c) where $u = \omega$)

$$\beta_\omega = \sqrt{\frac{\int_{-\infty}^{\infty} d\omega |dS(\omega)/d\omega|^2}{\int_{-\infty}^{\infty} d\omega |S(\omega)|^2}} \tag{12.30c}$$

Consider first the case of the rectangular pulse of duration τ_p, for which the integrals in (12.30b) are very easily evaluated, leading to

$$\beta_\omega = \sqrt{\frac{2\int_0^{\tau_p/2} dt\, t^2}{2\int_0^{\tau_p} dt}} = \frac{\tau_p}{\sqrt{12}} = 0.2887\tau_p \tag{12.31}$$

For the Gaussian-shaped pulse as given by (12.25a), Eq. (12.30b) yields

$$\beta_\omega = \sqrt{\frac{\int_{-\infty}^{\infty} dt\, t^2 e^{-\pi t^2/\tau_p^2}}{\int_{-\infty}^{\infty} dt\, e^{-\pi t^2/\tau_p^2}}} = \frac{\tau_p}{\sqrt{2\pi}} = 0.399\tau_p \tag{12.32}$$

For the pulse with linearly sloping edges as given by (12.26a)

$$\beta_\omega = \sqrt{\frac{\int_0^{(\tau_p-\delta)/2} dt\, t^2 + \int_{(\tau_p-\delta)/2}^{(\tau_p+\delta)/2} dt\, t^2\left\{\frac{1}{2} - \frac{1}{\delta}[t - (\tau_p/2)]^2\right\}^2}{\int_0^{(\tau_p-\delta)/2} dt(1) + \int_{(\tau_p-\delta)/2}^{(\tau_p+\delta)/2} dt\left\{\frac{1}{2} - \frac{1}{\delta}[t - (\tau_p/2)]^2\right\}^2}}$$

$$= 0.2887\tau_p \sqrt{\frac{1 - \xi + \xi^2 - \xi^3/5}{1 - \xi/3}} \tag{12.33}$$

where $\xi = \delta/\tau_p$.

For infinite slope, where $\xi = 0$, β_ω is that of the rectangular pulse as given by (12.31). If ξ has its largest possible value, 0.5, then (12.33) gives us $\beta_\omega = 0.2343\ \tau_p$.

From Eqs. (12.31) through (12.33)′ and Eq. (12.30a), the rms error in frequency measurement is always inversely proportional to the pulse duration and the voltage SNR and does not appear to be highly sensitive to the pulse shape, varying by a factor less than 2 over this limited set of examples.

12.2.3 Accuracy of Angle Measurements

Before details of angle measurement procedures are discussed, let us consider a uniformly illuminated aperture in the x–y plane, as discussed in Section 10.2. The aperture illumination in the x-direction is given by (see Eqs. (10.7a, b) and Fig. 10.2)

$$f(x) = u\left(x + \frac{a}{2}\right) - u\left(x - \frac{a}{2}\right) \tag{12.34a}$$

where a is the aperture dimension in the x-direction.

The pattern in the x–z principal plane resulting from this illumination function $f(x)$, neglecting the factor $(1 + \cos\theta)$ from (10.8), is

$$F(\theta) = \text{sinc}\left(\frac{\pi a}{\lambda}\right)\sin\theta \tag{12.34b}$$

The noise-induced error in measurement of the parameter $u = k_0 \sin\theta$, as given by (12.131), is

$$\epsilon_{k_0 \sin\theta} = k_0\sqrt{\langle[\Delta(\sin\theta)]^2\rangle} = \frac{1}{\sqrt{\rho_0}\beta_{k_0 \sin\theta}} \tag{12.35a}$$

where, with the aid of (12.34a)

$$\beta_{k_0 \sin\theta} = \sqrt{\frac{\displaystyle\int_{-a/2}^{a/2} dx\, x^2}{\displaystyle\int_{-a/2}^{a/2} dx}} = \sqrt{\frac{2(a/2)^3(1/3)}{a}} = \frac{a}{\sqrt{12}} = 0.2887a \tag{12.35b}$$

We note that, if $\Delta\theta = \theta_{\text{meas}} - \theta_{\text{true}}$

$$\Delta(\sin\theta) = \sin\theta_{\text{meas}} - \sin\theta_{\text{true}} = \sin\theta_{\text{true}}(\cos\Delta\theta - 1) \tag{12.35c}$$

Assuming the error to be sufficiently small to permit the small angle approximations for $\sin\Delta\theta$ and $\cos\Delta\theta$, (12.35c) can be approximated by

$$\begin{aligned}\Delta(\sin\theta) &= \cos\theta_{\text{true}}\Delta\theta - \sin\theta_{\text{true}}\frac{(\Delta\theta)^2}{2} \\ &\simeq \Delta\theta\cos\theta_{\text{true}} \text{ to first order}\end{aligned} \tag{12.35d}$$

Since the error is dependent on the true angle θ_{true}, we should account for the possibility of a different geometry for the aperture distribution. If the direction of illumination is along the z-axis (for example, a long dipole along the z-axis with a specified current distribution along that axis), then $\sin\theta$ is replaced by $\cos\theta$ in (12.33a). In that case, using the same small angle approximations, (12.35d) would be replaced by

$$\begin{aligned}\Delta(\cos\theta) &= \cos\theta_{\text{true}}(\cos\Delta\theta - 1) - \sin\theta_{\text{true}}\sin\Delta\theta \\ &= -\Delta\theta\sin\theta_{\text{true}} + \frac{(\Delta\theta)^2}{2}\cos\theta_{\text{true}} \simeq -\Delta\theta\sin\theta_{\text{true}} \text{ to first order}\end{aligned} \tag{12.35e}$$

We are interested in the error for the angle θ, which we can obtain from (12.35a) for the uniform x-directed aperture illumination with the aid of (12.35b, d). This error is

$$\epsilon_\theta = \sqrt{\langle(\Delta\theta)^2\rangle} = \frac{1}{k_0 \cos \theta_{\text{true}}} \epsilon_{k_0 \sin \theta} = \frac{1}{\sqrt{\rho_0}\beta_\theta} \tag{12.35f}$$

where

$$\beta_\theta = \frac{2\pi}{\lambda} \cos \theta_{\text{true}} \beta_{k_0 \sin \theta} = \frac{\pi a \cos \theta_{\text{true}}}{\lambda\sqrt{3}}$$

where λ is wavelength. If the illumination is in the z-direction it is evident that cos θ_{true} replaces sin θ_{true} in (12.35f).

Since there are an infinite variety of possible radar antenna geometries, and the error in θ depends on its true value θ_{true}, consider the ground-based radar situation depicted in Figure 12.5.

The target's elevation angle ϕ_{el}, as depicted in Figure 12.5a, is the angle between line of sight from radar to target and the x' axis in the local coordinate

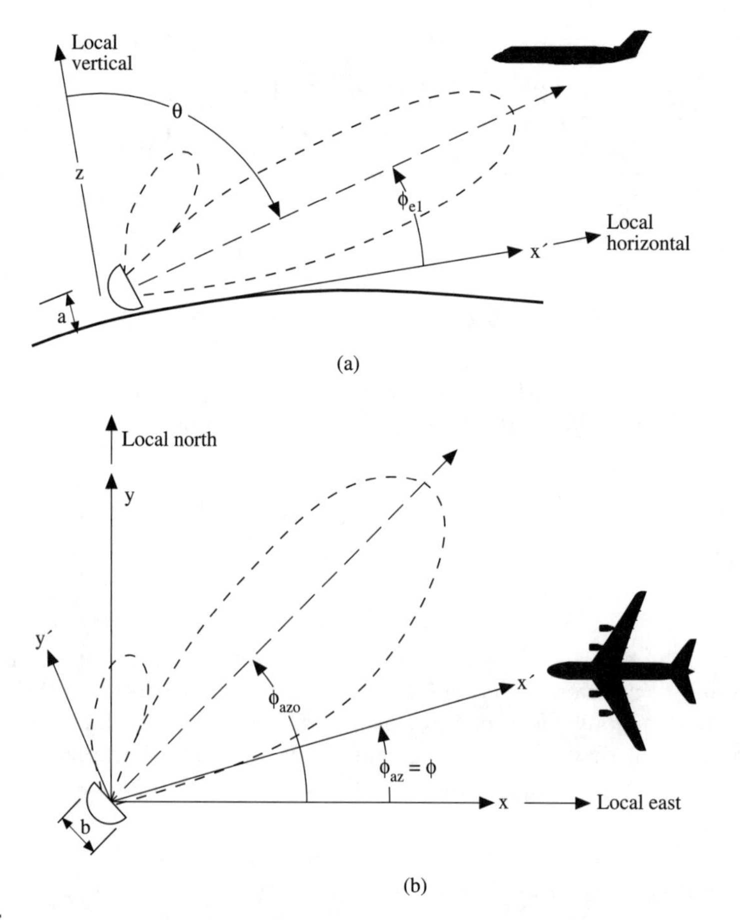

FIGURE 12.5
Elevation and azimuth angles. (a) View in vertical plane; ϕ_{el} = elevation angle of target. (b) Down view; ϕ_{az} = target azimuth, ϕ_{azo} = azimuth of peak of beam.

system of the radar, i.e., it is the angle in a locally vertical plane designated as the elevation plane or the x–z plane. Figure 12.5b shows ϕ_{az}, the azimuth angle of the target, which is defined in the radar's local horizontal plane. The positive local horizontal coordinate $+x$ is in the eastward direction and the positive local coordinate $+y$ is in the northward direction. The primed horizontal coordinates (x', y') are used to define the elevation plane. The relationship between the standard spherical coordinate angles (θ, ϕ) and the elevation and azimuth angles (ϕ_{el}, ϕ_{az}) in degrees are

$$\phi_{el} = 90 - \theta$$
$$\phi_{az} = \phi \tag{12.35g}$$

If, as shown in Figure 12.5, the antenna beam is pointed in the directions of the elevation and azimuth angles ϕ_{el0} and ϕ_{az0}, respectively, and the target is at angles (ϕ_{el}, ϕ_{az}), then the direction of aperture illumination for the elevation beam is $z \cos \phi_{el0}$ and that for the azimuth beam is $y\ \cos_{az0}$. If the principal plane patterns in elevation and azimuth planes both arise from uniform illumination in their respective directions, and if the dimensions of the apertures for elevation and azimuth pattern generation are a_{el} and a_{az} respectively, then the beam pattern can be inferred from (10.7b) and are given by (neglecting the factor $1 + \cos\theta$ and the angular dependences associated with the aperture field polarizations)

$$F(\phi_{el}, \phi_{az}) = \operatorname{sinc}\left(\frac{\pi a_{el}}{\lambda}\sin(\phi_{el} - \phi_{el0})\cos(\phi_{az} - \phi_{az0})\right)\cdots$$
$$\operatorname{sinc}\left(\frac{\pi a_{az}}{\lambda}\sin(\phi_{el} - \phi_{el0})\sin(\phi_{az} - \phi_{az0})\right) \tag{12.36}$$

Now we will apply the theory that led to (12.35f) for a general angle θ to the angles ϕ_{el} and ϕ_{az}. Measurements of these angles will be made when the target is very close to the peak of the beam. This applies especially to the tracking mode, where rough estimates of the angles are presumed to exist already and the measurement processes we are discussing here are designed to refine them. Another consideration is that the process requires a high SNR, a requirement that will be fulfilled only if the target is "in the beam," i.e., near the peak of the beam. In view of this restriction, we define the elevation plane as that for which $\phi_{az} = \phi_{az0}$ and the azimuth plane as that for which $\phi_{el} = \phi_{el0} + 90°$. Then from (12.36) the elevation pattern is

$$F_{el}(\phi_{el}) = \operatorname{sinc}\left[\frac{\pi a_{el}}{\lambda}\sin(\phi_{el} - \phi_{el0})\right] \tag{12.37a}$$

and the azimuth pattern is

$$F_{az}(\phi_{az}) = \operatorname{sinc}\left(\frac{\pi a_{az}}{\lambda}\sin(\phi_{az} - \phi_{az0})\right) \tag{12.37b}$$

Now we return to the discussion of accuracy in measurement of elevation and azimuth angles. The factor $\cos\theta_{true}$ that appears in (12.35f) can be approximated by 1 in first order (where θ_{true} is the true value of either ϕ_{el} or ϕ_{az}), where the error is assumed to be small due to high SNR, as discussed earlier.

Based on that assumption, the rms errors in elevation and azimuth angles can be inferred from (12.35f) and are given by

$$\epsilon_{\phi_{az}^{el}} = \sqrt{\langle(\Delta\phi_{az}^{el})^2\rangle} = \frac{1}{\sqrt{\rho_0}\beta_{\phi_{az}^{el}}} = \frac{\lambda\sqrt{3}}{\pi\sqrt{\rho_0}a_{az}^{el}} \tag{12.37c}$$

In Chapter 10, Section 10.5, we discussed the tapering of apertures to achieve sidelobe level reduction. It was pointed out that the price for sidelobe reduction is a wider mainbeam, caused by the reduction in the *effective* aperture length. It is instructive to consider the effect of this measure on angle measurement error.

Consider the linearly tapered aperture, as discussed in Section 10.5. The counterpart of (12.34a) is

$$f(x) = \left[u\left(x + \frac{a}{2}\right) - u\left(x - \frac{a}{2}\right)\right]\left[1 - \frac{2|x|}{a}\right] \tag{12.38a}$$

from which it follows that the numerator of $\beta^2_{k_0 \sin\theta}$ is

$$\beta^2_{num} = \int_{-a/2}^{a/2} dx\, x^2\left[1 - \frac{2|x|}{a}\right]^2 = 2\int_0^{a/2} dx\, x^2\left(1 - \frac{2x}{a}\right)^2 = \frac{a^3}{120} \tag{12.38b}$$

The denominator is

$$\beta^2_{dnom} = \int_{-a/2}^{a/2} dx\left(1 - \frac{2|x|}{a}\right)^2 = \frac{a}{3} \tag{12.38c}$$

From (12.38b, c)

$$\beta_{k_0 \sin\theta} = \frac{a}{2\sqrt{10}} \tag{12.38d}$$

from which, through application of (12.35a, f) to the linear tapered aperture case, approximating $\cos\theta_{true}$ as 1, and determining the counterpart of (12.37a, b) for this case, we obtain

$$\left(e_{\phi_{az}^{el}}\right)_{LTA} = \frac{\lambda}{a_{az}^{el}\sqrt{\rho_0}}\frac{\sqrt{10}}{\pi}\left(\epsilon_{\phi_{az}^{el}}\right)_{UA}\sqrt{\frac{10}{3}} = 1.826\left(\epsilon_{\phi_{az}^{el}}\right)_{UA} \tag{12.38e}$$

where LTA and UA refer to "linearly tapered aperture" and "uniform aperture," respectively. Equation (12.38e) indicates that the rms noise-induced error in measurement of angle for the linearly tapered aperture is nearly twice that for the uniform aperture if both have the same SNR.

To interpret this result, we first note that the rms error could also have been expressed in the equivalent form (for an x-directed aperture)

$$\epsilon_\theta = \sqrt{\frac{N_0}{\int_{-\infty}^{\infty} dx\, x^2|f(x)|^2}} \tag{12.38f}$$

where N_0 is the noise power density, which is independent of the antenna aperture illumination and hence can be considered as a fixed quantity. From (12.38b, c, d), the ratio of rms angle error for the linearly tapered aperture relative to that for the uniform aperture with the *same noise power density* is

$$\frac{(\epsilon_\theta)_{\mathrm{LTA}}}{(\epsilon_\theta)_{\mathrm{UA}}} = \sqrt{\frac{\int_{-a/2}^{a/2} dx\, x^2}{\int_{-a/2}^{a/2} dx\, x^2[1 - (2|x|/a)]^2}} = \sqrt{10} = 3.16 \tag{12.38g}$$

This is a greater difference in the two cases than indicated by (12.38e), because the noise power density rather than the SNR is the same for the two aperture distributions. The comparison indicated by (12.38g) rather than (12.38e) seems more valid because the signal energy, which is the numerator of the matched filter output SNR, is dependent on the aperture distribution while its denominator N_0 is not. Keeping the SNR constant while changing the aperture distribution, other parameters remaining the same, is not realistic.

To further investigate this issue, we define the *effective* aperture length as the average length over which power is distributed on the aperture, given by

$$a_{\mathrm{eff}} = \int_{-a/2}^{a/2} dx\, |f(x)|^2 \tag{12.39a}$$

From (12.38e, f)

$$\frac{(a_{\mathrm{eff}})_{\mathrm{UA}}}{(a_{\mathrm{eff}})_{\mathrm{LTA}}} = \frac{\int_{-a/2}^{a/2} dx}{\int_{-a/2}^{a/2} dx\left[1 - \frac{2|x|}{a}\right]^2} = 3 \tag{12.39b}$$

This is a little less than but quite close to the value 3.16 obtained in (12.38g), which points to the idea that the rms error is monotonically decreasing with the effective aperture dimension.

For further investigation consider the cosine and cosine-square tapers given by (10.35a) and (10.35b), respectively. The counterparts of (12.39b) for these two tapers are

$$\frac{(a_{\mathrm{eff}})_{\mathrm{UA}}}{(a_{\mathrm{eff}})_{\cos}} = \frac{a}{\int_{-a/2}^{a/2} dx\, \cos^2(\pi x/a)} = \frac{a}{(a/2)} = 2 \tag{12.40a}$$

$$\frac{(a_{\mathrm{eff}})_{\mathrm{UA}}}{(a_{\mathrm{eff}})_{\cos^2}} = \frac{a}{\int_{-a/2}^{a/2} dx\, \cos^4(\pi x/a)} = \frac{a}{(3a/8)} = \frac{8}{3} = 2.6667 \tag{12.40b}$$

From (12.40a, b), we see that the cosine taper halves the effective aperture dimension relative to the uniformly illuminated aperture and the cosine-square taper reduces it to about 37.5% of its value for uniform illumination.

To examine the rms angle error increases over the uniform case for each of these tapers, we invoke the counterparts of (12.38b, c), with the results (where subscripts cos, $\cos^2$, and UA are used to distinguish between the three aperture illumination functions).

Cosine taper:

$$(\beta^2_{\cos})_{\text{dnom}} = (a_{\text{eff}})_{\cos} = \frac{a}{2} \tag{12.41a}$$

$$(\beta^2_{\cos})_{\text{num}} = \int_{-a/2}^{a/2} dx\, x^2 \cos^2\left(\frac{\pi x}{a}\right) = \frac{a^3}{24}\left[1 - \frac{6}{\pi^2}\right] = 0.01633a^3 \tag{12.41b}$$

$$\beta_{\cos} = \frac{a}{\sqrt{12}}\sqrt{1 - \frac{6}{\pi^2}} = 0.1807a \tag{12.41c}$$

Cosine-square taper:

$$(\beta^2_{\cos^2})_{\text{dnom}} = (a_{\text{eff}})_{\cos^2} = \frac{3}{8}a \tag{12.42a}$$

$$(\beta^2_{\cos^2})_{\text{num}} = \int_{-a/2}^{a/2} dx\, x^2 \cos^4\left(\frac{\pi x}{a}\right) = \frac{a^3}{32}\left(1 - \frac{15}{2\pi^2}\right) = 0.0074999a^3 \tag{12.42b}$$

$$\beta_{\cos^2} = \frac{a}{\sqrt{12}}\sqrt{1 - \frac{15}{2\pi^2}} = 0.1414a \tag{12.41c}$$

From (12.35b), for the uniform aperture (where $a_{\text{eff}} = a$)

$$(\beta^2_{\text{UA}})_{\text{num}} = \frac{a^3}{12} = 0.08333a^3 \tag{12.43a}$$

$$\beta_{\text{UA}} = \frac{a}{\sqrt{12}} = 0.2887a \tag{12.43b}$$

For a constant SNR, the ratio of rms errors for the cosine taper relative to that for a uniform aperture can be obtained from (12.35f), (12.41c), and (12.43b). The result is

$$\left(\frac{(\epsilon_\theta)_{\cos}}{(\epsilon_\theta)_{\text{UA}}}\right)_{\text{fixed } \rho_0} = \frac{\beta_{\text{UA}}}{\beta_{\cos}} = \frac{0.2887a}{0.1807a} = 1.5977 \tag{12.44a}$$

For constant noise power density (implying that signal power depends on the aperture illumination function and therefore SNR is *not* constant), from (12.38f), (12.41b), and (12.43a), we obtain

$$\left(\frac{(\epsilon_\theta)_{\cos}}{(\epsilon_\theta)_{\text{UA}}}\right)_{\text{fixed } N_0} = \sqrt{\frac{(\beta^2_{\text{UA}})_{\text{num}}}{(\beta^2_{\cos})_{\text{num}}}} = \sqrt{\frac{2}{1 - (6/\pi^2)}} = \sqrt{5.102} = 2.2588 \tag{12.44b}$$

The counterparts of (12.44a) and (12.44b) for the cosine-square taper are [with the aid of (12.35f), (12.42c) and (12.43b)]

$$\left[\frac{(\epsilon_\theta)_{\cos^2}}{(\epsilon_\theta)_{UA}}\right]_{\text{fixed } \rho_0} = \frac{\beta_{UA}}{\beta_{\cos^2}} = \frac{0.2887a}{0.1414a} = 2.0417 \tag{12.45a}$$

for constant SNR. With the aid of (12.38f), (12.42b), and (12.43a), the same ratio for constant noise power density is

$$\left[\frac{(\epsilon_\theta)_{\cos^2}}{(\epsilon_\theta)_{UA}}\right]_{\text{fixed } N_0} = \sqrt{\frac{(\beta^2_{UA})_{\text{num}}}{(\beta^2_{\cos^2})_{\text{num}}}} = \sqrt{\frac{(8/3)}{1-(15/2\pi^2)}} = 3.333 \tag{12.45b}$$

The effect of the taper on the rms error in measurement of the angle is seen to be more pronounced when the noise power density remains constant while the SNR is allowed to vary as the taper is changed. The tapering reduces the peak antenna gain and thereby reduces the signal level with fixed transmitter power when the target is near boresight. From that viewpoint, the comparison of rms errors with fixed value of N_0, as given by (12.44b) or (12.45b), seems to be a more valid measure of the effect of the taper than that with fixed SNR as given by (12.44a) or (12.45a).

We can infer from (12.44b) and (12.45b) that the increased error when the taper is changed from $\cos(\pi x/a)$ to $\cos^2(\pi x/a)$ is $3.333/2.2588 = 1.476$, i.e., an increase of about 48%. From (12.40a, b), we note that the corresponding increase in effective aperture dimension is $2.667/2 = 1.334$, i.e., an increase of about 34%. If the comparison is made with fixed SNR through (12.44a) and (12.45a), then the ratio of rms error values is $2.0417/1.5977 = 1.278$, an increase of only about 28%.

If the comparisons are made between the two tapered apertures and the uniform illuminated aperture, then with fixed noise power density, the error increase for the cosine taper [Eq. (12.44b)] is about 126% while the increased effective aperture dimension [Eq. (12.40a)] is about 100%. For the cosine-square taper, the error increase [Eq. (12.45b)] is about 233%, when the increase in effective aperture dimension is about 167%.

These results do not enable us to make any precise quantitative inferences about the relationship between the increase of rms angle error with increasing effective aperture dimension, but they do point to a monotonically increasing variation between the error and the effective dimension. Moreover, the same relationships would hold if the effective aperture dimension were defined as the spacing between −3 dB points rather than the definition (12.39a). In the latter case, the ratios given by (12.40a) and (12.40b) are 2 and 2.746, respectively, so the conclusions discussed above change very little.

The basic idea emphasized here is the tradeoff between sidelobe reduction through tapering and increasing the effective aperture dimension to improve angle measurement accuracy. This would be a consideration in design of a radar antenna that has both a maximum allowable sidelobe level specification and a maximum allowable rms angle measurement error. It may not be possible to realize both specifications with a given limit in transmitted power.

12.3 ANGLE MEASUREMENT FOR TRACKING RADARS

There are two basically different generic ways in which target angle is measured in tracking radars, as follows:

1. The variation of amplitude of the target signal as a target changes its angular position in the radar's antenna beam, usually through scanning.
2. The use of phase-delay difference between signals arriving at two different antennas, known as phase comparison or radar interferometry. In this class of techniques, the antenna's amplitude pattern has little or no effect on the measurement.

In what follows, we will discuss some of the techniques that are customarily used in tracking radars. These are usually either "amplitude comparison" or "phase comparison" methods. The first is a subclass of category (1) above and the second is category (2). The standard texts on radar, e.g., Barton[8] and Skolnick,[6] have coverage on these topics that is more extensive on details of implementation than the coverage in this text, where the emphasis will be on some of the basic theoretical ideas behind these techniques.

12.3.1 Angle Measurement Through Amplitude Comparison

A class of measurement techniques commonly used for angle tracking uses the difference between two signals from a target received simultaneously on two angularly separated but otherwise identical beams, as illustrated in Figure 12.6 (based on Fig. 9–11 on p. 316 of Raemer[1]). The diagram shows three point targets at different positions relative to the pointing angles of the two beams. All three targets are assumed to be in the plane of the paper. The beams are symmetrical about the axis AA', which is called the boresight axis. The beams point in the directions of the dashed lines labeled L_1 and L_2. The angle between L_1 and AA' or equivalently that between AA' and L_2 is the "offset angle."

Target 1 is near the peak of the left beam and nearly out of the right beam and hence the differential amplitude (left beam amplitude minus right beam amplitude) is positive and nearly equal to that of the peak of the left beam. Target 3 has a negative differential amplitude and Target 2, at boresight, should return zero differential amplitude since its returns on both beams have the same amplitude.

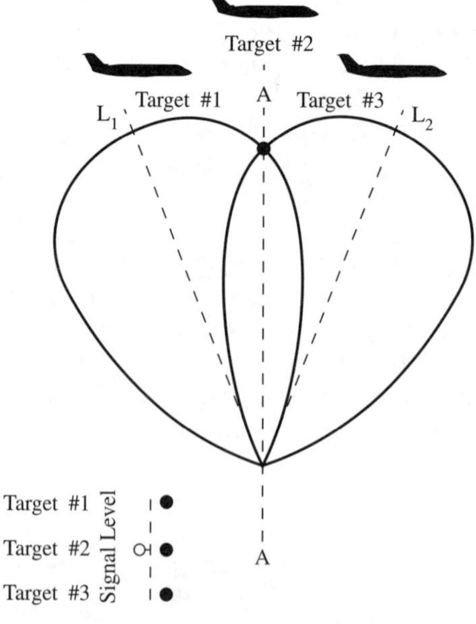

FIGURE 12.6
Amplitude comparison in a plane. From Raemer,[1] Figure 9-11, page 316.

To study this quantitatively, we will simplify the mathematical model of the antenna pattern to remove the effect of sidelobes, which greatly complicates the differential amplitude measurement process. Any situation in which the target is in the sidelobes of the left- or right-hand beam introduces an amplitude ambiguity. In fact, the target must be between the peaks at L_1 and L_2 to avoid ambiguity.

A convenient beam pattern model that brings out the important features of the process is one in which the pattern shape is assumed Gaussian, and in which it is assumed that the beam is so narrow and the target so far from the radar that we can think of the angular arc about the radar as a plane, as shown in the diagram. This is a small angle approximation. From the target antenna geometry shown in the diagram, we will indicate the output of an "amplitude comparator," which measures the difference between the amplitudes of the signal received on the two beams and divides by the sum of these two amplitudes. That output is

$$u_0(\theta_T) = \frac{|\hat{s}_{01}(\theta_T) + \hat{n}_{01}| - |\hat{s}_{02}(\theta_T) + \hat{n}_{02}|}{|\hat{s}_{01}(\theta_T) + \hat{n}_{01}| + |\hat{s}_{02}(\theta_T) + \hat{n}_{02}|} \tag{12.46}$$

where $\hat{s}_{01}$ and $\hat{s}_{02}$ are the complex envelopes of the signals on beam channels 1 and 2, respectively, given as functions of θ_T, the target angle in the plane of the diagram, and $\hat{n}_{01}$ and $\hat{n}_{02}$ are the complex envelopes of the additive noise waveforms for the two beam channels. The target angle θ_T is measured relative to the boresight axis.

If the SNR on both channels is assumed sufficiently high, we can express each amplitude in (12.46) in the form

$$\begin{aligned} |\hat{s}_0 + \hat{n}_0| &= \sqrt{|\hat{s}_0|^2 + |\hat{n}_0|^2 + 2Re(\hat{s}_0\hat{n}^*)} \\ &\simeq |\hat{s}_0| + \frac{Re(\hat{s}_0\hat{n}^{*0})}{|\hat{s}_0|} = |\hat{s}_0| + |\hat{n}_0|\cos(\psi_n - \psi_s) \end{aligned} \tag{12.47a}$$

where ψ_s and ψ_n are signal and noise phase, respectively.

Applying the approximation (12.47a) to (12.46), we obtain

$$u_0(\theta_T) = \frac{|\hat{s}_{01}(\theta_T)| - |\hat{s}_{02}(\theta_T)| + |\hat{n}_{01}|\cos(\psi_{n1} - \psi_{s1}) - |\hat{n}_{02}|\cos(\psi_{n2} - \psi_{s2})}{|\hat{s}_{01}(\theta_T)| + |\hat{s}_{02}(\theta_T)| + |\hat{n}_{01}|\cos(\psi_{n1} - \psi_{s1}) + |\hat{n}_{02}|\cos(\psi_{n2} - \psi_{s2})} \tag{12.47b}$$

We can assume the noise term in the denominator to be small relative to the signal term in the denominator (an assumption that is reasonable for the denominator but would not be valid in the numerator, whose signal term could be very small, e.g., zero at boresight). We can then factor out the signal term in the denominator and expand the remaining factor in a power series in the ratio of noise-to-signal terms, truncating after the linear terms, then invoke the approximation $1/(1 + x) \simeq 1 - x$ for $|x| << 1$, leading to the result

$$u_0(\theta_T) = u_{0s}(\theta_T) + u_{0n} \tag{12.47c}$$

where the signal component of the output is

$$u_{0s}(\theta_T) = K_0\left(\frac{|\hat{s}_{01}(\theta_T)| - |\hat{s}_{02}(\theta_T)|}{|\hat{s}_{01}(\theta_T)| + |\hat{s}_{02}(\theta_T)|}\right) \tag{12.47c)'}$$

where K_0 is a calibration constant and the noise component of the output is

$$u_{0n} = K_0\left(\frac{n_a}{s_b} - \frac{s_a n_b}{s_b^2} - \frac{n_a n_b}{s_b^2}\right) = K_0\left[\frac{n_a s_b - s_a n_b}{s_b^2}\right] \tag{12.47c)''}$$

where s_a is the signal component of the numerator, s_b that of the denominator, and n_a and n_b the noise components of numerator and denominator, respectively, and where the term $-(n_a n_b)/s_b^2$ can be neglected due to the high SNR assumption. The constituents of (12.47c)″ are

$$n_{\mathrm{b}}^{\mathrm{a}} = |\hat{n}_{01}|\cos(\psi_{n1} - \psi_{s1}) \mp |\hat{n}_{02}|\cos(\psi_{n2} - \psi_{s2})$$

$$s_{\mathrm{b}}^{\mathrm{a}} = |\hat{s}_{01}(\theta_T)| \mp |\hat{s}_{02}(\theta_T)| \tag{12.47c)'''}$$

Let us first concentrate on the noise-free case. We can assume that the two signals $\hat{s}_{01}$ and $\hat{s}_{02}$ depend only on the location of the target within the beam and are independent of range or other parameters; hence

$$|\hat{s}_{0\genfrac{}{}{0pt}{}{1}{2}}(\theta_T)| = f(\theta_T \pm \theta_0) \tag{12.47d}$$

where $f(\theta)$ is the one-way power (or equivalently two-way field amplitude) pattern function and where θ_0 is the offset angle, as shown in Figure 12.6 (the antenna pattern function of each beam is given in terms of the angle relative to its peak and the left and right beam pattern functions are identical except for the offset).

Assuming that the target is at a small angle off of the peak of each beam, we can expand $f(\theta)$ in a power series around its peak and truncate the series beyond quadratic terms, with the result [based on (12.47c)′ and (12.47d)]

$$\frac{u_{0s}(\theta_T)}{K_0} = \frac{[f_0^+ + (f'_{0+})\theta_T + \frac{1}{2}(f''_{0+})\theta_T^2] - [f_0^-(f'_{0-})\theta_T + \frac{1}{2}(f''_{0-})\theta_T^2]}{[f_0^+ + (f'_{0+})\theta_T + \frac{1}{2}(f''_{0+})\theta_T^2] + [f_0^- + (f'_{0-})\theta_T + \frac{1}{2}(f''_{0-})\theta_T^2]} \tag{12.47e}$$

where

$$f_0^{\pm} = f(\pm\theta_0); \qquad f'_{0\pm} = f'(\pm\theta_0)' f''_{0\pm} = f''(\pm\theta_0)$$

Assuming that $f(\theta)$ is an even function of θ, it follows that

$$f_0^{+\prime} = f_0^- = f_0; \qquad f'_{0+} = -f'_{0-} = f'_0; \qquad f''_{0+} = f''_{0-} = f''_0 \tag{12.47f}$$

From (12.47e, f), if we set K_0 equal to f_0/f'_0

$$u_{0s}(\theta_T) = \frac{K_0 f'_0 \theta_T}{f_0 + (f''_0/2)\theta_T^2} = \frac{\theta_T}{1 + (f''_0/2f_0)\theta_T^2} \tag{12.47g}$$

If $|(f_0''/2f_0)\theta_T^2| << 1$, then

$$u_{0s}(\theta_T) \approx \theta_T \tag{12.47h}$$

In the linear region of the noise-free comparator output, where (12.47h) holds, we can read the angle off of boresight directly from that output. The noise-free percentage error in the reading is determined by subtraction of (12.47h) from (12.47g), dividing by (12.47h) and multiplying by 100. The result is

$$|\epsilon_{nf}(\theta_T)| = \left|\frac{100(f_0''/2f_0)\theta_T^2}{1 + (f_0''/2f_0)\theta_T^2}\right| \tag{12.47i}$$

Consider the specialization of $f(\theta)$ to the Gaussian pattern, i.e.

$$f(\theta) = e^{-2\gamma(\theta/\theta_b)^2} \tag{12.48a}$$

where θ_b is the one-way -3 dB beamwidth and $\gamma = 1.386$.

From (12.47h)

$$f_0 = e^{-2\gamma(\theta_0/\theta_b)^2}; \qquad f_0' = -\frac{2\gamma\theta_0}{\theta_b^2}e^{-2\gamma(\theta_0/\theta_b)^2}; \qquad f_0'' = -\frac{2\gamma}{\theta_b^2}e^{-2\gamma(\theta_0/\theta_b)^2}\left(1 - \frac{2\gamma\theta_0^2}{\theta_b^2}\right) \tag{12.48b}$$

Using (12.47g) and (12.48b), we set the calibration constant to

$$K_0 = \left|\frac{f_0}{f_0'}\right| = \left|-\frac{\theta_b^2}{2\gamma\theta_0}\right| \tag{12.48c}$$

The condition for validity of (12.47h) is

$$\left|\frac{f_0''}{2f_0}\theta_T^2\right| = \frac{\gamma}{\theta_b^2}\left|1 - \frac{2\gamma\theta_0^2}{\theta_b^2}\right|\theta_T^2 << 1 \tag{12.48d}$$

From (12.47i) and (12.48b), the noise-free error is

$$|\epsilon_{nf}(\theta_T)| = \frac{100(\gamma/\theta_b^2)|1 - (2\gamma\theta_0^2/\theta_b^2)|\theta_T^2}{|1 - (\gamma/\theta_b^2)(1 - (2\gamma\theta_0^2/\theta_b^2))\theta_T^2|} \tag{12.48e}$$

The curve of Figure 12.7a shows the comparator output vs. $|\theta_T|$ from (12.47g) and (12.48b) for the Gaussian pattern case for $\theta_b = 6°$, $\theta_0 = 3°$, and values of θ_T between 0 and θ_b. In Figure 12.7b, a plot of the noise-free error as given by (12.48e) is shown for the same values of θ_0 and θ_b. The upper limit on θ_T for maintaining linearity could be estimated from curves like these for the specified values of beamwidth and offset angle.

The next consideration is the error due to additive noise. To determine that error, we invoke (12.47c) and (12.47c)″, square the latter, and take the square root of its ensemble average. From (12.47c)″

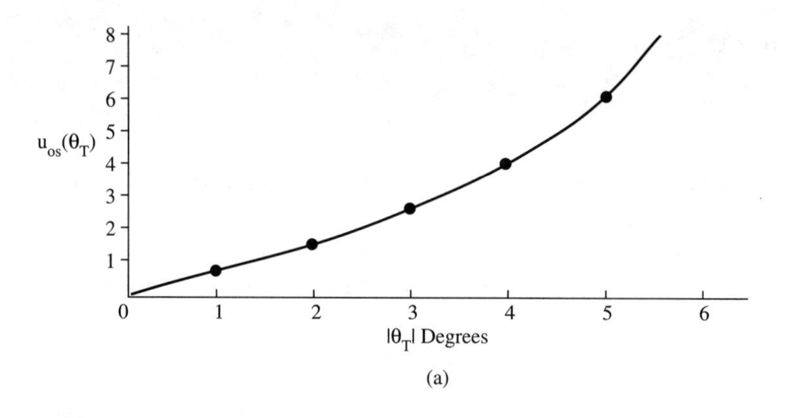

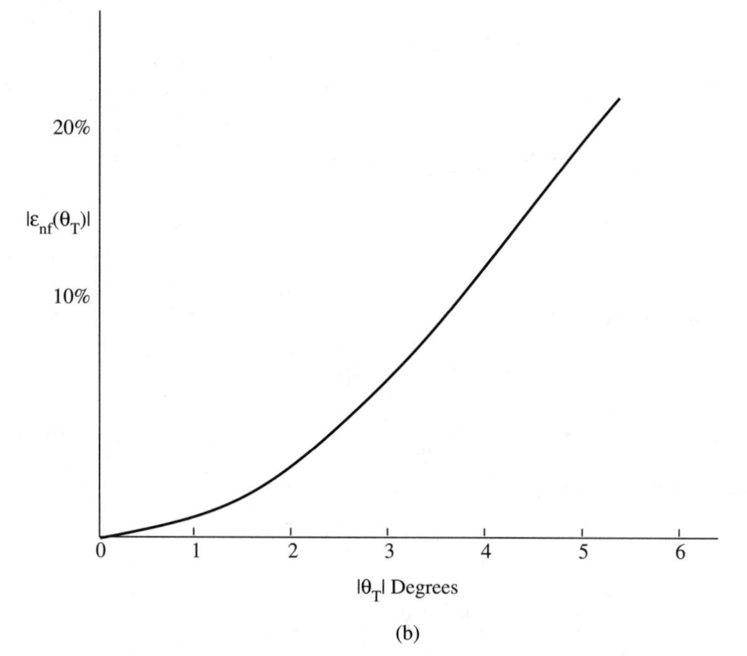

FIGURE 12.7
Amplitude comparison method of angle measurement. (a) Comparator output vs. angle off of boresight. (b) Noise-free error vs. angle.

$$\epsilon_n = \sqrt{\langle u_{on}^2 \rangle} = \frac{K_0}{s_b^2} \sqrt{s_b^2 \langle n_a^2 \rangle + s_a^2 \langle n_b^2 \rangle - 2 s_a s_b \langle n_a n_b \rangle} \tag{12.49a}$$

We rewrite the quantities $|\hat{n}_{0l}|\cos(\psi_{nl} - \psi_{sl})$ appearing in (12.47b) in the form

$$|\hat{n}_{0l}|\cos(\theta_{nl} - \psi_{sl}) = x_{nl} \cos \psi_{sl} + y_{nl} \sin \psi_{sl} = n_{0l} \tag{12.49b}$$

where x_{nl} and y_{nl} are inphase and quadrature components of the noise. The signal phase is arbitrary.

From (12.47c)‴ and (12.49b)

$$\langle n_{\substack{a\\b}}^2 \rangle = \langle (n_{01} \mp n_{02})^2 \rangle = \langle n_{01}^2 \rangle + \langle n_{02}^2 \rangle \mp 2 \langle n_{01} n_{02} \rangle \tag{12.49c}$$

$$\langle n_a n_b \rangle = \langle n_{01}^2 \rangle - \langle n_{02}^2 \rangle \tag{12.49d}$$

If we assume that both noises have the same rms values, then Eqs. (12.49c, d) imply that

$$\langle n_{\substack{a \\ b}}^2 \rangle = 2\sigma_n^2(1 \mp R_{12}) \tag{12.49e}$$

$$\langle n_a n_b \rangle = 0 \tag{12.49f}$$

where $\sigma_n^2 = \langle n_{01}^2 \rangle = \langle n_{02}^2 \rangle$, and $R_{12} = \langle n_{01} n_{02} \rangle / \sigma_n^2$ = cross-correlation function of n_{01} and n_{02} normalized to σ_n^2.

From (12.49a, e, f), the rms error due to noise is

$$\epsilon_n = \frac{K_0}{s_b^2} \sigma_n \sqrt{2(s_a^2 + s_b^2)} \sqrt{1 + [(s_a^2 - s_b^2)/(s_a^2 + s_b^2)] R_{12}} \tag{12.49g}$$

which becomes, with the aid of the second expression in (12.47c)‴ and Eq. (12.47d),

$$\epsilon_n = \frac{2K_0 \sigma_n \sqrt{f^2(\theta_T + \theta_0) + f^2(\theta_T - \theta_0)}}{[f(\theta_T + \theta_0) + f(\theta_T - \theta_0)]^2} \cdots$$

$$\sqrt{1 - \frac{2f(\theta_T + \theta_0) f(\theta_T - \theta_0)}{f^2(\theta_T + \theta_0) + f^2(\theta_T - \theta_0)} R_{12}} \tag{12.49h}$$

Applying the power series expansion used in (12.47e) to (12.49h) and truncating beyond terms of order θ_T^2, we obtain

$$\epsilon_n = \frac{K_0 \sigma_n \sqrt{2} \sqrt{1 + a\theta_T^2}}{2f_0[1 + (f_0''/f_0)\theta_T^2]} \sqrt{1 - R_{12} \frac{(1 + b\theta_T^2)}{(1 + a\theta_T^2)}} \tag{12.49i}$$

where

$$a = \frac{f_0''}{2f_0} + \left(\frac{f_0'}{f_0}\right)^2, \qquad b = \frac{f_0''}{2f_0} - \left(\frac{f_0'}{f_0}\right)^2, \qquad K_0 = \frac{f_0}{f_0'},$$

Specializing to the Gaussian beam as given by (12.48a) and invoking the expressions in (12.48b), we obtain

$$\begin{matrix} a \\ b \end{matrix} = -\frac{\gamma}{\theta_b^2}\left(1 - \frac{2\gamma\theta_0^2}{\theta_b^2}\right) \pm \frac{4\gamma^2\theta_0^2}{\theta_b^4} = \frac{\gamma}{\theta_b^2} + \begin{pmatrix} 6 \\ (-2) \end{pmatrix} \frac{\gamma^2\theta_0^2}{\theta_b^4} \tag{12.49j}$$

$$f_0 = e^{-2\gamma(\theta_0/\theta_b)^2}$$

When the target is on boresight, i.e., when $\theta_T = 0$, the noise-induced error as given by (12.49i) is

$$(\epsilon_n)_{\theta_T=0} = \frac{K_0 \sigma_n \sqrt{2}}{2f_0} \sqrt{1 - R_{12}} = \frac{\sigma_n}{\sqrt{2}|f_0'|} \sqrt{1 - R_{12}} \tag{12.49k}$$

Plots of ϵ_n as a function of R_{12} are shown in Figure 12.8 for $\theta_T = 0$ as given by (12.49k) and (from 12.49i) for a few other values of θ_T less than θ_0 (which is a requirement for viability of this method) for $\theta_0/\theta_b = 0.5$, $\theta_0 = 3°$, and $\sigma_n = \sqrt{2}\,|f_0'|$.

It is evident from the figure that the noise-induced error is greatest when the noises are uncorrelated ($R_{12} = 0$) and decreases with increasing values of R_{12}, becoming very small (zero at boresight) when they are completely correlated ($R_{12} = 1$). In the latter case, at boresight the differencing operation would subtract out the noise with the signal and reduce the effective noise level to zero.

12.3.2 Conical Scanning

Conical scanning is a technique wherein the antenna beam is offset from the boresight axis and scanned continuously around that axis in such a manner that the output signal is roughly proportional to the angle off of boresight. The technique is illustrated in Figure 12.9.

As noted in the figure, $\theta_T'(t)$ is the time-varying angle between 0–T, the vector from the origin to the target and the beam axis 0–B. The cosine of this angle is given by

$$\cos \theta_T'(t) = \cos \theta_0 \cos \theta_T + \sin \theta_0 \sin \theta_T \cos[\phi_0(t) - \phi_T] \tag{12.50a}$$

where θ_T is the target angle relative to the boresight axis, ϕ_T and $\phi_0(t)$ are the azimuth angles of the target and beam axis, respectively, in the coordinate frame whose z-axis is the boresight axis, θ_0 is the offset angle of the rotating beam axis relative to the boresight axis (sometimes called "squint angle"), and the time variation in $\phi_0(t)$ is linear, due to the fixed rate of rotation of the beam axis about the boresight axis.

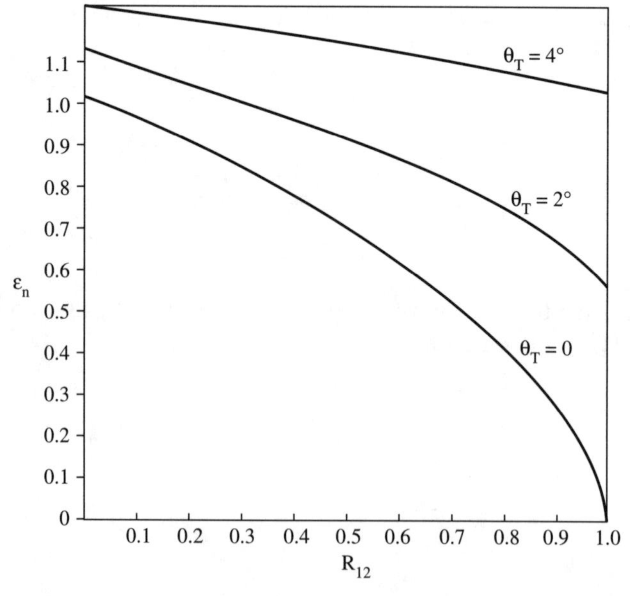

FIGURE 12.8
Noise-induced error in angle measurement by amplitude comparison.

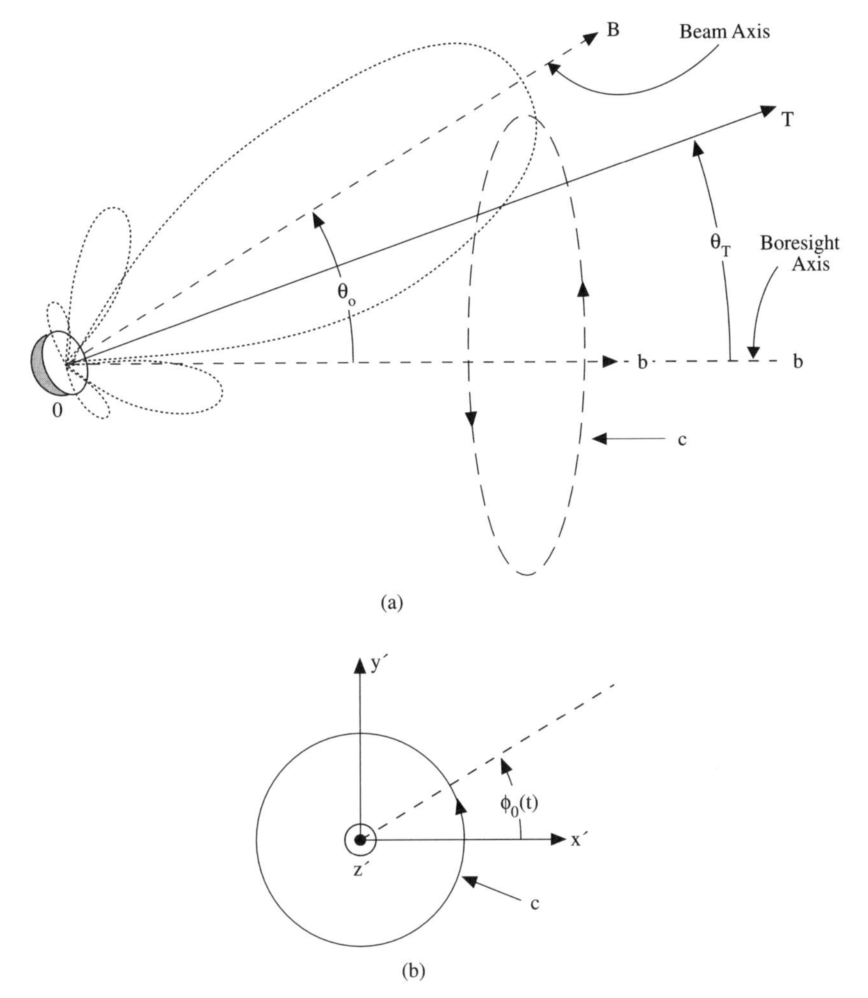

FIGURE 12.9
Illustration of conical scanning. (a) 3-dimensional view. (b) View along boresight axis.

The scanning rate is denoted by R_s and is in radians per second. The scanning function is

$$\phi_0(t) = \phi_{00} + R_s t \tag{12.50b}$$

where ϕ_{00} is the beam's azimuthal angle at $t = 0$.

The beam pattern, whatever its form, is a function of $\sin\theta'(t)$. If we specialize to the Gaussian pattern as given by (12.48a), it is useful to recognize that the function given in (12.48a) is an approximation for a small value of θ, which is $\theta'(t)$ in this case. To allow a larger value of $\theta'(t)$ in the analysis to follow we rewrite the pattern function (noting its time variation and its functional dependence on θ_T) in the form

$$f(\theta_T;\, t) = e^{-2\beta \sin^2[\theta'(t)]} \tag{12.50c}$$

where β is a positive constant.

Substituting (12.50a, b) into (12.50c), we obtain

$$f(\theta_T;\, t) = e^{-2\beta} e^{2\beta[\cos\theta_0 \cos\theta_T + \sin\theta_0 \sin\theta_T \cos(\phi_{00} - \phi_T + R_s t)]^2}$$

$$= A_0 e^{c_1 \cos(\phi_{00} - \phi_T + R_s t) + c_2 \cos[2(\phi_{00} - \phi_T + R_s t)]} \tag{12.50d}$$

where

$$A_0 = e^{-2\beta\{1 - \cos^2\theta_0 \cos^2\theta_T - (\sin^2\theta_0 \sin^2\theta_T/2)\}}$$

$$c_1 = 4\beta \cos\theta_0 \cos\theta_T \sin\theta_0 \sin\theta_T$$

$$c_2 = \beta \sin^2\theta_0 \sin^2\theta_T$$

We can construct a Fourier series from (12.50d) by expanding the exponential quantity in a Taylor series. That series is of the form

$$f(\theta_T;\, t) = A_0[1 + c_1 \cos\eta(t) + c_2 \cos 2\eta(t)$$
$$+ \frac{1}{2}\left(\frac{c_1^2}{2}[1 + \cos 2\eta(t)] + \frac{c_2^2}{2}[1 + \cos 4\eta(t)] + c_1 c_2[\cos\eta(t) + \cos 3\eta(t)]\right)$$
$$+ \cdots] \tag{12.50e}$$

where $\eta(t) = \phi_{00} - \phi_T + R_s t$

If the exponent is assumed to be small and we also assume $c_2 << c_1$ (based on the assumption of small θ_T, which implies small values of both c_1 and c_2 and that $c_2 << c_1$) then we can truncate the series beyond the terms containing cos $2\eta(t)$, with the result

$$f(\theta_T;\, t) = a_0 + a_1 \cos\eta(t) + a_2 \cos 2\eta(t) \tag{12.50f}$$

where

$$a_0 = A_0 + \frac{c_1^2}{4}$$

$$a_1 = A_0 c_1\left(1 + \frac{c_2}{4}\right) = A_0 c_1$$

$$a_2 = A_0\left(c_2 + \frac{c_1^2}{4}\right)$$

The quantity a_0 is the average value of $f(\theta_T; t)$ over the scan; a_1 is the amplitude of the first harmonic, which we will show to be roughly proportional to θ_T and hence acts as a measure of θ_T; a_2 is the amplitude of the second harmonic.

If $f(\theta_T; t)$ as given by (12.50e) is passed through a lowpass filter whose cutoff frequency is between R_s and $2R_s$ radians/second, then only the DC term a_0 and the fundamental term $a_1 \cos\eta(t)$ will be seen at the filter output. The term $a_2 \cos 2\eta(t)$ and higher harmonics will be diminished. For small values of both θ_0 and θ_T, we can invoke the approximations $\cos\theta_T \approx \cos\theta_0 \approx 1$, $\sin\theta_T \approx \theta_T$, $\sin\theta_0 \approx \theta_0$ from which the quantities defined in (12.50d) are approximated by

$$A_0 = e^{\beta\theta_0^2\theta_T^2}$$

$$c_1 = 4\beta\theta_0\theta_T$$

$$c_2 = \beta\theta_0^2\theta_T^2$$
$$c_1^2 = 16\beta^2\theta_0^2\theta_T^2 \tag{12.50g}$$

from which it follows that

$$\begin{aligned} a_0 &= e^{\beta\theta_0^2\theta_T^2} + 16\beta^2\theta_0^2\theta_T^2 = e^{\beta\theta_0^2\theta_T^2} \\ a_1 &= e^{\beta\theta_0^2\theta_T^2}(4\beta\theta_0\theta_T) \\ a_2 &= \beta e^{\beta\theta_0^2\theta_T^2}\theta_0^2\theta_T^2(1 + 4\beta) \end{aligned} \tag{12.50h}$$

If $f(\theta_T; t)$ is passed through the lowpass filter alluded to above, then with the aid of (12.50h)

$$\begin{aligned} f(\theta_T; t) &= a_0 + a_1 \cos \eta(t) \\ &= e^{\beta\theta_0^2\theta_T^2}[1 + (4\beta\theta_0)\theta_T \cos \eta(t)] \end{aligned} \tag{12.50i}$$

Division of a_1 by a_0 [accomplished through an automatic gain control ("AGC") loop] with a properly chosen value of the calibration constant will produce an output equal to the target angle

$$u_0(\theta_T) = K_0 \frac{a_1}{a_0} \approx \theta_T \tag{12.50j}$$

where $K_0 = 1/4\beta\theta_0$.

Linear rectification of $f(\theta_T; t)$ will give us the DC term a_0. Passing $f(\theta_T; t)$ through a bandpass filter centered at R_s and with bandwidth much less than R_s followed by linear rectification of the filter output will give us a_1. Taking the ratio of a_1 to a_0 and multiplying it by K_0 will result in the noise-free output given by (12.50j).

12.3.3 Sequential Lobing

Conical scan, (as illustrated in Fig. 12.9) is merely a particular technique for implementation of the amplitude comparison method discussed in Section 12.3.1. Another technique for that purpose is "sequential lobing." The discussion and analysis in Sections 12.3.1 and 12.3.2 can be used as a springboard to introduce the sequential lobing concept. Instead of being continuously rotated about the boresight axis, the beam is switched sequentially between four azimuthal angles 90° apart (Fig. 12.10). The amplitude comparison technique of Section 12.3.1 is applied between the first and third beam position to form sum and difference signals in a plane containing the boresight axis. Then the same is done between the second and fourth beam positions to form sum and difference signals in another plane containing the boresight axis but orthogonal to the first plane. Then the two outputs are combined to produce a signal proportional to θ_T, the angle off of boresight.

To relate the analysis to follow to the conical scan discussion in Section 12.3.2 we take two readings of $f(\theta_T; t)$ as given by (12.50i), one at $t = t_1$ and the other at $t = t_1 + \pi/R_s$. If we take the ratio of the difference to the sum of these two readings and multiply by a calibration constant K_0 we obtain

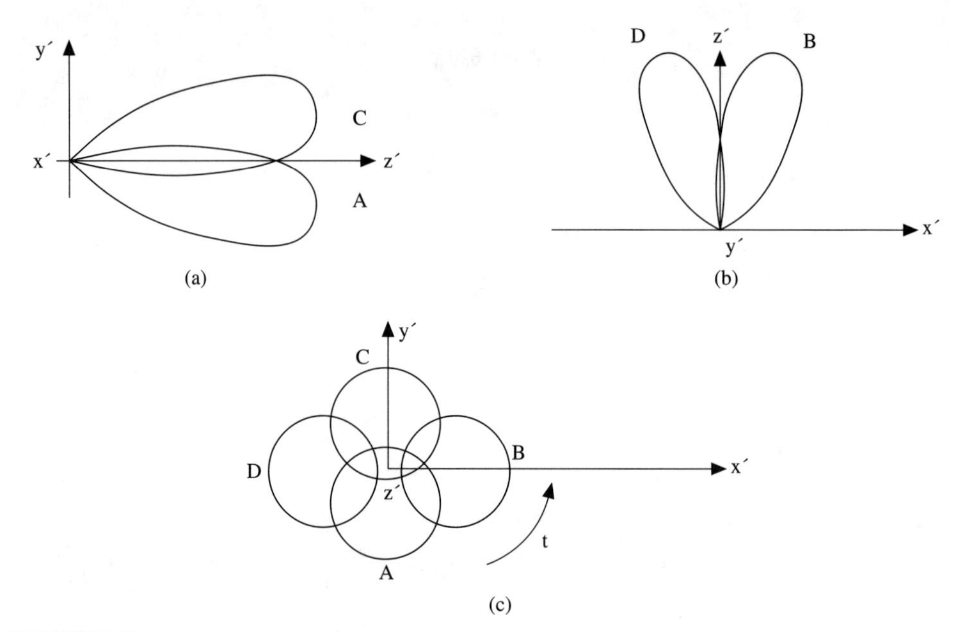

FIGURE 12.10
Illustration of sequential lobing. (a) Up-down beams. (b) Left-right beams. (c) View along boresight axis. A: $t = t_1$; B: $t = t_1 + \Delta t$; C: $t = t_1 + 2\Delta t$; D: $t = t_1 + 3\Delta t$; E: $t = t_1 + 4\Delta t$.

$$u_{01}(\theta_T) = K_0\left\{\frac{(a_0 + a_1\cos\eta_1) - [a_0 + a_1\cos(\eta_1 + \pi)]}{(a_0 + a_1\cos\eta_1) + [a_0 + a_1\cos(\eta_1 + \pi]}\right\} \tag{12.51a}$$

$$= K_0\frac{a_1}{a_0}\cos\eta_1$$

where $\eta_1 = \eta(t_1) = \phi_{00} - \phi_T + R_s t_1$.

This would give us the implementation of the differential amplitude method on a plane containing the boresight axis which we might choose to call the "left-right plane" in a geometry where the boresight axis is nearly horizontal. If we then took another two readings, at $t = t_1 - \pi/2R_s$ and $t = t_1 + \pi/2R_s$, and then performed the same operations, we would have

$$u_{02}(\theta_T) = K_0\left\{\frac{[a_0 + a_1\cos(\eta_1 - \frac{\pi}{2})] - [a_0 + a_1\cos(\eta_1 + \frac{\pi}{2})]}{[a_0 + a_1\cos(\eta_1 - \frac{\pi}{2})] + [a_0 + a_1\cos(\eta_1 + \frac{\pi}{2})]}\right\}$$

$$= K_0\frac{a_1}{a_0}\sin\eta_1 \tag{12.51b}$$

which gives us the implementation of the amplitude comparison method on another plane containing the boresight axis and orthogonal to the left–right plane, which we might call the "up–down plane."

The square root of the sum of the square of $u_{01}(\theta_T)$ and $u_{02}(\theta_T)$ yields

$$u_0(\theta_T) = \sqrt{u_{01}^2 + u_{02}^2} = K_0\frac{a_1}{a_0} \tag{12.51c}$$

which is the conical scan output as given by (12.50j).

The earlier analysis of the amplitude comparison technique in Section 12.3.1 was two-dimensional, and carried the tacit assumption that the target is in the plane in which the analysis is being done. The diagram of Figure 12.6 is based on the notion that the target is in the plane of the paper. In a three-dimensional world, taking the difference between the signals on the two beams in the same plane might produce a zero signal if the target were not in that plane. It may be at a spherical polar angle θ_T off boresight but at an azimuthal angle that is out of plane. This is the case, for example, if we perform the differencing operation given by (12.51a) in the case where $\phi_{00} = 0$, $\phi_T = \pi/2$, $t = 0$ in which case $\eta_1 = \pi/2$ and hence $u_{01} = 0$. This corresponds to the situation in which the target is in the up–down plane where the readings at $t = 0$ and $t = \pi/R_s$ are for the left–right plane. If we perform the differencing operation in the up–down plane in this case, the output $u_{02} = K_0(a_1/a_0)$. If the target is midway between up–down and left–right planes, i.e., $\eta_1 = \pi/2$, then both outputs u_{01} and u_{02} are $K_0(a_1/a_0\sqrt{2})$. The point is that the differencing in only one plane containing the boresight axis does not suffice to produce the desired signal proportional to θ_T. We must perform that operation in two mutually orthogonal planes ("up–down" and "left–right" planes as indicated above) and combine the two outputs as in (12.51c) to be assured that we will receive the signal from the target regardless of its azimuthal angles in the plane normal to the boresight axis. All of the variations of the amplitude comparison scheme are based on that idea.

The error analysis culminating in (12.48e) and (12.49g) will not be repeated here because it would be essentially the same for conical scan or sequential lobing as for the two-dimensional amplitude comparison technique discussed in Section 12.3.1. However, there is one point worth mentioning regarding the noise error as given by (12.49g). For conical scan the correlation between the two noise waveforms separated by 180° depends on the scan rate. If (π/R_s) is large compared to the noise correlation time (reciprocal noise bandwidth), implying a slow scan, then the noise waveforms in the two differencing channels are uncorrelated and $R_{12} = 0$. If (π/R_s) is small compared to noise correlation time, implying a very rapid scan, then the noises in the two channels are highly correlated, i.e., $R_{12} \approx 1$. We have discussed the effect of R_{12} on the noise error for the amplitude comparison technique.

In sequential lobing, the scan rate R_s is not really defined, although we have used it indirectly in this discussion. In that case, the parameter π/R_s should be designated as the "switching time" between samples separated by 180° in azimuth. It is the ratio of the switching time to the noise correlation time that determines the value of R_{12}.

12.3.4 Simultaneous Lobing or Monopulse

Another variation of the amplitude comparison method employs two antenna beams for the up–down plane and two antenna beams for the left–right plane (Fig. 12.11). Since the two signals whose amplitude sums and differences are taken for measurement of the angle off of boresight in the up–down plane are simultaneous, as are the two signals used for the left–right plane, this technique is called "simultaneous lobing." Because it can be done with a single pulse transmitted simultaneously from all four antennas it is often called "monopulse."

The discussion in Section 12.3.1 applies to monopulse in either the up–down or left–right plane. The combining of the outputs in the two orthogonal planes is the same as that described in Section 12.3.3 [Eqs. (12.51a, b, c)]. Hence these discus-

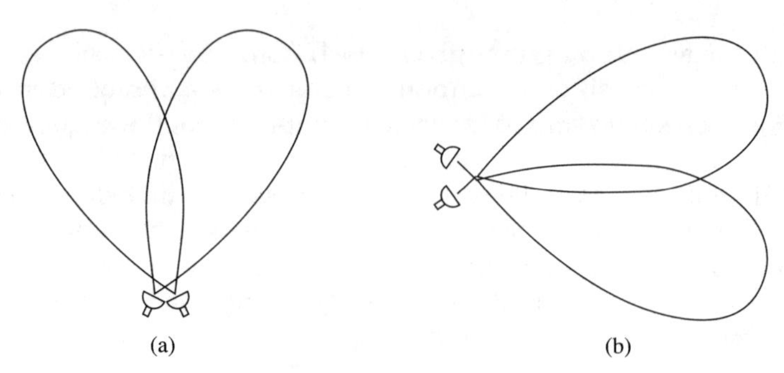

FIGURE 12.11
Simultaneous lobing or monopulse. (a) Left-right beams. (b) Up-down beams.

sions will not be repeated here. The principal differences between monopulse and (either) conical scan or sequential lobing are practical rather than conceptual. As pointed out by Skolnick,[14] monopulse allows faster processing than either of the other methods because it circumvents the problem of pulse-to-pulse amplitude fluctuations, especially where the fluctuations may contain amplitude modulation components near the conical scanning frequency or the sequential lobing rate. This could severely limit tracking accuracy in cases where many pulses must be processed in order to extract the angular error signal that activates the tracking loop.

12.3.5 Phase Comparison or Radar Interferometry*

The basic idea of phase comparison is illustrated in Figure 12.12. Two antennas A_1 and A_2 are separated by a distance d. Radar waves emitted simultaneously by these antennas both strike a target T at an angle θ relative to the normal to $A_1 - A_2$. The wave originating at A_1, striking T and returning to A_1 travels a distance $2r$ in its round trip to target and back. The wave from A_2 travels a greater distance $2r + 2d \sin\theta$ in its round trip to T and back. The electrical path length difference between the target returns at A_1 and A_2 is $(4\pi d/\lambda)\sin\theta$. The two signals are fed into a "phase comparator," a device whose output is proportional to the sine of the phase difference, i.e., to $\sin[(4\pi d/\lambda)\sin\theta]$.

There are certain practical limitations in this method that should be mentioned before we embark on a discussion of its statistical aspects.

If the absolute value of the angle θ is very small, e.g., less than 15°, it can be assumed that

$$\sin\theta \simeq \theta \tag{12.52a}$$

and the approximate comparator output is

$$u_0(\theta) \simeq \sin\left(\frac{4\pi d\theta}{\lambda}\right) \tag{12.52b}$$

* This subsection, including Figures 12.12 and 12.13, is taken nearly verbatim from one of the chapters on radar applications in Raemer,[1] a book published by the author in 1969.

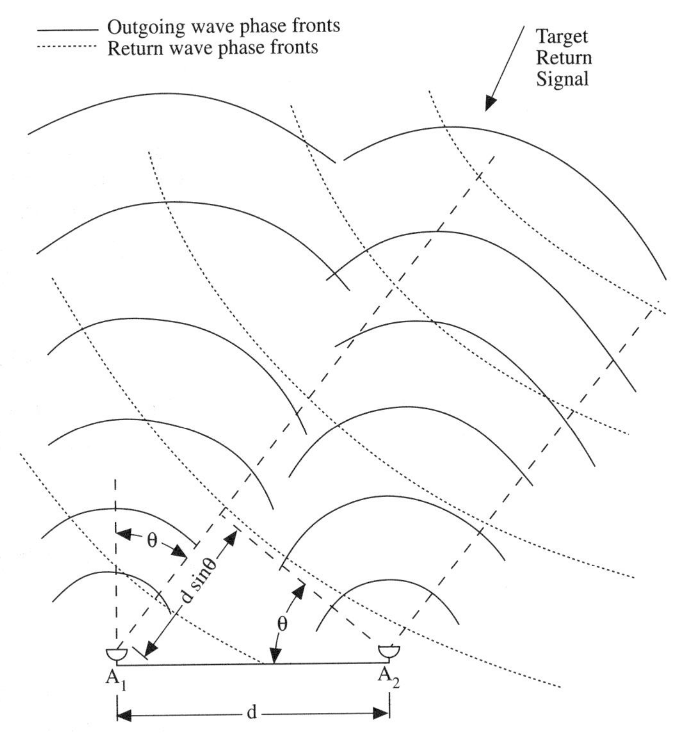

FIGURE 12.12
Basic idea of radar interferometry. Phase difference = $(2\pi d/\lambda)\sin\theta$. From Raemer,[1] Figure 9-8, page 309.

In general with a calibration constant of unity, the comparator output is

$$u_0(\theta) \approx \sin\left(\frac{4\pi d}{\lambda}\sin\theta\right) \tag{12.52c}$$

To illustrate some of the limitations of this technique, we present in Figure 12.13 the plots of $u_0(\theta)$ versus θ for two cases: (a) spacing small compared to wavelength ($d = 0.1\lambda$) and (b) spacing comparable to wavelength ($d = \lambda$). In the case $d = 0.1\lambda$, the variation with θ is smooth and monotonic increasing, and it is nearly linear for $\theta \leq 40°$.

The striking point about the curve corresponding to the case $d = \lambda$, obvious from Eq. (12.52a), is the oscillatory variation of $u_0(\theta)$ with θ: An ambiguity in angle measurement appears as soon as $(4\pi d/\lambda)\sin\theta$ exceeds $\pi/2$. A practical measurement technique requires that the variable being measured be a single-valued function of the output of the measuring instrument. This is not the case if $(4\pi d/\lambda)\sin\theta > \pi/2$. Thus, to preclude ambiguity in the measurement of angle, we must have the conditions

$$|\theta| < \frac{\pi}{2} \tag{12.52d}$$

$$|\sin\theta| < \frac{\lambda}{8d} \tag{12.52e}$$

For a given wavelength and a given separation the conditions (12.52d) and (12.53e) place limits on the size of the angle that can be measured.

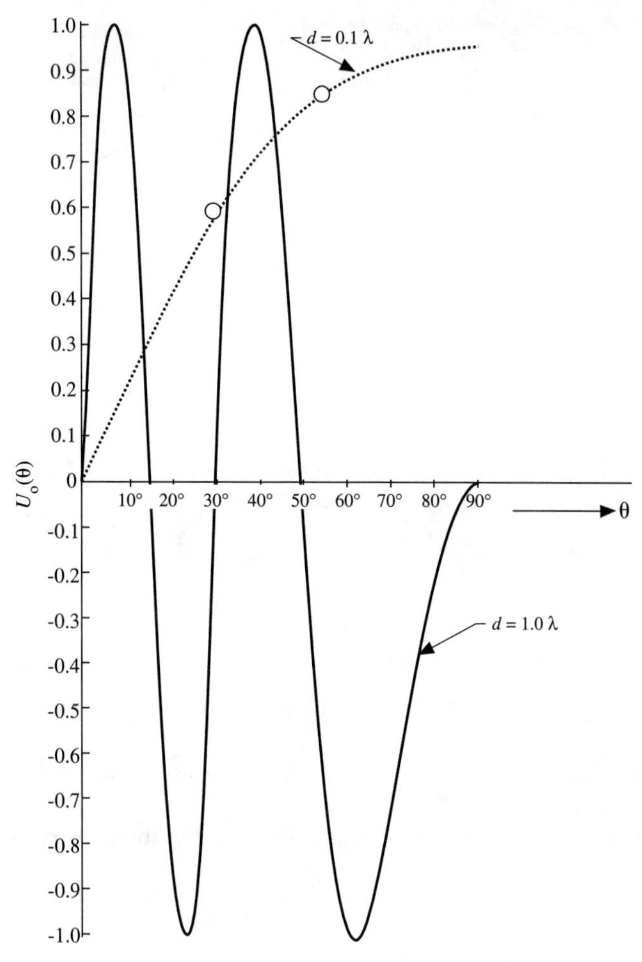

FIGURE 12.13
Phase comparator output versus target angle. From Raemer,[1] Figure 9-9, page 311.

Now let us study noise as a factor in limiting measurement accuracy. We will circumvent the precise application of maximum likelihood estimator theory to the problem of phase difference measurement, which has not been covered in this text. In lieu of processing for optimum phase difference measurement, we will base our discussion on measurement of the phase difference between two signals, each of which is processed for (approximately) optimum phase measurement, according to the discussion in Section 12.1.2. This implies that we would first do something as close as possible to matched filtering on the signals entering the two antenna channels (assuming that the two antennas are connected to the same RF head, so that the noise waveforms in the two channels are identical) and then take the phase difference between the two matched filter outputs. If the signal is a pure sinusoidal pulse of length τ_p, the signal-plus-noise matched filter output complex envelopes of the receiver channels corresponding to A_1 and A_2 are, respectively

$$[\hat{v}_0(t)]_{A_1} = [\hat{s}_0 + \hat{n}_0(t)]e^{j\omega_0 t} \tag{12.52f}$$

$$[\hat{v}_0(t)]_{A_2} = [\hat{s}_0 e^{j(-(\pi/2)+(4\pi d/\lambda)\sin\theta)} + \hat{n}_0(t)]e^{j\omega_0 t} \tag{12.52g}$$

where $\hat{s}_0 = a_0 e^{j\psi_0}$.

The output of an ideal phase comparator whose input complex envelopes are $[\hat{v}_0(t)]_{A_1}$ and $[\hat{v}_0(t)]_{A_2}$ is

$$u_0(\theta) = \frac{\lambda}{4\pi d|\hat{s}_0|^2\tau_p} Re\left\{\int_0^{\tau_p} dt'[\hat{v}_0(t)]_{A_1}[v_0(t)]_{A_2}\right\}$$

$$= \frac{\lambda}{4\pi d}\sin\left(\frac{4\pi d}{\lambda}\sin\theta\right) \tag{12.52h}$$

$$+ \frac{\lambda}{4\pi d|\hat{s}_0|\tau_p} Re\left[\int_0^{\tau_p} dt'\,(e^{j\psi_0}\hat{n}_0^*(t') + je^{-j\psi_0}\hat{n}_0(t')e^{(j4\pi d/\lambda)\sin\theta})\right.$$

$$\left. + \frac{1}{|\hat{s}_0|}\int_0^{\tau_p} dt'|\hat{n}_0(t')|^2\right]$$

If $|(4\pi d/\lambda)\sin\theta| << \pi/2$, and $\theta << \pi/2$, then (12.52h) takes the approximate form

$$u_0(\theta) \simeq \theta + \frac{\lambda}{4\pi d|\hat{s}_0|\tau_p} Re\left\{\int_0^{\tau_p} dt'[\hat{v}_0(t)]_{A_1}[\hat{v}_o(t)]_{A_2}\right\}$$

$$= \frac{\lambda}{4\pi d}\sin\left(\frac{4\pi d}{\lambda}\sin\theta\right) \tag{12.52i}$$

$$+ \frac{\lambda}{4\pi d|\hat{s}_0|\tau_p} Re\left[\int_0^{\tau_p} dt'\,(e^{j\psi_0}\hat{n}_0^*(t') + je^{-j\psi_0}\hat{n}_0(t')e^{(j4\pi d/\lambda)\sin\theta})\right.$$

$$\left. + \frac{1}{|\hat{s}_0|}\int_0^{\tau_p} dt'|\hat{n}_0(t')|^2\right]$$

where the nonlinear terms in θ can be neglected if the above assumptions hold. If the SNR is sufficiently high to justify neglect of the noise–noise term in (12.52i), then the rms error in the measurement is approximated by a quantity proportional to the square root of the ensemble average of the square of the quantity in braces. This calculation, which is somewhat cumbersome, is left as an exercise for the reader. For wideband noise and high SNR, the rms error is roughly proportional to the rms noise, as can be shown by detailed evaluation of (12.52i) with the noise–noise term neglected.

REFERENCES

1. Raemer, H.R., "Statistical Communication Theory and Applications." Prentice-Hall, Englewood Cliffs, NJ, 1969; Chapter 8, also Sections 9.5, pp. 295–318.
2. Davenport, W.B., and Root, W.L., "Theory of Random Signals and Noise." McGraw-Hill, New York, 1958, Chapter 14.
3. Van Trees, "Detection, Estimation and Modulation Theory," Part I. Wiley, New York, 1958, Chapter 2, or Part III, 1971, Chapter 14.
4. Helstrom, C.W., "Probability and Stochastic Processes for Engineers," 2nd ed. Prentice-Hall, Englewood Cliffs, NJ, 1991, Chapter 7, in particular Sections 7.1 through 7.4, pp. 480–560.
5. Scharf, L.L., "Statistical Signal Processing." Addison-Wesley, Reading, MA, 1991, Chapter 6.
6. Helstrom, C.W., "Elements of Signal Detection and Estimation." Prentice-Hall, Englewood Cliffs, NJ, 1995, Chapter 6.

6. Skolnick, M.I., "Introduction to Radar Systems," 2nd ed. McGraw-Hill, New York, 1980, Chapter 5 (on tracking radar).
7. Eaves, J.L. and Reedy, E.K. (eds.), "Principles of Modern Radar." Van Nostrand Reinhold, New York, 1987, Part 5 (on tracking radar); Chapter 17 (on range tracking) by J.A. Bruder; Chapter 18 (on angle tracking) by G.W. Ewell and N.T. Alexander; Chapter 19 (on Doppler tracking) by E.V. Morris.
8. Barton, D.K., "Modern Radar System Analysis." Artech, Norwood, MA, 1988, Chapter 8 (on angle measurement and tracking); Chapter 9 (on range and Doppler measurement); Chapter 10 (on tracking considerations).
9. Levanon, N., "Radar Principles." Wiley, New York, 1988, Chapter 9.
10. Carpentier, M.H., "Principles of Modern Radar Systems." Artech, Norwood, MA, 1988, Chapter 6 (on angle measurement).
11. Spiegel, M., "Mathematical Handbook." McGraw-Hill, New York, 1968.
12. Barton, D.K., "Modern Radar System Analysis." Artech, Norwood, MA, 1988, pp. 381–382; in particular, Table 8.1 on p. 381; also pp. 425–432.
13. Barton, D.K., "Modern Radar System Analysis." Artech, Norwood, MA, 1988, Figures 9.1.1 on p. 426, 9.1.2 on p. 427, and 9.1.3 on p. 428.
14. Skolnick, M.I, "Introduction to Radar Systems," 2nd ed. McGraw-Hill, New York, 1980, Section 5.4, pp. 160–167.

PROBLEMS

PROBLEMS FOR CHAPTER 1

Note: In all problem sets, radars are monostatic unless otherwise indicated. Transmit and receive antennas are the same unless otherwise indicated.

*Problem 1.1**

Calculate the round-trip travel time of a radar pulse originating on earth to the following bodies

(a) Moon (3.84×10^8 m)
(b) Mars (5.3×10^{10} m);
(c) Jupiter (5.8×10^{11} m)

Problem 1.2

(a) An airborne radar moving horizontally at 500 mph irradiates a target moving at 60° to the vertical at a speed of 400 mph. Both objects are moving in a plane parallel to the paper. If the radar frequency is 9 GHz (transmitted wave), find the frequency of the return wave.

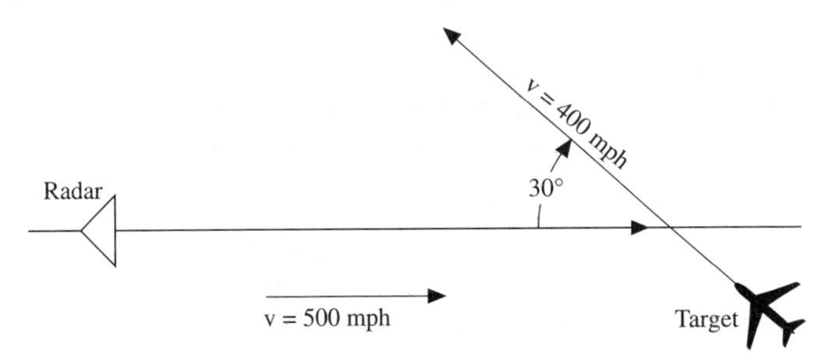

(b) Same as Part (a) but the target has now reversed its motion and is moving in the direction opposite to its original direction.

* From S. A. Hovanessian, "Radar System Design and Analysis," Artech, Norwood, MA, 1984, page 14.

Problem 1.3

The situation in Problem 1.2 occurs at $t = 0$. Redo Problems 1.2(a) and 1.2(b) 20 sec later, assuming the radar can still get return signals from the target. (For purposes of preservation of the validity of the problem at that time, assume isotropic antenna and scattering patterns so the target never gets to be "outside the beam." In this case we sacrifice realism to focus on a simple principle, as we often do in "academic exercises.") Assume the initial distance between radar and target to be 5 km.

Problem 1.4

A coherent monostatic airborne radar can measure range and Doppler of a target directly. Consider an airborne radar in horizontal flight at an altitude h_R illuminating a stationary target on the ground at a frequency f_0. The position vector of the radar as a function of time is

$$\mathbf{r}_R(t) = \hat{x}(x_{RO} + v_x t) + \hat{y}(y_R) + \hat{z}(h_R)$$

where $\hat{x}$, $\hat{y}$ and $\hat{z}$ are unit vectors in the eastward, northward, and upward vertical directions, respectively. The position vector of the target is

$$\mathbf{r}_T(t) = \hat{x}x_T + \hat{y}y_T + \hat{z}(0)$$

The curvature of the earth is neglected. Derive an algorithm to determine the unknown target coordinates (x_T, y_T) from the known parameters x_{RO}, v_x, y_R, h_R and f_0 and the measured range $R(t)$ and Doppler frequency $D(t)$ at an arbitrary instant of time.

Problem 1.5

Use the algorithm derived in Problem 1.4 to determine the location of the target if $f_0 = 1$ GHz, $x_{RO} = 0$, $v_x = 300$ mph, $h_R = 2$ miles, and $y_R = 1$ mile. The range and doppler readings at time $t = 0$ are $R(0) = 2.5$ miles and $D(0) = +100$ Hz. Does an answer to this problem necessarily exist and if so, is it unique? What does it mean if you pose the problem, specifying $R(0)$ and $D(0)$ arbitrarily and then find that no solution exists?

Problem 1.6

Generalize the algorithm to include the possibility that the target is at a known altitude h_T rather than at ground level and moving northward at constant speed v_{yT}, starting at a position (x_T, y_{TO}, h_T) at time $t = 0$. In this case could the algorithm be used to determine the target trajectory from the known parameters x_{RO}, v_x, y_R, h_R, v_{yT}, h_T, and f_0 and the range and Doppler measurements $R(t)$ and $D(t)$ as functions of time? Would this be possible if v_{yT} and h_T were not known quantities?

PROBLEMS FOR CHAPTER 2

*Problem 2.1**

A ground-based air-detection radar has the listed characteristics. Determine the received power and signal-to-noise ratio for a target with cross-section of 5 m^2 at distances of 50 and 200 km.

P_T = peak transmitted power = 1 MW;
pulse duration τ_p = 1 μsec; receiver bandwidth B_R = 1 MHz;
effective aperture area = 10 m^2; noise figure = 10 dB;
wavelength λ = 25 cm

Problem 2.2

If a radar of 1 MW peak power and peak antenna gain of 30 dB irradiates a target with RCS = 1 m^2 with a 10 μsec pulse at 1000 km range, what is the energy density (energy/area) of the return pulse? What is the elapsed time of back-and-forth pulse travel?

Problem 2.3

The radar receiver in Problem 2.2 now has an antenna different from the transmitting antenna and the receiving antenna has a circular aperture with an (effective) radius of 5 m. The "effective noise temperature" T_e is the product of the noise figure and the absolute temperature. If the receiver bandwidth is 1.5 times the reciprocal pulse duration, what value of T_e will give SNR = 1?

Problem 2.4

Find the maximum detection range of a particular pulsed radar for a target with RCS = 1 m^2. The radar characteristics are P_T = 1 kW, peak gain of antenna = 20 dB, effective noise figure = 8 dB, receiver bandwidth = reciprocal of pulse duration, the latter being 2 μsec. Radar frequency = 10 GHz. Successful detection requires SNR ≥ 3 dB.

Problem 2.5

Same parameter values as Problem 2.4, except that (1) the target range is now specified as 100 km; (2) successful detection now requires that the ratio of pulse energy (only 1 pulse) to noise power density must be at least 3 dB. Plot a curve (a *calibrated* rough sketch will do) showing the variation of peak power with pulse duration that will just barely meet this requirement.

Problem 2.6

Consider a monostatic radar in which a 3 dB SNR at the receiver is needed to detect a target. Target RCS = 0.5 m^2. Transmitting and receiving antennas are the same. The *power density* of the transmitted signal *at the target* is 1 μW/m^2 if the target is at 10 km range and at boresight. The peak antenna gain is 20 dB. Receiver band-

* From FT Ulaby, RK Moore and AK Fung, "Microwave Remote Sensing," Volume II, Artech House, Norwood, MA, 1982. Problem 7.1, p. 554, with some variations on problem statement and parameter values.

width = 300 MHz; radar frequency = 3 GHz. Noise figure = 8 dB. Find the range r_{max} at which the target is barely detectable.

Problem 2.7

In Problem 2.6, all parameters are the same except that the effective aperture area of the antenna is assumed fixed while the frequency increases by 50%. What is the effect on r_{max}?

Problem 2.8

A bistatic radar is illustrated below.

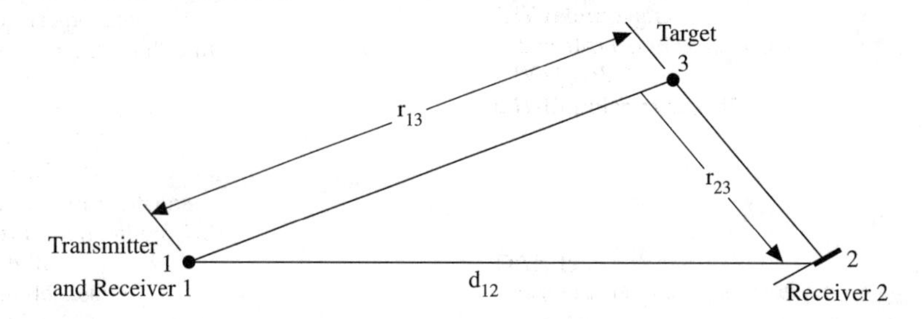

Note that there is also a receiver at the transmitter location, using the same antenna as the transmitter. Given the following parameters, find the relationship between r_{13}, r_{23}, and d_{12} that will result in the same power received at Points 1 and 2.

Transmitted power $P_T = 1$ kW
Both antennas have isotropic coverage (unrealistic but for ease of solution)
The target scatters isotropically
Radar wavelength = 3 cm.

Problem 2.9

What is the effect of doubling the transmitted power and halving the radar wavelength on the answer to Problem 2.8? Explain. All other conditions and parameters are the same.

Problem 2.10

Redo Problem 2.8 if the target still scatters isotropically but the antennas both have one-way power patterns of the forms $e^{-\beta(\theta/\Delta\theta)^2}$, where z_1 and z_2 axes are vertical as shown and where $\Delta\theta$ in each case is the angle θ at which the power drops to 1/2 of its peak value (half-power points or −3 dB points). Let $\Delta\theta_1 = \Delta\theta_2$; what is the value of β?

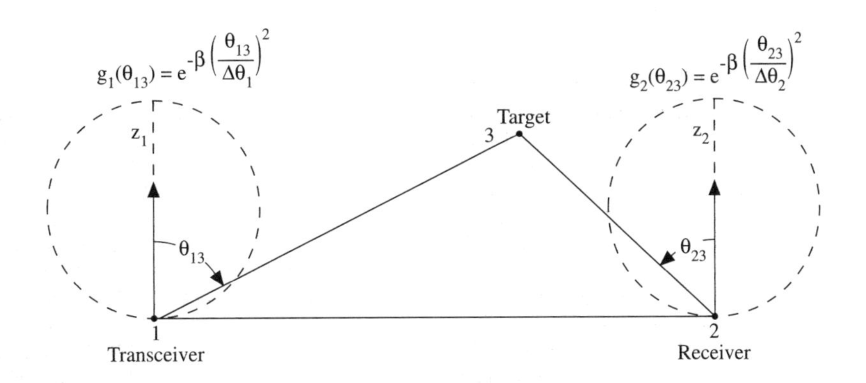

Problem 2.11

A ground-based monostatic pulsed radar has the following parameters and conditions:

Antenna same for transmit and receive
RCS = 0.5 m^2
A_{e0} = 100 cm^2
Transmitter at $x = y = z = 0$
Target at peak of beam and located at x = 20 km, y = 0, z = 5 km
f_0 = 30 GHz
Detection threshold = 0.2 micromicrowatt

(a) Target is barely detectable at receiver if it is at the peak of the antenna beam. Find the power density of the radiation incident on the target.
(b) Calculate and plot the maximum detection range vs. radar frequency from 500 MHz to 50 GHz (on log scale).

Problem 2.12

A particular search radar is required to have a SNR of 5 dB at the output of a matched filter in the receiver. (It was noted in Section 2.8 that this SNR is the ratio of signal energy to noise power density.) Find the minimum power aperture (see Chapter 2, Section 2.10) required to meet this specification for a 1-m^2 target at a range of 10 km if the noise figure (including system losses lumped into the noise figure) is 10 dB, the absolute temperature is 293°, peak antenna gain is 20 dB, and dwell time is 10 sec.

Problem 2.13

Given the radar system parameters of Problem 2.12, including the 6 dB required output SNR for detection, plot

(a) the detection range vs. log of the power aperture for a dwell time of 10 sec
(b) the detection range vs. log of the dwell time for the power aperture obtained in Problem 2.12.

PROBLEMS FOR CHAPTER 3

Problem 3.1

In a radar receiver, the mixer input–output characteristic has the form of Eq. (3.22), where $s_{RF}(t)$ is given by (3.20) and $s_{LO}(t)$ by (3.19). The parameter values are

In (3.19); $A_{LO} = 10$, $\omega_{RF} = 2\pi(10^9)$, $\omega_{IF} = 2\pi(10^6)$
In (3.20); $ap(t) = 1$, $\psi(t) = 0$ (CW waveform)
In (3.22);$a_0 = 0$, $a_1 = 1$, $a_2 = 1$, $a_3 = 0.3$, $a_4 = 0.2$, $a_5 = 0.1$, $a_k = 0$ for $k > 5$

The IF bandwidth is 30% of the IF (between -3 dB points).

Find the intermodulation products passed by the IF amplifier.

Problem 3.2

In Problem 3.1, all parameters are the same except that the noise waveform of the form (3.23) is added to the input [the input is now $(s_{RF} + n_{RF} + s_{LO})$]. Find the additional spurious signals in the IF amplifier output that are due to the input noise. Discuss *qualitively* the overall effect on the output SNR of the mixer if the input SNR is 6 dB.

PROBLEMS FOR CHAPTER 4

Problems 4.1, 4.2, 4.3, 4.4, 4.5

Consider a monostatic radar with the following parameters;

$$P_T = 10 \text{ kW}$$

$$G_0 = 15 \text{ dB}$$

$$f_0 = 6 \text{ GHz}$$

$$\sigma_0 = 2 \text{ m}^2$$

$$\tau_p = 10^{-6} = 1\ \mu\text{sec}$$

Transmitted signal—Train of 5000 equally spaced identical rectangular RF pulses; entire train transmitted in 10 sec
Noise—additive to signal; power $= 10^{-12}$ W
Noise bandwidth = 5 MHz

The receiver designer wants to construct a linear filter that maximizes the SNR at the input to the nonlinear stages (i.e. at the IF output). He or she assumes that the atmosphere, target scattering, and receiver do not distort the signal shape and designs the filter accordingly. Target detection requires SNR of 3 dB at the output of the filter.

Problem 4.1

The designer knows the target range (*if* the target is present) and the RF phase exactly and knows the target to be stationery and incorporates this knowledge into the filter design. What is the maximum achievable detection range r_{max}?

Problem 4.2

More realistically, the RF phase is unknown to the designer. What must be done to account for this? What is the effect on r_{max}? Explain your answer.

Problem 4.3

Conditions same as Problem 4.1, but *without the designer's knowledge*, the target has moved from range r_0 (the assumed range value) to range $r_0 + \Delta r$. If $r_0 =$ 10 km, what is the effect on the output SNR if

(a) $\Delta r = +10$ m
(b) $\Delta r = -20$ m
(c) $\Delta r = +100$ m
(d) $\Delta r = -300$ m

Problem 4.4

Again without the designer's knowledge, each return pulse has been (say, due to propagation effects)

(a) *reduced* in duration by 25%, its shape remaining intact and its amplitude remaining intact.

(b) *increased* in duration by 25%, its shape remaining intact and its amplitude being reduced by 50%. What is the effect of changes (a) and (b) on the SNR at the filter output and on the maximum attainable detection range r_{max}?

Problem 4.5

In Problem 4.1, the designer *thinks* the target is stationary, but it is *actually* approaching the radar at a relative speed $V_t > 0$. Redo Problem 4.1 for $V_t = 50$, 100, 200, 300, 400, and 500 miles/hour.

Problem 4.6

In Problem 4.5, what is the effect of the unknown target motion on the minimum required transmitted power $(P_T)_{min}$ and on the minimum detectable target cross-section $(\sigma_0)_{min}$ where in each case all *other* parameter values are those given for Problems 4.1 through 4.5? The relative speed values are those given in Problem 4.5. Let the range r be 50 km.

Problem 4.7

Plot rough-sketch curves (with some crude calibrations) of (a) r_{max} for fixed (P_T, σ_0); (b) $(P_T)_{min}$ for fixed (r, σ_0); and (c) $(\sigma_0)_{min}$ for fixed (r, P_T), all vs. V_t from V_t $= -500$ to $+500$ miles/hour; where in each case all fixed values are those given for Problems 4.1 through 4.5 and in cases (b) and (c), $r = 50$ km.

Problem 4.8

Extend Problem 4.5 to include the cases where (again *without* the designer's knowledge),

(a) the pulses have been distorted and now have a Gaussian shape with parameter values as shown. Other parameter values are as given for Problems 4.1 through 4.5.

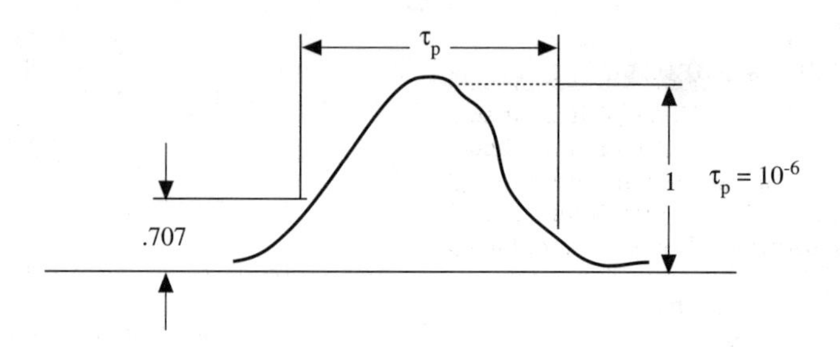

(b) Pulse shape and duration are as given for Problems 4.1 through 4.5 but the PRF has been decreased by 1% of its original value.

(c) Same as (b) but the decrease in PRF is now 2%.

Problem 4.9

Derive Eq. (4.88 g) from the rectangular pulse function $p(t) = u[t + (\tau_p/2)] - u[t - (\tau_p/2)]$, where $u(t)$ is the unit step function.

Problems 4.10, 4.11, 4.12, 4.13

Assume: monostatic radar, point target
Equivalent noise temperature $= T^0 F_N = 1200$
Radar frequency $= 10$ GHz
Peak gain of antenna $= 30$ dB

Peak transmitter power P_T—unspecified
Target RCS σ_0—will be specified

Transmitted signal—uniformly spaced train of rectangular RF pulses:
Pulse duration $\tau_p = 1$ μsec
PRF $= 1000$ pps
Pulse train "goes on forever"

No waveshape distortion

Angular patterns neglected, i.e., assume target at peak of beam and radar at peak of scattering pattern.

The objective is to detect a target with a given RCS. The *actual* target present is at a range of 100 km and its RCS is 1 m^2. It is *not* moving with respect to the radar. An attempt is made to design a matched filter at the IF output in order to maximize the SNR at the input to the detection operation. Successful detection requires an SNR of at least 6 dB at that point.

The designer of the detection scheme knows that (1) the range is between 85 and 115 km; (2) the line-of-sight velocity is between $-|V|_{max}$ km/sec and $+|V|_{max}$ km/sec. The strategy for detection can be

(a) design a bank of matched filters centered at particular ranges ("range gates") covering the possible ranges from 85 to 115 km.

or

(b) use one range gate and sweep it over the 85–115 km range spread searching for the target

or

(c) some combination of these strategies.

In general, doppler is not precisely known; therefore (a), (b), or (c) must be applied with the filter tuned to different frequencies. This means that *in general* techniques analogous to (a), (b), or (c) must also be applied with respect to doppler

Problem 4.10

$|V|_{max} = 0.001$; *RF phase precisely known*. Assume only 1 filter is allowed. The target must be detected within 2 minutes. How much transmitter power P_T is required? Explain how you got your answer.

Problem 4.11

Again $|V|_{max} = .001$, *RF phase precisely known*. Assume $P_T = 100$ kW. A maximum of 10 filters is allowed. How long will it take to detect the target? Explain.

Problem 4.12

Redo Problems 4.10 and 4.11 with $|V|_{max}$ increased to 0.3 km/sec.

Problem 4.13

Make the following changes in Problems 4.10 and 4.11, all other parameters and conditions remaining the same:

(1) RF phase now completely *unknown*
(2) In lieu of a pulse train, we have a *single* pulse of unspecified duration.
(3) $P_T = 10$ kW
(4) 20 filters are allowed.

How long must the pulse be for target detection?

Problem 4.14

(*Note:* This problem and the next one will be used as a basis for other problems in later chapters.)

A ground-based surveillance radar with antenna pattern coverage that is azimuthally omnidirectional looks out at an aggregate of aircraft targets at a fixed time t_0. The targets at $t = t_0$ are at coordinates (x, y, z) relative to the radar at the origin of coordinates. (Targets are labeled $T1, T2, \ldots$, etc. All coordinates are given in kilometers)

$$T1: (3, 2, 2),\quad T2: (1, 4, 3),$$
$$T3: (4, 6, 2),\quad T4: (5, -3, 1),\quad T5: (-3, -4, 1.5).$$

The radar parameters are: $P_T = 250$ kW, $G_0 = 10$ dB, $f_0 = 2$ GHz, $\tau_p = 3$ μsec, PRF = 400 Hz, $T^0F_N = 1300°$, losses = 8 dB. The RCS values of all of the targets are approximated as 2 m^2 and the elevation beam is wide enough to justify considering all of the targets to be at the peak of the elevation beam. Assume that no other targets are in the radar's field of view.

A bank of range gates all of the same width is set up to cover the ranges from 0.5 to 10 km. Note that a range gate is really a matched filter tuned to a time delay corresponding to the desired range. Set the range gate widths to maximize their output SNR for a single pulse. Assume coherent integration of 400 pulses. Plot the distribution of range gate output SNR values in dB from 0.5 to 10 km at time t_0 (like an A-scope display).

Problem 4.15

Assume same conditions and parameter values as in Problem 4.14 except that the PRF is increased to 8 kHz, while P_T is reduced to 1.25 kW. Assume that the targets are all in straight-and-level flight, with x and y velocity components (given in meters/second) as follows:

$T1$: (100, 0), $T2$: (50, 200),
$T3$: (−150, 30), $T4$: (−175, −90), $T5$: (−100, −100)

Set up a bank of Doppler filters all of the same bandwidth. The filter bank covers the Doppler frequency region between −3 and +3 kHz. The output of each range gate is driven through the entire bank of Doppler filters. Plot (from −5 to +5 kHz) the distribution of Doppler filter outputs for each range gate that contains target signals at time t_0.

PROBLEMS FOR CHAPTER 5

Problems 5.1, 5.2, 5.3, 5.4

Repeat Problems 4.10, 4.11, 4.12, 4.13 with the following changes:

Instead of requiring an SNR of at least 6 dB at the detection decision point, we now require an SNR large enough to attain the following detection probabilities and false alarm rates

(1) $P_d = 98\%$, $P_{fa} = 10^{-12}$

(2) $P_d = 90\%$, $P_{fa} = 10^{-8}$

(3) $P_d = 85\%$, $P_{fa} = 10^{-4}$

Otherwise all conditions are the same as in Problems 4.10 through 4.13.

Note: You *need not* redo Problems 4.10 through 4.13 to do the analogous Problems 5.1 through 5.4. All you need to do is (in each case) look at your answer and determine how that answer would change due to the change in required SNR incurred by each of the specifications (1), (2), or (3). You can get those new SNR's by examining the curves in Figures 5.12 and 5.13, so the amount of additional work required is minimal.

Problems 5.5, 5.6, 5.7

The following parameters are fixed

$$F_N = 6 \text{ dB (noise figure)}$$
$$k = 1.38(10^{-23})$$
$$T^0 = 290°\text{k}$$
$$c = 3(10^8)$$

Radar is monostatic *unless* otherwise indicated.

Problem 5.5

A CW radar has the following parameters:
$P_T = 100$ kW, $G_0 = 10$ dB; detection threshold $= 6$ dB SNR at IF output; $f = 3$ GHz, Receiver bandwidth $B = 30$ MHz

(a) Find the maximum detection range for a 0.2-m^2 target.
(b) How large a target (minimum required RCS) can be detected at a range of 20 km if the target is at the peak of the beam?
(c) What is the effect on the answer to (a) of
  (1) halving the receiver bandwidth
  (2) increasing the antenna's peak effective aperture area by a factor of 3
  (3) doubling the radar frequency.

All other parameters remain the same in each case (1), (2), or (3). You can give your answer in percentage change or fractional change.

Problem 5.6

Consider the bistatic radar shown in the diagram. Transmitter, receiver and target are at points 1, 2, 3, respectively

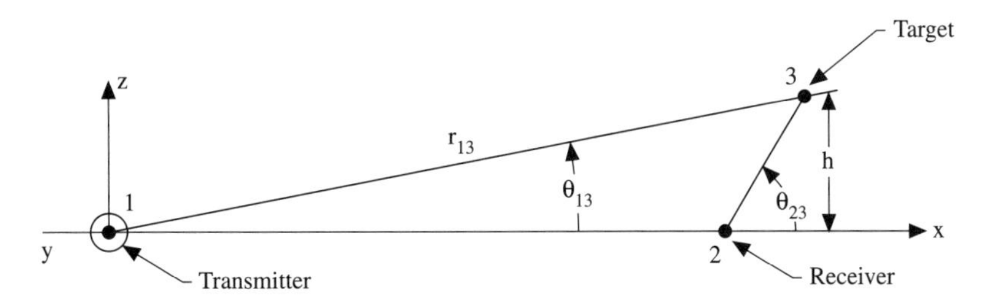

Transmitter at $x_1 = y_1 = z_1 = 0$
Receiver at $x_2 = d_{12}, y_2 = 0, z_2 = 0$
Target at $x_3 = d_{13}, y_3 = 0, z_3 = h$
The receiver's antenna pattern is approximated as *isotropic*. That of the transmitter has a (one-way power relative gain) pattern of the approximate form:

$$g(\theta) = 0.1 + 0.9\cos^2\theta; \quad 0 \le \theta \le \pi/2$$
$$= 0; \quad \pi/2 < \theta \le \pi$$

(The peak of the beam is in the *horizontal* direction, i.e., θ is the angle relative to the x-direction in the x–z plane.)

(a) Find the power density (power per unit area) incident on the target if $d_{12} = 10$ km, $d_{13} = 15$ km, $z_3 = 2$ km, and $P_T = 10$ kW, $f = 6$ GHz, $B = 300$ MHz.
(b) Same as (a) but in this case $z_3 = 5$ km.
(c) Find the SNR for the signal received at Point 2 with the parameter values given in Part (a) and $\sigma = 1\ \text{m}^2$ (isotropic target scattering).
  (1) at the IF output
  (2) at the output of a matched filter whose input is the IF output, given that the transmitted signal is a train of pulses with $\tau_p = 1\ \mu$sec, PRF $= 3$ kHz and total pulse train duration of 1 min.

Problem 5.7

The task is to detect a target with a given RCS, stationary at range $= 20$ km, and within the antenna beam. Prior knowledge about the target is:

The target is *known* to be stationary and within the beam (*if* it is present) and to be at a range between 18 and 22 km. We are trying to detect the target *optimally* (i.e. as close as possible to matched filtering). Fixed system parameters are

$$B = 10\% \text{ of radar frequency } f;$$

$$f = 10 \text{ GHz}, \ \tau_p = 3\ \mu\text{sec}, \ \text{PRF} = 600 \text{ Hz},$$

$$G_0 = 20 \text{ dB}, \ \text{RCS} = 3\ \text{m}^2$$

Only *one* pulse is used. Successful detection is defined as
P_d = detection probability $\ge 90\%$
P_{fa} = false alarm probability $\le 10^{-4}$
(a) If the phase is exactly known, find the *maximum attainable SNR* at the point where the detection decision is to be made, if $P_T = 2$ kW.
(b) Find the *minimum RCS* required to meet the detection specification, with a *1-min* time limit, if all parameter values (except RCS, of course) are the same as in Part (a) under the conditions:
  (1) Phase is exactly known
  (2) We are permitted to use a bank of range gates whose centers are separated by $c\tau_p/2$. How many range gates are needed?
(c) Same as Part (b) except
  (1) Phase is unknown
  (2) Target signal has Rayleigh fluctuations
  (3) A single pulse is used.
Again find the minimum required RCS.

Problems 5.8, 5.9, 5.10

Consider a target as moving so slowly that it can be considered as fixed during a time interval T. A monostatic radar antenna beam is scanning in azimuth at a rate of R_s degrees per second (continuous linear scanning).

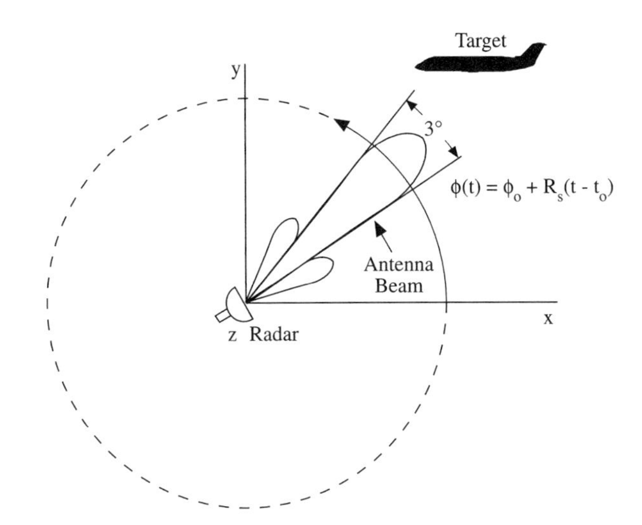

It is transmitting a train of uniformly spaced rectangular pulses with PRF $= fr$, duration $= \tau_p$. For simplification purposes, we model the antenna pattern as uniform over a 3 degree beamwidth in the azimuth angle ϕ, and zero outside that beamwidth. We also set up a parameter

$$K_0 = \frac{G_0^2 \lambda^2}{(4\pi)^3 (kT^0 F_N)} = 10^{18}$$

and write the SNR at IF (prior to matched filtering) as

$$\rho = \frac{P_R}{P_N} = \frac{P_T K_0 \sigma_0}{r^4 B_N}$$

and at the output of a matched filter at IF as

$$\rho_0^{(0)} = \frac{P_T K_0 \sigma_0}{r^4} N_p \tau_p$$

The objective is to detect the target within the time period T. Assume that there are an unlimited supply of range gates, so we need not worry about degradation due to range (or Doppler) mismatch. Just allow a 6 dB safety factor on the required SNR to account for any such effects that exist. Also assume that we do the best possible job of attempting to matched filter each pulse before the detection decision is made (i.e., to *optimize* the detection process). Unspecified variables are

P_T, σ_0, r, N_p, τ_p (also B_N, but you won't need it under the assumptions made), T, R_s.

In all these problems, assume that phase is a priori *unknown*.

Problem 5.8

Setting $P_{fa} = 10^{-4}$, determine the maximum range at which we could obtain a 90% probability of detection of a 1-m^2 target in *one* scan of the beam past the target, if $\tau_p = 1$ μsec, PRF = 500 Hz, $R_s = 1$ degree per second, and peak transmitted power is 10 kW, under conditions

(a) No target fluctuations
(b) Swerling 1
(c) Swerling 2
(d) Swerling 3
(e) Swerling 4

Problem 5.9

Same as Problem 5.8 *except* range has the value such that the required single-pulse SNR for the specified detection performance is 6dB." *How many scans past the target* are required for the specified detection performance?

(a), (b), (c), (d), (e)—same as Problem 5.8.

Problem 5.10

Same as Problem 5.9 *except* that the target must be detected in *3 scans and* the PRF is unspecified. Find the minimum PRF to meet detection specifications under the same conditions (a), (b), (c), (d), (e).

Problems 5.11, 5.12, 5.13

For this set of problems, set P_{fa} at $0.693/n'$ as on the DiFranco and Rubin curves. Now fix n' at 10^6. The objective of each problem will be to plot a curve of P_d vs. range for various combinations of other parameters; *then* to plot P_d vs. range for the same cases where the phase is exactly known (coherent detection). The plots should cover ranges from 1 to 1000 km. Since that would require 3 cycles on a logarithmic scale, you will need a 3-cycle log scale for range.

Problem 5.14

Try to demonstrate, using the Di-Franco and Rubin detection curves for multiple pulses ($N_p = 30, 100, 300, 1000, 3000$) and comparing them with noncoherent detection curves for $N_p = 1$, that the integration loss *for large* N_p obeys a simple rule (roughly) *like* ($10 \log_{10}\sqrt{N_p} - 5.5$) as indicated in 5.132, for the nonfluctuating signal case and for Swerling 1, 2, 3, and 4. You can do this by taking values of the curves for given detection performance levels [e.g. (P_d = 99%, $P_{fa} = 10^{-8}$) or (P_d = 90%, $P_{fa} = 10^{-6}$), etc.] and different values of N_p. The law you get may be different for different cases and performance levels, but it should be close to $\sqrt{N_p}$ for all cases if N_p is large enough. This would be an appropriate computer exercise and would enhance your sense of the importance of integration loss in pulsed radar.

PROBLEMS FOR CHAPTER 6

Problem 6.1

The analysis done to obtain the detection probabilities for the fluctuating models was given in some detail in Chapter 5, Section 5.4 for a single pulse or a coherently-summed pulse train. Show the steps in extending this procedure to the case of a Chi-square model with $k = 3$. (Remember that $k = 1$ is Rayleigh and $k = 2$ is one-dominant-plus-Rayleigh.) Go as far as you can toward evaluating P_d for this case.

Problems 6.2, 6.3, 6.4

Do some rough plots (you may want to use your PC to make accurate plots) of the PDF's of RCS for the following target models:

Confine this to the case of a single pulse or a coherently-summed pulse train.

Problem 6.2

(a), (b), (c) lognormal: Standard deviation of $\ln(\sigma/\sigma_m) = 1, 2, 3$

Problem 6.3

(a), (b), (c), (d)

Ricean {Ratio of mean of dominant signal power to mean power from the Rayleigh part = 1, 5, 10, 100}

Problem 6.4

Weibull: magnitude parameter $\lambda = 1/\langle\sigma\rangle^{\beta}$;

(a) Shape parameter $\beta = 2$; (b) $\beta = 3$; (c) $\beta = 4$; (d) What trends do you observe as β increases?

All of these should be plotted as a function of the *normalized* variable $\sigma/\langle\sigma\rangle$; except lognormal, where the normalization parameter is σ_m, the *median* value of σ.

Problems 6.5, 6.6, 6.7, 6.8, 6.9

For each of the target amplitude PDFs considered in Problems 6.1 through 6.4, solve the following target signal detection problems. The radar and target parameters are:

Peak transmitted power = 100 kW; peak antenna gain (same for transmitter and receiver) = 20 dB; radar frequency = 1 GHz; mean target RCS = 1.5 m^2; target distance from radar = 10 km; radar system losses = 5 dB; noise temperature = 1200°K; The noise is zero-mean Gaussian; target at peak of beam. You will need to do numerical integration to solve most of these problems. If you have software such as Matlab, Mathematica, or Mathcad on your PC these integrations can be accomplished easily.

Problem 6.5

Find the detection probability for a single pulse of 1 μsec duration for a target with chi-square amplitude distribution, as in Problem 6.1, where (a) $k = 4$; (b) $k = 6$ and compare with the case of the nonfluctuating target, where the detection threshold is set at the same level in both cases ($P_{fa} = 10^{-4}$).

Problem 6.6

Same as 6.5 but for a lognormal distribution of target amplitude as in Cases (a), (b), (c) of Problem 6.2.

Problem 6.7

Same as 6.5 or 6.6 but for a Ricean target amplitude PDF, where the ratios of mean of dominant signal amplitude to rms of power from the Rayleigh part = 1, 5, 10, and 100 in Cases (a), (b), (c), and (d), respectively as in Problem 6.3.

Problem 6.8

Same as 6.5, 6.6, or 6.7 but for a Weibull distribution of target amplitude where, as in Problem 6.4, $\lambda = 1/\langle\sigma\rangle^{\beta}$ and $\beta = 2$, 3 and 4 in Cases (a), (b), and (c), respectively.

Problem 6.9

Same as 6.5, 6.6, 6.7, or 6.8 but for a *K*-distribution of target amplitude where $\nu = -0.75, -0.5, 0, 0.5$ and infinity in Cases (a), (b), (c), (d), and (e), respectively (see Fig. 6.5).

Problem 6.10

What general qualitative conclusions can be drawn from the results obtained in Problem 6.5 through 6.10 relating to the effect of target amplitude distribution on detection probability?

PROBLEMS FOR CHAPTER 7

The range-Doppler ambiguity function, the time resolution constant, frequency resolution constant, time-span, frequency span, and area of ambiguity (AF, TRC, FRC, TS, FS, AA) were derived in Section 7.2 for three classes of single-pulse waveform; the rectangular pulse, Gaussian-shaped pulse, and linear FM (chirp) pulse. Also, the case of a train of identical, uniformly spaced pulses of arbitrary shape was covered in Section 7.3. Results were given in terms of both pulse duration and bandwidth for each case.

In Problems 7.1 through 7.5, you are asked to obtain the AF, TRC, FRC, TS, FS, AA, and the time-bandwidth product or compression ratio for the following waveforms and discuss the comparison of these cases with the case of a single rectangular pulse.

Problem 7.1

A chirp pulse with rectangular envelope.

Problem 7.2

A binary sequence 1101101 where the total duration of the sequence is T and each element is of duration $T/7$. Treat a "0" in this binary sequence as a "-1", since it represents a 180° phase shift.

Problem 7.3

A train of 6 pulses, each of Gaussian shape, and of the *same* amplitude and duration, but spaced as shown in the diagram (i.e., the PRF is increased for each pulse according to the rule $T_{m,m+1} = mT_{m-1,m}$; $m = 1, 2, 3, 4, 5$).

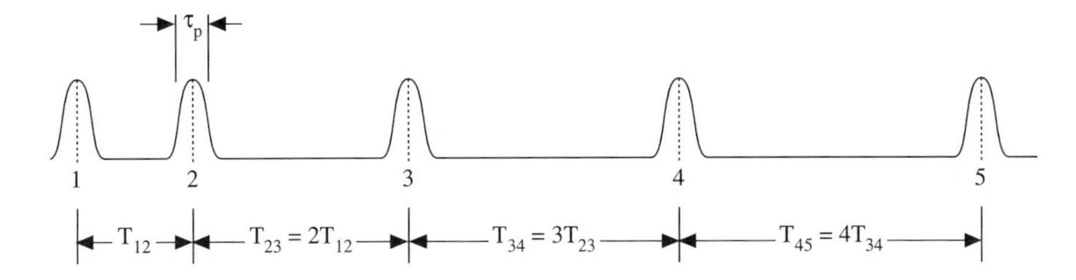

Problem 7.4

A pulse with frequency spectrum as shown below—flat from $\omega = \omega_0 - W/2$ to $\omega = \omega_0 + W/2$

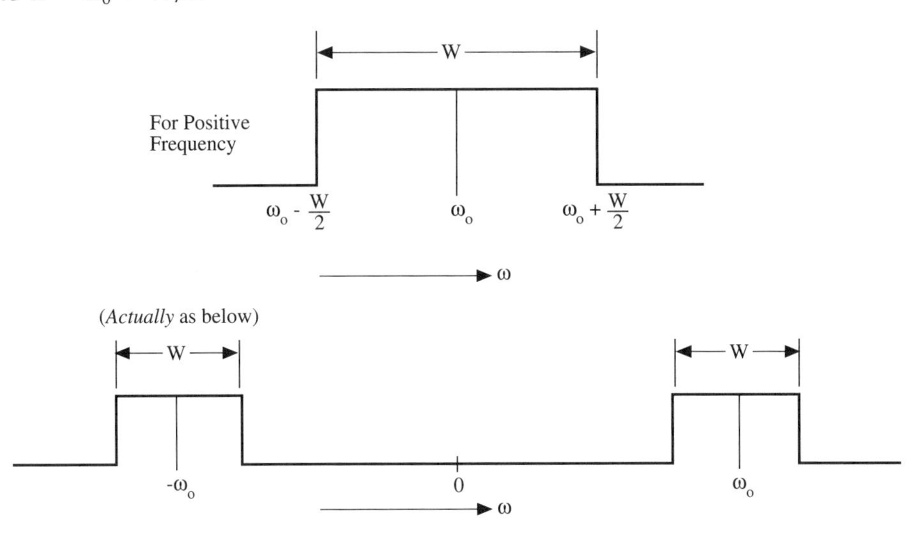

Problem 7.5

For each case above, translate into radar context by the following means:

In Problem 7.1:
Pulse duration $\tau_p = 5$ μsec
Frequency is varied over 1 GHz during the 5 μsec pulse interval

In Problem 7.2:

$$T = 7 \ \mu\text{sec}$$

In Problem 7.3:

$$T_{12} = 1 \ \text{msec}$$

$$\tau_p = 1 \ \mu\text{sec}$$

(where τ_p is defined as the width of a *rectangular* pulse of the same total energy as this Gaussian pulse)

In Problem 7.4:

$$\frac{W}{2\pi} = 1 \text{ GHz}$$

In each case (Problems 7.1, 7.2, 7.3, 7.4) estimate the *minimum resolvable range* and *minimum resolvable doppler speed* for monostatic radar.

Problem 7.6

Find the time ambiguity function $\Lambda(\tau, 0)$ with a flat spectrum like that of Problem 7.4 *except* for a weighting of the form

$$0.08 + 0.92 \cos^2\left(\frac{\pi\omega}{W}\right) \text{(called the "Hamming weighting")}$$

Sketch the AF vs. τ' from $\tau' = \tau/\tau_p = -1$ to $+1$, where τ_p is the pulse duration. Compare it with $\Lambda(\tau, 0)$ for a chirp pulse with rectangular envelope, as shown in Figure 7.21, which has a spectrum somewhat like that of Problem 7.4 for a sufficiently large compression ratio (see Fig. 7.20). A weighting like this can be used to reduce sidelobes in the output waveform of the radar receiver. These sidelobes can be falsely interpreted as targets or can mask true targets at the ranges corresponding to their position on the time-base.

Problem 7.7

Repeat Problem 4.14, with all conditions and parameter values remaining the same except as follows:

(1) Drop the condition that 400 pulses are coherently integrated and substitute the condition that the PRF is now increased and the dwell time is specified as 1 sec.
(2) Increase P_T to 10 Mw.
(3) Increase PRF to 15 kHz.
(4) Add two more aircraft targets *T6* and *T7*, at positions (43, 0, 175) and (60, 40, 0.75), respectively.
(a) Do the same plots as in Problem 4.14, under the new conditions, from range 0.5 to 10 km.
(b) Repeat Part (a) extending the upper limit of range on the plot to 20 km.
(c) Discuss the effect of the increased PRF in Parts (a) and (b).

Problem 7.8

Repeat Problem 4.15 with all conditions and parameter values remaining the same except as follows:

(1) Drop the condition that 400 pulses are integrated and substitute the condition that the PRF is now decreased and the dwell time is now specified as 1 second.

(2) Increase P_T to 6.25 kW.
(3) Decrease the PRF to 1600 Hz.

PROBLEMS FOR CHAPTER 8

Problems 8.1, 8.2, 8.3

These three problems, obtained from a reference on active remote sensing by radar,* relate to a radar illuminating a patch of earth's surface terrain. For radars seeking point targets, such as aircraft, the return from such a patch within the same range gate as the target constitutes the ground clutter.

Note that $\sigma^{(0)}$ means normalized radar cross-section ("NRCS") = RCS *per unit area* for a surface distribution of reflecting material, e.g., earth's surface. It is a pure number = (area/area) and is usually given in dB.

Problem 8.1

A scatterometer (radar mounted on an airborne vehicle, in this case a satellite, for radar remote sensing of earth terrain, e.g., terrain mapping by radar) has the following parameters:

Height = 435 km
Peak antenna gain = 45 dB
Transmitter power = 15 W
Bandwidth = 80 kHz
Noise temperature = 1300°K
Frequency = 11 GHz
Antenna pattern is circular
Neglect earth curvature
Assume CW transmission

Determine the preintegration SNR at an incidence angle of 35° (grazing angle of 55°) if $\sigma^{(0)} = -12$ dB. Find the beamwidth from the peak antenna gain

Problem 8.2

A given airborne radar illuminating the earth's surface has

Peak transmitted power $P_T = 30$ kW
Peak antenna effective aperture area = 7 m^2
Frequency = 10 GHz
Horizontal (azimuthal) beamwidth = 8 milliradians
Pulse duration = 90 nsec
Receiver bandwidth = 14 MHz
Noise figure = 6 dB
Neglect earth curvature

* From F.T. Ulaby, R.K. Moore, and A.K. Fung, "Microwave Remote Sensing," Volume II, Artech House, Norwood, MA, 1982. Problems 7.2, 7.3, 7.4, pp. 554–555, with some variations on problem statements and parameter values.

Determine the SNR for a single pulse, with $\sigma^{(0)} = -20$ dB at a radar height of 4 km and slant range of 12 km. Assume that the antenna beam is pointed directly at the surface patch.

Problem 8.3

A proposed spacecraft "synthetic aperture radar" illuminating the earth's surface has the following characteristics:

Average transmitted power = 80 watts
PRF = 2500 Hz
Antenna aperture = 8×1.25 m
Frequency = 5 GHz
Slant-range resolution = 30 m
Grazing angle extends from 25° to 30°
Height of radar = 400 km
Receiver bandwidth = reciprocal of effective pulse duration
Noise figure $F_N = 5$ dB
Neglect earth curvature
Determine single-pulse SNR for $\sigma^{(0)} = -30$ dB
(a) for total signal from the beam
(b) for the resolution cell with 30 km along-track resolution and then with 30 m along-track resolution.

*Problem 8.4**

Three targets scatter with equal amplitude. They are spaced a distance of 5 m from a to b and 5 m from b to c in a straight line.

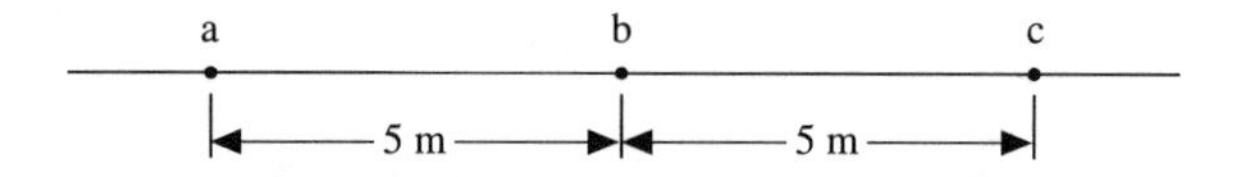

They are observed from *very far away* with a radar at a frequency of 6 GHz. (Whole assembly looks like a point target from the radar.)

(a) Find and sketch the received voltage pattern if the radar travels parallel to line of the targets.
(b) Same as (a) except radar moves perpendicular to line of targets.
(c) Describe the pattern characteristics. What is the nearest null-spacing? Do for both Parts (a) and (b).

Problem 8.5

An airborne radar is mounted on an aircraft at altitude 3 km in straight-and-level flight. The antenna is pointed at a fixed elevation angle and is scanning continuously in azimuth at a fixed rate. The antenna can be approximated as having

* From F.T. Ulaby, R.K. Moore, and A.K. Fung, "Microwave Remote Sensing," Volume II, Artech House, Norwood, MA, 1982. Problems 7.6, 7.7 on page 555 with some variations on problem statements.

Gaussian-shaped elevation and azimuth beam patterns, i.e., a one-way power pattern of the form

$$F(\theta, \phi) = \exp\left(-2.77\left[\left(\frac{\theta_{el} - \theta_{el0}}{W_{el}}\right)^2 + \left(\frac{\phi - \phi_0}{W_{az}}\right)^2\right]\right)$$

where θ_{el} and ϕ are elevation and azimuth angles, respectively, as measured in the earth-based frame, W_{el} and W_{az} are the -3 dB beamwidths in elevation and azimuth, respectively, and θ_{el0} and ϕ_0 are elevation and azimuth beam pointing angles. The peak gain of the antenna can be estimated from (2.29) where the solid angle of coverage is approximated as the product of the 3 dB beamwidths in radians.

The radar is trying to detect an aircraft flying parallel to the radar and *below* the radar's altitude. The system parameters are P_T = peak transmitted power = 10 MW, antenna beamwidths: $W_{el} = 40°$, $W_{az} = 5°$, pulse duration = 2 μsec (no pulse compression), frequency = 2.1 GHz; PRF = 100 Hz, azimuthal scan rate = 360°/sec, target RCS = 5 m^2. Radar system losses = 6 dB, Ground clutter NRCS vs. angles of incidence given in Figure 8.14 for all the illuminated terrain, $-\theta_{el0}$ = depression angle of peak of antenna beam = 20°, range-gate width = same as pulse duration in range units, receiver bandwidth = reciprocal pulse duration, noise temperature = 1100°K. *Note:* Neglect earth curvature and the effect of range and doppler ambiguities in this problem.

(a) Find the per-pulse SNR, CNR and SIR at the output of the range gate containing the target as the antenna beam scans past the target, given that the instantaneous position coordinates of radar (x_R, y_R, z_R) and target (x_T, y_T, z_T) are as follows: $x_R = y_R = 0$, $z_R = 2$ km, $x_T = 2.8$ km, $y_T = 0$, $z_T = 1$ km.

(b) What is the effect of a 10 dB increase in transmitted power on SNR, CNR,and SIR, all other parameters remaining the same?

(c) What is the effect of pointing the elevation beam in a horizontal direction on the answers to Part (a)?

(d) What is the effect on these answers of pointing the beam at an elevation angle of +30°?

(e) How would the changes in elevation beam pointing angle indicated in (c) and (d) affect the answer to Part (b)?

Problem 8.6

(a) In Problem 8.5, assume that both radar and target are moving so slowly that we can consider them stationary over a period of 10 sec. With the same parameter values as in Problem 8.5(a), determine the detection probability under the assumptions of a nonfluctuating target and noncoherent pulse integration. Also assume that the clutter statistics are the same as those of the noise, such that clutter-plus-noise voltage for a single pulse can be considered as a zero-mean Gaussian random process with rms value equal to the square root of the clutter-plus-noise power. The false alarm rate is $6.93(10^{-2})$.

(b) Find the minimum required target RCS for a detection probability of 90%, all other parameters and conditions being the same.

Problem 8.7

Parameter values are the same as those of Problem 8.5 except that the antenna pattern is now that of a rectangular aperture. [See Chapter 10, Sec. 10.2, Eqs. (10.7b) and (10.8).] In the context of this problem the one-way power pattern is

$$F(\theta, \phi) = \left|\text{sinc}\left[\frac{\pi a}{\lambda}\sin(\theta_{el} - \theta_{el0})\cos(\phi - \phi_0)\right]\text{sinc}\left[\frac{\pi b}{\lambda}\sin(\theta_{el} - \theta_{el0})\sin(\phi - \phi_0)\right]\right|^2.$$

where λ is the wavelength and a and b are the aperture dimensions in the elevation and azimuth planes, respectively. Let $a = 8\lambda$, $b = 5\lambda$. The peak gain of this antenna is $4\pi ab/\lambda^2$. In this problem, the beam pointing angle is fixed in azimuth at $\phi = \phi_0$ but is allowed to vary in elevation. The instantaneous target and radar coordinates in the earth frame are the same as those in Problem 8.5.

Draw a sketch of the elevation beam pattern, the radar, the target and the clutter patch in the same range gate as the target and calculate the target signal power, clutter power, SCR, and SIR for the following geometries:

(a) the beam points directly at the target, (b) the beam points directly at the clutter patch, (c) the beam depression angle is $-10°$ (meaning its elevation angle is $+10°$. (d) Indicate whether each of the cases (a), (b), and (c) are clutter limited, noise limited, or limited significantly by both clutter and noise.

Problem 8.8

Redo a modified version of Problem 8.5(a) accounting for range ambiguities, i.e., assume range gates of the same width as the pulse duration measured in range units. Fix the antenna beam pointing angles at $\theta_{el0} = -20°$, $\phi_0 = 0°$.

For the range gate containing the target, calculate contributions to received target signal power, clutter power, SCR, and SIR due to clutter at ambiguous ranges. All parameters are the same as given in Problem 8.5 *except* that the PRF and transmitted power are changed, $(x_T - x_R)$ is increased to 28 km, $z_R = 11$ km, and $(z_T - z_R)$ is changed to -10 km. Do this for (a) PRF = 800 Hz, $P_T = 1.25$ Mw, (b) PRF = 1200 Hz, $P_T = 833$ kW, (c) PRF = 3 kHz, $P_T = 333.2$ kW, (d) PRF = 20 kHz, $P_T = 50$ kW, (e) PRF = 50 kHz, $P_T = 20$ kW.

In all cases, determine the ambiguous ranges first. Note that it is assumed that no other targets are present in this scene and the terrain is assumed uniform throughout the scene. If your preliminary analysis tells you that ambiguous range contributions change the results by less than 1 dB then neglect them.

Problem 8.9

For this problem, we allow the radar and target to move and deal with instantaneous Doppler frequencies and doppler ambiguities. The radar-bearing aircraft is flying eastward (i.e., in the $+x$-direction in the coordinate system used here) at 500 mph and altitude 3 km and the target is at altitude 1 km and also flying eastward at 250 mph. The antenna pattern is that of Problem 8.5. The radar system parameters are those given in Problem 8.5 *except* where otherwise indicated.

(a) The radar is at $x = 0$ at $t = 0$. The target is at $x = 5$ km at $t = 0$. We are interested only in the period from $t = 0$ to $t = 10$ min. Find the minimum and maximum values of (unambiguous) target range and minimum and maximum (unambiguous) target Doppler we will have to deal with during that period. At which times do these maxima and minima occur? Explain.

(b) Find the (unambiguous) Doppler shift of the return from the ground clutter patch that is in the same range gate as the target signal at $t = 0$ and $t = 10$ min *if* such a clutter patch exists. If so, where is the clutter patch in each case?

(c) Considering only the terrain region illuminated by the antenna at $t = 0$ and within the target's range gate, at what ambiguous Doppler frequencies would we see extraneous clutter returns? Same question at $t = 10$ min.

(d) Calculate, at $t = 0$, the clutter power, the CNR and SIR including the *significant* contributions from the doppler ambiguities determined in (c). Define "significant" as changing the result by at least 1 dB.

Problem 8.10

In this problem, the conditions and target and radar parameter values of Problem 8.9 hold *except* where otherwise indicated. The modifications are (1) the radar is ground-based, on top of a hill at position $x_R = r_R = 0$, $z_R = 500$ m and the antenna beam is pointed in the eastward direction and 30° upward, such that the ground clutter is negligible; (2) a rainfall volume covers the entire region illuminated by the beam; (3) the frequency is changed to values to be indicated below; (4) the target is moving in the westward direction (i.e., toward the radar) in the y-z plane at altitude 3 km.

(a) At the instant when the target is at the peak of the beam, calculate the instantaneous target return, clutter return, and SIR at frequencies 9.4, 35, 70, and 95 GHz using Eq. (8.41) to obtain the clutter within the target's range gate. The rainfall rate is 1.25 mm/hr. Neglect path attenuation due to the rainfall (see Section 11.5).

(b) Repeat Part (a) at frequencies 9.4, 12, and 15 GHz using Eq. (8.39) [or equivalently (8.42)], applicable for $f \leq 15$ GHz.

Problem 8.11

(a) Conditions and parameter values of Problem 4.14 hold except where otherwise indicated. The radar is at an altitude of 1 km. Ground clutter is distributed uniformly within the radar's field of view. Consider all ground clutter returns to obey Eq. (8.20b) (i.e., the law for small grazing angle) and use Figure 8.14 to obtain the NRCS vs. grazing angle. Plot the distribution of range gate output CNR values in dB from 5 to 10 km.

(b) What is the effect on this plot of a 10-fold increase in PRF, other parameter values remaining intact?

(c) Extend the coverage of the plot to the region 0.5 to 5 km, replacing (8.20b) with (8.20c).

PROBLEMS FOR CHAPTER 9

Problem 9.1

Determine the blind speeds incurred through use of an N-pulse canceller for a radar with a transmitted frequency of 2 GHz for five different PRFs: (1) 100 Hz, (2) 300 Hz, (3) 500 Hz, (4) 1 kHz, (5) 5 kHz. For a target or clutter patch at a range of 50 km, what are the ambiguous ranges for each PRF? Which ambiguous ranges are below the true range for each of these PRFs?

Problem 9.2

Same as Problem 9.1 but this time the PRF is fixed at 300 Hz and there are five different frequencies spanning most of the commonly used radar bands, as follows:

(1) 500 MHz, (2) 2 GHz, (3) 10 GHz, (4) 20 GHz, (5) 30 GHz.
Find blind speeds and ambiguous ranges below the true range for the target or clutter patch in Problem 9.1

Problems 9.3, 9.4, 9.5

Suppose, in designing a ground-based radar that detects moving targets and discriminates against stationary clutter using an N-pulse canceller type of MTI, we know that moving targets to be detected could have any LOS velocity between −200 and +200 m/sec and hence we could not tolerate blind speeds below 200 m/sec except for the blind speed at zero. Also, targets could have any range between 25 and 50 km and hence we want to do our best to avoid ambiguous range returns from clutter within those limits.

The problems to follow do not necessarily have unique answers. Their purpose is to give you some practice in applying some of the ideas in this chapter to a problem involving the choice of design parameters.

Problem 9.3

(a) Assume you have to choose a frequency somewhere within S-band and you are limited to a single PRF. Choose a PRF that meets the goal of eliminating or at least minimizing blind speeds below 200 m/sec (except for the obvious one at zero) and minimizing clutter returns from ambiguous ranges that lie between 25 and 50 km. What are the ambiguous ranges, if any, within those limits for a true range of 25 km and for a true range of 50 km?

(b) Repeat for a chosen frequency within X-band.

(c) Discuss qualitatively the differences between Cases (a) and (b).

Problem 9.4

(a) Use a staggered PRF scheme to meet the goals indicated in Problem 9.3, where you are allowed to choose a maximum of 3 different PRFs to reduce the effect of blind speeds. Choose PRFs 1.027, 2.353, and 2.931 kHz. Plot a set of curves of response vs. f_d for each PRF like those of Figure 9.5a, b, and c, all on the same horizontal scale. Then, on that same scale, plot the composite curve for the staggered PRF scheme, like that of Figure 9.5d. Do this for a 3-pulse canceller (note that Fig. 9.5 is for a 2-pulse canceller).

(b) Add a fourth PRF, 564 Hz, to Part (a).

(c) Determine the ambiguous ranges below 200 km at which clutter returns might occur if the true range is 35 km. Do this for Part (a), then for Part (b).

Problem 9.5

(a) As an alternative to the staggered PRF system, use a staggered radar frequency scheme in a 3-pulse canceller, where the PRF is fixed at the value chosen in Problem 9.3 for X-band. Choose 3 X-band frequencies, 8.253, 9.708, and 11.231 GHz, and superpose the three returns as discussed in Section 9.2 for the staggered PRF system. Determine the blind speeds for each frequency. Plot a set of response curves like those of Figure 9.5a, b, c, and d where the horizontal axis is calibrated in meters per second (Doppler speed).

(b) Determine the ambiguous ranges below 200 km if the true range is 35 km.

Problem 9.6

(a) Assume the conditions and parameter values of Problem 8.9. For Problem 9.6, assume a two-pulse canceller at the output of the range gate containing the target. Calculate instantaneous per-pulse values of the SNR, CNR, and SIR in dB at the *output* of the canceller at $t = 0$. Assume the clutter spectrum given in 8.43(a), where the clutter Doppler spread Δf_c is determined from σ_v, the corresponding spread in clutter Doppler speed, assumed to be 2 m/sec. Assume that $\eta_v = 0$ in Eqs. (9.32g)′, and a large enough spread around the mean target Doppler speed to justify neglect of all terms but the first in the numerators of Eqs. (9.32g)′ (i.e., $e^{-\zeta_S^2} << 1$).

(b) Compare your answers to those of Problem 8.9(d). What are the effects of the two-pulse canceller on the SNR, CNR, and SIR?

(c) Suppose you replaced the two-pulse canceller with a three-pulse canceller. How much would that change the answers to (b)? Same question for a four-pulse canceller.

(d) Leave all conditions and parameter values the same *except* that the radar is now moving *northward* at the same speed and crosses the x-axis at $t = 0$, Recalculate the instantaneous per-pulse SNR, CNR, and SIR in dB at the canceller output (two-pulse, three-pulse, *and* four-pulse) at $t = 0$. Do the cancellers have a greater effect than in (c)? If so, why?

Problem 9.7

As the radar-target-clutter-patch geometry changes along the trajectory, the SIR, both with and without Doppler filtering, can be widely variable. To demonstrate that process, calculate the instantaneous SIR in Problem 9.6(a) for the following values of time.

(a) $t = 10$ sec; (b) $t = 20$ sec; (c) $t = 30$ sec; (d) $t = 45$ sec.

At each instant, compare your answers *with* and *without* the 2-pulse canceller, where $\sigma_v = 2$ m/sec and $\eta_v = 0$.

(e) In (a)–(d), does the MTI always increase the SIR? If not, why not?

(f) The pulse canceller type of MTI is specifically designed for moving targets and stationary clutter and works best for ground-based radar, except in cases where the target Doppler is near a blind speed. In the previous problems, we have been dealing with airborne MTI, where the clutter is almost never stationary with respect to the radar. Make the assumptions of (1) a stationary radar and (2) $e^{-\zeta_S^2} << 1$. How would these assumptions affect the answers to (a) and (b), all other parameters and conditions remaining the same?

Problem 9.8

For the scenario of Problem 9.6 (all parameters and conditions are the same as in Problem 9.6 except where otherwise indicated) you have already calculated the signal power and clutter power in the range gate containing the target and the SCR improvement attainable with pulse cancellers. In this problem we replace the canceller with a bank of digital Doppler filters at the output of each range gate, as discussed in Section 9.4.2. We know a priori that the target's range is between 1 and 10 km and its range-rate is between −1500 and +1500 mph. We must cover the corresponding Doppler spread with our bank of filters and the range spread with a bank of range gates whose width is the pulse duration in range units.

At the output of the range gate containing the target, the Doppler velocity spread of the target around its central Doppler velocity is assumed to be very small.

To be specific, say it is 0.2 m/sec. Assume the spread in Doppler velocity of the clutter around *its* central Doppler velocity to be entirely due to the geometry of the radar and the clutter patch corresponding to a given range gate.

(a) Design the width of each Doppler filter to correspond to the 0.2 m/sec spread alluded to above. Find the resulting improvement in SCR at the output of the Doppler filter containing the target at $t = 0$.

(b) Repeat Part (a) at $t = 10$ sec and at $t = 20$ sec. Are the results significantly different from those of Part (a)? Explain.

PROBLEMS FOR CHAPTER 10

Problem 10.1

An antenna with a rectangular aperture in the x–y plane whose illumination is uniform and whose x-dimension and y-dimension are 1.5 and 3 m, respectively, operates at a frequency of 3 GHz.

(a) Determine the values of θ at the first five sidelobe peaks in the (x–z) and (y–z) principal planes and plot those patterns vs. θ in dB from $\theta = 0$ to $\theta = 90°$.

(b) Find the peak gain and approximate 3 dB beamwidth of this antenna.

Problem 10.2

For the antenna in Problem 10.1, find the values of ϕ for the nulls in the one-way power patterns for (a) $\theta = 45°$, (b) $\theta = 90°$. How many nulls are within visible space in cases (a) and (b)?

Problem 10.3

In problems 10.1 and 10.2, no mention was made of polarization. In this problem, the electric field on the aperture is in the x-direction and has a peak magnitude of 1 V/m.

(a) Define and illustrate the E-plane and H-plane for this case.

(b) Find the magnitudes of the θ and ϕ components of the electric field vector in volts/meter and the power density in watts/m^2 at a distance of 500 m from the antenna at (1) $\theta = 30°$, $\phi = 30°$, (2) $\theta = 45°$, $\phi = 30°$, (3) $\theta = 60°$, $\phi = 45°$, (4) $\theta = 90°$, $\phi = 60°$.

Problem 10.4

Repeat Part (a) of Problem 10.3 for a circular aperture with uniform illumination, frequency = 2 GHz and aperture radius = 1 m, other parameters and conditions remaining the same. What do your results tell you about the effect of the circular symmetry of the aperture on the polarization of the radiated fields?

Problem 10.5

(a) Repeat Part (a) of Problem 10.1 for the linearly tapered aperture discussed in Section 10.5 [see Eq. (10.24)]. The taper is only in the x-direction and the aperture illumination is uniform in the y-direction.

(b) Same as Part (a) but for the cosine taper of Eq. (10.35a) in the x-direction and uniform in the y-direction.

(c) Same as Part (a) but for the cosine-square taper of Eq. (10.36a) in the x-direction and uniform in the y-direction.

Problem 10.6

An array of 7 isotropic antenna elements has its array axis in the x-direction. The amplitude of the excitation is the same on all the elements. Determine and sketch-plot the normalized array factor in the x–z plane as a function of β, the angle between the array axis and the direction of the radiation from $\beta = 0$ to $\beta = 180°$. The interelement spacing is a half-wavelength. The frequency is 1 GHz.

(a) The phase of the excitation is the same for all the elements.

(b) The phase of the excitation varies linearly at the rate of $+2(\pi/\lambda)$ radians per meter.

(c) The phase of the excitation varies linearly at the rate of $-(\pi/\lambda)$ radians per meter.

(d) Characterize the array as "broadside," "endfire," or somewhere inbetween in Cases (a), (b), and (c). In each case, specify the mainbeam pointing angle β_0 and the angle of the first sidelobe peak on either side of the mainbeam peak.

Problem 10.7

Redo Part (a) of Problem 10.6 for the case where the isotropic elements are replaced by half-wave dipoles oriented in the y-direction. How does that affect the peak gain and effective aperture area of the array?

Problem 10.8

Redo Part (a) of Problem 10.6 for the case where the element excitation amplitudes are (numbered in the $+x$-direction), the phases being the same for all the elements:

$$C_1 = 1, \quad C_2 = 4, \quad C_3 = 9,$$

$$C_4 = 16, \quad C_5 = 9, \quad C_6 = 4, \quad C_7 = 1$$

Problem 10.9

In problem 10.6, determine the extent of visible space in the parameter $w = kd \cos \beta + \alpha d$ in Cases (a), (b), and (c).

Problem 10.10

In a particular radar application, at S-band (specifically 2.7 GHz), an H-plane horn antenna is to be used. The largest permissible x and y dimensions of the cross-section of the waveguide feeding the horn are a half wavelength and a quarter wavelength respectively.

(a) What is the peak gain of an open rectangular waveguide aperture antenna using this waveguide without a horn?

(b) The design specification is for a peak gain of at least 12 dB. How long must the horn be to achieve that specification?

(c) If the maximum allowable horn length is 1 m, what is the highest achievable peak gain in dB?

(d) How high a peak gain could be achieved with a 1-m-long H-plane horn if the frequency and the waveguide dimensions are increased 10-fold?

(e) In Part (c), how large an E-plane flare angle would be required for a 12 dB gain if we were to use a 1-m-long pyramidal horn?

Problem 10.11

The antenna array in Problem 10.6 is electronically scanned by varying the phase rate linearly from $+1.8\pi/\lambda$ to $-1.8\pi/\lambda$ radians per meter in 10 sec. The center of the array is at $x = y = z = 0$.

(a) Calculate and plot the angle θ_m in the x–z plane vs. time t from $t = 0$ to $t = 10$ sec. The angle θ_m is defined in Section 10.9.2. The plot should be in degrees, not radians.

(b) If the linear scanning were done mechanically, the variation of θ_m vs. t would be linear. Plot the deviation from linearity in degrees from $t = 0$ to $t = 10$ seconds if $\theta_m(0) = -55°$ and $\theta_m(10) = +55°$. Over what time interval is the electronic scanning approximately linear?

Problem 10.12

The aperture of Problem 10.1 is electronically scanned by allowing the x-directed phase rate α_x and the y-directed phase rate α_y to both vary linearly from $-\pi/\lambda$ to $+\pi/\lambda$ radians/second between $t = 0$ and $t = 10$ sec.

(a) Calculate and plot $\dot{\theta}_m$ and $\dot{\phi}_m$, the rates of change of angles θ_m and ϕ_m vs. time t from $t = 0$ to $t = 10$ sec. The angles are defined in Section 10.9.2. The plots should be in degrees/second, not radians/second.

(b) Redo Part (a) where the only difference is that $\alpha_x = \pi/\lambda =$ constant and $\alpha_y = \alpha_x$ at $t = 0$ and $-\alpha_x$ at $t = 10$ sec.

Problems 10.13, 10.14

In this chapter there was considerable emphasis on the tapering of aperture illumination or array excitation in order to reduce sidelobe levels (see Sections 10.5 and 10.6). These two problems deal with that issue in the context of a radar scenario.

Problem 10.13

Redo part of Problem 8.7 with only *one* modification in parameter values or conditions. Change the one-way power antenna pattern function $F(\theta, \phi)$ to the form

$$F(\theta, \phi) = |F_x(\theta, \phi)F_y(\theta, \phi)|^2$$

where $F_y(\theta, \phi)$ is the sinc function on the *right-hand side* in Problem 8.7 and $F_x(\theta, \phi)$, which replaces the sinc function on the *left-hand side* in Problem 8.7, is

$$F_x(\theta, \phi) = \left\{ \text{sinc}\left[\frac{\pi a}{N\lambda} \sin(\theta_{el} - \theta_{el0})\cos(\phi - \phi_0) \right] \right\}^N$$

where N is a positive integer exceeding 1. [This is a generalization of the form (10.30) in Section 10.5.] With this modification, recalculate the SCR for the conditions of Parts (a), (b), and (c) of Problem 8.7. Do these calculations for (a) $N = 2$, (b) $N = 4$, (c) $N = 10$, (d) Compare your answers with those of Problem 8.7 and explain the difference.

Problem 10.14

Same as Problem 10.13, except that $F_x(\theta, \phi)$ is replaced by (a) cosine taper of Eq. (10.35a), (b) cosine-square taper of Eq. (10.35b).

PROBLEMS FOR CHAPTER 11

Problem 11.1

The detection range for a target with RCS = 1 m^2 of a particular ground-based radar is determined by the radar equation to be 200 km assuming a clear atmosphere and infinite free space (i.e., neglecting the presence of the earth). Given standard atmospheric conditions (see Section 11.1) and given that the radar antenna is 5 m above ground,

(a) Find the maximum slant range where a 1 m^2 target could be detected at ground level.
(b) How high would the antenna have to be before the radar could detect a 1 m^2 target at its full infinite free-space detection range? Would this be a practical height for a ground-based radar?
(c) In Eq. (11.5), the value of α is increased by a factor of 3. What is the effect on the answers to Parts (a) and (b)?

Problem 11.2

In this problem you will be asked to include the propagation factor as discussed in Section 11.2 as a part of the radar equation. We will start with a set of radar system and geometrical parameters.

Peak transmitter power = 1 MW
Radar frequency = 1 GHz
Radar and target are both in the x–z plane of the cartesian coordinate system
Transceiver antenna: rectangular aperture with uniform illumination: dimension of aperture in x–z plane = 1 m; aperture dimension in y-direction = 2 m.
Pulse duration = 1 μsec; no pulse compression
PRF = 300 Hz
Target RCS = 1 m^2 (approximate the target scattering pattern as isotropic)
Receiver losses = 3 dB
Transmitted radiation is horizontally polarized
The receiver accepts only horizontally polarized radiation.

Receiver noise temperature = 1200°K
Receiver bandwidth = reciprocal of pulse duration
Terrain within the radar-target environment for purposes of calculation of terrain reflected waves can be modelled as if it were perfectly conducting

(a) Determine the received SNR in the clear atmosphere neglecting the propagation factor and assuming radar is at coordinates $x_R = 0$, $y_R = 0$, $z_R = 1$ km and target is at $x_T = 50$ km, $y_T = 0$, $z_T = 2$ km.
The assumed geometry is based on a flat earth on the x–y plane of our coordinate system. The antenna beam points directly at the target.
(b) Find the earth-reflection point for the indirect rays in the multipath process (see Section 11.2.1).
(c) Extend Part (a) to include the pattern propagation factor for the indirect-direct path (R→ERP→T→R) and direct-indirect path (R→T→ERP→R).
(d) Enhance Part (c) by adding the indirect-indirect path (R→ERP→T→ERP→R) to the propagation factor.

Problem 11.3

Repeat Parts (c) and (d) of Problem 11.2 with all parameters and conditions the same except that the effect of the antenna elevation pattern is neglected, i.e., the elevation pattern is approximated as uniform at the earth reflection point. (In this case we would refer to the "propagation factor," not the "pattern propagation factor.") Does this make a significant difference in the result? Explain.

Problems 11.4, 11.5, 11.6, 11.7

Parameter values and conditions same as in Problem 11.2 except where otherwise indicated. These problems all involve the generalizations discussed in Sections 11.2.1, 11.2.2, and 11.2.3.

Problem 11.4

Repeat Problem 11.2 accounting more accurately for the properties of the terrain medium. The only changes in parameters are

(a) the medium is sea water: set relative permittivity $(\epsilon/\epsilon_0) = 80$, conductivity $\sigma = 4.5$ S/m.
(b) the medium is a type of soil with $\epsilon/\epsilon_0 = 2.5$, $\sigma/\omega\epsilon = 0.05$ at $f = 1$ GHz.

Problem 11.5

Redo Problem 11.4 with the following change:
The excitation field on the transmitting antenna aperture is vertically polarized and the receiver accepts only vertically polarized radiation.

Problem 11.6

Redo Problems 11.2 and 11.3 with inclusion of

(a) the effect of surface roughness as discussed in Section 11.2.3, with rms height of the surface near the ERP equal to 0.5 m
(b) same as (a) with rms surface height = 1.5 m

(c) What do your answers to (a) and (b) tell you about the effects of terrain-height fluctuation on the propagation factor?

Problem 11.7

Redo Problem 11.4 accounting for earth curvature (see Section 11.2.2). Note that the value of the grazing angle and the location of the ERP are affected by earth curvature. Both of these quantities are required in determination of the divergence factor and other constituents of the propagation factor and those of the return power without that factor. Use four-thirds of the true earth radius in your calculations. What is the percentage error in the return power due to neglect of earth curvature?

Problem 11.8

Use the same set of system parameters as in Problem 11.2 except where otherwise indicated. In Part (a) of Problem 11.2 the radar and target coordinates are given. Change these to $x_R = 0$, $y_R = 0$, $z_R = 5$ m (i.e., ground-based radar): $x_T = 10$ km, $y_T = 0$, z_T unspecified. Leave transmitted power unspecified. Place a hill along the path between radar and target, at the position $x_h = 6$ km, $y_h = 0$. The height of the hill $z_h = 50$ m. It is assumed that knifedge diffraction theory (see Section 11.3) is adequate to model the effect of the hill. As in Problem 11.3, the minimum SNR required for detection of the target is 6 dB.

Find the minimum transmitted power required for detection at

(a) $z_T = 0$, (b) $z_T = 20$ m, (c) $z_T = 50$ m, (d) $z_T = 100$ m.

Problem 11.9

Repeat Problem 11.4, Part (a), where all parameter values are the same except that the frequency is increased to

(a) 3 GHz (S-band)
(b) 10 GHz (X-band)
(c) 30 GHz (Ka-band)
(d) 60 GHz (V-band)

In Parts (a) through (d), account for all frequency-dependent effects along the propagation path including atmospheric attenuation as discussed in Section 11.4, but assume that it is *not raining*. If you include surface roughness with $\langle z^2 \rangle = 0.5$ m, as in Problem 11.6(a), how will that affect these results?

Problem 11.10

Repeat Parts (a) and (b) of Problem 11.9 in the case where it *is* raining along the entire propagation path, where the rainfall rate p is (see Fig. 11.4)

(a) 0.25 mm/hr (drizzle)
(b) 1.25 mm/hr (light rain)
(c) 12.5 mm/hr (moderately heavy rain)

PROBLEMS FOR CHAPTER 12

Problem 12.1

A particular monostatic radar system has the following parameter values: peak transmitter power = 100 kW, frequency = 3 GHz, noise temperature = 1200°K, transmitter and receiver losses = 6 dB, pulse duration = 2 μsec, no pulse compression, PRF = 300 Hz; antenna: same for transmission and reception, rectangular aperture with uniform illumination; x dimension = 1 m, y dimension = 50 cm; receiver noise bandwidth at IF stage prior to matched filtering = 300 MHz.

Consider a signal from a target with RCS = 1 m^2 at the peak of the antenna beam. Assume (realistically) that phase is completely unknown so that coherent matched filtering of the entire pulse train received during the dwell time Δt can be accomplished through quadrature filtering, where cosine and sine filter outputs are squared and summed to accomplish a close approximation to the optimal amplitude measurement discussed in Section 12.1.1.

(a) How large a dwell time must we have to achieve a fractional rms error of 0.01 or below in amplitude measurement at a target range of 25 km?
(b) Compare this with the minimum dwell time required for target detection with P_d = 95% and $P_{fa} = 10^{-6}$ at that same range. Assume that the target is nonfluctuating.
(c) In optimal phase measurement as discussed in Section 12.1.2 the quadrature technique alluded to above is an inherent part of the process. Find the minimum dwell time required for a phase measurement accurate to within 3 degrees. Also find the phase error in degrees for the dwell time determined in Part (a).

Problem 12.2

(a) Determine β_τ as defined below Eq. (12.11j) [or alternatively in (12.20c) where $x \to \tau$] for a train of uniformly-spaced identical pulses.
(b) Same as (a) for β_ω as defined below Eq. (12.16).

Problem 12.3

Conditions and parameters same as in Problem 12.1, but add the condition that the pulses have a Gaussian shape where pulse duration τ_p has the meaning previously assigned to it for that case, the duration of a rectangular pulse with the same total energy as the Gaussian pulse (see Section 7.5, Fig. 7.5).

(a) Find the rms error in range measurement for a single pulse.
(b) Determine the minimum required dwell time for a range measurement accurate to within 5 m (i.e. rms error = 5 m).
(c) Same as (a) (i.e., measurement with a single pulse) but this time use the pulse shape with sloping edges as depicted in Section 12.2.1, Eqs. (12.26a) and Figure 12.3. Compare your answers with those of Part (a) and comment on the interpretation of that comparison.
(d) If you doubled the PRF and halved the pulse duration, what would be the effect on the answer to Part (b)? Also, how would halving the PRF and doubling the pulse duration affect that answer? Explain.

Problem 12.4

Conditions and parameter values same as in Problem 12.1. This time the measurement is of range-rate as determined from Doppler shift, for a 1 m^2 target.

(a) Find the rms error in Doppler speed measurement for a single rectangular pulse.
(b) Using Gaussian pulses of the specified duration, find the minimum required dwell time for a rms error in range-rate measurement of 2 m/sec or less. How would use of a Gaussian pulse affect your answer?
(c) Repeat Part (b) maintaining all parameter values intact except frequency, which is doubled, then reduce frequency by a factor of two. What is the effect of doubling the frequency on the answer to (b)? of halving the frequency?

Problem 12.5

Consider a ground-based radar with the parameters and conditions given for Problem 12.3, except where indicated otherwise. The antenna is scanning linearly and continuously in azimuth at the rate of 10°/sec. The target, with RCS = 1 m^2, is moving so slowly that it can be considered stationary within a single scan. Its position, relative to the radar at $x_R = y_R = z_R = 0$, is $x_T = 10$ km, $y_T = 5$ km, $z_T = 5$ km. The antenna beam is pointed at an elevation of 24° above the horizontal. The y direction on the antenna aperture is the horizontal (i.e., azimuthal) direction. It is required to measure the target's azimuth angle within a single scan.

(a) What is the best attainable measurement accuracy in degrees?
(b) What is the effect on this answer of doubling the target altitude if the elevation beam angle and all other parameters remain the same?
(c) All parameters except frequency remain constant. What is the effect of halving the radar frequency? of doubling the frequency?

Problem 12.6

All parameters and conditions are the same as in Problem 12.5 except where otherwise indicated (e.g., a single target at $x_T = 10$ km, $y_T = z_T = 5$ km is within the radar's field of view). The radar is searching for the target in range, elevation angle, and azimuth angle. A single range gate whose width is the pulse duration in range units is stepped through range space from 5 to 25 km, remaining at a particular range for a period T_A sec. During that period, the elevation beam is step-scanned from +10° to +70° in increments equal to the −3 dB elevation beamwidth θ_B. While the range gate and elevation beam are at fixed settings, the azimuthal beam scans continuously and linearly over 360°. Eventually, the target appears in the range gate and in the elevation and azimuth beams. During that short interval, the measurement of target position parameters is made.

(a) What is the time interval during which the target appears in the radar return (i.e., the "dwell time") if $T_A = 10$ sec?
(b) What are the rms errors in measurement of range, elevation angle, and azimuth angle? The transmitted pulses are assumed to have a Gaussian shape.
(c) What are the rms errors in measurement of x_T, y_T and z_T?

Note: In calculating SNR, account should be taken of losses due to the fact that the target is not necessarily at the *peak* of the range gate or antenna beam during the dwell.

Problem 12.7

This is a continuation of Problem 12.6, where we now take account of the target's motion.

(a) The target is flying westward (i.e., in the $-x$-direction) at 400 mph. During the small time interval when the target is in the beam and in the range gate, what is the doppler frequency shift in the received signal?
(b) Given $T_A = 10$ sec as in Problem 12.6, determine the rms error in measurement of Doppler frequency. What is the corresponding rms error in the line-of-sight velocity component?
(c) The choice of $T_A = 10$ sec in Problem 12.6 is not realistic for measuring parameters for a target moving at 400 mph. The range, angles, and Doppler are changing too fast to allow that much search time. How much change would occur in range, direction angles and Doppler in 10 sec? 1 sec? 0.1 sec? 0.01 sec?
(d) From your answers to (c), choose a value of T_A during which the range does not change by more than 150 m, the direction angles do not change by more than 1°, and the Doppler frequency does not change by more than 50 Hz. For that value of T_A calculate the rms errors in all four parameters. Comment on the difference between these results and the corresponding results for Problem 12.6 and Part (b) of this problem.

Problem 12.8

(a) Calculate the error with the noise-free comparator output for the amplitude comparison method of angle measurement as discussed in Section 12.3.1, using Eq. (12.47i) but changing from the Gaussian pattern of Eq. (12.48a) to the more realistic sinc function pattern of Eq. (10.8). Obtain the counterpart of (12.48e) and plot the curve in Figure 12.7(b) for the sinc function pattern, where $\theta_0 = 3°$ and the value of (d/λ_0) in (10.8) is chosen so that $\theta_b = 6°$.
(b) Calculate the rms error due to noise based on (12.49i) for the sinc function pattern used in Part (a). Plot the counterpart of the curves on Figure 12.8 for this pattern.
(c) Discuss the comparison between your results and the results given in the text for the Gaussian pattern.

INDEX

H

I

K

L

M

N

O

P

Q

R

X

Z